PATHOPHYSIOLOGY OF CARDIOVASCULAR DISEASE

PROGRESS IN EXPERIMENTAL CARDIOLOGY

Edited by Naranjan S. Dhalla, Ph.D., M.D. (Hon.), D. Sc. (Hon.)

1. S. Mochizuki, N. Takeda, M. Nagano, N.S. Dhalla (eds.): *The Ischemic Heart*. 1998
   ISBN 0-7923-8105-X
2. N.S. Dhalla, P. Zahradka, I. Dixon, R. Beamish (eds.): *Angiotensin II Receptor Blockade: Physiological and Clinical Implications*. 1998.
   ISBN 0-7923-8147-5
3. N. Takeda, M. Nagano, N.S. Dhalla (eds.): *The Hypertrophied Heart*. 2000.
   ISBN 0-7923-7741-9
4. B. Ostadal, M. Nagano, N.S. Dhalla (eds.): *Cardiac Development*. 2002.
   ISBN 1-4020-7052-7
5. P. Singal, I. Dixon, L. Kirshenbaum, N.S. Dhalla (eds.): *Cardiac Remodeling and Failure*. 2002.
   ISBN 1-4020-7177-9
6. N.S. Dhalla, N. Takeda, M. Singh, A. Lukas (eds.): *Myocardial Ischemia and Preconditioning*. 2002.
   ISBN 1-4020-7195-7
7. N.S. Dhalla, L. Hryshko, E. Kardami, P.K. Singal (eds.): *Signal Transduction and Cardiac Hypertrophy*. 2003. ISBN 1-4020-7218-X
8. G.N. Pierce, M. Nagano, P. Zahradka, N.S. Dhalla (eds.): *Atherosclerosis, Hypertension and Diabetes*. 2003. ISBN 1-4020-7311-9
9. N.S. Dhalla, A. Chockalingam, H.I. Berkowitz, P.K. Singal (eds.): *Frontiers in Cardiovascular Health*. 2003. ISBN 1-4020-7451-4
10. N.S. Dhalla, H. Rupp, A. Angel, G.N. Pierce (eds.): *Pathophysiology of Cardiovascular Disease*. 2004.
    ISBN 1-4020-7452-2

PATHOPHYSIOLOGY OF CARDIOVASCULAR DISEASE

Editors
NARANJAN S. DHALLA, PhD, MD (Hon), DSc (Hon)
Distinguished Professor and Director
Institute of Cardiovascular Sciences
St. Boniface General Hospital Research Centre
Faculty of Medicine, University of Manitoba
Winnipeg, Canada

HEINZ RUPP, PhD
Professor of Medicine
Philipps University of Marburg
Marburg, Germany

AUBIE ANGEL, MD
Professor of Internal Medicine
Faculty of Medicine, University of Manitoba
Winnipeg, Canada

GRANT N. PIERCE, PhD, FACC
Professor & Director
Division of Stroke & Vascular Disease
St. Boniface General Hospital Research Centre
Faculty of Medicine, University of Manitoba
Winnipeg, Canada

SPRINGER SCIENCE+BUSINESS MEDIA, LLC

Library of Congress Cataloging-in-Publication Data
International Society for Heart Research. Congress (17th : 2001 : Winnipeg, Man.)
Pathophysiology of cardiovascular diseases / editors, Naranjan S. Dhalla ...[et al.].
p. ; cm. – (Progress in experimental cardiology ; 10)
Includes bibliographical references and index.

ISBN 978-1-4613-5084-2 ISBN 978-1-4615-0453-5 (eBook)
DOI 10.1007/978-1-4615-0453-5

Cardiovascular system –Pathophysiology—Congresses. 2.
Heart—Diseases—Treatment—Congresses. I. Dhalla, Naranjan S. II. Title. III. Series.
[DNLM: 1. Cardiovascular Diseases—physiopathology—Congresses. 2. Cardiovascular Diseases—therapy—Congresses. WG 120 I617p 2003]
RC669.9I57 2001
616.1'07—dc21 2003048839

Printed on acid-free paper.

Contents

PROFESSOR HORACIO CINGOLANI, MD
La Plata, Argentina

A Tribute to Professor Horacio Cingolani, MD

This book is dedicated to Professor Horacio Cingolani to recognize his outstanding achievements in the area of Cardiovascular Science and Medicine. Dr. Horacio E. Cingolani is the Director of the Cardiovascular Research Center in La Plata School of Medicine, La Plata, Argentina. He is the founder of the Latin American Section of the ISHR and an established investigator of the Consejo Nacional de Investigaciones Científicas y Técnicas (Argentina) with the highest category (Investigador Superior).

He got his medical degree in La Plata in 1961, where he reached the academic position of Professor and Chairman of Physiology after all the steps of the career structure. Dr. Cingolani did his postdoctoral training in Cardiovascular Physiology at National Heart Institute, NIH, Bethesda MD, USA under the direction of Stanley Sarnoff.

He was Associate Professor in Vanderbilt University Medical Center, Nashville, TN, USA and Visiting Professor in the University of Utrecht, Holland, R.W. Johnson School of Medicine, NJ, USA and Johns Hopkins University, Baltimore, USA. He was President of the Argentine Society for Clinical Investigation and of the Argentine Federation of Cardiology.

Dr. Cingolani has authored/co-authored over 150 research papers, many of which have been published in international prestigious scientific journals. He has served on the editorial board of Canadian Journal of Cardiology, Journal of Pharmacological and Toxicological Methods, European Journal of Cardiology, Circulation Research, Journal Molecular and Cellular Cardiology, Experimental and Clinical Cardiology. He is co-director and one of the authors of the textbook of Physiology "Bernardo Houssay Medical Physiology".

Dr. Cingolani is a member of the International Academy of Cardiovascular Sciences and during his career he has been the recipient of several awards. These include Argentine Society of Cardiology Award, Argentine Federal of Cardiology Award, Argentine Society of Hypertension Award and Hippocrates Award of the National Academy of Medicine.

Dr Horacio E. Cingolani is married to Dr Gladys Chiappe de Cingolani and has three children: Oscar, Celia and Eugenio, all of them are MD.

Preface

This volume of *"PATHOPHYSIOLOGY OF CARDIOVASCULAR DISEASE"* was compiled as a result of invitations given to participants from the XVII World Congress of the International Society for Heart Research. The World Congress was held in Winnipeg, Canada July 7–11, 2001. The meeting was an outstanding event attracting approximately 2000 attendees with nearly 500 invited speakers and chairpersons from all over the world. The excitement and enthusiasm generated from these research presentations and the ensuing discussions surrounding the data stimulated the Editors of this volume to create a compendium of excellent research manuscripts on the topic of the pathophysiology of cardiovascular disease.

This volume has been subdivided into 4 sections. The 4 sections deal with four specific issues involved in the pathophysiology of cardiovascular disease. These include, heart dysfunction and its associated characteristics (hypertrophy, cardiomyopathy and failure), vascular dysfunction and disease, ischemic heart disease, and novel therapeutic interventions to address the problem of cardiovascular disease. These four broad categories have led to the submission of a wide variety of different research topics within each category. The volume is particularly valuable because it discusses so many aspects of cardiovascular disease but within each paper it focuses in great depth on specific aspects of the disease process. We have also selected articles which describe a biochemical or functional process that occurs in the normal heart or vessel but can participate in the cardiovascular disease process when abnormalities in these pathways occur.

The first section in this volume describes the cellular processes involved in heart dysfunction. A variety of cardiac diseases are discussed. These include septic cardiomyopathy, heart failure and the involvement of mitochondria in heart failure, reactive oxygen species, calcium homeostasis, the extracellular matrix in cardiomyopathies, signalling pathways in cardiac hypertrophy and preconditioning, and signal transduction pathways in the heart and in cardiac fibroblasts. This section also deals with adrenergic stimulation of the ventricle, ion channels, ventricular fibrillations and the measurement of cytosolic pH and isometric tension in cardiac muscle preparations.

The second section of this volume concentrates on ischemic heart disease. Excellent manuscripts have been written on the topics of the involvement of reactive oxygen species in ischemic heart disease, the general cell biology of acute reperfusion injury, the involvement of reactive oxygen species and ion movements (in particular calcium in ischemic reperfusion injury) and the use of optical mapping techniques to understand calcium signalling in myocardial ischemia. In addition, intriguing data has been included here to discuss intercellular signalling during ischemia through an examination of connexin phosphorylation. Two papers examine ischemic pre-conditioning as a mode to protect the heart and a third paper examines a variety of cardioprotective interventions. Two additional papers examine unique approaches to ischemia/reperfusion-induced myocardial injury by presenting data on the involvement of mast cells in this process and the use of a novel drug to scavenge oxygen derived free radicals to alter calcium homeostasis in the reperfused myocardium.

The third section of this volume investigates the involvement of the vascular systems in cardiovascular disease. Papers discuss the modulation of fatty acid metabolism, the identification of endothelium derived hyperpolarism factors and their involvement in the regulation of blood flow, cytokinesis in smooth muscle and how this may play an important role in smooth muscle cell hyperplasia and hypertrophy, and the importance of ADP-ribosylation in the cardiovascular system under both healthy and diseased conditions. Two other interesting papers examine the interactions between the heart and vessels during disease. Specifically, the influence of cardiac contractility on coronary venous flow and the influence of vascular factors on aortic stenosis and heart transplantation is discussed. The cell signalling pathways induced by oxidative stress in vascular smooth muscle cells is also an important subject for evaluation in this text. An important interrelationship of a preconditioning phenomena with myocardial angiogenesis is another topic for discussion. The final paper in this section involves an examination of the arachidonic acid cascade in experimental myocardial infarctions.

Perhaps the most interesting and novel section within this text is the discussion of the new therapeutic approaches to cardiovascular disease. Of particular interest is the lead article which examines the potential for using the nuclear envelope itself as a new target for drug action. There appears to be more and more data to suggest that the nucleus may exist as a cell within a cell. This paper by Bkaily and coworkers essentially addresses this possibility and extends it to address the intriguing potential of using the nuclear envelope membrane as a new target for therapeutic drug action. We are becoming increasingly aware of the relationship of inflammation to atherosclerotic disease. However, less appreciated is the role of inflammation in cardiac function and repair. The incidence of myocardial infarctions continues to rise in North America today. Thus, the need to understand how to repair the heart following an infarction is critical. Two papers examine that, one by focusing on a particular group of proteins (the Smad family) as targets for therapeutic action. Estrogen therapy has also received a great deal of attention as a therapeutic mode of altering cardiovascular disease, particularly in women. An interesting article is included in this text on phytoestrogens and their association with cardiovascular disorders. A unique method for cardioprotection has been introduced in one of the paper by the use of chronic exposure to long lasting ambient heat. Perhaps one of the most exciting new developments in the treatment of heart disease is the use of cell transplantation into infarcted tissue to restore cardiac function. Several papers use different cells (bone marrow stromal cells, cardiomyocytes and tissue transplantation) to treat a heart that has experienced a myocardial infarction. The use of magnetic resonance imaging in the assessment of both myocardial injury and brain injury is also evaluated. Finally, the potential for using mechanical circulatory devices in the treatment of heart failure is discussed by an expert panel of international experts.

Cardiovascular disease is the leading cause of death in the world today. In the industrialized countries, cardiovascular disease is also the most expensive disease in terms of both direct and indirect costs to the health care system. It is not well appreciated that despite this economic cost, the amount of research that is dedicated to understanding and treating cardiovascular disease is relatively small. In North America, the investment in research on cardiovascular disease remains far below the investment in other disease entities when expressed relative to the economic costs of each of the diseases. It is clear that if we are to make an impact in our understanding and our treatment of cardiovascular disease and if we truly wish to lower the economic cost of this disease and the mortality that results from this disease, we will have to invest a great deal more time, energy and financial resources to research cardiovascular disease.

This volume represents a compendium of different approaches to understanding cardiovascular disease, and identifying the proteins, pathways and processes involved in cardiovascular disease. This exciting volume of research is as a result of an association of scientists at a world meeting held in Canada. It is clear evidence that international gatherings of medical scientists can generate friendly, prosperous and successful interactions in times when the world may appear quite hostile in view of recent international developments. It is refreshing and encouraging to note that medical research is still productive and successful even on an international stage of collaboration.

Naranjan S. Dhalla
Heinz Rupp
Aubie Angel
Grant N. Pierce

Acknowledgments

We are grateful to the following corporations and granting agencies for their generous donations in support of the XVII World Heart Congress of the International Society for Heart Research, the first Public Heart Health Forum as well as publication of this book:

PATRONS:

Government of Canada (Dept. of Western Diversification)
Government of Manitoba (Depts. of Industry Trade and Mines; Health; Post-Secondary Education; Culture Heritage and Tourism)
Merck Frosst Canada, Ltd.
Mitsubishi-Tokyo Pharmaceuticals Inc.

PARTNERS:

American Section of the International Society for Heart Research
AstraZeneca
Aventis Pharmaceuticals Inc.
Bayer Canada, Inc.
City of Winnipeg
International Academy of Cardiovascular Sciences
International Society for Heart Research (Kaito Fund, Bayer Yakuhin Fund and Canon Fund)
Kowa Pharmaceuticals
Pfizer Canada
St. Boniface General Hospital Research Foundation

COLLABORATORS:

CanWest Global Foundation
CIHR Institute of Circulatory and Respiratory Health
Eli Lilly
Great West Life and London Life
Manitoba Liquor Control Commission
Mars Incorporated
Medicure, Inc.

Myles Robinson Memorial Heart Fund
Safeway Food and Drug
University of Manitoba (Faculty of Medicine; Departments of Physiology and Human Anatomy & Cell Science)

BENEFACTORS:
ATL Canada
Beckman Coulter Canada Inc.
Canadian Cardiovascular Society
Canadian Institutes of Health Research
Cardiovascular Solutions, Inc.
Dairy Farmers of Canada
De Fehr Foundation
Faculty of Health Sciences, University of Western Ontario
Heart and Stroke Foundation of Manitoba
Institute of Biodiagnostics, National Research Council of Canada
Japanese Working Group on Cardiac Structure and Metabolism
Manitoba Hydro
Merck KGaA (Germany)
Pulsus Group Inc.
St. Boniface General Hospital Research Centre
Wawanesa Mutual Insurance Company
World Heart Corporation

The collaboration of Ms. Eva Little, Ms. Janet Labarre, Ms. Diane Stowe, Ms. Florence Willerton and Ms. Susan Zettler in coordinating diverse editorial activities associated with this book is gratefully acknowledged. Special thanks are due to Mr. Zachary Rolnik, Ms. Mimi T. Breed, Ms. Melissa Ramondetta and their editorial staff at Kluwer Academic Publishers for their patience, interest and hard work in assembling this volume.

PATHOPHYSIOLOGY OF CARDIOVASCULAR DISEASE

I. Cellular Processes Involved in Heart Dysfunction

Pathophysiology of Cardiovascular Disease,
edited by N.S. Dhalla, H. Rupp,
A. Angel & G.N. Pierce
Kluwer Academic Publishers, Boston, 2004

Septic Cardiomyopathy

Karl Werdan, Hendrik B. Schmidt, Frithjof Schlegel,
Anja Herklotz, and Ursula Mueller-Werdan

Department of Medicine III, Martin-Luther-University of Halle-Wittenberg, Halle(Saale), Germany

Summary. The existence of human septic myocardial depression was only unequivocally proven in the 1980s by the group of Parrillo, utilizing a nuclear imaging technique in intensive care patients. Heart failure in sepsis is frequently masked by a seemingly normal cardiac output. However, relative to the lowered systemic vascular resistance—resulting in a reduced afterload, cardiac outputs and ventricular ejection fractions of septic patients are often not adequately enhanced. This septic cardiomyopathy involves both the right and the left ventricle and is potentially reversible. In response to volume substitution, the hearts can be considerably enlarged. The disease is not primarily hypoxic in nature, but may be aggravated by ischemia. Autonomic dysfunction, documented by a reduced heart rate variability and impaired baro- and chemoreflex sensitivities, forms part of the disease entity. The severity of myocardial depression correlates with a poor prognosis. Non-infectious systemic inflammatory response syndrome (SIRS) can give rise to an analogous disease entity, namely SIRS cardiomyopathy.

The etiology of the disease is multifactorial. Several candidates with potential pathogenic impact on the heart were identified: bacterial toxins, cytokines and mediators including tumor necrosis factor α, interleukin-1 and nitric oxide, cardiodepressant factors, oxygen reactive species, catecholamines. Symptomatic treatment consists of volume substitution and of catecholamine support; causal therapeutic approaches aiming at an interruption of the proinflammatory mediator cascades are being tested.

Key words: Cardiomyopathy, septic—Sepsis—Autonomic dysfunction—Shock, septic—Heart failure, septic.

Corresponding author: Prof. Dr. med. Karl Werdan, Universitaetsklinik und Poliklinik fuer Innere Medizin III, Klinikum Kroellwitz der Martin-Luther-Universitaet Halle-Wittenberg, Ernst-Grube-Strasse 40, D-06097 Halle(Saale), Germany. Tel.: +49-345-557-2601/4545, Fax: +49-345-557-2072/4546, E-mail: karl.werdan@medizin.uni-halle.de.

INTRODUCTION

Acute septic myocarditis in the preantibiotic era (1) was a purulent disease of the heart. Nowadays, merely nonspecific pathomorphological and pathohistological alterations characterize the myocardium of patients whose hearts had failed in septic shock (2). Septic myocardial depression for decades has been attributed to the release of cardiodepressant factors into the blood stream, while the existence of human septic myocardial depression was only unequivocally proven in the early 1980s by the group of Parrillo (3), who had examined patients in the ICU with nuclear imaging techniques. In recent years, the concept of a septic cardiomyopathy was proposed (4,5), which emphasizes alterations of cardiac cellular phenotype as a basis of organopathy in response to a variety of agents acting on heart cells, like bacterial toxins and endogenous cytokines, hormones, mediators and cardiodepressant factors.

The intention of this review is to highlight newer aspects of cardiac involvement in systemic inflammation and especially in sepsis. Not organ-related infectious heart diseases like viral myocarditis or bacterial endocarditis are in the focus of this paper, but the uniform reaction pattern of the heart to generalised inflammatory processes like sepsis and other systemic inflammatory response syndromes (4–8).

These heart diseases of infectious or systemic inflammatory origin are increasingly paid attention to beyond the field of intensive care medicine: The pathogenic mechanisms involved in these disorders may also apply to common chronic heart disorders like coronary artery disease or heart failure (9), which go along with a moderate inflammatory response.

SEPTIC CARDIOMYOPATHY—A SECONDARY CARDIOMYOPATHY IN THE SCOPE OF A SYSTEMIC DISEASE

The present concept of the pathogenesis of bacterial sepsis is centered around the idea that numerous and very heterogeneous stimuli—including all classes of microorganisms or even isolated bacterial toxins like endotoxin or superantigens—uniformally evoke an activation of mediator cells with secondary release of cytokines and proinflammatory mediators (Figures 1, 2). These primarily protective mechanisms are directed against invading microbes and aim at inactivating toxins. The inflammatory response may, however, take an exaggerated course and thereby be complicated by an impairment of the patients' organs. An overwhelming activation of mediator cascades presents clinically as sepsis (Figure 2). In addition to the detrimental action of endogenous mediators, a direct toxicity of the bacterial toxins—like endotoxin (4,5) and Pseudomonas exotoxin A (10)—potentially may further aggravate organ malfunctioning.

In recent years, it became evident that not only infectious, but also non-infectious stimuli can trigger these defense mechanisms and thus give rise to a clinical picture that may be indiscriminable from bacterial sepsis (Figure 1); examples of such non-infectious aggressions are major trauma, surgery with the assistance of the cardiopulmonary bypass, pancreatitis, ischemia-reperfusion injuries or allograft rejection. Whatever the initial trigger may have been—infectious or non-infectious—, the sequence of events following adheres to a uniform pattern: there is a massive

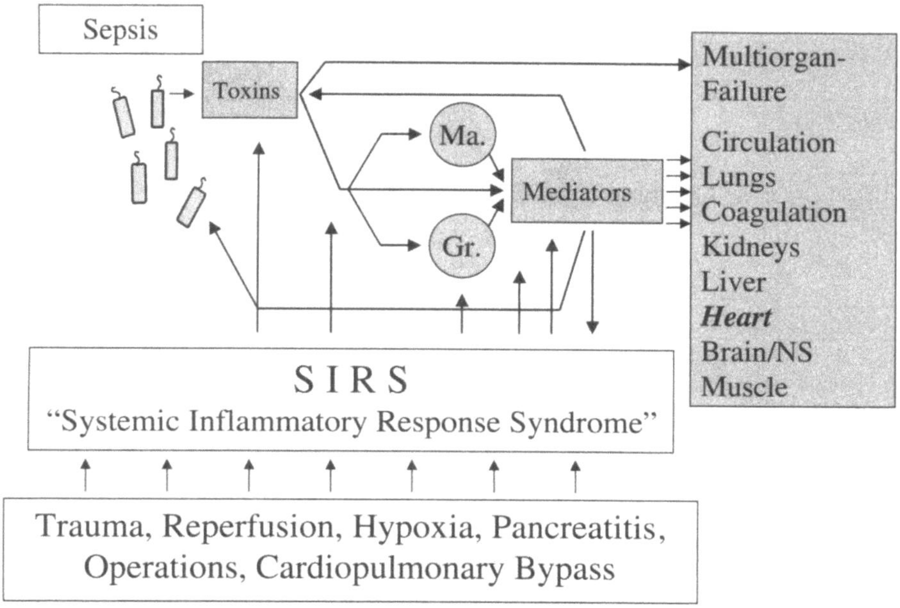

Figure 1. Pathophysiology of sepsis and escalating SIRS. Bacterial toxins like endotoxin from Gram-negative bacteria damage the patient's organs and also trigger the release of mediators from activated macrophages (Ma) and granulocytes (Gr). These mediators—released to kill bacteria and inactivate toxins—also add to the patient's damage in the sense of a multiple organ dysfunction syndrome or multiorgan failure. Not only infections but also noninfectious triggers can induce the deleterious mediator network termed systemic inflammatory response syndrome (SIRS).

production and release of mediators interwoven in a complex network resulting in a mediator-induced systemic inflammatory response syndrome (SIRS) and multiple organ dysfunction syndrome (MODS), which includes the heart (4,5,11–14).

It has long been denied that a cardiac involvement forms part of septic MODS, as cardiac output values of septic patients usually are seemingly normal or may even be enhanced in comparison to the physiological range (Figure 3). However, heart failure becomes evident, when cardiac output is considered in relation to the systemic vascular resistance, which is severely lowered due to sepsis-induced vasodilation (Figure 4): a healthy heart could compensate for the pathological fall in afterload down to 1/3 or 1/4 of the normal value by an up to three- or fourfold increase in cardiac output (Figure 3), while—very often—the observed values in our septic patients are considerably lower: the compensatory increase in pump activity is not high enough to stabilize blood pressure. This is the consequence of septic cardiomyopathy (Table 1), a disease entity characterised by a dominant left ventricular failure, which is not primarily hypoxic in nature, as coronary arteries are dilated and coronary blood flow is high (15,16). However, septic cardiomyopathy can be aggra-

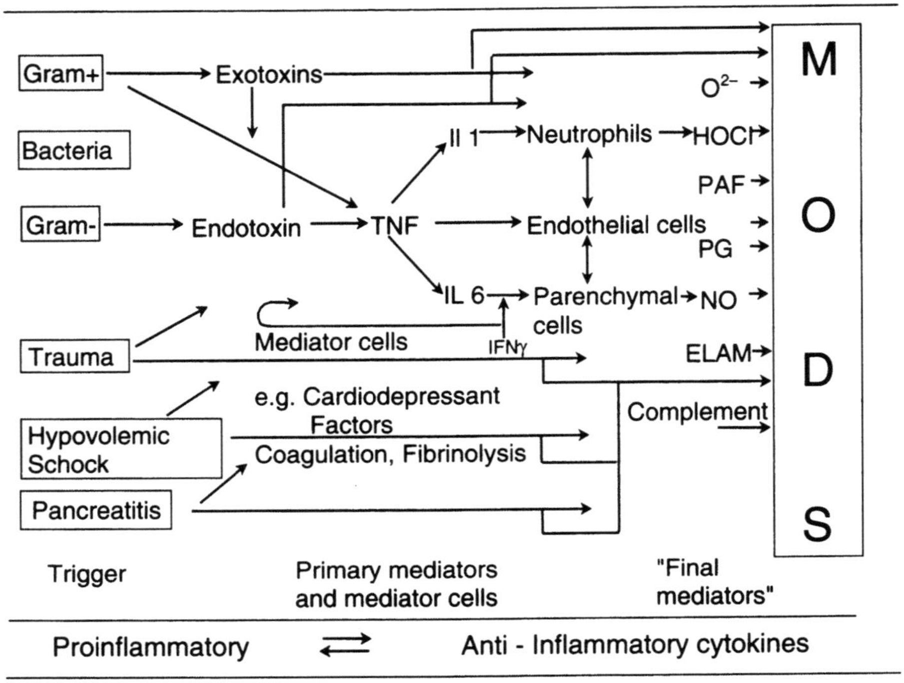

Figure 2. Toxin- and mediator-network in sepsis and SIRS. For explanation see legend to figure 1. ELAM = endothelial leukocyte adhesion molecule as representative of adhesion molecules; Gram+ = Gram-positive bacteria; Gram− = Gram-negative bacteria; HOCl = hypochloric acid; IFNγ = interferon γ; IL = interleukin; MODS = multiple organ dysfunction syndrome; NO = nitric oxide; O^{2-} = superoxide anion as representative of reactive oxygen species including peroxinitrite; PAF = platelet activator factor; PG = prostaglandins; TNF = tumor necrosis factor-α.

vated by myocardial ischemia (5,12,17), particularly in patients with preexisting coronary artery disease, as the increased coronary blood flow in sepsis narrows the coronary reserve. Left ventricular stroke work indices were found to be reduced to a similar degree (Figure 5; (14,18)) in patients with various forms of Gram-negative, Gram-positive or fungal sepsis indicating that not such much the bacterial virulence factors but rather the common mediator network determines the occurrence and severity of the disease. Additional right ventricular dysfunction essentially belongs to septic myocardial depression and can be accentuated in the presence of pulmonary hypertension due to ARDS (19); right ventricular dilatation and a reduced right ventricular ejection fraction can further impair left ventricular performance by a fall in left ventricular filling pressure and a mechanical compromise of the left ventricle by a septum shift. Although septic cardiomyopathy is potentially completely reversible, it still is a condition of high prognostic importance: it accounts for about 10% of fatalities witnessed in sepsis and septic shock (3), with non-survivors having more depressed afterload-related myocardial depres-

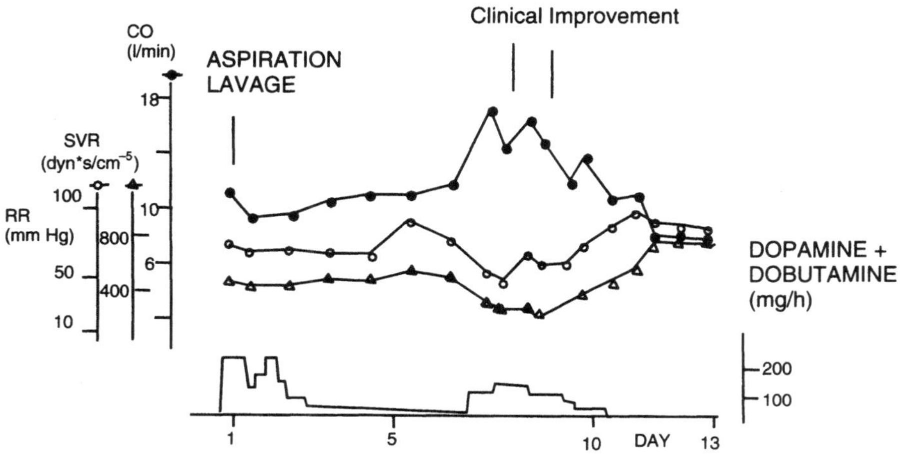

from: Müller-Werdan et al. (1996) Cytokines and the Heart. Landes Company

Figure 3. Case report: Cardiovascular changes in Pseudomonas sepsis. The graph shows the time course of the hemodynamic changes seen in a patient with Pseudomonas sepsis: Due to the toxin- and mediator-induced toxic vasodilatation—as measured by a fall in the systemic vascular resistance (SVR; normal value 1100 ± 200 dynes × sec × cm^{-5})—mean blood pressure (RR) falls. In a compensatory manner, cardiac output (CO) increases up to values of about 18 l/min. During recovery SVR rises to normal, blood pressure stabilizes and cardiac output returns to values found in healthy persons. In this patient, heart function is not severely impaired by sepsis. However, in most septic patients cardiac output does not rise up to the values seen in this patient, thereby indicating myocardial depression in sepsis (see figure 4).

sion than survivors (K. Werdan, A. Herzklotz, U. Mueller-Werdan: unpublished results).

In view of all these findings, cardiac impairment in sepsis can be classified as "septic cardiomyopathy", as organ involvement of the heart within the scope of the systemic disease "sepsis".

Diagnosis of cardiac impairment in sepsis

In clinical practice, the diagnosis of septic cardiomyopathy is frequently hampered by the fact that all reference values for cardiac function parameters are normalised to an afterload of 1100 dynes × cm^{-5} × sec, but that reference values for a systemic vascular resistance of 300 dynes × cm^{-5} × sec have never been established. According to Figure 4, an inverse correlation between afterload—represented best, though a calculated and not measured value, by the systemic vascular resistance—and cardiac performance could be demonstrated and quantitated (K. Werdan, A. Herklotz, U. Mueller-Werdan: unpublished results). If this inverse correlation is not taken into account, data from echocardiographical or hemodynamic measurements (20,21) may be misinterpreted.

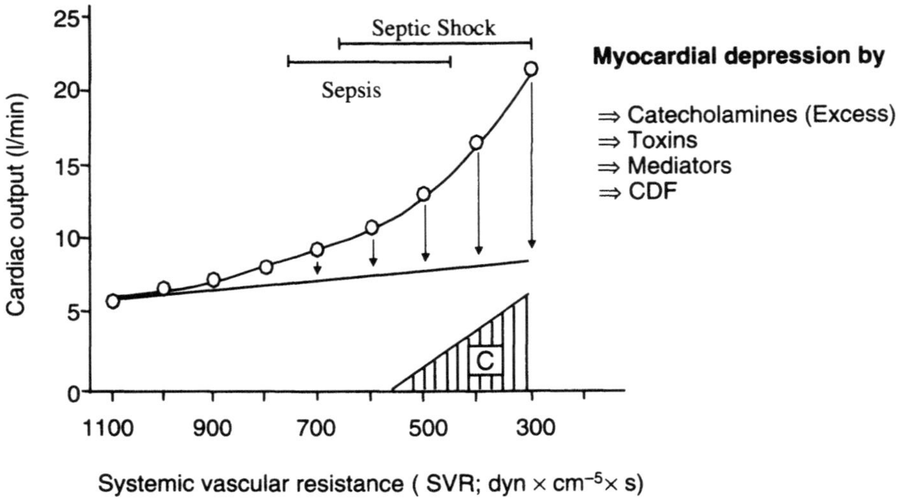

Figure 4. Afterload dependance of heart function in sepsis. To compensate for a progressive decrease in systemic vascular resistance due to vasodilatation, the calculated rise in cardiac output (O-O-O) should be necessary to maintain a mean arterial pressure of 90 mm Hg (according to: [mean arterial pressure—mean right atrial pressure (10 mm Hg)] = systemic vascular resistance (dynes $\times$ cm^{-5} $\times$ sec) $\times$ cardiac output (l/min): 80). The cardiac outputs measured in sepsis and septic shock (——), however, are usually lower than the calculated values, as a result of septic cardiomyopathy. "C": catecholamines applied to stabilize blood pressure; "CDF": cardiodepressant factor. Modified from (5).

Table 1. Features of acute septic cardiomyopathy

Myocardial depression:
- In relation to the lowered systemic vascular resistance
  - ⇒ cardiac index not adequately raised
  - ⇒ right and left ventricular ejection fractions and stroke work indices not enhanced or even decreased
  - ⇒ pump volumes inadequately low
- Global and regional contractile disturbances
- Contraction and relaxation abnormalities
- Increase in ventricular compliance (shift of the pressure-volume-curves to the left)
- Considerable dilatation of the heart is possible
- Coronary arteries dilated, high coronary blood flow

Right ventricular compromise and dilatation due to ARDS

Superimposed hypoxic heart damage

Accentuated in preexisting coronary artery disease, particularly in manifest shock

Autonomic dysfunction

Narrowed range of heart rate variability, related to severity of MODS

Incidence of arrhythmias?

Modified from (4,5).

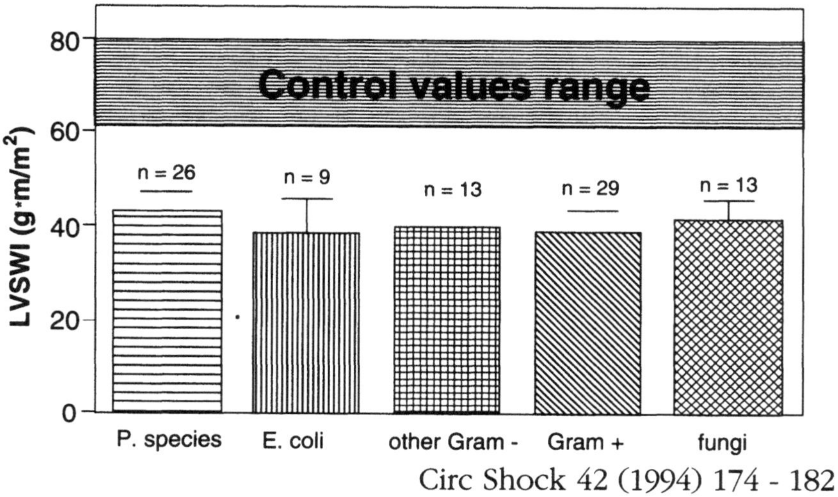

Figure 5. Myocardial depression in sepsis. In patients with septic shock due to infection with Pseudomonas (P.), E. coli, other Gram-negative and Gram-positive pathogens as well as due to fungi, left ventricular stroke work index (LVSWI) as clinical measure of inotropy was determined. All patients independent of the infectious agent showed a marked depression of LVSWI. Modified from (18).

Myocardial depression in sepsis—triggers and mechanisms

Pleiotropy and redundancy: Endotoxin, tumor necrosis factor alpha and interleukin-1 as triggers

In the majority of cases numerous bacterial toxins and primary, secondary and final mediators are involved in the pathogenesis of systemic inflammation (Figure 2). However, the sole administration of endotoxin or TNF-α is sufficient to induce the hemodynamic alterations of sepsis, including myocardial depression. Several cardiodepressant factors of not fully unravelled chemical composition have been described (4,5,22), with one myocardial depressant substance identified as the combination of TNF-α and IL-1β (23).

TNF-α is regarded as the pertinent cardiodepressant cytokine in sepsis, its negative inotropic impact frequently being ascribed to the induction of inducible nitric oxide (NO) synthase (iNOS) and an enhanced production of NO in the heart (4). Although iNOS induction was found in some myocardial specimen from patients with heart failure or sepsis, this was not a uniform finding (24). Results from *in vitro* studies (Figure 6) rather suggest a complex interaction of TNF-α with the heart, with pleiotropic effects on cardiomyocyte performance (Figure 7) due to an interference with several inotropic pathways as well as TNF-α-like cardiodepressive effects of other cytokines like interleukin-1. The experimentally supported concepts of TNF-α cardiodepression include an induction of iNOS and inhibition of constitutive NOS

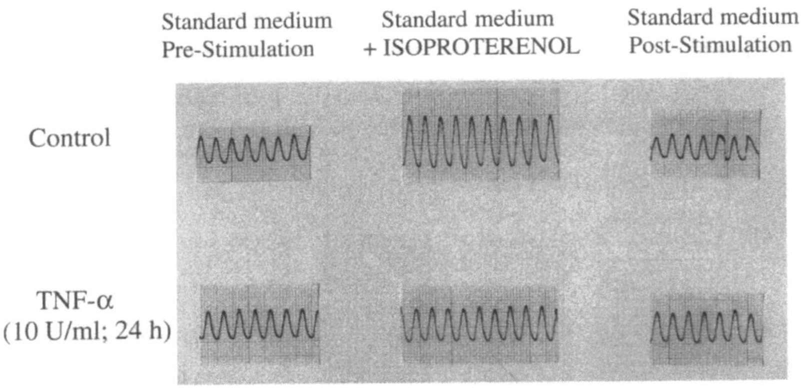

Figure 6. Preincubation of Cardiomyocytes with TNF-α blocks β-Adrenoceptor-mediated Increase in Pulsation Amplitude. Neonatal rat cardiomyocytes were cultured in serum free culture medium for 24 hours, in the absence (Control) and in the presence of a non-cytotoxic concentration of tumor necrosis factor-α (TNF-α). Thereafter, cells were perfused with the β-adrenergic agent isoproterenol, and pulsation amplitude of the electrically driven cells were monitored by an electrooptical system. Pretreatment of the cells with TNF-α suppresses the isoproterenol-induced increase in pulsation amplitude. For methods and further details see (30).

(cNOS or eNOS) at high TNF-α concentrations, but also NO-independent cardiodepression at low, pathophysiologically more relevant concentrations (25). TNF-α effects on the heart also vary with regard to the kinetics of the process (26): rapidly occuring cardiodepressant effects include a release of sphingosine (27) and a suppression of the calcium transient, while chronic administration of TNF-α was shown to depress the synthesis of precursors for the phosphoinositide pathway (28) and inhibit pyruvate dehydrogenase activity and mitochondrial function (Figure 9; (29)). Cardiomyocytes are not only effector cells for cytokines, but upon stimulation produce cytokines, which may be secreted (IL-6) (30) or retained intracellularly (IL-1) (30): it seems likely that secondary cytokines induced in cardiomyocytes by TNF-α or endotoxin (30,31)—via a CD14-dependent pathway (31)—contribute to cardiodepression. Whether apoptotic signals activated by TNF-α (32,33) are involved in the negative inotropic pathways is presently unknown.

Conclusive evidence for a pathogenetic impact of chronic TNF-α elevation on the heart comes from *in vivo* studies (34) and transgenic mice (35–37). TNF-α and interleukin-1 may, however, not only have detrimental effects on the heart, but in myocytes also confer resistance to hypoxic injury (38) or α-adrenoceptor-induced arrhythmias (Figure 10, (4,25,28)).

Nitric oxide and reactive oxygen species: not only impairment of inotropic cascades, but also of mitochondrial function and cellular oxygen utilization

In the hierarchy of the proinflammatory mediator cascades, TNF-α and IL-1 (39) belong to the primary players, while nitric oxide (NO) (40) and oxygen free

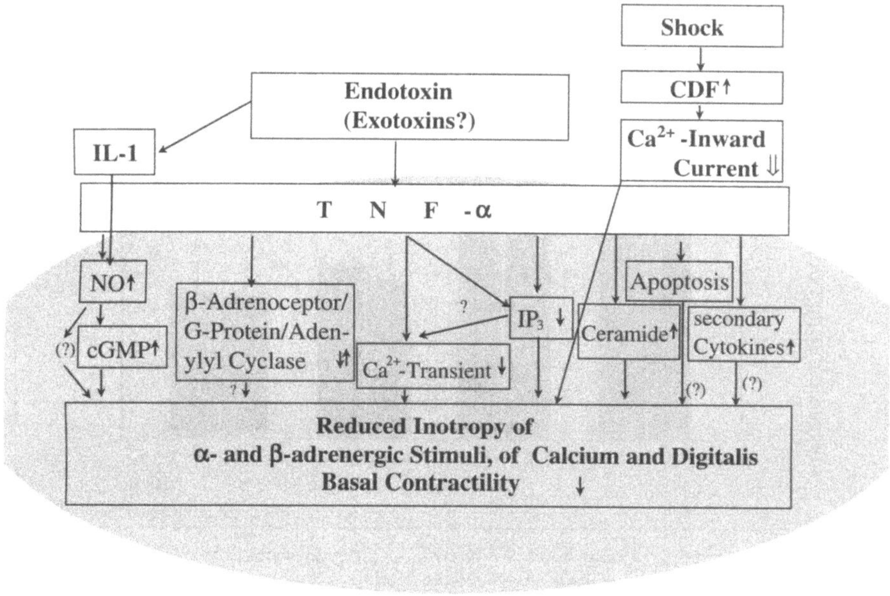

Figure 7. Cardiodepression in sepsis—experimentally supported concepts. The mechanisms of contractile compromise of the agents given in the figure have been elucidated in in vivo or in vitro experimental settings. IL-1 = Interleukin-1; IFN-γ = Interferon-γ; TNF-α = Tumor necrosis factor α; CDF = cardiodepressant factor; NO = nitric oxide; cGMP = cyclic guanosine monophosphate; β-AR = β-adrenoceptor; G Prot = guanine nucleotide-binding protein; AC = adenylate cyclase; IP_3 = inositol triphosphate. Modified from (4,5).

radicals (41) are well-known final effectors in the scenario of SIRS cardiodepression. Endogenous peroxynitrite formation from NO and superoxide may be the most toxic compound—both with respect to contractility and respiration (see below)—in the concert of free radicals in systemic inflammation (42,43). The detection of an abundant expression of the recently discovered cyclooxygenase-2 in the myocardium in sepsis (44) insinuates a possible importance of prostanoids as final mediators of septic heart failure.

High levels of nitrite and nitrate, the stable end products of NO metabolism, are found in patients with severe sepsis, and relate inversely to systemic vascular resistance indices and global oxygen extraction rates, but directly to blood lactate levels (45). It is well-established that the reduced afterload in sepsis is determined by NO; inhibitors of NOS unmask a tonic pressor response of the counterregulatory endothelin-1 (46), the blood levels of which are elevated in human sepsis (45,46). The role of NO for septic myocardial depression is, however, less lucid. Striking similarities between the cardiac modifications in liver cirrhosis, which are related to NO (47) in an animal model, and sepsis have been emphasized.

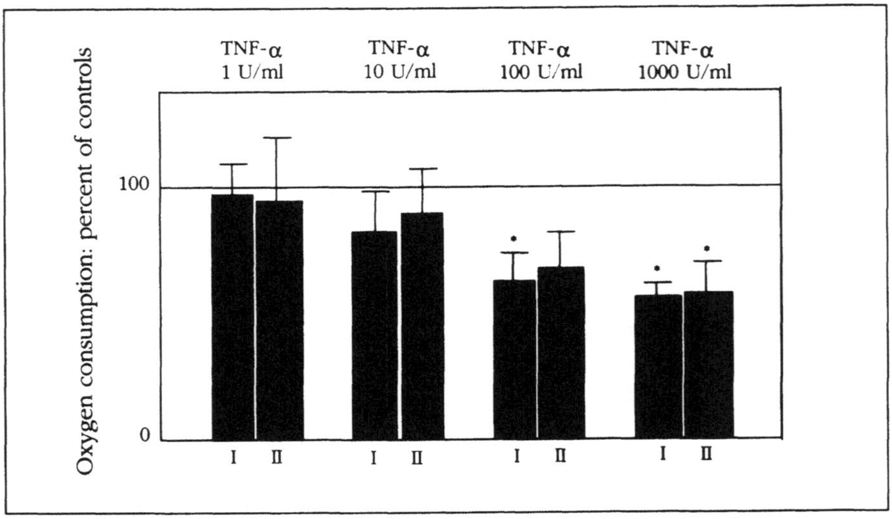

Figure 8. O_2 consumption of mitochondria isolated from cardiomyocytes: Effect of a 24-Hour-Preincubation of the Cells with TNf-α. Spontaneously beating neonatal rat cardiomyocytes were cultured for 24 hours in serum-free medium in the absence or in the presence of non-cytotoxic concentrations of tumor necrosis factor-α (TNF-α). Thereafter, mitochondria were isolated, and oxygen consumption due to complex I and complex II of the respiratory chain were measured. TNF-α concentration-dependant decreases oxygen consumption due to complex I as well as due to complex II respiration. For further details and for methods see (29).

A pertinent intracellular target for NO with respect to contractile properties is guanylate cylase. While a low increase in cyclic guanosyl monophosphate (cGMP) by low doses of NO was reported to improve contractile response of myocytes (48), its activation by excessive amounts of NO results in negative inotropy, mainly due to an interference with compounds of the β-adrenoceptor pathway (49). Additional cGMP-related cardiac NO effects include the regulation of coronary flow (50), which however is maintained in eNOS deficient mice (51) suggesting important compensatory mechanisms, and an enhanced cardiac adenosine production (52). However, cellular targets of NO other than guanylate cyclase may be pathogenically relevant, particularly an interaction of NO with energy metabolism was found to be related with contractile dysfunction (53). Consistently, in the hearts of baboons with experimentally induced E. coli sepsis, diminished activities of the complexes I and II of the mitochondrial respiratory chain were found (Table 2; (54) which might be due to detrimental effects of sepsis mediators like NO (53), TNF-α (Figure 8), Interleukin-1 (29) and others. Interestingly, in patients with sepsis, skeletal partial oxygen pressure is not lowered but raised in the very severe state of sepsis (55)),

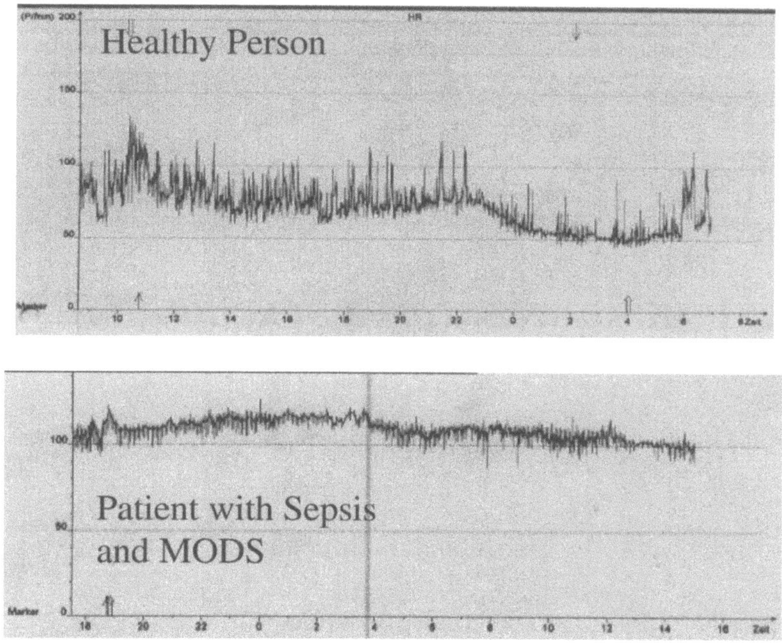

Figure 9. 24-h-Heart Rate-Monitoring in a healthy person and in a patient with sepsis and multiple organ dysfunction syndrome (MODS). Continuous 24 h-monitoring of heart rate clearly demonstrates a narrowing of the heart rate variability in the patient with sepsis and MODS in comparison to the heart rate variability in the healthy person. For further details see (68).

being well in agreement with a toxin- and mediator-induced impairment of mitochondrial function leading to a disturbance of cellular oxygen utilisation.

Several studies have undermined a possible role of myocytes' NOS in non-septic human heart failure (most recently: (56,57)), while another investigation confirmed the expression of constitutive and inducible NOS in failing human cardiac myocytes but attributed inflammatory myocyte injury rather to abundant iNOS expression within infiltrating macrophages (58). A recent paper questions an essential role of NO for contractile depression or β-adrenoceptor desensitisation in myocytes from failing human ventricle, as pharmacological inhibition of NOS did not increase contraction or β-adrenoceptor sensitivity (59). During severe sepsis in humans, both β-adrenergic receptor-dependent and -independent stimulation of adenylate cyclase were shown to be impaired, suggesting heterologous desensitisation (60). Evidence is at hand for an interplay of NO with sympathetic and parasympathetic signaling in the heart (49,61), but data from eNOS deficient mice refute the idea that eNOS is obligatory for the normal autonomic control of cardiac muscle function (62). In summary, a very complex picture of the contribution of NO and NOS to septic

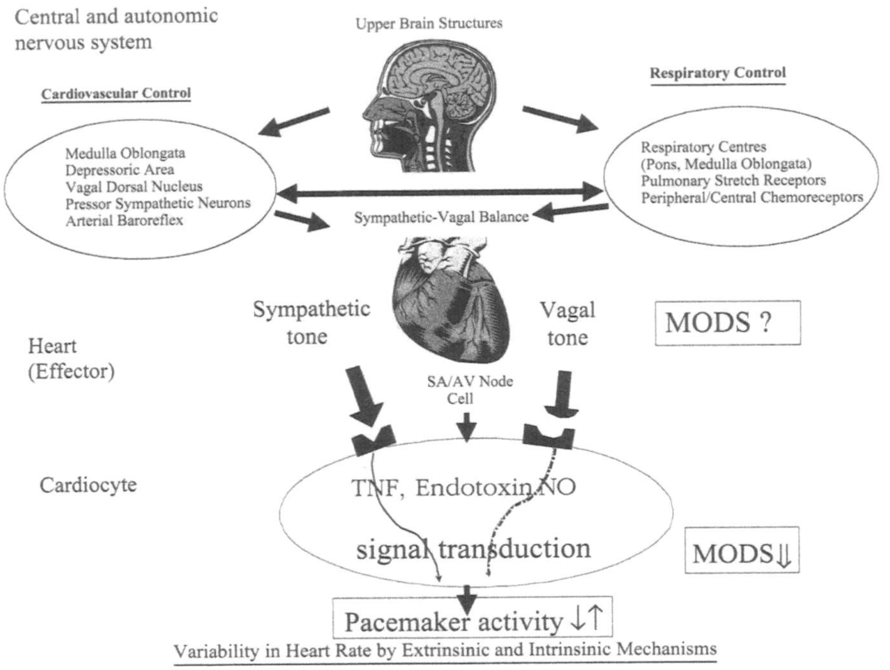

Figure 10. Assessment of extrinsic and intrinsic mechanisms underlying the depressed heart rate variability in patients with multiple organ dysfunction syndrome (MODS). The sympathetic-vagal balance modulating the heart rate at the level of the sinuatrial node is the result of several reflex mechanisms. In addition to the probable alterations of the sympathteic and vagal tone in MODS, toxin- and mediator-induced impairments at the target cells (pacemaker cell) of adrenergic and muscarinergic receptors, intracellular signal transduction pathways and the pacemaker ionic channels attenuate the nervous signals leading to the observed depression of both, the sympathetic as well as the vagal activity in MODS. For further details see (68).

and non-septic heart failure is evolving and some more pieces of the puzzle are expected to emerge in the future.

The complex clinical spectrum of septic cardiomyopathy: Myocardial depression is not all!

Arrhythmias

In contrast to the well-documented myocardial depression in sepsis, only few experimental and clinical studies on the occurrence of arrhythmias in sepsis have been reported. A case of the coincidence of long QT syndrome and torsade de pointes with septic cardiomyopathy was reported (63). In a retrospective analysis of score evaluated patients, no increased incidence of ventricular and supraventricular arrhythmias was documented in 25 septic MODS patients compared to 15 non-

Table 2. Septic cardiomyopathy in E. coli sepsis in baboons—myocardial respiratory chain and glycolysis enzymes

Respiratory chain/Enzymes	Controls (n = 3)	*E. coli* Sepsis: Survivors (n = 26)	*E. coli* Sepsis: Non-Survivors (n = 20)
Complex I/III	5.1 ± 4.1	3.4 ± 2.9	0.6 ± 0.6*
Complex II/III	9.7 ± 2.9	4.4 ± 4.6	1.3 ± 2.4*
Complex III	153 ± 57	140 ± 114	90 ± 56
Cytochrome-c-oxidase	77 ± 11	61 ± 42	39 ± 29*
Citrate synthase	424 ± 81	584 ± 451	507 ± 307
Succinate dehydrogenase	25 ± 6.4	22 ± 16	15 ± 13
Phosphofructokinase	288 ± 168	312 ± 175	47 ± 32*
Glucosephosphate-isomerase	1768 ± 214	1628 ± 494	1082 ± 480
Lactate dehydrogenase	1165 ± 501	960 ± 312	709 ± 28
Noncollagen protein	164 ± 26	159 ± 60	138 ± 43

Dimensions: Units/gram noncollagen protein; *$p < 0.05$.
Data from (54).

septic MODS patients (64). At present no specific arrhythmogenic sepsis profile has been identified.

On the other hand, in postoperative ICU patients, sepsis and SIRS have been identified as main triggers of tachyarrhythmias, with an adjusted odd's ratio of 36,48 (65)!

Autonomic dysfunction

A narrowed heart rate variability essentially belongs to critical illness (Figure 8) and was found to be a marker of poor prognosis—associated with a severalfold increased mortality (66). A reduction in heart rate variability is a measure of autonomic dysfunction, reflecting a loss of the balance between sympathetic and vagal tone (67). Sepsis and multiple organ dysfunction syndrome (MODS) in particular are coupled with an impairment of autonomic dysfunction (68), a finding that culminated in the hypothesis of a loss of the neuro-humoral organ interaction ("uncoupling of biological oscillators") as major pathogenic factor in MODS (69–71). Interestingly, the administration of endotoxin to healthy humans reversibly induces narrowed heart rate variability (72). A study investigating heart rate variability in septic versus non-septic MODS could demonstrate that MODS more than sepis is the major determinant of reduced heart rate variability (73).

Not only heart rate variability is reduced in patients with MODS, but also baroreflex sensitivity and chemoreflex sensitivity are impaired, all dysfunctions correlating with an unfavourable prognosie (68). The correlation between the severity of autonomic dysfunction and severity of septic cardiomyopathy, however, remains to be determined.

An interesting difference of the quality of autonomic dysfunction in MODS patients in comparison to cardiac patients is evident: while in cardiac patients the sympathetic activity dominates over the vagal tone, in MODS patients both the sympathetic as well as the vagal tone are attenuated (Figure 10). This lead us to the hypothesis that in MODS patients not only an impairment of the autonomous nervous system is evident but also an attenuation of the sympathetic as well as of the vagal signal in the target cell, in case of heart rate variability narrowing at the level of the pacemaker cell in the sinus node of the heart (68). This attenuation of the nervous signals might be due either to a toxin- and/or mediator-induced alteration of the intracellular signal transduction pathways or to the pacemaker ion channel activity itself. In agreement with this hypothesis, preincubation of spontaneously beating neonatal rat cardiomyocytes in culture with endotoxin results in a narrowing of the beating rate variability of these pacemaker cells (74).

SIRS cardiomyopathy

While compromise of heart function due to bacterial sepsis has been thoroughly studied, few groups have clearly documented the presence of SIRS cardiomyopathy from non-infectious disease. Severe trauma goes along with an abnormal left ventricular function (75). Prolonged hemorrhagic shock resulted in contractile impairment in an animal model (76). Extracorporeal circulation unequivocally induces an inflammatory response. It has by now been well documented by several groups that the impact of the cardio-pulmonary bypass on cytokine release is very drastic (77) and relates to myocardial dysfunction (77,78). An escalating SIRS as opposed to a non-escalating SIRS in patients after cardiopulmonary bypass-assisted cardiac surgery is associated with an increased mortality, with exaggerated TNF-α and TNF receptor blood levels and lowered left ventricular stroke work indices (79), the latter not being due to different afterloads, but rather an intrinsic myocardial depression, namely SIRS cardiomyopathy (5).

Causal therapeutic approaches towards the treatment of acute septic cardiomyopathy

Current consensus guidelines to the therapy of sepsis give detailed recommendations for fluid resuscitation and vasopressor and inotropic therapy (80,81). NO inhalation therapy does not seem to have depressant effects on heart function (82,83). A modulation of cardiac inflammatory reactions by phosphodiesterase inhibitors (84) or adenosine (85) was reported. Causal therapeutic strategies—aiming at an interruption of mediator cascades—to alleviate septic cardiomyopathy and vasculopathy are being tested, but do not form part of the standard regimens. Table 3 summarizes data, which convey the notion that by antagonising and eliminating pertinent proinflammatory mediators septic vasculopathy is better approachable than septic cardiomyopathy (hemofiltration (86–88), hydrocortisone (89)). Still, none of the listed agents is as effective as noradrenaline in raising systemic vascular resistance. The options for a causal therapy of septic cardiovascular damage presently at hand thus have to be rated as very limited.

Table 3. Treatment of acute septic cardiomyopathy and vasculopathy—causal therapeutic approaches

	CI	LVSWI	RVSWI	SVR	PVR
Endotoxin antibody (Centoxin®)	±	±/↓		±	
TNF-α antibody	±	±/↑		±/↑	
Hemofiltration	±	±	±	↑	±
Plasmapheresis	±	±		±	
PAF antagonist	↑	↑		↑ (MAP)	↓ (PAP)
Hydrocortisone	±	±		↑	
NOS inhibitors	±/↓	±/↓/↑ (transiently)	±	↑	↑
Methyleneblue	±	↑		↑	↑
Pentoxifylline	±	±	±	±	±
Hemoperfusion with endotoxin adsorption	↓	±	±	↑	↑
In comparison:					
Noradrenaline	↑	↑	↑	↑	±

CI cardiac index; **LVSWI** left ventricular stroke work index; **RVSWI** right ventricular stroke work index; **SVR** systemic vascular resistsnace; **PVR** pulmonary vascular resistance; **MAP** mean arterial pressure; **PAP** mean pulmonary arterial pressure.
Modified from (4,5), substituted by (89).

Synopsis

The original concept of septic myocardial depression proposing a negative inotropic blood factor has undergone an evolution in recent years. Heart failure in severe sepsis and SIRS nowadays is regarded as a symptom of a secondary cardiomyopathy, characterised by an altered cellular phenotype due to the impact of multiple mediators and toxins. Considerable progress has been made in characterising the clinical spectrum—including autonomic dysfunction—and in elucidating the pathogenetic mechanisms underlying septic and SIRS cardiomyopathy. However trials to establish causal therapies for the disease have so far not yielded convincing results and necessitate continuing efforts in both basic and clinical research. The pathogenic mechanisms underlying septic heart failure may also be operative in non-septic heart failure, which goes along with a measurable systemic inflammatory response.

REFERENCES

1. Romberg E. 1921. Die septische akute Myokarditis ("The septic acute myocarditis"). In: Lehrbuch der Krankheiten des Herzens und der Blutgefäße ("Textbook of diseases of the heart and veseels"). Ed. E Romberg, p. 494. Stuttgart, Germany: Enke-Verlag.
2. Mueller-Hoecker J, Haerty W. 1986. Pathomorphological aspects of the hearts in septic patients. In: Pathophysiology of Shock, Sepsis, and Organ Failure. Ed. G Schlag and H Redl, 18–54. Berlin, Heidelberg, New York, Oxford: Elsevier.
3. Parrillo JE. 1989. The cardiovascular pathophysiology of sepsis. Ann Rev Med 40:469–485.
4. Mueller-Werdan U, Werdan K. 1999. Septic cardiomyopathy. Current Opinion in Critical Care 5:415–420.
5. Mueller-Werdan U, Reithmann C, Werdan K (Eds.). 1996. Cytokines and the heart—Molecular mechanisms of septic cardiomyopathy. Landes Company, Austin, USA/Springer Heidelberg New York Berlin.

6. Parker MM. 1998. Pathophysiology of cardiovascular dysfunction in septic shock. New Horiz 6:130–138.
7. Grocott-Mason RM, Shah AM. 1998. Cardiac dysfunction in sepsis: new theories and clinical implications. Intensive Care Med 24:286–295.
8. Suffredini AF. 1998. Myocardial dysfunction in sepsis: clinical and experimental investigations. Schweiz Med Wochenschr 128:1444–1452.
9. Werdan K. 1998. Activated immune system in congestive heart failure—From dropsy to the cytokine paradigm. J Int Med 243:87–92.
10 Müller-Werdan U, Pfeifer A, Hübner G, Seliger C, Reithmann C, Rupp H, Werdan K. 1997. Partial inhibition of protein synthesis by Pseudomonas exotoxin A deranges catecholamine sensitivity of cultured rat heart myocytes. J Mol Cell Cardiol 29:799–811.
11. Guest TM, Ramanathan AV, Tuteur PG, Schechtman KB, Ladenson JH, Jaffe AS. 1995. Myocardial injury in critically ill patients—A frequently unrecognized complication. JAMA 273:1945–1949.
12. Spies C, Haude V, Fitzneer R, Schröder K, Overbeck M, Runkel N, Schaffartzik W. 1998. Serum cardiac troponin T as a prognostic marker in early sepsis. Chest 113:1055–1063.
13. Kuhn C, Mueller-Werdan U, Schmitt DV, Lange H, Pilz G, Kreuzer E, Mohr FW, Zerkowski H-R, Werdan K. 2000. Improved outcome of APACHE II score-defined escalating systemic inflammatory response syndrome in patients post cardiac surgery in 1996 compared to 1988–1999: the ESSICS-study pilot project. Eur J Cardio-thoracic Surg 17:30–37.
14. Briassoulis G, Narlioglou M, Zavras N, Hatzis T. 2001. Myocardial injury in meningococcus-induced purpura fulminans in children. Intensive Care Med 27:1073–1082.
15. Cunnion RE, Schaer GL, Parker MM, et al. 1986. The coronary circulation in human septic shock. Circulation 73:637–644.
16. Dhainaut JF, Hughebaert M-F, Monsallier JF, et al. 1987. Coronary hemodynamics and myocardial metabolism of lactate, free fatty acids, glucose, and ketones in patients with septic shock. Circulation 75:533–541.
17. Wu AHB. 2001. Increased troponin in patients with sepsis and septic shock: myocardial necrosis or reversible myocardial depression? Intensive Care Med 27:959–961.
18. Pilz G, McGinn P, Boekstegers P, Kääb S, Weidenhöfer S, Werdan K. 1994. Pseudomonas sepsis does not cause more severe cardiovascular dysfunction in patients than Non-Pseudomonas sepsis. Circ Shock 42:174–182.
19. Dhainaut J-F, Pinsky MR, Nouria S, Slomka F, Brunet F. 1997. Right ventricular function in human sepsis—A thermodilution study. Chest 112:1043–1049.
20. Poortmans G, Poelaert J. 1999. Transesophageal echocardiographic evaluation of left ventricular function. In: Yearbook of Intensive Care and Emergency Medicine 1999. Ed. Vincent J-L, 468–481. Heidelberg New York Berlin: Springer.
21. Porembka DT. 1998. Use of transesophageal and transthoracic echocardiography for monitoring and diagnosis of critical illness. Curr Opin Crit Care 4:195–207.
22. Hoffmann JN, Werdan K, Hartl WH, et al. 1999. Hemofiltrate from patients with severe sepsis and depressed left ventricular contractility contains cardiotoxic compounds. Shock 12:174–180.
23. Kumar A, Thota V, Dee L, Olson J, Uretz E, Parrillo JE. 1996. Tumor necrosis factor α and interleukin 1β are responsible for in vitro myocardial cell depression induced by human septic shock serum. J Exp Med 183:949.
24. Thoenes M, Förstermann U, Tracey WR, Bleese NM, Nüssler AK, Scholz H, Stein B. 1996. Expression of inducible nitric oxide synthase in failing and non-failing human heart. J Mol Cell Cardiol 28:165–169.
25. Müller-Werdan U, Schumann H, Fuchs R, Reithmann C, Loppnow H, Koch S, Zimny-Arndt U, Chang He, Darmer D, Jungblut P, Stadler J, Holtz J, Werdan K. 1997. Tumor necrosis factor α (TNFα) is cardiodepressant in pathophysiologically relevant concentrations without inducing inducible nitric oxide-(NO)-synthase (iNOS) or triggering serious cytotoxicity. J Mol Cell Cardiol 29:2915–2923.
26. Müller-Werdan U, Engelmann H, Werdan K. 1998. Cardiodepression by tumor necrosis factor α. Eur Cytokine Netw 9:689–691.
27. Oral H, Dorn GW, Mann DL. 1997. Sphingosine mediates the immediate negative inotropic effects of tumor necrosis factor-α in the adult mammalian cardiac myocyte. J Biol Chem 272:4836.
28. Reithmann C, Werdan K. 1994. Tumor necrosis factor α decreases inositol phosphate formation and phosphatidylinositol-bisphosphate (PIP_2) synthesis in rat cardiomyocytes. Naunyn-Schmiedeberg's Arch Pharmacol 349:175–182.

29. Zell R, Geck P, Werdan K, Boekstegers P. 1997. TNF-α and IL-1β inhibit both pyruvate dehydrogenase activity and mitochondrial function in cardiomyocytes: evidence for primary impairment of mitochondrial function. Mol Cell Biochem 177:61–67.
30. Müller-Werdan U, Schumann H, Loppnow H, Fuchs R, Darmer D, Stadler J, Holtz J, Werdan K. 1998. Endotoxin and tumor necrosis factor α exert a similar proinflammatory effect in neonatal rat cardiomyocytes, but have different cardiodepressant profiles. J Mol Cell Cardiol 30:1027–1036.
31. Comstock KL, Krown KA, Page MT, Martin D, Ho P, Pedraza M, Castro EN, Nakajima N, Glembotsik CC, Quintana PJE, Sabbadini RA. 1998. LPS-induced TNF-α release from and apoptosis in rat cardiomyocytes: obligatory role for CD14 in mediating the LPS response. J Mol Cell Cardiol 30:2761–2775.
32. Krown KA, Page MT, Nguyen C, Zechner D, Gutierrez V, Comstock KL, Glembotski CC, Quintana PJE, Sabbadini RA. 1996. Tumor necrosis factor alpha-induced apoptosis in cardiac myocytes. J Clin Invest 98:2854.
33. Natoli G, Costanzo A, Guido F, Moretti F, Levrero M. 1998. Apoptotic, non-apoptotic, and anti-apoptotic pathways of tumor necrosis factor signalling. Biochem Pharmacol 56:915–920.
34. Bozkurt B, Kribbs SB, Clubb FJ Jr, Michael LH, Didenko VV, Hornsby PJ, Seta Y, Oral H, Spinale FG, Mann DL. 1998. Pathophysiologically relevant concentrations of tumor necrosis factor-α promote progressive left ventricular dysfunction and remodelling in rats. Circulation 97:1382–1391.
35. Kubota T, McTiernan CF, Frye CS, Demetris AJ, Feldman AM. 1997. Cardiac-specific over-expression of tumor necrosis factor-α causes lethal myocarditis in transgenic mice. J Card Fail 3:117–124.
36. Kubota T, McTiernan CF, Frye CS, Slawson SE, Lemster BH, Koretsky AP, Demetris AJ, Feldman AM. 1997. Dilated cardiomyopathy in transgenic mice with cardiac-specific overexpression of tumor necrosis factor-α. Circ Res 81:627–635.
37. Bryant D, Becker L, Richardson J, Shelton J, Franco F, Franco F, Peshock R, Thompson M, Giroir B. 1998. Cardiac failure in transgenic mice with myocardial expression of tumor necrosis factor-α. Circulation 97:1375–1381.
38. Nakano M, Knowlton AA, Dibbs Z, Mann DL. 1998. Tumor necrosis factor-α confers resistance to hypoxic injury in the adult mammalian cardiac myocyte. Circulation 97:1392–1400.
39. Loppnow H, Werdan K, Reuter G, Flad H-D. 1998. The interleukin-1 and interleukin-1 converting enzyme families in the cardiovascular system. Eur Cytokine Netw 9:675–680.
40. Kelly RA, Balligand J-L, Smith TW. 1996. Nitric oxide and cardiac function. Circ Res 79:363–380.
41. Singal PK, Khaper N, Palace V, Kumar D. 1998. The role of oxidative stress in the genesis of heart disease. Cardiovascular Research 40:426–432.
42. Xie Y-W, Kaminski PM, Wolin MS. 1998. Inhibition of rat cardiac muscle contraction and mitochondrial respiration by endogenous peroxynitrie formation during posthypoxic reoxygenation. Circ Res 82:891–897.
43. Szabo C. 1996. The pathophysiological role of peroxynitrite in shock, inflammation, and ischemia-reperfusion injury. Shock 6:79–88.
44. Wong SCY, Fukuchi M, Melnyk P, Rodger I, Giaid A. 1998. Induction of cyclooxygenase-2 and activation of nuclear factor-κB in myocardium of patients with congestive heart failure. Circulation 98:100–103.
45. Groeneveld ABJ, Hartemink KJ, de Groot MCM, Visser J, Thijs LG. 1999. Circulating endothelin and nitrate-nitrite relate to hemodynamic and metabolic variables in human septic shock. Shock 11:160–166.
46. Avontuur JAM, Boomsma F, van den Meiracker AH, de Jong FH, Bruining HA. 1999. Endothelin-1 and blood pressure after inhibition of nitric oxide synthesis in human septic shock. Circulation 99:271–275.
47. Van Obbergh L, Vallieres Y, Blaise G. 1996. Cardiac modifications occurring in the ascitic rat with biliary cirrhosis are nitric oxide related. J Hepatol 24:747–752.
48. Kojda G, Kottenberg K, Nix P, Schluter KD, Piper HM, Noack E. 1996. Low increase in cGMP induced by organic nitrates and nitrovasodilators improves contractile response of rat ventricular myocytes. Circ Res 78:91–101.
49. Balligand J-L. 1998. Molecular mechanisms of action of nitric oxide. In: Yearbook of Intensive Care and Emergency Medicine 1998. Ed. J-L Vincent, 107–124. Heidelberg New York Berlin: Springer.
50. Kostic MM, Petronijevic MR, Jakovljevic VLJ. 1996. Role of nitric oxide (NO) in the regulation of coronary circulation. Physiol Res 45:273–278.
51. Gödecke A, Decking UKM, Ding Z, Hirchenhain J, Bidmon H-J, Gödecke S, Schrader J. 1998. Coronary hemodynamics in endothelial NO synthase knockout mice. Circ Res 82:186–194.

52. Obata T, Sato T, Yamanaka Y, Arita M. 1998. NO and cGMP facilitate adenosine production in rat hearts via activation of ecto-5′-nucleotidase. Pflügers Arch—Eur J Physiol 436:984–990.
53. Kelm M, Schäfer S, Dahmann R, Dolu B, Perings S, Decking UKM, Schrader J, Strauer B. 1997. Nitric oxide induced contractile dysfunction is related to a reduction in myocardial energy generation. Cardiovascular Research 36:185–194.
54. Gellerich FN, Trumbeckaite S, Hertel K, Zierz S, Müller-Werdan U, Werdan K, Redl H, Schlag G. 1999. Impaired energy metabolism in hearts of septic baboons: diminished activities of complex I and complex II of the mitochondrial respiratory chain. Shock 11:336–341.
55. Boekstegers P, Weidenhöfer S, Kapsner T, Werdan K. 1994. Skeletal muscle partial pressure of oxygen in patients with sepsis. Crit Care Med 22:640–650.
56. Drexler H, Kastner S, Strobel A, Studer R, Brodde OE, Hasenfuss G. 1998. Expression, activity and functional significance of inducible nitric oxide synthase in the failing human heart. J Am Coll Cardiol 32:955–963.
57. Paulus WJ. 1999. Myocardial inducible nitric oxide synthase and left ventricular performance in the human heart. In: Yearbook of Intensive Care and Emergency Medicine 1999. Ed. J-L Vincent, 497–503. Heidelberg New York Berlin: Springer.
58. Fukuchi M, Hussain SNA, Giaid A. 1998. Heterogenous expression and activity of endothelial and inducible nitric oxide synthases in end-stage human heart failure—Their relation to lesion site and β-adrenergic receptor therapy. Circulation 98:132–139.
59. Harding SE, Davies CH, Money-Kyrle AM, Poole-Wilson PA. 1998. An inhibitor of nitric oxide synthase does not increase contraction or β-adrenoceptor sensitivity of ventricular myocytes from failing human heart. Cardiovascular Research 40:523–529.
60. Bernardin G, Strosberg, Bernard A, Mattei M, Marullo S. 1998. β-adrenergic receptor-dependent and -independent stimulation of adenylate cyclase is impaired during severe sepsis in humans. Intensive Care Med 24:1315–1322.
61. Scrogin KE, Hatton DC, Chi Y, Luft FC. 1998. Chronic nitric oxide inhibition with L-NAME: effects on autonomic control of the cardiovascular system. Am J Physiol 274:R367–R374.
62. Vandecastelle G, Eschenhagen T, Scholz H, Stein B, Verde I, Fischmeister R. 1999. Muscarinic and β-adrenergic regulation of heart rate, force of contraction and calcium current is preserved in mice lacking endothelial nitric oxide synthase. Nature Medicine 5:331–334.
63. Varriale P, Ramaprasad S. 19995. Septic Cardiomyopathy as a Cause of Long QT Syndrome. J Electrocardiology 28:327–329.
64. Prondzinsky R, Stache N, Witthaut R, Winkler M, Fraunberger P, Walli AK, Seidel D, Werdan K. 1997. Multiorgan-failure (MOF) with and without sepsis: differences in incidence and pattern of detected arrhythmias. Crit Care 1 (Suppl 1): P30.
65. Knotzer H, Mayra, Ulmer H, Lederer W, Schobersberger W, Mutz N, Hasibeder W. 2000. Tachyarrhythmias in a Surgical Intensive Care Unit: A case-controlled epidemiologic study. Intensive Care Med 26:908–914.
66. Winchell RJ, Hoyt DB. 1996. Spectral analysis of heart rate variability in the ICU. J Surg Res 63: 11–16.
67. Stein PK, Kleiger RE. 1999. Insights from the study of heart rate variability. Annu Rev Med 50: 249–261.
68. Schmidt H, Werdan K, Mueller-Werdan U. 2001. Autonomic dysfunction in the ICU patient. Current Opinion in Critical Care 7:314–322.
69. Godin PJ, Buchman TG. 1996. Uncoupling of biological oscillators. A complementary hypothesis concerning the pathogenesis of multiple organ dysfunction syndrome. Crit Care Med 24:1107–1116.
70. Seely AJE, Christou NV. 2000. Multiple organ dysfunction syndrome: exploring the paradigm of complex nonlinear systems. Crit Care Med 28:2193–2200.
71. Marshal JC. 2000. Complexity, chaos, and incomprehensibility: parsing the biology of critical illness. Crit Care Med 28:2646–2648.
72. Godin PJ, Fleisher LA, Eidsath A, Vandivier RW, Preas HL, Banks SM, Buchman TG, Suffredini AF. 1996. Experimental human endotoxemia increases cardiac regularity: Results from a prospective, randomized, crossover trial. Crit Care Med 24:1117–1124.
73. Heinroth KM, Kuhn C, Stache N, Witthaut R, Müller-Werdan U, Werdan K, Prondzinsky R. 1999. Reduced heart rate variability in patients with septic and non-septic multiple organ dysfunction syndrome. Intensivmed 36:••.
74. Schmidt HB, Mueller-Werdan U, Saworski J, Werdan K. 2000. Can LPS of TNF-α narrow the beating rate variabiltiy of contracting cardiomyocytes? Shock 13:79.

75. Smail N, Descorps Declere A, Duranteau J, Vigue B, Samii K. 1996. Left ventricular function after severe trauma. Intensive Care Med 22:439–442.
76. McDonough KH, Giaimo M, Quinn M, Miller H. 1999. Intrinsic myocardial function in hemorrhagic shock. Shock 11:205–210.
77. Menasché P. 1995. The inflammatory response to cardiopulmonary bypass and its impact on postoperative myocardial function. Current Opinion in Cardiology 10:597–604.
78. Hennein HA, Ebba H, Rodriguez JL, Merrick SH, Keith FM, Bronstein MH, Leung JM, Mangano DT, Greenfield LJ, Rankin JS. 1994. Relationship of the proinflammatory cytokines to myocardial ischemia and dysfunction after uncomplicated coronary revascularization. J Thorac Cardiovasc Surg 108:626–635.
79. Pilz G, Kääb S, Kreuzer E, Werdan K. 1994. Evaluation of definitions and parameters for sepsis assessment in patients after cardiac surgery. Infection 22:8–17.
80. Task Force of the American College of Critical Care Medicine, Society of Critical Care Medicine. 1999. Practice parameters for hemodynamic support of sepsis in adult patients in sepsis. Crit Care Med 27:639–660.
81. Marik PE, Varon J. 1998. The hemodynamic derangements in sepsis—Implications for treatment strategies. Chest 114:854–860.
82. Hayward CS, Kalnins WV, Rogers P, Feneley MP, MacDonald PS, Kelly RP. 1997. Effect of inhaled nitric oxide on normal human left ventricular function. J Am Coll Cardiol 30:49–56.
83. Argenziano M, Dean DA, Moazami N, Goldstein DJ, Rose EA, Spotnitz HM, Burkhoff D, Oz MC, Dickstein ML. 1998. Inhaled nitric oxide is not a myocardial depressant in a porcine model of heart failure. J Thorac Cardiovasc Surg 115:700–708.
84. Takeuchi K, del Nido PJ, Ibrahim AE, Cao Danh H, Friehs I, Glynn P, Poutias D, Cowan DB, McGowan FX. 1999. Vesnarinone and amrinone reduce the systemic inflammatory response syndrome. J Thorac Cardiovasc Surg 117:375–381.
85. Wagner DR, McTiernan C, Sanders VJ, Feldman AM. 1998. Adenosine inhibits lipopolysaccharide-induced secretion of tumor necrosis factor-α in the failing human heart. Circulation 97:521–524.
86. Dhainaut J-F A, Vinsonneau C, Journois D. 1999. Hemofiltration and left ventricular function in sepsis: Mechanisms and clinical implications. Crit Care Med 27:473–474.
87. Hoffmann JN, Hartl WH, Deppisch R, Faist E, Jochum M, Inthorn D. 1996. Effect of hemofiltration on hemodynamics and systemic concentrations of anaphylatoxins and cytokines in human sepsis. Intensive Care Med 22:1360–1367.
88. Iwama H, Komatsu T. 1998. Effect of an endotoxin-removing column containing immobilized polymyxin B fiber in a patient with septic shock from Gram-positive infection. Acta Anaesthesiol Scand 42:590–593.
89. Briegel J, Forst H, Haller M, Schelling G, Kilger E, Kuprat G, Hemmer B, Hummel T, Lenhart A, Heyduck ••, Stoll C, Peter K. 1999. Stress doses of hydrocortisone reverse hyperdynamic septic shock: A prospective, randomized, double-blind, single center study. Crit Care Med ••.

Pathophysiology of Cardiovascular Disease,
edited by N.S. Dhalla, H. Rupp,
A. Angel & G.N. Pierce
Kluwer Academic Publishers, Boston, 2004

Mitochondrial Function—A Limiting Factor in Heart Failure?

Anne-Marie L. Seymour

Department of Biological Sciences, University of Hull, Hull UK

Summary. Mitochondria are the major sources of energy for contraction in the heart, generating ATP from oxidative processes. The metabolic adaptations that occur in cardiac hypertrophy cause a switch in substrate provision from fatty acid to glucose oxidation. These modifications, together with depletion in critical cofactors and altered gene expression, may jeopardise mitochondrial function, resulting in apoptosis and cell loss. The progressive nature of these adaptations may underlie the transition from hypertrophy to heart failure.

Key words: Mitochondria, Magnetic Resonance Spectroscopy, Metabolic Remodelling, Cardiac Hypertrophy, Heart Failure.

INTRODUCTION

The heart has an enormous requirement for oxygen in order to generate sufficient energy to carry out its function to pump blood around the body (1). This is reflected in the large number of mitochondria within heart cells together with a high density of contractile elements (myofibrils). These significant intracellular structures make up greater than 80% of the myocardial cell volume. The mitochondria can be considered the powerhouses within the cell, generating energy in the form of ATP from the oxidation of nutrients. The enzymes of the tricarboxylic acid (TCA) cycle are contained within the mitochondrial matrix and produce NADH and $FADH_2$ for oxidation by the electron transport chain. This series of complexes is situated in the inner mitochondrial membrane and is tightly coupled to the ATP synthesising unit, the F_0F_1-ATPase. The ATP synthesised within the mitochondria is then

Dr Anne-Marie L. Seymour, Department of Biological Sciences, University of Hull, Cottingham Road, Hull HU6 7RX UK. Tel: 44-1482-465517, Fax: 44-1482-465458, E-mail: a.m.seymour@hull.ac.uk.

transported via the adenine nucleotide translocase and the mitochondrial isoform of creatine kinase, into the cytosol for contraction, biosynthesis and maintenance of ionic gradients.

As the heart has a continuous requirement for energy, the balance between energy production and energy utilisation has to be carefully regulated (2). This is achieved through a number of different mechanisms, each playing a role at different stages of maturation. In the neonatal heart, the primary means of control of energy production is through the availability of ADP (3), whereas, in the mature heart, this is achieved by provision of sufficient reducing equivalents via calcium activated dehydrogenases (4). Overall, the amount of ATP remains constant in the heart but its rate of turnover can change dramatically in response to altered workload. Acute increases in workload can modulate key enzyme activities and fluxes through metabolic pathways, whereas chronic alterations can lead to changes in phenotype. The mitochondria thus play a critical role in ensuring that the heart continues to function and it is expected that any dysfunction within this organelle will be a major component in the progression towards heart failure.

This chapter highlights the application of magnetic resonance spectroscopy as a non-invasive probe of myocardial oxidative metabolism. Investigations of adaptation in fuel selection and oxidation in cardiac hypertrophy are described, demonstrating the progressive nature of such adaptations. It is suggested that these metabolic consequences of hypertrophy can become the causes of deterioration into heart failure, jeopardising mitochondrial function and resulting in loss of myocytes.

MAGNETIC RESONANCE SPECTROSCOPY

Over the past 20 years, the powerful technique of magnetic resonance spectroscopy has become an increasingly important tool with which to investigate cardiac energy metabolism in situ. Phosphorus-31 MRS has provided a unique non-invasive and non-destructive way to visualise cardiac energetics and mitochondrial function (5–7). From the relative simple spectrum (Figure 1), the major energy components can be identified—phosphocreatine, the energy back-up or reserve, the three phosphates of ATP, the energy currency in the cell, and inorganic phosphate, from which the intracellular pH can be determined. This type of spectrum whether obtained from a perfused heart in vitro or an in vivo heart provides a gold standard with which alterations that occur in diseased states can be monitored.

More recently, the application of ^{13}C MRS techniques has allowed re-examination of intracellular metabolism in situ, determining fluxes through key metabolic pathways simultaneously (8). Carbon-13 is a stable isotope with a natural abundance of only 1.1% and is thus ideally suited as an isotopic tracer. In spectral terms, carbon-13 has a wide chemical shift range or "window" which allows each labelled carbon atom within a metabolite to be resolved. The major strength of this technique is its ability to determine, without separation and isolation, the enrichment and concentration of a wide range of metabolic intermediates (9). This method represents a substantial advance over radioisotope techniques, which focus on the measurement of end products and thus limit the number of pathways that can be investigated.

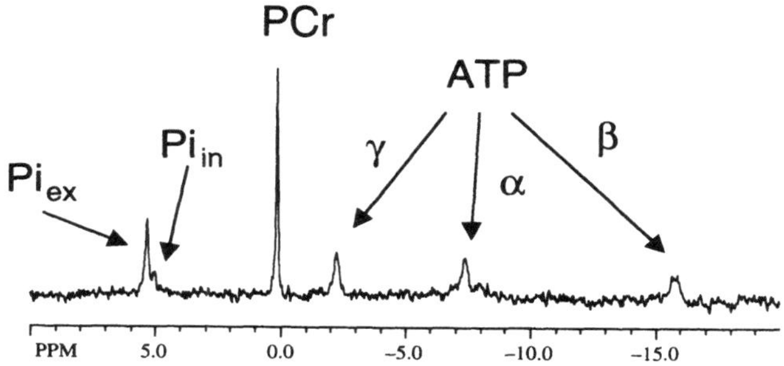

Figure 1. ^{31}Phosphorus magnetic resonance spectrum of a perfused rat heart, showing the major energy containing metabolites: phosphocreatine (PCr), the three phosphates of ATP (α, β and γ) and inorganic phosphate (both intracellular Pi_{in} and extracellular Pi_{ex}).

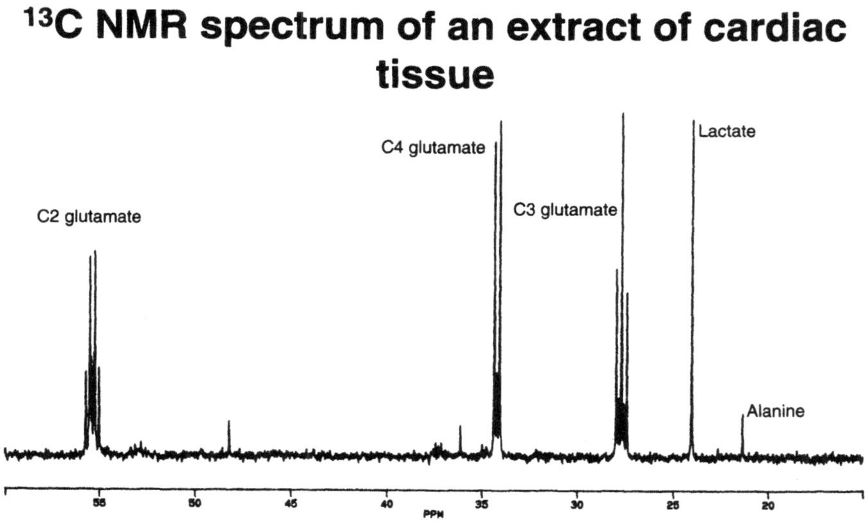

Figure 2. ^{13}C MR spectrum of a heart with 5 mM 1-^{13}C labelled glucose, highlighting the ^{13}C labelling of the carbon 2, 3 and 4 positions of glutamate.

One potential limitation, however, is the insensitivity of the ^{13}C technique, which means that only intermediates, present in sufficient concentration, can be detected by MRS. Although the majority of TCA cycle intermediates are present at concentrations below the level of detection, fortuitously, 2-oxoglutarate is in exchange with glutamate—a component present in sufficient amounts (Figure 2). Thus metabolic fluxes in the heart can be assessed from the ^{13}C spectral pattern of glutamate (10).

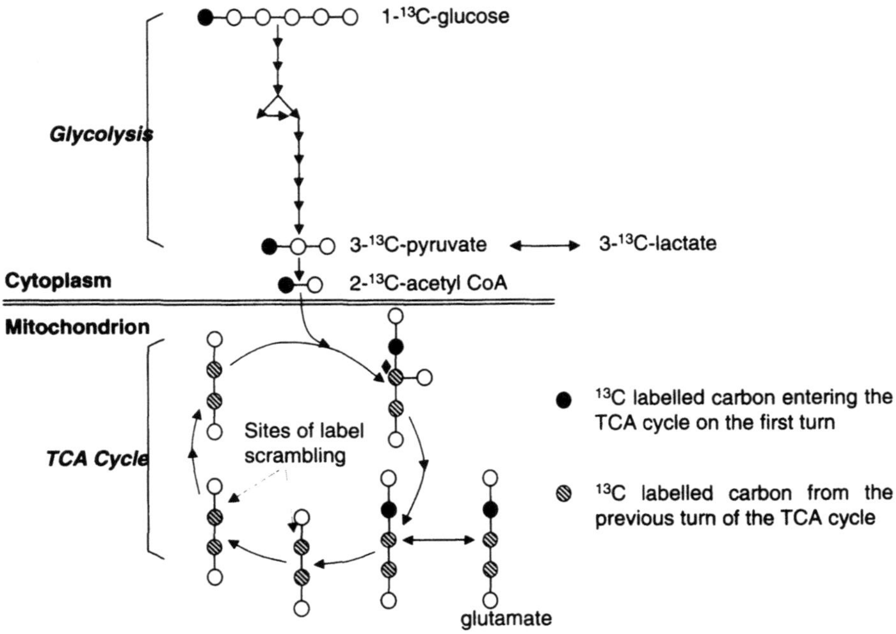

Figure 3. Schematic representation of the sequential labelling of TCA cycle intermediates and glutamate, following administration of 1-^{13}C glucose.

Administration of ^{13}C enriched substrates, such as 1-^{13}C labelled glucose or uniformly labelled (U-^{13}C) palmitate, provide a direct way to investigate the contributions of different substrates to oxidative pathways. Analysis of the ^{13}C isotopomer patterns arising from glutamate yields the relative contribution of both ^{13}C labelled and unlabelled (i.e. ^{12}C) substrates to oxidative pathways and, together with a measure of oxygen consumption, can determine fluxes through metabolic pathways. For example when 1-^{13}C glucose is oxidised by the heart, the ^{13}C label will be metabolised via glycolysis, resulting in labelling of the pyruvate pool in the carbon 2 position and, subsequently, in the carbon 2 position of acetyl CoA. As this intermediate enters the TCA cycle, in the first turn of the cycle, the label will be transferred to the carbon 4 position of citrate and subsequently to the carbon 4 position of 2-oxoglutarate and glutamate. As the ^{13}C-labelled oxoglutarate is further metabolised within the TCA cycle, the label will be randomised between the carbon 2 and carbon 3 positions (Figure 3). Thus on the second turn of the cycle, ^{13}C label will enter from acetyl CoA again and from the carbon 2 or 3 position of oxaloacetate. Label will then build up in the different positions of glutamate, giving rise to multiplets within the individual ^{13}C resonances (Figure 4). With time, the ratio of different multiplets will alter, corresponding to the number of turns through the

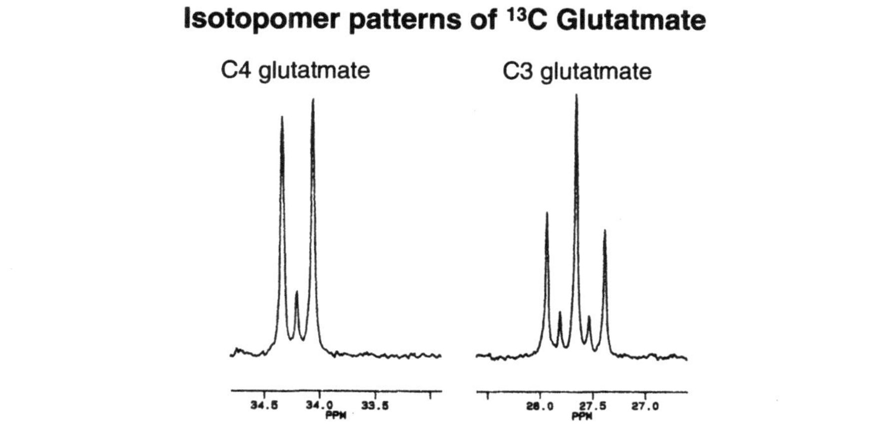

Figure 4. ^{13}C isotopomer patterns in glutatmate carbon 3 and 4 positions, following administration of 1-^{13}C glucose.

TCA cycle. This approach has been developed by Jeffrey and colleagues (10) and has been used to address questions of substrate competition in the experimental setting in the normal heart (11,12) and more recently in the hypertrophied (13) and the diabetic heart (14).

CARDIAC HYPERTROPHY

Cardiac hypertrophy is recognised as one of the important independent risk factor in heart failure (15). This enlargement or "remodelling" is the adaptive response of the heart to chronic pressure or volume overload and occurs at the cellular and organ level, The trigger of stretch or neurohumoral stimulation induces expression of proto-oncogenes within the myocyte and in turn alterations in gene transcription. In consequence, protein expression changes both at a quantitative and qualitative level, resulting in modulation of enzyme activities, increased or decreased expression of protein and isoform switching (16). Many of the changes are characteristic of the foetal heart (17), leading to the suggestion that hypertrophy leads to a re-expression of the foetal phenotype. More recent evidence has shown that a combined response occurs with a specific down regulation of isoforms expressed in the adult phenotype, as well (18). Overall, the hypertrophied heart does not merely represent an enlarged normal heart but a separate hypertrophied phenotype.

In addition to the increase in myocyte size, there are a number of structural alterations, which contribute to modifications in metabolism and function of the hypertrophied heart. Inter-capillary distance is augmented as a result of hypertrophy, whereas capillary density is reduced giving rise to an overall increased diffusion distance from vessels to myocytes (19). Although protein synthesis is enhanced, there is an imbalance between mitochondrial proliferation and the increase in

Table 1. Metabolic consequences in cardiac hypertrophy

• Decreased energy reserve
• Enhanced glucose utilisation
• Reduced fatty acid oxidation
• Depletion of essential cofactors

myofibrillar mass. Thus many of the consequences of the onset of hypertrophy—including metabolic factors (Table 1), may themselves become important elements in the transition from compensated hypertrophy to decompensated heart failure.

METABOLIC ADAPTATION IN CARDIAC HYPERTROPHY

Many studies have shown that cardiac hypertrophy is associated with impaired contractile function and an enhanced susceptibility to ischaemic injury (20–22). The mechanisms underlying these effects are not clearly understood. One important hypothesis proposes that the hypertrophied heart is in an energy-deprived state resulting from an imbalance between energy production and utilisation (23,24). As substrate oxidation (energy provision) is tightly coupled to cardiac work (energy utilisation), changes in the profile of myocardial energy substrate preference in the hypertrophied heart will play a crucial role in the development of pathophysiological hypertrophy and heart failure.

We and others have investigated myocardial energy metabolism in animal models of cardiac hypertrophy and failure (7,25–27) for a number of years. Interestingly, we always observe a significant decrease in the concentration of myocardial phosphocreatine, the energy reserve of the heart, irrespective of the model under examination. At the same time, we have demonstrated that cardiac function is markedly impaired. In parallel with our studies, other groups have observed similar alterations in phosphocreatine in experimental models of hypertrophy (27,28) and, more recently, in studies of patients with heart failure (29–31). Overall these results highlight the fact that the hypertrophied failing heart is in an energy-deprived state. We postulate that this situation arises as a consequence of inadequate provision of substrate to the mitochondrion.

It is clear, from experimental studies, that the hypertrophied myocardium is markedly different from the normal heart in both its biochemistry and function (32). In the normal heart, energy is derived predominantly from fatty acid and lactate oxidation with a small but consistent use of glucose (1,33). The exact profile of substrate selection, however, will depend upon a number of factors including oxygen supply, workload and hormonal influences. Indeed, experimental studies have shown that substrate limitation can lead to impaired cardiac function in the normal heart. Taegtmeyer and co-workers (34) demonstrated that the use of the ketone body, acetoacetate, as the sole substrate, results in deterioration of contractile function in the isolated perfused rat heart. Inclusion of anaplerotic substrates such as lactate overcomes this contractile impairment. Taegtmeyer has postulated that sequestration of co-enzyme A (CoA) by acetoactetate inhibits one of the critical rate limiting steps of the tricarboxylic acid cycle (35), which, in turn, is responsible for the dysfunction of the heart.

Over recent years, the profile of substrate utilisation and the relative contributions of different metabolic pathways towards the provision of ATP in the failing heart have proved to be an explosive area of research (see 36 for review). Early studies identified that glycolytic capacity was enhanced in cardiac hypertrophy (25). Subsequently we demonstrated that glucose transport was increased in experimental cardiac hypertrophy (37) and Zhang and colleagues observed similar findings in vivo (38). These data are consistent with the proposal that myocardial substrate selection switches from fatty acid oxidation towards glucose metabolism in hypertrophy (39–41), a more efficient process in terms of the amount of ATP derived per atom of oxygen consumed (42). More recently, using carbon-13 MRS, we have investigated the profile of substrate provision in a model of compensated cardiac hypertrophy (13). Under conditions where function is normal, we still observe a greater oxidative use of glucose over the control heart, but with a similar contribution from fatty acids (Figure 5). These observations imply that the reliance on glucose

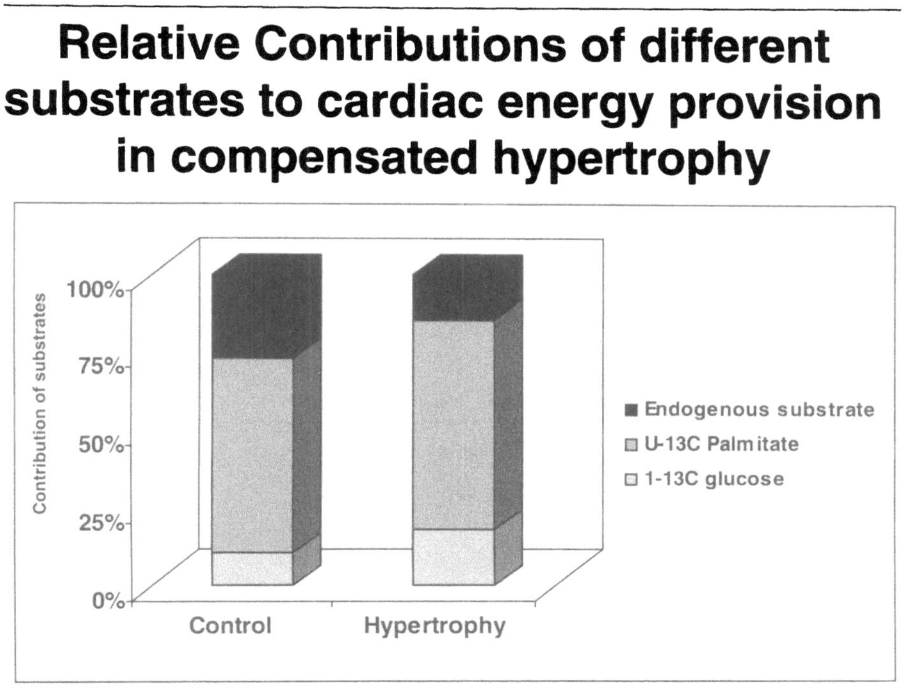

Figure 5. The relative contributions of glucose (light grey), palmitate (dark grey) and endogenous substrate (black) oxidation in control and hypertrophied rat hearts.

precedes other metabolic changes and may reflect a generalised protective response (43) rather than a true re-expression of the foetal phenotype.

In parallel with investigations on glucose metabolism, other workers have shown that there is a marked reduction in oxidation of long chain fatty acids such as palmitate in the hypertrophied myocardium (44,45). This alteration occurs in part as a result of a decrease in the myocardial carnitine content (46,47). Carnitine is an essential component for the transport of activated long chain fatty acids (acyl CoA) from the cytosol to the mitochondrion, where β-oxidation and subsequent ATP production occurs. Any reduction in this cofactor will limit entry of fatty acids into the mitochondria and their subsequent oxidation. In addition, the accumulation of fatty acyl components will tend to sequester any available CoA and thus reduce the amount of free CoA available, which may, in turn, affect the rate of TCA cycle turnover and contractile function (35). Alterations in the amount of available CoA will also have consequences for the oxidation of glucose (48). A decrease in available CoA will reduce the acetyl CoA to CoA ratio, an important factor in the regulation of pyruvate dehydrogenase (PDH), the key enzymatic step in the oxidative use of glucose. In consequence, despite an increased reliance on glucose (or, more generally, carbohydrate metabolism), the hypertrophied heart may have a limited capacity to use both glucose and fatty acids (49) and becomes energy depleted and unable to function.

These metabolic adaptations can build up progressively to render individual myocytes susceptible to injury (50). Under such conditions, the coupling between oxidation and ATP synthesis may be jeopardised, with the mitochondria becoming de-energised, resulting in a loss of membrane potential. At this point, we can speculate that a vicious cycle starts, with deterioration in the integrated functioning of the mitochondria. Alterations in membrane potential (which may result from formation of the membrane permeability transition, 51) will allow loss of cytochrome c, which then can act as an activator of caspase 3 and the apoptotic cascade (52). Individual myocytes would then be lost through apoptosis (53), causing the remaining cardiac cells to bear a greater workload. These cells may hypertrophy further

Table 2. Progression to Heart Failure

Adaptation to hypertrophic response
↓
Re-expression of fetal phenotype
↓
Changes in energy provision
↓
Limited delivery of substrate to mitochondria
↓
Mismatch between supply and demand
↓
Deterioration in mitochondrial function
↓
Apoptosis with hypertrophy of remaining cells

and, in turn, through impaired provision of substrate for oxidation, suffer the same fate and undergo apoptosis (Table 2). Ultimately, the contractile performance of the heart will deteriorate into failure. There is increasing evidence for mitochondrial-induced apoptosis in heart failure and one of the potential mechanisms for this is through abnormal oxidative metabolism.

CONCLUSION

With an increased understanding of the metabolic alterations that occur in cardiac hypertrophy, it is clear that these modifications can endanger mitochondrial function and potentially lead to heart failure. Focus on metabolic therapies (54,55) which slow or prevent some of these modifications is now timely and offers hope for future treatment of heart failure.

ACKNOWLEDGEMENTS

The author would like to thank the British Heart Foundation for financial support and Dr John Chatham, Dr David Ashford and Dr John Greenman for constructive and stimulating discussions.

REFERENCES

1. Taegtmeyer H. 1994. Energy metabolism of the heart: from basic concepts to clinical applications. Curr Probl Cardiol 19:62–113.
2. Harmsen E, Seymour A-ML. 1988. The importance of the determination of ATP and catabolites. In "Myocardial Energy Metabolism" Eds de Jong JW. Publ: Martinus Nijhoff Dordrecht, Netherlands.
3. Balaban RS, Heinmann F. 1992. In "The Heart and the Cardiovascular System" Eds Fozzard H et al. Publ.
4. McCormack JG, Denton RM. 1989. Influence of calcium ions on mammalian intramitochondrial dehydrogenases. Meth Enzymol 174:95–118.
5. Chatham JC, Blackband SJ. 2001. NMR spectroscopy and imaging in animal research, IlAR Journal 42:189–208.
6. Rajagopalan B, Blackledge MJ, McKenna WJ, Bolas N, Radda GK. 1987. Measurement of phosphocreatine to ATP ratio in normal and diseased human heart by ^{31}P magnetic resonance spectroscopy using the rotating frame-depth selection technique. Ann NY Acad Sci 508:321–332.
7. Raine AEG, Seymour A-ML, Roberts AFC, Radda GK, Ledingham JGG. 1993. Impairment of cardiac function and energetics in experimental renal failure. J Clin Invest 92:2934–2940.
8. Shulman RG, Rothman DL. 2001. ^{13}C NMR of intermediary metabolism: Implications for Systemic Physiology. Annu Rev Physiol 63:15–48.
9. Chatham JC, Forder J, Glickson JD, Chance EM. 1995. Calculation of absolute flux and the elucidation of pathways of glutamate labelling in perfused rat hearts by ^{13}C NMR spectroscopy. J Biol Chem 270:7999–8008.
10. Malloy CR, Sherry AD, Jeffrey FMH. 1990. Analysis of tricarboxylic acid cycle of the heart using ^{13}C isotope isomers. Am J Physiol 259:H987–H995.
11. Jeffrey FM, Dickzu V, Sherry AD, Malloy CR. 1995. Substrate selection in the isolated working rat heart. Basic Res Cardiol 90:38–396.
12. Weiss RG, Chacko VP, Gerstenblith G. 1989. Fatty acid regulation of glucose metabolism in the intact beating rat heart assessed by carbon-13 NMR spectroscopy J Mol Cell Cardiol 21:469–478.
13. Clarke SJ. 2001. PhD Thesis, University of Hull.
14. Chatham JC, Seymour A-ML. 2002. Changes in cardiac metabolism precede development of contractile function in Type 2 diabetes. Cardiovas Res 55:104–112.
15. Braunwald E, Bristow M. 2000. Congestive heart failure—fifty years of progress Circulation 102(suppl IV):14–23.

16. Boheler KR, Schwartz K. 1992. Gene expression in cardiac hypertrophy. Trends in Cardiovasc Med 2:172–178.
17. Depre C, Shipley Gl, Chen W, Han Q, Doenst T, Moore ML, Stepkowski S, Davies PJA, Taegtmeyer H. 1998. Unloaded heart in vivo replicates foetal gene expression of cardiac hypertrophy. Nature Med 4:1269–1275.
18. Rosenblatt-Velin N, Montessuit C, Papageoriou I, Terrand J, Lerch R. 2001. Postinfarction heart failure is associated with an upregulation of GLUT 1 and downregulation of genes of fatty acid metabolism. Cardiovasc Res 52:407–416.
19. Anveras P, Capasso JM. 1991. Loss of intermediary sized coronary arteries and capillary proliferation after left ventricular failure in rats. Am J Physiol 260:H1552–H2560.
20. Gaasch WH, Zie MR, Hoshino PK, Weinberg EO, Rhodes DR, Apstein CS. 1990. Tolerance of the hypertrophied heart to ischaemia. Circulation 81:1644–1653.
21. Schonekess BO, Allard MF, Lopaschuk GD. 1996. Recovery of glycolysis and oxidative metabolism during postischemic reperfusion of hypertrophied hearts. Am J Physiol 271:H798–H805.
22. Wambolt RB, Henning SL, English DR, Dyachkova Y, Lopaschuk GD, Allard M. 1999. Glucose utilisation and glycogen turnover are accelerated in hypertrophied hearts during severe low-flow ischemia. J Mol Cell Cardiol 31:493–502.
23. Ingwall JS. 1993. Is cardiac failure a consequence of decreased energy reserve? Circulation 87(sVII): 58–62.
24. Katz AM. 2001. Heart Failure in 2001—a prophecy revisited. Am J Cardiol 87:1383–1389.
25. Seymour A-ML, Eldar H, Radda GK. 1990. Hyperthyroidism results in increased glycolytic capacity in the rat heart—a 31-P NMR study. Biochim Biophys Acta 1055:107–116.
26. Smolenski RT, Jayakumar J, Seymour A-ML, Yacoub MH. 1998. Energy metabolism and mechanical recovery after cardioplegia in moderately hypertrophied rats. Molec Cell Biochem 180: 137–143.
27. Neubauer S, Horn M, Naumann A, Tian R, Hu K, Laser M, Friedrich J, Gaudron P, Schnackerz K, Ingwall JS, Ertl G. 1995. Impairment of energy metabolism in intact residual myocardium of rat hearts with chronic myocardial infarction. J Clin Invest 95:1092–1100.
28. Tian R, Ingwall JS. 1996. Energetic basis for reduced contractile reserve in isolated hearts. Am J Physiol 270:H1207–H1216.
29. Conway MA, Allis J, Ouwerkerk R, Niioka T, Rajagopalan B, Radda GK. 1991. Detection of low phosphocreatine/ATP ratio in failing hyprtrophied human myocardium by P-31 MRS. Lancet 338: 973–976.
30. Neubauer S, Krahe T, Schindler R, Horn M, Hillenbrand H, Enzeroth C, Mader H, Kromer E, Riegger G, Lackner K, Ertl G. 1992. ^{31}P MRS in dilated cardiomyopathy and coronary heart disease. Circulation 86:1810—1818.
31. Bottomley PA. 1989. Human in vivo NMR spectroscopy in diagnostic medicine: clinical tool or research probe? Radiol 170:1–15.
32. Swynghedauw B. 1999. Molecular mechanisms of myocardial remodelling. Physiol Rev 79:215–262.
33. Lopaschuk GD, Belke DD, Gamble J, Ito T, Schonekess BO. 1994. Regulation of fatty acid oxidation in the mammalian heart in health and disease. Biochim Biophys Acta 1213:263–276.
34. Russell RR, Taegtmeyer H. 1991. Changes in citric acid cycle flux and anaplerosis antedate the functional decline in isolated rat hearts utilising acetoacetate. J Clin Invest 87:384–390.
35. Russell RR, Taegtmeyer H. 1992. Coenzyme A sequestration in rat hearts oxidising ketone bodies. J Clin Invest 89:968–973.
36. Stanley WC. (ed) 2002. Special Issue on Myocardial energy metabolism in heart failure. Heart Failure Revs 7:113—219.
37. Paternostro G, Clarke K, Heath J, Seymour A-M, Radda GK. 1995. Decreased GLUT-4 mRNA content and insulin-sensitive deoxyglucose uptake show insulin resistance in the hypertensive rat heart. Cardiovasc Res 30:205–211.
38. Zhang J, Duncker DJ, Ya X, Zhang Y, Pavek T, Wei H, Merkle H, Ugurbil K, From AHL, Bache RJ. 1995. Effect of left ventricular hypertrophy secondary to chronic pressure overload on transmural myocardial 2-deoxyglucose uptake—A ^{31}P NMR spectroscopic study. Circulation 92:1274–1283.
39. Allard MF, Schonekess BO, Henning SL, English DR, Lopaschuk GD. 1994. Contribution of oxidative metabolism and glycolysis to ATP production in hypertrophied hearts. Am J Physiol 267:H742–H750.
40. Young ME, Laws FA, Goodwin G, Taegtmeyer H. 2001. Reactivation of PPARα is associated with contractile dysfunction in the hypertrophied rat heart. J Biol Chem 276:44390–44395.

41. Davila-Roman VG, Vedala G, Herrero P, de las Fuentes L, Rogers JG, Kelly DP, Gropler RJ. 2002. Altered myocardial fatty acid and glucose metabolism in idiopathic dilated cardiomyopathy. J Am Coll Cardiol 40:271–277.
42. Collins-Nakai RL, Noseworthy D, Lopaschuk GD. 1994. Epinephrine increases ATP production in hearts by preferentially increasing glucose metabolism. Am J Physiol 267:H1862–H1871.
43. Dispersyn GD, Borgers M. 2001. Apoptosis in the heart: about programmed cell death and survival. News Physiol Sci 16:41–47.
44. El Alaoui-Talibi Z, Landormy S, Loireau A, Moravec J. 1992. Fatty acid oxidation and mechanical performance of volume-overloaded rat hearts. Am J Physiol 262:H1068–H1074.
45. Ben Cheikh R, Guendouz A, Moravec J. 1994. Control of oxidative metabolism in volume overloaded rat hearts: effects of different substrates. Am J Physiol 266:H2090–H2097.
46. Reibel DK, Uboh CE, Kent RL. 1983. Altered coenzyme A and carnitine metabolism in pressure-overloaded hypertrophied hearts. Am J Physiol 244:H839–H843.
47. El Alaoui-Talibi Z, Guendouz A, Moravec M, Moravec J. 1997. Control of oxidative metabolism in volume-overloaded rat hearts. Am J Physiol 272:H1615–H1624.
48. Timmons JA, Poucher SM, Constantin-Teodosiu D, Worrall V, MacDonald I, Greenhaff PL. 1996. Increased acetyl groups availability enhances contractile function of canine skeltal muscle during ischemia. J Clin Invest 97:879–883.
49. Seymour A-ML, Chatham JC. 1997. The effects of hypertrophy and diabetes on cardiac pyruvate dehydrogenase activity. J Mol Cell Cardiol 29:2771–2778.
50. Lesnefsky EJ, Moghaddas S, Tandler B, Kerner J, Hoppel CL. 2001. Mitochondrial dysfunction in cardiac disease: Ischemia-reperfusion, Aging and Heart failure. J Mol Cell Cardiol 33:1065–1089.
51. Halestrap A. 2000. Mitochondria and cell death. Biochemist 22(2):19–24.
52. Anversa P, Leri A, Kajstura J, Nadal-Ginard B. 2002. Myocyte growth and cardiac repair. J Mol Cell Cardiol 34:91–105.
53. Green DR, Reed JC. 1998. Mitochondria and apoptosis. Science 281:1309–1316.
54. Kantor PF, Lucien A, Kozak R, Lopaschuk GD. 2000. Anti-anginal drug, trimetazidine, shifts cardiac energy metabolism from fatty acid oxidation to glucose oxidation. Circ Res 86:580–588.
55. Stanley WC, Hoppel CL. 2000. Mitochondrial dysfunction in heart failure: potential for therapeutic interventions? Cardiovasc Res 45:805–806.

Pathophysiology of Cardiovascular Disease,
edited by N.S. Dhalla, H. Rupp,
A. Angel & G.N. Pierce
Kluwer Academic Publishers, Boston, 2004

Myosin Heavy Chain Isoforms Modulate Motor Function of Cardiac Myosin by Changing Crossbridge Kinetics

Hiroshi Yamashita,[1] Seiryo Sugiura,[1] Hideo Fujita,[1] Soichiro Yasuda,[1] Ryozo Nagai,[1] Yasutake Saeki,[2] and Haruo Sugi[3]

[1] *Department of Cardiovascular Medicine, Graduate School of Medicine, University of Tokyo, 7-3-1 Hongo, Bunkyo-ku, Tokyo 113-8655, Japan*
[2] *Department of Physiology, School of Dental Medicine, Tsurumi University, 2-1-3 Tsurumi, Tsurumi-ku, Yokohama 230-8501, Japan*
[3] *Department of Physiology, School of Medicine, Teikyo University, 2-11-1 Kaga, Itabashi-ku, Tokyo 173-8605, Japan*

Summary. In mammalian hearts, there are two myosin heavy chain (MHC) isoforms, fast-type α-MHC and slow-type β-MHC, and the isoform composition changes in response to mechanical load. In human atria and rodent ventricles, isoform composition shifts from α- to β-MHC under pressure overload. Since α-MHC is related with higher ATPase activity and higher velocity of muscle shortening compared to β-MHC, the redistribution towards β-MHC could be adaptive response to increased mechanical load for better economy. To dissect the molecular mechanism underlying the higher economy of β-MHC, we purified V1 (homodimer of α-MHC) and V3 (homodimer of β-MHC) myosins from rat ventricles and characterized the functional differences in vitro. Both Ca^{2+}-activated ATPase activity and actin filament velocity (VEL) in the in vitro motility assay were twice higher in V1 compared to V3. Ensemble force generated by several myosin molecules was measured with a laser trap. Isometric force per unit length of actin filament, an index of average force of several myosin molecules and time-averaged force of an individual molecule (F_{ave}), was similar between V1 and V3. Displacement under low load and isometric force generated by a single (or a very small number of) myosin molecule(s) were determined by reducing the number of myosin molecules interacting with actin filaments. Both unitary displacement and unitary force were equal between these myosins. Notably, however, the event duration was significantly longer in V3 under both low load and isometric condition compared to V1.

Corresponding author: Hiroshi Yamashita, M.D., Ph.D., Department of Cardiovascular Medicine, Graduate School of Medicine, University of Tokyo, 7-3-1 Hongo, Bunkyo-ku, Tokyo 113-8655, Japan. Phone: 81-3-3815-5411 (ext.33072), Fax: 81-3-3814-0021, E-mail: hiroyama-tky@umin.ac.jp.

Assuming the two-state model of crossbridge, these results suggest that V3 may have proportionally slower kinetics in both attachment and detachment of the crossbridge, which can explain the similar F_{ave}. Importantly, however, to maintain equal level of isometric force, V3 may hydrolyze less number of ATP molecules and thus have better economy than V1.

Key words: myosin isoform, in vitro motility assay, unitary displacement, unitary force, duty ratio.

INTRODUCTION

Cardiovascular remodeling is structural reconstruction in response to various mechanical or neurohumoral stimuli. The remodeling process includes gross morphological changes in the ventricle such as chamber dilatation and compensatory hypertrophy accompanied by alterations in protein expression in myocardial cells, the molecular remodeling. Increased mechanical stress imposed on cardiac muscles by cardiovascular diseases, such as hypertension, myocardial infarction and valvular diseases, induces myocardial hypertrophy. In earlier stages of hypertrophy, mechanical stress is normalized because of the increased wall thickness and size of individual cardiomyocytes, and cardiac output is maintained. Even in compensated hypertrophy, however, the rate of muscle shortening and relaxation decreases, indicating impaired contractile and diastolic function. These changes are associated with altered protein expression in EC coupling and contractile apparatus, which generally reduces ATP consumption rate and speed of muscle shortening and relaxation. In this respect, these forms of molecular remodeling can be adaptive response to an increased mechanical load for better economy of force generation at the expense of shortening and relaxation speed.

Myosin is a motor protein transforming the chemical energy derived from ATP hydrolysis into mechanical work while cyclically interacting with actin. The cardiac myosin is a hexamer consisting of 2 heavy chains and two pairs of light chains, the essential light chain (LC1 or ELC) and the regulatory light chain (LC2 or RLC). The heavy chain contains sites for actin-binding, ATP hydrolysis and light chain-binding, all of which are essential for catalytic and motor function of the molecule. In mammalian cardiac myosin, there are two types of isoforms in the heavy chain and each of the light chains, and expression of each isoform is regulated in tissue-specific and developmental stage-specific manners. The isoform composition is also known to change in response to mechanical load or various hormonal stimuli.

There are two types of myosin heavy chain isoforms expressed in mammalian hearts, α- and β-heavy chains (MHC), which are encoded by closely-related two genes positioned sequentially on chromosome 14. Initially, the myosin isoforms in the ventricles were separated on pyrophosphate gel electrophoresis into three isoforms, termed V1, V2, and V3 according to the mobility on the native condition (1), which are consisted of αα, αβ, and ββ MHC dimmers, respectively (Table 1). In adult rat ventricles, V1 is the predominant isoform, and the composition shifts from V1 to V3, i.e., α- to β-MHC, under pressure overload. The MHC isoforms also changes in response to thyroid status: the composition redistributes towards V1 in

Table 1. Myosin heavy chain (MHC) isoforms associated with distinct enzymatic and motor function

	V1		V2		V3
MHC isoform	αα		αβ		ββ
ATPase activity		>		>	
Shortening velocity		>		>	
Economy		<		<	
Force		=	?	=	

hyperthyroidism and towards V3 in hypothyroidism. Distinct MHC isoforms are related with differences in both enzymatic and mechanical function of cardiac muscles (Table 1). Cardiac muscle fibers with higher content of V1 shorten faster than those containing predominantly V3, and the maximum shortening velocity (V_{max}) correlated with V1 content of cardiac myosin (2,3). As Barany initially demonstrated in various types of muscles (4), there is a good correlation between Vmax and ATPase activity in cardiac muscles (2,3).

From an energetic aspect, Alpert and Mulieri determined thermomechanical economy (tension dependent heat divided by force-time integral) in hypertrophied rabbit cardiac muscles induced by pressure overload and revealed that hypertrophied muscles with increased content of V3 had higher economy compared to the control subjects (5) (Table 1). They proposed that different crossbridge cycling rates between these isoforms could account for the difference in both mechanical and enzymatic properties, i.e., V1 has faster crossbridge cycling rate and higher ATPase activity and shortening velocity with lower economy of force generation compared to V3 (6). In a thermomechanical study, Hasenfuss et al. estimated average force-time integral per crossbridge cycle and showed that it correlated with V3 content (7). These results supported the idea that the isoform shift from V1 to V3 could be adaptive response to an increased hemodynamic load, in view of increased economy of force generation.

In these studies with muscle preparations, however, confounding effects of EC coupling and cooperative activation process of the thin filament could not be eliminated. To circumvent these problems, we adopted the in vitro assay systems, where actin-myosin interaction was reconstituted from isolated proteins in vitro, and characterized the functional differences between these isoforms. Recent progress in molecular physiology enabled us to measure distance and force in nanometer and piconewton resolution and analyze the motor function of myosin at the molecular level.

MATERIALS AND METHODS

Proteins

V1 and V3 myosins were obtained from the ventricles of 12-week-old male rats and hypothyroid rats treated with 1-methyl-2-mercaptoimidazole in drinking water, respectively (8). Purity of each myosin isoform was confirmed by pyrophosphate

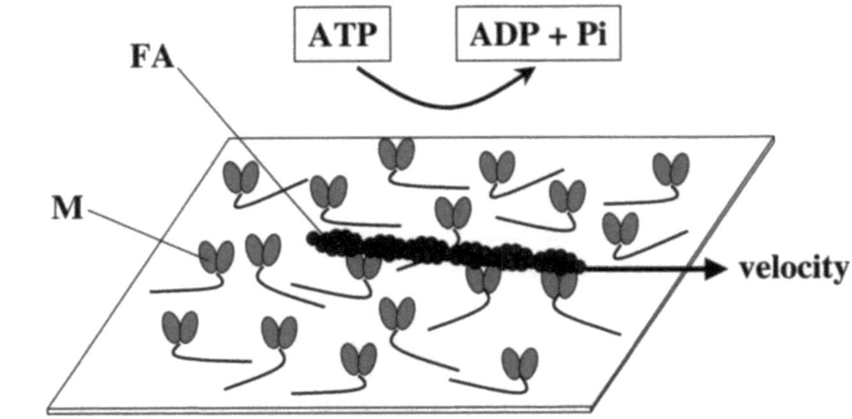

Figure 1A. The in vitro motility assay system. Fluorescently labeled actin filaments (FA) slide over myosin molecules (M) fixed on a glass cover slip. The velocity of sliding movement of the filaments was determined under a fluorescence microscope.

gel electrophoresis (data not shown). Actin and plasma gelsolin were prepared from rabbit back muscle and bovine plasma, respectively, as described (9).

In vitro motility assay system

The sliding actin filament assay was done by the methods of Kron and Spudich (10) (Figure 1A). Briefly, fluorescently labeled actin filaments slid over myosin molecules fixed on a glass coverslip, and the sliding velocity of the filaments was measured under fluorescence microscope. Since the load imposed on the actin filaments was considered to be negligible, the sliding velocity of the filaments indicated the rate of unloaded sliding movement of actin and myosin in vitro and was comparable to the values estimated from unloaded muscle shortening (10).

Measuring ensemble force of crossbridges

To measure force produced by several myosin molecules, we utilized laser trap techniques (optical tweezers) (9). A strongly focused laser beam produced force pulling a small particle towards the high focus of the beam and the force amplitude was proportional to the distance from the focus like a Hookian spring (11–13) (Figure 2A–C). Briefly, a small polystylene bead (1 µm in diameter) was crosslinked with gelsolin, which specifically bound to the barbed end of an actin filament, and a single fluorescent actin filament was bound to the bead via gelsolin (9). The bead was captured by a laser trap and manipulated so that the actin filament could interact with the myosin molecules fixed on a glass surface. Myosin molecules interacted with actin and pulled the filament towards the pointed end. Since the force was proportional to the distance between the bead position and the trap center, the force

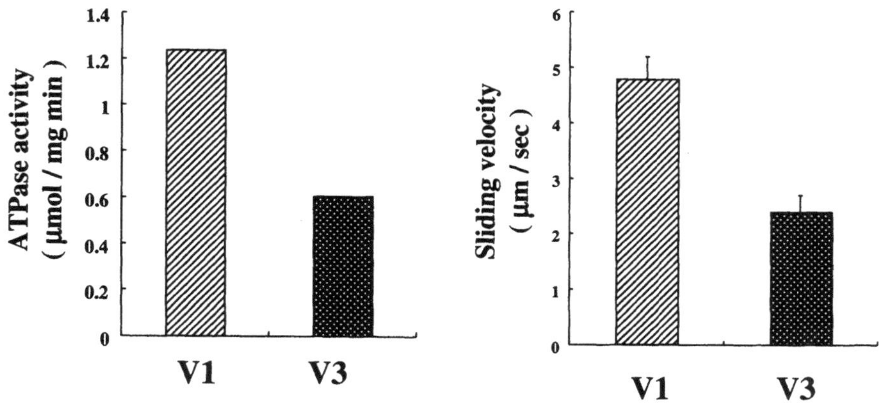

Figure 1B. Ca^{2+}-activated ATPase activity (left) and sliding velocity (right) of V1 and V3. Both ATPase activity and sliding velocity of V1 were approximately twice higher than those of V3 (1.23 vs. 0.60 µmol/mg min, 4.8 vs. 2.4 µm/sec), which showed significant correlation between these parameters.

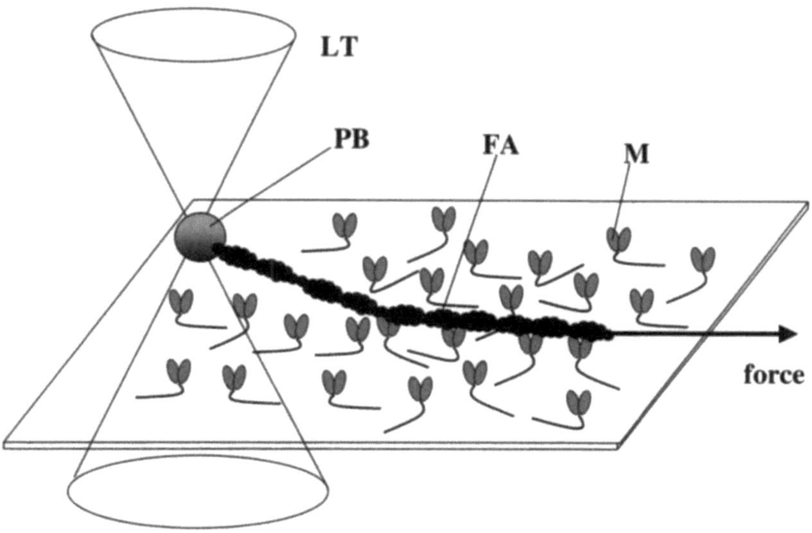

Figure 2A. Measuring ensemble force of several myosin molecules with a laser trap. A small polystylene bead (PB) attached to a single fluorescently labeled actin filament (FA) was manipulated with a laser trap (LT) so that FA could interact with myosin molecules (M) on a glass coverslip.

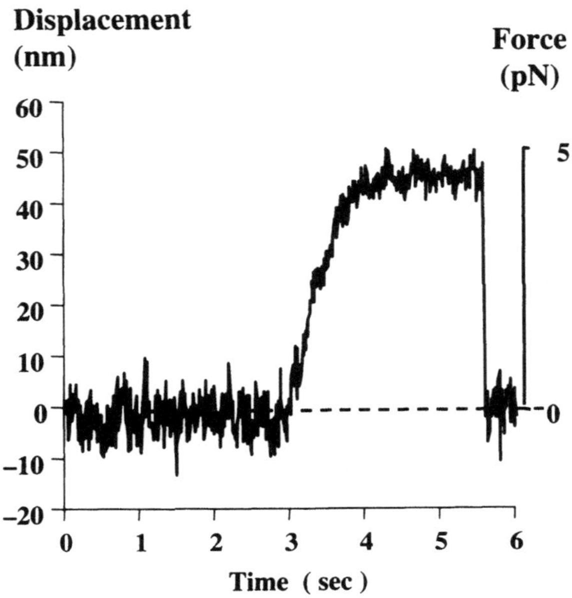

Figure 2B. A record of time course of the bead position. As the actin filament slid over myosin, the bead was pulled away from the trap center and finally stalled, when the trap force was balanced with the isometric force generated by the myosin molecules (stall force).

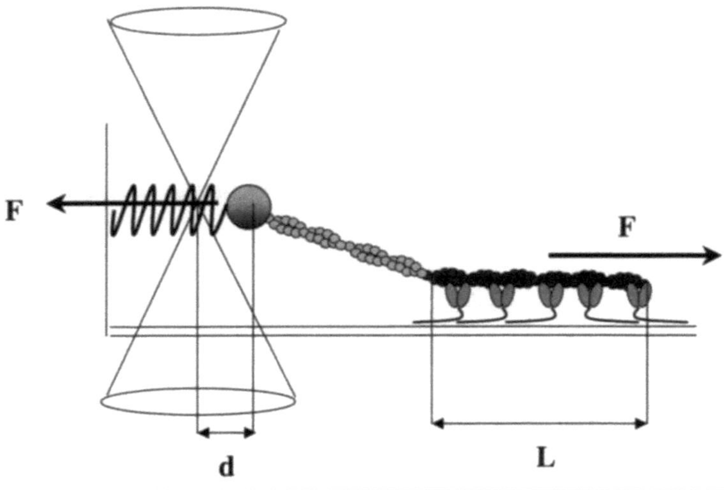

Figure 2C. Since the laser trap acts as a Hookian spring, the force exerted on the bead (F) equals to the product of the trap stiffness (k) and the distance between the bead and the trap center (d), i.e., $F = k \times d$. Since the number of myosin molecules pulling the bead is proportional to the length of actin filament interacting with the myosin layer (L), the stall force (F) linearly correlates with L.

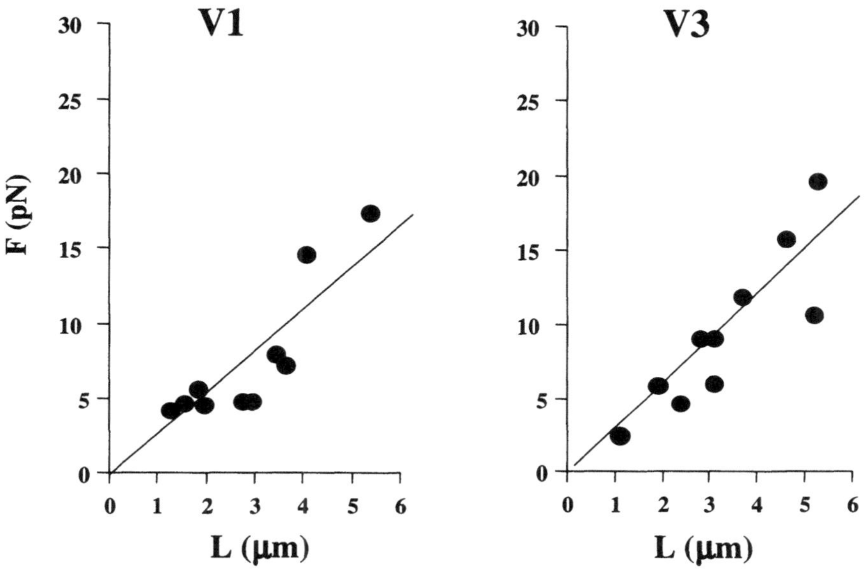

Figure 2D. The relationship between the stall force (F) and the filament length (L) in V1 and V3. F linearly correlated with L in both myosins with similar slopes.

could be calculated as the product of the displacement from the trap center and the trap stiffness. The length of filament interacting with the myosin molecules was measured on the focal plane under the fluorescence microscope (Figure 2C).

Measuring unitary displacement and force

To analyze motor function of a myosin molecule, we recorded single events of actin-myosin interaction in vitro and determined force and displacement generated by a myosin molecule (14). By adsorbing myosin molecules on the glass surface very sparsely and improving time and spatial resolution with a quadrant photodiode detector, we succeeded in recording single events (Figure 3A). When a single (or a very small number of) myosin molecule(s) was interacting with the actin filament, the position of the bead exhibited discrete step-wise displacements at low trap stiffness (Figure 3B). To measure isometric force generated by a single myosin molecule, the bead position was feedback controlled to be held stationary in position (14) (Figure 3C). Discrete steps were scored in the records by eye according to criteria similar to those of Finer et al. (12).

Statistical analysis

Data were expressed as mean ± SD. A 2-tailed unpaired Student t-test was used to determine significant differences between group means and fit parameters. A value of $p < 0.05$ was considered significant.

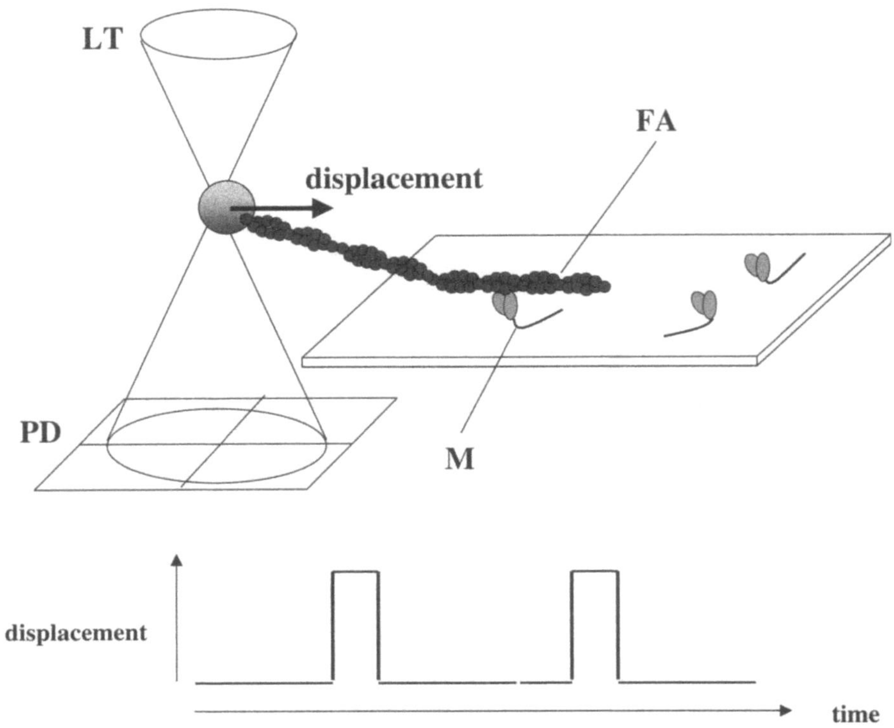

Figure 3A. Recording of single events of actin-myosin interaction by a laser trap (LT). Myosin molecules (M) were adsorbed very sparsely by diluting the myosin solution to 0.5–1.0 µg/mL before adsorbing on the glass surface. The bead movement was recorded by projecting the bright-field image of the bead onto a quadrant photodiode detector (PD) and the position was detected in nanometer resolution. Discrete events of interaction of a single actin filament (FA) and a single (or a very small number of) myosin molecule(s) were recorded as step-wise displacements of the bead position (bottom).

RESULTS

Actin filament velocity and Ca^{2+}-activated ATPase activity

The actin filament velocity and Ca^{2+}-activated ATPase activity of V1 were twice higher than those of V3, showing correlation between these parameters (8) (Figure 1B). These results were comparable to the values estimated from unloaded muscle shortening (2,3) and those of similar assays by other researchers (15,16).

Ensemble force of crossbridges

The stall force, which was balanced with the trap force, was considered to be ensemble force generated by several myosin molecules and correlated with the length of

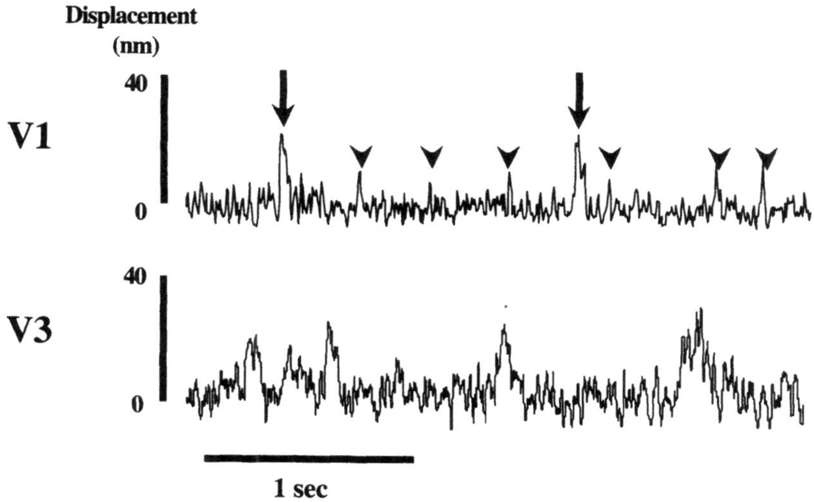

Figure 3B. Displacement records at low trap stiffness (0.03–0.06 pN/nm) and very low MgATP concentration (0.5 μmol/L). Step displacements were composed of higher (arrows) and lower (arrow heads) amplitudes. The mean amplitudes of discrete steps were not different between V1 and V3, while the duration of each step (T_{low}) was longer in V3 than V1 (Table 2).

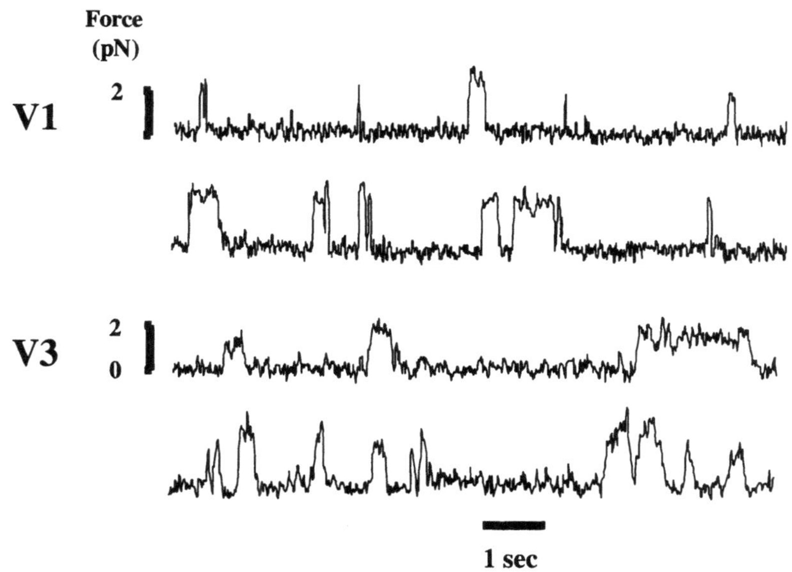

Figure 3C. Isometric force records with feed-back control of the trap position. The output of the quadrant photodiode detector (PD) was fed to a feed-back circuit to control the trap position by driving piezoelectric mirrors rapidly so that the bead was held stationary in position. Vertical deflections indicate movement of the trap position, representing force transients. The mean amplitudes of force transients were not different between V1 and V3, while the duration of each step (T_{iso}) was longer in V3 than V1 (Table 2).

filament on the myosin layer. When both V1 and V3 were absorbed on the surface at the same concentration (250 μg/ml), the slopes of the relationship between the force and filament length were similar between these myosins (2.8 vs. 3.1 pN/μm) (Figure 2D). Since the densities of myosin molecules on the glass surface estimated by measuring ATPase activity per unit surface area of the myosin layer were similar between these myosins (1.47 ± 0.19 vs. 1.57 ± 0.13 ng/mm^2), the number of myosin molecules interacting with unit length of actin filament was considered to be similar. Therefore, average force generated by several myosin molecules could be similar between V1 and V3.

Unitary displacement and unitary force

Typical recordings of the displacement events were shown in Figure 3B. Mean amplitudes of discrete steps were not different between these myosins (15.3 ± 14.6 vs. 14.9 ± 6.7 nm, N.S.). Histogram of the displacement amplitude exhibited a broad distribution with two peaks, which correspond to the displacement steps with higher and lower amplitudes (arrows and arrow heads in Figure 3B). If we assume that each step is a sum-up of unitary displacements (D_{uni}) of single events of actin-myosin interaction, the amplitude distribution consisted of two gaussian curves with peaks at around 9 nm and 18 nm in both myosins. The results suggested that the amplitude of D_{uni} was equal between V1 and V3, being around 9 nm. The mean duration of displacement events (T_{low}), however, was 38% longer in V3 compared to V1 (Table 2).

Figure 3C shows typical recordings of force transients. The mean amplitudes of force transients were not significantly different between these myosins (1.2 ± 0.6 vs. 1.6 ± 0.9 pN, N.S.). In both myosins, the force amplitudes distributed with a single peak at around 1.8 pN, suggesting that unitary force (F_{uni}) was similar between V1 and V3, being around 1.8 pN. However, the mean duration of force transients (T_{iso}) was 50% longer in V3 than in V1 (Table 2). The results were basically consistent with those by Palmiter et al., who also reported equal D_{uni} (~7 nm) and F_{uni} (~0.8 pN) between V1 and V3 and longer duration of displacement events for V3 (17). In summary, the amplitudes of D_{uni} and F_{uni} were not different between V1 and V3, while the event duration was significantly longer in V3 than V1.

Table 2. Summarized data of single molecule experiments under low load and isometric conditions

	Low load		Isometric	
	Duni (nm)	Tlow (ms)	Funi (pN)	Tiso (ms)
V1	~9	205 ± 5	~1.8	333 ± 14
V3	~9	**283 ± 20	~1.8	**488 ± 82

Duni, unitary displacement; Tlow, duration of the displacement events; Funi, unitary force; Tiso, duration of the isometric events. **p < 0.01 vc. V1.

DISCUSSION

V1 and V3 have equal force-generating ability

The long-lasting question has been why and how V3 has higher economy of force generation compared to V1. Does V3 molecule produce higher level of force per crossbridge, or similar level of force compared to V1? If V1 and V3 generate similar level of force, why V3 has higher economy of force generation? Although there were a number of studies measuring force of muscle strips or single cardiomyocytes containing predominantly V1 or V3, their results were still controversial on this point (18–24). In the literature, the number of studies reporting similar level of force between V1 and V3 was larger than that of studies concluding higher force level for V3 (18–20,23,24). Hypertrophied cardiac muscles with higher V3 content induced by pressure overload produced similar level of maximum force compared to V1-dominant control subjects (18,19). Shibata et al. also reported similar force-generating ability between hyperthyroid and euthyroid rabbits (V1 and V3-dominant, respectively) (20). On the other hand, some researchers reported higher force-generating ability in V3-dominant cardiac muscles (21,22). More recently, Fitzsimons et al. and Metzger et al. utilizing skinned single cardiomyocytes reached the same conclusion: similar maximal tension between control (V1-dominant) and hypothyroid (100% V3) rats (23,24). Metzger et al. also measured instantaneous stiffness to estimate the number of crossbridges interacting with actin and concluded that force production per crossbridge was not different between V1 and V3 (24). The present results were consistent with these recent results. It is noteworthy, however, that VanBuren et al., using a glass microneedle to measure ensemble force generated by myosin molecules, reported twice higher average force in V3 compared to V1 (16). The cause of these discrepant results is not clear, but different species of the animal model (young rats vs. hyperthyroid rabbits for V1; hypothyroid rats vs. hypothyroid rabbits for V3), methods in measuring force (laser trap vs. glass microneedle), and methylcellulose concentrations in the assay buffers (0.2% vs. 0.375%) might explain the discrepancy.

Two-state model of crossbridge

Assuming that V1 and V3 have similar force-generating ability as shown above, why V3 has higher economy of force generation than V1? To answer the question, we interpreted the present results in the two-state model of crossbridge, originally proposed by A. F. Huxley (25). It is believed that cyclic interaction of actin and myosin is a multistep enzymatic reaction coupled with ATP hydrolysis. In this model, multiple intermediate states of the crossbridge cycle were grouped into two states, the attached state (or force-generating state) and the detached state (or non force-generating state) (Figure 4). If we assume that myosin molecules interact with actin filament randomly, the average force of several myosin molecules equals to the time-averaged force of an individual molecule (F_{ave}), which is given as the product of the unitary force (F_{uni}) and the duty ratio (f), where the unitary force is the amplitude of force generated by a single myosin molecule in the attached state, and the duty

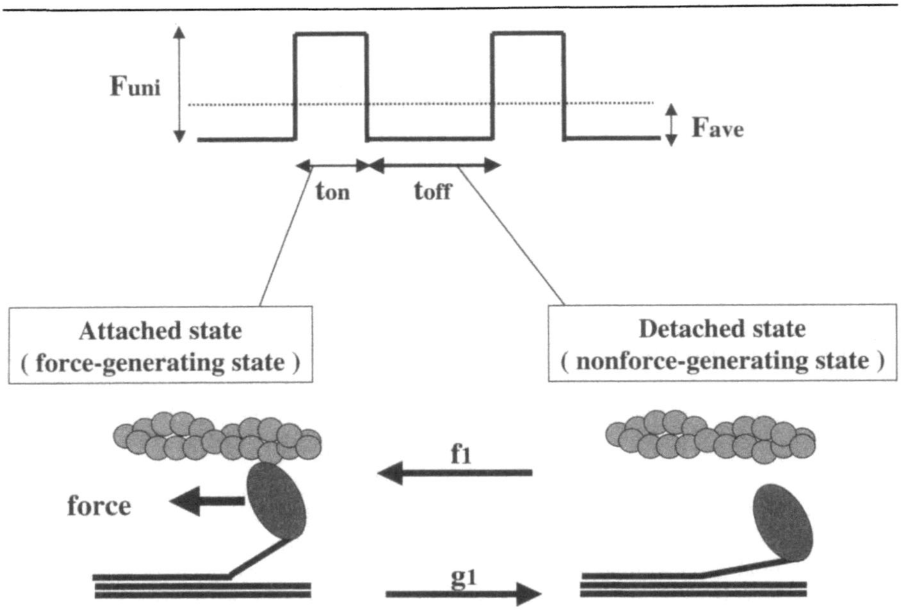

Figure 4. The two-state model of crossbridge. F_{uni}, unitary force of a single crossbridge; F_{ave}, time-averaged force of a single crossbridge; t_{on}, duration of the attached state; t_{off}, duration of the detached state; f_1, rate constant of the transition from the detached state to the attached state; g_1, rate constant of the transition from the attached state to the detached state.

ratio (f) is the fraction of time in the attached state (t_{on}) against the total cycling time (t_{on} + t_{off}) under isometric condition.

$$Fave = Funi \times f$$

$$f = ton/ton + toff$$

Similar F_{ave} and F_{uni} between V1 and V3 in our results suggest that f can be also similar between these myosins. In this model, if f_1 and g_1 are defined as the rate constants of transitions from the detached state to the attached state and the attached state to the detached state, respectively, t_{on} and t_{off} will be given as inverses of g_1 and f_1, respectively.

$$ton = 1/g1$$

$$toff = 1/f1$$

Therefore, duty ratio (f) will be given as $f_1/(f_1 + g_1)$ and a function of g_1/f_1 ratio

$$f = f1/(f1 + g1) = 1/(1 + g1/f1)$$

Since T_{iso} in the our results is the duration of isometric events in vitro and can be an index of t_{on} in the model, the longer T_{iso} in V3 suggests that t_{on} may also be longer and thus g_1 may be smaller in V3 than V1. ATPase rate under isometric condition (ATPase) of a single crossbridge is given as an inverse of the total cycling time (t_{on} + t_{off}).

$$\text{ATPase} = 1/f1 + 1/g1 = (f1 + g1)/(f1 \times g1)$$

If we define the economy of force production of a single crossbridge (E_{co}) as the time-averaged force (F_{ave}) divided by the ATPase rate, Eco equals to F_{uni}/g_1.

$$\begin{aligned} \text{Eco} &= F_{ave}/\text{ATPase} \\ &= F_{uni}/g_1 \end{aligned}$$

Since F_{uni} is similar between V1 and V3, E_{co} will be inversely correlated with g_1. Thus, the better economy in force generation reported by researchers can be explained by the longer t_{on} and smaller g_1 in the crossbridge kinetics in V3 compared to V1. If we assume that f_1 is proportionally smaller in V3 than V1 and g_1/f_1 ratio is equivalent between these myosins, we can explain the similar duty ratio (f) and time-averaged force (F_{ave}). Importantly, while V1 and V3 have similar force-generating ability, V3 consumes less number of ATP molecules to maintain the same level of isometric force and thus may have better economy in force maintenance compared to V1.

Functional significance of MHC isoforms in humans

In human atria, under pressure or volume overload, the MHC isoforms shift from α- to β-MHC, and the fraction of β-MHC correlates with the severity of the pressure overload (26). In human ventricles, however, it is controversial whether MHC isoforms also change under mechanical overload, because β-MHC is the predominant isoform in normal subjects. Formerly, Mercadier et al. reported that human ventricles contained up to 15% of α-MHC and ATPase activity of myosin correlated with the α-MHC content (27). Recently, it was reported that relative amount of α-MHC mRNA to the total MHC mRNA in the ventricles was much higher in normal subjects (33.3%) than reported in previous studies and significantly decreased in chronic end-stage heart failure (2.2%) (28,29). Also, at the protein level, ventricular tissues from nonfailing humans contained up to 10% of α-MHC in the total MHC, and the relative content of α-MHC significantly decreased in failing ventricles (30,31). Although the net change of MHC isoform content in failing human ventricles was much smaller than that observed in rodent hypertrophy models, even a small shift in the MHC isoform could have significant physiological relevance, because transgenic expression of β-MHC up to 12% of the total MHC had dominant-negative functional consequences in mice (32).

CONCLUSION

Two MHC isoforms, α- and β-MHC, have distinct catalytic and motor function. The β-MHC may have slower kinetics of crossbridge attachment and detachment, while the unitary displacement and force of a single crossbridge were indistinguishable between these isoforms. Although force-generating ability, estimated from the average force of several myosin molecules, was similar between these isoforms, the β-MHC could hydrolyze less number of ATP molecules to maintain equal level of force and have higher economy of force generation than V1.

ACKNOWLEDGMENTS

This study was supported in part by a Grant-in-Aid for Scientific Research on Priority Areas (Differentiation and Regulation of Cardiac Cells), a Grant-in-Aid for Scientific Research (C-13670689) from the Ministry of Education, Culture, Sports, Science and Technology, and the Vehicle Racing Commemorative Foundation.

REFERENCES

1. Hoh JFY, McGrath PA, Hale PT. 1978. Electrophoretic analysis of multiple forms of rat cardiac myosin: effects of hypophysectomy and thyroxine replacement. J Mol Cell Cardiol 10: 1053–1076.
2. Schwartz K, Lecarpentier Y, Martin JL, Lompre AM, Mercadier JJ, Swynghedauw B. 1981. Myosin isozymic distribution correlates with speed of myocardial contraction. J Mol Cell Cardiol 13:1071–1075.
3. Pagani ED, Julian FJ. 1984. Rabbit papillary muscle myosin isozymes and the velocity of muscle shortening. Circ Res 54:586–594.
4. Barany M. 1967. ATPase activity associated with the speed of muscle shortening. J Gen Physiol 50S:197–218.
5. Alpert NR, Mulieri LA. 1982. Increased myothermal economy of isometric force generation in compensated cardiac hypertrophy induced by pulmonary artery constriction in the rabbit. Circ Res 50:491–500.
6. Holubarsch CH, Goulette RP, Litten RZ, Martin BJ, Mulieri LA, Alpert NR. 1985. The economy of force development, myosin isoenzyme pattern and myofibrillar ATPase activity in normal and hyperthyroid rat myocardium. Circ Res 56:78–86.
7. Hasenfuss G, Mulieri LA, Blanchard EM, Holubarsch C, Leavitt BJ, Ittleman F, Alpert NR. 1991. Energetics of isometric force development in control and volume-overload human myocardium: comparison with animal species. Circ Res 68:836–846.
8. Sata M, Sugiura S, Yamashita H, Momomura S, Serizawa T. 1993. Dynamic interaction between cardiac myosin isoforms modifies velocity of actomyosin sliding in vitro. Circ Res 73:696–704.
9. Sugiura S, Kobayakawa N, Momomura S, Chaen S, Omata M, Sugi H. 1996. Different cardiac myosin isoforms exhibit equal force-generating ability in vitro. Biochim Biophys Acta 1273:73–76.
10. Kron SJ, Spudich JA. 1986. Fluorescent actin filaments move on myosin fixed to a glass surface. Proc Natl Acad Sci U S A 83:6272–6276.
11. Ashkin A. 1992. Forces of a single-beam gradient laser trap on a dielectric sphere in the ray optics regime. Biophys J 61:569–582.
12. Finer JT, Simmons RM, Spudich JA. 1994. Single myosin molecule mechanics: piconewton forces and nanometre steps. Nature 368:113–119.
13. Guilford WH, Dupuis DE, Kennedy G, Wu J, Patlak JB, Warshaw DM. 1997. Smooth muscle and skeletal muscle myosins produce similar unitary forces and displacement in the laser trap. Biophys J 72:1006–1021.
14. Sugiura S, Kobayakawa N, Fujita H, Yamashita H, Momomura S, Chaen S, Omata M, Sugi H. 1998. Comparison of unitary displacements and forces between 2 cardiac myosin isoforms by the optical trap technique: molecular basis for cardiac adaptation. Circ Res 82:1029–1034.

15. Harris DE, Work SS, Wright RK, Alpert NR, Warshaw DM. 1994. Smooth, cardiac, and skeletal muscle myosin force and motion generation assessed by cross-bridge mechanical interaction in vitro. J Muscle Res Cell Motil 15:11–19.
16. VanBuren P, Harris DE, Alpert NR, Warshaw DM. 1995. Cardiac V1 and V3 myosins differ in their hydrolytic and mechanical activities in vitro. Circ Res 77:439–444.
17. Palmiter KA, Tyska MJ, Dupuis DE, Alpert NR, Warshaw DM. 1999. Kinetic differences at the single molecule level account for the functional diversity of rabbit cardiac myosin isoforms. J Physiol 519: 669–678.
18. Maughan D, Low E, Litten R III, Brayden J, Alpert N. 1979. Calcium-activated muscle from hypertrophied rabbit hearts: mechanical and correlated biochemical changes. Circ Res 44:279–287.
19. Rupp H. 1981. The adaptive changes in the isoenzyme pattern of myosin from hypertrophied rat myocardium as a result of pressure overload and physiological training. Basic Res Cardiol 76:79–88.
20. Shibata T, Hunter WC, Sagawa K. 1987. Dynamic stiffness of barium-contractured cardiac muscle with different speeds of contraction. Circ Res 60:770–779.
21. Loiselle DS, Wendt IR, Hoh JFY. 1982. Energetic consequences of thyroid-modulated shifts in ventricular isomyosin distribution in the rat. J Muscle Res Cell Motil 3:5–23.
22. Saeki Y, Kako C, Totsuka T, Yanagisawa K. 1987. Mechanical properties and ATPase activity in glycerinated cardiac muscle of hyperthyroid rabbit. Pflugers Arch 408:578–583.
23. Fitzsimons DP, Patel JR, Moss RL. 1998. Role of myosin heavy chain composition in kinetics of force development and relaxation in rat myocardium. J Physiol 513:171–183.
24. Metzger JM, Wahr PA, Michele DE, Albayya F, Westfall MV. 1999. Effects of myosin heavy chain isoform switching on Ca^{2+}-activated tension development in single adult cardiac myocytes. Circ Res 84:1310–1317.
25. Huxley AF. 1957. Muscle structure and theories of contraction. Prog Biophys Biophys Chem 7: 255–317.
26. Tsuchimochi H, Sugi M, Kuro-o S, Ueda F, Takaku F, Furuta S, Shirai T, Yazaki Y. 1984. Isozymic change in myosin of human atrial myocardium induced by overload: immunohistochemical study using monoclonal antibodies. J Clin Invest 74:662–665.
27. Mercadier JJ, Bouveret P, Gorza L, Schiaffino S, Clark WA, Zak R, Swynghedauw B, Schwartz K. 1983. Myosin isoenzymes in normal and hypertrophied human ventricular myocardium. Circ Res 53:52–62.
28. Nakao K, Minobe W, Roden R, Bristow MR, Leinwand LA. 1997. Myosin heavy chain gene expression in human heart failure. J Clin Invest 100:2362–2370.
29. Lowes BD, Minobe W, Abraham WT, Rizeq MN, Bohlmeyer TJ, Quaife RA, Roden RL, Dutcher DL, Robertson AD, Voelkel NF, Badesch DB, Groves BM, Gilbert EM, Bristow MR. 1997. Changes in gene expression in the intact human heart: downregulation of α-myosin heavy chain in hypertrophied, failing ventricular myocardium. J Clin Invest 100:2315–2324.
30. Miyata S, Minobe W, Bristow MR, Leinwand LA. 2000. Myosin heavy chain isoform expression in the failing and nonfailing human heart. Circ Res 86:386–390.
31. Reiser PJ, Portman MA, Ning XH, Schomisch Moravec C. 2001. Human cardiac myosin heavy chain isoforms in fetal and failing adult atria and ventricles. Am J Physiol 280:H1814–H1820.
32. Tardiff JC, Hewett TE, Factor SM, Vikstrom KL, Robbins J, Leinwand LA. 2000. Expression of the β(slow)-isoform of the MHC in the adult mouse heart causes dominant-negative functional effects. Am J Physiol 278:H412–H419.

Pathophysiology of Cardiovascular Disease,
edited by N.S. Dhalla, H. Rupp,
A. Angel & G.N. Pierce
Kluwer Academic Publishers, Boston, 2004

Role of Reactive Oxygen Species in the Regulation of Cardiac Myocyte Phenotype

Melanie Maytin, Douglas B. Sawyer, and Wilson S. Colucci

Cardiovascular Medicine Section, Department of Medicine, Boston University Medical Center and the Myocardial Biology Unit, Boston University School of Medicine, Boston, MA

Summary. In cardiac myocytes *in vitro*, ROS can cause either hypertrophy or apoptosis in a concentration-dependent manner with hypertrophy in response to low levels of ROS and apoptosis in response to higher levels. Likewise, there is evidence that ROS mediate the hypertrophic effects of α-adrenergic receptor stimulation and low-level mechanical strain, and the apoptotic effects of β-adrenergic receptor stimulation or higher amplitude mechanical strain. The MAPK signaling pathway appears to mediate several of these effects of ROS. Erk is involved in the hypertrophic response to low levels of ROS, low amplitude mechanical strain and α-adrenergic receptor stimulation. In contrast, JNK is involved in the apoptotic effect of higher levels of ROS, high amplitude mechanical strain and β-adrenergic receptor stimulation. These observations indicate that ROS play a critical role in the determination of myocyte phenotype in response to a variety of stimuli associated with cellular growth or death.

Key words: cardiac myocytes, phenotype, apoptosis, hypertrophy, reactive oxygen species.

INTRODUCTION

There is mounting evidence that oxidant stress is increased in the failing heart and may contribute to myocardial remodeling at the level of the cardiac myocyte (1) (2–4). Myocardial remodeling involves a number of structural and functional changes, characterized at the molecular and cellular level by myocyte hypertrophy (5), cell loss via apoptosis or necrosis (6), and changes in the quantity and quality

Address for Correspondence: Wilson S. Colucci, M.D., Cardiovascular Medicine, Boston University Medical Center, 88 East Newton Street, Boston, MA 02118. Tel: 617-638-8706, Fax: 617-638-8712, E-mail: wilson.colucci@bmc.org.

of the interstitial matrix (7,8). As our understanding of the mechanisms that underlie remodeling grows, reactive oxygen species (ROS) and oxidant stress increasingly appear to play a central role in regulating the phenotype of cardiac myocytes. Cardiac myocytes may express multiple phenotypes as related to cell structure and function. At least two phenotypes, hypertrophy and apoptosis, have been implicated in the pathophysiology of myocardial remodeling and failure.

MYOCYTE HYPERTROPHY

ROS appear to exert graded, concentration-dependent effects on myocyte phenotype in vitro. A small increase in oxidant stress in cultured cardiac myocytes stimulates myocyte growth, whereas larger increases stimulate apoptosis. We increased O_2^- cellular superoxide (O_2^-) levels by inhibiting copper-zinc superoxide dismutase (CuZnSOD) with the copper chelator diethyldithiocarbamic acid (DDC). Exposure of neonatal rat cardiac myocytes in primary culture to low concentrations of DDC (1–10 μM) for 24 hours caused a small increase in ROS levels and was associated with increases in protein synthesis, cell size and ANF mRNA expression (9). These cellular events were prevented by treatment with antioxidants, supporting the conclusion that oxidant stress stimulates myocyte hypertrophy.

ROS-dependent hypertrophic stimuli

Other stimuli for myocyte hypertrophy appear to also be ROS-mediated. For example, we found that low amplitude mechanical strain (5% nominal stretch) increases ROS formation in cultured cardiac myocytes and is associated with increases in protein synthesis, cell size and fetal gene expression (10). This hypertrophic effect of mechanical strain was inhibited by the SOD/catalase mimetic Mn(II/III)tetrakis(1-methyl-4-peridyl) (MnTMPyP), thus implicating the role of ROS. TNF-α, angiotensin II and α-adrenergic receptor stimulation all cause ROS-mediated myocyte hypertrophy which can be inhibited by concurrent treatment with soluble antioxidants (11,12). Similarly, endothelin modifies early response gene expression via ROS-dependent activation of *ras* (13) and ouabain causes myocyte hypertrophy through ROS-dependent activation of a pathway involving *ras* and mitogen-activated protein kinases (MAPK) (14).

Source of hypertrophic ROS

The mechanism by which these hypertrophic stimuli increase myocyte ROS formation is unclear. However, there is evidence that NAD(P)H oxidase may play a role for at least some of the stimuli (Figure 1). In adult rat ventricular myocytes, we found that α_1-adrenergic receptor-stimulated activation of the extracellular-signal regulated kinase 1 and 2 (ERK) is completely prevented by the NAD(P)H oxidase inhibitor, diphenylene iodonium, whereas inhibition of the mitochondrial respiratory chain with rotenone had no effect (15,16). The neutrophil NAD(P)H oxidase is a complex consisting of at least four subunits, including two membrane-spanning components (p22phox, gp91phox) and two cytosolic components (p67phox,

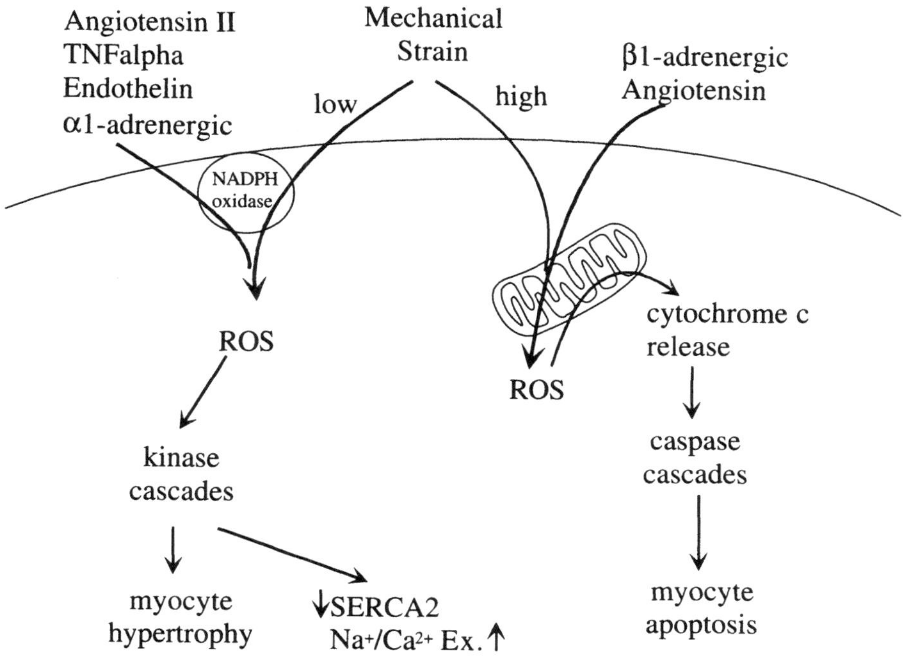

Figure 1. Proposed central role of ROS in determining cardiac myocyte phenotype. Several stimuli which cause myocyte hypertrophy, including angiotensin, tumor necrosis factor-α, endothelin, α_1-adrenergic receptor agonists and low amplitude mechanical strain, activate growth promoting kinase pathways in an ROS-dependent manner possibly involving an NAD(P)H oxidase. Other stimuli including β_1-adrenergic receptor stimulation and higher amplitude mechanical strain, cause myocyte apoptosis which may be mediated by an ROS-dependent activation of the mitochondrial death pathway. Many aspects of this model remain to be elucidated including the precise sources of ROS for each type of remodeling stimulus, and the role of the type, location and intensity of ROS in determining the net effect of a particular stimulus on distal signaling pathways and myocyte phenotype. Reproduced with permission (37).

p47phox) (17). In the cardiovascular system, a similar oxidase system has been identified in several cell types, albeit with different component structures and biochemical characteristics (17). The expression of the four major subunits (p22phox, gp91phox, p67phox, p47phox) of NAD(P)H oxidase has been demonstrated in adult rat ventricular myocytes by reverse transcriptase-polymerase chain reactions (RT-PCR) and northern blot analysis (16). The relevance of NAD(P)H oxidase in the myocardium has been suggested by in vivo experiments with the gp91phox knockout mouse (18,19). In this mouse, subpressor doses of angiotensin II caused cardiac hypertrophy associated with increased expression of a fetal gene program, increased myocyte cross-sectional area and decreased myocardial collagen content as compared with wild-type mice.

MYOCYTE APOPTOSIS

While low levels of ROS cause myocyte growth, higher levels are capable of activating the apoptotic death program that has been implicated in the pathophysiology of heart failure (9,10,20). As discussed above, in neonatal rat ventricular myocytes we found that low levels of ROS caused by low concentrations of DDC stimulate hypertrophy. In contrast, higher concentrations of DDC, which resulted in higher levels of ROS, stimulate myocyte apoptosis as assessed by nuclear morphology and TUNEL staining for DNA fragmentation (9). This effect of DDC was associated with increased expression of Bax, a pro-apoptotic member of the Bcl-2 family, whereas lower, hypertrophic concentrations of DDC did not increase Bax. At still higher levels, ROS can result in necrotic cell death. The apoptotic effect of ROS described above occurs at sub-necrotic concentrations.

Mechanical strain

Apoptosis induced by remodeling stimuli such as mechanical strain may also be mediated by ROS. Cheng and colleagues demonstrated an increase in $O_2^{\bullet-}$ production and myocyte apoptosis in papillary muscles subjected to a static mechanical load for four hours (21). The apoptosis was inhibited by addition of a source of NO, which may have acted to decrease $O_2^{\bullet-}$ levels by combing to form $^-$ONOO. Further evidence for the role of ROS in stretch-induced cell death comes from studies of isolated ventricular myocytes. High amplitude strain (25% nominal stretch) in isolated cardiac myocytes increased the frequency of apoptosis as measured by TUNEL or DNA laddering, with an associated increase in Bax expression (10). The stretch-induced apoptosis was associated with increased cellular ROS and was inhibited by treatment with an SOD mimetic, thus implicating ROS in the mediation of stretch-induced apoptosis.

β-Adrenergic receptor-stimulated apoptosis

β-adrenergic receptor (βAR) stimulated-apoptosis appears to be mediated by ROS. We (22) and others (23) have shown that βAR stimulation for 24 hours causes an increase in the frequency of apoptosis in adult rat ventricular myocytes. In this model system, apoptosis is increased by stimulation of β_1AR coupled to G_s/adenylyl cyclase. In contrast, β_2AR stimulation exerts an anti-apoptotic effect via coupling to G_i (24). βAR-stimulated apoptosis can be inhibited by treatment with an SOD/catalase-mimetic, further implicating a role of ROS (25). Similar to other models of ROS-mediated apoptosis, βAR stimulation is associated with increased Bax and decreased Bcl-XL and Bcl-2 expression, thus implicating a mitochondrial pathway (23). Moreover, βAR stimulation increases the translocation of cytochrome c from the mitochondria to the cytosol, as measured by protein levels and increased caspase-3 activity (25). Further support for the involvement of mitochondrial pathways in βAR stimulated-apoptosis comes from Zaugg and colleagues who demonstrated an increase in caspase 9, but not caspase 8, activity with β_1AR stimulation (23).

Anthracyclines

Anthracyclines can induce cardiac myocyte apoptosis and necrosis and has been associated with ROS generation (26). In vitro, we have shown that the iron chelator, dexrazoxane, inhibits anthracycline-stimulated myocyte apoptosis, but not necrosis (27). Although there was no measurable increase in H_2O_2 with anthracycline, the effect of dexrazoxane suggests that anthracycline-induced apoptosis involves the formation of hydroxyl radicals via Fenton chemistry, a reaction that is favored by increased superoxide levels (27,28).

ROS SIGNALING

Direct effects of ROS

There are several potential downstream targets for ROS signaling in cardiac myocytes. The mitogen-activated protein kinase (MAPK) superfamily is known to be responsive to ROS and has been shown to be involved in the determination of myocyte hypertrophy and apoptosis (29–31). In adult rat ventricular myocytes in primary culture, we (S. Kwon, unpublished data) and others (32) have found that direct application of ROS in the form of H_2O_2 results in activation of all members of the MAPK family—ERK, JNK and p38 kinase, albeit at different concentrations. ERK is activated at low concentrations, whereas JNK and p38 require higher concentrations. We have further found that the ERK inhibitor PD98059 prevents H_2O_2-stimulated hypertrophy, whereas the JNK inhibitor SP600125 prevents apoptosis. This later finding is consistent with the observation in the H9C2 cardiac cell line that a dominant-negative JNK diminishes oxidant-stress induced apoptosis (33). Relevant to this, it has been shown that ischemia/reperfusion-induced myocyte apoptosis, which is believed to be mediated by ROS, is associated with JNK activation that is inhibited by carvedilol, a β-blocker with antioxidant properties (34).

Remodeling stimuli

α-adrenergic receptor stimulation causes ROS-dependent myocyte hypertrophy that is associated with activation of ERK, and prevented by the ERK inhibitor PD98059 (35,36). It has also been shown that inhibition of ERK increases the magnitude of ROS-stimulated apoptosis, suggesting that ERK has both hypertrophic and pro-survival effects (28). Likewise, we have found that ERK inhibition prevents the hypertrophic effects of low level mechanical strain (10). In contrast, we have found that the apoptotic effect of β-adrenergic receptor stimulation is prevented by the JNK inhibitor SP600125 (A. Remondino, unpublished data).

SUMMARY

Myocardial remodeling and failure are associated with increased levels of ROS. In cardiac myocytes *in vitro,* ROS can cause either hypertrophy or apoptosis in a concentration-dependent manner with hypertrophy in response to low levels of ROS and apoptosis in response to higher levels. Likewise, there is evidence that

ROS mediate the hypertrophic effects of α-adrenergic receptor stimulation and low-level mechanical strain, and the apoptotic effects of β-adrenergic receptor stimulation or higher amplitude mechanical strain. The MAPK signaling pathway appears to mediate several of these effects of ROS. ERK is involved in the hypertrophic response to low levels of ROS, low amplitude mechanical strain and α-adrenergic receptor stimulation. In contrast, JNK is involved in the apoptotic effect of higher levels of ROS, high amplitude mechanical strain and β-adrenergic receptor stimulation.

These observations indicate that ROS play a critical role in the determination of myocyte phenotype in response to a variety of stimuli associated with cellular growth or death, and suggest that a better understanding of the mechanisms that regulate ROS and mediate their effects in cardiac myocytes will lead to new approaches to the treatment of cardiac remodeling and failure.

REFERENCES

1. Ide T, Tsutsui H, Kinugawa S, Suematsu N, Hayashidani S, Ichikawa K, Utsumi H, Machida Y, Egashira K, Takeshita A. 2000. Direct evidence for increased hydroxyl radicals originating from superoxide in the failing myocardium. Circ Res 86:152–157.
2. Ghatak A, Brar MJ, Agarwal A, Goel N, Rastogi AK, Vaish AK, Sircar AR, Chandra M. 1996. Oxy free radical system in heart failure and therapeutic role of oral vitamin E. Int J Cardiol 57:119–127.
3. Diaz-Velez CR, Garcia-Castineiras S, Mendoza-Ramos E, Hernandez-Lopez E. 1996. Increased malondialdehyde in peripheral blood of patients with congestive heart failure. Am Heart J 131:146–152.
4. Mallat Z, Philip I, Lebret M, Chatel D, Maclouf J, Tedgui A. 1998. Elevated levels of 8-iso-prostaglandin F2alpha in pericardial fluid of patients with heart failure: a potential role for in vivo oxidant stress in ventricular dilatation and progression to heart failure. Circulation 97:1536–1539.
5. Gerdes AM, Liu Z, Zimmer HG. 1994. Changes in nuclear size of cardiac myocytes during the development and progression of hypertrophy in rats. Cardioscience 5:203–208.
6. MacLellan WR, Schneider MD. 1997. Death by design. Programmed cell death in cardiovascular biology and disease. Circ Res 81:137–144.
7. Weber KT, Anversa P, Armstrong PW, Brilla CG, Burnett JC, Jr, Cruickshank JM, Devereux RB, Giles TD, Korsgaard N, Leier CV, et al. 1992. Remodeling and reparation of the cardiovascular system. J Am Coll Cardiol 20:3–16.
8. Mann DL, Spinale FG. 1998. Activation of matrix metalloproteinases in the failing human heart: breaking the tie that binds. Circulation 98:1699–1702.
9. Siwik DA, Tzortzis JD, Pimental DR, Chang DL, Pagano PJ, Singh K, Sawyer DB, Colucci WS. 1999. Inhibition of copper-zinc superoxide dismutase induces cell growth, hypertrophic phenotype, and apoptosis in neonatal rat cardiac myocytes in vitro. Circ Res 85:147–153.
10. Pimentel DR, Amin JK, Xiao L, Miller T, Viereck J, Oliver-Krasinski J, Baliga R, Wang J, Siwik DA, Singh K, Pagano P, Colucci WS, Sawyer DB. 2001. Reactive oxygen species mediate amplitude-dependent hypertrophic and apoptotic responses to mechanical stretch in cardiac myocytes. Circ Res 89:453–460.
11. Nakamura K, Fushimi K, Kouchi H, Mihara K, Miyazaki M, Ohe T, Namba M. 1998. Inhibitory effects of antioxidants on neonatal rat cardiac myocyte hypertrophy induced by tumor necrosis factor-alpha and angiotensin II. Circulation 98:794–799.
12. Amin JK, Xiao L, Pimental DR, Pagano PJ, Singh K, Sawyer DB, Colucci WS. 2001. Reactive oxygen species mediate alpha-adrenergic receptor-stimulated hypertrophy in adult rat ventricular myocytes. J Mol Cell Cardiol 33:131–139.
13. Cheng TH, Shih NL, Chen SY, Wang DL, Chen JJ. 1999. Reactive oxygen species modulate endothelin-I-induced c-fos gene expression in cardiomyocytes. Cardiovasc Res 41:654–662.
14. Kometiani P, Li J, Gnudi L, Kahn BB, Askari A, Xie Z. 1998. Multiple signal transduction pathways link Na+/K+-ATPase to growth-related genes in cardiac myocytes. The roles of Ras and mitogen-activated protein kinases. J Biol Chem 273:15249–15256.
15. Tanaka K, Honda M, Takabatake T. 2001. Redox regulation of MAPK pathways and cardiac hypertrophy in adult rat cardiac myocyte. J Am Coll Cardiol 37:676–685.

16. Xiao L, Pimentel DR, Wang J, Singh K, Colucci WS, Sawyer DB. 2002. Role of reactive oxygen species and NAD(P)H oxidase in alpha(1)-adrenoceptor signaling in adult rat cardiac myocytes. Am J Physiol Cell Physiol 282:C926–C934.
17. Griendling KK, Sorescu D, Ushio-Fukai M. 2000. NAD(P)H oxidase: role in cardiovascular biology and disease. Circ Res 86:494–501.
18. Bendall JK, Cave AC, Heymes C, Gall N, Shah AM. 2002. Pivotal role of a gp91(phox)-containing NADPH oxidase in angiotensin II-induced cardiac hypertrophy in mice. Circulation 105:293–296.
19. Wang HD, Xu S, Johns DG, Du Y, Quinn MT, Cayatte AJ, Cohen RA. 2001. Role of NADPH oxidase in the vascular hypertrophic and oxidative stress response to angiotensin II in mice. Circ Res 88:947–953.
20. von Harsdorf R, Li PF, Dietz R. 1999. Signaling pathways in reactive oxygen species-induced cardiomyocyte apoptosis. Circulation 99:2934–2941.
21. Cheng W, Li B, Kajstura J, Li P, Wolin MS, Sonnenblick EH, Hintze TH, Olivetti G, Anversa P. 1995. Stretch-induced programmed myocyte cell death. J Clin Invest 96:2247–2259.
22. Communal C, Singh K, Pimentel DR, Colucci WS. 1998. Norepinephrine stimulates apoptosis in adult rat ventricular myocytes by activation of the beta-adrenergic pathway. Circulation 98: 1329–1334.
23. Zaugg M, Xu W, Lucchinetti E, Shafiq SA, Jamali NZ, Siddiqui MA. 2000. Beta-adrenergic receptor subtypes differentially affect apoptosis in adult rat ventricular myocytes. Circulation 102:344–350.
24. Communal C, Singh K, Sawyer DB, Colucci WS. 1999. Opposing effects of beta(1)- and beta(2)-adrenergic receptors on cardiac myocyte apoptosis: role of a pertussis toxin-sensitive G protein. Circulation 100:2210–2212.
25. Communal C, Xie X, Sawyer DB, Singh K, Colucci WS. 2000. Beta-adrenergic receptor-stimulated apoptosis involves mitochondrial pathways and reactive oxygen species. Circulation 102:II-9 (Abstract).
26. Rajagopalan S, Politi PM, Sinha BK, Myers CE. 1988. Adriamycin-induced free radical formation in the perfused rat heart: implications for cardiotoxicity. Cancer Res 48:4766–4769.
27. Sawyer DB, Fukazawa R, Arstall MA, Kelly RA. 1999. Daunorubicin-induced apoptosis in rat cardiac myocytes is inhibited by dexrazoxane. Circ Res %19;84:257–265.
28. Aikawa R, Komuro I, Yamazaki T, Zou Y, Kudoh S, Tanaka M, Shiojima I, Hiroi Y, Yazaki Y. 1997. Oxidative stress activates extracellular signal-regulated kinases through Src and Ras in cultured cardiac myocytes of neonatal rats. J Clin Invest 100:1813–1821.
29. Sugden PH Clerk A. 1998. "Stress-responsive" mitogen-activated protein kinases (c-Jun N-terminal kinases and p38 mitogen-activated protein kinases) in the myocardium. Circ Res 83:345–352.
30. Wang Y, Su B, Sah VP, Brown JH, Han J, Chien KR. 1998. Cardiac hypertrophy induced by mitogen-activated protein kinase kinase 7, a specific activator for c-Jun NH2-terminal kinase in ventricular muscle cells. J Biol Chem 273:5423–5426.
31. Jiang Y, Gram H, Zhao M, New L, Gu J, Feng L, Di Padova F, Ulevitch RJ, Han J. 1997. Characterization of the structure and function of the fourth member of p38 group mitogen-activated protein kinases, p38delta. J Biol Chem 272:30122–30128.
32. Wei S, Rothstein EC, Fliegel L, Dell'Italia LJ, Lucchesi PA. 2001. Differential MAP kinase activation and Na(+)/H(+) exchanger phosphorylation by H(2)O(2) in rat cardiac myocytes. Am J Physiol Cell Physiol 281:C1542–C1550.
33. Turner NA, Xia F, Azhar G, Zhang X, Liu L, Wei JY. 1998. Oxidative stress induces DNA fragmentation and caspase activation via the c-Jun NH2-terminal kinase pathway in H9c2 cardiac muscle cells. J Mol Cell Cardiol 30:1789–1801.
34. Yue TL, Ma XL, Gu JL, Ruffolo RR, Jr, Feuerstein GZ. 1998. Carvedilol inhibits activation of stress-activated protein kinase and reduces reperfusion injury in perfused rabbit heart. Eur J Pharmacol 345:61–65.
35. Xiao L, Pimental DR, Amin JK, Singh K, Sawyer DB, Colucci WS. 2001. MEK1/2-ERK1/2 mediates alpha1-adrenergic receptor-stimulated hypertrophy in adult rat ventricular myocytes. J Mol Cell Cardiol 33:779–787.
36. Reed JC. 1997. Cytochrome c: can't live with it—can't live without it. Cell 91:559–562.
37. Sawyer DB, Siwik DA, Xiao L, Pimentel DR, Singh K, Colucci WS. 2002. Role of oxidative stress in myocardial hypertrophy and failure. J Mol Cell Cardiol 34:379–388.

Pathophysiology of Cardiovascular Disease,
edited by N.S. Dhalla, H. Rupp,
A. Angel & G.N. Pierce
Kluwer Academic Publishers, Boston, 2004

Functional Significance of Ryanodine Receptor-Mediated Calcium Leaks in Sarcoplasmic Reticulum Membranes

Penelope Rampersad, Mark Mutawe, Bernard Abrenica, Tom Cook, and James Gilchrist

Departments of Oral Biology and Physiology, University of Manitoba, Division of Stroke and Vascular Disease, 4024 St. Boniface General Hospital Research Centre, 351 Taché Avenue, Winnipeg, Manitoba, Canada R2H 2A6

Summary. Since their original purification and identification as the main Ca^{2+} efflux pathway from sarcoplasmic reticulum (SR) membranes, ryanodine receptors (RyRs) have attracted an enormous level of research interest. This is not surprising for several reasons. First, they sit at the epicentre of regulatory mechanisms involved in excitation-contraction coupling. Second, RyRs represent the main building blocks of a huge multimeric protein assembly with a molecular mass estimated to be around 2.3 mega Daltons. Third, RyRs are able to bind an unprecedented number of regulatory proteins, toxins and other molecules producing, in many cases, quite complex effects upon its Ca^{2+} ion channel properties. Much remains to be determined about how *in vitro* kinetic and features of RyR activation and inactivation account for observed Ca^{2+} release properties *in vivo*. Some studies favor the idea that in intact SR membranes ATP-dependent Ca^{2+} pumps (SERCA) can also influence both activation and inactivation of Ca^{2+} release. More recently, however, we have come to appreciate the significant leakiness of SR membranes to Ca^{2+} at rest and the possibility that this occurs primarily through activation of RyRs. A number of more recent lines of evidence suggest this leakiness has some physiological significance in activating RyRs to initiate Ca^{2+} sparks and trigger Ca^{2+} release. Because of the potential importance of this to RyR function a great deal of interest has been generated in an attempt to reveal underlying regulatory mechanism of RyR leakiness. A class of immunophillin binding proteins, known as FKBP12/12.6, is now thought to play a very important role in determining RyR leaks in both normal and patho-

Corresponding author: James Gilchrist, Ph.D., Division of Stroke and Vascular Disease, 4024 St. Boniface General Hospital Research Centre, 351 Taché Avenue, Winnipeg, Manitoba, Canada R2H 2A6. Tel: 204 235 3048, Fax: 204 231 1151, E-mail: James_Gilchrist@SBRC.ca.

logical settings. However, species and tissue differences in FKBP12/12.6 distribution make it hard to make any firm generalizations, at this point, regarding in vivo RyR leak regulation.

Key words: Calcium Leaks, Ryanodine Receptors, FKBP, Sarcoplasmic Reticulum, SERCA.

INTRODUCTION

It is now widely appreciated that the ability of cells to mobilize, distribute and store intracellular Ca^{2+} in a highly regulated fashion is of paramount importance to survival because of the quintessential role of this cation as a second messenger (1,2). Cells achieve this through the coordinated actions of Ca^{2+} pumps and channels both at the cell surface and on various internal membrane sites including (but not limited to) mitochondria, nuclei and sarco/endoplasmic reticulum. Our earlier understanding of this was based on observations from spectroscopic studies of Ca^{2+}-sensitive probes in a variety of model systems (3). However, the advent of fluorescence confocal microscopy has allowed us to more clearly "see" these Ca^{2+} movements and, thus, readily appreciate the enormously varied and complex (4–7), but ultimately exquisite (8,9), ways different cells are able to produce them. Spatial distribution of pumps and channels is clearly a key factor in determining patterns of intracellular Ca^{2+} activity but one interesting way cells may also fine-tune their Ca^{2+} movements is through splice variant selection of gene transcripts encoding multimeric transporters (10–12). This tells us that the assembly of different pump and channel combinations is an important feature of how the various roles of Ca^{2+} are subserved.

The specific repertoire and spatial arrangement of pumps and channels in striated muscle serves the requirement for contraction and relaxation to be regulated by relatively fast release and re-uptake of sarcoplasmic reticulum (SR) Ca^{2+}. These SR "Ca^{2+} transients" are triggered from resting cytosolic free Ca^{2+} levels at ~100 nM up to 10 fold higher peak levels within, in some cases, a few milliseconds (13). In all striated muscle this fast initial release of SR Ca^{2+} occurs through activation of a large multimeric protein channel complex known as the ryanodine receptor (RyR) (14–16). All three RyR isoforms (RyR1–3) are known to regulate SR Ca^{2+} release in striated muscle although selection appears not only based upon Ca^{2+} conductance properties but also on curious differences in structural domains potentially important in Ca^{2+}-dependent regulation of Ca^{2+} release. Subsequent relaxation then occurs by actively reducing free cytosolic Ca^{2+} back to its resting level within a similar time frame. This is due to the action of the SR Ca^{2+} ATPase (SERCA) and here, also, isoform selection markedly impacts upon kinetics of relaxation (17–22). Essential mechanistic details for SERCA pumps were worked out over two decades ago (23,24) but since then a vast accumulation of information about its molecular and cellular physiology shows it has a major impact upon cell function and phenotype in both physiological and pathophysiological settings.

Much is now known about mechanisms governing the activation and inactivation of both SERCA and RyR proteins (barring some debate) and we also now have a much clearer picture about which part of the molecule can do what (25–27). SERCA pumps have attracted much attention as key regulators of cell survival with *in vitro* observations, for the most part, being well-reconciled with those in vivo. Although potentially quite important mechanistic questions about SERCA pump reversal and pump "slippage" (28,29) remain unanswered, there appear to be vastly more questions about diverse modulatory effects from an array of RyR ligands. Indeed, it's complex multimeric (~2.3 mega Dalton) structure and its unprecedented ability to interact with and be regulated by a wide range of proteins, drugs, toxins and other molecules have made RyRs the object of great research interest over the past decade. One of its more curious properties is its leakiness (30) and, again, why this occurs is not quite clear. RyR Ca^{2+} leaks have been linked to the mechanism of activation, but we remain far from certain about how somewhat controversial mechanistic fine-detail obtained from single-channel studies fits with observation of Ca^{2+} release in situ or in vivo. There are many excellent reviews on the mechanism and molecular physiology of RyR Ca^{2+} release and the reader is encouraged to refer to several of these for more in depth material (26,31–33). One purpose of this chapter is to discuss what our current understanding of RyR Ca^{2+} release mechanisms tells us about new observations concerning SR Ca^{2+} leaks. The other is to highlight evidence suggesting some involvement of SERCA pumps in this process.

MECHANISMS OF TRIGGERED SR Ca^{2+} RELEASE

In both cardiac and skeletal muscle, SR Ca^{2+} release occurs from specialized regions known as terminal cisternae (31,32), the junctional face membranes of which are closely apposed to cytoplasmic membrane leaflets of t-tubules (34,35). Triggering of skeletal muscle Ca^{2+} release is mediated through intimate associations of RyR tetrameric complexes, known as "feet", on SR junctional surfaces with t-tubule dihydropyridine receptor (DHPR) complexes, known as "tetrads" (34). Ultimately, t-tubule membrane depolarization also initiates RyR activation in cardiac muscle (36–39). In this case, however, depolarization activates voltage-dependent DHPR Ca^{2+} channels to locally increase cytosolic Ca^{2+} within the vicinity of RyR2 Ca^{2+} channels (37,40). Ca^{2+} binding to RyR2 high-affinity cytosolic Ca^{2+} activation sites then stimulates SR Ca^{2+} efflux (31,40) by "Calcium-Induced Ca^{2+} release" (CICR). In contrast, skeletal muscle SR Ca^{2+} release is initiated via charge transfer mediated by the physical coupling of RyR1 "feet" and DHPR "tetrads" at SR junctional face membranes (36). However, up to 50 percent of the total RyR1 channel population may be distributed on lateral non-junctional surfaces of terminal cisternae membranes (41). In this case, the initial stimulus for Ca^{2+} release may be propagated from junctional regions to activate non-junctional RyR1 channels as indicated from rapid scanning confocal methods. In cardiac muscle propagation of RyR2-mediated Ca^{2+} release at these non-junctional sites (37,42) possibly occurs via a regenerative (37) or "cluster bomb" (42) mechanism of Ca^{2+}-dependent activation as illustrated in Figure 1.

Ca^{2+}-dependent regulation of SR calcium release

All RyR Ca^{2+} channel isoforms are complexly regulated by Ca^{2+} ions upon both luminal and cytoplasmic faces (31,43–51). In striated muscle much more is known

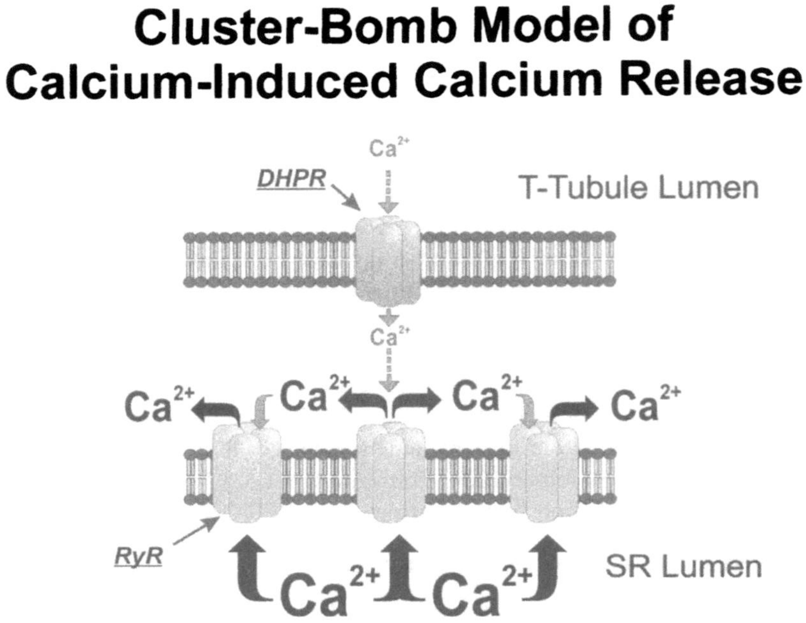

Figure 1. A scheme to illustrate the principle of "Cluster-Bomb" activation of RyR-mediated Ca^{2+} release via CICR (42). The model suggests single DHPR molecules on the t-tubule membrane produces a small flux of trigger Ca^{2+} which binds to the cytoplasmic face of a single RyR molecule in close proximity on the sarcoplasmic reticulum membrane. Local Ca^{2+} release from this molecule then activates neighboring RyR molecules in a regenerative feed-forward manner as a means of amplifying the Ca^{2+} original signal.

about RyR1 and RyR2 and the role that they play in E-C coupling although more recent evidence suggests RyR3 may be involved in subunit assembly and in conferring some form of functional modulation (10). Cytoplasmic Ca^{2+} concentrations up to about 100 μM maximally activate RyR1 channels (31,51), whereas cytoplasmic Ca^{2+} concentrations in the millimolar range (above 1 mM) are inhibitory (31,51,52). This bimodal regulation is generally attributed to separate activatory and inhibitory Ca^{2+} binding sites accessible from the cytoplasmic side. The latter is not the case with cardiac RyR2 which does not appear to inactivate at these cytosolic Ca^{2+} levels where it also shows no loss of [^{3}H]-ryanodine binding (53). In addition, both RyR channel isoforms are regulated by Ca^{2+} ions on its luminal face (47–49,51). Here, however, much less is known and the opinion is somewhat divided as to exactly where luminal Ca^{2+} acts. Both (a) vesicle studies of Ca^{2+} release and (b) bilayer studies of single RyR1 and RyR2 channels reveal a bimodal mechanism of Ca^{2+}-dependent regulation in which Ca^{2+} may cause both activation and inhibition of channels (51,54,55). Vesicle ion flux studies have also produced conflicting results regarding the influence of Ca^{2+} loads upon Ca^{2+} release rates from passively loaded SR vesicles. Earlier studies revealed Ca^{2+} release was steeply dependent upon increases in total luminal Ca^{2+} loads up to ~110 μM in skeletal muscle triads and progressively declined to a plateau following further filling to millimolar levels (54). Essentially similar results were obtained with active SR Ca^{2+} loading in skinned frog skeletal muscle fibres, in situ (55). However, others have reported progressive increases in Ca^{2+} release rates from passively loaded frog and rabbit skeletal muscle HSR vesicles with attainment of luminal Ca^{2+} loads in the low mM range (56).

In contrast to studies with passively loaded vesicles, studies of active Ca^{2+} transport have revealed an apparent luminal Ca^{2+} threshold dependence of Ca^{2+} release (57–65). It has been postulated that calsequestrin may exert direct luminal modulation of RyR1 and that luminal Ca^{2+} binding to calsequestrin may be involved in the expression of the threshold dependence of Ca^{2+} release (61,66). One other contributory factor to this apparent threshold dependence of HSR Ca^{2+} release may be back-inhibition of SERCA1 pumps during active Ca^{2+} sequestration, particularly since it commonly occurs at millimolar luminal Ca^{2+} loads (67). However, it is far from clear exactly how luminal Ca^{2+} is compartmentalized in HSR membranes. Previous estimates using $^{45}Ca^{2+}$ radiotracer methods suggest approximately 90% of luminal $[Ca^{2+}]$ in skeletal muscle triads is bound, presumably to calsequestrin (54). Estimates from ^{19}F NMR studies suggest free luminal Ca^{2+} concentration in cardiac SR around 1.5 mM (68). These values agree reasonably well with values of 700 μM reported in cardiac SR vesicle studies using luminally-trapped Mag-Fura-2 (69) as do similar values obtained with Mag-Fura-2 trapped in skeletal muscle triads (70).

Effects of luminal Ca^{2+} upon RyR conductance properties has been explored in bilayer studies of single channels although this work has generated further conflicting ideas about Ca^{2+}-dependent inhibitory mechanisms. In one study, luminal free Ca^{2+} concentrations up to 250 μM reportedly activated RyR Ca^{2+} channels and further Ca^{2+} increases above 1 mM were found to be inhibitory (51). However, this observation was controversial because others have shown that RyR channels are fully active at millimolar luminal Ca^{2+} (47,48). Whether these luminal Ca^{2+} actions arise from binding to specific sites on the luminal face of RyR channels (47–49,71,72) or from complex effects of Ca^{2+} permeation through RyR channels (51) is not clear. Recent mutagenesis studies identified the M2 putative transmembrane domain in RyR3 as a candidate region involved in Ca^{2+}-dependent activation (52). A single E3885A point mutation in this domain produced a 10,000 fold loss in Ca^{2+} sensitivity (52). This region is conserved in all other RyR isoforms suggesting a Ca^{2+} sensor located in the channel domain. Other domain switching studies with RyR chimeras expressed in HEK-293 cells have also localized Ca^{2+} activation and inactivation sites to the COOH terminal region which forms much of the Ca^{2+} permeable pore (45).

Recent models of Ca^{2+}-dependent control: adaptation or permeation?

It is now quite clear that Ca^{2+}-dependent regulation of RyR channels both *in situ* and *in vivo* is not readily explained by cytosolic Ca^{2+} regulation alone as invoked by many models of CICR. For the cell, the challenge is to coordinate the bulk

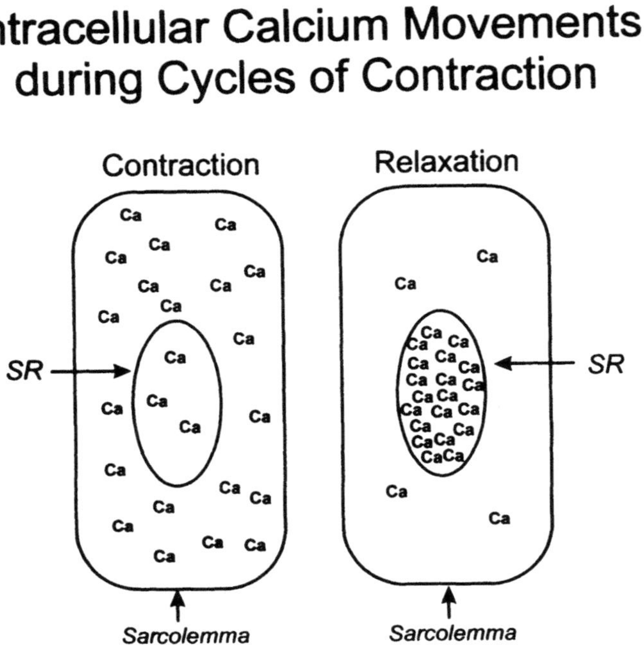

Figure 2. A simplified scheme of intracellular Ca^{2+} movements between SR and cytosol illustrating that each cycle of relaxation and contraction attends inverse changes in the concentration of Ca^{2+} ions in the SR lumen and the cytosol.

movement of Ca^{2+} between one compartment and another while having to deal with the consequences of a constantly changing Ca^{2+} environment from a resting state to an active state as shown simply in Figure 2. This sets up the possibility of complex and non-linear RyR regulation by both cytosolic and luminal Ca^{2+} because as cytosolic Ca^{2+} is reduced, luminal Ca^{2+} increases and vice versa. This behavior is hard to capture and describe under steady state conditions, *in vitro*, as evidenced from the fact that modeling studies of CICR, while very good at describing its feed-forward properties, poorly account for its feed-back characteristics (42). RyR "adaptation" has been proposed as a theory to explain Ca^{2+} dependent control (73–80). Adaptation was thought to involve Ca^{2+}-dependent allosteric interactions between RyR subunits which countered inherently unstable feed-forward properties of CICR. Adaptation was initially proposed following the observation that RyR Ca^{2+} channels appeared capable of re-activation without the requirement for complete reduction of cytosolic Ca^{2+}. This property was thought to explain both discrepant findings amongst single channel studies of Ca^{2+} dependent activation/inactivation and the observed mechanistic features of CICR and ligand-induced Ca^{2+} release. While attractive, acceptance of this putative mechanism has met with considerable criticism concerning both methodological artifact (81,82) and

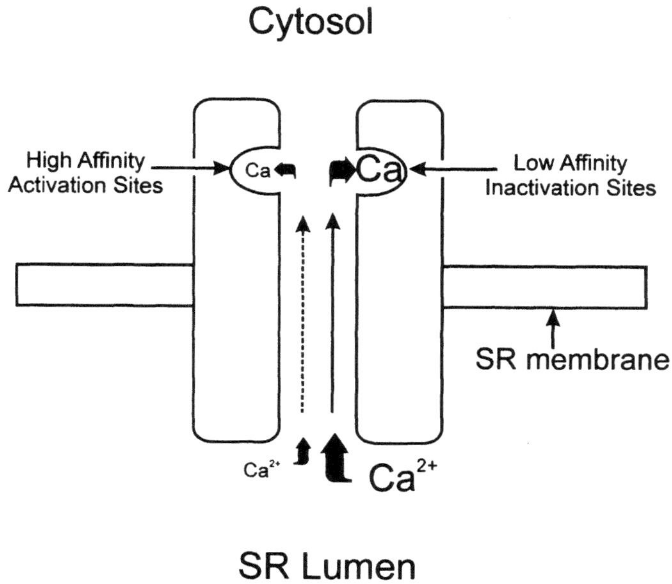

Figure 3. A highly simplified scheme illustrating how Ca^{2+} gradients across luminal and cytosolic faces of RyR channels might regulate its activation and inactivation through rate-dependent binding of Ca^{2+} ions at high and low affinity sites in its Ca^{2+}-permeable pore (see text).

its thermodynamic feasibility (83). In essence, adaptation theorists propose that the affinity of cytosolic Ca^{2+} activation sites decreases with progressive Ca^{2+} occupancy at activation sites. Thus, RyR channels are thought to adapt to a given level of cytosolic Ca^{2+} rather than inactivate. However, for this to occur there must be energy input and since adaptation has been putatively demonstrated in bilayer studies in the absence of ATP, the theory becomes considerably weakened. Currently, there is growing opposition for this initially appealing model. More recent evidence from rapid line-scan confocal studies in intact cells suggest that termination of Ca^{2+} release is largely determined by use-dependent RyR inactivation and not adaptation (84).

One intriguing proposal that Ca^{2+} ions on the luminal side of RyR channels have access to Ca^{2+} activation and inactivation sites on the cytosolic side (51). Tripathy and Meissner (51) demonstrated that the rate of luminal to cytosolic Ca^{2+} flux exerted a bimodal influence such that ion fluxes up to 2 pA activated the channel but increasingly faster Ca^{2+} fluxes up to 12 pA inactivated it (51). Figure 3 schematically illustrates the principle of this proposed mechanism. High Ca^{2+} concentrations in the vicinity of low-affinity inactivation sites are required for RyR channel closure and this occurs when fast outward Ca^{2+} fluxes are driven by elevated luminal Ca^{2+} concentrations. In contrast, only relatively low luminal-to-cytosolic Ca^{2+} fluxes and, therefore, low luminal Ca^{2+} concentrations are required for binding to high affinity Ca^{2+} activation sites. This suggests that the rate at which Ca^{2+} ions flow through the channel provides an important means of regulating RyR channel opening and closing via a Ca^{2+} gradient-sensitive mechanism. An important corollary to this idea is that SERCA pumps, through their influence on Ca^{2+} gradient formation, play an important role in determining Ca^{2+} dependent activation and inactivation of RyR channels in intact membranes.

Functional interactions between SERCA pumps and RyR channels

It has long been suspected that SERCA pumps participate in regulation of SR Ca^{2+} release but few have little idea about how this might occur and to what extent this may be brought about by SERCA-mediated (a) reduction of cytosolic Ca^{2+}, (b) elevation of luminal Ca^{2+}, or (c) both (60,66,70,85–92). This question is especially relevant when considering Ca^{2+} exchanges in non-junctional cisternal SR and corbular SR regions where substantial co-localization of SERCA and RyR proteins exist (35,41). The problem surrounds the fact that RyR channels and SERCA pumps exhibit similar cytosolic Ca^{2+} dependency for activation. In principle, this could produce futile Ca^{2+} cycles in the absence of coordinated pump and channel activity. It has been suggested that RyR channels are sensitive to trans SR Ca^{2+} gradients which, in intact membranes, are largely determined by SERCA activity (67,92). What is mechanistically at the heart of Ca^{2+} gradient sensitivity is not clear although one intriguing possibility is that it arises from a dependence of RyR activation/inactivation upon the rate of Ca^{2+} ion permeation through the channel itself as described above (51). This model predicts that the speed with which SERCA pumps are able to generate high outward luminal-to-cytosolic Ca^{2+} gradients may then be an important determinant of RyR channel inactivation. It also predicts that when SERCA pumps fail, RyR channels become leaky (93–96). Our recent demonstration that formation of ryanodine-inactivated RyR2 states required active SERCA2 pumps is consistent with this (67). On the other hand, loss of SERCA pump activity may promote leaky RyR channels and imbalances in intracellular Ca^{2+} homeostasis.

Several vesicle studies have hinted at a complex involvement of SERCA1 pumps in RyR1-mediated Ca^{2+} release regulation but mechanistic details of this remain obscure (60–62,88,90). Data from rapid kinetic experiments suggested, for example, that early conformational changes and complex oscillations in SERCA1 intermediary kinetic states may have a causative role in Ca^{2+} release (88,90). This was determined from transient increases in *EP* formation and intrinsic tryptophan fluorescence seen to precede Ca^{2+} release induced by caffeine, ADP and a Ca^{2+} jump. SERCA1 pump reversal, *per se*, appears not to be a realistic pathway for physiological Ca^{2+} efflux, as earlier proposed (97) although more recent evidence suggests it may be an important source of SR Ca^{2+} leaks in the failing heart (98). More recently, it has also been suggested that rates of SR Ca^{2+} release and Ca^{2+} re-uptake are coordinated at critical luminal Ca^{2+} loads (70,92). However, the physiological significance of this is somewhat obscure.

One of the most striking observations to indicate some involvement of SERCA pumping in SR Ca^{2+} release comes from the observation that SERCA1 overexpression in transgenic mice (a) elevated rates of myocardial pressure development, (b) shortened time to peak pressure and (c) doubled the amplitude of cytosolic Ca^{2+} transients (85,86). These enhanced Ca^{2+} release dynamics were not due to altered RyR2 protein expression suggesting SERCA1 activity indirectly impacts on RyR-mediated Ca^{2+} release. Why this occurs is not clear. One obvious possibility is that increasing the ability to pump Ca^{2+} into the SR simply loads the SR with more releasable Ca^{2+}. However, several other studies suggest that it is not a simple relationship as indicated from studies of SERCA1 and RyR1 mutant overexpression in HEK-293 cells (50). Here, different combinations of SERCA1 and mutant/wild type RyR1 channel expression produced varying effects upon (a) transmembrane Ca^{2+} distributions and (b) responses to RyR channel agonists that defied explanation by any simple Ca^{2+} handling paradigm.

Effects of SERCA pump inhibition on SR Ca^{2+} release

Although potential mechanistic links between RyR1 activation and SERCA1 activity have so far not been described, the idea that SERCA pumps are involved in regulating SR Ca^{2+} efflux is supported by results of studies showing that SERCA1 inhibitors, thapsigargin and cyclopiazonic acid, induce RyR1-mediated Ca^{2+} release from isolated HSR membranes (96,99,100). Because thapsigargin and cyclopiazonic acid solely target SERCA1 pumps in HSR membranes (101–108) they are useful tools for probing potential involvement of SERCA pumps in Ca^{2+} release. A curious observation was the fact that the efficacy of both compounds upon activation of release from HSR vesicles was dependent upon the extent of luminal Ca^{2+} loading (96). Thus, it is worth examining how these compounds are known to act. Thapsigargin is a naturally occurring sesquiterpine lactone and, in "light" SR membrane preparations primarily enriched in SERCA pumps, its principle effects are inhibition of (a) SERCA mediated catalysis, (b) EP formation and, (c) Ca^{2+} uptake (101,105–107). Thapsigargin is known to bind irreversibly with low picomolar affinity to the M3 and M4 transmembrane stalk region of SERCA1 Ca^{2+} pumps (109,110) in a 1:1 molar ratio (106,107,111,112). A simplified scheme of the SERCA pump cycle is $E_1 \rightarrow E_1 \sim P \rightarrow E_2\text{-P} \rightarrow E_2$ where E_1 is the outwardly-directed high affinity Ca^{2+} bound state, $E_1 \sim P$ is the high energy Ca^{2+}-occluded phosphoenzyme, E_2-P is low energy phosphoenzyme with low Ca^{2+} binding affinity, E_2 is the dephosphoenzyme which undergoes conformational change to return SERCA to the E_1 form. The preferred SERCA1 conformation favoring thapsigargin binding is the E_2 state and thapsigargin reduces or inhibits the binding of Ca^{2+} and ATP to the ATPase in the E_2 state (112). The fact then that elevated luminal Ca^{2+} favors formation of the E_2 state (113) suggests it (luminal Ca^{2+}) is an important factor in predisposing SR membranes to Ca^{2+} release. In contrast to thapsigargin, less is known about the specific actions of cyclopiazonic acid. It is an indole-based metabolite of *Aspergillus* and *Penicillium* with an apparent Ki for SERCA1 catalytic activity estimated at 162 nM (104). Cyclopiazonic acid is thought to competitively inhibit ATP binding to SERCA1 and appears to (a) stabilize SERCA1 in the E_2 conformation and (b) inhibit high affinity Ca^{2+} binding (104,108). This again reinforces the above idea that luminal Ca^{2+} accumulation, in favoring the E_2 state may mediate this apparent role of SERCA pumps in SR Ca^{2+} release.

The importance of rapid RyR1 channel closure during initial phases of SR Ca^{2+} sequestration is that it promotes efficient luminal Ca^{2+} filling and high affinity Ca^{2+} binding to calsequestrin (61,94). As calsequestrin is progressively bound up with Ca^{2+}, RyR channels then become progressively more sensitive to cytosolic ligands, as our previous studies suggest (93). As calsequestrin becomes saturated at a total physiological luminal Ca^{2+} load of approximately $100\,nmol{\cdot}mg^{-1}$ luminal free Ca^{2+} would be expected to increase to levels sufficient to impact upon SERCA activity. Even at a relatively low peak value of approximately $200\,\mu M$, a predicted 30% reduction in SERCA catalytic potential (113) would arise from back-inhibition of SERCA pumps. This form of autogenic inhibition of SERCA pumps may explain supralinear feedback of ER/SR Ca^{2+} transport (114) and may also account for the markedly non-linear gain in Ca^{2+} release observed in isolated cardiomyocytes at threshold levels of luminal SR Ca^{2+} filling (115). The significance is that luminal Ca^{2+} loads may determine whether cytosolic Ca^{2+} is rapidly accumulated or remains in the cytosol to trigger RyR activation. Thus, some level of SERCA/RyR "Cross-Talk" may exist in which SERCA pumps influence RyR inactivation/activation through its determination of Ca^{2+} distributions across SR membranes.

The nature of any functional interaction between SERCA pumps and RyR channels is likely to be very complex given that each protein is bimodally regulated within similar concentration ranges of luminal and cytosolic Ca^{2+}. We imagine that a higher level of complexity then arises from the fact that, *in situ*, Ca^{2+} exchanges are dynamic. That is, filling of one compartment (i.e. SR lumen) occurs at the expense of synchronous depletion of the other (i.e. cytosol). However, currently available techniques which offer some hope for describing mechanisms are often limited to more steady-state analyses of RyR function and lack the ability to track the constantly fluctuating Ca^{2+} environment. To explore these complexities it seems necessary in the future to develop novel methods in order to track the dynamics of SERCA catalytic activity and transmembrane Ca^{2+} distribution in both free and bound compartments under physiological conditions of dynamic Ca^{2+} exchange (i.e. in the absence of cytosolic or luminal Ca^{2+} buffering).

ORIGIN AND REGULATION OF SR MEMBRANE CA^{2+} LEAKS

The presence of SR membrane Ca^{2+} leaks from resting striated muscle is now well recognized. However, their physiological significance (if any) is unclear. In addition, the complete identity of all possible ion channel pathways responsible for these leaks under normal and pathophysiological conditions remains to be determined. That Ca^{2+} leaks were present in resting cardiomyocytes was initially inferred from studies reporting time-dependent depletion of SR luminal Ca^{2+} (116–119). However, these studies provided little direct information about the rate of unidirectional SR Ca^{2+} efflux because of unknown contributions of SERCA pumps to reloading of SR Ca^{2+} stores. This contribution was later revealed in rest-decay studies of SR luminal Ca^{2+} in which the ability of SERCA pumps to reload leaked Ca^{2+} is usurped by artificially activating Na/Ca exchanger activity (118). These observations served to confirm the idea that, in the resting cardiomyocyte, SR Ca^{2+} leaks are well compensated by SERCA-mediated Ca^{2+} re-loading. Bassani and Bers were the first to measure the actual magnitude of unidirectional SR efflux from resting cardiomyocytes. This was determined from the rate of post-rest depletion of SR luminal Ca^{2+} estimated from decline of caffeine-induced Ca^{2+} release under conditions where (a) Na/Ca exchanger was activated and (b) SERCA pumps were thapsigargin-inhibited (30). Their results showed that the unidirectional Ca^{2+} efflux rate is relatively small (0.27–$0.32\,\mu M \cdot s^{-1}$), being about four orders of magnitude lower than the maximum rates of SR Ca^{2+} efflux following physiological activation of SR Ca^{2+} release during excitation-contraction coupling. Calculations revealed this unidirectional Ca^{2+} efflux rate was also much less than the maximum rate of SERCA-mediated Ca^{2+} uptake but consistent with rates of SR membrane Ca^{2+} efflux associated with "Ca^{2+} sparks" (0.2–$0.8\,\mu M \cdot s^{-1}$). Thus, these results also suggested that a major fraction of the resting SR Ca^{2+} leak was due to a spontaneous RyR activation.

"Ca^{2+} sparks", luminal Ca^{2+} and SERCA pumps

The fact that the SR Ca^{2+} leak rate compares favorably with Ca^{2+} efflux arising from "Ca^{2+} sparks" suggests the former may trigger the latter. Ca^{2+} sparks were first described in cardiomyocytes by Cheng et al. (120) who observed local spontaneous cytosolic Ca^{2+} transients in resting rat heart cells. Sparks were ryanodine-sensitive but exhibited slow Ca^{2+} release kinetics. It was concluded that sparks represented an elementary Ca^{2+} release event during excitation contraction coupling. This study helped to explain the occurrence of both spontaneous and triggered Ca^{2+} release. Subsequent studies have shown that opening of a single voltage-gated L-type Ca^{2+} channel (dihydropyridine receptor or DHPR) on the surface of cardiomyocytes can activate between four and six RyR2 multimers in close proximity on the sarcoplasmic reticulum membrane (121). These triggering DHPR Ca^{2+} fluxes were referred to as "Ca^{2+} sparklets". As in the earlier Cheng study (120), subsequent studies have identified a major effect of luminal Ca^{2+} upon the activation of spark frequency (122). The frequency of sparks reportedly declines with reduction in luminal Ca^{2+} and a similar effect arises when SERCA pumps are inhibited by thapsigargin (123,124). This latter effect of pump inhibition is thought to arise as a result of its effect to deplete luminal Ca^{2+}. The study of Lukyanenko et al. (123) proposed the existence of a luminal Ca^{2+} sensor linking RyR functional activity to the luminal Ca^{2+} loading level as part of an autoregulatory mechanism for adjusting RyRs activation and therefore the gain of CICR.

A particularly significant observation in this regard comes from reported effects of low-affinity Ca^{2+} buffers (ADA, citrate and maleate) incorporated into SR of permeabilized rat cardiomyocytes (125). Terentyev et al. (125) showed these buffers were effective in dramatically increasing total luminal Ca^{2+} and this produced 2–3 fold increases in spark amplitude, time-to-peak amplitude, and duration of sparks. However, the frequency of sparks from the same site was markedly decreased. It was concluded, that local pools of luminal Ca^{2+} were important regulators of spark termination. Curiously, Chu et al. (126,127), nearly two decades ago, observed comparable effects of dicarboxylate anions (maleate and succinate) to promote Ca^{2+} uptake into isolated SR vesicle but at the same time enhance subsequent spontaneous Ca^{2+} release and SERCA mediated catalytic activity. Thus, these experiments have helped to define a critical role for luminal Ca^{2+} in promoting membrane Ca^{2+} leaks. Exactly how this occurs is not known because there is considerable disagreement and uncertainty about how Ca^{2+} stores (free versus bound) are organized in SR membranes and how these contribute to both the activation and source of released Ca^{2+}.

Numerous groups have since joined in the investigation of these phenomena because of their potential role as initiators of CICR. Indeed, rapid line scanning confocal techniques required to analyse sparks have become de rigueur within the last decade. Despite all of this interest, relatively little is known about (a) how sparks are themselves initiated and (b) what our current understanding of sparks tells us about mechanisms of triggered Ca^{2+} release. Anywhere between one and fifteen RyR Ca^{2+} channels have been reported to participate in the spontaneous appearance of a Ca^{2+} spark (128,129). The latter number is thought to be a more accurate estimate because previous estimates were based upon the use of erroneous flux reconstruction algorithms. However, regardless of the eventually determined number of RyRs involved in spontaneous sparks, the observation and description of spontaneous SR Ca^{2+} release is not new. Nearly two decades ago, Ca^{2+} transport studies had reported slow spontaneous Ca^{2+} release from SR vesicles actively loaded with Ca^{2+} (130–133). This appeared not to be a consequence of impaired SERCA pump function since this released Ca^{2+} was then re-accumulated in a subsequent phase of Ca^{2+} uptake. This suggests there may be a variable leak in SR membranes. However, this observation was made all the more obscure by the fact that these events were observed in regions of the SR not expected to contain Ca^{2+} release sites (130). This raises the possibility that that there may be additional leaks in SR membranes. More recently, it has been discovered that preceding and following many sparks is the appearance of what have been described as "Ca^{2+} embers" (134,135). Embers are seen as small, transient elevations in cytoplasmic Ca^{2+} and the fact that they are temporarily and spatially associated with sparks suggests that they may reveal events involved in the activation and termination of sparks. However, the true relationship between embers and sparks remains very much a black box. Calculated Ca^{2+} release flux associated with embers is consistent with the opening of a single RyR channel. But how this then leads to the activation of an apparent minimum cluster of 15 or more other channels remains unknown. One of the most striking effects upon Ca^{2+} spark frequency is the influence of mitochondrial function in intact cells. After metabolic poisoning of skeletal muscle single fibres with oligomycin, FCCP or antimycin, Ca^{2+} spark frequency was found to markedly decline (136). It was concluded that in intact cells, mitochondria were important downregulators of SR Ca^{2+} spark frequency because of the importance of ATP supply in promoting SERCA-mediated Ca^{2+} uptake. Similar results have been recently obtained in permeabilized cardiomyocytes following phosphocreatine withdrawal (137). Thus, it seems likely that the primary role for SERCA pumps in Ca^{2+} sparks is in reloading of Ca^{2+} stores for subsequent release.

Ca^{2+} leaks and RyR-associated immunophillin binding proteins

Within the last decade, a separate area of investigation into RyR regulation has revealed an important level of control attained through the association of a class of intracellular rapamycin and FK506 (tacrolimus)-binding proteins, known as FKBPs, with RyR Ca^{2+} channels. Interest in the role of FKBP proteins in ryanodine receptor regulation began with the finding that FKBP12 was tightly associated following purification of RyRs from CHAPS-solubilized skeletal and cardiac sarcoplasmic reticulum membranes (138). Early three-dimensional reconstructions of cryo-electronmicrograph sections reveal CHAPS-solubilized RyRs possess a 4-fold symmetry (139–141). Similar studies have since revealed FKBP12 and FKBP 12.6, respectively, bind to domain 6 within the cytosolic assembly region of skeletal and cardiac muscle RyR channels (14,142). It is now known 4 FKBP molecules associate with one RyR channel resulting in formation of a functional heterotetrameric complex (143). Both rapamycin and FK506 are highly effective in physically removing FKBPs from skeletal and cardiac RyRs (144,145). This suggests that these immunophillins reduce the binding affinity of FKBPs for RyR Ca^{2+} channel complexes. Bilayer studies of single RyR channels have shown one of the major effects of FKBP is to regulate the stability of the open and closed states of the ryanodine receptor Ca^{2+} channel (146). FKBP removal from RyR channel complexes is seen to produce 4 subconductance states of 25%, 50%, 75% and 100% maximal (147). Because the channel multimer is made up of 4 individual RyR monomers this data suggests that the extent of RyR subunit association is reflected in the individual conductance states. Thus, FKBPs appear to have an important role in coordinating the proper gating of the channel and stabilizing it between fully open and fully closed configurations.

What is the physiological function of FKBP12/12.6?

Consistent with this regulatory action of FKBPs upon stability of RyR channel states, functional studies of Ca^{2+} transport in isolated sarcoplasmic reticulum membranes have also shown reduction in rates of Ca^{2+} uptake in FKBP-denuded vesicles and FK506-treated vesicles (143). In FKBP-denuded vesicles normal transport efficiency could then be restored upon re-addition of purified FKBPs (143). Thus, one major functional effect of FKBPs upon SR membranes appears to be in the regulation of RyR-mediated Ca^{2+} leaks. Exactly how this is achieved is uncertain. FKBP proteins may assist in the formation of functional multimeric RyR channels possibly by increasing subunit cooperativity (148,149). However, FKBP knockout studies do not indicate removal of immunophillin binding proteins produces any aberrant alignment and structural organization of RyR channel multimers. Indeed, in a recent study which proposed the existence of coupled RyR gating in closely packed RyR arrays it was shown that FKBP12.6 removal functionally uncouples this gating without affecting the physical association at cytosolic contacts (149). Interestingly, PKA phosphorylation of FKBP12.6 was also found to promote its dissociation from RyR2 in cardiac SR (150). PKA phosphorylation was found to increase RyR2 open probability resulting in a defective channel. Such effects are thought to explain how, in the failing heart, PKA hyperphosphorylation leads to leaky RyR2 Ca^{2+} channels thus contributing to intracellular Ca^{2+} overload and contractile dysfunction. Consistent with this, recent studies of tachycardia-induced heart failure showed that low-dose propranolol administration (to block β-adrenergic stimulation and cyclic AMP production) restored normal both RyR2 function and normal FKBP12.6/RyR2 stoichiometry in the left ventricular muscle of dog hearts (151). However, in other studies a role for FKBPs has been proposed in regulating the physical coupling of RyRs with the II-III loop domain of DHPRs (152). In intact RyR multimers peptide sequences corresponding to regions of this loop domain were shown to activate RyRs (147). However, in FKBP-denuded RyRs this activatory effect of DHPR loop peptides was absent (147).

FKBP12/12.6 transgenic studies

Although the collective evidence from various studies, *in vitro*, indicates an important role for FKBP proteins in regulating RyR leak states, there remain significant questions regarding the physiological significance of these proteins for E-C coupling, *in vivo*. Part of the problem surrounds the fact that in FKBP knockout studies, where the attempt is to reveal mechanisms through loss of function experiments, we remain uncertain about the extent to which cells compensate for untranscribed genes. In this respect FKBP12 knockout mice develop severe defects in heart structure and function and embryos, in almost all cases, die close to birth (153). Although this is not particularly surprising given the presumed role of the FKBPs upon RyR function, the results are hard to reconcile with the finding that FKBP12.6 knockout mice do not die at birth (154,155) especially since FKBP12.6 is the preferred cardiac isoform for binding to RyR2 (156,157). Even more intriguing, is the finding of gender differences in FKBP12.6 knockout mice. As shown by Xin et al. (154) only male mice exhibited cardiac hypertrophy. Female mice did not exhibit this physical defect suggesting possible hormonally-related differences as an explanation for this discrepancy. In keeping with this idea, Xin et al. (154) tested the effects of Tamoxifen administration in female knockout mice to inhibit estrogen production. The results clearly showed that Tamoxifen treatment promoted the same degree of hypertrophy seen in male knockout mice suggesting that, at some point, estrogen interposes itself within second messenger pathways linked to FKBP12.6-dependent mechanisms. One possibility is that estrogen acts at the level of NFAT which is dependent upon calcineurin and FKBP12 activity. These important observations clearly reveal a further level of complexity in the cellular physiology of FKBPs within striated muscle and seem certain to attract a greater interest in this protein.

The uncertainties regarding the physiological role of FKBP12.6 and its involvement in excitation-contraction coupling are deepened by the observation that, in some animals species, (rabbits in particular) the intracellular content of this protein is much lower than that expected in the case where every RyR2 multimer possessed 4 copies of FKBP proteins. In earlier studies, Timerman et al. (157) showed that nearly 20% of the total RyR2 pool in cardiomyocytes was unoccupied by the available level of FKBP12.6 molecules. Why this should be the case is not clear although it does suggest that not all RyR2 populations require FKBPs proteins for function. Indeed, the effects of rapamycin are quite divergent and seem to depend on species and cell type. Despite the fact that rapamycin-induced removal of FKBPs from SR membranes is effective in disrupting normal RyR2 gating, it remains to be established whether this is strictly a function of the loss of FKBP or some effect of residual rapamycin used to engineer protein dissociation in the first place. Rapamycin is highly hydrophobic and it is possible that it may stably associate with SR membranes to independently modify conductance properties of RyR2. In a recent study by Prestle et al. (158) markedly lower levels of FKBP12.6 versus FKBP12 were observed in rabbit cardiomyocytes. In that study, rapamycin (which in bilayer and vesicle studies has marked effects upon RyR2 activity) had barely any effect upon Ca^{2+} transport properties in intact cells. This, again, is consistent with the idea that not all RyRs require large endogenous levels of FKBPs for function. However, in studies of adenovirally-mediated FKBP12.6 transfection in cardiomyocytes marked improvements in Ca^{2+} transport efficiency and Ca^{2+} loading were observed suggesting that if this protein is present it acts to stabilize RyR-dependent leaks through SR membranes, *in vivo*, as well as *in vitro* (158).

While the above Prestle study provides solid evidence of major FKBP12.6 effects upon RyR2, *in vivo*, Valdivia (159) noted that this study tells us, at best, only what FKBP12.6 can do under these circumstances rather than what the FKBP12.6 actually does. Perhaps the most curious observation from the Prestle study was the fact that, although larger Ca^{2+} releases were observed from FKBP12.6 transfected cardiomyocytes, the kinetics of cell shortening was decreased even though the amplitude of Ca^{2+} transients and total cell contractions were increased. These observations suggest FKBP12.6 may modulate a slow component of RyR-mediated Ca^{2+} release rather than the fast initial component. Thus, there may be some kinetic component to FKBP12.6 regulation of Ca^{2+} release and the extent to which it is involved in cellular control of SR Ca^{2+} fluxes may depend on the specific contribution of SR membranes to intracellular Ca^{2+} regulation during E-C coupling.

ACKNOWLEDGEMENTS

The authors would like to express their gratitude to Dr. G.N. Pierce for helpful discussion during the preparation of the manuscript. The authors also wish to thank the Canadian Institutes for Health Research for their support of this work.

REFERENCES

1. Carafoli E, Santella L, Branca D, Brini M. 2001. Generation, control, and processing of cellular calcium signals. Crit Rev Biochem Mol Biol 36:107–260.
2. Carafoli E. 2002. Calcium signaling: a tale for all seasons. Proc Natl Acad Sci U S A 99:1115–1122.
3. Tsien RY. 1986. New tetracarboxylate chelators for fluorescence measurement and photochemical manipulation of cytosolic free calcium concentrations. Soc Gen Physiol Ser 40:327–345.
4. Wier WG, Lopez-Lopez JR, Shacklock PS, Balke CW. 1995. Calcium signalling in cardiac muscle cells. Ciba Found Symp 188:146–160; discussion 160–144.
5. Schneider MF, Klein MG. 1996. Sarcomeric calcium sparks activated by fiber depolarization and by cytosolic Ca2+ in skeletal muscle. Cell-Calcium 20:123–128.
6. Cannell MB. 1997. New insights into cardiac excitation-contraction coupling in normal and hypertension/failure animal models. J Hum Hypertens 11:555–558.
7. Kraus D, Khoury S, Fendyur A, Kachalsky SG, Abu-Hatoum T, Rahamimoff R. 2000. Intracellular calcium dynamics–sparks of insight. J Basic Clin Physiol Pharmacol 11:331–365.
8. Bkaily G, Jacques D, Pothier P. 1999. Use of confocal microscopy to investigate cell structure and function. Meth Enzymol 307:119–135.
9. Bkaily G, Pothier P, D'Orleans-Juste P, Simaan M, Jacques D, Jaalouk D, Belzile F, Hassan G, Boutin C, Haddad G, Neugebauer W. 1997. The use of confocal microscopy in the investigation of cell structure and function in the heart, vascular endothelium and smooth muscle cells. Mol Cell Biochem 172:171–194.
10. Protasi F, Takekura H, Wang Y, Chen SR, Meissner G, Allen PD, Franzini-Armstrong C. 2000. RYR1 and RYR3 have different roles in the assembly of calcium release units of skeletal muscle. Biophys J 79:2494–2508.
11. Fessenden JD, Wang Y, Moore RA, Chen SR, Allen PD, Pessah IN. 2000. Divergent functional properties of ryanodine receptor types 1 and 3 expressed in a myogenic cell line. Biophys J 79:2509–2525.
12. Ward CW, Schneider MF, Castillo D, Protasi F, Wang Y, Chen SR, Allen PD. 2000. Expression of ryanodine receptor RyR3 produces Ca2+ sparks in dyspedic myotubes. J Physiol 525(Pt 1):91–103.
13. Ebashi S, Endo M. 1968. Calcium ion and muscle contraction. Prog Biophys Mol Biol 18:123–183.
14. Wagenknecht T, Radermacher M, Grassucci R, Berkowitz J, Xin HB, Fleischer S. 1997. Locations of calmodulin and FK506-binding protein on the three-dimensional architecture of the skeletal muscle ryanodine receptor. J Biol Chem 272:32463–32471.
15. Zhang L, Kelley J, Schmeisser G, Kobayashi YM, Jones LR. 1997. Complex formation between junctin, triadin, calsequestrin, and the ryanodine receptor. Proteins of the cardiac junctional sarcoplasmic reticulum membrane. J Biol Chem 272:23389–23397.
16. Campbell KP, Knudson CM, Imagawa T, Leung AT, Sutko JL, Kahl SD, Raab CR, Madson L. 1987. Identification and characterization of the high affinity [3H]ryanodine receptor of the junctional sarcoplasmic reticulum Ca2+ release channel. J Biol Chem 262:6460–6463.
17. Lytton J, Westlin M, Hanley MR. 1991. Thapsigargin inhibits the sarcoplasmic or endoplasmic reticulum Ca-ATPase family of calcium pumps. J Biol Chem 266:17067–17071.
18. Lytton J, Westlin M, Burk SE, Shull GE, MacLennan DH. 1992. Functional comparisons between isoforms of the sarcoplasmic or endoplasmic reticulum family of calcium pumps. J Biol Chem 267:14483–14489.
19. Wu KD, Lytton J. 1993. Molecular cloning and quantification of sarcoplasmic reticulum Ca(2+)-ATPase isoforms in rat muscles. Am J Physiol 264:C333–341.
20. Wu KD, Lee WS, Wey J, Bungard D, Lytton J. 1995. Localization and quantification of endoplasmic reticulum Ca(2+)-ATPase isoform transcripts. Am J Physiol 269:C775–784.
21. Poch E, Leach S, Snape S, Cacic T, MacLennan DH, Lytton J. 1998. Functional characterization of alternatively spliced human SERCA3 transcripts. Am J Physiol 275:C1449–1458.
22. Wu KD, Bungard D, Lytton J. 2001. Regulation of SERCA Ca2+ pump expression by cytoplasmic Ca2+ in vascular smooth muscle cells. Am J Physiol Cell Physiol 280:C843–851.
23. de Meis L, Vianna AL. 1979. Energy interconversion by the Ca2+-dependent ATPase of the sarcoplasmic reticulum. Ann Rev Biochem 48:275–292.
24. Ribeiro JM, Aragao ES, Vianna AL. 1980. Steady state kinetics of the (Ca2+ + Mg2+)-dependent P-nitrophenylphosphatase activity of sarcoplasmic reticulum vessicles. An Acad Bras Cienc 52: 403–409.
25. East JM. 2000. Sarco(endo)plasmic reticulum calcium pumps: recent advances in our understanding of structure/function and biology (review). Mol Membr Biol 17:189–200.

26. Stamler JS, Meissner G. 2001. Physiology of nitric oxide in skeletal muscle. Physiol Rev 81:209–237.
27. Ikemoto N, Yamamoto T. 2002. Regulation of calcium release by interdomain interaction within ryanodine receptors. Front Biosci 7:d671–683.
28. Berman MC. 2001. Slippage and uncoupling in P-type cation pumps; implications for energy transduction mechanisms and regulation of metabolism. Biochimica Et Biophysica Acta 1513: 95–121.
29. Berman MC, McIntosh DB, Davidson GA, Ross DC, Seebregts CJ. 1987. Mechanism of action of the calcium pump of sarcoplasmic reticulum of skeletal muscle. S Afr Med J 72:777–780.
30. Bassani RA, Bers DM. 1995. Rate of diastolic Ca release from the sarcoplasmic reticulum of intact rabbit and rat ventricular myocytes. Biophys J 68:2015–2022.
31. Meissner G. 1994. Ryanodine receptor/Ca2+ release channels and their regulation by endogenous effectors. Ann Rev Physiol 56:485–508.
32. Martonosi AN. 1984. Mechanisms of Ca2+ release from sarcoplasmic reticulum of skeletal muscle. Physiol Rev 64:1240–1320.
33. Fill M, Zahradnikova A, Villalba-Galea CA, Zahradnik I, Escobar AL, Gyorke S. 2000. Ryanodine receptor adaptation. J Gen Physiol 116:873–882.
34. Block BA, Imagawa T, Campbell KP, Franzini Armstrong C. 1988. Structural evidence for direct interaction between the molecular components of the transverse tubule/sarcoplasmic reticulum junction in skeletal muscle. J Cell Biol 107:2587–2600.
35. Franzini Armstrong C, Jorgensen AO. 1994. Structure and development of E-C coupling units in skeletal muscle. Ann Rev Physiol 56:509–534.
36. Chandler WK, Schneider MF, Rakowski RF, Adrian RH. 1975. Charge movements in skeletal muscle. Phil Trans R Soc Lond Biol Sci 270:501–505.
37. Endo M. 1977. Calcium release from the sarcoplasmic reticulum. Physiol Rev 57:71–108.
38. Schneider MF, Chandler WK. 1973. Voltage dependent charge movement of skeletal muscle: a possible step in excitation-contraction coupling. Nature 242:244–246.
39. Schneider MF. 1981. Membrane charge movement and depolarization-contraction coupling. Ann Rev Physiol 43:507–517.
40. Fabiato A. 1985. Time and calcium dependence of activation and inactivation of calcium-induced release of calcium from the sarcoplasmic reticulum of a skinned canine cardiac Purkinje cell. J Gen Physiol 85:247–289.
41. Dulhunty AF, Junankar PR, Stanhope C. 1992. Extra-junctional ryanodine receptors in the terminal cisternae of mammalian skeletal muscle fibres. Proc R Soc Lond B Biol Sci 247:69–75.
42. Stern MD. 1992. Theory of excitation-contraction coupling in cardiac muscle. Biophys J 63:497–517.
43. Du GG, Imredy JP, MacLennan DH. 1998. Characterization of recombinant rabbit cardiac and skeletal muscle Ca2+ release channels (ryanodine receptors) with a novel [3H]ryanodine binding assay. J Biol Chem 273:33259–33266.
44. Du GG, MacLennan DH. 1998. Functional consequences of mutations of conserved, polar amino acids in transmembrane sequences of the Ca2+ release channel (ryanodine receptor) of rabbit skeletal muscle sarcoplasmic reticulum. J Biol Chem 273:31867–31872.
45. Du GG, MacLennan DH. 1999. Ca(2+) inactivation sites are located in the COOH-terminal quarter of recombinant rabbit skeletal muscle Ca(2+) release channels (ryanodine receptors). J Biol Chem 274:26120–26126.
46. Lynch PJ, Tong J, Lehane M, Mallet A, Giblin L, Heffron JJ, Vaughan P, Zafra G, MacLennan DH, McCarthy TV. 1999. A mutation in the transmembrane/luminal domain of the ryanodine receptor is associated with abnormal Ca2+ release channel function and severe central core disease [see comments)]. Proc Natl Acad Sci U S A 96:4164–4169.
47. Sitsapesan R, Williams AJ. 1994. Regulation of the gating of the sheep cardiac sarcoplasmic reticulum Ca(2+)-release channel by luminal Ca2+. J Memb Biol 137:215–226.
48. Sitsapesan R, Williams AJ. 1995. The gating of the sheep skeletal sarcoplasmic reticulum Ca(2+)-release channel is regulated by luminal Ca2+. J Memb Biol 146:133–144.
49. Sitsapesan R, Williams AJ. 1997. Regulation of current flow through ryanodine receptors by luminal Ca2+. J Memb Biol 159:179–185.
50. Tong J, McCarthy TV, MacLennan DH. 1999. Measurement of resting cytosolic Ca2+ concentrations and Ca2+ store size in HEK-293 cells transfected with malignant hyperthermia or central core disease mutant Ca2+ release channels. J Biol Chem 274:693–702.

51. Tripathy A, Meissner G. 1996. Sarcoplasmic Reticulum Lumenal Ca2+ Has Access to Cytosolic Activation and Inactivation Sites of Skeletal Muscle Ca2+ Release Channel. Biophys J 70:2600–2615.
52. Chen SR, Ebisawa K, Li X, Zhang L. 1998. Molecular identification of the ryanodine receptor Ca2+ sensor. J Biol Chem 273:14675–14678.
53. Du GG, Guo X, Khanna VK, MacLennan DH. 2001. Functional characterization of mutants in the predicted pore region of the rabbit cardiac muscle Ca(2+) release channel (ryanodine receptor isoform 2). J Biol Chem 276:31760–31771.
54. Ikemoto N, Ronjat M, Mészáros LG, Koshita M. 1989. Postulated role of calsequestrin in the regulation of calcium release from sarcoplasmic reticulum. Biochemistry 28:6764–6771.
55. Kurebayashi N, Ogawa Y. 1998. Effect of luminal calcium on Ca2+ release channel activity of sarcoplasmic reticulum in situ. Biophys J 74:1795–1807.
56. Donoso P, Prieto H, Hidalgo C. 1995. Luminal calcium regulates calcium release in triads isolated from frog and rabbit skeletal muscle. Biophys J 68:507–515.
57. Kim DH, Ohnishi ST, Ikemoto N. 1983. Kinetic studies of calcium release from sarcoplasmic reticulum in vitro. J Biol Chem 258:9662–9668.
58. Morii H, Tonomura Y. 1983. The gating behavior of a channel for Ca2+-induced Ca2+ release in fragmented sarcoplasmic reticulum. J Biochem 93:1271–1285.
59. Morii H, Takisawa H, Yamamoto T. 1985. Inactivation of a Ca2+-induced Ca2+ release channel from skeletal muscle sarcoplasmic reticulum during active Ca2+ transport. J Biol Chem 260:11536–11541.
60. Kim DH, Sreter FA, Ohnishi ST, Ryan JF, Roberts J, Allen PD, Meszaros LG, Antoniu B, Ikemoto N. 1984. Kinetic studies of Ca2+ release from sarcoplasmic reticulum of normal and malignant hyperthermia susceptible pig muscles. Biochimica et Biophysica Acta 775:320–327.
61. Gilchrist JS, Belcastro AN, Katz S. 1992. Intraluminal Ca2+ dependence of Ca2+ and ryanodine-mediated regulation of skeletal muscle sarcoplasmic reticulum Ca2+ release. J Biol Chem 267: 20850–20856.
62. Nelson TE, Nelson KE. 1990. Intra- and extraluminal sarcoplasmic reticulum membrane regulatory sites for Ca2(+)-induced Ca2+ release. Febs Letters 263:292–294.
63. Nelson TE, Lin M, Volpe P. 1991. Evidence for intraluminal Ca++ regulatory site defect in sarcoplasmic reticulum from malignant hyperthermia pig muscle. J Pharmacol Exp Therap 256:645–649.
64. Ohnishi ST. 1979. Calcium-induced calcium release from fragmented sarcoplasmic reticulum. J Biochem 86:1147–1150.
65. Ohnishi ST. 1979. Interaction of metallochromic indicators with calcium sequestering organelles. Biochimica Et Biophysica Acta 585:315–319.
66. Ikemoto N, Antoniu B, Mészáros LG. 1985. Rapid flow chemical quench studies of calcium release from isolated sarcoplasmic reticulum. J Biol Chem 260:14096–14100.
67. Abrenica B, Gilchrist JSC. 2000. Nucleoplasmic calcium loading regulated by mobilization of perinuclear calcium. Cell-Calcium 28:127–136.
68. Chen W, Steenbergen C, Levy LA, Vance J, London RE, Murphy E. 1996. Measurement of free Ca2+ in sarcoplasmic reticulum in perfused rabbit heart loaded with 1,2-bis(2-amino-5,6-difluorophenoxy)ethane-N,N,N′,N′-tetraacetic acid by 19F NMR. J Biol Chem 271:7398–7403.
69. Shannon TR, Bers DM. 1997. Assessment of intra-SR free [Ca] and buffering in rat heart. Biophys J 73:1524–1531.
70. Saiki Y, Ikemoto N. 1997. Fluorescence probe study of the lumenal Ca2+ of the sarcoplasmic reticulum vesicles during Ca2+ uptake and Ca2+ release. Biochem Biophys Res Comm 241: 181–186.
71. Lukyanenko V, Györke I, Györke S. 1996. Regulation of calcium release by calcium inside the sarcoplasmic reticulum in ventricular myocytes. Pflugers Archiv. European J Physiol 432:1047–1054.
72. Györke I, Györke S. 1998. Regulation of the cardiac ryanodine receptor channel by luminal Ca2+ involves luminal Ca2+ sensing sites. Biophys J 75:2801–2810.
73. Dettbarn C, Gyorke S, Palade P. 1994. Many agonists induce "Quantal" Ca2+ release or adaptive behavior in muscle ryanodine receptors. Mol Pharmacol 46:502–507.
74. Gyorke S, Fill M. 1993. Ryanodine receptor adaptation: control mechanism of Ca(2+)-induced Ca2+ release in heart [see comments]. Science 260:807–809.
75. Gyorke S. 1999. Ca2+ spark termination: inactivation and adaptation may be manifestations of the same mechanism [letter]. J Gen Physiol 114:163–166.

76. Jafri MS, Rice JJ, Winslow RL. 1998. Cardiac Ca2+ dynamics: the roles of ryanodine receptor adaptation and sarcoplasmic reticulum load [published erratum appears in iophys 1998 Jun;74(6):3313]. Biophys J 74:1149–1168.
77. Zahradnikova A, Zahradnik I. 1995. Description of modal gating of the cardiac calcium release channel in planar lipid membranes. Biophys J 69:1780–1788.
78. Valdivia HH, Kaplan JH, Ellis-Davies GC, Lederer WJ. 1995. Rapid adaptation of cardiac ryanodine receptors: modulation by Mg2+ and phosphorylation. Science 267:1997–2000.
79. Sachs F, Qin F, Palade P. 1995. Models of Ca2+ release channel adaptation [letter; comment]. Science 267:2010–2011.
80. Yasui K, Palade P, Gyorke S. 1994. Negative control mechanism with features of adaptation controls Ca2+ release in cardiac myocytes [see comments]. Biophys J 67:457–460.
81. Lamb GD, Stephenson DG. 1995. Activation of ryanodine receptors by flash photolysis of caged Ca2+. Biophys J 68:946–948.
82. Lamb GD. 1997. Ryanodine receptor "adaptation": a flash in the pan? J Mus Res Cell Motil 18:611–616.
83. Stern MD. 1996. "Adaptive" behavior of ligand-gated ion channels: constraints by thermodynamics. Biophys J 70:2100–2109.
84. Sham JS, Song LS, Chen Y, Deng LH, Stern MD, Lakatta EG, Cheng H. 1998. Termination of Ca2+ release by a local inactivation of ryanodine receptors in cardiac myocytes. Proc Natl Acad Sci U S A 95:15096–15101.
85. Baker DL, Hashimoto K, Grupp IL, Ji Y, Reed T, Loukianov E, Grupp G, Bhagwhat A, Hoit B, Walsh R, Marban E, Periasamy M. 1998. Targeted overexpression of the sarcoplasmic reticulum Ca2+-ATPase increases cardiac contractility in transgenic mouse hearts. Circ Res 83:1205–1214.
86. Ji Y, Loukianov E, Loukianova T, Jones LR, Periasamy M. 1999. SERCA1a can functionally substitute for SERCA2a in the heart. Am J Physiol 276:H89–97.
87. Matsui H, MacLennan DH, Alpert NR, Periasamy M. 1995. Sarcoplasmic reticulum gene expression in pressure overload-induced cardiac hypertrophy in rabbit. Am J Physiol 268:C252–258.
88. Mészáros LG, Ikemoto N. 1985. Conformational changes of the Ca2+-ATPase as early events of Ca2+ release from sarcoplasmic reticulum. J Biol Chem 260:16076–16079.
89. Mészáros LG, Ikemoto N. 1985. Ruthenium red and caffeine affect the Ca2+-ATPase of the sarcoplasmic reticulum. Biochem Biophys Res Comm 127:836–842.
90. Mészáros LG, Brown KL, Ikemoto N. 1987. 4′,6-Diamidino-2-phenylindole, a novel conformational probe of the sarcoplasmic reticulum Ca2+ pump, and its effect on Ca2+ release. J Biol Chem 262:11553–11558.
91. Periasamy M, Reed TD, Liu LH, Ji Y, Loukianov E, Paul RJ, Nieman ML, Riddle T, Duffy JJ, Doetschman T, Lorenz JN, Shull GE. 1999. Impaired cardiac performance in heterozygous mice with a null mutation in the sarco(endo)plasmic reticulum Ca2+-ATPase isoform 2 (SERCA2) gene. J Biol Chem 274:2556–2562.
92. Saiki Y, Ikemoto N. 1999. Coordination between Ca2+ release and subsequent re-uptake in the sarcoplasmic reticulum. Biochemistry 38:3112–3119.
93. Gilchrist JSC, Palahniuk C, Abrenica B, Cook T. 2002. RyR1/SERCA1 Cross-Talk Regulation of Calcium Transport in Heavy Sarcoplasmic Reticulum Vesicles. J Biol Chem (under revision).
94. Gilchrist JSC, Palahniuk C, Abrenica B, Cook T. 2002. Mag-Fura-2 and Chlorotetracycline Fluorescence Reveal Two Distinct HSR Luminal Calcium Pools. J Biol Chem (under revision).
95. Palahniuk CP, Abrenica B, Cook T, Pierce GN, Gilchrist JSC. 2002. Thapsigargin and cyclopiazonic acid activate ryanodine-inhibitable RyR1 calcium release from heavy sarcoplasmic reticulum vesicles. under revision.
96. Palahniuk C, Gilchrist JSC. 1996. Thapsigarigin induces a rapid, ryanodine-inhibtable calcium release from skeletal muscle sarcoplasmic reiculum. Biophys J 70:A285.
97. Gould GW, McWhirter JM, East JM, Lee AG. 1987. A model for the uptake and release of Ca2+ by sarcoplasmic reticulum. Biochemical J 245:739–749.
98. Smith GL, Duncan AM, Neary P, Bruce L, Burton FL. 2000. P(i) inhibits the SR Ca(2+) pump and stimulates pump-mediated Ca(2+) leak in rabbit cardiac myocytes. Am J Physiol Heart Circ Physiol 279:H577–585.
99. Favero TG, Abramson JJ. 1994. Thapsigargin-induced Ca2+ release from sarcoplasmic reticulum and asolectin vesicles. Cell-Calcium 15:183–189.

100. Dettbarn C, Palade P. 1998. Effects of three sarcoplasmic/endoplasmic reticulum Ca++ pump inhibitors on release channels of intracellular stores. J Pharmacol Exp Therap 285:739–745.
101. Kirby MS, Sagara Y, Gaa S, Inesi G, Lederer WJ, Rogers TB. 1992. Thapsigargin inhibits contraction and Ca2+ transient in cardiac cells by specific inhibition of the sarcoplasmic reticulum Ca2+ pump. J Biol Chem 267:12545–12551.
102. Thastrup O, Cullen PJ, Drobak BK, Hanley MR, Dawson AP. 1990. Thapsigargin, a tumor promoter, discharges intracellular Ca2+ stores by specific inhibition of the endoplasmic reticulum Ca2(+)-ATPase. Proc Natl Acad Sci USA 87:2466–2470.
103. Inesi G, Sagara Y. 1992. Thapsigargin, a high affinity and global inhibitor of intracellular Ca2+ transport ATPases. Arch Biochem Biophys 298:313–317.
104. Karon BS, Mahaney JE, Thomas DD. 1994. Halothane and cyclopiazonic acid modulate Ca-ATPase oligomeric state and function in sarcoplasmic reticulum. Biochemistry 33:13928–13937.
105. Sagara Y, Inesi G. 1991. Inhibition of the sarcoplasmic reticulum Ca2+ transport ATPase by thapsigargin at subnanomolar concentrations. J Biol Chem 266:13503–13506.
106. Sagara Y, Fernandez-Belda F, de Meis L, Inesi G. 1992. Characterization of the inhibition of intracellular Ca2+ transport ATPases by thapsigargin. J Biol Chem 267:12606–12613.
107. Sagara Y, Wade JB, Inesi G. 1992. A conformational mechanism for formation of a dead-end complex by the sarcoplasmic reticulum ATPase with thapsigargin. J Biol Chem 267:1286–1292.
108. Thomas D, Hanley MR. 1994. Pharmacological tools for perturbing intracellular calcium storage. Meth Cell Biol 40:65–89.
109. Nørregaard A, Vilsen B, Andersen JP. 1993. Chimeric Ca(2+)-ATPase/Na+,K(+)-ATPase molecules. Their phosphoenzyme intermediates and sensitivity to Ca2+ and thapsigargin. Febs Letters 336:248–254.
110. Davidson GA, Varhol RJ. 1995. Kinetics of thapsigargin-Ca(2+)-ATPase (sarcoplasmic reticulum) interaction reveals a two-step binding mechanism and picomolar inhibition. J Biol Chem 270:11731–11734.
111. Wictome M, Henderson I, Lee AG, East JM. 1992. Mechanism of inhibition of the calcium pump of sarcoplasmic reticulum by thapsigargin. Biochem-J 283:525–529.
112. Kijima Y, Ogunbunmi E, Fleischer S. 1991. Drug action of thapsigargin on the Ca2+ pump protein of sarcoplasmic reticulum. J Biol Chem 266:22912–22918.
113. Inesi G, de Meis L. 1989. Regulation of steady state filling in sarcoplasmic reticulum. Roles of back-inhibition, leakage, and slippage of the calcium pump. J Biol Chem 264:5929–5936.
114. Favre CJ, Schrenzel J, Jacquet J, Lew DP, Krause KH. 1996. Highly supralinear feedback inhibition of Ca2+ uptake by the Ca2+ load of intracellular stores. J Biol Chem 271:14925–14930.
115. Shannon TR, Ginsburg KS, Bers DM. 2000. Reverse Mode of the Sarcoplasmic Reticulum Calcium Pump and Load-Dependent Cytosolic Calcium Decline in Voltage-Clamped Cardiac Ventricular Myocytes. Biophys J 78:322–333.
116. Bridge JH. 1986. Relationships between the sarcoplasmic reticulum and sarcolemmal calcium transport revealed by rapidly cooling rabbit ventricular muscle. J Gen Physiol 88:437–473.
117. Bers DM, Bassani RA, Bassani JW, Baudet S, Hryshko LV. 1993. Paradoxical twitch potentiation after rest in cardiac muscle: increased fractional release of SR calcium. J Mol Cell Cardiol 25: 1047–1057.
118. Bassani RA, Bers DM. 1994. Na-Ca exchange is required for rest-decay but not for rest-potentiation of twitches in rabbit and rat ventricular myocytes. J Mol Cell Cardiol 26:1335–1347.
119. Banijamali HS, Gao WD, MacIntosh BR, ter Keurs HE. 1991. Force-interval relations of twitches and cold contractures in rat cardiac trabeculae. Effect of ryanodine. Circ Res 69:937–948.
120. Cheng H, Lederer WJ, Cannell MB. 1993. Calcium sparks: elementary events underlying excitation-contraction coupling in heart muscle. Science 262:740–744.
121. Wang SQ, Song LS, Lakatta EG, Cheng H. 2001. Ca2+ signalling between single L-type Ca2+ channels and ryanodine receptors in heart cells. Nature 410:592–596.
122. Lukyanenko V, Viatchenko-Karpinski S, Smirnov A, Wiesner TF, Gyorke S. 2001. Dynamic regulation of sarcoplasmic reticulum Ca(2+) content and release by luminal Ca(2+)-sensitive leak in rat ventricular myocytes. Biophys J 81:785–798.
123. Lukyanenko V, Subramanian S, Gyorke I, Wiesner TF, Gyorke S. 1999. The role of luminal Ca2+ in the generation of Ca2+ waves in rat ventricular myocytes. J Physiol 518(Pt 1):173–186.
124. Lukyanenko V, Gyorke I, Wiesner TF, Gyorke S. 2001. Potentiation of Ca(2+) release by cADP-ribose in the heart is mediated by enhanced SR Ca(2+) uptake into the sarcoplasmic reticulum. Circ Res 89:614–622.

125. Terentyev DA, Viatchenko-Karpinski S, Valdivia HH, Escobar A, Gyorke S. 2002. Luminal Ca Determines Termination of Cardiac Ca Sparks. Biophys J 82:511.
126. Chu A, Bick RJ, Tate CA, Van Winkle WB, Entman ML. 1983. Anion effects on in vitro sarcoplasmic reticulum function. Co-transport of anions with calcium. J Biol Chem 258:10543–10550.
127. Chu A, Tate CA, Bick RJ, Van Winkle WB, Entman ML. 1983. Anion effects on in vitro sarcoplasmic reticulum function. The relationship between anions and calcium flux. J Biol Chem 258:1656–1664.
128. Bridge JH, Ershler PR, Cannell MB. 1999. Properties of Ca2+ sparks evoked by action potentials in mouse ventricular myocytes. J Physiol 518 (Pt 2):469–478.
129. Soeller C, Cannell MB. 2002. Estimation of the sarcoplasmic reticulum ca(2+) release flux underlying ca(2+) sparks. Biophys J 82:2396–2414.
130. Palade P, Mitchell RD, Fleischer S. 1983. Spontaneous calcium release from sarcoplasmic reticulum. General description and effects of calcium. J Biol Chem 258:8098–8107.
131. Mitchell RD, Palade P, Fleischer S. 1984. Spontaneous calcium release from sarcoplasmic reticulum. Assessment of other ionic influences. J Biol Chem 259:1073–1081.
132. Volpe P, Mrak RE, Costello B, Fleischer S. 1984. Calcium release from sarcoplasmic reticulum of normal and dystrophic mice. Biochimica et Biophysica Acta 769:67–78.
133. Volpe P, Palade P, Costello B, Mitchell RD, Fleischer S. 1983. Spontaneous calcium release from sarcoplasmic reticulum. Effect of local anesthetics. J Biol Chem 258:12434–12442.
134. Rios E, Brum G. 2002. Ca2+ release flux underlying Ca2+ transients and Ca2+ sparks in skeletal muscle. Front Biosci 7:D1195–1211.
135. Gonzalez A, Zhou J, Kirsch W, Uttenweiler D, Fink R, Rios E, Brum E. 2002. Morphology of Ca2+ Sparks of Mammalian Muscle. Biophys J 82:510a.
136. Isaeva EV, Shirokova N. 2002. Metabolic Control of Local Calcium Signaling in Mammalian Skeletal Muscle. Biophys J 82:510a.
137. Yang Z, Steele DS. 2002. Effects of phosphocreatine on SR Ca(2+) regulation in isolated saponin-permeabilized rat cardiac myocytes. J Physiol 539:767–777.
138. Jayaraman T, Brillantes AM, Timerman AP, Fleischer S, Erdjument-Bromage H, Tempst P, Marks AR. 1992. FK506 binding protein associated with the calcium release channel (ryanodine receptor). J Biol Chem 267:9474–9477.
139. Radermacher M, Wagenknecht T, Grassucci R, Frank J, Inui M, Chadwick C, Fleischer S. 1992. Cryo-EM of the native structure of the calcium release channel/ryanodine receptor from sarcoplasmic reticulum. Biophys J 61:936–940.
140. Saito A, Inui M, Radermacher M, Frank J, Fleischer S. 1988. Ultrastructure of the calcium release channel of sarcoplasmic reticulum. J Cell Biol 107:211–219.
141. Marks AR, Fleischer S, Tempst P. 1990. Surface topography analysis of the ryanodine receptor/junctional channel complex based on proteolysis sensitivity mapping. J Biol Chem 265:13143–13149.
142. Sharma MR, Penczek P, Grassucci R, Xin HB, Fleischer S, Wagenknecht T. 1998. Cryoelectron microscopy and image analysis of the cardiac ryanodine receptor. J Biol Chem 273:18429–18434.
143. Qi Y, Ogunbunmi EM, Freund EA, Timerman AP, Fleischer S. 1998. FK-binding protein is associated with the ryanodine receptor of skeletal muscle in vertebrate animals. J Biol Chem 273:34813–34819.
144. Cameron AM, Steiner JP, Sabatini DM, Kaplin AI, Walensky LD, Snyder SH. 1995. Immunophilin FK506 binding protein associated with inositol 1,4,5-trisphosphate receptor modulates calcium flux. Proc Natl Acad Sci U S A 92:1784–1788.
145. Brillantes AB, Ondrias K, Scott A, Kobrinsky E, Ondriasova E, Moschella MC, Jayaraman T, Landers M, Ehrlich BE, Marks AR. 1994. Stabilization of calcium release channel (ryanodine receptor) function by FK506-binding protein. Cell 77:513–523.
146. Ondrias K, Marx SO, Gaburjakova M, Marks AR. 1998. FKBP12 modulates gating of the ryanodine receptor/calcium release channel. Ann N Y Acad Sci 853:149–156.
147. O'Reilly FM, Robert M, Jona I, Szegedi C, Albrieux M, Geib S, De Waard M, Villaz M, Ronjat M. 2002. FKBP12 modulation of the binding of the skeletal ryanodine receptor onto the II-III loop of the dihydropyridine receptor. Biophys J 82:145–155.
148. Gaburjakova M, Gaburjakova J, Reiken S, Huang F, Marx SO, Rosemblit N, Marks AR. 2001. FKBP12 binding modulates ryanodine receptor channel gating. J Biol Chem 276:16931–16935.
149. Marx SO, Gaburjakova J, Gaburjakova M, Henrikson C, Ondrias K, Marks AR. 2001. Coupled gating between cardiac calcium release channels (ryanodine receptors). Circ Res 88:1151–1158.

150. Marx SO, Reiken S, Hisamatsu Y, Jayaraman T, Burkhoff D, Rosemblit N, Marks AR. 2000. PKA phosphorylation dissociates FKBP12.6 from the calcium release channel (ryanodine receptor): defective regulation in failing hearts. Cell 101:365–376.
151. Doi M, Yano M, Kobayashi S, Kohno M, Tokuhisa T, Okuda S, Suetsugu M, Hisamatsu Y, Ohkusa T, Matsuzaki M. 2002. Propranolol prevents the development of heart failure by restoring FKBP12.6-mediated stabilization of ryanodine receptor. Circulation 105:1374–1379.
152. Dulhunty AF, Laver DR, Gallant EM, Casarotto MG, Pace SM, Curtis S. 1999. Activation and inhibition of skeletal RyR channels by a part of the skeletal DHPR II-III loop: effects of DHPR Ser687 and FKBP12. Biophys J 77:189–203.
153. Shou W, Aghdasi B, Armstrong DL, Guo Q, Bao S, Charng MJ, Mathews LM, Schneider MD, Hamilton SL, Matzuk MM. 1998. Cardiac defects and altered ryanodine receptor function in mice lacking FKBP12. Nature 391:489–492.
154. Xin HB, Senbonmatsu T, Cheng DS, Wang YX, Copello JA, Ji GJ, Collier ML, Deng KY, Jeyakumar LH, Magnuson MA, Inagami T, Kotlikoff MI, Fleischer S. 2002. Oestrogen protects FKBP12.6 null mice from cardiac hypertrophy. Nature 416:334–338.
155. Nelson MT, Herrera GM. 2002. Molecular physiology: protecting the heart. Nature 416:273–274.
156. Barg S, Copello JA, Fleischer S. 1997. Different interactions of cardiac and skeletal muscle ryanodine receptors with FK-506 binding protein isoforms. Am J Physiol 272:C1726–1733.
157. Timerman AP, Onoue H, Xin HB, Barg S, Copello J, Wiederrecht G, Fleischer S. 1996. Selective binding of FKBP12.6 by the cardiac ryanodine receptor. J Biol Chem 271:20385–20391.
158. Prestle J, Janssen PM, Janssen AP, Zeitz O, Lehnart SE, Bruce L, Smith GL, Hasenfuss G. 2001. Overexpression of FK506-binding protein FKBP12.6 in cardiomyocytes reduces ryanodine receptor-mediated Ca(2+) leak from the sarcoplasmic reticulum and increases contractility. Circ Res 88:188–194.
159. Valdivia HH. 2001. Cardiac ryanodine receptors and accessory proteins: augmented expression does not necessarily mean big function. Circ Res 88:134–136.

Pathophysiology of Cardiovascular Disease,
edited by N.S. Dhalla, H. Rupp,
A. Angel & G.N. Pierce
Kluwer Academic Publishers, Boston, 2004

Cell-Cell and Cell-Extracellular Matrix Interactions in Cardiomyopathy

Shunzo Onishi,[1,2] Naomasa Kawaguchi,[1] Soiji Mori[1] and Nariaki Matsuura[1]

[1] *Department of Pathology, School of Allied Health Sciences, Faculty of Medicine, Osaka University, Suita, Osaka, Japan*
[2] *Laboratory Center, Tane General Hospital, Osaka, Japan*

Summary. An immunohistochemical study was performed to examine the morphological changes of cell-cell and cell-extracellular matrix adhesion molecules in the heart of 300-day-old (heart failure stage) male BIO-14.6 cardiomyopathic hamsters. Age- and sex-matched BIO-F1B were used as normal control. Confocal laser microscopy using monoclonal antibodies against pan-cadherin (adherens junctions), desmosomal protein (desmosomes), and α- and β-dystroglycan were employed. In control myocardium, N-cadherin and desmosomal protein showed an irregular immmunolabelling pattern at the intercalated disks, and α- and β-dystroglycan were observed along the myocyte surface. In cardiomyopathic myocardium, immunolabelling of N-cadherin, desmosomal protein and β-dystroglycan were not altered, whereas immunostaining of α-dystroglycan in the sarcolemma of cardiomyopathic myocardium was significantly reduced as compared with that of control myocardium. We conclude that a disruption of cell-extracellular matrix interaction resulting from loss of α-dystroglycan in myocardium may relate to decreased myocardial function in cardiomyopathic hamster failing hearts.

Key words: Adherens junction, Desmosome, Dystroglycan, Cardiomyopathy, Immunohistochemistry.

INTRODUCTION

The cardiomyopathic BIO-14.6 Syrian hamster has provided the unique possibilities of studying the frequent involvement of the myocardium in human primary muscle disorders (1–3). Cardiac insufficiency develops progressively in cardiomyo-

Correspondence: to Shunzo Onishi, MD, PhD, Laboratory Center, Tane General Hospital, 1-2-31 Sakaigawa, Nishi-ku, Osaka 550-0024, Japan. Phone & Fax: +81-6-6581-3412, E-mail: onishi@abox6.so-net.ne.jp.

pathic BIO-14.6 hamster through three distinct stages, that is, stages of myolysis (30–90 days of age), hypertrophy (90–270 days of age), and insufficiency or heart failure (over 270 days of age) (3). Many BIO-14.6 hamsters die from congestive heart failure (2).

The sarcolemma contains three types of intercellular junctions: gap junctions, adherens junctions, and desmosomes, that are frequently combined in the intercalated disks. N-cadherin is a member of the transmembrane Ca + 2-dependent glycoprotein cell adhesion molecules and is localized in the adherens junctions (4). Desmosomes mediate cell-cell adhesion through the cadherin-family proteins desmogleins and desmocolins.

Dystroglycan was identified in skeletal muscle as a component of the dystrophin-glycoprotein complex and consists of α- and β-dystroglycan; the one is a 156 kDa extracellular laminin-binding glycoprotein and the other is a 43 kDa transmembrane glycoprotein (5).

In the present study, we investigated the spatiotemporal changes of cell-cell and cell-extracellular matrix adhesion molecules in myocardium of the BIO-14.6 cardiomyopathic hamster at the congestive heart failure stages by an immunohistochemical method using monoclonal antibodies against desmosomal protein, cadherin, and α- and β-dystroglycan.

MATERIALS AND METHODS

Animal treatment and histology

Experiments were performed on male cardiomyopathic BIO-14.6 hamsters at the congestive heart failure (300 days of age, n = 5) stage. Age- and sex-matched BIO-F1B hamsters (n = 5) were used as normal controls in this study. The animals were slightly anesthetized with pentobarbital, and the hearts were rapidly removed and weighted. Cardiac tissue samples were rapidly embedded with O.C.T. compound in dry ice-aceton. In addition, the hearts were fixed with 10% buffered-formalin and embedded in paraffin. Paraffin sections were stained with hematoxylin-eosin or with picrosirius red for connective tissue.

Immunohistochemistry

Monoclonal antibodies desmosomal protein (Sigma, USA), pan-cadherin (Sigma, USA), α-dystroglycan (Upstate Biotechnology, USA) or β-dystroglycan (Novocastra, UK) were used as the primary antibodies. The cryosections were fixed in ice-cold acetone (for desmosomal protein and pan-cadherin) for 5 minutes at −20°C and 2% paraformaldehyde (for α- and β-dystroglycan) for 10 minutes at room temperature. All primary antibodies were used at a 1:100 dilution with phosphate buffer saline (PBS). The sections were incubated with primary antibodies for 14°C in a moist chamber. After rinsing in PBS, the sections were incubated with FITC-conjugated secondary antibody (Amersham, USA) at a 1:50 dilution with PBS for 1 hour at room temperature. The nuclei were stained with propidium iodide (Sigma, USA).

Immunohistochemical control sections were incubated as above with deletion of the primary antibody. All sections were examined by confocal laser microscope Radiance 2000 KR-3 (Bio-Rad, USA).

Statistics

Statistical analysis was performed using unpaired Student's *t*-test. A p-value <0.05 was considered a significant difference. Data are expressed as mean ± S.D.

RESULTS

Gross and light microscopic observation

The heart weights and the heart weight to body weight ratio were significantly increased in the heart of BIO-14.6 hamster as compared with that of control hamster (Figure 1). In light microscopic observation, the left ventricular (LV) myocardium of BIO-14.6 hamsters showed hypertrophy and atrophy of the cardiomyocytes and severe interstitial fibrosis. However, necrosis and granulation tissue are not found.

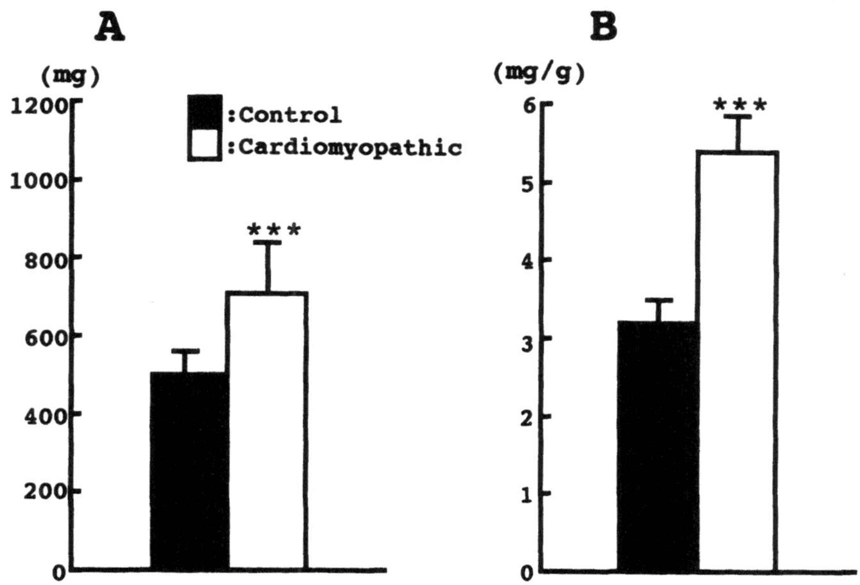

Figure 1. The heart weights (A) and heart weight to body weight ratio (B) of control and cardiomyopathic hamsters. The heart weights and the heart weight to body weight ratio were significantly increased in the heart of cardiomyopathic hamster as compared with that of control hamster. Values are mean ± S.D. ***; $p < 0.001$ vs control

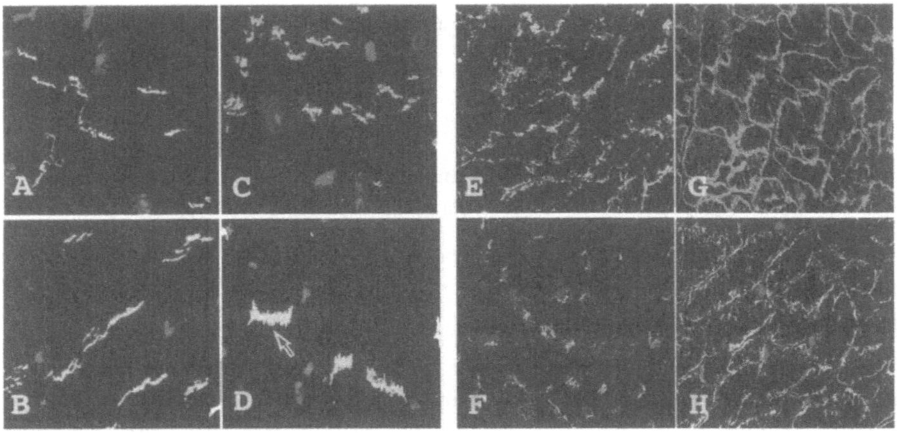

Figure 2. Confocal microscopic images of immunolabeled desmosomal protein (A and B), N-cadherin (C and D), α-dystroglycan (E and F) and β-dystroglycan (G and H) in left ventricular myocardium. Panels A, C, E and G: control myocardium (original magnification ×600) Panels B, D, F and H: cardiomyopathic myocardium (original magnification ×600). Immunostaining against N-cadherin in cardiomyopathic hamster myocardium shows an increase on fluorescence intensity at the intercalated disks (arrow) in hypertrophied cardiomyocytes (D). Immunoreactivity for α-dystroglycan in cardiomyopathic myocardium is markedly reduced (F). Immunostaining pattern of desmosomal protein (B) and β-dystroglycan (H) in cardiomyopathic myocardium is similar to that of control myocardium (A and G).

Immunohistochemistry

N-Cadherin and desmosomal protein

In normal LV myocardium, immunolabelling of desmosomal protein and N-cadherin showed distinct fluorescence at the intercalated disks (Figures 2A and 2C). The arrangement and distribution of immunostaining pattern of desmosomal protein and N-cadherin in the BIO-14.6 myocardium were similar to that of control myocardium. However, immunostaining against N-cadherin showed an increase in fluorescence intensity at the intercalated disks associated with hypertrophy of the cardiomyocytes (Figures 2B and 2D).

α-Dystroglycan and β-dystroglycan

In control LV myocardium, both α- and β-dystroglycans fluorescent signals were observed along the cardiac cell surface (Figures 2E and 2G). In cardiomyopathic LV myocardium, immunoreactivity for α-dystroglycan was substantially reduced as compared to that of control LV myocardium (Figure 2F). In contrast, immuno-fluorescence against β-dystroglycan in cardiomyopathic LV myocardium was not significantly altered (Figure 2H).

DISCUSSION

Cardiac myocytes are jointed end-to-end by means of the intercalated disks. The adherens junctions and desmosomes in the intercalated disks are thought to be the mechanical junctions in cardiac myocytes. N-cadherin is localized in the adherens junctions of myocytes. Fujio et al. (6) indicated that downregulation of N-cadherin was found in cardiomyopathic BIO-14.6 hamster at congestive heart failure stage. In the present study, the arrangement and distribution of immunofluorescence pattern of N-cadherin was not altered significantly in cardiomyopathic myocardium. However, an increase of immunofluorescence intensity for N-cadherin was observed in cardiomyopathic myocardium. Since N-cadherin of the adherens junctions links to the end of actin filaments of sarcomeres through α-, β- and γ-catenins, vinculin, and α-actinin (7), this augmentation of immunofluorescence intensity for N-cadherin may be associated with an increase in actin filament as a consequence of hypertrophic cardiomyocytes. Moreover, an increase of immunoreactivities for vinculin at the intercalated disks in hypertrophied cardiomyocytes have been reported in cardiomyopathic hamster failing hearts (8).

Desmosomes are button-like points of cell-cell contact that rivet together cells and serve as anchoring sites for intermediate filaments, which form a network in sarcoplasm and provide tensile strength (9). Wang et al. (4) demonstrated that the distribution of immunostaining pattern and fluorescence intensity for desmosones at the intercalated disks in left ventricular myocytes of a chronic aortic stenosis guinea pig model was not altered by quantitative confocal microscopy at the congestive heart failure stage. These results are in agreement with our observation.

The dystrophin-glycoprotein complex comprises dystrophin, α- and β-dystroglycan, and α-, β-, γ- and δ-sarcoglycan (10). β-Dystroglycan is connected with α-dystroglycan, which extracellularly binds to laminin in the basement membrane of the skeletal and cardiac muscle. Moreover, β-Dystroglycan binds to dystrophin on the inside of the muscle cells (11). Dystroglycan is a rod like cytoskeletal protein, and the NH_2-terminal domain of dystrophin binds to actin filament. Dystroglycan is thought to contributes to structural integrity of the sarcolemma (5). Disruption of the α-dystroglycan-extracellular matrix interaction causes muscular dystrophy (12). In this study, immunoreactivity for α-dystroglycan was substantially reduced in cardiomyopathic LV myocardium. We suggest that a downregulation of α-dystroglycan may be associated with a remodelling of extracellular matrix as a result of severe interstitial fibrosis in cardiomyopathic hamster myocardium.

It is concluded that a disruption of cell-extracellular matrix interaction resulting from loss of α-dystroglycan in myocardium may relate to decreased myocardial function in cardiomyopathic hamster failing hearts.

REFERENCES

1. Bajusz E, Baker JR, Nixon C W, Homburger F. 1969. Spontaneous hereditary myocardial degeneration and congestion heart failure in a strain of Syrian hamster. Ann NY Acad Sci 156:59–88.
2. Buchner F, Onishi S, Wada A. 1978. Cardiomyopathy associated with systemic myopathy: Genetic defect of actomyosin influencing muscular structure and function. 1–99 Baltimore-Munich: U & S.
3. Strobeck JE, Factor SM, Bhan A, Sole M, Liew CC, Fein F, Sonnenblick EH. 1979. Hereditary and acquired cardiomyopathies in experimental animals: Mechanical, biochemical, and structural features. Ann NY Acad Sci 317:59–88.
4. Wang X, Gerdes AM. 1999. Chronic pressure overload cardiac hypertrophy and failure in guinea pigs: III. Intercalated disc remodeling. J Moll Cell Cardiol 31:333–343.

5. Henry MD, Campbell KP. 1996. Dystroglycan: an extracellular matrix receptor linked to the cytoskeleton. Curr Opin Cell Biol 8:602–607.
6. Fujio Y, Yamada-Honda F, Sato N, Funai H, Wada A, Awata N, Shibata N. 1995. Disruption of cell-cell adhesion in an inbred strain of hereditary cardiomyopathic hamsters (Bio 14.6). Cardiovasc Res 30:899–904.
7. Barth AIM, Nathke IS, Nelson WJ. 1997. Cadherin, catenins and APC protein: interplay between cytoskeleton comlpexes and signaling pathways. Curr Opin Cell Biol 6:683–690.
8. Kawaguchi N, Fujitani N, Schaper J, Onishi S. 1995. Pathological changes of myocardial cytoskeleton in cardiomyopathic hamster. Mol Cell Biochem 144:75–80.
9. Alberts B, Bray D, Lewis J, Raff M, Roberts K, Watson JD. 1989. Cell adhesion, cell junctions, and the extracellular matrix. In: molecular biology of the cell. Ed. M Robertson, 797–798. New York and London: Garland Publishing.
10. Lee GH, Badorff C, Knowlton KU. 2000. Dissociation of sarcoglycans and the dystrophin carboxyl terminal from the sarcolemma in enteroviral cardiomyopathy. Circ Res 87:489–495.
11. Stevemson S, Rothery S, Cullen MJ, Severs NJ. 1998. Spatial relationship of the C-terminal domains of dystrophin and β-Dystroglycan in cardiac muscle support a direct molecular interaction at the plasma membrane interface. Circ Res 82:82–93.
12. Sunada Y, Bernier SM, Kozak CA, Yamada Y, Campbell KP. 1994. Deficiency of merosin in dystrophic *dy* mice and genetic linkage of laminin M chain gene to *dy* locus. J Biol Chem 269:13729–13732.

Pathophysiology of Cardiovascular Disease,
edited by N.S. Dhalla, H. Rupp,
A. Angel & G.N. Pierce
Kluwer Academic Publishers, Boston, 2004

Role of the Phosphoinositide 3-Kinase Pathway in Cardiac Hypertrophy

Tetsuo Shioi, Julie R. McMullen, and Seigo Izumo

Cardiovascular Division, Department of Medicine
Beth Israel Deaconess Medical Center and Harvard Medical School
Boston, MA

Summary. Phosphoinositide 3-kinase (PI3K) is a lipid kinase that catalyzes the addition of a phosphate molecule specifically to the 3-position of the inositol ring of phosphoinositides. Biochemical studies using cultured mammalian cells show that PI3K is activated by receptor tyrosine kinases that include insulin or insulin-like growth factors, and PI3K regulates the activities of several downstream molecules such as Akt, p70 ribosomal S6 kinase, and glycogen synthase kinase-3β. The mammalian target of rapamycin, an intracellular target of the immunosuppressant rapamycin, interacts with the PI3K pathway. PI3K regulates diverse cellular activities, including growth, DNA synthesis, apoptosis, and cytoskeletal organization in cultured cells. Recent genetic studies in *Drosophila* reveal that the PI3K pathway regulates organ and body size in the developmental process in flies. Various agonists or stimuli activate PI3K or its downstream effectors in cultured myocytes and heart tissue. Perturbation of the PI3K pathway significantly modulates the phenotype of cultured myocytes and the intact heart. Since agonists that activate PI3K, such as growth hormone, have been implicated in the treatment of heart failure, further studies elucidating the role of PI3K and its downstream effectors in the heart may be useful for the development of new strategies to treat cardiac hypertrophy and failure.

Key words: Heart failure, Cardiac hypertrophy, Signal transduction.

Cardiac hypertrophy, an increase in heart size, is induced by several primary cardiovascular diseases. Cardiac hypertrophy initially serves to maintain the cardiac output to meet the increased work load. However, prolonged hypertrophy is

Address for Correspondence: Seigo Izumo, MD, Cardiovascular Division, Beth Israel Deaconess Medical Center, 330 Brookline Avenue, Boston, MA 02215. Tel: 617-667-4858, Fax 617-975-5268, E-mail: sizumo@caregroup.harvard.edu.

associated with increases in cardiovascular morbidity and mortality (1). The roles of intracellular signaling molecules in the development of cardiac hypertrophy have been intensively studied (2,3). Phosphoinositide 3-kinase is a lipid kinase that is activated by tyrosine kinases such as insulin or insulin-like growth factors. Phosphoinositide 3-kinase has been shown to regulate diverse cellular activities including growth, DNA synthesis, apoptosis, and cytoskeletal organization in cultured cells. In this review, we will summarize recent findings about the role of phosphoinositide 3-kinase and its downstream effectors in cardiac hypertrophy.

1. PHOSPHOINOSITIDE 3-KINASE AND ITS DOWNSTREAM EFFECTORS

Phosphoinositide 3-kinase

Phosphatidylinositol (PtdIns) is a component of eukaryotic cell membranes that is unique among phospholipids in that its inositol head group can be phosphorylated at multiple free hydroxyls (4,5). Several phosphorylated derivatives of phosphatidylinositol, collectively termed phosphoinositides, have been identified in eukaryotic cells from yeast to mammals. The enzymes that phosphorylate phosphatidylinositol and its derivatives are termed phosphoinositide kinases. Phosphoinositide 3-kinases (PI3Ks) are a subfamily of lipid kinases that catalyze the addition of a phosphate molecule specifically to the 3-position of the inositol ring of phosphoinositides. There are multiple isoforms of PI3Ks which are divided into three classes according to their structure and substrate specificity (6). Class I PI3Ks are the best characterized class. Class I PI3Ks are subdivided into class IA and IB enzymes. Class IA PI3Ks utilize PtdIns, PtdIns(4)P, and PtdIns(4,5)P_2 as substrates in vitro, and class IA PI3Ks appear to prefer PtdIns(4,5)P_2 as a substrate in cells. Class IA PI3Ks are activated by tyrosine kinases, and the representative tyrosine kinases are insulin and insulin-like growth factor (IGF) receptors. Class IB PI3Ks are activated by heterotrimeric G protein-coupled receptors. Class II PI3Ks phosphorylate PtdIns and PtdIns(4)P in vitro, but not PtdIns(4,5)P_2. Class III PI3Ks only use PtdIns as a substrate. Class III PI3Ks appear to have a housekeeping role in constitutive membrane trafficking and vesicle morphogenesis. In this article, we will review the role of class IA PI3K in cardiac hypertrophy and refer to class IA PI3K as PI3K. The role of class IB PI3K in cardiac hypertrophy has been studied previously (7,8).

The p110 subunit of class IA PI3Ks exists in a complex with an adaptor protein that has two Src homology 2 (SH2) domains. The SH2 domains bind to phosphorylated tyrosine residues that are generated by activated tyrosine kinases in receptors and various receptor-associated adaptor proteins. Phosphotyrosine binding is thought to allow translocation of the cytosolic PI3Ks to the membranes, where their lipid substrates reside. The lipid products of PI3K regulate several downstream effectors. A lipid phosphatase PTEN (phosphatase and tensin homologue on chromosome ten) removes the phosphate from the 3-position of the inositol ring and negatively regulates PI3K activity (9). Biochemical analysis using cultured mammalian cells showed that PI3K regulated activities of several downstream molecules

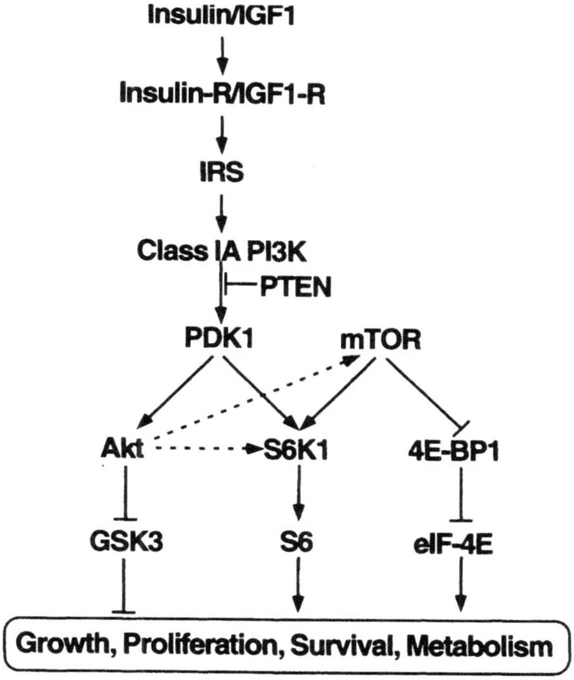

Figure 1. Phosphoinositide 3-kinase and its downstream effectors. Phosphoinositide 3-kinase (PI3K) and its downstream effectors regulate body or organ size by possibly regulating growth, proliferation, survival and metabolism of cells. IGF1: insulin-like growth factor 1, Insulin-R: insulin receptor, IGF1-R: IGF1 receptor, IRS: insulin receptor substrate, PI3K: phosphoinositide 3-kinase, PTEN: phosphatase and tensin homologue on chromosome ten, PDK1: phosphoinositide dependent protein kinase 1, mTOR: mammalian target of rapamycin, S6K1: p70 S6 kinase 1, 4E-BP1: eukaryotic translation initiation factor 4E binding protein 1, GSK3: glycogen synthase kinase 3, S6: 40S ribosomal S6 protein, eIF4E: eukaryotic translation initiation factor 4E.

such as Akt, p70 ribosomal S6 kinase, and glycogen synthase kinase-3β (Figure 1). The mammalian target of rapamycin, an intracellular target of the immunosuppressant rapamycin, interacts with the PI3K pathway. PI3K is known to regulate diverse cellular activities including growth, DNA synthesis, apoptosis, and cytoskeletal organization in cultured cells (5).

Effectors of PI3K

Akt

Akt, also known as RAC-PK (protein kinase related to A and C kinases) (10) or PKB (protein kinase B) (11), is a well characterized downstream target of PI3K (12,13). Mammals have three closely related Akt genes, encoding the isoforms Akt1, Akt2 and Akt3 (13). Akt2 and Akt3 show 81% and 83% amino acid identity respectively with Akt1. All Akt isoforms show a broad tissue distribution. Akt is activated by a multistep process via a variety of signals (12). In the early steps of this process, phosphoinositides generated by PI3K bind the pleckstrin homology domain of Akt and translocate the kinase to the membrane. On the membrane, Thr308 of Akt is phosphorylated by phosphoinositide dependent protein kinase 1 (PDK1) (14) and Ser 473 is phosphorylated by a complex mechanism that may involve autophosphorylation (15). Akt is reported to be activated in a PI3K independent manner. Ca^{2+}/calmodulin-dependent protein kinase kinase directly activates Akt, and is likely to promote survival of some cultured neurons in a PI3K independent manner (16,17). Akt transduces signals that regulate multiple biological processes including apoptosis, gene expression, and cellular proliferation (12).

Glycogen synthase kinase-3β

Glycogen synthase kinase-3 (GSK3) is a ubiquitously expressed serine/threonine kinase (18). In humans there are two isoforms of GSK3, namely GSK3α (51 kDa) and GSK3β (47 kDa). GSK3α and GSK3β have 97% sequence homology within their kinase domains, and GSK3α has an extended N-terminal glycine-rich tail. GSK3β is enzymatically active in unstimulated cells. Inactivation of GSK3β is induced by phosphorylation of Ser9 by upstream kinases (19). Akt is one of the important physiological kinases of GSK3β (20). However, GSK3β is also phosphorylated in a PI3K or Akt independent manner (21,22). GSK3β plays important roles in diverse biological processes such as protein synthesis, cell differentiation, cell proliferation, microtubule dynamics, cell motility and/or apoptosis (18). GSK3β phosphorylates the GTP-GDP exchange factor eIF2B and negatively regulates initiation factor eIF2-directed methionyl-tRNA binding to the 40S ribosome (23). Inactivation of GSK3β by Akt might increase protein synthesis by increasing the efficiency of translation.

p70 Ribosomal S6 kinase

p70 ribosomal S6 kinase 1 (S6K1) is a physiological serine/threonine kinase for the ribosomal S6 protein (24). A highly homologous gene, termed S6K2, has also been identified (25). In the case of S6K1, at least eight phosphorylation sites are believed to mediate kinase activation in a hierarchical fashion (24). Akt alone is sufficient to activate S6K1 in some cultured cell lines (26), although it is likely that activation of S6K1 is not solely dependent on Akt (27). Protein kinase C or calcium also activates S6K1 in a PI3K independent mechanism (28,29). S6K1 is involved in the selective translation of a unique family of mRNAs, by mediating the multiple phosphorylation of 40S ribosomal protein S6 (24). The mRNAs are characterized by an oligopyrimidine tract at their 5′ transcription start site. These mRNAs encode for components of the translational apparatus, including ribosomal proteins and translational elongation factors.

Mammalian target of rapamycin

Rapamycin, a lipophilic macrolide, has been used as a potent immunosuppressant (30). In search for molecules affected by rapamycin, TOR (target of rapamycin) was identified in yeast (reviewed in Gingras et al. (31)). Purification and molecular cloning of the mammalian TOR (mTOR) revealed a 290 kDa protein, which is highly related to yeast TOR. The mTOR protein contains a carboxy-terminal domain with homology to phosphatidylinositol kinases. Rapamycin binds with high affinity to FKBP12 (FK506-binding protein, MW of 12 kD) and this complex then binds to mTOR, inhibiting the function of mTOR. Although overexpression of a constitutively active Akt mutant lead to a modest increase in mTOR kinase activity, and moderately increased mTOR phosphorylation (32–34), an mTOR mutant protein possessing an alanine substitution at a putative Akt consensus phosphorylation site retained its ability to activate S6K1 after growth factor stimulation (34). Thus, the role of Akt in the regulation of mTOR kinase activity is not yet clear. More recent studies have shown that mTOR activity can be modulated in cells by levels of small molecules such as branched-chain amino acids, ATP and phosphatidic acid (35,36). These results suggest that mTOR may work as the link between mitogen pathways and nutrient signals. The mammalian TOR controls a diverse set of downstream effectors which are important for cellular growth (37). These effectors include S6K1 and eukaryotic translation initiation factor 4E binding proteins (4E-BPs), which are also potential downstream effectors of Akt (26,38).

2. ROLE OF THE PI3K PATHWAY IN BODY OR ORGAN SIZE REGULATION

Recent genetic studies in *Drosophila* reveal that several molecules that are activated by insulin or IGFs in mammals regulate organ and body size in the fly (39). Hypomorphic mutations in *Inr*, which encodes the insulin/IGF receptor, results in smaller flies (40). Overexpression of activated or dominant-negative PI3K, a target of the insulin/IGF receptor, increases and decreases the size of fly wings respectively (41). Null mutation in *chico*, which encodes the insulin receptor substrate protein in *Drosophila*, causes a reduction in size and the number of cells in each organ (42). In contrast, mutation in ribosomal S6 kinase, a downstream effector of PI3K, results in flies which have a normal number of cells, but cells are smaller than wild type (43). Deletion of genes for IGFs or their receptors (44–46), insulin receptor substrates (47–49) or S6K1 (25) results in dwarfism in mice, suggesting the importance of the conserved insulin/IGF-PI3K pathway in body size determination in mammals. However, the role of PI3K in relation to organ size regulation in mammals had not been well studied.

3. ROLE OF THE PI3K PATHWAY IN CARDIAC HYPERTROPHY

Activation of the PI3K pathway in the heart

Various agonists or stimuli have been shown to activate PI3K or its downstream effectors in cultured myocytes and in the intact heart.

Cultured myocytes

Insulin (50), IGF1 (51), phenylephrine (52), leukemia inhibitory factor (LIF) (53), and β_2-adrenergic receptor stimulation (54) induced PI3K activation in cultured cardiac myocytes. Insulin (50), IGF1 (55), LIF (53), cardiotrophin-1 (56), tumor necrosis factor (TNF)-α (57), Fas ligand (58), β_2-adrenergic receptor (54,59), and stretching of isolated adult cardiac myocytes (60) have all been shown to activate Akt in cultured myocytes. Angiotensin II (61), phenylephrine (62), neuregulin-1 (63), β-adrenoreceptor agonists (64), and LIF (53) were reported to induce S6K1 activation. GSK3β is phosphorylated (inactivated) by endothelin (58,65), IGF1 (58,65), Fas ligand (58) or isoproterenol (66). Finally, insulin (67) or electrically stimulated contraction (68) induced phosphorylation of 4E-BP1.

Heart tissue

IGF1 mRNA was increased in the hearts of rats with cardiac hypertrophy induced by hypertension (69). Increased cardiac IGF1 formation was associated with the physiological hypertrophy seen in athletes (70). IGF1 mRNA was increased in hearts from patients with aortic stenosis or regurgitation (71). Akt phosphorylation was enhanced in the heart of transgenic mice expressing IGF1 (72). Akt was activated in hypertrophied hearts from patients with heart failure (73). Ischemia or volume overload increased Akt phosphorylation in rat hearts (74). PI3K or Akt was not activated in pressure overloaded hearts (75). In another report, class IB PI3K and Akt were activated, but class IA PI3K was not activated in response to pressure overload (7). Pressure overload activated S6K1 in heart tissue (7,75,76). Pressure overload (65) induced GSK3β phosphorylation in mouse hearts (58,65). GSK3β was phosphorylated in hypertrophied hearts from patients with heart failure (73).

Modulation of the PI3K pzathway in the heart

Cultured myocytes

Perturbation of the PI3K pathway has been shown to modulate the phenotype of cultured myocytes. IGF1 induced hypertrophy of cultured myocytes (77). Overexpression of wild type PTEN in cultured myocytes induced myocyte apoptosis, whereas expression of a catalytically inactive PTEN induced cardiac myocyte hypertrophy (78). TNFα-induced hypertrophy of cultured myocytes was suppressed by adenoviral gene transfer of dominant negative PI3K or Akt (57). LY294002 (an inhibitor of PI3K) or rapamycin (an inhibitor of mTOR) attenuated phenylephrine induced myocyte hypertrophy (62). Phenylephrine increased PI3K activity, and wortmannin attenuated phenylephrine induced myocyte hypertrophy (52). Expression of constitutively active Akt induced myocyte hypertrophy in culture (66). Expression of constitutively active PI3K or Akt increased protein synthesis in cultured myocytes (65). Overexpression of GSK3β inhibited cardiac protein synthesis and hypertrophic gene expression (65,66,79). Effects of IGF1 on the gene expression profile in cultured myocytes was examined, and genes involved in intracellular signaling, cell cycle, transcription/translation, mitochondrial function, cell survival, ion channels, and humoral factors were modulated by IGF1 (80).

Animal models

The role of the PI3K pathway in the intact heart has been examined using genetically engineered mouse models. Overexpression of IGF1 in the heart induced cardiac hypertrophy in transgenic mice (81). In cardiomyocyte-selective insulin receptor knockout mice, hearts were reduced in size by 22–28% (82). We have perturbed the activity of PI3K specifically in cardiac myocytes by expressing a constitutively active PI3K (caPI3K) mutant or a dominant-negative PI3K (dnPI3K) mutant in transgenic mice (83). The caPI3K mutant (iSH2p110) is a chimeric molecule that contains the iSH2 domain of the p85 regulatory subunit fused to the N-terminus of bovine p110α by a flexible glycine-linker (84). The dnPI3K mutant (p110Δkinase) is a truncated p110 mutant which contains the p85 binding domains, but lacks the kinase domain. The caPI3K transgene resulted in a larger heart associated with an increase in myocyte size, whereas expression of the dnPI3K mutant resulted in a smaller heart associated with smaller myocytes (Figure 2). The proportion of the chamber size, architecture of the myocardium or cardiac function were not disturbed by the modulation of PI3K activity, suggesting that PI3K specif-

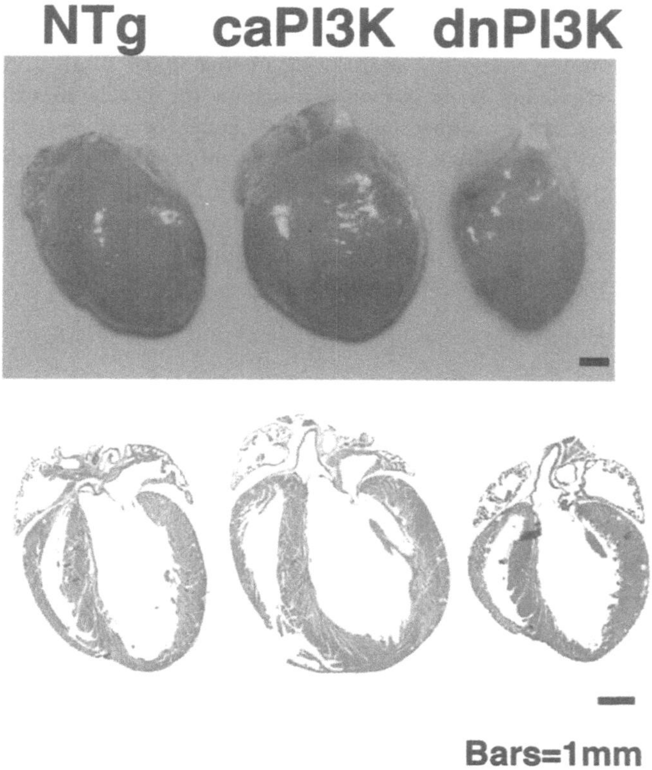

Figure 2. PI3K is necessary and sufficient to promote heart growth. In transgenic mice expressing a constitutively active PI3K (caPI3K) mutant in the heart, there was a proportional increase in the size of all chambers and ventricular wall thickness. In transgenic mice expressing a dominant-negative PI3K (dnPI3K) mutant, the heart was smaller than non-transgenic mice. Bars represent 1 mm.

ically regulates heart size. Furthermore, we have generated and characterized transgenic mice expressing a constitutively active Akt1 gene (caAkt) or a kinase-deficient Akt gene (kdAkt) specifically in the heart (85). For the caAkt mutant (Akt(T308D/S473D)), 2 residues critical for activation of Akt were replaced by negative charged amino acids (T308D/S473D). This mutant has been shown to have comparable activity to endogenous Akt stimulated with insulin (27). For the kdAkt mutant (Akt(K179M)), the critical ATP binding site was mutated. The heart weight of caAkt transgenic mice was increased by 2.0 fold compared with that of non-transgenic mice (Figure 3A). Expression of the kdAkt transgene did not profoundly decrease the basal activity of endogenous Akt, and the heart size of kdAkt transgenic mice was normal. To determine whether Akt is genetically downstream of PI3K in the intact mammalian heart, Akt transgenic mice were crossed with PI3K transgenic mice. The kdAkt mutant attenuated the caPI3K induced overgrowth of the heart (Figure 3B), and the caAkt mutant circumvented cardiac growth retardation induced by the dnPI3K mutant (Figure 3B). These data suggest Akt is genetically downstream of PI3K. In addition, rapamycin attenuated caAkt induced overgrowth of the heart, suggesting that mTOR or effectors of mTOR mediated the caAkt induced heart growth.

These studies using transgenic animals suggest that IGF1, PI3K and Akt play a critical role in regulating heart growth throughout the developmental stage. The role of these molecules in cardiac hypertrophy induced by pathological stress, such as pressure overload, needs to be determined in future studies. Load-induced cardiac hypertrophy occurred normally in mice with severe IGF1 deficiency (86). In contrast, cardiac specific overexpression of an unphosphorylatable from of GSK3β attenuated developmental heart growth as well as cardiac hypertrophy in response to aortic banding and isoproterenol stimulation (87).

4. DISCUSSION

How does the PI3K pathway regulate cardiac hypertrophy? Cardiac myocyte hypertrophy is associated with increased efficiency and capacity for protein synthesis (88). Stretching papillary muscles (89) or increasing the intraventricular pressure in Langendorff preparations (90) results in increased protein synthesis. Stretch (91) as well as several growth factors, such as angiotensin II (61) or phenylephrine (62), induce protein synthesis in neonatal cultured myocytes. Since it has been shown that PI3K or its downstream effectors are involved in protein synthesis, it is likely that the PI3K pathway induces cardiac hypertrophy by increasing the efficiency of protein synthesis. In addition, PI3K, Akt and mTOR have been shown to regulate cytoskeletal organization and gene expression in cultured cell lines. Hypertrophy of myocytes also requires changes in this aspect. Thus, it is likely that the PI3K pathway induces cardiac hypertrophy by regulating protein synthesis, cytoskeletal organization and gene expression in an integrated manner.

The cardiovascular effects of growth hormone have been extensively examined (92). Biventricular hypertrophy was observed in patients with acromegaly (93).

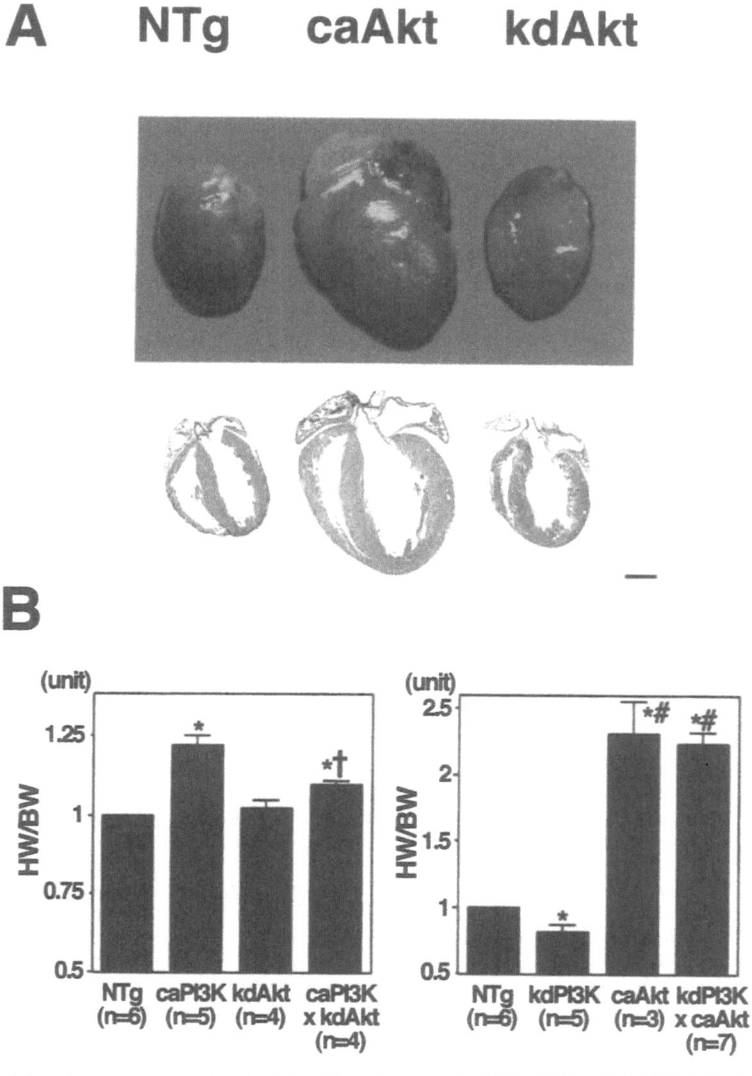

Figure 3. Akt is sufficient to promote heart growth. (A) In transgenic mice expressing constitutively active Akt (caAkt) there was a marked increase in the size of all chambers and in ventricular wall thickness. The proportion of the size of the chambers was maintained. Overexpression of kdAkt did not inhibit the activity of endogenous Akt profoundly, and the heart size of kdAkt transgenic mice was normal. Bars represent 1 mm. **(B)** Heart weight/body weight (HW/BW) ratio of double transgenic mice expressing both caPI3K and kdAkt transgenes was significantly lower than that of caPI3K mice (left panel, 1.10 ± 0.01 vs 1.22 ± 0.03, $p = 0.0096$, NTg = 1.00). HW/BW ratio of double transgenic mice having both dnPI3K and caAkt was similar to that of caAkt (right panel, 2.24 ± 0.08 vs 2.31 ± 0.25, $p = 0.7438$, NTg = 1.00). These results indicate that Akt is genetically downstream of PI3K in relation to heart size regulation. n, number of mice. $^{*}p < 0.05$ vs NTg, $^{\dagger}p < 0.05$ vs caPI3K, $^{\#}p < 0.05$ vs dnPI3K.

Growth hormone is reported to increase left ventricular wall thickness, reduce chamber size and improve clinical symptoms (94). However, in a larger clinical trial, growth hormone increased left ventricular mass in patients with dilated cardiomyopathy, but did not improve clinical symptoms or exercise tolerance (95). Since the biological effects of growth hormone are sometimes different from those of IGF1, a lack of therapeutic efficiency with growth hormone does not necessarily attest to an ineffectiveness of IGF1 therapy for cardiomyopathy (96). Furthermore, IGF1 activates signaling pathways other than the PI3K pathway (51). Thus, further studies regarding the roles of PI3K and its downstream effectors in the heart may be useful for the development of new strategies to treat cardiac hypertrophy and failure.

ACKNOWLEDGEMENT

This study was supported by NIH grants HL65742 and AG61716 to SI.

REFERENCES

1. Levy D, Garrison RJ, Savage DD, Kannel WB, Castelli WP. 1990. Prognostic implications of echocardiographically determined left ventricular mass in the Framingham Heart Study. N Engl J Med 322:1561–1566.
2. Molkentin JD, Dorn IG, 2nd. 2001. Cytoplasmic signaling pathways that regulate cardiac hypertrophy. Annu Rev Physiol 63:391–426.
3. Sadoshima J, Izumo S. 1997. The cellular and molecular response of cardiac myocytes to mechanical stress. Annu Rev Physiol 59:551–571.
4. Fruman DA, Meyers RE, Cantley LC. 1998. Phosphoinositide kinases. Annu Rev Biochem 67:481–507.
5. Vanhaesebroeck B, Leevers SJ, Ahmadi K, Timms J, Katso R, Driscoll PC, Woscholski R, Parker PJ, Waterfield MD. 2001. Synthesis and function of 3-phosphorylated inositol lipids. Annu Rev Biochem 70:535–602.
6. Vanhaesebroeck B, Leevers SJ, Panayotou G, Waterfield MD. 1997. Phosphoinositide 3-kinases: a conserved family of signal transducers. Trends Biochem Sci 22:267–272.
7. Naga Prasad SV, Esposito G, Mao L, Koch WJ, Rockman HA. 2000. Gbetagamma-dependent phosphoinositide 3-kinase activation in hearts with in vivo pressure overload hypertrophy. J Biol Chem 275:4693–4698.
8. Naga Prasad SV, Barak LS, Rapacciuolo A, Caron MG, Rockman HA. 2001. Agonist-dependent recruitment of phosphoinositide 3-kinase to the membrane by beta-adrenergic receptor kinase 1. A role in receptor sequestration. J Biol Chem 276:18953–18959.
9. Maehama T, Dixon JE. 1998. The tumor suppressor, PTEN/MMAC1, dephosphorylates the lipid second messenger, phosphatidylinositol 3,4,5-trisphosphate. J Biol Chem 273:13375–13378.
10. Jones PF, Jakubowicz T, Pitossi FJ, Maurer F, Hemmings BA. 1991. Molecular cloning and identification of a serine/threonine protein kinase of the second-messenger subfamily. Proc Natl Acad Sci USA 88:4171–4175.
11. Coffer PJ, Woodgett JR. 1991. Molecular cloning and characterisation of a novel putative protein-serine kinase related to the cAMP-dependent and protein kinase C families [published erratum appears in Eur J Biochem 1992 May 1;205(3):1217]. Eur J Biochem 201:475–481.
12. Chan TO, Rittenhouse SE, Tsichlis PN. 1999. Akt/PKB other D3 phosphoinositide-regulated kinases: kinase activation by phosphoinositide-dependent phosphorylation. Annu Rev Biochem 68:965–1014.
13. Vanhaesebroeck B, Alessi DR. 2000. The PI3K-PDK1 connection: more than just a road to PKB. Biochem J 346:561–576.
14. Alessi DR, James SR, Downes CP, Holmes AB, Gaffney PR, Reese CB, Cohen P. 1997. Characterization of a 3-phosphoinositide-dependent protein kinase which phosphorylates and activates protein kinase Balpha. Curr Biol 7:261–269.
15. Toker A, Newton AC. 2000. Akt/Protein kinase B is regulated by autophosphorylation at the hypothetical PDK-2 site. J Biol Chem 275:8271–8274.

16. Yano S, Tokumitsu H, Soderling TR. 1998. Calcium promotes cell survival through CaM-K kinase activation of the protein-kinase-B pathway. Nature 396:584–587.
17. Soler RM, Egea J, Mintenig GM, Sanz-Rodriguez C, Iglesias M, Comella JX. 1998. Calmodulin is involved in membrane depolarization-mediated survival of motoneurons by phosphatidylinositol-3 kinase- and MAPK-independent pathways. J Neurosci 18:1230–1239.
18. Frame S, Cohen P. 2001. GSK3 takes centre stage more than 20 years after its discovery. Biochem J 359:1–16.
19. Hardt SE, Sadoshima J. 2002. Glycogen synthase kinase-3beta: a novel regulator of cardiac hypertrophy and development. Circ Res 90:1055–1063.
20. Cross DA, Alessi DR, Cohen P, Andjelkovich M, Hemmings BA. 1995. Inhibition of glycogen synthase kinase-3 by insulin mediated by protein kinase B. Nature 378:785–789.
21. Ballou LM, Tian PY, Lin HY, Jiang YP, Lin RZ. 2001. Dual regulation of glycogen synthase kinase-3beta by the alpha1A-adrenergic receptor. J Biol Chem 276:40910–40916.
22. Fang X, Yu SX, Lu Y, Bast RC, Jr., Woodgett JR, Mills GB. 2000. Phosphorylation and inactivation of glycogen synthase kinase 3 by protein kinase A. Proc Natl Acad Sci USA 97:11960–11965.
23. Clemens MG. 1996. Protein kinases that phosphorylate eIF2 and eIF2B, and their role in eukaryotic cell translational control. Cold Spring Harbor Laboratory Press, NY.
24. Dufner A, Thomas G. 1999. Ribosomal S6 kinase signaling and the control of translation. Exp Cell Res 253:100–109.
25. Shima H, Pende M, Chen Y, Fumagalli S, Thomas G, Kozma SC. 1998. Disruption of the p70(s6k)/p85(s6k) gene reveals a small mouse phenotype and a new functional S6 kinase. Embo J 17:6649–6659.
26. Burgering BM, Coffer PJ. 1995. Protein kinase B (c-Akt) in phosphatidylinositol-3-OH kinase signal transduction. Nature 376:599–602.
27. Dufner A, Andjelkovic M, Burgering BM, Hemmings BA, Thomas G. 1999. Protein kinase B localization and activation differentially affect S6 kinase 1 activity and eukaryotic translation initiation factor 4E- binding protein 1 phosphorylation. Mol Cell Biol 19:4525–4534.
28. Conus NM, Hemmings BA, Pearson RB. 1998. Differential regulation by calcium reveals distinct signaling requirements for the activation of Akt and p70S6k. J Biol Chem 273:4776–4782.
29. Ballou LM, Cross ME, Huang S, McReynolds EM, Zhang BX, Lin RZ. 2000. Differential regulation of the phosphatidylinositol 3-kinase/Akt and p70 S6 kinase pathways by the alpha(1A)-adrenergic receptor in rat-1 fibroblasts. J Biol Chem 275:4803–4809.
30. Saunders RN, Metcalfe MS, Nicholson ML. 2001. Rapamycin in transplantation: a review of the evidence. Kidney Int 59:3–16.
31. Gingras A-C, Raught B, Sonenberg N. 2001. Regulation of translation initiation by FRAP/mTOR. Genes and Development 15:807–826.
32. Scott PH, Brunn GJ, Kohn AD, Roth RA, Lawrence JC, Jr. 1998. Evidence of insulin-stimulated phosphorylation and activation of the mammalian target of rapamycin mediated by a protein kinase B signaling pathway. Proc Natl Acad Sci U S A 95:7772–7777.
33. Peterson RT, Beal PA, Comb MJ, Schreiber SL. 2000. FKBP12-rapamycin-associated protein (FRAP) autophosphorylates at serine 2481 under translationally repressive conditions. J Biol Chem 275:7416–7423.
34. Sekulic A, Hudson CC, Homme JL, Yin P, Otterness DM, Karnitz LM, Abraham RT. 2000. A direct linkage between the phosphoinositide 3-kinase-AKT signaling pathway and the mammalian target of rapamycin in mitogen-stimulated and transformed cells. Cancer Res 60:3504–3513.
35. Fang Y, Vilella-Bach M, Bachmann R, Flanigan A, Chen J. 2001. Phosphatidic acid-mediated mitogenic activation of mTOR signaling. Science 294:1942–1945.
36. Dennis PB, Jaeschke A, Saitoh M, Fowler B, Kozma SC, Thomas G. 2001. Mammalian TOR: a homeostatic ATP sensor. Science 294:1102–1105.
37. Schmelzle T, Hall MN. 2000. TOR, a central controller of cell growth. Cell 103:253–262.
38. Gingras AC, Kennedy SG, O'Leary MA, Sonenberg N, Hay N. 1998. 4E-BP1, a repressor of mRNA translation, is phosphorylated and inactivated by the Akt(PKB) signaling pathway. Genes Dev 12:502–513.
39. Weinkove D, Leevers SJ. 2000. The genetic control of organ growth: insights from Drosophila. Curr Opin Genet Dev 10:75–80.
40. Chen C, Jack J, Garofalo RS. 1996. The Drosophila insulin receptor is required for normal growth. Endocrinology 137:846–856.
41. Leevers SJ, Weinkove D, MacDougall LK, Hafen E, Waterfield MD. 1996. The Drosophila phosphoinositide 3-kinase Dp110 promotes cell growth. Embo J 15:6584–6594.

42. Bohni R, Riesgo-Escovar J, Oldham S, Brogiolo W, Stocker H, Andruss BF, Beckingham K, Hafen E. 1999. Autonomous control of cell and organ size by CHICO, a Drosophila homolog of vertebrate IRS1-4. Cell 97:865–875.
43. Montagne J, Stewart MJ, Stocker H, Hafen E, Kozuma SC, Thomas G. 1999. Drosophila S6 kinase: a regulator of cell size. Science 285:2126–2129.
44. Baker J, Liu JP, Robertson EJ, Efstratiadis A. 1993. Role of insulin-like growth factors in embryonic and postnatal growth. Cell 75:73–82.
45. DeChiara TM, Efstratiadis A, Robertson EJ. 1990. A growth-deficiency phenotype in heterozygous mice carrying an insulin-like growth factor II gene disrupted by targeting. Nature 345:78–80.
46. Liu JP, Baker J, Perkins AS, Robertson EJ, Efstratiadis A. 1993. Mice carrying null mutations of the genes encoding insulin-like growth factor I (Igf-1) and type 1 IGF receptor (Igf1r). Cell 75:59–72.
47. Araki E, Lipes MA, Patti ME, Bruning JC, Haag Br, Johnson RS, Kahn CR. 1994. Alternative pathway of insulin signalling in mice with targeted disruption of the IRS-1 gene. Nature 372:186–190.
48. Tamemoto H, Kadowaki T, Tobe K, Yagi T, Sakura H, Hayakawa T, Terauchi Y, Ueki K, Kaburagi Y, Satoh S, et al. 1994. Insulin resistance and growth retardation in mice lacking insulin receptor substrate-1 [see comments]. Nature 372:182–186.
49. Withers DJ, Gutierrez JS, Towery H, Burks DJ, Ren JM, Previs S, Zhang Y, Bernal D, Pons S, Shulman GI, et al. 1998. Disruption of IRS-2 causes type 2 diabetes in mice. Nature 391:900–904.
50. Aikawa R, Nawano M, Gu Y, Katagiri H, Asano T, Zhu W, Nagai R, Komuro I. 2000. Insulin prevents cardiomyocytes from oxidative stress-induced apoptosis through activation of PI3 kinase/Akt. Circulation 102:2873–2879.
51. Foncea R, Andersson M, Ketterman A, Blakesley V, Sapag-Hagar M, Sugden PH, LeRoith D, Lavandero S. 1997. Insulin-like growth factor-I rapidly activates multiple signal transduction pathways in cultured rat cardiac myocytes. J Biol Chem 272:19115–19124.
52. Schluter KD, Simm A, Schafer M, Taimor G, Piper HM. 1999. Early response kinase and PI 3-kinase activation in adult cardiomyocytes and their role in hypertrophy. Am J Physiol 276:H1655–1663.
53. Oh H, Fujio Y, Kunisada K, Hirota H, Matsui H, Kishimoto T, Yamauchi-Takihara K. 1998. Activation of phosphatidylinositol 3-kinase through glycoprotein 130 induces protein kinase B and p70 S6 kinase phosphorylation in cardiac myocytes. J Biol Chem 273:9703–9710.
54. Chesley A, Lundberg MS, Asai T, Xiao RP, Ohtani S, Lakatta EG, Crow MT. 2000. The beta(2)-adrenergic receptor delivers an antiapoptotic signal to cardiac myocytes through G(i)-dependent coupling to phosphatidylinositol 3′-kinase. Circ Res 87:1172–1179.
55. Wu W, Lee WL, Wu YY, Chen D, Liu TJ, Jang A, Sharma PM, Wang PH. 2000. Expression of constitutively active phosphatidylinositol 3-kinase inhibits activation of caspase 3 and apoptosis of cardiac muscle cells. J Biol Chem 275:40113–40119.
56. Craig R, Wagner M, McCardle T, Craig AG, Glembotski CC. 2001. The cytoprotective effects of the glycoprotein 130 receptor-coupled cytokine, cardiotrophin-1, require activation of NF-kappa B. J Biol Chem 276:37621–37629.
57. Hiraoka E, Kawashima S, Takahashi T, Rikitake Y, Kitamura T, Ogawa W, Yokoyama M. 2001. TNF-alpha induces protein synthesis through PI3-kinase-Akt/PKB pathway in cardiac myocytes. Am J Physiol Heart Circ Physiol 280:H1861–1868.
58. Badorff C, Ruetten H, Mueller S, Stahmer M, Gehring D, Jung F, Ihling C, Zeiher AM, Dimmeler S. 2002. Fas receptor signaling inhibits glycogen synthase kinase 3 beta and induces cardiac hypertrophy following pressure overload. J Clin Invest 109:373–381.
59. Zhu WZ, Zheng M, Koch WJ, Lefkowitz RJ, Kobilka BK, Xiao RP. 2001. Dual modulation of cell survival and cell death by beta(2)-adrenergic signaling in adult mouse cardiac myocytes. Proc Natl Acad Sci U S A 98:1607–1612.
60. Petroff MG, Kim SH, Pepe S, Dessy C, Marban E, Balligand JL, Sollott SJ. 2001. Endogenous nitric oxide mechanisms mediate the stretch dependence of Ca2+ release in cardiomyocytes. Nat Cell Biol 3:867–873.
61. Sadoshima J, Izumo S. 1995. Rapamycin selectively inhibits angiotensin II-induced increase in protein synthesis in cardiac myocytes in vitro. Potential role of 70-kD S6 kinase in angiotensin II-induced cardiac hypertrophy. Circ Res 77:1040–1052.
62. Boluyt MO, Zheng JS, Younes A, Long X, O'Neill L, Silverman H, Lakatta EG, Crow MT. 1997. Rapamycin inhibits alpha 1-adrenergic receptor-stimulated cardiac myocyte hypertrophy but not activation of hypertrophy-associated genes. Evidence for involvement of p70 S6 kinase. Circ Res 81:176–186.

63. Baliga RR, Pimental DR, Zhao YY, Simmons WW, Marchionni MA, Sawyer DB, Kelly RA. 1999. NRG-1-induced cardiomyocyte hypertrophy. Role of PI-3-kinase, p70(S6K), and MEK-MAPK-RSK. Am J Physiol 277:H2026–2037.
64. Simm A, Schluter K, Diez C, Piper HM, Hoppe J. 1998. Activation of p70(S6) kinase by beta-adrenoceptor agonists on adult cardiomyocytes. J Mol Cell Cardiol 30:2059–2067.
65. Haq S, Choukroun G, Kang ZB, Ranu H, Matsui T, Rosenzweig A, Molkentin JD, Alessandrini A, Woodgett J, Hajjar R, et al. 2000. Glycogen synthase kinase-3beta is a negative regulator of cardiomyocyte hypertrophy. J Cell Biol 151:117–130.
66. Morisco C, Zebrowski D, Condorelli G, Tsichlis P, Vatner SF, Sadoshima J. 2000. The Akt-glycogen synthase kinase 3beta pathway regulates transcription of atrial natriuretic factor induced by beta-adrenergic receptor stimulation in cardiac myocytes. J Biol Chem 275:14466–14475.
67. Pham FH, Sugden PH, Clerk A. 2000. Regulation of protein kinase B and 4E-BP1 by oxidative stress in cardiac myocytes. Circ Res 86:1252–1258.
68. Tuxworth WJ, Jr., Wada H, Ishibashi Y, McDermott PJ. 1999. Role of load in regulating eIF-4F complex formation in adult feline cardiocytes. Am J Physiol 277:H1273–1282.
69. Donohue TJ, Dworkin LD, Lango MN, Fliegner K, Lango RP, Benstein JA, Slater WR, Catanese VM. 1994. Induction of myocardial insulin-like growth factor-I gene expression in left ventricular hypertrophy. Circulation 89:799–809.
70. Neri Serneri GG, Boddi M, Modesti PA, Cecioni I, Coppo M, Padeletti L, Michelucci A, Colella A, Galanti G. 2001. Increased cardiac sympathetic activity and insulin-like growth factor-I formation are associated with physiological hypertrophy in athletes. Circ Res 89:977–982.
71. Serneri GG, Modesti PA, Boddi M, Cecioni I, Paniccia R, Coppo M, Galanti G, Simonetti I, Vanni S, Papa L, et al. 1999. Cardiac growth factors in human hypertrophy. Relations with myocardial contractility and wall stress. Circ Res 85:57–67.
72. Yamashita K, Kajstura J, Discher DJ, Wasserlauf BJ, Bishopric NH, Anversa P, Webster KA. 2001. Reperfusion-activated Akt kinase prevents apoptosis in transgenic mouse hearts overexpressing insulin-like growth factor-1. Circ Res 88:609–614.
73. Haq S, Choukroun G, Lim H, Tymitz KM, del Monte F, Gwathmey J, Grazette L, Michael A, Hajjar R, Force T, Molkentin JD. 2001. Differential activation of signal transduction pathways in human hearts with hypertrophy versus advanced heart failure. Circulation 103:670–677.
74. Kim CH, Cho YS, Chun YS, Park JW, Kim MS. 2002. Early expression of myocardial HIF-1alpha in response to mechanical stresses: regulation by stretch-activated channels and the phosphatidylinositol 3-kinase signaling pathway. Circ Res 90:E25–33.
75. Laser M, Kasi VS, Hamawaki M, Cooper Gt, Kerr CM, Kuppuswamy D. 1998. Differential activation of p70 and p85 S6 kinase isoforms during cardiac hypertrophy in the adult mammal. J Biol Chem 273:24610–24619.
76. Sanada S, Kitakaze M, Node K, Takashima S, Ogai A, Asanuma H, Sakata Y, Asakura M, Ogita H, Liao Y, et al. 2001. Differential subcellular actions of ACE inhibitors and AT(1) receptor antagonists on cardiac remodeling induced by chronic inhibition of NO synthesis in rats. Hypertension 38:404–411.
77. Ito H, Hiroe M, Hirata Y, Tsujino M, Adachi S, Shichiri M, Koike A, Nogami A, Marumo F. 1993. Insulin-like growth factor-I induces hypertrophy with enhanced expression of muscle specific genes in cultured rat cardiomyocytes. Circulation 87:1715–1721.
78. Schwartzbauer G, Robbins J. 2001. The tumor suppressor gene PTEN can regulate cardiac hypertrophy and survival. J Biol Chem 276:35786–35793.
79. Morisco C, Seta K, Hardt SE, Lee Y, Vatner SF, Sadoshima J. 2001. Glycogen synthase kinase 3beta regulates GATA4 in cardiac myocytes. J Biol Chem 276:28586–28597.
80. Liu T, Lai H, Wu W, Chinn S, Wang PH. 2001. Developing a strategy to define the effects of insulin-like growth factor-1 on gene expression profile in cardiomyocytes. Circ Res 88:1231–1238.
81. Reiss K, Cheng W, Ferber A, Kajstura J, Li P, Li B, Olivetti G, Homcy CJ, Baserga R, Anversa P. 1996. Overexpression of insulin-like growth factor-1 in the heart is coupled with myocyte proliferation in transgenic mice. Proc Natl Acad Sci U S A 93:8630–8635.
82. Belke DD, Betuing S, Tuttle MJ, Graveleau C, Young ME, Pham M, Zhang D, Cooksey RC, McClain DA, Litwin SE, et al. 2002. Insulin signaling coordinately regulates cardiac size, metabolism, and contractile protein isoform expression. J Clin Invest 109:629–639.
83. Shioi T, Kang PM, Douglas PS, Hampe J, Yballe CM, Lawitts J, Cantley LC, Izumo S. 2000. The conserved phosphoinositide 3-kinase pathway determines heart size in mice. EMBO J 19:2537–2548.
84. Franke TF, Kaplan DR, Cantley LC, Toker A. 1997. Direct regulation of the Akt proto-oncogene product by phosphatidylinositol-3,4-bisphosphate. Science 275:665–668.

85. Shioi T, McMullen JR, Kang PM, Douglas PS, Obata T, Franke TF, Cantley LC, Izumo S. 2002. Akt/Protein kinase B promotes organ growth in transgenic mice. Mol Cell Biol 22:2799–2809.
86. Lembo G, Rockman HA, Hunter JJ, Steinmetz H, Koch WJ, Ma L, Prinz MP, Ross J, Jr., Chien KR, Powell-Braxton L. 1996. Elevated blood pressure and enhanced myocardial contractility in mice with severe IGF-1 deficiency. J Clin Invest 98:2648–2655.
87. Antos CL, McKinsey TA, Frey N, Kutschke W, McAnally J, Shelton JM, Richardson JA, Hill JA, Olson EN. 2002. Activated glycogen synthase-3 beta suppresses cardiac hypertrophy in vivo. Proc Natl Acad Sci U S A 99:907–912.
88. Morgan HE, Gordon EE, Kira Y, Chua HL, Russo LA, Peterson CJ, McDermott PJ, Watson PA. 1987. Biochemical mechanisms of cardiac hypertrophy. Annu Rev Physiol 49:533–543.
89. Peterson MB, Lesch M. 1972. Protein synthesis and amino acid transport in the isolated rabbit right ventricular papillary muscle. Effect of isometric tension development. Circ Res 31:317–327.
90. Xenophontos XP, Gordon EE, Morgan HE. 1986. Effect of intraventricular pressure on protein synthesis in arrested rat hearts. Am J Physiol 251:C95–98.
91. Sadoshima J, Jahn L, Takahashi T, Kulik TJ, Izumo S. 1992. Molecular characterization of the stretch-induced adaptation of cultured cardiac cells. An in vitro model of load-induced cardiac hypertrophy. J Biol Chem 267:10551–10560.
92. Colao A, Marzullo P, Di Somma C, Lombardi G. 2001. Growth hormone and the heart. Clin Endocrinol (Oxf) 54:137–154.
93. Fazio S, Cittadini A, Sabatini D, Merola B, Colao AM, Biondi B, Lombardi G, Sacca L. 1993. Evidence for biventricular involvement in acromegaly: a Doppler echocardiographic study. Eur Heart J 14:26–33.
94. Fazio S, Sabatini D, Capaldo B, Vigorito C, Giordano A, Guida R, Pardo F, Biondi B, Sacca L. 1996. A preliminary study of growth hormone in the treatment of dilated cardiomyopathy [see comments]. N Engl J Med 334:809–814.
95. Osterziel KJ, Strohm O, Schuler J, Friedrich M, Hanlein D, Willenbrock R, Anker SD, Poole-Wilson PA, Ranke MB, Dietz R. 1998. Randomised, double-blind, placebo-controlled trial of human recombinant growth hormone in patients with chronic heart failure due to dilated cardiomyopathy. Lancet 351:1233–1237.
96. Butler AA, Le Roith D. 2001. Control of growth by the somatropic axis: growth hormone and the insulin-like growth factors have related and independent roles. Annu Rev Physiol 63:141–164.

Pathophysiology of Cardiovascular Disease,
edited by N.S. Dhalla, H. Rupp,
A. Angel & G.N. Pierce
Kluwer Academic Publishers, Boston, 2004

Role of Ras/GTPase and Calcium/Calmodulin-dependent Protein Kinase II in the Signal Transduction Mechanisms Involved in Hyperthermic Preconditioning

Ibrahim F. Benter,[1] Jasbir S. Juggi,[2] and Saghir Akhtar[2]

Departments of Pharmacology[1] and Physiology[2], Faculty of Medicine, Kuwait University, P.O. Box: 24923, Safat 13110, Kuwait

Summary. The signaling pathways involved in ischemic heart disease are not well understood. In this study, the role of Ras/GTPase, and the serine threonine Calcium/Calmodulin-dependent protein kinase II (CaMK II) in mediating global ischemia and hyperthermic preconditioning (PC) in a perfused rat heart model was investigated. A 60 minute episode of global ischemia in perfused rat hearts produced significantly impaired cardiac function, measured as left ventricular developed pressure (Pmax) and left ventricular end-diastolic pressure (LVEDP), and impaired coronary hemodynamics, measured as coronary flow (CF) and coronary vascular resistance (CVR). PC and/or pre-treatment with Ras/GTPase inhibitor FPT III (232 ng/min for 6 days) significantly enhanced cardiac recovery. Combination of PC and FPT III treatment did not produce any additive benefits as compared to either treatment alone. In contrast, pre-treatment of hearts with CamKII inhibitor, KN-93 (578 ng/min for 6 days), produced detrimental effects on recovery of cardiac function and coronary hemodynamics after 30 min ischemia. PC did not produce any improvement in cardiac function in animals that were pretreated with KN-93. However, PC did produce partial improvement in function in animals that were pretreated with both FPT III and KN-93. These observations suggest that beneficial effects produced by hyperthermic preconditioning are mediated by inhibition of signal transduction pathways involving Ras/GTPase. These results also indicate that CaMK II-mediated regulation of calcium homeostasis is part of the hyperthermic preconditioning mechanisms that improve recovery after ischemia/reperfusion. For reasons that are not clear inhibition of Ras/GTPase attenuated the detrimental effects of CaMK II

Address for Correspondence: Dr. I.F. Benter, Chairman, Department of Pharmacology & Toxicology, Faculty of Medicine, Kuwait University, P.O. Box 24923, Safat, 13110, Kuwait. Telephone: 965 5312300 (Ext.) 6307/6344, Fax: 965 5342583, E-mail: ibenter@hsc.kuniv.edu.kw.

inhibition suggesting that removal of Ras/GTPase input improves recovery either by improving calcium regulation and/or some other mechanism.

Key words: Cardiac ischemia/reperfusion, preconditioning, heart, sodium-hydrogen exchanger, ischemia, Ras/GTPase, CaMK II.

INTRODUCTION

Cardiac ischemia arising from coronary heart disease or open-heart surgery can lead to cardiac failure. The development of rationally-designed therapeutic agents will depend on a thorough understanding of the underlying signalling mechanisms involved in the development of cardiac ischemia as these are poorly understood at present. Several G-protein-coupled receptors (GPCR) agonists such as norepinephrine, angiotensin and thrombin increase activity of cardiac NHE-1 activity by increasing sensitivity of exchanger to H^+ (1). GPCRs constitute the largest family of cell-surface molecules involved in signal transmission and play key physiological roles, with their dysfunction resulting in several disease states. Stimulation of GPCRs results in activation of several molecular routes that includes Ras/GTPase family of molecules. Activation of the Ras/GTPase signalling pathway by GTP loading of Ras/GTPase proteins is the critical step that connects the events at the plasma membrane with cellular and nuclear responses. Farnesylation of p21Ras/GTPase is catalysed by the enzyme farnesyltransferase and is an obligatory initial step that targets Ras proteins to the plasma membrane for future activation.

We previously showed that Ras/GTPase activity in the heart is increased during the development of hypertension (2,3). Further we showed that normalization of Ras/GTPase activity by the selective inhibition with FPT III prevented elevation of blood pressure and hypertension-related end organ damage in several models of hypertension (2). Abe et al. (4,5) suggested a role for Ras/GTPase in reactive oxygen species-mediated cardiovascular disease by demonstrating that Ras/GTPase is essential for both p90RSK and ERK1/2 activation by H_2O_2 which is increased in hypertension, inflammation, ischemia-reperfusion injury, and atherosclerosis (6). These observations led to our hypothesis that Ras/GTPase activity may play an important role in cardiac dysfunction due to ischemia-reperfusion damage and cardioprotection by preconditioning.

Ischemia/reperfusion (I/R) can cause calcium overload which may lead to impaired myocardial contraction. It has been suggested that PC attenuates reduction in cardiac function due to I/R by improving the mechanisms involved in handling of excessive calcium due to I/P. Dhalla and colleagues showed that attenuation of PC-mediated protection of changes in cardiac function by an inhibitor of CaMK II correlated linearly with reduced sarcoplasmic reticulum calcium uptake, which in turn correlated well with attenuated CaMK II activity (7). However, the contribution of CaMK II to hyperthermic preconditioning and the effect of Ras/GTPase-mediated signal transduction on CaMK II-mediated calcium regulation in a model of global ischemia have not been elucidated.

The aim of the present study was to examine the role of the Ras/GTPase and CaMK II in mediating events during cardiac ischemia/reperfusion and hyperthermic preconditioning, and subsequent recovery of function in a clinically relevant model of sustained global cardiac ischemia.

METHODS

Experimental protocol

To study the effects of inhibitors of Ras farnesyl transferase (FPT III) and CaMK II (KN-93) on hyperthermic preconditioning, rats were divided into various groups. FPT III and KN-93 were administered using osmotic minipumps (model 2ML1). The osmotic minipumps were filled with FPT III, KN-93 or vehicle and placed in the abdominal cavity. The osmotic minipumps delivered FPT III (138 ng/min) or KN-93 (578 ng/min) over a period of one week. The doses of FPT III and KN-93 used in these studies were determined by our previous studies where we showed normalization of hypertension-related elevation of Ras/GTPase and CaMK II levels in the heart (2,3).

The harvested hearts from FPT III-, KN-93 or vehicle-treated rats were perfused with the perfusion medium for 30 minutes before subjecting them to 60 minutes of unprotected ischemia at 34°C or hyperthermic (42°C) preconditioned ischemia (60 min at 34°C). At the end of the ischemic period all the hearts were reperfused for 30 min at 34°C when the post-ischemic left ventricular contractility was recorded and compared. Our previous studies indicated that hyperthermic (42°C) preconditioning provided optimal protection to left ventricular function (8) and therefore this model of preconditioning was selected for this study.

Heart perfusion

Rats were lightly anaesthetized with Intraval Sodium (40 mg/kg body weight) and hearts were rapidly removed after intravenous heparinization (1000 U/kg body weight). This investigation conforms to the guide for the care and use of Laboratory animals published by the US National Institute of Health (NIH publication No 85–23, revised 1985). The excised hearts were immediately mounted on the Langendorff perfusion assembly (Hugo Sachs Electronics (HSE) Freiburg, Germany), and were perfused with a constant pressure perfusion of 50 mm Hg with oxygenated (95% O_2 + 5% CO_2) Krebs-Henselit buffer (34°C) of the following composition (in mM): NaCl 117; KCl 4.39; $CaCl_2$ 2.5; $NaHCO_3$ 20.0; KH_2PO_4 1.21; $MgCl_2.6H_2O$ 1.2; Glucose 12.0; osmolarity 300 mOsm/l, pH 7.35. A water-filled balloon was introduced into the left ventricle and connected to a Statham pressure transducer (P23Db) and balloon volume was adjusted to give the baseline end-diastolic pressure of 5 mmHg. Left ventricular developed pressure (P_{max}) and left ventricular end-diastolic pressure (LVEDP), were continuously monitored and computed for statistical significance by paired students t-test.

RESULTS

Table 1 shows Pmax values at the end of 30 minutes of reperfusion following 1 hour of ischemia. Recovery of Pmax was very poor in hearts that underwent ischemia alone (46 ± 7 mmHg). PC and FPT-III pretreatment significantly ($p <$

Table 1. Effect of FPT III on the left ventricular functions in ischemic-reperfused and hyperthermic (42°C) preconditioned hearts

	P_{max} (mm Hg)		LVEDP (mm Hg)	
	Con	Rep	Con	Rep
Ischemia 34°C (n = 7)	111 ± 6	46 ± 7*	5 ± 0.1	43 ± 4*
PC + Ischemia 34°C (n = 7)	124 ± 6	80 ± 6*†	5 ± 0.3	18 ± 2*†
FPT + Ischemia 34°C (n = 6)	119 ± 8	97 ± 5*†	6 ± 0.2	20 ± 2*†
FPT + PC + Ischemia 34°C (n = 7)	109 ± 7	88 ± 6*†	5 ± 0.1	14 ± 1*†

The data was computed at 30 min reperfusion and expressed as mean ± SEM.
Con, Control; Rep, Reperfusion; P_{max}, Left ventricular developed pressure; LVEDP, Left ventricular end-diastolic pressure.
*, $p < 0.05$ compared with respective controls; †, $p < 0.05$ compared to Ischemia 34°C.

0.05) improved the post-reperfusion recovery in the left ventricular Pmax (80 ± 6 and 97 ± 5 mmHg, respectively, and significantly ($p < 0.05$) reduced the excessive rise of LVEDP (18 ± 2 and 20 ± 2 mmHg, respectively, when compared with vehicle-treated ischemia-reperfusion LVEDP: (43 ± 5 mmHg). FPT-III did not show any additive protection when used in combination with PC. Reperfusion recovery in Pmax and LVEDP did not significantly differ between the hyperthermic (42°C) preconditioned hearts; FPT-III pretreated ischemic hearts and FPT-III pretreated preconditioned (42°C) hearts (Table 1).

In contrast, pretreatment with CaMK II inhibitor KN-93 resulted in opposite results as compared to pretreatment with FPT III (Table 2). As shown before, recovery of Pmax was very poor in hearts that underwent ischemia alone (46 ± 7 mmHg) and recovery was even worse with KN-93 pretreatment (22 ± 3 mmHg, $p < 0.05$). As expected, Pmax values were much higher in hearts that were preconditioned before ischemia (80 ± 6 mmHg). However, pretreatment with CaMK II inhibitor KN-93 before PC prevented the beneficial effect on Pmax observed in PC hearts (23 ± 6 mmHg). Interestingly, inhibition of Ras/GTPase with FPT III with simultaneous inhibition of CaMK II attenuated the detrimental effects of KN-93 on PC-induced improvements. (Pmax: 55 ± 5 mmHg). Similar pattern of effects were observed on LVEDP values (Table 2).

DISCUSSION

In ischemic heart disease, the heart is exposed to many cell stresses including the increased production of ROS, ionic imbalances, metabolic deprivation, and osmotic and mechanical stresses (6). As a result, such ischemic triggers are known to initiate release of mediators that activate numerous receptors including G-protein coupled receptors and receptor tyrosine kinases. For example, ROS stimulates activation of small G-proteins and kinases such as Ras/GTPase and MAP Kinases that can ultimately lead to calcium overload and cardiac dysfunction (6). However, there is considerable controversy in the exact role of specific signalling molecules involved in both ischemia-reperfusion and preconditioning (9,10). For example, the inhibi-

Table 2. Effect of KN93 on the left ventricular functions in ischemic-reperfused and hyperthermic (42°C) preconditioned hearts. Influence of combined protection with FPT III

	P_{max} (mm Hg)		LVEDP (mm Hg)	
	Con	Rep	Con	Rep
Ischemia 34°C (n = 7)	111 ± 6	46 ± 7*	5 ± 0.1	43 ± 4*
PC + Ischemia 34°C (n = 7)	124 ± 6	80 ± 6*†	5 ± 0.3	18 ± 2*†
KN93 + Ischemia 34°C (n = 3)	123 ± 5.4	22 ± 3*†‡	5.5 ± 0.2	41 ± 7*‡
KN93 + PC + Ishemia 34°C (n = 5)	99 ± 5	23 ± 6*†	5.3 ± 0.1	30.3 ± 3*†
FPT + KN93 + PC + Isch. 34°C (n = 3)	95 ± 8	55 ± 5*†$#	5.3 ± 0.5	13.7 ± 4*†$#

The data was computed at 30 min reperfusion and expressed as mean ± SEM.
Con, Control; Rep, Reperfusion; P_{max}, Left ventricular developed pressure; LVEDP, Left ventricular end-diastolic pressure.
*, $p < 0.05$ compared with respective controls; †, $p < 0.05$ compared to Ischemia 34°C; ‡, $p < 0.05$ compared to PC + Ischemia and $, $p < 0.05$ compared to KN93 + Ischemia; #, $p < 0.05$ compared to KN93 + PC + Ischemia.

tion of p38 MAP kinase has been purported to both improve and exacerbate recovery from ischemia-reperfusion and preconditioning (9,10). For example, the inhibition of p38 MAP kinase has been purported to both improve and exacerbate recovery from ischemia-reperfusion implying opposing roles for this MAP kinase in the development of ischemia (5,11).

The isolated perfused rat heart exhibited poor recovery in cardiac function and coronary hemodymanics when exposed to a 60 minutes episode of ischemia followed by 30 minutes reperfusion. Hyperthermic PC produced significant improvement in recovery from ischemia as compared to ischemia alone and was consistent with previous findings (8).

Ras/GTPase inhibition prevents cardiac ventricular dysfunction due to ischemia-reperfusion

We have previously shown that Ras/GTPase activity in the heart is increased during the development of hypertension, and that normalization of Ras/GTPase activity, by selective inhibition with FPT III, prevented elevation of blood pressure and hypertension-related end organ damage in several models of hypertension (2,3). This, and the previous finding that Ras/GTPase can be activated by reactive oxygen species, led to our hypothesis that Ras/GTPase activity may play an important role in cardiac dysfunction due to ischemia-reperfusion. This study shows for the first time that inhibition of Ras/GTPase, by the use of an inhibitor of Ras/GTPase farnesylation, significantly improves cardiac function in a manner comparable to PC. FPT III pretreatment significantly improved left ventricular contractility (Pmax and LVEDP) to values similar to those observed with PC.

Inhibition of CaMK II exacerbates left-ventricular contractile dysfunction due to ischemia-reperfusion

CaMK II is a multifunctional serine threonine kinase responsible for regulating the activity of several Ca^{2+}-handling proteins that play important roles in cardiac and vascular functions. We previously showed that CaMK II activation represents an important component of the mechanisms initaiting vascular smooth muscle cell proliferation and development of hypertension (2,3). Since calcium overload has been implicated as an important intermediatory step in cardiac dysfunction, in the present study we examined the role of CaMK II in recovery of cardiac function following ischemia/reperfusion. Inhibition of CaMK II by KN93 produced detrimental effects on recovery of cardiac functions. These findings contribute to the understanding of signalling mechanisms involved in ischemia/reperfusion. Our data suggest that CaMK II is involved in signalling pathways leading to recovery from cardiac ischemia whereas Ras/GTPase signalling pathways are critical in the development of cardiac dysfunction due to ischemia.

Our present study and that of others suggest the schematic scheme represented as Figure 1 for signalling molecules involved in cardiac dysfunction due to ischemia/reperfusion. It suggests that release of various mediators by ischemia leads to activation of RTK- and GPCR-mediated signal transduction pathways. The

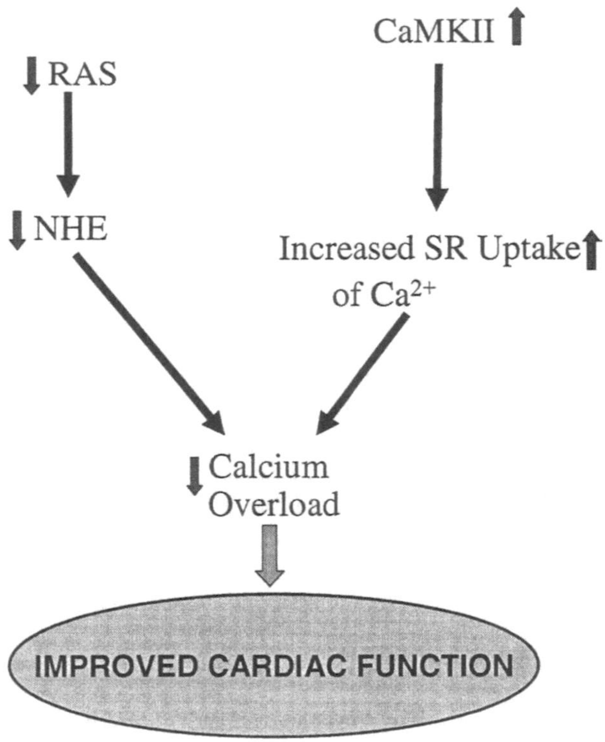

Figure 1. Schematic representation for the signaling molecules involved in Cardiac dysfunction due to Ischemia-reperfusion.

Ras/GTPase pathway leads to activation of NHE-1 which produces calcium overload and cardiac dysfunction. On the other hand, the data from our present study and that of Dhalla and colleagues suggest that activation of the CaMK II pathway prevents development of calcium overload and thereby plays a protective role during ischemia-reperfusion.

In conclusion, this study has demonstrated that inhibition of Ras/GTPase significantly improved recovery of left ventricular function after global ischemia implicating Ras/GTPase-activated signal transduction as a mediator of ischemia-reperfusion injury. In contrast, inhibition of CaMK II produced detrimental effects on recovery of cardiac function in conditions with or without PC suggesting that CaMK II is part of the protective mechanisms activated during ischemia-reperfusion or activated due to PC.

REFERENCES

1. Avkiran M, Gross G, Karmazyn M, Klein H, Murphy E, Ytrehus K. 2001. Na+/H+ exchange in ischemia, reperfusion and preconditioning. Cardiovasc Res 50:162–166.
2. Muthalif MM, Benter IF, Khandekar Z, Gaber L, Estes A, Malik S, Parmentier JH, Manne V, Malik KU. 2000. Contribution of Ras/GTPase and cytochrome P450 metabolites to deoxycorticosterone-salt-induced hypertension. Hypertension 35:457–463.
3. Muthalif MM, Karzoun NA, Benter IF, Gaber L, Ljuca F, Uddin MR, Khandekar Z, Estes A, Malik KU. 2002. Functional Significance of Activation of Calcium/Calmodulin–Dependent Protein Kinase II in Angiotensin II–Induced Vascular Hyperplasia and Hypertension. Hypertension 39[part 2]: 704–709.
4. Abe J, Okuda M, Huang Q, Yoshizumi M, Berk BC. 2000. Reactive oxygen species activate p90 ribosomal S6 kinase via Fyn and Ras. J Biol Chem 275:1739–1748.
5. Abe J, Baines PC, Berk CB. 2000. Role of Mitogen-Activated Protein Kinases in Ischemia and Reperfusion Injury: The Good and the Bad. Circ Res 86:607–609.
6. Takeishi Y, Abe J, Lee JD, Kawakatsu H, Walsh RA, Berk BC. 1999. Differential Regulation of p90 Ribosomal S6 Kinase and Big Mitogen-Activated Protein Kinase 1 by Ischemia/Reperfusion and Oxidative Stress in Perfused Guinea Pig Hearts. Circ Res 85:1164–1172.
7. Osada M, Netticadan T, Kawabata K, Tamura K, Dhalla NS. 2000. Ischemic preconditioning prevents I/R-induced alterations in SR calcium-calmodulin protein kinase II. Am J Physiol 278:H1791–H1798.
8. Juggi JS, Al-Awadi F, Joseph S, Telehoun G, Prahash A. 1997. Ischemic preconditioning is not additive to preservation with hypothermia or crystalloid cardioplegia in the globally ischemic rat heart. Mol Cell Biochemistry 176:303–313.
9. Schneider S, Chen W, Hou J, Steenbergen C, Murphy E. 2001. Inhibition of p38 MAPK α/β reduces ischemic injury and does not block protective effects of preconditioning. Am J Physiol 280: H499–H508.
10. Schulz R, Cohen MV, Behrends M, Downey JM, Heusch G. 2001. Signal transduction of ischemic preconditioning. Cardiovas Res 52:181–198.
11. Ping P, Murphy E. 2000. Role of p38 Mitogen-Activated Protein Kinases in Preconditioning. A Detrimental Factor or a Protective Kinase? Circ Res 86:921–922.

Pathophysiology of Cardiovascular Disease,
edited by N.S. Dhalla, H. Rupp,
A. Angel & G.N. Pierce
Kluwer Academic Publishers, Boston, 2004

Interactions of the Cyclic AMP and Nitric Oxide Pathways in Cardiac Fibroblasts

Åsa B. Gustafsson and Laurence L. Brunton

Department of Pharmacology, 0636
School of Medicine
University of California, San Diego
La Jolla, CA 92093
U.S.A.

Summary. We have studied NO and cyclic AMP signaling in freshly isolated rat cardiac fibroblasts in primary culture. Although these cells express eNOS, basal production of NO is low and is only slightly stimulated by treatments that elevate intracellular Ca^{++}. However, iNOS is readily induced in these cells by IL-1β and by the combination of TNFα and IFNγ, the latter causing cell death that L-NAME inhibits. Activation of β_2-adrenergic receptors enhances iNOS-induction and NO production, an effect mimicked by forskolin and dibutyryl cyclic AMP and antagonized by inhibition of PKA. The effect of cAMP-PKA to enhance iNOS induction is due to message stabilization that results in a doubling of the half-life of iNOS mRNA. Conversely, NO (from iNOS induction or from pharmacologic NO donors) reduces cyclic AMP accumulation in rat cardiac fibroblasts. This cyclic AMP lowering effect of NO does not result from inhibition of cyclic AMP production but rather from stimulation of cyclic AMP degradation, an effect of NO that depends on soluble guanylyl cyclase activity, is independent of PKG activation, and results from the stimulation of PDE2 by cyclic GMP. Experiments with PDE inhibitors indicate that PDE4 is the main PDE activity in these cells (62%), with PDE2 (8%) and PDE3 (1%) making smaller contributions in isoproterenol-stimulated cells. PDE1 seems to be absent and an additional PDE activity is implied by inhibitor data.

Key words: nitric oxide (NO), cyclic AMP, IL-1β, cyclic GMP, PDE2.

Correspondent: Laurence L. Brunton, Ph.D., Department of Pharmacology 0636, UCSD School of Medicine, La Jolla, CA 92093-0636, USA. Phone: 858-534-4667, Fax: 858-534-6833, E-mail: lbrunton@ucsd.edu.

INTRODUCTION

We have recently been defining the pharmacology of second messenger generation in ventricular fibroblasts isolated from adult rat heart. The rationale for this work is that fibroblasts comprise 65% of the heart by cell number and about 16% of the heart by cell mass. Furthermore, these cells are responsible for the deposition and maintenance of the cardiac extracellular matrix, for healing in damaged areas of myocardium, and for some adverse contributions to cardiac hypertrophy. It seemed likely to us that these aspects of fibroblast function might be usefully regulated by hormones and drugs and that fibroblasts might produce diffusible messengers, such as nitric oxide (NO), that could affect the functions of both fibroblasts and myocytes.

In the course of this work, we have defined conditions under which fibroblasts produce substantial quantities of NO. Although cardiac fibroblasts express eNOS (the Ca^{++}/calmodulin calmodulin activated, *endothelial* form of NO synthase), they make little NO basally or when cell Ca^{++} is elevated (1). However, certain cytokines cause a prominent induction of iNOS (the *inducible* NO synthase), with a concomitant production of substantial NO (2).

This paper summarizes both published and unpublished work that demonstrates several interactions between the iNOS/NO/cyclic GMP pathway and the cyclic AMP pathway in cardiac fibroblasts. The data indicate that IL-1β induces iNOS expression and that activation of the cyclic AMP/PKA pathway stabilizes iNOS mRNA, thereby enhancing induction of iNOS and the production of NO. Conversely, NO activates the soluble guanylyl cyclase and the resulting cyclic GMP activates PDE2, the cyclic GMP-stimulated isoform of cyclic AMP phosphodiesterase. Thus, cardiac fibroblasts can produce large quantities of NO in response to IL-1β; furthermore, immune activation and β_2-adrenergic stimulation can interact to affect cellular responses to these stimuli, enhancing the NO pathway and diminishing the cyclic AMP pathway.

MATERIALS AND METHODS

Materials

Rat recombinant IL-1β, Sigma-Aldrich (St. Louis, MO). Collagenase I and trypsin, Worthington (Freehold, NJ). SNAP, Alexis Biochemicals (San Diego, CA). Monoclonal antibody to iNOS, Transduction Laboratories (Lexington, KY). α-[^{32}P]ATP and [^{3}H]cAMP, Perkin Elmer Lifesciences (Boston, MA). All other reagents and chemicals were of reagent grade from Sigma-Aldrich or Calbiochem-Novabiochem (La Jolla, CA).

Isolation of adult ventricular fibroblasts

Rat cardiac fibroblasts were isolated from adult male Sprague-Dawley rats (250–275 g) as recently described (2,3), using a trypsin/collagenase digestion. Cells were cultured in Dulbecco's modified Eagle's medium (DMEM) supplemented with 10% fetal bovine serum and penicillin/streptomycin (100 U/ml). All cells used in experiments were from passages 2 through 4. The purity of these cultures was greater than 95% as determined by immuno-staining.

Western blot analysis

Fibroblasts (on 60 mm culture dishes) were lysed at 4°C in a buffer containing 50 mM β-glycerolphosphate (pH 7.5 at 4°C), 1 mM EGTA, 10 mM $MgCl_2$, 1% TritonX-100, 1 mM PMSF and 10 μg/ml leupeptin. Equal amounts of total protein/lane were loaded and separated on a 7.5% SDS-polyacrylamide gel, then transferred to an Immobilon-P membrane (Millipore, Bedford, MA). After blocking in 5% non-fat milk, the membrane was incubated with a monoclonal iNOS antibody overnight at 4°C, followed by a series of washes and incubation with a secondary antibody coupled to horseradish peroxidase for 1 h at 20°C. iNOS was detected using enhanced chemiluminescence (Amersham Life Science).

Measurement of nitrite levels

NO production was estimated by measuring nitrite accumulation in the culture medium. Fibroblasts were plated on 60 mm culture dishes and grown to 80–90% confluency. For experiments, cells were incubated in phenol-red free DMEM supplemented with 1.5 mM L-arginine, 0.1 mg/ml BSA, 10 μg/ml leupeptin and 100 U/ml penicillin/streptomycin with vehicle or drug added for 24 h. Nitrite in media was assayed by the method of Griess (4).

Assay of cyclic GMP accumulation

In experiments with IL-1β, cells were treated for 24 h with the cytokine, then 1 mM isobutylmethylxanthine (IBMX) was added to each plate 45 min before terminating the experiment. In experiments with an NO donor, the cells (after serum-free incubation) were treated for 15 min with 1 mM IBMX and then stimulated with 1 mM sodium nitroprusside (SNP) for 45 min. All experiments were terminated by aspiration of the media and addition of ice-cold 5% trichloroacetic acid (TCA). TCA extracts were extracted four times with water-saturated ether and the cGMP content was determined by RIA (5).

Assay of cyclic AMP accumulation

Cardiac fibroblasts were incubated with DMEM without serum for 2 h and then treated as described in Results, after which the medium was aspirated and ice-cold 5% TCA was added. The extracts were purified over Dowex-50 and the cyclic AMP content was determined by the method of Gilman (6).

Adenylyl cyclase assay

Membranes were prepared by Dounce homogenization of the cells (in 50 mM HEPES pH 7.6, 5 mM MgCl2, 1 mM EGTA, 0.3 mM PMSF, and 10 μg/ml leupeptin). Membranes were collected by centrifugation (3000 × g for 5 min) and suspended in homogenization buffer. 40 μg of membrane protein was incubated with 1 mM SNP or vehicle on ice for 30 min and adenylyl cyclase activity was assayed by monitoring the conversion of [α-^{32}P]ATP to [^{32}P]cAMP at 30°C for 10 min (7).

Reverse transcription-polymerase chain reaction

Total RNA was prepared from isolated cardiac fibroblasts and ventricular tissue using Trizol (Gibco BRL) and subjected to RT-PCR using the SUPERSCRIPT Pre-amplification System (Gibco BRL). Specific sense and antisense oligonucleotide primers were used to detect expression of the different PDE transcripts (8). The samples were subjected to 35 cycles of 95°C for 1 min, 57°C for 1 min, and 72°C for 1 min, except for the primers for PDE3B where an annealing temperature of 50°C was used. The products were separated by electrophoresis through a 1% agarose gel and visualized by UV light.

Protein determination

Protein content was estimated by the method of Bradford (9) using bovine serum albumin as a standard.

Analysis of data

Analysis and graphing of data were performed with Prism 2.0 (GraphPad Software, San Diego, CA). Data are expressed as mean ± SE. Statistical analysis was performed by Student's t-test. P values < 0.05 were considered to indicate significance.

RESULTS

iNOS induction

Among common cytokines, IL-1β alone stimulates iNOS induction and NO production by cardiac fibroblasts; the effect is potent (half maximally effective concentration, ≈3 ng/ml in a 24 hour assay of NO accumulation) and the production of NO is quite large (2). Indeed, in response to TNFα plus IFNγ, iNOS induction and NO production are so large and persistent that cell death results. Since the NOS inhibitor, L-NAME, prevents cell death, the cells are apparently dying by some NOS-NO-dependent mechanism. From these studies, we conclude that cardiac fibroblasts, under the influence of cytokines, can become a significant source of NO and that NO may affect the viability of the fibroblast itself. Details of some of this work have been recently published (2,10). Having established that IL-1β reproducibly induced iNOS in these cells, we have studied the influence of the cyclic AMP pathway on iNOS induction.

Effect of the cyclic AMP/PKA pathway on iNOS induction

There are reports in the literature of both stimulation and inhibition of iNOS induction by cyclic AMP. In cardiac fibroblasts, the results are striking and clear: isoproterenol doubles iNOS induction and consequent NO production in response

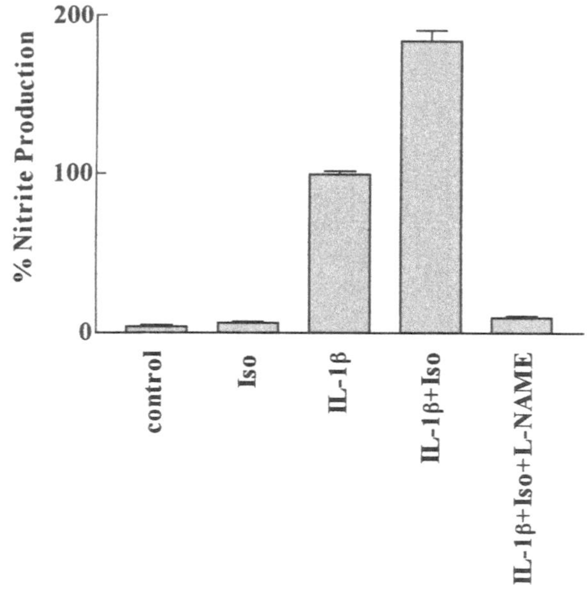

Figure 1. Isoproterenol enhances NO production in IL-1β-stimulated adult rat cardiac fibroblasts. Cells were incubated with vehicle, Iso (10 μM) Iso, IL-1β (10 ng/ml), IL-1β + Iso, and IL-1β + Iso + L-NAME (1 mM) for 24 h, then nitrite accumulation in the culture medium was measured. Effects of IL-1β and IL-1β + Iso are significant ($P < 0.0001$). Data are mean ± SEM, $n = 4$. Adapted from reference 2.

to IL-1β (Figure 1 and reference 2). This effect can also be produced by dibutyryl cyclic AMP, forskolin and high concentrations of phosphodiesterase inhibitors. H-89, an inhibitor of PKA (the cyclic AMP-dependent protein kinase) blocks this enhanced induction of iNOS activity. The enhanced iNOS activity parallels an increase in iNOS protein and mirrors an increase in iNOS mRNA. Thus, we focused on whether the production or stability of the mRNA for iNOS was altered by β-adrenergic stimulation (cyclic AMP/PKA). Nuclear run-on experiments with nuclei isolated from control and isoproterenol-treated fibroblasts show no difference in message production (Figure 2). However, experiments on IL-1β-induced cells in which message half-life was assessed indicate that isoproterenol causes nearly a doubling of the half-life of iNOS mRNA, from 1 hr to 1.9 hr (Figure 3). This alteration is sufficient to account for the observed increase in iNOS protein and NO production (2).

Effect of NO on cyclic AMP production

Following induction of iNOS by IL-1β treatment (10 ng/ml, 24 hours), accumulation of cyclic AMP in response to isoproterenol and forskolin is roughly halved (Figure 4). Addition of the NOS inhibitor L-NMMA fully restores the cyclic AMP

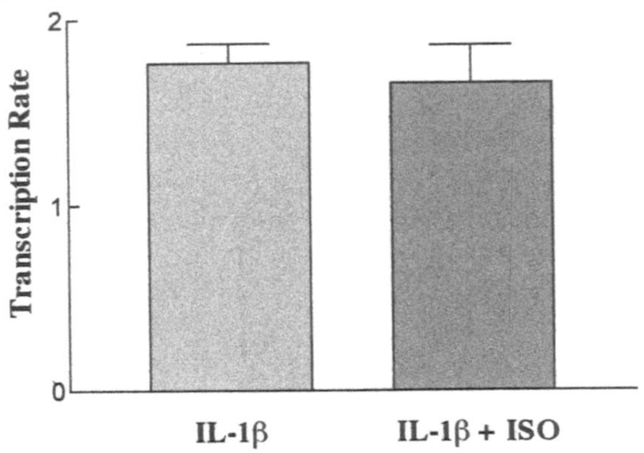

Figure 2. Effects of isoproterenol on iNOS gene transcription.
Cardiac fibroblasts were treated with diluent, 10 μM Iso, 10 ng/ml IL-1β, or Iso + IL-1β. After 20 h, nuclei were prepared and subjected to nuclear run-off assay. The data from the nuclear run-off experiments were quantified and iNOS signals were normalized to β-actin signals, this unitless ratio being termed the "transcription rate" (ordinate). IL-1β treatment stimulated transcription of the iNOS gene; Iso did not significantly enhance the rate of transcription in IL-1β-stimulated fibroblasts ($n = 3$, $P > 0.5$). Adapted from reference 2.

response, indicating that the observed decrease results from enzymatically produced NO. We reproduced these effects of iNOS induction pharmacologically by short-term (45 min) treatment of cells with NO donors [sodium nitro prusside (SNP) and S-nitroso-N-acetyl penicillamine (SNAP)]. Subsequent experiments use this pharmacologic mimicry of iNOS induction.

These are a number of testable pathways by which NO could modulate cyclic AMP production (Figure 5): direct effects of NO on components of hormonally-responsive adenylyl cyclase; actions of NO mediated by cyclic GMP; actions of cyclic GMP mediated by PKG (cyclic GMP-dependent protein kinase) or by the cyclic GMP-stimulated cyclic nucleotide phosphodisterase (PDE2).

With respect to direct inhibitory effects of NO on adenylyl cyclase, we find that the lowering of cAMP accumulation by NO is readily and rapidly reversible ($t_{1/2} \approx$ 5 min), suggesting that long-lived covalent interactions are probably not involved and that reversible interactions such as those with heme groups may account for the effect. Experiments with pertussis toxin show that the inhibitory actions of NO are not mediated by activation of the G_i pathway.

Recent studies (11,12) indicate that NO interacts directly to inhibit isoforms 5 and 6 of adenylyl cyclase (AC5 and AC6). Using the same protocol employed by McVey et al. (12), we find only a slight (12%) inhibition of AC activity in cardiac fibroblast membranes, an effect that we believe is too small to account for our observations in whole cells. To the extent that the effect of NO described by McVey et al. (12) is diagnostic of the presence at AC5 and AC6, we conclude that AC5 and AC6 are minor contributors to cyclic AMP production by rat cardiac fibroblasts.

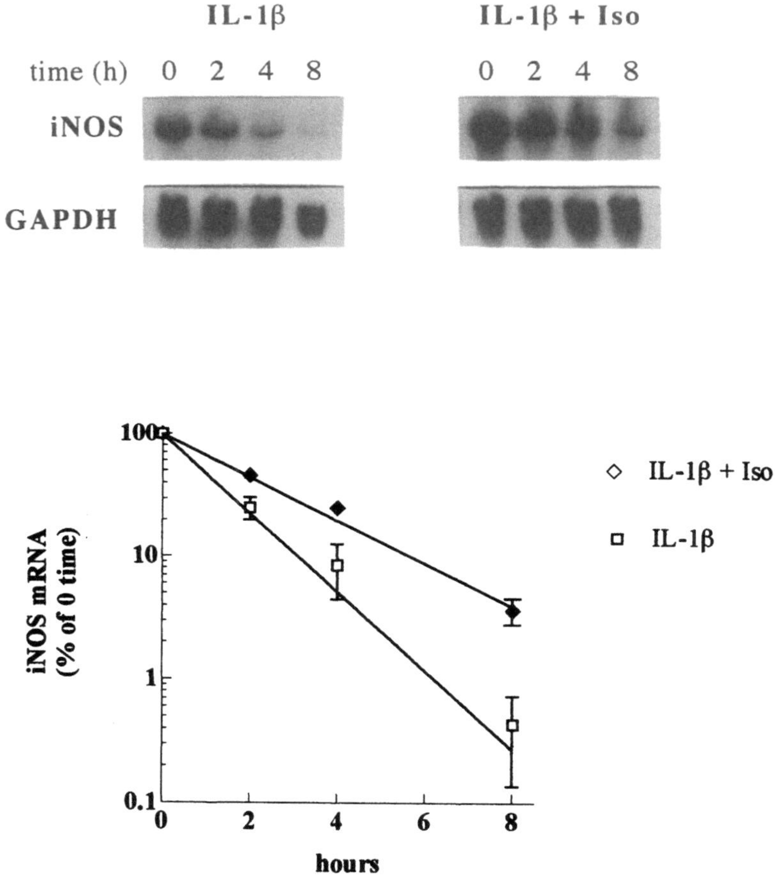

Figure 3. Isoproterenol stabilizes iNOS mRNA in IL-1β-stimulated cardiac fibroblasts. Cells were stimulated (24 h) with IL-1β (10 ng/ml) or IL-1β + Iso (10 μM), then 5,6-dichloro-1-β-ribofuranosylbenzimidazole (DRB, 65 μM) was added and total RNA was extracted at the indicated times after the administration of DRB. Northern analyses were performed (top panel is representative). Band densities of iNOS mRNA were normalized to GAPDH mRNA values and plotted as a percentage of zero time values (bottom panel). iNOS mRNA half-lives are 1.0 h for IL-1β and 1.9 h for IL-1β Iso; the effect of isoproterenol is significant, $P < 0.05$. Each data point represents the mean ± S.E. of three independent experiments. Adapted from reference 2.

Our data indicate that the lowering of cyclic AMP accumulation by NO is mediated by cyclic GMP: NO donors increase cyclic GMP content (Figure 6A), 10 μM ODQ (an inhibitor of the soluble guanylyl cyclase) inhibits cyclic GMP production and restores hormone-sensitive cyclic AMP production (Figure 6B). Even though many effects of cyclic GMP are mediated by PKG, the PKG inhibitor KT5823 (at 1 μM) has no effect on the action of NO to reduce cyclic AMP accumulation.

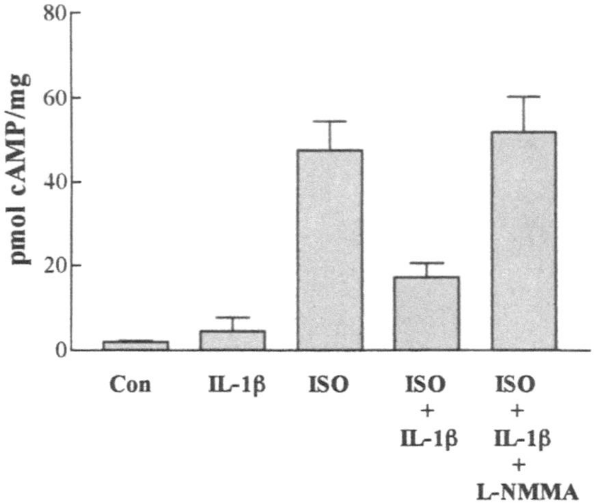

Figure 4. IL-1β treatment inhibits cyclic AMP accumulation.
Cardiac fibroblasts were treated with IL-1β (10 ng/ml, 24 h) and then stimulated for 5 min with 1 μM isoproterenol and the cAMP content determined. IL-1β treatment significantly decreases cAMP accumulation in response to isoproterenol ($P < 0.01$, $n = 4$). L-NMMA (1 mM) completely restores isoproterenol-stimulated cAMP accumulation. Data are mean ± SE, $n = 3$. Additional supportive data appear in reference 2.

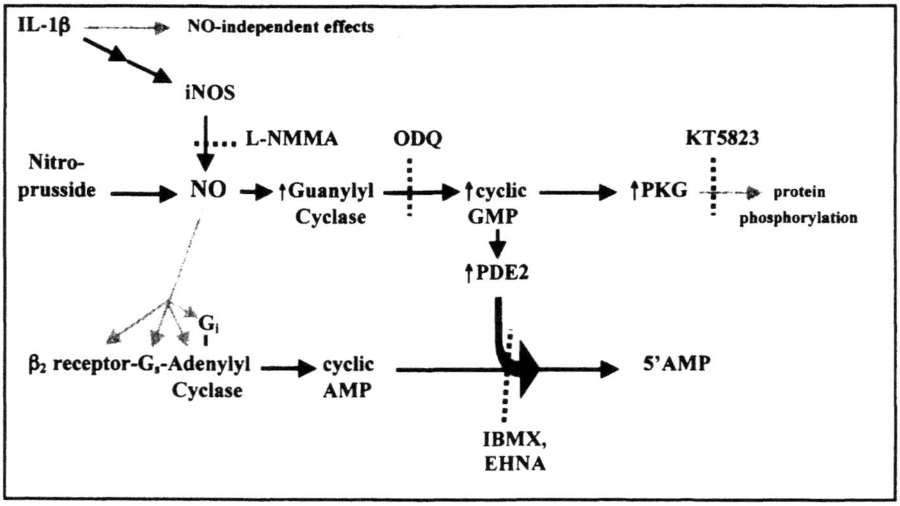

Figure 5. Testable mechanisms by which IL-1β, iNOS and NO could affect cyclic AMP accumulation.
Black lines indicate pathways involved in NO action in cardiac fibroblasts. Gray lines are pathways tested and found not to explain the actions of NO. Dashed lines indicate sites of action of pharmacologic inhibitors used diagnostically in our experiments.

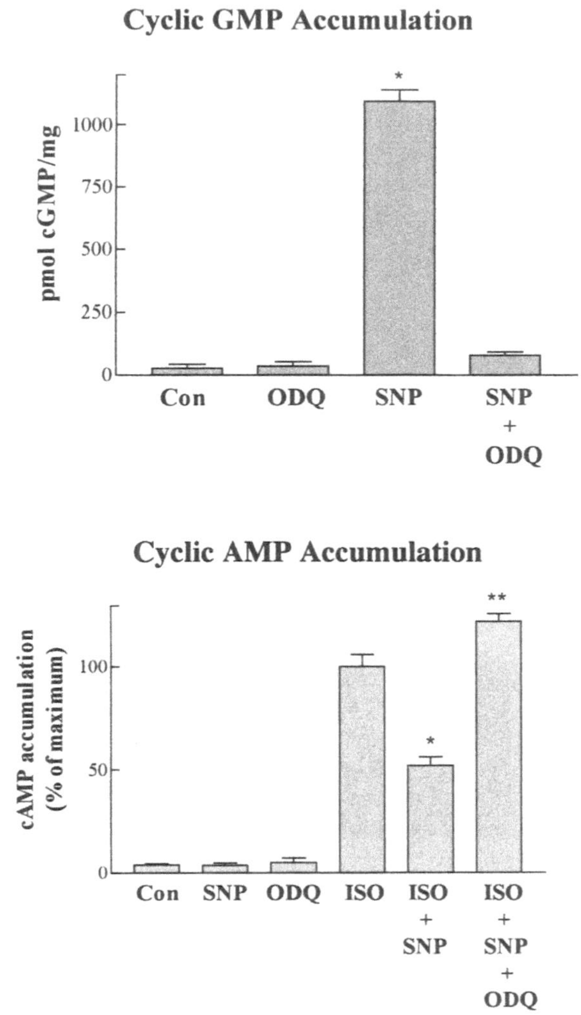

Figure 6. Effects of NO are mediated via the soluble guanylyl cyclase.
Top panel: Cardiac fibroblasts were incubated with or without 10 μM ODQ, an inhibitor of the soluble guanylyl cyclase, for 1 h, then stimulated with 1 mM SNP for 45 min and the cGMP content determined. SNP treatment significantly stimulated cGMP accumulation over control (*, $P < 0.001$); ODQ inhibited SNP-stimulated cGMP accumulation. **Bottom panel:** Following treatments with ODQ (10 μM) and SNP (1 mM), as above, cells were stimulated with 1 μM isoproterenol for 5 min and the cAMP content was determined. ODQ completely restored the β-adrenergic response in SNP-treated fibroblasts (* vs. **, $P < 0.001$). Data are mean ± SE, n = 3; see reference 3 for details.

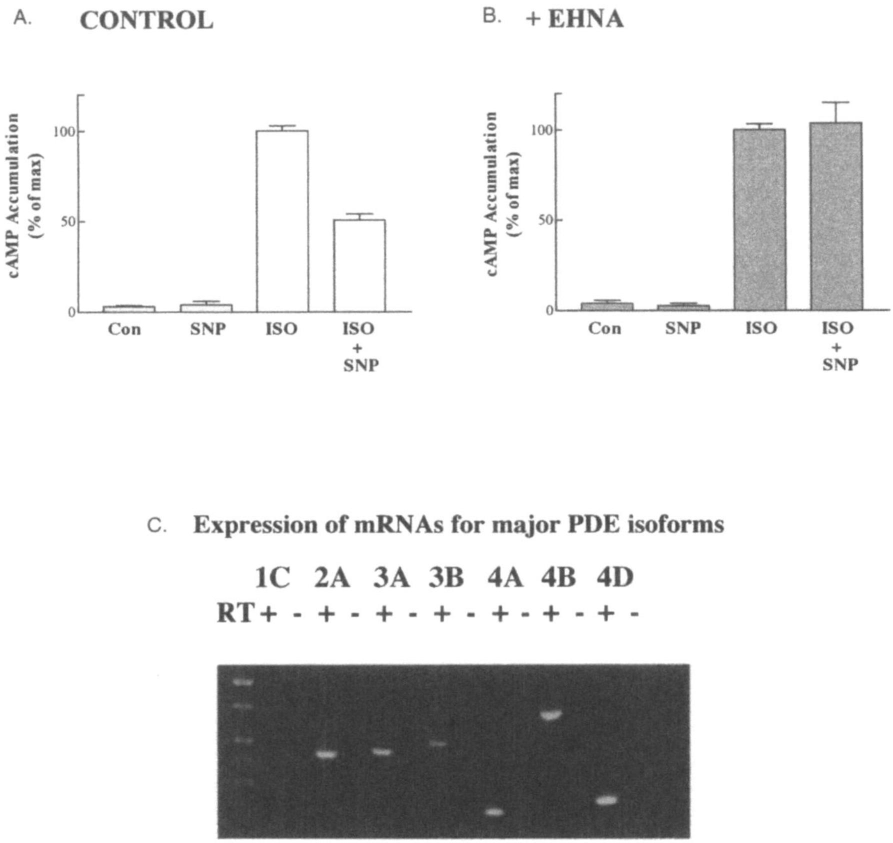

Figure 7. PDE2 inhibitor restores isoproterenol-stimulated cAMP accumulation in SNP-treated cardiac fibroblasts.
Cardiac fibroblasts were incubated with diluent (**top left**), or 0.2 mM EHNA, a PDE2-specific inhibitor (**top right**), for 15 min prior to SNP treatment (1 mM, 45 min), after which cells were stimulated with 1 μM isoproterenol for 5 min and the cAMP content determined. Data are mean ± SE, n = 3; see reference 10. **Bottom panel:** RT-PCR data from cardiac fibroblasts using primers for common PDE isoforms.

Thus, we turned our attention to the remaining pathway, the cyclic nucleotide phosphodiesterase activity that cyclic GMP stimulates (see scheme, Figure 5).

When NO donors are added to cells to reduce cyclic AMP accumulation, addition of the non-specific PDE inhibitor IBMX (1 mM) fully restores the β-adrenergic response (Figure 7A). Furthermore, addition of the specific inhibitor of PDE2, EHNA, likewise restores cAMP accumulation (Figure 7B), whereas inhibitors of PDEs 1, 3, and 4 (vinpocetine, milrinone, rolipram) do not (see reference 10). RT-PCR analysis confirms that rat cardiac fibroblasts express PDE2A, a cGMP-stimulated form (Figure 7C).

Relative PDE Activities in Cardiac Fibroblasts
(as judged by efficacy of isoform-specific inhibitors)

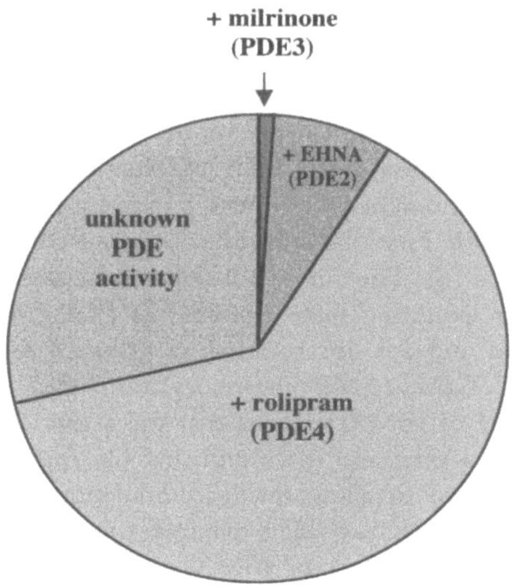

100% = 1 μM isoproterenol + 1 mM IBMX

Figure 8. Relative efficacies of PDE inhibitors compared to IBMX.
Monolayer cultures were incubated with diluent, 1 mM IBMX, 200 μM EHNA, or 10 μM rolipram for 15 min, stimulated with 1 μM isoproterenol for 5 min and then the cAMP content was determined. The whole circle represents cAMP accumulation in response to Iso + 1 mM IBMX. Relative to IBMX, accumulation to Iso + milrinone = 1%, Iso + EHNA = 8% and to Iso + rolipram = 62%. Data are mean of 3–4 replicate experiments.

The effect of 1 mM IBMX is large—a 26-fold increase in cAMP accumulation in response to isoproterenol—demonstrating the presence of abundant PDE activity in the fibroblasts. The PDE2 inhibitor EHNA increases isoproterenol-stimulated cyclic AMP levels about 2-fold, demonstrating the presence of basal PDE2 activity even in cells in which cGMP has not been elevated. Using fully effective concentrations of isoform specific PDE inhibitors (13–16) and comparing their effects to that of 1 mM IBMX (a non-specific inhibitor), we have tried to judge the relative contributions of PDE isoforms to cyclic AMP metabolism. PDE2 appears to be a minor player. PDE4, the cyclic AMP-specific PDE, seems to be the main isoform of PDE involved in cyclic AMP metabolism in cardiac fibroblasts. Indeed, 10 μM rolipram, a PDE4 specific inhibitor, increases isoproterenol-stimulated cyclic AMP accumulation to 62% of the maximal accumulation observed in the presence of 1 mM IBMX (Figure 8). To the extent that we may rely on the full efficacy and specificity of these inhibitors, these data suggest that cyclic AMP metabolism is additionally regulated by another, as yet unidentified, isoform of PDE, since a major portion (~30%) of the cyclic AMP accumulation in the presence of 1 mM IBMX cannot be accounted for with any of the isoform-specific PDE inhibitors that we employed.

DISCUSSION

The maximal capacity of cardiac fibroblasts to destroy cyclic AMP is very large. The evidence for this is the 26-fold enhancement in the cyclic AMP response by a maximally effective concentration of the PDE inhibitor IBMX. This result does not imply that the fibroblast's moment to moment metabolism of cyclic AMP exceeds the synthesis of cyclic AMP. First, the affinities of these PDEs vary widely, ranging from 0.2 to 150 μM (13–16). Thus, it is possible to accumulate modest levels of cAMP and not be near saturating these isoforms of PDE. In addition, the PDEs themselves are regulated and are not running at maximal capacity except under certain conditions (Ca^{2+}/CaM, cGMP, phosphorylation, etc.). The efficacy of isoproterenol in the absence of IBMX suggests that the activities of adenylyl cyclases and PDEs are separated in time and space and that the capacity of the stimulated adenylyl cyclases is sufficient to allow small accumulations of cyclic AMP (4–10 fold), probably more than enough to elicit maximal physiological responses.

Another inference from the effect of IBMX is that in the absence of a PDE inhibitor most of the cyclic AMP that the fibroblasts produce is destroyed. From the 26-fold enhancement of the β-adrenergic response by IBMX, we estimate that 96% of the cyclic AMP in isoproterenol-stimulated fibroblasts is normally destroyed and converted to 5′-AMP in the absence of IBMX. Thus, there is a flux of cyclic AMP → 5′-AMP and this 5′-AMP could be active as an intracellular messenger, itself, causing, for instance, activation of the AMP-activated kinase (AMPK). AMPK is present in cardiac tissue and is activated during ischemia (17), with an EC_{50} for AMP of 20 μM (18). Our estimations suggest that hormonal activation of adenylyl cyclase in cardiac fibroblasts could cause AMPK activation. Since fibroblasts accumulate about 1250 pmol cAMP/mg of protein in the presence of IBMX, let us assume that the cells could maximally produce that amount of 5′-AMP from cyclic AMP in the absence of a PDE inhibitor during a short burst of adenylyl cyclase stimulation. Further, let us assume that the cell is about 10% protein and 70% water; thus, 1250 pmol/mg protein ≈ 1250 pmol/7 mg water ≈ 180 pmol/μl or about 180 μM of 5′-AMP could be achieved during a maximal cyclic AMP response, assuming that the cyclic AMP and 5′-AMP are distributed uniformly in total cell water. Surely, cellular 5′-AMP is, itself, not metabolically stable. Nevertheless, half-maximal activation of AMPK would require that only about 10% of the cyclic AMP be converted and preserved as AMP.

This analysis, albeit overly simplified, suggests that there is abundant AMP produced whenever cyclic AMP synthesis is active, and that cyclic AMP degradation could be a source of AMP that could activate AMPK. While it is not clear exactly what AMPK accomplishes in these cells, AMPK can phosphorylate eNOS and seems to be activated as a part of a stress response to ischemia and hypoxia (17). The idea that the adenylyl cyclase-cyclic AMP-PDE pathway represents an hormonal way of activating AMPK has not been much considered in the literature. Such a scheme could provide a novel mechanism for PKA-independent effects of cyclic AMP.

While our analysis indicates that the PDE2 isoform largely accounts for the effects of NO on cyclic AMP metabolism, PDE2 is a small component of the total PDE activity in cardiac fibroblasts. In these cells, PDE4 is largely responsible for cyclic AMP degradation: rolipram produces a large increase in isoproterenol-stimulated cAMP accumulation (to 62% of that with 1 mM IBMX). Inhibition of PDE1 (for which no mRNA is detected) and PDE5 does not enhance the response to isoproterenol; inhibition of PDE3 with milrinone increases the isoproterenol response only slightly (~25%). The pie chart (Figure 8) attempts to summarize these data and to point out that a PDE isoform(s) sensitive to IBMX and that is not PDEs 1–5 may account for about 30% of IBMX-sensitive PDE activity in these cells. PDE7, exhibiting a high affinity for cyclic AMP (0.2 μM), low affinity for IBMX and for which no specific inhibitors are yet available, is found in the heart and is insensitive to rolipram (18,19), as required by the putative, unknown fibroblast isoform. It is also possible, of course, that lack of specificity and full efficacy of the isoform-specific PDE inhibitors that we employed (Figure 8) could contribute to such a result.

How this distribution of PDE activities changes when cyclic GMP is elevated following NO production is difficult to quantify, since cyclic GMP is both an activator and, at higher concentrations, a competitive substrate for PDE2 (13). Under the conditions of our experiments, PDE2 is activated by NO/cyclic GMP by roughly 2-fold; thus, PDE2 becomes a larger fractional player in the presence of NO. Simultaneously, PDE3 is inhibited by cyclic GMP and its influence decreases. Clearly, PDE4 continues to be the dominant hydrolyzer of cyclic AMP. However, since PDE activities, like adenylyl cyclases and protein kinases, can be confined to discrete subcellular regions in cardiac cells (20), the shift toward PDE2 and away from PDE3 when NO production is elevated could have profound local effects within the cell.

We may compare these data from adult rat cardiac fibroblasts with similar data from adult rat ventricular myocytes. In assaying HPLC-resolved PDE activities, we found essentially no PDE1 activity in myocytes (Sprague-Dawley), where PDE2, 3 and 4 predominate (21). Measuring cyclic AMP-sensitive I_{Ca} in adult ventricular myocytes (Wistar), Verde and colleagues (22) found essentially no effect of a PDE1 inhibitor (8-methoxymethyl-3-isobutyl-1-methylxanthine) and concluded that PDE4 and PDE3 activities predominate with PDE2 being a modest contributor. Thus, both fibroblasts and myocytes exhibit prominent PDE4 activity, but lack significant PDE1 activity. Yet extracts of whole ventricle display a prominent Ca^{++}/calmodulin-dependent PDE activity (PDE1) (21). This suggests that PDE1 is expressed in endothelial or smooth muscle cells and that perturbations that would selectively modulate PDE1 would not alter cyclic AMP-dependent events in myocytes and fibroblasts. As our knowledge increases of the cyclic nucleotide-regulated functions of the non-myocyte/non-fibroblast component of heart tissue (presumably endothelial and smooth muscle cells), the capability to selectively alter a major PDE isoform in this cellular compartment of the heart may become experimentally or even therapeutically useful.

In conclusion, we have demonstrated that cardiac fibroblasts can produce large quantities of NO as a consequence of the induction of iNOS by cytokines, an induction that the cyclic AMP/PKA pathway enhances by causing stabilization of iNOS message. The resulting NO depresses cyclic AMP accumulation. The mechanism of this effect does not involve inhibition of adenylyl cyclase by NO; rather, the effect depends on the stimulation of the soluble guanylyl cyclase by NO, is independent of the stimulation of PKG by the resultant cyclic GMP, and occurs via the stimulation of PDE2 by the elevated cyclic GMP. The data also indicate that cyclic AMP metabolism in cardiac fibroblasts is dominated by PDE4 and by an unknown PDE, possibly PDE7, and that PDE1 activity is lacking in these cells. Further, the action of PDEs could supply large quantities of AMP and activate AMPK, giving rise to cyclic AMP-dependent but PKA-independent protein phosphorylation.

REFERENCES

1. Villegas S, Brunton LL. 1996. Cellular localization and characterization of cyclic GMP production in rat heart. Cardiovasc Pathobiol 1:5–12.
2. Gustafsson ÅB, Brunton LL. 2000. Regulation of the Expression of Inducible Nitric Oxide Synthase in Adult Rat Cardiac Fibroblasts by IL-1β and β-Adrenergic Stimulation. Mol Pharmacol 58:1470–1478.
3. Meszaros JG, Gonzalez AM, Endo-Mochizuki Y, Villegas S, Villarreal F, Brunton LL. 2000. Identification of G protein-coupled signaling pathways in cardiac fibroblasts: crosstalk between G_q and G_s. Am J Physiol Cell Physiol 278:C154–C162.
4. Green LC, Wagner DA, Glogowski J, Skipper PL, Wishnok JS, Tannenbaum SR. 1982. Analysis of nitrate, nitrite, and [^{15}N]nitrate in biological fluids. Anal Biochem 126:131–138.
5. Harper JF, Brooker G. 1975. Femtomole sensitive radioimmunoassay for cyclic AMP and cyclic GMP after 2′0 acetylation by acetic anhydride in aqueous solution. J Cyclic Nucleotide Res 1:207–218.
6. Gilman AG. 1970. A protein binding assay for adenosine 3′:5′-cyclic monophosphate. Proc Natl Acad Sci U S A 67:305–312.
7. Salomon Y, Londos C, Rodbell M. 1974. A highly sensitive adenylate cyclase assay. Anal Biochem 58:541–548.
8. Kostic MM, Erdogan S, Rena G, Borchert G, Hoch B, Bartel S, Scotland G, Huston E, Houslay MD, Krause EG. 1997. Altered expression of PDE1 and PDE4 cyclic nucleotide phosphodiesterase isoforms in 7-oxo-prostacyclin-preconditioned rat heart. J Mol Cell Cardiol 29:3135–3146.
9. Bradford MM. 1976. A rapid and sensitive method for the quantitation of microgram quantities of protein utilizing the principle of protein-dye binding. Anal Biochem 72:248–254.
10. Gustafsson ÅB, Brunton LL. 2002. Attenuation of Cyclic AMP Accumulation in Adult Rat Cardiac Fibroblasts by IL-1β and Nitric Oxide: Role of Cyclic GMP-stimulated Phosphodiesterase (PDE2). Am J Physiol Cell Physiol, 283:C463–C471.
11. Hill J, Howlett A, Klein C. 2000. Nitric oxide selectively inhibits adenylyl cyclase isoforms 5 and 6. Cell Signal 12:233–237.
12. McVey M, Hill J, Howlett A, Klein C. 1999. Adenylyl cyclase, a coincidence detector for nitric oxide. J Biol Chem 274:18887–18892.
13. Michie AM, Lobban M, Muller T, Harnett MM, Houslay MD. 1996. Rapid regulation of PDE-2 and PDE-4 cyclic AMP phosphodiesterase activity following ligation of the T cell antigen receptor on thymocytes: analysis using the selective inhibitors erythro-9-(2-hydroxy-3-nonyl)-adenine (EHNA) and rolipram. Cell Signal 8:97–110.
14. Nicholson CD, Challiss RA, Shahid M. 1991. Differential modulation of tissue function and therapeutic potential of selective inhibitors of cyclic nucleotide phosphodiesterase isoenzymes. Trends Pharmacol Sci 12:19–27.

15. Pyne NJ, Cooper ME, Houslay MD. 1987. The insulin- and glucagon-stimulated "dense-vesicle" high-affinity cyclic AMP phosphodiesterase from rat liver. Purification, characterization and inhibitor sensitivity. Biochem J 242:33–42.
16. Torphy TJ, Cieslinski LB. 1990. Characterization and selective inhibition of cyclic nucleotide phosphodiesterase isozymes in canine tracheal smooth muscle. Mol Pharmacol 37:206–214.
17. Kemp BE, Mitchelhill KI, Stapleton D, Michell BJ, Chen ZP, Witters LA. 1999. Dealing with energy demand: the AMP-activated protein kinase. Trends Biochem Sci 24:22–25.
18. Michaeli T, Bloom TJ, Martins T, Loughney K, Ferguson K, Riggs M, Rodgers L, Beavo JA, Wigler M. 1993. Isolation and characterization of a previously undetected human cAMP phosphodiesterase by complementation of cAMP phosphodiesterase-deficien Saccharomyces cerevisiae. J Biol Chem 268:12925–12932.
19. Essayan DM. 1999. Cyclic nucleotide phosphodiesterase (PDE) inhibitors and immunomodulation. Biochem Pharmacol 57:965–973.
20. Steinberg SF, Brunton LL. 2001. Compartmentation of G protein-coupled signaling pathways in cardiac myocytes. Ann Rev Pharmacol Toxicol 41:751–773.
21. Bode DC, Kanter JR, Brunton LL. 1991. Cellular distribution of phosphodiesterase isoforms in rat cardiac tissue. Circ Res 68:1070–1079.
22. Verde I, Vandecasteele G, Lezoualc'h F, Fischmeister R. 1999. Characterization of the cyclic nucleotide phosphodiesterase subtypes involved in the regulation of the L-type Ca^{2+} current in rat ventricular myocytes. Br J Pharmacol 127:65–74.

Pathophysiology of Cardiovascular Disease,
edited by N.S. Dhalla, H. Rupp,
A. Angel & G.N. Pierce
Kluwer Academic Publishers, Boston, 2004

Do Humoral and Mechanical Stimulation of Cardiomyocytes Share Common Transduction Pathways?

Arié Pinson and Régine Tirosh

The Hebrew University-Hadassah Medical School
P.O. Box 12272, 91120 Jerusalem, ISRAEL
FAX 972-2-6784010
E-mail: pinson@post.tau.ac.il

Summary. Physical forces are the most basic and non-specific forces to which living organisms were exposed during evolution, whereas humoral effectors developed in more complex organisms. We tested the hypothesis that both adrenergic (humoral) and hypergravity (mechanical) stimulation may share common signal transduction pathways. Mechanical loading remains the most critical effector in the development of cardiac hypertrophy, while hypergravity is currently used to train aircraft and space crews and may lead to aberrations in cardiac function. In this study, we used cultured cardiomyoctes submitted to 10 min hypergravity or to adrenostimulation. Specific inhibitors of adrenoreceptors and of protein kinases were used to identify cell reactivity. The mRNA of *c-fos* and creatine kinase isoenzymes was determined by Northern blot. We showed that with respect to *c-fos* induction, although both effectors are initiated at different transduction levels, apparently share a common transduction pathway—Protein kinase C—but in reality they are transduced via different isoenzymes of the PKC complex. Creatine kinase is expressed through two independent pathways. We suggest that studying the opposite effects of hypo- and hypergravity could add new insights into how cells sense mechanical stimulation that might lead to identification of the cellular messengers participating in the mechanosensitive pathways.

Key words: Adrenostimulation, cardiomyocytes, gene expression, hypergravity, signal transduction.

INTRODUCTION

Cardiac hypertrophy is a response to an elevated work load on the heart, which is expressed as an increase in cardiomyocyte size (1). However, hypertrophy is not the result of mechanical overload alone, but is also associated with complex interactions between mechanical stress and release of neurotransmitters and hormones.

Already in 1982, Simpson et al. (2) showed that adrenostimulation induced hypertrophy in cultured cardiomyocytes mediated via the α_1-adrenoreceptor (3). Rupp et al. (4) showed a reprogramming of myosin by β-adrenostimulation, while Hannan et al. (5) found that both α- and β-receptors increased the levels of various transcriptional factors, including phosphoprotein UBF and ribosomal DNA.

Using the so-called "redifferentiated" adult cardiomyocytes, Pinson et al. (6) found that both α- and β-adrenostimulation increased cellular mass. However, only the α_1-agonist phenylephrine increased uridine incorporation into total RNA and stimulated the expression of the CK-B isoform ("fetal shift").

Hypertrophy, however, is mainly mediated by physical forces, such as hemodynamic overload, which, at the level of the heart, is translated into stretching of the ventricular wall. Over 25 years ago, Peterson and Lesch (7) showed increased protein synthesis in response to stretching an isolated papillary muscle. In extensive research using cultured cardiomyocytes by Yazaki's group (for reviews, see 8–10), it was found that stretching cultured cardiomyocytes increased cellular mass, induced transient expression of protooncogenes and activated a new gene expression program ("fetal shift"), which clearly plays an important role in the adaptation of the heart to changing conditions.

Gravity is one of the physical forces the heart works against (11). Hypergravity is currently used to train aircraft and space crews. In the framework of space research, the effects of hypogravity changes have been studied in in vivo and in vitro models, while, although hypergravity is part of the training program, much less attention has been focused on it. However, serious transient and sometimes persistent cardiac dysrhythmias have been reported during such training (12). Frey et al. (13) reported 15% and 27% increases in heart mass and mitochondrial volume, respectively, in hypergravity-exposed mice, while Tran et al. (9) did not find myocardial hypertrophy or transition in myosin isoforms in hypergravity-exposed rats. Surprisingly few studies have been published on the effect of hypergravity on mammalian cells.

Kumei et al. (14) showed that hypergravity enhances the expression of the *c-myc* oncogene. De Groot et al. (15) found that EGF-induced expression of *c-fos* in human epidermal carcinoma cells is enhanced by hypergravity and decreased by microgravity and is affected by changes in the gravity conditions (16). In cultured osteoblastic cells, while PKC inhibition or exhaustion by phorbol esters completely repressed *c-fos* expression, it had no apparent effect on *erg-1* expression (17). Hypergravity also stimulated inositol 1,4,5 triphosphate production and phosphorylation of a microtubule-associated protein (18).

All the findings to date indicate that, like other physical forces, hypergravity activates certain signal transduction pathways involved in growth. For this reason, we were interested in comparing the effects of hypergravity and adrenoreceptor stimulation on the expression of both early and late responding genes, and, in particular, whether hypergravity and adrenostimulation utilize the same or distinct signal transduction pathways.

MATERIALS AND METHODS

Cultured ventricular myocytes were obtained as previously described (19). The final cell suspension was seeded on Thermanox coverslips, 25 mm in diameter placed into 6-well multidishes (Nunc, Denmark). The medium was changed on days 2 and 4. Six h after the last medium change, the coverslips were transferred into sterile, covered plastic buckets (Weltplast, Holon, Israel) and kept overnight in serum-free M199 medium (Biological Industries, Beit Haemek, Israel).

Hypergravity. To prevent shear stress caused by whirlpool formation, the buckets containing the coverslips were filled with serum-free M199, and were centrifuged for 10 min at 40 g. After exposure to hypergravity, the cells were maintained in an incubator at 37°C for various time periods. The control cells were not exposed to hypergravity, but were otherwise similarly treated.

Total RNA and total protein synthesis were determined by adding 2 μCi [^{14}C]uridine or L[U-^{14}C]phenylalanine (Amersham) to the medium before subjecting the cells to hypergravity treatment for 10 min, and post-hypergravity incubation for 6 h. The radioactivity of the TCA-insoluble fractions was determined as previously described (6).

RNA preparation and Northern blot analysis. Total RNA was extracted by TriReagent M.R.C. (Cincinnati, Ohio, USA) according to the manufacturer's instructions.

Total RNA (15 ng) was fractionated in 6.3% and 1.5% formaldehyde/agarose gels and transferred to Hybond N-membrane (Amersham) by the capillary blot procedure.

cDNA was labeled by random oligonucleotide priming with ^{32}P αdCTP (Amersham). The blots were hybridized overnight at 42°C (NE Blot kit, New England Bio Lab. MA USA).

The membranes were washed and then autoradiographed with Fuji Rx film. The radioactivity of the membrane was quantified by phosphostimulated luminescence autoradiography using a Bas 1000 (Fujix) bio-imaging analyzer. Quantification was expressed in PSL which is a unit of radiation dose. It is proportional to standard units such as becquerel or dpm. When figures are given in a graphic presentation the mRNA was quantified by normalization using 18S rRNA labeling.

c-fos was determined immediately on exposure to hypergravity (0 min) and then after 15, 30 and 60 min.

Creatine kinase isoforms were determined at 3.6 and 22 h after hypergravity.

Phenylephrine (Sigma, Israel) was added to the cells at a concentration of 10 μM.

Chelerytrine (Alomon Labs. Jerusalem Israel), a PKC inhibitor, was added to the cells at a concentration of 0.7 μM 30 min before application of hypergravity (20).

H89 (Calbiochem), a PKA inhibitor, was added to the cells at a concentration of 20 μM 1 h before exposure to hypergravity (21).

Go 6976, a PKCα and β inhibitor was added to the cells at a concentration of 2.3 nM (22).

Calphostine, a Ca independent PKC inhibitor was added to the cells at a concentration of 50 nM (22).

TPA, a PKCα and PKCd4ε stimulator was added at a concentration of 10 ng/ml.

RESULTS

To determine whether hypergravity, like other physical forces, affects protein synthesis in cardiomyocytes, we used [2-^{14}C]-uridine and L-[U-^{14}C]-phenylalanine as markers of total RNA and protein synthesis, respectively. In cells exposed to hypergravity for 10 min, followed by 6 h incubation, the incorporation of uridine was increased by 93% and of phenylalanine into TCA-insoluble proteins by 42%, as compared to controls (Figure 1).

The subsequent figures show representative data of one out of 3-to-5 experiments, all of which presented similar results.

To further probe the effect of hypergravity on the expression of specific genes, we chose *c-fos* as a typical early oncogene and creatine phosphokinase isoforms (CK-M and -B) as examples of late gene expression.

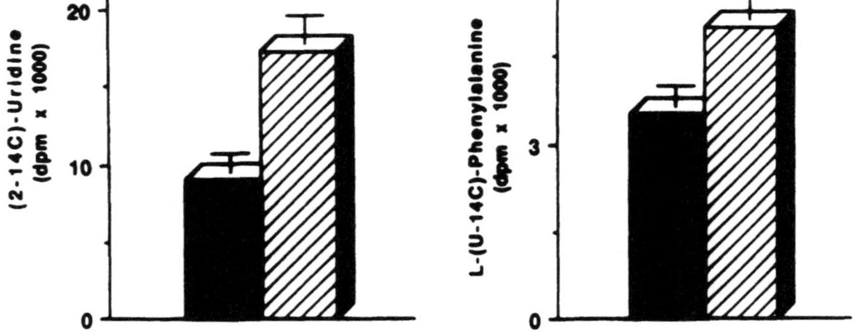

Figure 1. [U-^{14}C]Phenylalanine and [2-^{14}C]Uridine incorporation into TCA-insoluble fractions for 6 h following hypergravity stress. Uridine incorporation increased by 93% ± 24 and phenylalanine by 42.6% ± 16.4 (n = 7). Control ■ Hypergravity ▨

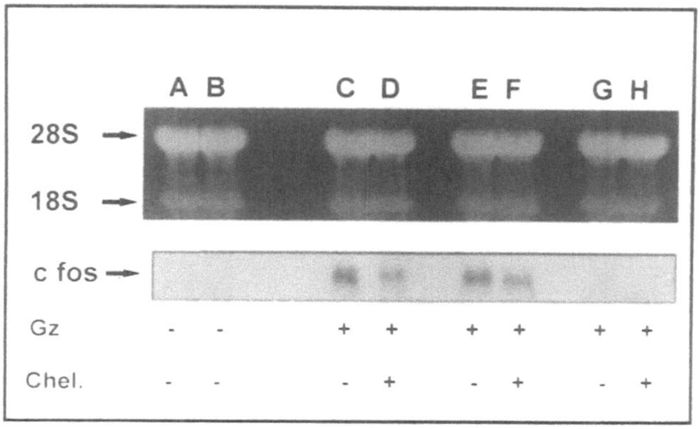

Figure 2. Expression of *c-fos* and effect of Chelerytrine in cells exposed to hypergravity. Chelerytrine (0.7 M), a PKC inhibitor, was added to M199 medium immediately before hypergravity. A and B—Controls; C, E and G—15, 30 and 60 min after hypergravity, respectively (without Chelerytrine). D, F and H are 15, 30 and 60 min after hypergravity with Chelerytrine. D and F showed 30% decreased expression as compared to C and E. Upper panel, Etbr staining of RNA.

The expression of *c-fos* (Figure 2) was undetectable in both control (untreated) cells or in cells immediately after exposure to 10 min hypergravity (lanes A and B), but was clearly expressed at 15 and 30 min after hypergravity exposure (lanes C and E), becoming undetectable again after 60 min (lane G).

Addition of the PKC inhibitor chelerytrine (0.7 μM) to the culture medium led to a 30% inhibition of *c-fos* expression (Figure 2). The PKA inhibitor H89 did not affect *c-fos* expression (not shown).

Of the various adrenoagonists (Figure 3), only the α_1-adrenoreceptor agonist phenylephrine markedly stimulated *c-fos* expression. Various α_2, β and β_2 agonists had little or no effect.

With concomitant application of hypergravity and phenylephrine the increase of *c*-fos expression was additive (Figure 4). The α_1-adrenoreceptor antagonist prazosine inhibited exclusively the phenylephrine stimulatory effect, whereas the PKC inhibitor chelerytrine (Figures 2 and 4) blocked *c-fos* induction by both, phenylephrine and hypergravity. Taken together, these findings indicate that while hypergravity and phenylephrine are not mediated via the same receptor, they share a common signal transduction pathway.

Five-day-old cultured cardiomyocytes are differentiated cells in which both CK isoforms are expressed with predominance of the M-isoform. Hypergravity exposure for 10 min increased the expression of both CK isoforms (Figure 5). The CKB isoform increased by 75% and 190% after 6 and 24 h, respectively, while CKM only increased by 25% after 24 h. To determine which protein kinase plays a central role in the signal transduction pathway induced by hypergravity stimulation, the cells were treated with either chelerytrine or H89, the respective inhibitors of

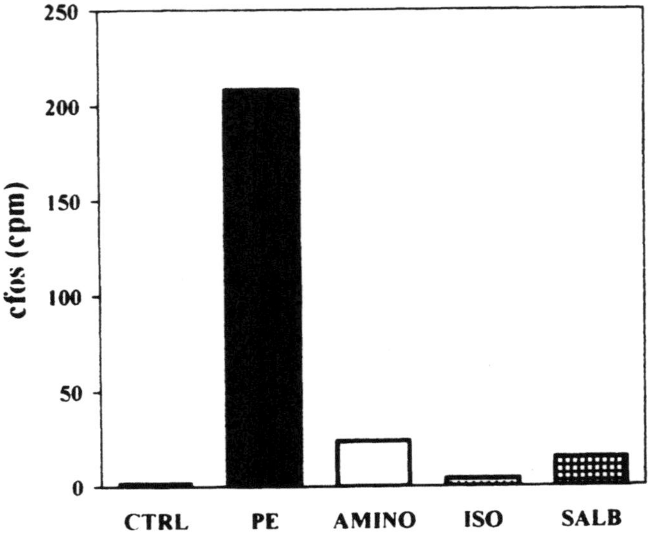

Figure 3. Effect of various types of adrenostimulation for 30 min on *c-fos* expression. The blots were quantified by phospholuminescence scanning using control hybridization with 18S RNA.
CRTL—Control; PE—Phenylephrine (α1-agonist); AMINO—p Aminoclonidine (α2-agonist); ISO—isoproterenol (β-agonist); SALB—Salbutamol (β2-agonist). Chelerytrine inhibited *c-fos* expression by PE (not shown).

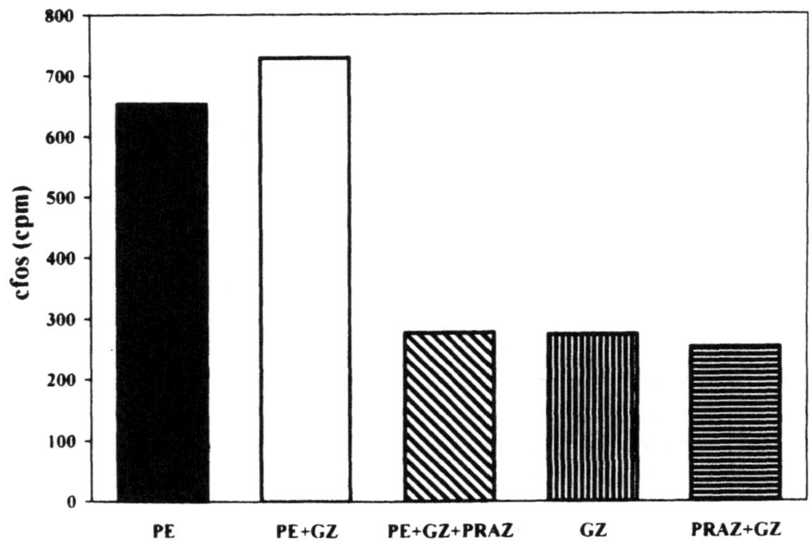

Figure 4. Combined effect of hypergravity and Phenylephrine on *c-fos* expression (with or without α1-receptor inhibitor).
Blots were quantified as in Figure 3. PE—Phenylephrine; Gz—hypergravity; PRAZ—Prazosine (α1-antagonist).

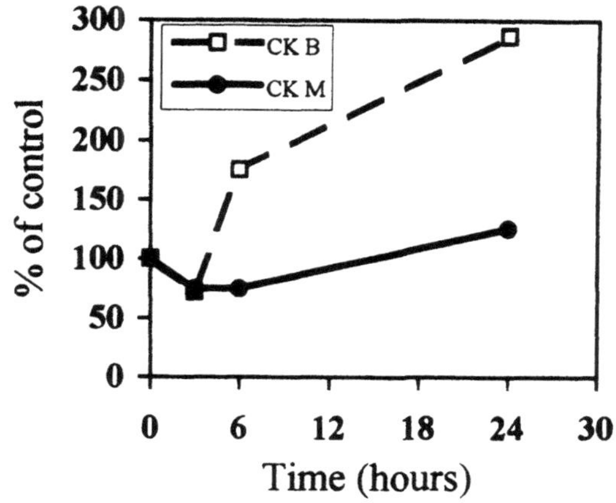

Figure 5. Effect of hypergravity on CK isoenzyme expression 3, 6 and 24h following 10min hypergravity. (Quantification as in Figure 3).

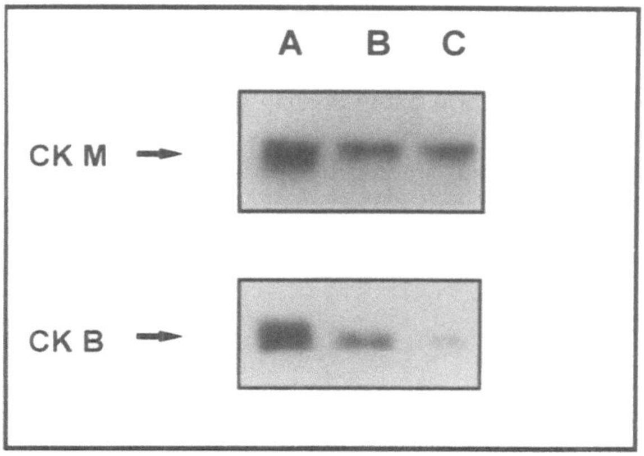

Figure 6. Effect of H89, a PKA inhibitor, on CK isoenzyme expression.
H89 (20 µM) was added before hypergravity. 6a) A—control cells (not exposed to hypergravity). B—control cells treated 24h with H-89 (not exposed to hypergravity); C—CK-M expression 22h after hypergravity; D and E—H-89 treated cells 10 and 22h after hypergravity, respectively. H89 inhibited CK expression by 7% in controls (B) and by 55% 22h after hypergravity (E). 6b) A, B, C, D and E as in 6a). CK-B was inhibited by 66% in controls (B) and by 86% 22h after hypergravity (E).

PKC and PKA. As shown in Figure 6, in cells incubated with H89 and exposed to hypergravity, both CK isozymes, but predominantly CKB (Figure 6), whereas chelerytrine had no inhibitory effect on either isozyme. These findings indicate that during hypergravity stimulation, CK signaling expression is transduced through PKA.

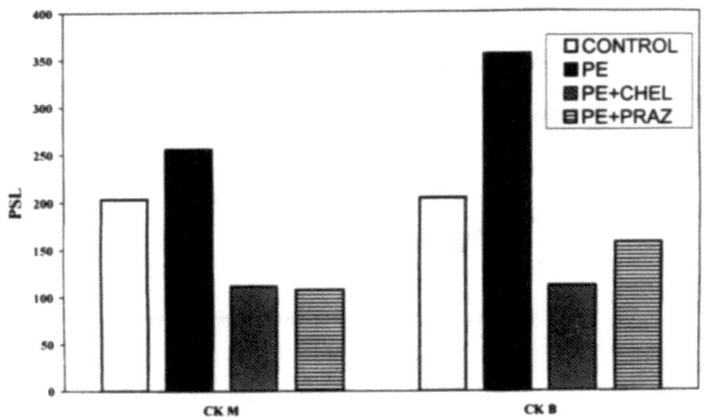

Figure 7. Combined effect of phenylephrine, Chelerytrine and Prazosine for 24 h on CK M and CK B expression (quantification as in Figure 3).

Phenylephrine, as shown in Figure 7, also stimulated the expression of both isoforms with predominance of the B-isoform. After 24 h of incubation, CKB was increased by 75% and CKM by 25%. Both prazosine and chelerytrine inhibited the phenylephrine effect, indicating that PE stimulates CK expression via a completely different transducer/transduction pathway than that induced by hypergravity.

The question of whether adrenostimulation and hypergravity use the same or different PKC isoenzymes, was studied by specific inhibitors. Figure 8 shows that

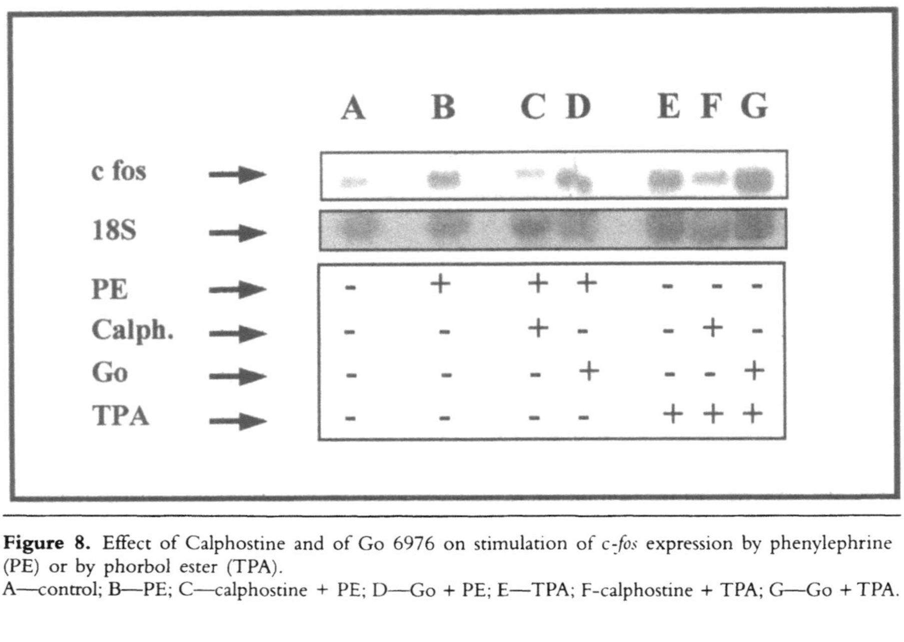

Figure 8. Effect of Calphostine and of Go 6976 on stimulation of *c-fos* expression by phenylephrine (PE) or by phorbol ester (TPA).
A—control; B—PE; C—calphostine + PE; D—Go + PE; E—TPA; F-calphostine + TPA; G—Go + TPA.

calphostine, which inhibits the activity of the Ca-independent isoenzymes, totally inhibited c-fos expression mediated by adrenostimulation but not that mediated via TPA.

DISCUSSION

Physical factors are the most primary and non-specific forces to which microorganisms and multicellular organisms were exposed during evolution. This implies that mechano–transducers and mechanogenic signal transduction mechanisms were the first to develop in primitive organisms. With the emergence of more complex organisms, paracrine and autocrine factors became essential for the transmission of information. Thus certain mechanogenic informational pathways could have been adapted for this purpose (23). Indeed, mechanical loading remains the most effective stimulus for trophic and hypertrophic responses in living organisms. It is the most critical factor in cardiac hypertrophy, in which the heart works against mechanical forces, including gravity (11,23).

This study raised the question of whether humoral effectors such as adreno–stimulation and physical forces such as hypergravity share the same signal transduction pathways. After showing that hypergravity promotes cell growth as it increases total RNA and protein neosynthesis, we chose to study *c-fos*, as member of the immediate early response genes, and the CK isozymes, as representatives of the adaptational expression of cardiac specific genes. Both the α_1-adrenoreceptor agonist and hypergravity stimulated *c-fos* expression via PKC, since inhibition of this enzyme by chelerytrine reduced the effect of both stimuli. However, inhibition of the α_1-adrenoreceptor by prazosine had no effect on the stimulation of *c-fos* expression by mechanical force. These findings indicate that, although initiated by different transducers, both effectors share a common signal transduction pathway. It may well be, however, that mechano- and adreno- stimulation are mediated via either different PKC subfamilies or the selective activation of various compartmentalized PKC pools. PKC is a complex isoenzyme system. Four PKC isoenzymes are expressed in neonatal cardiomyocytes: the α-isoenzyme which is Ca-dependent, and the ε-, δ-, and ζ-isoenzymes, which are Ca-independent. We clearly show that calphostine, which inhibits the activity of Ca-independent isoenzymes, also totally inhibited *c-fos* expression mediated by adreno–stimulation, but not that mediated via hypergravity or TPA. At this stage preliminary findings would suggest that *c-fos* expression by hypergravity is mediated via the PKCα isoenzyme. This might explain how various effectors utilize an apparently universal pathway while each of them is associated with specific phosphorylation events.

Gene expression reprogramming in the heart in response to hemodynamic overload is characterized by a reexpression of protein isoforms to a pattern usually expressed in the fetal heart, known as the "fetal shift" (1).

Stretching (8,10) or adrenostimulation (3,6,24) of cultured ventricular myocytes induced the reexpression of skeletal α-actin, ANF, β-MHC and CKB isoforms.

Phenylephrine stimulated both CKB and CKM expression, with predominance of the CKB isoform. This stimulation, mediated by α_1-adrenoreceptors, is transduced through PKC. Hypergravity also stimulated expression of both CK isozymes with predominance of the CKB isoform; however, this mechanostimulation was transduced via PKA, as shown by the inhibitory effect of H89. This signal transduction pathway was not affected by the inhibition of the phenylephrine pathway. Taken together, our data show that, in the case of CK, these α_1 adrenoreceptor stimulation and stimulation by hypergravity are mediated via completely separate pathways.

In this paper, we have shown that cells exposed to hypergravity for 10 min underwent, at least during the subsequent 24 h, a "fetal" reprogramming of CK isoforms. In view of the higher substrate affinity of the B-isoform (25), this may be interpreted as an adaptational response to increase the efficiency of the CK-ATP shuttle in cells undergoing hypertrophic stimulation.

The way how cells sense mechanical forces is still unclear. It is very unlikely that primitive organisms had specialized receptors when they were first exposed to mechanical forces. Several researchers proposed that the signal for mechanical stimulation may be mediated via deformation of the extracellular matrix and the cytoskeleton (26), or by mechanosensitive ion channel activation, which does not however, induce reprogramming of gene expression (27,28). Anyhow, a clear mechanism for such an effect has not been demonstrated. Indeed, any "positive" mechanical force (23), including hypertension, stretching, shear forces, osmolarity and viscosity, may initiate the release of a cascade of signaling compounds leading to cellular trophic or hypertrophic events via a mechanotransducing system. Anyhow, the nature and size of these mechanical deformations are certainly not similar and inhibitors of mechanical forces have not been found.

Hyper- and hypogravity were shown to be the only mechanical effectors acting in opposite directions: hypergravity as a stimulating effector and hypogravity as an inhibitory one (16). In this respect, the answer to the question of how cells sense mechanical forces may possibly be resolved by this fact, and since hypogravity is also deformative (29,30). Ingber (31) adapted the architectural principle of tensegrity (tensional integrity) as reflecting the cytoskeleton structure to explain how locally applied mechanical forces are integrated into a whole-cell response. This occurs via transmembrane adhesion receptors, such as integrins, anchored to the extracellular matrix, and are connected to the cytoskeleton network or to other cells. As such, integrins would function as receptors for mechanical stimulations. This interesting model might not only explain how mechanical forces are sensed, but may also suggest an explanation for the opposite directions in cellular response to hyper- versus hypo-gravity, or for example atrophy versus hypertrophy.

In conclusion, although adrenostimulation and hypergravity stimulation are initiated respectively via sarcolemmal receptors or an unknown mechanism, we have shown that with respect to early oncogene expression, they apparently share the same signal transduction pathway but in reality are transduced via different isoenzymes. As for the expression of the late protein CK, the signal transduction pathways are completely separate.

The question of how cells sense mechanical forces—tensegrity—and the nature of the second messengers of the mechanosensitive pathways is still unresolved.

In our view, research comparing the events of hypo- and hypergravity could provide new insights into this problem.

ACKNOWLEDGMENTS

This study was supported by a grant from the Chief Scientist, Ministry of Health, Israel, and a grant in memory of Rachel and Nissim Hasday.

REFERENCES

1. Morgan HE, Baker KM. 1991. Cardiac hypertrophy: Mechanical, neural and endocrine dependence. Circulation 83:13–25.
2. Simpson P, McGrath A, Savion Sh. 1982. Myocyte hypertrophy in neonatal rat heart cultures and its regulation by serum and catecholamines. Circ Res 51:787–801.
3. Long CS, Kariya K, Simpson PC. 1990. Trophic factors for cardiac myocytes. J Hypertensios 8: S219–S224.
4. Rupp H, Berger HJ, Pfeiffer A, Werdan K. 1991. Effect of positive ionotropic agents on myosin isozyme population and mechanical activity of cultured rat heart myocytes. Circ Res 68:1164–1173.
5. Hannan RD, Lukyen J, Rothblum LI. 1995. Regulation of rDNA transcription factors during cardiomyocyte hypertrophy induced by adrenergic agents. J Biol Chem 270:8290–8297.
6. Pinson A, Schluter KD, Zhou XJ, Schwartz P, Kessler-Icekson G, Piper HM. 1993. Regulation of protein synthesis of redifferentiated adult rat heart muscle cells by catecholamines. J Mol Cell Cardiol 25:477–490.
7. Peterson MB, Lesch M. 1972. Protein synthesis and amino acid transport in isolated rabbit right ventricular muscle. Circ Res 31:317–327.
8. Komuro I, Yazaki Y. 1993. Control of cardiac gene expression by mechanical stress. Ann Rev Physiol 55:55–75.
9. Tran CC, Stephane J, Serrurier B, Bigard AX, Guezennec CY. 1996. G-induced myocardium functional alterations are not related to structural changes in the heart. Aviat Space Environ Med 62:1170–1173.
10. Yamazaki T, Komuro I, Yazaki Y. 1995. Molecular mechanism of cardiac cellular hypertrophy by mechanical stress. J Mol Cell Cardiol 27:133–140.
11. Seymour RS, Hargens AR, Pedley TJ. 1993. The heart works against gravity. Am J Physiol 265:R715–R720.
12. McKenzie I, Gillingham KK. 1993. Incidence of cardiac dysrhythmias occurring during centrifuge training. Aviat Space Environ Med 64:687–691.
13. Frey M, von Kanel-Christen R, Stalder-Navarro V, Duke PJ, Wiebel ER, Hoppeler H. 1997. Effects of long-term hypergravity on muscle heart and lung structure in mice. J Com Physiol 167: 494–501.
14. Kumei Y, Nakajima T, Sato A, Kamata N, Enomoto S. 1989. Reduction of G_1 phase duration and enhancement of *c-myc* gene expression in HeLa cells at hypergravity. J Cell Sci 93:221–226.
15. De Groot RP, Rijken PJ, Boonst J, Verklej AJ, De Laat SW, Kruijer W. 1991. Experimental growth factor-induced expression of *c-fos* is influenced by altered gravity conditions. Aviat Space Environ Med 62:37–40.
16. De Groot R, Rijken PJ, Den Hertog J, Boonstra J, Verklej AJ, De Laat SW, Kruijer W. 1991a. Nuclear response to protein kinase signal transduction are sensitive to gravity changes. Exptl Cell Res 97: 87–90.
17. Nose K, Shibanuma M. 1994. Induction of early response genes by hypergravity in cultured osteoblastic cells (MC 3T3-E1). Exptl Cell Res 211:168–170.
18. Kumei Y, Whitson PA, Sato A, Cintron NM. 1991. Hypergravity signal transduction in HeLa cells with concomitant phosphorylation of proteins immunoprecipitated with anti microtubule-associated protein antibodies. Exptl Cell Res 192:492–496.
19. Pinson A, Padieu P, Harary I. 1987. Techniques for culturing heart cells. In Pinson A. (ed.) *The Heart Cell in Culture*, CRC Press Inc., Boca Raton Vol. 1, pp. 7–22.
20. Herbert JM, Augereau JM, Gleye J, Maffrand JP. 1990. Chelerytrine is a potent specific inhibitor of protein kinase C. Biochem Biophys Res Commun 172:993–999.

21. Chijiwa T, Mishima A, Hagiwara M, Sano M, Hayashi K, Inoue T, Naito K, Toshioka T, Hidaka H. 1990. Inhibition of a forskolin-induced neurite outgrowth and protein phosphorylation by a newly synthesized selective inhibitor of cyclic AMP-dependent protein kinase N-[2-(p-bromocinnamylamino) ethyl]-5-isoquinoline sulfonamide (H-89) of PC12D pheochromocytoma. J Biol Chem 265:5267–5272.
22. Gschwendt M, Dieterich S, Rennecke J, Kittstein W, Mueller H-J, Johannes F-J. 1996. Inhibition of protein kinase C by various inhibitors. Differentiation from Protein Kinase C isoenzymes. FEBS Lett 392:77–80.
23. Vanderburgh HH. 1992. Mechanical forces and their second cell messengers in stimulating cell growth in vitro. Am J Physiol 262:R350–R355.
24. Simpson PC, Kariya K, Krans LR, Long CS, Karliner JS. 1991. Adrenergic hormones and control of cardiac myocyte growth. Mol Cell Biochem 104:35–43.
25. Szasz G, Gruber W. 1978. CK in the serum. 4. Differences in substrate affinity among the isozymes. Clin Chem 24:245–249.
26. Ingber DE. 1990. Fibronectin controls capillary endothelial cell growth by modulation of cell shape. Proc Natl Acad Sci USA 87:3579–3583.
27. Bustamante JO, Ruknudin, Sachs F. 1991. Stretch-activated channels I heart cells: Relevance to cardiac hypertrophy. J Cardiovasc Pharmacol 17(Suppl. 2):S110–113.
28. Morris CE. 1990. Mechanosensitive ion channels. J Membr Biol 113:93–107.
29. Gruener R, Hoeger G. 1990. Vector-free gravity disrupts synapse formation in cell culture. Am J Physiol 27:C489–C494.
30. Gruener R, Hoeger G. 1991. Vector-averaged gravity alters myocyte and neuron properties in cell culture. Aviat Space Environ Med 62:1159–1165.
31. Ingber DE. 1997. Tensegrity: The architectural basis of cellular mechanotransduction. Annu Rev Physiol 59:575–599.

Pathophysiology of Cardiovascular Disease,
edited by N.S. Dhalla, H. Rupp,
A. Angel & G.N. Pierce
Kluwer Academic Publishers, Boston, 2004

Differential Inhibition by the Phosphatase Inhibitor Cantharidin of the Anti-adrenergic Effect of Endothelin-1 and Carbachol in the Canine Ventricular Myocardium

Masao Endoh, Li Chu, Ikuo Norota, and Kuniaki Ishii

Department of Pharmacology, Yamagata University School of Medicine, Yamagata 990-9585, Japan

Summary. Endothelin-1 (ET-1) and the muscarinic agonist carbachol do not affect the basal force of contraction, but induce a pronounced negative inotropic effect in the presence of β-adrenoceptor stimulation in the canine ventricular myocardium. We studied the influence of the phosphatase inhibitor cantharidin on the negative inotropic effect of ET-1 and carbachol in isolated canine right ventricular trabeculae and single ventricular myocytes. In the presence of 100 nM norepinephrine (NE), 10 nM ET-1 and 30 nM carbachol induced a negative inotropic effect of an identical extent. Cantharidin at 10 μM did not affect the basal force of contraction and the positive inotropic effect of NE mediated by β-adrenoceptor stimulation. Cantharidin (10 μM) markedly attenuated the negative inotropic effect of ET-1, but it did not affect the negative inotropic effect of carbachol. At 30 μM, cantharidin induced a positive inotropic effect and enhanced the positive inotropic effect of NE. Cantharidin (30 μM) suppressed significantly the negative inotropic effect of carbachol (30 nM and 100 nM). In canine single ventricular myocytes, ET-1 (10 nM) or carbachol (30 nM) did not affect the baseline level of cell shortening and the amplitude of intracellular Ca^{2+} transients, while they inhibited the NE (30 nM)-induced increases in cell shortening and Ca^{2+} transients. Cantharidin (10 μM) attenuated the inhibitory action of ET-1, but did not affect the effects of carbachol in the presence of NE. These results indicate that the activation of phosphatase

Corresponding author: Department of Pharmacology, Yamagata University School of Medicine, Yamagata 990-9585, Japan. Telephone: +81-23-628-5234, Fax: +81-23-628-5235, E-mail: mendou@med.id.yamagata-u.ac.jp.

that is susceptible to cantharidin is involved in the negative inotropic effect of both ET-1 and carbachol. The extent of contribution of phosphatase activation appears to be greater in the ET-1-induced negative inotropic effect than in the effect of the muscarinic receptor agonist carbachol in the canine ventricular myocardium.

Key words: Cantharidin, Endothelin-1, Carbachol, Ventricular contractility, Ca^{2+} transient.

In mammalian ventricular myocardium, muscarinic cholinergic receptor agonists do not affect the basal force of contraction, but elicit a pronounced negative inotropic effect when they are applied during induction of the positive inotropic effect by activation of β-adrenoceptors. This phenomenon has been termed the anti-adrenergic effect, indirect inhibitory action of muscarinic receptor agonists or "accentuated antagonism" (1), for which suppression of cyclic AMP-mediated signaling through the activation of pertussis toxin-sensitive G_i proteins may be responsible (1–3). While a decrease in β-adrenoceptor-mediated cyclic AMP generation via activation of G_i proteins subsequent to stimulation of muscarinic receptors may play a crucial role in "accentuated antagonism", it has recently been reported that the muscarinic receptor stimulation may counteract directly the effect of PKA activation and/or elicit activation of protein phosphatases (4). In vitro biochemical studies show that muscarinic agonists such as ACh and carbachol produce stimulation of protein phosphatase activity by inhibition of protein phosphatase inhibitor-1 and dephosphorylate a variety of functional proteins that have been phosphorylated by activation of PKA subsequent to β-adrenoceptor-mediated cyclic AMP accumulation (5–8). These findings in vitro could explain the anti-adrenergic effects of muscarinic stimulation that are not accompanied by a concomitant decrease in cyclic AMP levels or in PKA activation (6,7). Actually, Narayan and coworkers have recently shown that the phosphatase inhibitor cantharidin blocks the anti-adrenergic effect of adenosine A_1 receptor stimulation in rat cardiac myocytes (9).

Endothelin-1 (ET-1), 21-amino acid peptide originally isolated from vascular endothelium, has a potent vasoconstrictor effect and a variety of direct effects on the myocardium (10). ET-1 has a positive inotropic effect in most mammalian species, including rat, guinea pig, ferret, rabbit and human, does not have any inotropic effect in the dog (11,12), and has a negative inotropic effect in mouse (13) and neonatal rat ventricular myocytes (14). ET-1 interacts with cardiac signaling systems that act not only via G_q but also $G_{i/o}$ proteins (15,16), so as to elicit a positive inotropic effect or negative inotropic effect depending on species of experimental animals and experimental conditions.

In the canine ventricular myocardium, ET-1 alone does not produce a direct inotropic effect (11,17), but induces a negative inotropic effect (18–21) that is associated with a decrease in L-type Ca^{2+} currents (17,22) in the presence of β-adrenoceptor stimulation. The negative inotropic effect of ET-1 is susceptible to pertussis toxin, but the signaling process downstream to activation of the pertussis toxin-sensitive $G_{i/o}$ proteins remains unclear.

It has recently been shown in canine and human ventricular muscle that the negative inotropic effect of ET-1 is not accompanied by a significant reduction in accumulation of cyclic AMP (19,23) and that the nitric oxide (NO)-mediated signaling pathway may not be involved in the regulation (18). ET-1 inhibits the increase in L-type Ca^{2+} currents induced by 8-bromo-cyclic AMP in mammalian ventricular myocytes (24,25). It appears therefore highly likely that the signaling process subsequent to generation of cyclic AMP is involved in the ET-1-induced negative inotropic effect.

The purpose of the present study was to examine whether the phosphatase inhibitor cantharidin blocks the inhibitory action of ET-1 and carbachol on contractility and intracellular Ca^{2+} signaling in the presence of norepinephrine (NE) in the canine ventricular myocardium.

METHODS

The study involving treatment of experimental animals was carried out in accordance with *Guide for Animal Experimentation*, Yamagata University School of Medicine and Japanese Governmental Law (No. 105). The approval for the animal experiments was obtained from the Committee for Animal Experimentation, Yamagata University School of Medicine prior to the experiments, and the study was carried out also in accordance with the *Declaration of Helsinki*. Mongrel dogs (7–10 kg) of either sex were used in these experiments. Preparation of canine ventricular trabeculae, isolation of canine ventricular myocytes and loading of myocytes with indo-1/AM will be briefly described in this section, which has been described elsewhere in more detail (19,26).

Preparation of isolated right ventricular trabeculae

Two to four ventricular trabeculae were excised from canine right ventricle, and they were mounted in 20-ml organ baths that contained Krebs-Henseleit solution. The ventricular trabeculae were electrically stimulated with a pulse of 5-ms duration and a voltage 20% above the threshold (approximately 0.4 V) at a frequency of 0.5 Hz. The preparations of canine ventricular trabeculae had the average dimension of 6.38 ± 0.93 mm in length and 1.45 ± 0.19 mm^2 in cross-sectional area (n = 42, from 16 dogs).

Measurements of inotropic effects in ventricular trabeculae

Prazosin (300 nM) was allowed to act for 30 min before the addition of NE and present in organ baths throughout the experiments. NE (100 nM) was allowed to act for 60 min and the first response to NE was determined as a control in individual preparations. After NE was washed out for 60 min, NE (100 nM) was added again and allowed to act for 15 min before the addition of ET-1 or carbachol. ET-1 (10 nM) or carbachol (30 or 100 nM) was administered at a single concentration to each preparation. Phosphatase inhibitor cantharidin (10 or 30 μM) was administered 30 min before the addition of NE and was present in organ baths throughout the experiments.

Isolation of canine ventricular myocytes

A portion of the left ventricular free wall that is supplied with the branch of left anterior descending artery was excised, and the artery was then cannulated and perfused with Tyrode's solution that contained 1.0 mg/ml collagenase and 0.1 mg/ml protease by use of a recirculating system. Finally the muscle piece was perfused with Tyrode's solution that contained 0.2 mM $CaCl_2$ and cut into small pieces with a scalpel. The cell suspension was rinsed several times with the solution with gradual increases in Ca^{2+} concentration up to 1.8 mM.

Loading of myocytes with indo-1/AM

Myocytes were loaded with a fluorescent dye acetoxymethylester of indo-1 (indo-1/AM) by incubating them in 5 μM indo-1 solution for about 3 min at room temperature (24°C). After loading, they were centrifuged at 10×g for 1 min. The pellet was resuspended in HEPES-Tyrode solution. The myocytes were then laid in the chamber superfused with bicarbonate buffer for about 10 min.

Simultaneous measurements of cell shortening and Ca^{2+} transients

The myocytes were laid in a perfusion chamber placed on the stage of an inverted microscope (Diaphot TMD 300; Nikon, Tokyo, Japan). The perfusion was started with bicarbonate buffer containing 1.8 mM $CaCl_2$ at a rate of 3 ml/min at room temperature (24°C) and the cells were stimulated electrically by square-wave pulses with voltage about 30–40% above the threshold at a frequency of 0.5 Hz.

Fluorescence of indo-1 was excited with the light from a xenon lamp (150 W) at a wavelength of 355 nm, reflected by a 380 nm long-pass dichroic mirror, and detected by a fluorescence spectrophotometer (CAM-230; Japan Spectroscopic Co., Tokyo, Japan). Excitation light was applied to myocytes intermittently through a neutral density filter to minimize the photobleaching of indo-1. The emitted fluorescence was collected by an objective lens (CF Fluor DL40, Nikon, Japan) and then separated by a 580 nm long-pass dichroic mirror to permit simultaneous measurements of light at both 405 nm and 500 nm wavelengths through band-pass filters. A bright-field cell image was projected onto a photodiode array of the edge detector (C6294-01, Hamamatsu Photonics K.K., Hamamatsu, Japan) with 5 ms temporal resolution and the cell length was monitored simultaneously with indo-1 fluorescence.

Myocytes were perfused with the solution containing the agent examined. Prazosin (300 nM) was allowed to act for 20 min before the application of NE. After the effect of NE reached a steady state for 10 min, ET-1 or carbachol was administered at a single concentration to each preparation. The protein phosphatase inhibitor cantharidin was administered 20 min before the addition of NE and was present throughout the experiments. When the response of myocytes to the agent applied reached a steady state, the indo-1 fluorescence was measured and then the perfusion was switched to the solution that contained an additional agent. Cell length was continuously measured throughout the experiments, while the fluorescence of indo-1 was monitored intermittently to reduce the quenching. Cell length and Ca^{2+} transients were simultaneously recorded in the baseline state and in the presence of the agent when the response reached a steady state.

Data recording and analysis

Cell length and fluorescence of indo-1 were stored and displayed by means of a computer (Power Macintosh 8100/100AV, Apple Computer Inc., Cupertino, CA, U.S.A.) equipped with an A/D converter (MP-100A, BIOPAC Systems Inc., Santa Barbara, CA, U.S.A.) at 200 Hz and analyzed after low-pass filtering (cutoff frequency of 20 Hz). Data used for statistical analysis were obtained by signal-averaging of five successive tracings of cell shortening and Ca^{2+} transients. In analysis of data, the diastolic cell length and indo-1 fluorescence ratio prior to the first application of the agent examined in individual experiments were regarded as baseline values in each myocyte, which are assigned to 100%, respectively, and all the data are expressed as percentages of baseline values.

Drugs

Drugs used were ET-1 (Peptide Institute, Osaka, Japan); prazosin hydrochloride (Pfizer Taito, Tokyo, Japan); norepinephrine hydrochloride (NE; Nakarai Chemicals LTD, Kyoto, Japan); cantharidin, carbamylcholine chloride (carbachol) and protease type XIV (Sigma, Chemical Co., St. Louis, MO, U.S.A); pentobarbital sodium (Tokyo Kasei, Tokyo, Japan); indo-1/AM (Dojindo Chemical, Kumamoto, Japan); collagenase type II (Worthington Biochemical, Freehold, NJ, U.S.A.). Other reagents used were of the highest grade in purity that was commercially available.

Statistical analysis

Experimental values are presented as means ± SE. Significant differences between mean values were estimated by a repeated-measures analysis of variance and/or by Student's *t*-test with analytic software STATVIEW J-4.5 (Abacus Concepts, Berkeley, CA, U.S.A.). P values less than 0.05 were judged to represent significant differences.

RESULTS

Inotropic effects of cantharidin and its influence on the positive inotropic effect of NE

Inotropic effects of cantharidin and its influence on the positive inotropic effect of 100 nM NE were investigated in the presence of prazosin (300 nM) in isolated canine ventricular trabeculae. Cantharidin at 10 μM did not affect the basal force of contraction, and at 30 μM, it induced a long-lasting positive inotropic effect (118.0 ± 4.2% of the basal force; $n = 4$, $P < 0.05$). As shown in Figure 1, cantharidin at 10 μM did not affect the positive inotropic effect induced by 100 nM NE (95.4 ± 4.2% of the control response; $n = 7$). At a higher concentration (30 μM), cantharidin significantly enhanced the positive inotropic effect of NE to 159.2 ± 9.6% of the control response ($n = 6$; $P < 0.001$).

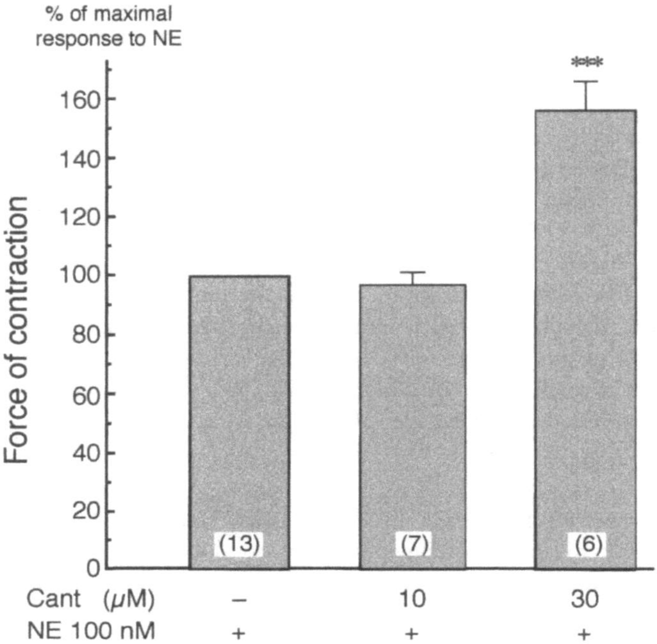

Figure 1. Influence of cantharidin (Cant) on the positive inotropic effect of norepinephrine (NE) in the presence of prazosin (300 nM) in isolated canine ventricular trabeculae. The maximum response to 100 nM NE in the absence of cantharidin was assigned to 100% for each preparation, and the maximum response to 100 nM NE in the presence of cantharidin are expressed as a percentage of the control response. Values are means ± SE. Numbers in parentheses indicate the numbers of muscle preparations examined. ***$P < 0.001$ vs. the response to 100 nM NE alone.

Influence of cantharidin on the negative inotropic effect of ET-1 and carbachol

Neither ET-1 (10 nM) nor carbachol (30 nM and 100 nM) significantly affected the basal force of contraction in isolated canine ventricular trabeculae (data not shown), but in the presence of NE (100 nM), they produced a definite negative inotropic effect. Figures 2 and 3 show the influence of cantharidin on the negative inotropic effect of ET-1 and carbachol. NE (100 nM) elicited a positive inotropic effect that developed gradually to achieve an elevated steady level at 10–15 min after the addition and was sustained essentially at a stable level for 60 min. After 60 min the positive inotropic effect of NE declined spontaneously to 88.2 ± 3.8% of the maximal level ($P < 0.05$). The positive inotropic effect of NE was reproducible when NE was applied again after washout for 60 min. ET-1 (10 nM) or carbachol (30 nM) elicited a negative inotropic effect of an identical extent in the presence of 100 nM NE when they were administered 15 min after the application of NE. As shown in Figure 2, cantharidin at 10 μM significantly suppressed the negative inotropic effect

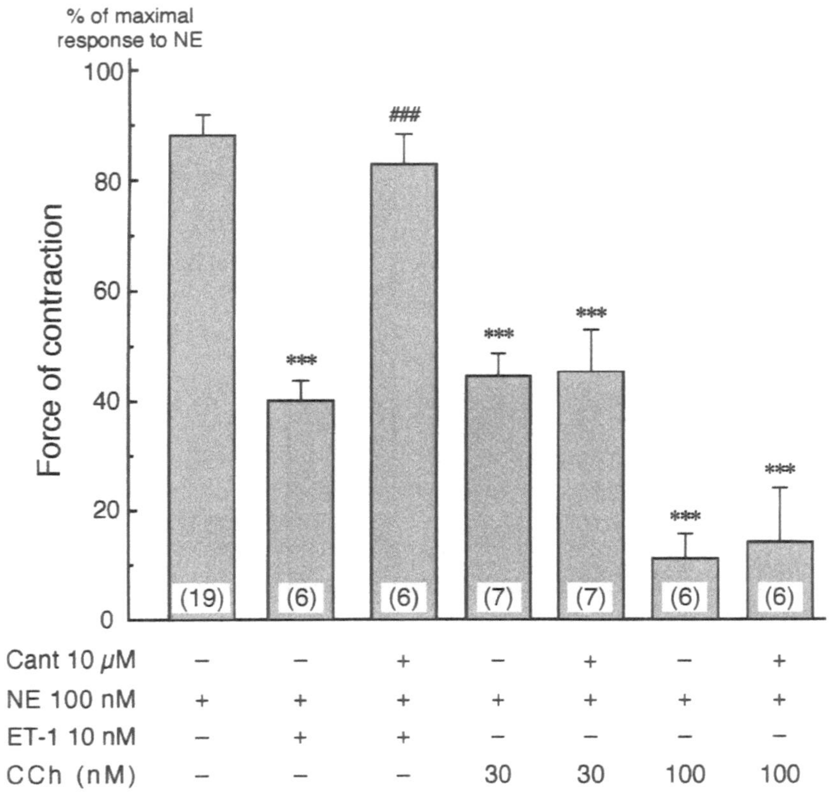

Figure 2. Influence of 10 μM cantharidin (Cant) on the negative inotropic effect of 10 nM ET-1 and carbachol (cch) in the presence of 100 nM NE in isolated canine ventricular trabeculae. The maximum response to 100 nM NE before the addition of ET-1 was assigned to 100% for each preparation, and the change in force of contraction is expressed as a percentage of the control response. Values are means ± SE. Numbers in parentheses indicate the numbers of muscle preparations examined. ***P < 0.001 vs. the response to NE alone; ###P < 0.001 vs. the response to 100 nM NE plus 10 nM ET-1.

of ET-1, but it did not affect significantly the negative inotropic effect of carbachol. In the presence of cantharidin (10 μM), the positive inotropic effect of NE measured at 45 min after the application of ET-1 (10 nM) was 82.9 ± 6.0% of the maximal response to NE, which was significantly different from the control response (without cantharidin) of 39.8 ± 4.0% (n = 6 each; P < 0.001). By contrast, the positive inotropic effect of NE after the application of carbachol (30 or 100 nM) in the presence of cantharidin was 45.1 ± 7.9% or 19.0 ± 10.5% of the maximal response to NE, which was not significantly different from the respective control response of 44.7 ± 4.0% or 13.8 ± 5.6% (n = 7 each; P > 0.05).

Cantharidin at a higher concentration (30 μM) suppressed the negative inotropic effect of carbachol (Figure 3). The positive inotropic effect of NE at 45 min after

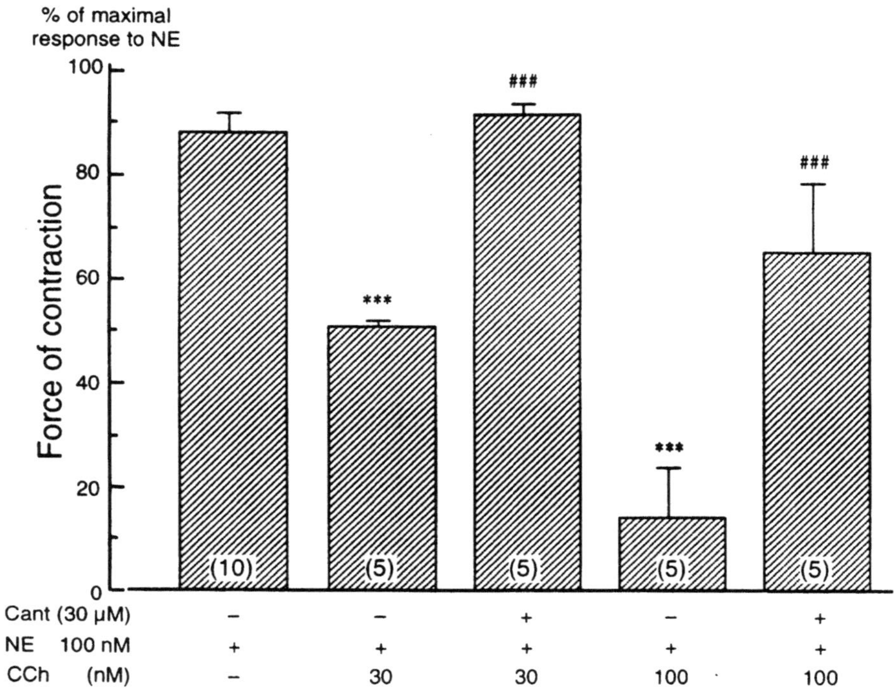

Figure 3. Influence of 30 μM cantharidin (Cant) on the negative inotropic effect of carbachol (cch, 30 or 100 nM) in the presence of 100 nM NE in isolated canine ventricular trabeculae. The maximum response to 100 nM NE before the addition of cch was assigned to 100% for each preparation, and the change in force of contraction was expressed as a percentage of the control response. Values are means ± SE. Numbers in parentheses indicate the numbers of muscle preparations examined. ***P < 0.001 vs. the response to NE alone; ###P < 0.001 vs. the response to 100 nM NE plus 30 or 100 nM carbachol.

the addition of carbachol (30 or 100 nM) was 91.7 ± 1.7% or 64.8 ± 12.8% of the maximal response to NE, which was significantly different from the respective control response (without cantharidin) of 50.1 ± 1.4% (n = 5 each; P < 0.001) or 14.7 ± 8.8% (n = 5 each; P < 0.001).

Influence of cantharidin on the effects of ET-1 and carbachol in canine single ventricular myocytes

A series of experiments was performed in canine ventricular myocytes to examine the influence of cantharidin on alterations of Ca^{2+} signaling and cell shortening induced by ET-1 and carbachol. As shown in Figures 4, 5 and 6, NE at 30 nM induced a pronounced increase in peak cell shortening associated with a remarkable elevation of the amplitude of Ca^{2+} transients. In addition, the positive inotropic effect of NE was associated with a marked abbreviation of the duration of cell short-

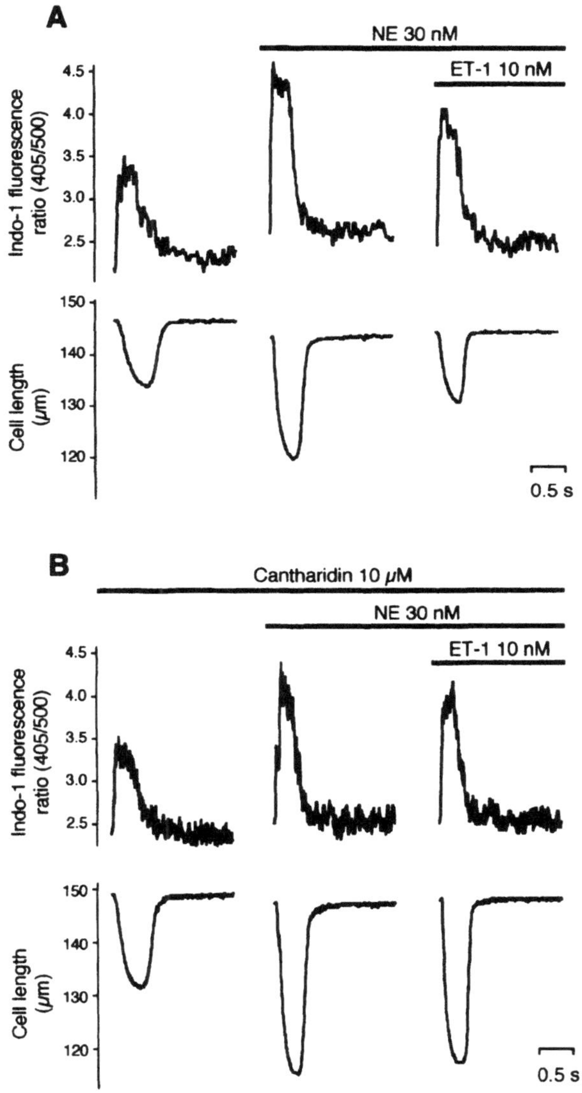

Figure 4. Influence of 10 μM cantharidin on the effects of 10 nM ET-1 on the indo-1 fluorescence ratio (upper traces) and cell shortening (lower traces) in the presence of 30 nM NE in a single canine ventricular myocyte. A: in the absence of cantharidin; B: in the presence of cantharidin. Individual tracings were obtained by means of signal averaging of five successive signals.

ening and Ca^{2+} transients. Both ET-1 and carbachol inhibited the NE-induced increase in cell shortening and the amplitude of Ca^{2+} transients to a similar extent, and antagonized also the NE-induced abbreviation of the duration of cell shortening and Ca^{2+} transients (data not shown).

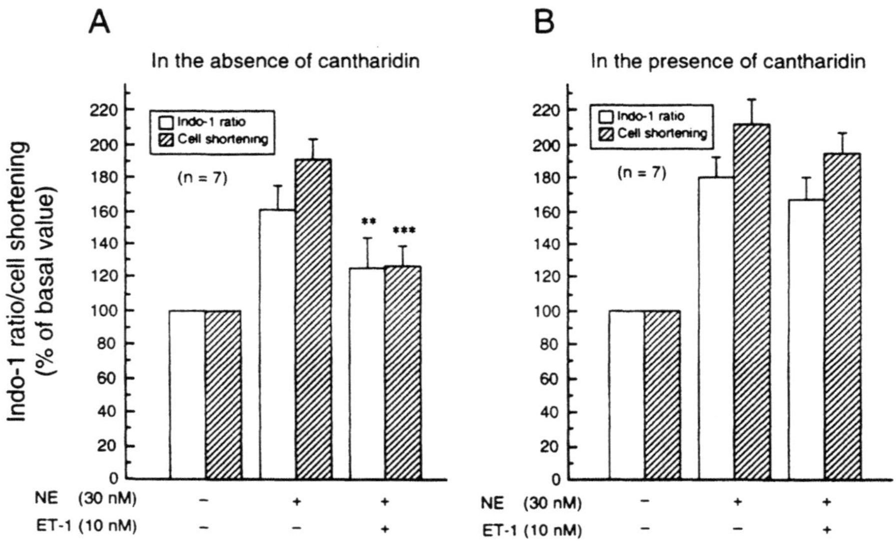

Figure 5. Influence of 10 μM cantharidin on the effects of 10 nM ET-1 on the indo-1 fluorescence ratio and cell shortening in the presence of 30 nM NE in single canine ventricular myocytes. A: in the absence of cantharidin; B: in the presence of cantharidin. Numbers in parentheses indicate the numbers of myocytes examined. $^{**}P < 0.01$; $^{***}P < 0.001$ vs. the response to 30 nM NE.

Cantharidin at 10 μM did not significantly affect the baseline indo-1 ratio and cell shortening by itself: on average, the indo-1 ratio and cell shortening were 98.3 ± 5.7% and 105.0 ± 4.6% of the basal values, respectively (n = 7). Cantharidin at 10 μM did not affect the NE-induced increase in cell shortening and Ca^{2+} transients. However, cantharidin at 10 μM significantly attenuated the ET-1-induced decrease in cell shortening and Ca^{2+} transients in the presence of NE (Figure 4 and Figure 5). Cantharidin at 10 μM had no effect on the carbachol-induced inhibitory action (Figure 6).

DISCUSSION

In canine ventricular myocardium, ET-1 elicits a negative inotropic effect in the presence of β-adrenoceptor stimulation induced by isoproterenol (19) or NE. As ET-1 does not affect the basal force of contraction and induces a negative inotropic effect only in the presence of β-adrenoceptor stimulation, the negative inotropic effect of ET-1 may be exerted by modulation of cyclic AMP-mediated signal pathway, including cyclic AMP generation and/or breakdown, PKA activation and phosphorylation of functional proteins.

It is likely that ET-1 produces the negative inotropic effect by activation of G_i proteins to decrease the β-adrenoceptor-mediated cyclic AMP generation, because the ET-1-induced negative inotropic effect is susceptible to pertussis toxin (19).

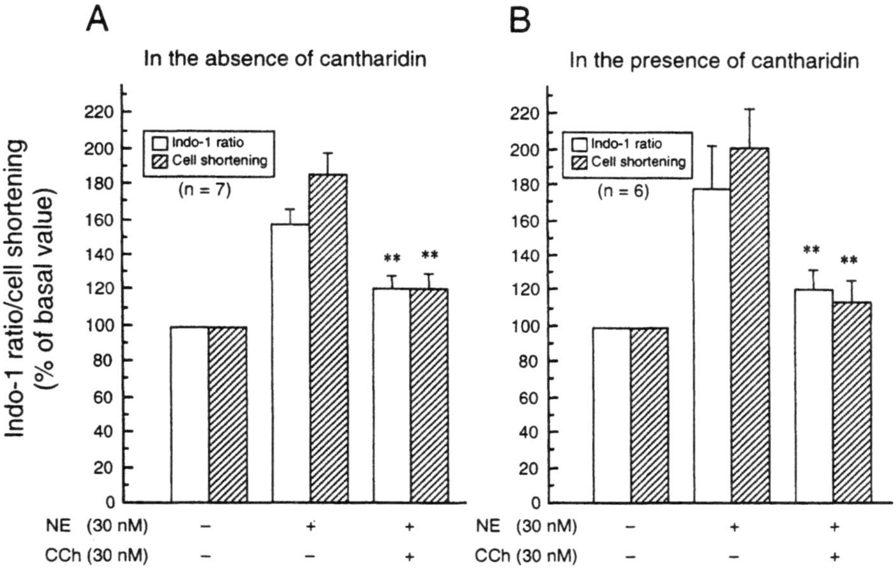

Figure 6. Influence of 10 μM cantharidin on the effects of 30 nM carbachol (cch) on the indo-1 fluorescence ratio and cell shortening in the presence of 30 nM NE in single canine ventricular myocytes. A: in the absence of cantharidin; B: in the presence of cantharidin. Numbers in parentheses indicate the numbers of myocytes examined. ** $P < 0.01$ vs. the response to 30 nM NE.

However, the following pieces of evidence imply that the regulation of cyclic AMP generation and/or breakdown may be of less importance for the ET-1-induced negative inotropic effect in canine ventricular myocardium. First, ET-1 induced a negative inotropic effect without detectable lowering of tissue cyclic AMP levels in canine ventricular myocardium (19). Similar findings have been reported also in the human myocardium (23). While it has been reported that ET-1 may activate cyclic AMP phosphodiesterase and thereby inhibit the β-adrenoceptor-mediated positive inotropic effect in the rat (15,16), such a signal pathway would not play an important role in the canine ventricular myocardium, since the cyclic AMP level was not altered by ET-1. Although these observations imply that cyclic AMP generation and/or hydrolysis may not be involved in the ET-1-induced negative inotropic effect, the tissue cyclic AMP level is a less sensitive indicator, so that the potential involvement of cyclic AMP lowering in functionally relevant compartments cannot be completely excluded. Nonetheless, by contrast to ET-1, the muscarinic receptor agonist carbachol decreased the cyclic AMP level that had been elevated by isoproterenol under the same experimental condition in canine ventricular myocardium (2,19). Second, ET-1 inhibited also the effect of 8-bromo cyclic AMP (24,25) that activates PKA directly beyond the process of cyclic AMP generation/breakdown. These observations indicate that the mechanism other than cyclic AMP generation/breakdown may be responsible for the ET-1-induced negative inotropic effect in canine ventricular myocardium, as postulated previously for the "accentuated antagonism" in ventricular myocardium of certain species of animals, including the guinea-pig ventricular myocardium (18–20,23,24).

Cantharidin, an inhibitor of phosphatase 1 and phosphatase 2A, has been shown to increase the phosphorylation state of cardiac regulatory proteins such as phospholamban, troponin I, and C protein without increasing the intracellular cyclic AMP level (27) and to enhance phosphorylation of regulatory proteins phosphorylated by β-adrenoceptor agonists (28). The current findings that cantharidin at 10 μM suppressed the negative inotropic effect of ET-1 support the view that activation of phosphatase plays a crucial role in the negative inotropic effect of ET-1 in canine ventricular myocardium. The concentration of cantharidin employed in the current study is compatible with that over which the compound affects the contractile force and induces phosphorylation of phospholamban and troponin I in isolated guinea pig papillary muscles (29). In guinea pig papillary muscles, cantharidin at 10 μM started to increase phosphorylation of Ser-16 of phospholamban and to induce a positive inotropic effect, an indication that the compound may penetrate effectively the cell membrane and modulate contractility in relation to phosphorylation of functional proteins (29).

We have recently found that the negative inotropic effect of ET-1 was also inhibited by the guanylyl cyclase inhibitor and the protein kinase G inhibitor (30). These observations together indicate that ET-1 may activate the cyclic GMP-mediated signaling process ultimately leading to the activation of phosphatase that may play a physiologically important role in termination of cyclic AMP-mediated signal transduction by dephosphorylation of the functionally relevant proteins that had been phosphorylated via activation of PKA.

Gupta and coworkers have recently found that muscarinic agonists reduce the β-adrenoceptor-mediated phosphorylation of regulatory proteins without decreasing cyclic AMP levels or the PKA activity (4,31). In addition, the inhibitory action of muscarinic agonists on the β-adrenoceptor-mediated protein phosphorylation was blocked by the phosphatase inhibitor, while the *functional* relevance of these biochemical findings has remained to be determined. Muscarinic agonists diminished the effect of cyclic AMP applied intracellularly (32), which implies that the muscarinic inhibition occurs at the process subsequent to cyclic AMP generation. In the pertinent study in isolated perfused heart (33) and cardiac myocytes (9,31), cantharidin blocked the adenosine A_1 receptor-mediated anti-adrenergic effect, which is consistent with the current finding with ET-1 in the canine ventricular myocardium.

The present results support the view that the activation of phosphatase activity may likewise play a role in signal transduction cascade triggered by muscarinic receptor agonists, as in the ET-1-induced regulation. Zhang and Macleod have recently reported that the negative inotropic effect of carbachol is dissociated from the changes in cyclic AMP and PKA in perfused rat hearts (34), which implies also the role of phosphatase activation in the muscarinic regulation. It is noteworthy, however, that cantharidin at 10 μM that suppressed significantly the ET-1-mediated negative inotropic effect did not affect the negative inotropic effect of carbachol. Cantharidin attenuated the carbachol-induced negative inotropic effect first at 30 μM, an indication that the anti-adrenergic effect of carbachol is less sensitive to cantharidin than the effect of ET-1. Since cantharidin at 30 μM enhanced the positive inotropic effect of NE, the apparent antagonistic action of cantharidin at 30 μM on the carbachol-induced negative inotropic effect could be partially due to the enhancement of NE-induced positive inotropic effect produced by cantharidin. Supporting this postulate, it has been shown that the anti-adrenergic effect of muscarinic and adenosine receptor agonists is markedly affected by the magnitude of the β-adrenoceptor-mediated positive inotropic effect and the excess stimulation of cyclic AMP-mediated signaling process counteracts the anti-adrenergic effect of these agonists (7,20,21).

In single canine ventricular cardiomyocytes loaded with indo-1, ET-1 and carbachol decreased cell shortening in association with suppression of the amplitude of Ca^{2+} transients, which indicates that the inhibitory action of these agents may be mainly due to a decrease in intracellular Ca^{2+} mobilization, i.e., being induced via the upstream mechanism of cardiac E-C coupling (35). As a mechanism of inhibition of Ca^{2+} transients, we have recently found that ET-1 inhibits the augmentation of L-type Ca^{2+} current induced by β-adrenoceptor stimulation through G_i protein-dependent pathway in canine ventricular myocytes (22). By contrast, ET-1 neither induced a negative inotropic effect (36) nor decreased L-type Ca^{2+} currents (37) when Bay K 8644, a Ca^{2+} channel agonist, had been employed to induce a positive inotropic effect or to increase Ca^{2+} currents.

In conclusion, the current findings indicate that cantharidin inhibits the negative inotropic effect of ET-1 and carbachol, an indication that both ET-1 and carbachol may induce a negative inotropic effect by activation of the protein phosphatase. The observation that the negative inotropic effect of ET-1 is inhibited by cantharidin at 10 μM that does not affect the negative inotropic effect of carbachol supports the view that the extent of contribution of phosphatase activation may be higher in the ET-1-induced negative inotropic effect than in the carbachol-induced negative inotropic effect in the canine ventricular myocardium. Further study is required with other phosphatase inhibitors to establish the different role of phosphatase activation in the negative inotropic effect induced by ET-1 and carbachol.

ACKNOWLEDGMENTS

This work was supported in part by Grant-in-Aid for Scientific Research (B) (nos. 11470021 and 11557203) from the Ministry of Education, Culture, Sports, Science, and technology, Japan, by the Research Grant for Cardiovascular Disease (11-1) from the Ministry of Health and Welfare, Japan, and the Research Grant from Sharyo-zaidan Foundation, Japan.

REFERENCES

1. Endoh M. 1999. Muscarinic regulation of Ca^{2+} signaling in mammalian atrial and ventricular myocardium. Eur J Pharmacol 375:177–196.
2. Endoh M. 1979. Correlation of cyclic AMP and cyclic GMP levels with changes in contractile force

of dog ventricular myocardium during cholinergic antagonism of positive inotropic actions of histamine, glucagon, theophyline and papaverine. Jpn J Pharmacol 29:855–864.
3. Endoh M, Honma M. 1979. Effects of papaverine and its interaction with isoprenaline and carbachol on the contractile force and cyclic nucleotide levels of the canine ventricular myocardium. Naunyn-Schmiedeberg's Arch Pharmacol 306:241–248.
4. Gupta RC, Neumann J, Boknik P, et al. 1994. M_2-specific muscarinic cholinergic receptor-mediated inhibition of cardiac regulatory protein phosphorylation. Am J Physiol 266:H1138–H1144.
5. Ahmad Z, Green F, Subuhi HS, Watanabe AM. 1989. Autonomic regulation of type 1 protein phosphatase in cardiac muscle. J Biol Chem 264:3859–3863.
6. Gupta RC, Neumann J, Watanabe AM. 1993. Comparison of adenosine and muscarinic receptor-mediated effects on protein phosphatase inhibitor-1 activity in the heart. J Pharmacol Exp Ther 266:16–22.
7. Sakai R, Shen JB, Pappano AJ. 1999. Elevated cAMP suppresses muscarinic inhibition of L-type calcium current in guinea pig ventricular myocytes. J Cardiovasc Pharmacol 34:304–315.
8. Herzig S, Meier A, Pfeiffer M, Neumann J. 1995. Stimulation of protein phosphatases as a mechanism of the muscarinic-receptor-mediated inhibition of the cardiac L-type Ca^{2+} channels. Pflügers Arch 429:531–538.
9. Narayan P, Mentzer RM Jr, Lasley RD. 2000. Phosphatase inhibitor cantharidin blocks adenosine A_1 receptor anti-adrenergic effect in rat cardiac myocytes. Am J Physiol 278:H1–H7.
10. Yanagisawa M, Kurihara H, Kimura S, Tomobe Y, Kobayashi M, Mitsui Y, Yazaki Y, Goto K, Masaki T. 1988. A novel potent vasoconstrictor peptide produced by vascular endothelial cells. Nature 332:411–415.
11. Takanashi M, Endoh M. 1991. Characterization of the positive inotropic effect of endothelin on mammalian ventricular myocardium. Am J Physiol 261:H611–H619.
12. Moravec CS, Reynolds EE, Stewart RW, Bond M. 1989. Endothelin is a positive inotropic agent in human and rat heart in vitro. Biochem Biophys Res Commun 159:14–18.
13. Izumi M, Miyamoto S, Hori M, Ozaki H, Karaki H. 2000. Negative inotropic effect of endothelin-1 in the mouse right ventricle. Eur J Pharmacol 393:109–117.
14. Kohmoto O, Ikenouchi H, Hirata Y, Momomura S, Serizawa T, Barry WH. 1993. Variable effects of endothelin-1 on $[Ca^{2+}]_i$ transients, pH_i and contraction in ventricular myocytes. Am J Physiol 265:H793–H800.
15. Hilal-Dandan R, Urasawa K, Brunton LL. 1992. Endothelin inhibits adenylate cyclase and stimulates phosphoinositide hydrolysis in adult cardiac myocytes. J Biol Chem 267:10620–10624.
16. Hilal-Dandan R, Merck D, Lujan J, Brunton LL. 1994. Coupling of the type A endothelin receptor to multiple responses in adult rat myocytes. Mol Pharmacol 45:1183–1190.
17. Banyasz T, Magyar J, Kortvely A, Szigeti G, Szigligeti P, Papp Z, Mohaesi A, Kovacs L, Nanasi PP. 2001. Different effects of endothelin-1 on calcium and potassium currents in canine ventricular cells. Naunyn-Schmiedeberg's Arch Pharmacol 363:383–390.
18. Zhu Y, Yang HT, Endoh M. 1997. Does nitric oxide contribute to the negative chronotropic and inotropic effects of endothelin-1 in the heart? Eur J Pharmacol 332:195–199.
19. Zhu Y, Yang HT, Endoh M. 1997. Negative chronotropic and inotropic effects of endothelin isopeptides in mammalian cardiac muscle. Am J Physiol 273:H119–H127.
20. Chu L, Endoh M. 2000. Biphasic inotropic effects of endothelin-1 in the presence of sympathomimetic drugs at different concentrations in canine ventricular myocardium (abstract). Jpn J Pharmacol 82 (suppl I):198P.
21. Chu L, Endoh M. 2000. Biphasic inotropic response to endothelin-1 in the presence of various concentrations of norepinephrine in dog ventricular myocardium. J Cardiovasc Pharmacol 36 (Suppl. 2):S9–S14.
22. Watanabe T, Endoh M. 2000. Antiadrenergic effects of endothelin-1 on L-type Ca^{2+} current in canine ventricular myocytes. J Cardiovasc Pharmacol 36:344–350.
23. Walker CA, Ergul A, Zile MR, et al. 2001. β-Adrenergic and endothelin receptor interaction in dilated human cardiomyopathic myocardium. J Cardiac Fail 7:129–137.
24. Watanabe T, Endoh M. 1999. Characterization of the endothelin-1 induced regulation of L-type Ca^{2+} current in rabbit ventricular myocytes. Naunyn-Schmiedeberg's Arch Pharmacol 360:654–664.
25. Reid JJ, Lieu AT, Rand MJ. 1991. Interactions between endothelin-1 and other chronotropic agents in rat isolated atria. Eur J Pharmacol 194:173–181.
26. Yang HT, Sakurai K, Sugawara H, Watanabe T, Norota I, Endoh M. 1999. Role of Na^+/Ca^{2+} exchange in endothelin-1-induced increases in Ca^{2+} transient and contractility in rabbit ventricular myocytes: pharmacological analysis with KB-R7943. Br J Pharmacol 126:1785–1795.

27. Neumann J, Herzig S, Boknik P, Apel M, Kaspareit G, Schmitz W, Scholz H, Tepel M, Zimmermann N. 1995. On the cardiac contractile, biochemical and electrophysiological effects of cantharidin, a phosphatase inhibitor. J Pharmacol Exp Ther 274:530–539.
28. Sulakhe PV, Vo XT. 1995. Regulation of phospholamban and troponin-I phosphorylation in the intact rat cardiomyocytes by adrenergic and cholinergic stimuli: roles of cyclic nucleotides, calcium, protein kinases and phosphatases and depolarization. Mol Cell Biochem 149–150:103–126.
29. Boknik P, Khorchidi S, Bodor GS, Huke S, Knapp J, Linck B, Luss A, Muller FU, Schnitz W, Neumann J. 2001. Role of protein phosphatases in regulation of cardiac inotropy and relaxation. Am J Physiol 280:H786–H794.
30. Endoh M, Chu L, Takahashi R, Norata I. 2001. Regulation of cardiac Ca^{2+} signaling by cross talk of endothelin-1 and norepinephrine (abstract). Jpn J Pharmacol. 85 (suppl I):20p.
31. Gupta RC, Neumann J, Durant P, Watanabe AM. 1993. A_1-adenosine receptor-mediated inhibition of isoproterenol-stimulated protein phosphorylation in ventricular myocytes. Evidence against a cAMP-dependent effect. Circ Res 72:65–74.
32. Levi RC, Alloati G, Penna C, Gallo MP. 1994. Guanylate-cyclase-mediated inhibition of cardiac I_{Ca} by carbachol and sodium nitroprusside. Pflügers Arch 426:419–426.
33. Sawmiller DR, Fenton RA, Dobson JG Jr. 1998. Myocardial adenosine A_1-receptor sensitivity during juvenile and adult stages of maturation. Am J Physiol 274:H627–H635.
34. Zhang JZ, Macleod KM. 1996. Dissociation of negative inotropic effect of carbachol from changes in cAMP and PKA in perfused rat hearts. Am J Physiol 271:H404–H409.
35. Blinks JR, Endoh M. 1986. Modification of myofibrillar responsiveness to Ca^{++} as an inotropic mechanism. Circulation 73 (suppl III):III-85–III-98.
36. Chu L and Endoh M. 2001. Differential inotropic effects of endothelin-1, angiotensin II and phenylephrine via α-adrenoceptor stimulation in dog ventricular myocardium (abstract). Folia Pharmacol Jpn 117:11P.
37. Delpech N, Soustre H, Putreau D. 1997. Endothelin-1 inhibits L-type Ca^{2+} current enhanced by isoprenaline in rat atrial myocytes. J Cardiovasc Pharmacol 29:136–143.

Pathophysiology of Cardiovascular Disease,
edited by N.S. Dhalla, H. Rupp,
A. Angel & G.N. Pierce
Kluwer Academic Publishers, Boston, 2004

Factors Involved in Self and Drug-Induced Spontaneous Ventricular Defibrillation: Intra and Inter Species Variations

Mordechai Manoach, Narcis Tribulova,[1] and Thomas Podzuweit[2]

Dept. Physiology, Tel Aviv Univ. Med. School, Tel Aviv, Israel
[1] Institute for Heart Research, SAS, Bratislava, Slovakia and
[2] Dept. Exp. Cardiology, Max- Planck-Institute Bad Nauheim, Germany

Summary. Ventricular fibrillation (VF) was thought to be a life threatening arrhythmia that requires immediate defibrillation. Recently it became evident that VF can be a transient one (TVF) that reverts spontaneously into a sinus rhythm and that compounds and hormones can facilitate this process. Various theories related this ability to the ventricular muscle mass and/or to different electrophysiological properties like: fibrillating and defibrillating thresholds, action potential duration and wavelength of reentrant impulse. However, these hypotheses could not explain the transient type of VF.

On the basis of results obtained in various mammals at different ages we have found that TVF is species and age-related and requires good intercellular coupling and synchronization that can be enhanced by catecholamines and cAMP. As to the mechanisms involved, we hypothesized that the defibrillating ability and its age and species variations can be related to the state of the cardiac myosin heavy chain (MHC) isoenzymes. Mammalian cardiomyocytes express functionally different isoforms of MHC designated as α and β (and V1 and V3 respectively). The former exhibits high ATPase activity, higher intrinsic contractility, faster velocity of shortening and is accompanied by increased sarcoplasmic reticulum (SR) Ca^{2+} uptake, while the latter exhibits low ATPase activity and is associated with greater economy of force maintenance. The relative expression of these isoforms is age and species specific and influenced by physiological conditions and pathological states. Catecholamines, thyroid hormones and cAMP up-regulate V1 and down-regulate V3, while hypertrophy, energy deprivation and hemodynamic overload have opposite

Mordechai Manoach Ph.D., Department of Physiology and Pharmacology, Tel Aviv University Medical School. Tel Aviv, ISRAEL. Phone No.: 972-3-6409108, Fax No.: 972-3-6409113, E-mail: manoach@imap.tau.ac.il.

effects. Comparing the incidence of TVF with the αMHC expression revealed very close correlation that can explain well the intra (age) and inter species differences (rat, GP, and rabbit versus dog and pig) and lead us to hypothesize that preferential expression of the cardiac V1 isoenzyme may facilitate TVF.

Following the data in the literature that V1 appears in newborn mammals few days after birth, we examined the 7days-old pig to electrically induced VF. The results showed well synchronized but still sustained VF that suggested that at this age it is a low and sub-sufficient amount of V1. Therefore, we treated the piglet, according to our previous experience, with maprotiline that increases catecholamine level and thereby cAMP and SR activity. The results revealed that this pretreatment transformed the VF into a TVF. However, it did not lead to TVF in old pig that has no V1 isoenzyme.

Following the information that thyroid hormones (T3 and T4) can increase the amount of V1 and gap junctional coupling, both requested for spontaneous defibrillation, we treated young pig with T3. This treatment lead to a synchronized but still sustained VF. Administration of maprotiline to the T3-pretreated pig transformed the sustained synchronized VF into a self-terminating ventricular fibrillo-flutter.

These results support our hypothesis and suggest that preferential expression of cardiac V1 isoenzyme predisposes to spontaneous defibrillation. In case that the amount of V1 and SR activity are sub-sufficient, a support of catecholamines and/or thyroid hormone that increased SR activity is required.

Key words: Ventricular fibrillation, Transient ventricular fibrillation, Myocardial heavy chain isoenzymes, Calcium activated sarcoplasmic reticulum ATPase, Thyroid hormones.

INTRODUCTION

Ventricular fibrillation (VF) is an uncoordinated electrical activity that leads to pump failure and, when it occurs, is usually fatal. VF is one of the major causes of sudden death in human. However, the exact mechanisms of initiation of VF are not fully understood and effective strategies for its prevention are still being sought.

Although it was commonly believed that VF never terminates spontaneously, there are observations that proved the contrary. For instance, Wiggers showed in 1929 (1) in animals that there are two types of VF: a sustained VF (SVF) that requires artificial defibrillation and a transient form (TVF) that spontaneously reverts into sinus rhythm within seconds after its occurrence. Since the latter was regularly observed in small animals such as rats and guinea pigs, the investigators concluded that TVF is an intrinsic property of small hearts (2). Following the assumption that VF was due to reentrant activity and muscle mass appeared to be important for the maintenance of reentrant circuits it was assumed that small hearts have not enough muscle mass to sustain fibrillation, whereas "larger hearts" like those of humans and dogs, have sufficient mass for continuing fibrillation (2,3).

This generally accepted hypothesis was questioned by more recent findings on two grounds. First, SVF was shown to occur in animals with small hearts, such as pigeons (4) and old rodents (5). Secondly, contrary to expectations, TVF occurred

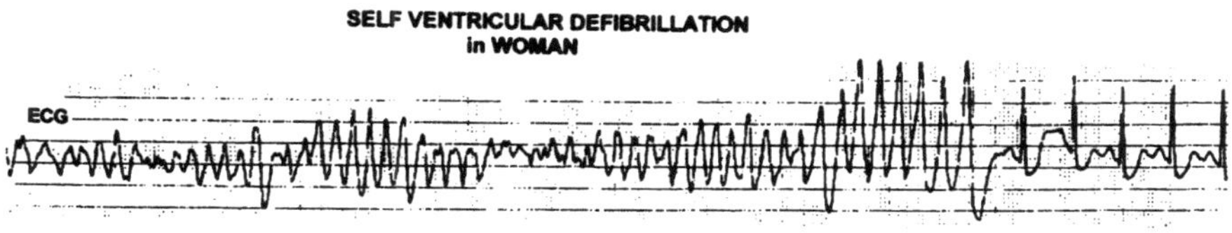

Figure 1. ECG record of spontaneous ventricular defibrillation in woman recorded in clinic.

in considerably "larger hearts" like those of cats (5) and young puppies (6). The existing evidence shows that transient VF also occurs in humans even not so rare (7) (Figure 1).

Following Schwartz et al. (8) that first described TVF in human subjects in 1949, transient life-threatening arrhythmias have been widely reported and designated by various names like torsade de pointes (9), idiopathic recurrent VF (10), paroxysmal VF with spontaneous reversion to sinus rhythm (11), polymorphous VT (12), fibrilo-flutter (13) etc. Since all of these arrhythmias are in fact transient life-threatening events, that resemble VF, we suggest that as a common name, Schwartz's term—TVF (8)—can be more adequate.

The main factors involved in the initiation and propagation of VF are thought to be both, structural and functional.

From structural point of view, inhomogeneities of myocardium that lead to dispersion of repolarization, refractory period and excitability, as well as differences in conductivity and unidirectional block, are classic requirements of reentry (14–18).

As far as functional properties are concerned, initiation and maintenance of reentrant arrhythmias were believed to depend on a critical relationship between wavelength (WL) of the reentrant impulse, conduction velocity (CV) and tissue refractoriness (ERP) (3,19–24).

THE WAVELENGTH THEORY

The wavelength is a product of the functional refractory period and conduction velocity (WL = FRP × CV) (19–21). The stability of reentrant circuits was thought to depend on the relationship between the wavelength and the physical length of the reentrant pathway. Consequently, attempts aimed at prolonging action potential duration (APD), EFR and WL seemed the most promising strategies for the prevention of VF. Accordingly, K^+ channel blockers known as Class III drugs that prolong APD and ERP, were developed as antiarrhythmic and antibrillatory compounds (25,26).

However, by searching the literature, we did not find any study showing how prolongation of APD and/or ERP prevented or terminated VF (27,28). In contrast, in many instances, prolongation of APD was proarrhythmic (29) and triggered VF, as it predisposes to early or delayed after depolarizations (30). Moreover, hearts of hypertensive (31), or hypokalaemic (32) animals as well as hypertrophic (33) hearts that exhibit prolonged APD and ERP are known to be more vulnerable to VF, whereas administration of adrenaline and other compounds that shorten APD and ERP and reentrant wavelength, proved potent defibrillating properties (34,35).

In the contrary to the defibrillating hypothesis, based on prolongation of APD, Johansen (36) (in 1985) summarized that prolongation of Q–T interval (which is related to prolongation of APD) is associated with electrical instability of the heart.

Today it is generally accepted that prolongation of APD has arrhythmogenic effect (28,29).

THE SYMPATHOMIMETIC HYPOTHESIS

Bacaner (37) and colleagues showed that pretreatment with bretylium tosylate facilitated conversion of VF from sustained to transient VF, in both, dogs (37) and humans (38,39). Initially, Bacaner linked the defibrillating ability of bretylium to its guanethidine like effect of chemical sympathectomy (40) as well to increase the fibrillation threshold. With time, Bacaner changed the explanation of the defibrillating effect of bretylium, relating it to a K^+ channel blocking effect i.e. to a class III effect (41).

The fact that in its first stage bretylium increases the cathecholamine level and the fact that the defibrillating activity of bretylium tosylate was blocked by prior reserpinization of the animals (40) lead us to suggest that the defibrillating activity of bretylium tosylate could or even should be related to its sympathomimetic properties.

This suggestion was strengths by our finding that the self-defibrillating capability of animals was age-dependent (5,6,42). We found that in young mammals, which have a predominant sympathetic auto-regulation, electrically-induced VF ceased spontaneously, whereas in older mammals (with vagal predominance) it was sustained and required artificial electrical defibrillation. The age range, during which spontaneous defibrillation occurred in young animals varied between species. In rats, which are highly sympathetic, transient VF occurred even in aged animals (older than 2 years), whereas in dogs with their strong vagal disposition, TVF was observed only in very young puppies (up to 5 weeks). Accordingly we showed that administration of agents that inhibit neuronal uptake of catecholamines such as tricyclic antidepressants (6,43,44) and thereby increased interstitial catecholamine content, facilitated transformation of SVF into TVF in adult mammals. In contrast, vagal stimulation or administration of either metacholine or beta adrenergic blocking agents prevented spontaneous defibrillation in young mammals (6,45).

THE MECHANISMS INVOLVED IN TVF

In an attempt to define the mechanisms involved in spontaneous cessation of VF, we examined electrophysiological data obtained from mammals of the same species that exhibit transient or sustained VF. No differences were found between the cardiac muscle mass, heart rate or APD in animals that exhibited transient or sustained VF (5).

Using floating microelectrodes for simultaneous recording of membrane electrical activity from two or more cardiomyocytes in vivo from different parts of the left ventricle, we found well synchronized, slow rate electrical fibrillating activity in TVF and a less synchronized electrical activity of higher frequency in SVF (46–48). "Synchronized" TVF occurred in hearts with functional intercellular coupling, which ensured proper propagation of electrical signals through gap junctions and allowed the cardiomyocytes to act in synchrony. Thus, preservation or enhancement of gap junctional coupling can facilitate conversion of SVF into TVF by decreasing the number of reentrant circuits. Functional gap junctional coupling reduces the fibrillating rate and transformes small local circuits into bigger ones, until the entire viable ventricular myocardium acts as a functional syncytium.

Intermyocyte communication mediated by gap junctions is a dynamic process (49), known to be modulated by intracellular Ca^{2+} and H^{+} ion concentration as well as by a variety of endogenous and exogenous factors (50–52). For instance, agents that increase cAMP up-regulated gap junctional channels, whereas channels can be down-regulated by high levels of cGMP (51). Alterations in the concentration of these mediators might affect intercellular electrical coupling. Decreased and/or abnormal intercellular coupling leads to a fractionated conduction (53), disturbances in synchronization and arrhythmogenesis (49,54). Moreover, excess of diastolic free Ca^{2+} concentration $[Ca^{2+}]_i$ inhibited gap junctional channels and decreased intercellular coupling most likely by increasing junctional and internal longitudinal resistance (53,55). Deterioration of myocardial intercellular synchronization increases the number of fibrillating micro-areas that allows VF to prevail and decreases the ability of the heart to defibrillate spontaneously (56–59).

The finding that pharmacological enhancement of gap junctional coupling and myocardial intercellular synchronization (with or without changes in APD) (60) facilitates transformation of SVF into TVF implies that neither muscle mass nor the duration of APD determine the transient nature of VF. Thus, we hypothesized, that the potency of defibrillating drugs is determined by their ability to enhance or recover intercellular coupling or prevent intercellular electrical uncoupling. These effects are most likely brought about by decreasing $[Ca^{2+}]_I$ and preventing Ca^{2+} overload, which may occur secondarily to an increase of cAMP (61).

In an attempt to test this hypothesis in detail, a series of experiments (28,61–66) was performed that examined the effects of various compounds possessing defibrillating ability (like adrenaline, dbcAMP, tricyclic antidepressants, phenotiazines, and some class III antiarrhythmics), on intercellular myocardial coupling and synchronization as well as cytoplasmic free Ca^{2+} concentration.

The following results were obtained:

1. Pretreatment with the named compounds prevented Ca^{2+} overload and maintained diastolic intracellular Ca^{2+} ($[Ca^{2+}]_i$) at its normal level;
2. All compounds tested decreased previously elevated $[Ca^{2+}]_i$ (induced by increase of extracellular Ca^{2+} concentration ($[Ca^{2+}]_o$) toward its basal level, whereas there was no effect on normal $[Ca^{2+}]_i$;
3. These compounds prevented the high $[Ca^{2+}]_i$-induced uncoupling and desynchronization and restored the hypoxia-induced electrical or dye uncoupling;

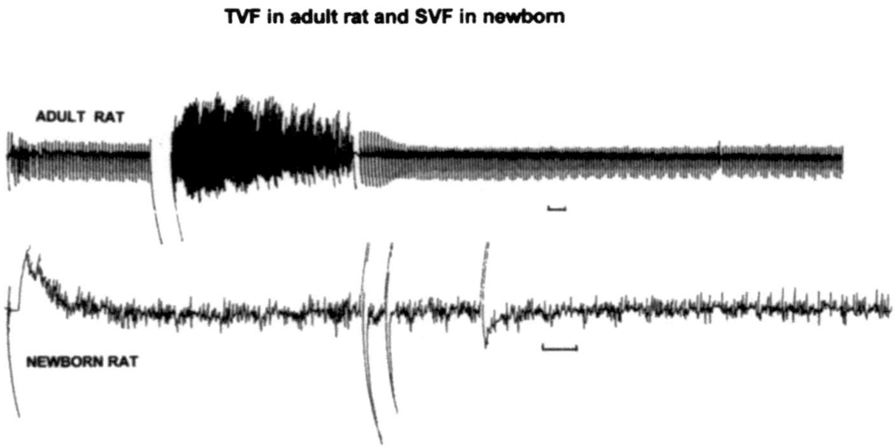

Figure 2. ECG record (LII) of VF electrically induced in adult (upper line) and 2 days old (lower line) rats. In adult rat VF terminated spontaneously in 12 sec, while in newborn on it sustained for >70 sec.

4. Several of these drugs were shown to increased myocardial cAMP levels (67), accelerated Ca^{2+} uptake by the sarcoplasmic reticulum (SR) (68) and exerted positive inotropic effects.
5. The named compounds increased intercellular synchronization and enabled the myocardium to act as a syncytium, to prevent spontaneous eruption of reentering wavelets and abolishing the continuity of reentry and fibrillation.

Although these sympathomimetic effects can explain many of the observed age and species differences, at least two questions remain to be answered:

First, since muscle mass and changes in APD can apparently be neglected as cause of spontaneous defibrillation, can we explain why dogs, pigs, and birds display sustained VF while rats, rabbits, guinea pigs and other species react with transient VF?

The second difficulty is to explain why new born rats and guinea pigs exhibit prolonged VF of about 90 sec. at the first day after birth, which decreases within few days (about 70 sec. during the second day) and TVF (shorter than 12 sec) after a week, although hearts of newborn animals are much smaller and have prolonged APD in comparison to more developed hearts which exhibit TVF (Figure 2).

THE NOVEL HYPOTHESIS

Following the assumption that TVF requires high level of sarcoplasmic (SR) Ca^{2+}-ATPase activity and gap junctional conductivity mediated by connexcin 43, we looked for a correlation between TVF, SR activity and myocardial structural variations. Searching the literature we found good correlation between SR Ca^{2+}-ATPase activity, cardiac myosin heavy chain (MHC) isoforms and the type of VF.

Ventricles of mammalian hearts contain mainly few types of myosin that differ structurally in their heavy chains and functionally in the Ca^{2+}-activated ATPase activity (69). According to Hoh et al (69) myosin from ventricular muscle contains in general three isoforms (V1–V3). These isoforms differ in their MHC composition. The V1 form is composed of two alpha MHC, V3 has two beta MHC and V2 has one alpha and one beta MHC. The V1 isomyosin that moves most rapidly during gel electrophoresis has a highest ATPase activity, whereas the slower moving one, V3, has the lowest ATPase activity (70,71). The high Ca^{2+} ATPase type V1 isomyosin and/or the lower ATPase type V3 occur in most of the species hearts while their relative amounts differ among mammals (72,73). Changes in the relative amount of myosin isozymes are responsible for alterations in the spread of muscle shortening (74). The ratio of these isomyosins is not static and varies according to age and species, as well as to physiological and pharmacological conditions (75). For example V3 is predominant during fetal life in all mammalian species (76). At or around birth there is a progressive appearance of V1, whereas its postnatal development varied with species (77). In rodent ventricles, V1 shifts with age toward V3. The range of age and amount in which V1 is observed (young age) vary between species. This range is prolonged in rats and very short, if any, in big animals. In adult big mammals (like dog, beef, pig and baboon) and in chicken only V3 was observed (72,78), while in humans V1 was found to range from 0 till 15% of total myosin and to vary from one heart to the other (75,78,79). There is a linear correlation between the myofibrillar myosin calcium-stimulated ATPase and percent V1 of total isoenzyme in normal hearts (80).

Increased expression of V3 was observed in various models of cardiac hypertrophy (81,82) as well as in hypothyroidism (83), whereas expression of V1 was predominant in hyperthyroidism, and could be increased in hypothyroid animals by treatment with thyroid hormones (84). An increased expression of V1 and concomitant decrease of V3 is associated with enhanced SR function.

Species	Age	Ventricular muscle mass	Type of VF	Heavy chain Isoenzymes*
Rat	young	small	TVF	only V1
	adult			V1 > V3
Guinea pig	young	small	TVF	V1 > V3
	adult		SVF	only V3
Rabbit	young	medium	TVF	V1 > V3
	adult		SVF	only V3
Cat	young	medium	TVF	no data
	adult		SVF	
Dog	young	big	TVF	V1 > V3
	adult		SVF	only V3
Pig	adult	big	SVF	only V3
Pigeon		small	SVF	only V3
Chicken		small	SVF	only V3

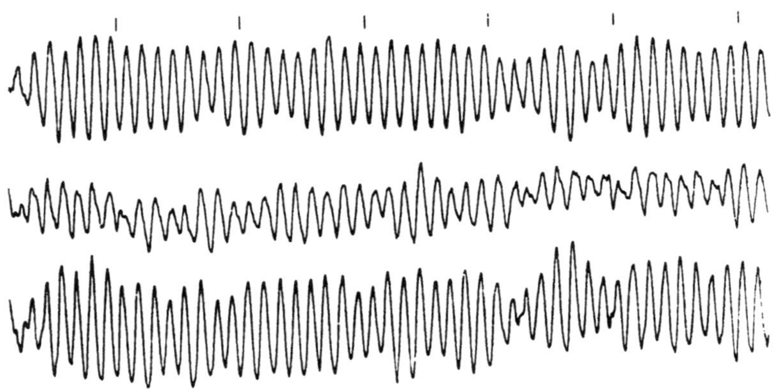

Figure 3. ECG (L I–III) recorded during electrically induced VF in untreated anaesthetized 7 day old pig.

The data depicted in Table 1 shows that TVF correlates well with the predominance of V1 and thereby with higher SR Ca^{2+} ATPase activity, whereas SVF correlates with the predominance of V3 and lower SR activity.

The observed correlations with the MHC isoforms can explain well the interspecies differences (rats, guinea pigs and rabbits that exhibit TVF versus dogs, pigs and pigeons that exhibit SVF). It can also explain the intra (age) differences and the variations in the length of the "young age". The fact that V3 is the only one isoform observed in mammals at birth can explain the previous mentioned unexpected finding that neonate rat (Figure 3) and guinea pig exhibit sustained or very prolonged VF despite of the newborn's heart is smaller in comparison to older rat and guinea pig that exhibit short TVF.

In view of this correlation it seems important to examine whether transformation of V3 into V1 or at least increase of alpha-MHC-induced SR activity can support self-ventricular defibrillation in animals like pig that normally have only V3 (78) and exhibit SVF.

Following Lompre et al. findings (76) that V1 appears in all mammals after birth (during the period of "young age"), we tried to examine the possibility for spontaneous defibrillation in very young pigs. Lompre (76) showed that V1 appears in very young pigs in a low percentage. Electrically induction of VF in young pigs at age of 7 and 12 days still exhibited SVF but with well synchronized ECG pattern (Figure 3) (in preparation). Our assumption was that V1 appears in pig at this age but probably in an insufficient SR activity like in adult cats, rabbits and guinea pigs. Therefore, following our previous findings (45) that pretreatment of adult cats, rabbits and guinea pigs with antidepressants that elevate cardiac catecholamine level exhibit TVF in spite of their low V1, we examined whether similar procedure would

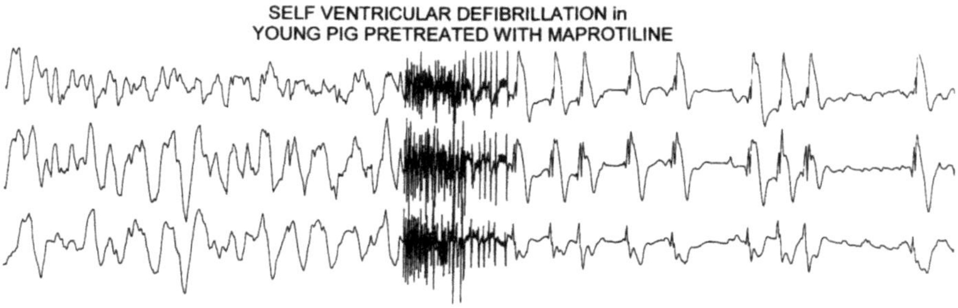

Figure 4. ECG (L I–III) recorded during spontaneous ventricular defibrillation in anaesthetized 7 days old pig pretreated with maprotiline (3 mg/kg body weight). The VF was electrically induced like in Figure 3.

support self-defibrillation in young pigs. To address this suggestion, we pretreated 7 days old pigs with maprotiline, an antidepressant compound known to inhibit catecholamine reuptake that thereby increases catecholamine level, cAMP and SR activity (45,56,60). The experiments revealed (in preparation) that this pretreatment transformed SVF into TVF (Figure 4) like in the adult mammals (6,45). Similar administration of the defibrillating drugs did not lead to a self-ventricular defibrillation in older pigs in which V1 probably does not exist.

The fact that treatment of animals with thyroid hormones T3 and T4 increased the expression of V1 or transforms a part of myocardial content of V3 into V1 (78,85) and increase the SR activity and gap junction conductivity (86) led us to explore whether administration of T3 and/or T4 can support self defibrillation in animals, like dogs and pigs, known to have only V3. The results of recent pilot experiment showed that pretreatment of pig (in preparation) with T3 or T4 did not lead to self defibrillation. However, administration of maprotiline to pigs pretreated with T3 or T4 supported the appearance of TVF or at least of self terminating torsade de points (Figure 5). The results fit well with the known interaction between thyroid hormones and sympathetic activity (87–90) associated with increased cAMP induced SR activity.

Our preliminary results suggest that V1 isoenzyme, associated with high SR and sympathomimetic activities, is necessary for ventricular synchronization, for prevention of VF and for self-ventricular defibrillation. In case that V1 amount and SR activity are sub-sufficient, a support of defibrillating compounds, cathecholamines and/or thyroid hormones, that increase SR activity are strongly required.

CONCLUSION

Based on the presently available evidence, we consider activity of SR Ca-ATPase as one of the main factors involved in spontaneous ventricular defibrillation. For this reason the present study was focused on the influence of the myosin isoenzyme V1 as a main predictor of the transient type ofVF. However, the correlation between

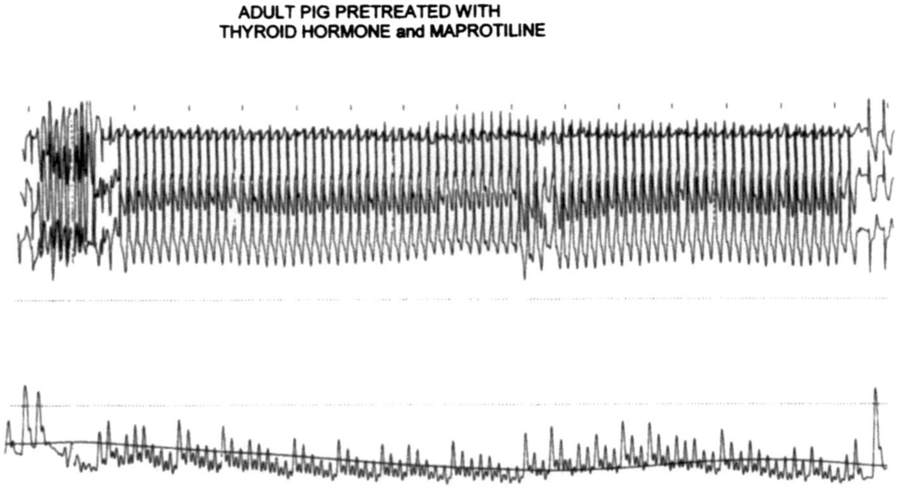

Figure 5. ECG (L I–III) recorded during spontaneous termination of ventricular fibrillo-flutter in 40 kg pig, pretreated with T3 (2 mg/kg p.o.) for 7 days and 5 mg/kg i.v. maprotiline 5 min before the induction of the arrhythmia. The fibrillo-flutter was induced like in Figure 3.

TVF and V1 expression does not exclude the possibility that other cellular components play a role in the ability of the heart to defibrillate spontaneously. As was pointed out before, the intercellular coupling-synchronization as well as dispersion of repolarization has great influence on spontaneous reversal of VF.

As the available evidence supports the correlation between TVF, amount of V1, SR activity and gap junctional conductivity C×43, it is important to address the question, whether or not it is possible to increase simultaneously V1 in heart with high risk for VF. Is it possible to achieve it in acute cases during arrhythmia and VF? Could it lead to alternative treatment for prevention of the fatal event of SVF?

These questions are now the base of our present studies. Future will tell whether this is fact or an epiphenomenon.

REFERENCES

1. Wiggers CJ. 1929. Studying of ventricular fibrillation caused by electrical shock. Am Heart J 5:351–365.
2. West TC, Laud JF. 1962. Minimal mass required for induction of a sustained arrhythmia in isolated atrial segment. Am J Physiol 202:232–236.
3. Moe GK, Rheinboldt WC, Abildskov JA. 1964. A computer model of atrial fibrillation. Am Heart J 67:200–220.
4. Kobrin VI, Kudinova ED. 1979. Electrical activity of intact myocardial cellsin different animals during normal activity and ventricular fibrillation. In: Proceedings of the International Symposium on Problems in Comperative Electrophysiology, abstract 36. Syktivkar, USSR.
5. Manoach M, Netz H, Erez M, Weinstock M. 1980. Ventricular defibrillation in mammals: age and drug dependence. Age Ageing 9:112–116.
6. Manoach M, Varon D, Neuman M, Netz H. 1987. Spontaneous termination and initiation of ventricular fibrillation as a function of heart size, age, autonomic autoregulation and drugs: A comparative study on different species of different age. Heart Vessels 2 (Suppl):56–68.

7. Clayton R, Higham D, Murray A, Campell R. 1992. Self terminating ventricular fibrillation. J Am Coll Cardiol (abst) 19:265A.
8. Shwartz SP, Orloff J, Fox C. 1949. Transient ventricular fibrillation. Am Heart J 37:21–35.
9. Evans TR, Curry PV, Fitchett DH, Krikler DM. 1976. Torsades de pointes initiated by electricular stimulation. J Electrocardiol 9:255–258.
10. Ledwich JR, Faye JE. 1969. Idiopathic recurrent ventricular fibrillation. Am J Cardiol 24:255–257.
11. Goble AJ. 1965. Paroxysmal ventricular fibrillation with spontaneous reversion to sinus rhythm. Brt Heart J. 27:62–68.
12. Sclarovsky S, Strasberg B, Lewin R, Agmon J. 1979. Polymorphus ventricular tachycardia: clinical features and treatment. Am J Cardiol 44:339–344.
13. Ranquin R, Parizel G. 1977. Ventricular fibrillo-flutter ('Torsade de pointes'): an established electrocardiographic and clinical entity, Angiology 28:115–118.
14. Kuo CS, Reddy CP, Munakata K, Surawicz B. 1985. Mechanism of electrical arrhythmias caused by dispersion of repolarization. Eur Heart J 6:63–70.
15. Spach MS, Kootsey JM, Sloan JD. 1982. Active modulation of electrical coupling between cardiac cells of the dog. A mechanism for transient and steady state variations in conduction velocity. Circ Res 51:347–362.
16. Spach MS, Dolber PC. 1986. Relating extracellular potentials and their derivates to anisotropic propagation at a microscopic level in human cardiac muscle: evidence for electrical uncoupliof side-to-side fiber connections with increasing age. Circ Res 58:356–371.
17. Disertori M, Inama G, Vergara G, Guarner M, Del-Favero A, Furlanello. 1983. Evidence of a reentry in the common type of atrial flutter in man. Circulation 67:434–440.
18. Mines GR. 1913. On dynamic equilibrium of the heart. J Physiol (Lon) 46:349–382.
20. Mines GR. 1914. On circulating excitations in heart muscle and their possible relation to tachycardia and fibrillation. Trans R Soc Can [Section IV], pp 43–52.
21. Wiener N, Rosenblueeth A. 1946. The mathematical formulation of the problem of conduction of impulses in a network of connected excitable elements, specifically in cardiac muscle. Arch Inst Cardiol Mex 16:205–265.
22. Moe GK. 1962. On the multiple wavelet hypothesis of atrial fibrillation. Arch Int Pharmacodyn Ther 140:183–188.
23. Rensma PL, Allssie MA, Laqmmers WJEP, Bonke FIM, Schalij MA. 1988. Length of excitationwave and susceptibility to reentrant atrial arrhythmias in normal conscious dogs. Cir Res 62:395–410.
24. Smeeth JLRM, Allessie MA, Lammers WJEP, Bonke FIM, Hollen J. 1986. The wavelength of the cardiac impulse and rarrhythmiasin isolated rabbit atrium: The role of heart rate, autonomic transmitters, temperature and potassium. Cir Res 58:96–108.
25. Singh BN. 1990. Historical development of the concept of controlling cardiac arrhythmias by lengthening repolarization: Particular reference to sotalol. Am J Cardiol 65:3A–11A.
26. Curtis MJ. 1994. Chemical Defibrillation in acute myocardial ischemia: A hypothesis. J Basical Clin Physiol Pharmacol 5:1–17.
27. Tribulova N, M. Manoach M. 1999. Is the Antiarrhythmic Effects of PA Related to Wavelength? Circulation 99:2220–2221.
28. Manoach M, Tribulova N, Varon D, Zinman T, Issack A, Shainberg A, Imanaga I. 1999. Are the antiarrhythmic defibrillating effects of d-sotalol due to or despite of prolongation of the action potential duration? Life Sci 65:273–278.
29. Hohnloser SH, Singh BN. 1995. Proarrhythmia with class III antiarrhythmicdrugs: definition, factors and clinical implication. J Cardiovasc Electrophysiol 6(10 Pt 2):920–936.
30. Wier G. 1986. Mechanisms of arrhythmogenic delayed and early afterdepolarization in ferret ventricular muscle. J Clin Invest 78:1185–1192.
31. Versailles JT, Verscheure Y, Le Kim A, Pourrias B. 1982. Comparison between the ventricular fibrillation threshold of spontaneously hypertensive and normotensive rats: investigation of antidysrhythmic drugs. J Cardiovasc. Pharmacol 4:430–435.
32. Tribulova N, Manoach M, Varon D, Okruhlicova L, Slobodova Z. 1999. Hypokalemia-induced ultrastructural, histochemical and connexin-43 alterations resulting in atrial and ventricular fibrillations. Gen. Physiol. Biophys 18(suppl.1):15–18.
33. Kowey PR, Friehling TD, Sewter J, Wu Y, Sokil A, Paul J, Nocella J. 1991. Electrophysiological effects of left ventricular hypertrophy: Effect of calcium and potassium blockade. Circulation 83:2067–2075.
34. Manoach M, Varon D, Erez M. 1995. The role of catecholamines on the intercellular coupling, myocardial cell synchronization and self vetricular defibrillation. Mol Cell Biochem 147:181–185.

35. Manoach M, Watanabe Y. 1995. How can we facilitate spontaneous termination of ventricular fibrillation and prevent sudden cardiac death? J Cardiovasc Electrophysiol 6:584–590.
36. Johanson BW. 1985. Ventricular repolarization and fibrillation threshold in hibernating species. Eur Heart J 6(Suppl D):53–62.
37. Bacaner M. 1966. Bretylium tosylate for suppression of induced ventricular fibrillation. Am J Cardiol 17:528–534.
38. Sana G, Arcidiacono R. 1973. Chemical ventricular defibrillation of the human heart with bretylium tosylate. Am J Cardiol 32:982–986.
39. Bacaner M. 1968. Treatment of ventricular fibrillation and acute arrhythmias with bretyliu tosylate. Am J Cardiol 21:530–544.
40. Bacaner M. 1968. Quantitative comparison of bretylium with other antifibrillatory drugs. Am J Cardiol 21:504–512.
41. Bekaily G, Payet MD, Benabderrazik M, Sauve R, Renand JF, Bacaner M, Sperelakis N. 1988. Intracellular bretylium blocks Na+ and K+ currents in heart cells. Eur J Pharmacol 151:389–397.
42. Manoach M, Beker B, Erez M, Varon D, Netz H. 1979. Spontaneous termination of electrical induced ventricular fibrillation. In: Progress in Electrocardiology. Ed McFarlain PW, 361–365. Pitman Med Pub, England.
43. Manoach M, Tager S, Erez M, Varon D, Vaugham-Williams EM. 1992. The defibrillating effect of high catecholamine level. J Mol Cell Cardiol (abst) 24: suppl V: S 69.
44. Manoach M, Kauli N, Netz H, Beker B, Assael M. 1979. Dibenzepin as an antifibrillatory agent for spontaneous termunating electrically induced ventricular fibrillation. Isr J Med Sci 15:443–447.
45. Manoach M, Varon D, Neuman M, Erez M. 1988. The cardioprotective features of tricyclic antidepressants. Gen Pharmacol 20:269–275.
46. Manoach M, Wyatt RF. 1985. Intracellular myocardial recordings in-vivo during sustained and transient ventricular fibrillation. In Advances in electrocardiology. Ed d'Alche P, 401–404. Univ Caen, France.
47. Manoach M, Netz H, Beker B, Kauli N. 1981. Vectorcardiographic discrimination between sustained and transient ventricular fibrillation. In New frontiers in electrocardiology, Eds De Padua F & McFarlane PW, 139–143. Wiley, New York.
48. Manoach M, Erez M, Varon D. 1993. Properties required for self ventricular defibrillation: Influence of age and drugs. Cardiol Elderly 1:337–344.
49. Safitz JE, Hoyt RH, Luke RA, Kanter HL, Beyer EC. 1992. Cardiac myocyte interconnections at gap junction. Trends Cardiovasc Med 2:56–60.
50. DeMello WC. 1975. Effect of intracellular injection of calcium and strontium on cell communication in heart. J Physiol (Lond) 250:231–245.
51. Burt JM. 1987. Block of intercellular communication: interaction of intracellular H^+ and Ca^+. Am J Physiol (Cell Physiol22) C607–C612.
52. Noma A, Tsuboi N. 1987. Dependence of junctional conductance on proton, calcium and magnesium ions in cardiac paired cells of guinea pig. J Physiol 382:193–211.
53. Spach MS, Miller WT, Geselowitz, Barr RC, Kootsey JM, Johnson EA. 1981. The discontinuous nature of propagation in normal canine cardiac muscle: evidence for reccurent discontinuites of intracellular resistance that affect the membrane currents. Circ Res 48:39–54.
54. Trosko JE, Madhukar BV, Chang CC. 1993. Endogenous and exogenous modulation of gap junctional intercellular communication: toxicological and pharmacological implications. Life Sci (1993) 53:1–19.
55. Spach MS, Dolber PC, Heidlage JF. 1998. Influence of the passive anisotropic properties on directional differences in propagation following modification of the sodium conductance in human atrial muscle. Evidence for reccurent discontinuities of intracellular resistance that affect the membrane currents. Circ Res 63:182–206.
56. Manoach M, Ben-Bassat M. 1990. The structural-functional basis of spontaneous ventricular defibrillation. Int J Cardiol 26:129–136.
57. Peters NS, Green CR, Poole-Wilson PA, Severs NJ. 1995. Cardiac arrhythmogenesis and the gap junction. J Mol Cell Cardiol 27:37–44.
58. Peters NS, Coromilas J, Severs NJ, Wit AI. 1997. Disturbed connexin 43 gap junction distribution correlates with location of circuits in the epicardial border zone of healing canine infarct that cause ventricular tachycardia. Circulation 95:988–996.
59. Peters NS, Coromilas J, Hanna MS, Josefson ME, Costaes C, Wit AI. 1998. Characteristics of temporal and spacial excitable gap in anisotropic reentrant circuits causing sustained ventricular tachycardia. Cir Res 82:279–297.

60. Manoach M, Varon D, Neuman M, Erez M. 1988. A new group of defibrillatory drugs in the classification of antiarrhythmic agents. Intern J Cardiol 21:211–217.
61. Miyachi EI, Manoach M, Uchiyama H, Watanabe Y. 1995. Is cyclic AMP involved in the defibrillating effect of sotalol? Life Sci 57:PL 393–399.
62. Manoach M, Uchiyama H, Miyachi EL, Thormahlen D. 1997. Tedisamil enhances intercellular coupling: A new explanation of its antiarrhythmic defibrillating effects. Exp.Clin Cardiol 2:47–50.
63. Uchiyama H, Manoach M, Miyachi EI, Watanabe Y. 1995. Sotalol facilitates spontaneous ventricular defibrillation by enhancing intercellular coupling. An entirely new mechanism for its antiarrhythmic action. Heart Vessels 10:185–189.
64. Manoach M, Tribulova N, Imanaga I. 1996. The protective effect of d-sotalol against hypoxia induced myocardial uncoupling. Heart Vessels 11:281–288.
65. Manoach M, Varon D, Shainberg A, Zinman T, Isaack A, Halili-Rutman I, Kaplan D, Tribulova N. 1997. The protective effect of clss III antiarrhythmic agents against calcium overload in cultured myocytes.Life Sci 61:PL 227–234.
66. Manoach M, Varon D, Shainberg A, Zinman T, Issac A, Tribulova N. 1997. The antiarrhythmic defibrillatory effect of d-sotalol in monolayer of rat cardio myocytes. In: Electrocardiology '97. Ed. L Bacharova and PW Macfarlane, 8–13. World Scientific Publ London.
67. Tribulova N, Ravingerova T, Okruhlicova L, Gabauer I, Fickova M, Manoach M. 1998. Tedisamile modulates cAMP level in normal and ischemic guinea pig hearts. J Mol Cell Cardiol 30:A 68 (abs).
68. Manoach M, Varon D, Shaiberg A, Zinman T, Kaplan D, Khananshvili D, Tribulova N. 1999. The role of sarcoplasmic reticulum in the protective effect of class III drugs against Ca2+ overload. Gen Physiol Biophys. 18(suppl 1):14–24.
69. Hoh JFY, McGiath PA, Hale RT. 1977. Electrophoretic analysis of multiple forms of cardiac myosin: Effect of lipopolysectomy and thyroxine replacement. J Mol Cell Cardiology 10:1053–1076.
70. Schwartz K, Lecarpentier Y, Martin JL, Lompre AM, Mercandier JJ, Swynghedauw B. 1981. Myosin isoenzyme distribution correlates with speed of myocardial contraction. J Mol Cell Cardiol 13:1071–1075.
71. Pope B, Hoh JFY, Weeds A. 1980. The ATPase activity of rat cardiac myosin isoenzymes FEBS Letters 118:205–208.
72. Clark WA, Chizzonite RA, Everett AW, Rabinowitz M, Zak R. 1982. Species correlation between cardiac isomyosins. J Biol Chem 257:5449–5454.
73. Carey R, Bove A, Coulson R, Spann J. 1979. Correlation between cardiac myosin ATPase activity and velocity of muscle shortening. Biochem Med 21:235–245.
74. Pagani ED, Julian FJ. 1984. Rabbit papillary muscle myosin isoenzymes and the velocity of muscle shortening. Cir Res 54:586–594.
75. Mercadier JJ, Bouveret P, Gorza L, Schiaffino S, Clark WA, Zak R, Swynghedauw B, Schwartz K. 1983. Myosin isoenzymes in normal and hypertrophied human ventricular myocardium. Circ Res 53:52–62.
76. Lompre AM, Mercadier JJ, Wisnewsky C, Bouveret P, Pantaloni C, D'Albis A, Schwartz K. 1981. Species and age dependent changes in the relative amounts of cardiac myosin isoenzymes in mammals. Developmental Biol 84:286–290.
77. Chizzonite RA, Everett AW, Clark WA, Jakovcic S, Rabinowitz M, Zak R. 1982. Isolation and characterization of two molecular variants of myosin heavy chain from rabbit ventricles. Change in their content during normal growth and after treatment with thyroid hormone. J Biol Chem 257:2056–2065.
78. Morkin E. 1993. Regulation of myosin heavy chain genes in the heart. Circulation 87:1451–1460.
79. Miyata S, Minobe W, Bristow MR, Leinwand LA. 2000. Myosin heavy chain isoform expression in the failing and nonfailing human heart. Cir Res 86:386–390.
80. Litten RZ, Martin BJ, Buchthal RH, Nagani R, Low RB, Alpert NR. 1985. Heterogeneity of myosin isozyme content of rabbit heart. Cir Res 57:406–414.
81. Alpert NR, Mulieri LA. 1981. Heat, mechanics and myosin ATPase in normal and hypertrophied heart muscle. Fed Proc 41:192–198.
82. Lompre AM, Schwartz K, d'Albis A, Lacombe G, Van Thiem N, Swynghedaauw B. 1979. Myisin Isoenzyme redistribution in chronic hear overload. Nature 282:105–107.
83. Schwartz K, Lecarpentier Y, Martin JL, Lompre AM, Mercadier JJ, Swynghedauw B. 1981. Myosin isoenzymic distribution correlates with speed of myocardial contraction. J Mol Cell Cardiol 13:1071–1075.

84. Cappelli V, Moggio R, Polla B, Bottinelli R, Poggesi C, Reggiani C. 1988. The dual effect of thyroid hormones on contractile properties of the rat myocardium. Pflugers Arch 411:620–627.
85. Ojamaa K, Klemperer JD, McGilvray SS, Klein I, Samarel A. 1996. Thyroid hormone and hemodinamic regulation of beta-myosin heavy chain promoter in the heart. Endocrinology 137:802–808.
86. Deker RS. 1976. Hormonal regulation of gap junction differentiation. J cell Biol 69:669–685.
87. Waldstein SS. 1966. Thyroid-catecholamine interactions. Ann Rev Med 17:123–132.
88. Levey GS. 1971. Catecholamine sensitivity, thyroid hormone and the heart. Am J Med 50:413–420.
89. Levey GS, Klein I. 1990. Catecholamine-thyroid hormone interactions and the cardiovascular manifestations of hyperthyroidism. Am J Med 88:642–646.
90. Polikar R, Burger AG, Scherrer U, Nicod P. 1993. The thyroid and the heart. Circulation 87:1435–1441.

Pathophysiology of Cardiovascular Disease,
edited by N.S. Dhalla, H. Rupp,
A. Angel & G.N. Pierce
Kluwer Academic Publishers, Boston, 2004

Cardiac ATP-Sensitive Potassium Channel: A Bi-Functional Channel/Enzyme Multimer

Alexey E. Alekseev, Martin Bienengraeber, Leonid V. Zingman, Petras P. Dzeja, and Andre Terzic

Division of Cardiovascular Diseases
Departments of Medicine, Molecular Pharmacology and Experimental Therapeutics
Mayo Clinic, Mayo Foundation
Rochester, MN
USA

Summary. Maintenance of myocardial homeostasis critically depends on the ability of cardiac cells to adapt energy-dependent processes in response to metabolic challenge. This requires efficient monitoring of the cellular energetic status, secure delivery of information to energetic sensors and accurate translation of metabolic signals into cellular response. Although advances have been made in resolving the molecular identity of energy-response elements, mechanisms integrating metabolic sensor function with cellular metabolism are largely unknown. Sarcolemmal ATP-sensitive K^+ (K_{ATP}) channels are prototypic membrane metabolic sensors, which act as bi-functional protein multimers combining catalytic with ion conduction properties. In the hetero-octameric K_{ATP} channel complex, the sulfonylurea receptor (SUR2A) harbors an intrinsic ATPase activity and confers fine nucleotide modulation of K^+ permeation through the pore-forming Kir6.2 subunit. The metabolic sensor role of K_{ATP} channels stems from the non-equivalent properties of nucleotide binding domains (NBD1 and NBD2) in SUR2A. NBD1 binds nucleotides whereas NBD2 hydrolyzes ATP, yet cooperative interaction, rather than the independent contribution of each NBD, is critical for K_{ATP} channel regulation. The ATP hydrolysis cycle at SUR2A drives conformational transitions with distinct outcomes on channel gating to adjust membrane potential in response to intracellular metabolic oscillations. Nucleotide exchange between the channel ATPase and the cellular phosphotransfer network provides a mechanistic basis for coupling cell energetics with K_{ATP} channel gating. Disruption of phosphotransfer genes produces aberrant metabolic sensing

Correspondence should be addressed to Andre Terzic, MD, PhD, Mayo Clinic, Guggenheim 7, 200, First Street SW, Rochester, MN 55905, USA, Tel.: (507) 284-2747, Fax: (507) 284-9111, E-mail: terzic.andre@mayo.edu.

by K_{ATP} channels, and generates a cellular phenotype with electrical vulnerability to metabolic challenge. Assigning to the channel catalytic module a role in integrating ion permeation with intracellular enzymatic pathways identifies a novel principle in the regulation of cellular excitability, and defines an operational paradigm for channel/enzyme multimers.

***Key words*:** K_{ATP} channel, ATPase, nucleotide binding domains, metabolic sensor, phosphotransfer enzymes.

INTRODUCTION TO ATP-SENSITIVE K^+ CHANNELS

Plasmalemmal ATP-sensitive K^+ (K_{ATP}) channels are membrane metabolic sensors, which act as alarm systems to adjust cellular electrical activity in response to oscillations in cellular metabolism (1–9). K_{ATP} channels are expressed in high density in metabolically active tissues, in particular heart muscle where they were originally discovered two decades ago (10). Central to metabolic sensing is their tetradimeric composition, where four ion conducting subunits physically associate with respective regulatory modules into a functional hetero-octameric K_{ATP} channel complex (11–13; Figure 1).

The inwardly-rectifying K^+ channel, Kir6.2, serves most commonly as the pore-forming core for K_{ATP} channels (14). The Kir6.2 protein contains two transmembrane domains which flank the pore region (Figure 1). Although deletion of the very carboxy-terminus of Kir6.2 (13,15) or its over-expression (16) may give rise to measurable K_{ATP} channel current, it is through physical assembly with the regulatory module, recognized as the sulfonylurea receptor or SUR (Figure 1), that the full properties of tissue-specific channel phenotypes are generated (14,17–19). It is in this way that Kir6.2 and SUR1 genes, which are clustered on chromosome 11 (11p15.1), are cotranscribed and cotranslated to produce K_{ATP} channels in insulin-secreting pancreatic β-cells (14,17,20). Although mapped to a chromosome (12p11.12) distinct from that of Kir6.2 (21), SUR2A, an isoform which shares a 68% identity with SUR1 (18,22), when physically associated with Kir6.2 forms the cardiac K_{ATP} channel complex (23). The splice variant, SUR2B, pairs with Kir6.1, an inwardly-rectifying potassium channel related to Kir6.2, to generate vascular K_{ATP} channels (19,24,25).

In the cardiovascular system, K_{ATP} channels contribute to the cardioprotective mechanism of ischemic preconditioning and to regulation of vascular tone (26–29). Ischemic preconditioning reduces the infarct size in wild-type, but not in Kir6.2-knockout mice lacking functional cardiac K_{ATP} channels (27). In fact, following global ischemia/reperfusion, the increase of left ventricular end-diastolic pressure during ischemia is more marked, and the recovery of contractile function worse in Kir6.2-knockout than wild-type hearts underscoring the role of sarcolemmal K_{ATP} channels in modulating ischemia/reperfusion injury (27). It has now been established that opening of sarcolemmal K_{ATP} channels underlies ST elevation during ischemia, which is lost in hearts lacking Kir6.2 (28). Moreover, mice lacking the gene encoding Kir6.1 have a high rate of sudden death associated with spontaneous ST elevation followed by atrioventricular block, and display a phenotype

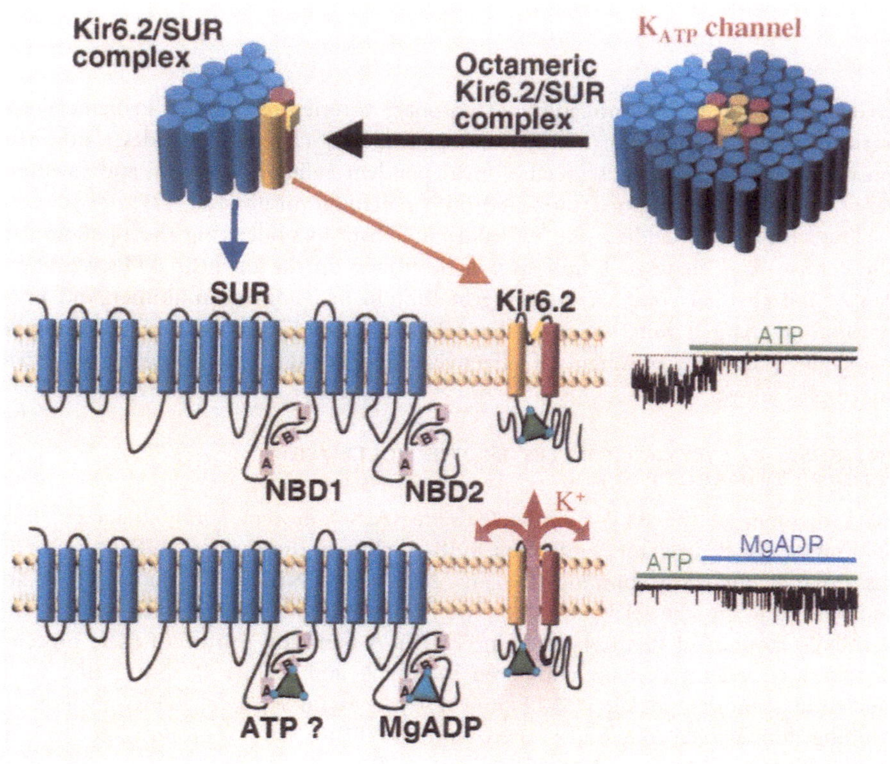

Figure 1. The K_{ATP} channel complex is an octamer composed of four pore-forming Kir6.2 subunits and four associated regulatory SUR subunits. Kir6.2 is formed by two transmembrane domains flanking a pore region. SUR is characterized by two cytosolic nucleotide-binding domains (NBD1 and NBD2) which contain highly-conserved Walker motifs (A and B) and a linker (L) region. Interaction of ATP (green triangle) with Kir6.2 induces pore closure. Upper channel record demonstrates ATP-induced K_{ATP} channel inhibition. MgADP (blue triangle) at NBD2 antagonizes ATP-induced pore inhibition, with an apparent requirement of ATP at NBD1. Lower channel record demonstrates MgADP-induced reduction of ATP-induced K_{ATP} channel inhibition leading to channel opening.

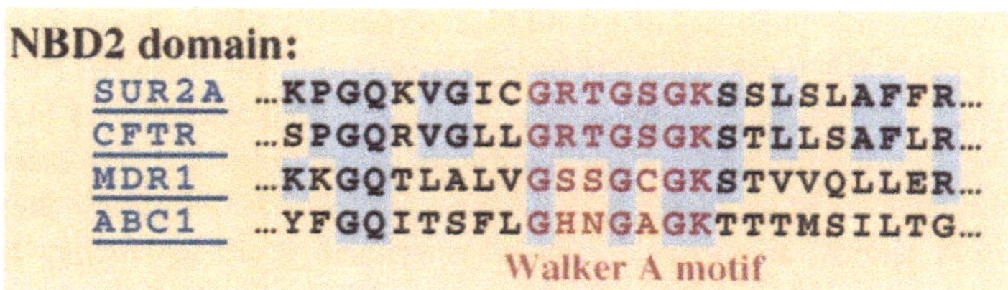

Figure 2. The cardiac SUR isoform, SUR2A, contains sequences with high homology within the nucleotide-binding domains of related ATP-binding cassette proteins, such as the cystic fibrosis transmembrane regulator (CFTR), the multi-drug resistance protein (MDR1) and the lipid transorting ATP-binding cassette transporter 1 (ABC1). Shown is the conserved Walker A motif (in bold) of the second nucleotide-binding domain of CFTR, MDR1 and ABC1 aligned with the corresponding region of SUR2A. Sequences are from human cDNAs.

characterized by hypercontractility of coronary arteries resembling Prinzmetal (or variant) angina in humans (29). Thus, targeting K_{ATP} channels provides a powerful mean to regulate numerous metabolism-dependent cellular functions, and to interfere with disease conditions associated with metabolic insults (30–34).

This chapter will address the molecular mechanism underlying the operation of the cardiac K_{ATP} channel. Emphasis will be placed on the intrinsic ATPase properties of the channel complex, signaling in the channel/enzyme multimer, and integration of channel function into the cellular metabolic network. The presented principles provide a paradigm of effective energetic signal decoding by a cardiac metabolic sensor.

INTRINSIC CATALYTIC ACTIVITY IN THE NUCLEOTIDE BINDING DOMAIN OF SUR

As a member of the ATP-Binding Cassette (ABC) protein family, the regulatory subunit of the K_{ATP} channel complex SUR, including the cardiac SUR2A isoform, possess two nucleotide binding domains, NBD1 and NBD2 (Figure 1). These are located between the eleventh and twelfth transmembrane regions and at the carboxy-terminus of the SUR protein (14,17). A common feature of ABC proteins is that such NBDs contain conserved Walker A and Walker B motifs that form nucleotide-binding pockets (35; Figure 1). In certain ABC proteins, nucleotide-binding domains have been associated with an ATPase activity proposed to serve as a switch between ATP- and ADP-liganded conformations, implicated in supporting transport function, ion conductance or DNA repair (36,37; Figure 2). In the K_{ATP} channel complex, interaction of ATP with Kir6.2 prevents ion permeation through the pore, whereas metabolic gating proceeds through interactions of MgATP/MgADP with NBDs of SUR modulating the ATP-sensitivity of Kir6.2 (5,6,15,38–40; Figure 1). However, the similar affinity of NBD2 for MgATP and MgADP (41) suggests that nucleotide competition at the binding site is not solely responsible for K_{ATP} channel regulation in a cellular environment with exceeding ATP over ADP levels. Therefore, rather than ligand competition *per se*, an ATPase-driven transition between nucleotide-liganded states of SUR may provide a more effective governance of channel gating (4,39,42,43).

In this regard, the presence of an ATPase activity in SUR stems from findings that α- but not γ-labeled ATP could be detected bound to the NBD2 fragment of SUR (42). In addition, mutations of key residues in Walker motifs of SUR, which precluded nucleotide binding and/or hydrolysis in other ABC transporters, do alter the responsiveness of K_{ATP} channels to adenine nucleotides (44,45), further suggesting that SUR harbors an intrinsic catalytic activity involved in channel regulation. In fact, exposed to γ-labeled [^{32}P]ATP, purified NBDs of SUR2A display an ATPase activity directly measured by monitoring the generation of [^{32}P]P_i (43; Figure 3A). The catalytic activity is primarily identified in NBD2, and to a lesser extent if at all in NBD1 (43; Figure 3A and 3B). The rate of the ATPase reaction in NBD2 of SUR2A is within the range of values reported for the ATPase activity of other ABC proteins (43). The ATPase activity of NBD2 is insensitive to inhibitors of

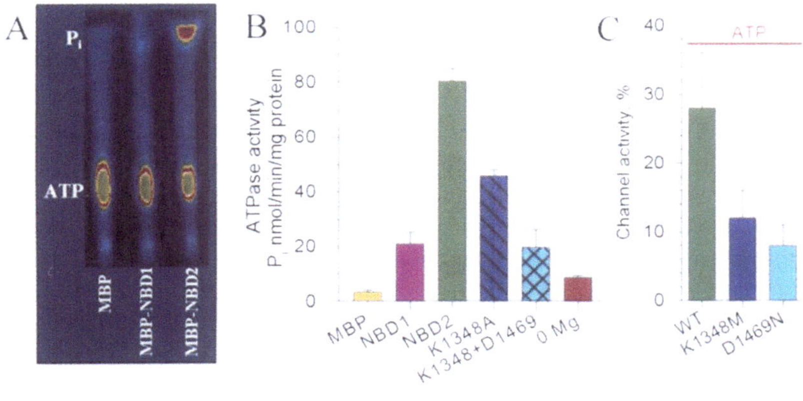

Figure 3. ATPase activity harbored in SUR regulates ATP-dependent K_{ATP} channel gating. **(A)** NBD1 (Ser[684]-Ser[884]) and NBD2 (Gly[1306]-Thr[1498]) of SUR2A were subcloned into pMal-c2 plasmid in-frame with maltose binding protein (MBP). Generation of $[^{32}P]P_i$ by purified NBD constructs from $[\gamma\text{-}^{32}P]$ATP was analyzed by thin-layer chromatography, followed by autoradiography with an image analyzer. Lane 1: MBP alone; lane 2: MBP-NBD1; lane 3: MBP-NBD2. **(B)** ATPase activity in NDB2 was significantly higher than the activity of MBP alone or NBD1. Lack of Mg^{2+} (0 Mg) as well as single (K1348A) or double mutations (K1348A + D1469N) in the Walker B motif of NBD2 significantly reduced the ATPase activity. **(C)** Sensitivity of recombinant wild-type and mutant cardiac K_{ATP} channels toward ATP. At 100 µM ATP, wild-type Kir6.2/SUR2A was inhibited by ~70%, whereas Kir6.2 co-expressed with mutant K1348M or D1469N SUR2A was inhibited by ~90%. Currents in each condition were obtained in inside-out patches, following expression in oocytes. Adapted from ref. 43.

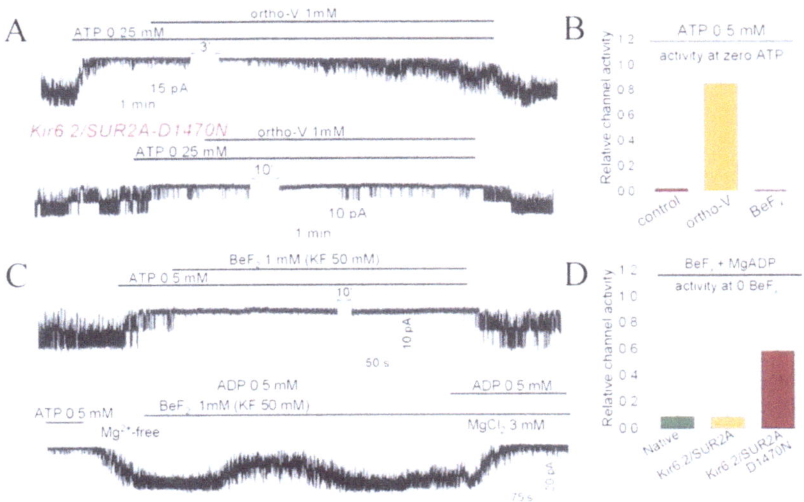

Figure 4. Regulation of the K_{ATP} channels by γ-phosphate analogs. **(A)** Orthovanadate (ortho-V) activated recombinant wild-type Kir6.2/SUR2A channels expressed in COS-1 cells (*upper trace*). Mutation of aspartate to asparagine (D1470N) in the NBD2 Walker B motif of SUR2A prevented the effect of ortho-V on K_{ATP} channel opening (*lower trace*). **(B)** Average effect of ortho-V and beryllium-fluoride (BeF_x) on ATP-inhibited K_{ATP} channel activity relative to the value obtained in the absence of ATP. Records are from native cardiac K_{ATP} channels in the open cell-attached configuration. **(C)** Beryllium-fluoride (BeF_2 + KF 50 mM = BeF_x) did not activate native K_{ATP} channels in cardiomyocytes (*upper trace*). BeF_x plus ADP, in the presence but not in the absence of magnesium, inhibited native K_{ATP} channels (*lower trace*). **(D)** The D1470N mutation in SUR2A prevented the effect of MgADP·BeF_x. Channel activity is presented relative to that measured in the absence of BeF_x and nucleotides. Adapted from ref. 49.

conventional ATPases and phosphatases, ouabain, oligomycin, thapsigargin and/or levamisole, but removal of Mg^{2+}, a co-factor in the ATPase reaction, reduces the hydrolytic activity (43; Figure 3B). Site-directed mutagenesis of the lysine residue (K1348A), in the signature Walker A motif of NBD2, along with mutation in the Walker B aspartate[1469] (D1469N) suppresses the ATPase activity (43; Figure 3B). These mutations in Walker A and B motifs produce K_{ATP} channels with a higher sensitivity to ATP compared to wild-type Kir6.2/SUR2A suggesting that transition from MgATP- to MgADP-liganded states (from high to low ATP sensitivity of Kir6.2, respectively) is catalyzed by an ATPase inherent to the channel complex (43; Figure 3C). Thus, the discovery of an intrinsic ATPase activty in the regulatory SUR module qualifies the K_{ATP} channel as a member of bi-functional protein complexes, recently recognized to combine enzymatic and ion conductance properties (36,43,46–49).

SIGNALING IN THE CHANNEL/ENZYME MULTIMER: ATPase TRANSITIONS IN SUR GATE K_{ATP} CHANNELS

Most recently, linkage between catalysis at SUR and ion permeation through the K_{ATP} channel pore was defined (49). Coupling of discrete conformations in the SUR ATPase cycle with channel gating was solved using nucleotide trapping procedures in conjunction with current recording to monitor in real-time the outcome of transitions during ATP hydrolysis on channel behavior (49). Capture of intermediates was achieved using γ-phosphate (P_i) analogs, orthovanadate (V_i) and beryllium-fluoride (BeF_x), which readily undergo changes in coordination geometry and stabilize the ATPase cycle in distinct conformations (36,50,51):

$$E + \text{MgATP} \leftrightarrow E^{\star}\cdot\text{MgATP} \leftrightarrow E^{\star\star}\cdot\text{MgADP}\cdot\text{P}_i \leftrightarrow E^{\star\star\star}\cdot\text{MgADP} + \text{P}_i \leftrightarrow E + \text{MgADP} + \text{P}_i$$

where E is the ATPase molecule and stars indicate transitional states during ATP hydrolysis (52). Orthovanadate forms a pentavalent pyramidal structure, like γ-phosphate in ATP undergoing hydrolysis ($E^{\star\star}\cdot\text{MgADP}\cdot\text{P}_i = E^{\star\star}\cdot\text{MgADP}\cdot\text{V}_i$), and stabilizes posthydrolytic states (53). Prehydrolytic states are distinguished using beryllium-fluoride which forms a tetrahedral structure, and mimics γ-phosphate in ATP prior to hydrolysis ($E^{\star}\cdot\text{MgATP} = E^{\star}\cdot\text{MgADP}\cdot\text{BeF}_x$; 54).

Accordingly, both γ-phosphate analogs suppress the ATPase activity in the NBD2 of SUR2A (49). Arrest of the ATPase cycle with orthovanadate is associated with reversal of ATP-induced K_{ATP} channel inhibition leading to vigorous channel opening (49; Figure 4A and 4B). Orthovanadate fails to activate K_{ATP} channels in the absence of divalent cations, essential co-factors in the ATPase reaction necessary for coordination of trapped nucleotides at the catalytic site (49). Moreover, a point mutation in the Walker B motif of NBD2, D1470N, which prevents coordination of magnesium at the catalytic site and decreases ATPase activity in NBD2 (43), abolishes the effect of orthovanadate on K_{ATP} channel gating (49; Figure 4A). Preventing intrinsic ATPase cycling, by replacement of ATP with non-hydrolyzable analogs, diadenosine pentaphosphate or AMP-PNP, also precludes orthovanadate-induced K_{ATP} channel opening (49). In contrast to orthovanadate, beryllium-fluoride which arrests the ATPase cycle in a prehydrolytic conformation does not reduce the sensitivity of K_{ATP} channels to ATP (49; Figure 4B and 4C). Rather, channels that are active in the presence of ADP are inhibited by complexing of beryllium-fluoride with MgADP at SUR (49; Figure 4C). The inhibitory effect of MgADP·BeF_x requires not only the divalent cation, but also an intact NBD2 ATPase and is abolished by the D1470N mutation in SUR (49; Figure 4C, 4D). Thus, formation of SUR★★·MgADP·V_i, a posthydrolytic transition state, translates into positive gating characterized by a reduced sensitivity of K_{ATP} channels to ATP (49). Stabilization of SUR★·MgADP·BeF_x, a prehydrolytic conformation (SUR★·MgATP), enhances closure of the K_{ATP} channel defining negative channel gating (49; Figure 6A).

Therefore, an intact NBD2 ATPase function is required for recruitment of transitional conformations driving ATP-dependent gating of the K_{ATP} channel complex (Figure 6A). In addition, an intact NBD1 is mandatory for NBD2 ATPase-dependent K_{ATP} channel gating (40). Stabilization of ATP at NBD1 depends on, and simultaneously promotes engagement of NBD2 into a MgADP-bound conformation required to counteract ATP-induced pore closure (40). Thus, rather than individual components of the regulatory subunit, it is the functional tandem formed by NBD1 and NBD2 that drives SUR-mediated nucleotide-dependent gating of the K_{ATP} channel complex (40).

In contrast to transporters that spend the energy of hydrolysis to transport compounds against a chemical gradient (55), ion channels allow movement of ions down an electrochemical gradient. As this does not require energy input, the ATPase in channel proteins may not fuel the process of ion permeation, but rather serve as an internal signaling system contributing to channel gating (49). In fact, when MgATP is continuously hydrolyzed at NBD2 of SUR, ATP interacting with Kir6.2 in a magnesium-independent manner keeps the pore closed (49). It is only arrest of the SUR ATPase in a posthydrolytic conformation that antagonizes ATP-induced pore closure (49). This may indicate that the energy of hydrolysis is utilized to support ATP-induced K_{ATP} channel inhibition. However, this is not the case as channel closure can be induced by arrest of the ATPase cycle in a prehydrolytic conformation (49). Rather, the ATPase supports transit through a spectrum of conformations required for transmission of different signals to the K_{ATP} channel pore *via* a single transduction pathway mediated by the SUR unit. Selection of a specific conformation from the spectrum at the regulatory unit would ultimately define channel behavior. Thus, the spectrum of ATPase-driven conformational intermediates, with different potential outcomes on channel gating, endows SUR with positive or negative regulation of ion permeation through K_{ATP} channels.

ATPase CYCLE AND K_{ATP} CHANNEL REGULATION BY POTASSIUM CHANNEL OPENERS

The ability of SUR to secure pluripotent gating of the K_{ATP} channel pore can be exploited by modulating the lifetime of specific conformations during the ATPase

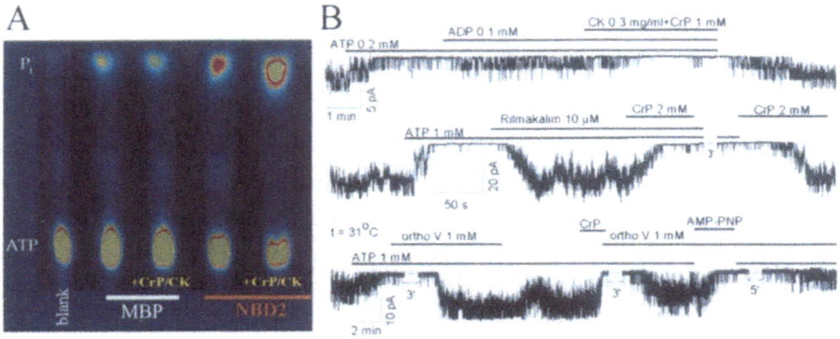

Figure 5. Modulation of SUR ATPase activity by the creatine kinase reaction (ADP + CrP ↔ ATP + Cr). **(A)** Creatine phosphate/creatine kinase (+CrP/CK) promoted hydrolysis of γ-labeled [^{32}P]ATP in NBD2 constructs, but not in MBP alone. **(B)** *Upper trace*: In inside-out patches, ADP-induced opening of cardiac K_{ATP} channels was reversed following application of exogenous creatine kinase activated by creatine phosphate. *Middle trace*: The potassium channel opener, rilmakalim, vigorously activates ATP-inhibited K_{ATP} channels, an effect reversed through activation of the ADP-scavenging creatine kinase system following application of the reaction substrate, creatine phosphate (CrP) in open-cell attached patches. Under this condition, endogenous creatine kinase remains within the cell amenable to regulation. *Lower trace*: Ortho-V (1 mM) induced opening of K_{ATP} channels (in the presence of 1 mM ATP) in open-cell attached patches was abolished on removal of the ATPase product, ADP, by intracellular creatine kinase activated using 2 mM creatine phosphate (CrP). Preventing ATPase-induced ADP formation by replacement of ATP with the non-hydrolyzable analog AMP-PNP also antagonized ortho-V induced channel opening. Adapted from ref. 43 and 49.

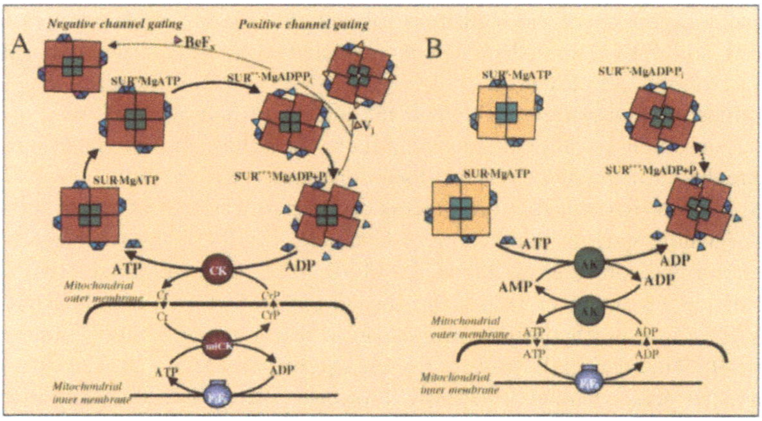

Figure 6. Conformations in the SUR ATPase cycle associated with closure and opening of K_{ATP} channels. A "bird's eye view" of the hetero-octameric K_{ATP} channel complex, composed of four Kir6.2 pore-forming and four SUR regulatory subunits, undergoing conformational transitions during the ATPase cycle. **(A)** Distinct transitional states during ATP hydrolysis can be captured by γ-phosphate analogs in the presence of MgADP. Beryllium-fluoride (BeF_x) stabilizes a prehydrolytic conformation associated with negative channel gating. Orthovanadate (V_i) stabilizes a posthydrolytic conformation associated with positive channel gating. Removal of ADP by the creatine kinase system (cytosolic CK and mitochondrial miCK) facilitates ATPase cycling, providing cross-talk between the intrinsic channel catalytic activity and cellular enzymatic pathways. High intracellular levels of creatine phosphate support ADP to ATP conversion at the channel site, and promote initiation of the ATPase cycle keeping K_{ATP} channels closed. A drop of intracellular creatine phosphate, as it occurs under metabolic stress, reduces the ability of creatine kinase to scavenge ADP. This would facilitate stabilization of the MgADP-bound posthydrolytic conformation and K_{ATP} channel opening. For simplicity of this scheme, the ATPase activity is presented as a collective property of the SUR module without ascribing specific functions to individual nucleotide binding domains. **(B)** Under metabolic stress, uncoupling of mitochondria promotes the F_1F_o-ATPase activity which through adenylate kinase (AK) favors ADP accumulation at the channel site increasing the lifetime of the SUR posthydyrolytic conformation associated with K_{ATP} channel opening.

cycle. Potassium channel openers (KCO) are powerful K_{ATP} channel activators that bind to SUR (56–59), and modulate channel ATPase activity (43). Mutations in Walker motifs of NBD2, that reduce the ATPase activity, diminish the efficacy of potassium channel openers to activate K_{ATP} channels (39,43,58,59). In fact, it has been proposed that KCO stabilize SUR in a MgADP bound state (39,44).

Using γ-phosphate analogs to dissect transitions associated with K_{ATP} channel gating, the molecular pharmacodynamics of pinacidil, a prototype KCO (60) was demonstrated (49). Arrest of the ATPase cycle, by beryllium-fluoride in a prehydrolytic transition state, prevents pinacidil-induced K_{ATP} channel activation. Trapping the ATPase cycle in a posthydrolytic conformation by orthovanadate does not prevent but rather potentiates the effect of the KCO (49). The efficacy of pinacidil, lost during blockade of the ATPase cycle by the non-hydrolyzable AMP-PNP, is rescued with bypass of inhibited hydrolysis by supplying the ATPase product, MgADP (49). Thus, entry of SUR into posthydrolytic conformations plays a permissive role in mediating the action of potassium channel openers. In a normally operating ATPase cycle, transit through the posthydrolytic state allows potassium channel openers to target this conformation. In turn, conformational changes during ATP hydrolysis may define the affinity of ligands for SUR, as proposed for the ATPase cycle in other ABC proteins (61,62). Specifically, the slow off-rate of openers in the presence of magnesium nucleotides (59) suggests that engagement in the posthydrolytic conformation may stabilize potassium channel opener binding. Thus, KCO-induced K_{ATP} channel opening relies on the lifetime of posthydrolytic conformations in the SUR ATPase cycle.

CELLULAR METABOLISM REGULATES ATPase CYCLING AT SUR AND ASSOCIATED K_{ATP} CHANNEL GATING

Identification of conformations during ATP hydrolysis associated with K_{ATP} channel gating raises the question of integration of this catalytic cycle in cellular metabolic sensing. The product of ATP hydrolysis, MgADP, suppresses ATPase activity at NBD2 (49). In principle, cytosolic ADP, which reaches the hundred-micromolar range under metabolic stress (63,64), would be sufficient to decelerate the ATPase cycle prolonging the lifetime of posthydrolytic states (49). However, in the highly compartmentalized intracellular environment of the cardiomyocyte, membrane ATPases constantly reduce local ATP concentration setting the submembrane ATP/ADP ratio distinct from that of the "bulk" cytosol (1,63–65). Such compartmentalized nucleotide fluctuations (64,66) hamper proper recognition of cellular signals. The extremely low diffusional flux of nucleotides in the subsarcolemmal space, due to molecular crowding and reduction in the free diffusional space (64,66–69), virtually secludes K_{ATP} channels from cellular "bulk" nucleotide oscillations (69–71). Due to such "fuzzy space" in the submembrane (66), channel gating would be relegated to local fluctuations of nucleotides ("metabolic background noise"), independent of the actual cellular metabolic status (71).

Since random fluctuations in adenine nucleotides distort energetic signaling, catalyzed phosphotransfer circuits are required to control exchange of nucleotides between the subsarcolemmal and cytosolic compartments securing proper signal processing. Cells with high and fluctuating energy demands, such as cardiomyocytes, possess well-defined phosphotransfer systems that facilitate energetic signaling between sites of ATP-production and ATP-utilization/sensing (72,73). One of the most vigorous ATP regenerating system in heart muscle is mediated through creatine kinase phosphotransfer (71,72). Activation of creatine kinase promotes the ATPase reaction at NBD2 of SUR (49; Figure 5A). In this way, the intrinsic ATPase activity of the K_{ATP} channel can be regulated through nucleotide exchange with cellular phosphotransfer reactions (43,71). Moreover, creatine kinase, activated by creatine phosphate, efficiently reverses K_{ATP} channel opening induced by stabilization of the MgADP-bound conformation (43,49,71; Figure 5B). Thus, in the cell, nucleotide flux through creatine kinase sets the turnover rate of the intrinsic ATPase cycle and thereby regulates K_{ATP} channel gating (Figure 6A).

Indeed, the dynamics of creatine kinase flux closely follow total ATP turnover, in line with tight coupling of creatine kinase phosphotransfer and the cellular metabolic state (71,74,75). Present throughout the cell, creatine kinase reactions form a phosphotransfer relay able to rapidly respond to changes in the cellular metabolic state and efficiently propagate metabolic waves between cellular compartments (9,71–76). Such catalyzed phosphotransfer delivers metabolic signals at a rate exceeding simple diffusion (9,74,76). Accordingly, under the normal status of cell metabolism, vigorous creatine kinase phosphotransfer dissipates local nucleotide gradients created by membrane ATPases and diffusional restrictions in the channel environment, facilitating ATPase cycling and keeping K_{ATP} channels predominantly closed (9,71,72). Under metabolic stress, however, reduced creatine kinase phosphotransfer is not sufficient to maintain continuous ADP to ATP regeneration at the channel site leading to stabilization of the MgADP-bound conformation, K_{ATP} channel opening, and adjustment of membrane excitability (71,72).

Reduced creatine kinase flux under metabolic insult is associated with upregulation of adenylate kinase phosphotransfer, which catalyzes conversion of ATP to ADP at the channel site (9,74; Figure 6B). Such interplay between phosphotransfer pathways amplifies the metabolic signal, translating into K_{ATP} channel opening and ultimately into shortening of the cardiac action potential under stress (8,9,28,71). Efficient integration of the SUR ATPase cycle with cellular metabolic oscillations is facilitated as phosphotransfer enzymes are found anchored in the intimate environment of the K_{ATP} channel complex (9,49,71,77,78). Genetic deletion of creatine kinase or adenylate kinase pathways generates a phenotype with compromised signal communication to K_{ATP} channels leading to increased electrical vulnerability and/or disrupted adequate metabolic sensing (9,71). Thus, an intact phosphotransfer network is a prerequisite for optimal decoding of energetic signals and the integration of the metabolic sensor with cellular metabolism.

ACKNOWLEDGMENTS

Work in the authors laboratory was supported by the National Institutes of Health, American Heart Association, Marriott Foundation, Miami Heart Research Institute,

Bruce and Ruth Rappaport Program in Vascular Biology and Gene Delivery, and American Physicians Fellowship for Medicine in Israel. AT is an Established Investigator of the American Heart Association.

REFERENCES

1. Weiss JN, Lamp ST. 1987. Glycolysis preferentially inhibits ATP-sensitive K^+ channels in isolated guinea pig cardiac myocytes. Science 238:67–69.
2. O'Rourke B, Ramza BM, Marban E. 1994. Oscillations of membrane current and excitability driven by metabolic oscillations in heart cells. Science 265:962–966.
3. Nichols CG, Shyng SL, Nestorowicz A, Glaser B, Clement JP, Gonzalez G, Aguilar-Bryan L, Permutt MA, Bryan J. 1996. Adenosine diphosphate as an intracellular regulator of insulin secretion. Science 272:1785–1787.
4. Ashcroft FM, Gribble FM. 1998. Correlating structure and function in ATP-sensitive K^+ channels. Trends Neurosci 21:288–294.
5. Seino S. 1999. ATP-sensitive potassium channels: A model of heteromultimeric potassium channel/receptor assemblies. Annu Rev Physiol 61:337–362.
6. Aguilar-Bryan L, Bryan J, Nakazaki M. 2001. Of mice and men: K_{ATP} channels and insulin secretion. Recent Prog Horm Res 56:47–68.
7. Miki T, Liss B, Minami K, Shiuchi T, Saraya A, Kashima Y, Horiuchi M, Ashcroft F, Minokoshi Y, Roeper J, Seino S. 2001. ATP-sensitive K^+ channels in the hypothalamus are essential for the maintenance of glucose homeostasis. Nature Neurosci 4:507–512.
8. Yamada K, Ji JJ, Yuan H, Miki T, Sato S, Horimoto N, Shimizu T, Seino S, Inagaki N. Protective role of ATP-sensitive potassium channels in hypoxia-induced generalized seizure. 2001. Science 292:1543–1546.
9. Carassco AJ, Dzeja PP, Alekseev AE, Pucar D, Zingman LV, Abraham MR, Hodgson D, Bienengraeber M, Puceat M, Janssen E, Wieringa B, Terzic A. 2001. Adenylate kinase phosphotransfer communicates cellular energetic signals to ATP-sensitive potassium channels. Proc Natl Acad Sci USA 98:7623–7628.
10. Noma A. 1983. ATP-regulated K^+ channels in cardiac muscle. Nature 305:147–148.
11. Clement JP, Kunjilwar K, Gonzalez G, Schwanstecher M, Panten U, Aguilar-Bryan L, Bryan J. 1997. Association and stoichiometry of K_{ATP} channel subunits. Neuron 18:827–838.
12. Shyng S, Nichols CG. 1997. Octameric stoichiometry of the K_{ATP} channel complex. J Gen Physiol 110:655–664.
13. Lorenz E, Alekseev AE, Krapivinsky GP, Carrasco AJ, Clapham DE, Terzic A. 1998. Evidence for direct physical interaction between a K_{ATP} channel (Kir6.2) and an ATP-binding cassette protein (SUR1) which affects cellular distribution and kinetic behavior of an ATP-sensitive K^+ channel. Mol Cell Biol 18:1652–1659.
14. Inagaki N, Gonoi T, Clement JP, Namba N, Inazawa J, Gonzalez G, Aguilar-Bryan L, Seino S, Bryan J. 1995. Reconstitution of I_{KATP}: An inward rectifier subunit plus the sulfonylurea receptor. Science 270:1166–1170.
15. Tucker SJ, Gribble FM, Zhao C, Trapp S, Ashcroft FM. 1997. Truncation of Kir6.2 produces ATP-sensitive K^+ channels in the absence of the sulphonylurea receptor. Nature 387:179–183.
16. John SA, Monck JR, Weiss JN, Ribalet B. 1998. The sulphonylurea receptor SUR1 regulates ATP-sensitive mouse Kir6.2 K^+ channels linked to the green fluorescent protein in human embryonic kidney cells (HEK 293). J Physiol 510:333–345.
17. Aguilar-Bryan L, Nichols CG, Wechsler S, Clement JP, Boyd A, Gonzalez G, Herrerasosa H, Nguy K, Bryan J, Nelson D. 1995. Cloning of the beta cell high-affinity sulfonylurea receptor: A regulator of insulin secretion. Science 268:423–426.
18. Inagaki N, Gonoi T, Clement JP, Wang CZ, Aguilar-Bryan L, Seino S, Bryan J. 1996. A family of sulfonylurea receptors determines the pharmacological properties of ATP-sensitive K^+ channels. Neuron 16:1011–1017.
19. Isomoto S, Kondo C, Yamada M, Matsumoto S, Higashiguchi O, Horio Y, Matsuzawa Y, Kurachi Y. 1996. A novel sulfonylurea receptor forms with BIR (Kir6.2) a smooth muscle type ATP-sensitive K^+ channel. J Biol Chem 271:24321–24324.
20. Bryan J, Aguilar-Bryan L. 1997. The ABCs of ATP-sensitive potassium channels: more pieces of the puzzle. Curr Opin Cell Biol 9:553–559.

21. Chutkow WA, Simon MC, Le Beau MM, Burant CF. 1996. Cloning, tissue expression, and chromosomal localization of SUR2, the putative drug-binding subunit of cardiac, skeletal muscle, and vascular K_{ATP} channels. Diabetes 45:1439–1445.
22. Babenko AP, Gonzalez G, Aguilar-Bryan L, Bryan J. 1998. Reconstituted human cardiac K_{ATP} channels: functional identity with the native channels from the sarcolemma of human ventricular cells. Circ Res 83:1132–1143.
23. Lorenz E, Terzic A. 1999. Physical association between recombinant cardiac ATP-sensitive K^+ channel subunits Kir6.2 and SUR2A. J Mol Cell Cardiol 31:425–434.
24. Yamada M, Isomoto S, Matsumoto S, Kondo C, Shindo T, Horio Y, Kurachi Y. 1997. Sulphonylurea receptor 2B and Kir6.1 form a sulphonylurea-sensitive but ATP-insensitive K^+ channel. J Physiol 499:715–720.
25. Repunte VP, Nakamura H, Fujita A, Horio Y, Findlay I, Pott L, Kurachi Y. 1999. Extracellular links in Kir subunits control the unitary conductance of SUR/Kir6.0 ion channels. EMBO J 18:3317–3324.
26. Gross GJ, Fryer RM. 1999. Sarcolemmal versus mitochondrial ATP-sensitive K^+ channels and myocardial preconditioning. Circ Res 84:973–979.
27. Suzuki M, Sasaki N, Miki T, Sakamoto N, Ohmoto-Sekine Y, Tamagawa M, Seino S, Marban E, Nakaya H. 2002. Role of sarcolemmal K_{ATP} channels in cardioprotection against ischemia/reperfusion injury in mice. J Clin Invest 109:509–516.
28. Li RA, Leppo M, Miki T, Seino S, Marban E. 2000. Molecular basis of electrocardiographic ST-segment elevation. Circ Res 87:837–839.
29. Miki T, Suzuki M, Shibasaki T, Uemura H, Sato T, Yamaguchi K, Koseki H, Seino S. 2002. Mouse model of Prinzmetal angina by disruption of the inward rectifier Kir6.1. Nature Med 8:466–472.
30. Abraham MR, Jahangir A, Alekseev AE, Terzic A. 1999. Channelopathies of inwardly rectifying potassium channels. FASEB J 13:1901–1910.
31. Koster JC, Marshall BA, Ensor N, Corbett JA, Nichols CG. 2000. Targeted overactivity of beta cell K_{ATP} channels induces profound neonatal diabetes. Cell 100:645–654.
32. Lawson K, Dunne MJ. 2001. Peripheral channelopathies as targets for potassium channel openers. Expert Opin Investig Drugs 10:1345–1359.
33. Terzic A. 1999. New frontiers of cardioprotection. Clin Pharmacol Ther 66:105–109.
34. Jahangir A, Terzic A, Shen W. 2001. Potassium channel openers: therapeutic potential in cardiology and medicine. Expert Opin Pharmacother 2:1995–2010.
35. Higgins CF. 1992. ABC-transporters: from microorganism to man. Annu Rev Cell Biol 8:67–113.
36. Senior AE, Gadsby DC. 1997. ATP hydrolysis cycles and mechanism in P-glycoprotein and CFTR. Semin Cancer Biol 8:143–150.
37. Hopfner KP, Karcher A, Shin DS, Craig L, Arthur LM, Carney JP, Tainer JA. 2000. Structural biology of Rad50 ATPase: ATP-driven conformational control in DNA double-strand break repair and the ABC-ATPase superfamily. Cell 101:789–800.
38. Drain P, Li L, Wang J. 1998. K_{ATP} channel inhibition by ATP requires distinct functional domains of the cytoplasmic C terminus of the pore-forming subunit. Proc Natl Acad Sci USA 95:13953–13958.
39. Shyng SL, Ferrigni T, Nichols CG. 1997. Regulation of K_{ATP} channel activity by diazoxide and MgADP: Distinct functions of the two nucleotide binding folds of the sulfonylurea receptor. J Gen Physiol 110:643–654.
40. Zingman LV, Hodgson DM, Bienengraeber M, Karger AB, Kathmann EC, Alekseev AE, Terzic A. 2002. Tandem function of nucleotide binding domains confers competence to sulfonylurea receptor in gating ATP-sensitive K^+ channels. J Biol Chem 277:14206–14210.
41. Matsuo M, Tanabe K, Kioka N, Amachi T, Ueda K. 2000. Different binding properties and affinities for ATP and ADP among sulfonylurea receptor subtypes, SUR1, SUR2A, and SUR2B. J Biol Chem 275:28757–28763.
42. Matsuo M, Kioka N, Amachi T, Ueda K. 1999. ATP binding properties of the nucleotide-binding folds of SUR1. J Biol Chem 274:37479–37482.
43. Bienengraeber M, Alekseev AE, Abraham MR, Carrasco AJ, Moreau C, Vivaudou M, Dzeja PP, Terzic A. ATPase activity of the sulfonylurea receptor: a catalytic function for the K-ATP channel complex. FASEB J 14:1943–1952.
44. Gribble FM, Tucker SJ, Haug T, Ashcroft FM. 1998. MgATP activates the β cell K_{ATP} channel by interaction with its SUR1 subunit. Proc Natl Acad Sci USA 95:7185–7190.

45. D'hahan N, Moreau C, Prost AL, Jacquet H, Alekseev AE, Terzic A, Vivaudou M. 1999. Pharmacological plasticity of cardiac ATP-sensitive potassium channels toward diazoxide revealed by ADP. Proc Natl Acad Sci USA 12:12162–12167.
46. Banfi B, Maturana A, Jaconi S, Arnaudeau S, Laforge T, Sinha B, Ligeti E, Demaurex N, Krause KH. 2000. A mammalian H^+ channel generated through alternative splicing of the NADPH oxidase homolog NOH-1. Science 287:138–142.
47. Gulbis JM, Zhou M, Mann S, MacKinnon R. 2000. Structure of the cytoplasmic beta subunit—T1 assembly of voltage-dependent K^+ channels. Science 289:123–127.
48. Runnels LW, Yue L, Clapham DE. 2001. TRP-PLIK, a bifunctional protein with kinase and ion channel activities. Science 291:1043–1047.
49. Zingman LV, Alekseev AE, Bienengraeber M, Hodgson D, Karger AB, Dzeja PP, Terzic A. 2001. Signaling in channel/enzyme multimers: ATPase transitions in SUR module gate ATP-sensitive K^+ conductance. Neuron 31:233–245.
50. Ajtai K, Peyser YM, Park S, Burghardt TP, Muhlrad A. 1999. Trinitrophenylated reactive lysine residue in myosin detects lever arm movement during the consecutive steps of ATP hydrolysis. Biochemistry 38:6428–6440.
51. Kerr KM, Sauna ZE, Ambudkar SV. 2001. Correlation between steady-state ATP hydrolysis and vanadate-induced ADP trapping in human P-glycoprotein: Evidence for ADP release as the rate limiting step in the limiting step in the catalytic cycle and its modulation by substrates. J Biol Chem 276:8657–8664.
52. Chabre M. 1990. Aluminofluoride and beryllofluoride complexes: new phosphate analogs in enzymology. Trends Biochem Sci 15:6–10.
53. Smith CA, Rayment I. 1996. X-ray structure of the magnesium(II) ADP vanadate complex of the *Dictyostelium discoideum* myosin motor domain to 1.9 Å resolution. Biochemistry 35:5404–5417.
54. Sankaran B, Bhagat S, Senior AE. 1997. Inhibition of P-glycoprotein ATPase activity by beryllium fluoride. Biochemistry 36:6847–6853.
55. Holland IB, Blight MA. 1999. ABC-ATPases, adaptable energy generators fuelling transmembrane movement of a variety of molecules in organisms from bacteria to humans. J Mol Biol 293:381–399.
56. Schwanstecher M, Sieverding C, Dorschner H, Gross I, Aguilar-Bryan L, Schwanstecher C, Bryan J. 1998. Potassium channel openers require ATP to bind to and act through sulfonylurea receptors. EMBO J 17:5529–5535.
57. Hambrock A, Loffler-Walz C, Kloor D, Delabar U, Horio Y, Kurachi Y, Quast U. 1999. ATP-Sensitive K^+ channel modulator binding to sulfonylurea receptors SUR2A and SUR2B: opposite effects of MgADP. Mol Pharmacol 55:832–840.
58. Moreau C, Jacquet H, Prost AL, D'hahan N, Vivaudou M. 2000. The molecular basis of the specificity of action of K_{ATP} channel openers. EMBO J 19:6644–6651.
59. Ashcroft FM, Gribble FM. 2000. New windows on the mechanism of action of K_{ATP} channel openers. Trends Pharmacol Sci 21:439–445.
60. Terzic A, Jahangir A, Kurachi Y. 1995. Cardiac ATP-sensitive K^+ channels: regulation by intracellular nucleotides and K^+ channel opening drugs. Am J Physiol 38:C525–C545.
61. Nagata K, Nishitani M, Matsuo M, Kioka N, Amachi T, Ueda K. 2000. Nonequivalent nucleotide trapping in the two nucleotide binding folds of the human multidrug resistance protein MRP1. J Biol Chem 275:17626–17630.
62. Sauna ZE, Ambudkar SV. 2000. Evidence for a requirement for ATP hydrolysis at two distinct steps during a single turnover of the catalytic cycle of human P-glycoprotein. Proc Natl Acad Sci USA 97:2515–2520.
63. Weiss JN, Venkatesh N. 1993. Metabolic regulation of cardiac ATP-sensitive K^+ channels. Cardiovasc Drugs Ther 7:499–505.
64. Weiss JN, Korge P. 2001. The cytoplasm:no longer a well-mixed bag. Circ Res 89:108–110.
65. Lederer WJ, Nichols CG. 1989. Nucleotide modulation of the activity of rat heart ATP-sensitive K^+ channels in isolated membrane patches. J Physiol 419:193–211.
66. Lederer WJ, Niggli E, Hadley RW. 1990. Sodium-calcium exchange in excitable cells: fuzzy space. Science 248:283.
67. Kaasik A, Veksler V, Boehm E, Novotova M, Minajeva A, Ventura-Clapier R. 2001. Energetic crosstalk between organelles: architectural integration of energy production and utilization. Circ Res 89: 153–159.
68. Saks VA, Kaambre T, Sikk P, Eimre M, Orlova E, Paju K, Piirsoo A, Appaix F, Kay L, Regitz-Zagrosek V, Fleck E, Seppet E. 2001. Intracellular energetic units in red muscle cells. Biochem J 356:643–657.

69. Rich TC, Fagan KA, Nakata H, Schaack J, Cooper DM, Karpen JW. 2000. Cyclic nucleotide-gated channels colocalize with adenylyl cyclase in regions of restricted cAMP diffusion. J Gen Physiol 116:147–161.
70. Kabakov AY. 1998. Activation of K_{ATP} channels by Na/K pump in isolated cardiac myocytes and giant membrane patches. Biophys J 75:2858–2867.
71. Abraham MR, Selivanov VA, Hodgson DM, Pucar D, Zingman LV, Wieringa B, Dzeja PP, Alekseev AE, Terzic A. 2002. Coupling of cell energetics with membrane metabolic sensing: Integrative signaling through creatine kinase phosphotransfer disrupted by M-CK gene knockout. J Biol Chem 277:24427–24434.
72. Dzeja PP, Terzic A. 1998. Phosphotransfer reactions in the regulation of ATP-sensitive K^+ channels. FASEB J 12:523–529.
73. Janssen E, Dzeja PP, Oerlemans F, Simonetti AW, Heerschap A, de Haan A, Rush PS, Terjung RR, Wieringa B, Terzic A. 2000. Adenylate kinase 1 gene deletion disrupts muscle energetic economy despite metabolic rearrangement. EMBO J 19:6371–6638.
74. Dzeja PP, Vitkevicius KT, Redfield MM, Burnett JC, Terzic A. 1999. Adenylate kinase-catalyzed phosphotransfer in the myocardium: Increased contribution in heart failure. Circ Res 84:1137–1143.
75. Pucar D, Dzeja PP, Bast P, Juranic N, Macura S, Terzic A. 2001. Cellular energetics in the preconditioned state: Protective role for phosphotransfer reactions captured by ^{18}O-assisted ^{31}P NMR. J Biol Chem 276:44812–44819.
76. Saks VA, Khuchua ZA, Vasilyeva EV, Belikova OYu, Kuznetsov AV. 1994. Metabolic compartmentation and substrate channelling in muscle cells. Role of coupled creatine kinases in in vivo regulation of cellular respiration—a synthesis. Mol Cell Biochem 133–134:155–192.
77. Elvir-Mairena JR, Jovanovic A, Gomez LA, Alekseev AE, Terzic A. 1996. Reversal of the ATP-liganded state of ATP-sensitive K^+ channels by adenylate kinase activity. J Biol Chem 271: 31903–31908.
78. Crawford RM, Ranki HJ, Botting CH, Budas GR, Jovanovic A. 2002. Creatine kinase is physically associated with the cardiac ATP-sensitive K^+ channel in vivo. FASEB J 16:102–104.

Pathophysiology of Cardiovascular Disease,
edited by N.S. Dhalla, H. Rupp,
A. Angel & G.N. Pierce
Kluwer Academic Publishers, Boston, 2004

Measurement of Cytosolic pH Simultaneouly with Isometric Tension in Canine Trabeculae

António Guia* and Ratna Bose

Dept. of Pharmacology and Therapeutics
Univ. of Manitoba, Winnipeg, Manitoba
Canada R3E 0W3

Summary. Dye leakage is a major problem for experiments used to measure cytosolic pH that last over several hours. The ratio of epifluorescence (FI_{500}/FI_{440}, 540 nm emission) of the pH indicator 2′,7′-bis(carboxyethyl)-5,(6)-carboxyfluorescein (BCECF) changes with time when it is loaded in the tissue. This change is enhanced by experimental protocols with large changes in pH and is probably due to dye leakage from the cells. Measurements of cytosolic pH (pH_i) which are performed long before the calibrations, are hence rendered unreliable. We describe a calibration technique for measuring pH_i during long experimental protocols performed on canine right ventricular trabeculae loaded with BCECF. The equation was modified to calculate pH_i according to the intracellular concentration of dye present during each measurement. This concentration-based correction compensates for dye leakage during long experiments. Fluorescence was measured simultaneously with isometric twitch tension. With this technique we show that prolonged acidification (45 minutes) in the presence or absence of bicarbonate in the buffer medium did not produce significant irreversible damage to the ventricular trabeculae as judged by developed isometric contractile tension, which was measured simultaneously.

Key words: heart, BCECF, acidification, lactate, methods.

Address all correspondence to: Dr. Ratna Bose, Dept. of Pharmacology and Therapeutics, Univ. of Manitoba, #311-753 McDermot Avenue, Winnipeg, Manitoba. Canada R3E 0W3. (204) 789-3562, (204) 789 3932 FAX, E-mail: rbose@ms.umanitoba.ca.
*Current address: Aviva Biosciences Corp., 11568-9 Sorrento Valley Rd, San Diego, CA 92121-1313, USA.

INTRODUCTION

Intracellular pH affects enzymes and membrane proteins and can profoundly modulate cellular function (1,2,3). Many techniques are available to measure pH_i, some only provide data on steady state events due to long equilibration periods. Where fast time-resolution is needed, such as with reperfusion-induced damage that occurs within minutes after reperfusion is started, pH-sensitive dyes (of which fluorescent dyes are the most sensitive) and microelectrodes are used. Nuclear magnetic resonance of 31-phosphate is useful for larger sample sizes.

pH microelectrodes have been used to measure pH_i in single cardiac myocytes and to measure pH_i simultaneously with contractile tension in multicellular preparations (4,5). Measurement of cytosolic pH of rapidly contracting tissues, with microelectrodes, is complicated by tissue movement causing the cell to move away from the tip, or breaking the fragile tip. The use of fluorescent dyes, such as BCECF, SNARF-1 provides a means of measuring pH_i in contracting tissue without the risk of breaking or plugging the tip of a microelectrode. The dye BCECF (2′,7′-bis(carboxyethyl)-5,(6)-carboxyfluorescein) has been used to measure cellular pH simultaneously with cell shortening (6) or tissue contraction (7).

BCECF may be loaded into a single cell via an injection microelectrode, or into many cells at once via the permeant acetoxy methyl ester form of the dye, BCECF-AM. Compartmentalization of BCECF has been shown not to occur (8). Although measurements with BCECF are stable in vitro and were originally assumed to be stable in the intracellular environment (9), there are indications that photobleaching and leakage of the dye from the cells can occur (10,11). This decrease in fluorescence over time makes quantitative experimental protocols of longer duration difficult to perform. A calibration method that is dependent on the amount of dye left in the cells has been described for smooth muscle (12). However, this approach does not work reliably with cardiac muscle and had to be modified. In other cell types, an acid transporter has been found to be responsible for the extrusion of the dye from the cytosol (13). Although there is no direct evidence for such a transporter in cardiac myocytes, a blocker of the transporter has been used with some success (7). Such a blocker could affect the observed pH_i. The objective of this study was therefore to develop a technique to assess cellular pH_i that compensates for dye leakage seen in cardiac tissue during long experimental protocols. This was accomplished by adjusting the in situ calibration curves of BCECF in cardiac tissue at the time of measurement. This adjustment was based on the amount of dye present inside the cells at the time of each measurement. The method was developed with tissues made permeable to H^+ with nigericin in a high-potassium solution (pH-clamped).

The method was verified with intact, non-permeablized canine ventricular trabeculae. Ischemia is commonly simulated in vitro by slowing or stopping perfusion thereby producing hypoxia and acidosis by modifying many parameters at once. Measurements of intracellular pH during ischemia-reperfusion have led to controversy over the mediators of cardiac tissue damage in ischemia-reperfusion experiments (5,14,15,16,17). One of the possible reasons for discrepancies in the literature is the use of different buffering systems in the experimental protocols. We measured the effect of acidification on pH_i and tension recovery in the presence and absence of bicarbonate in the buffer medium. The results show that in either buffer medium, prolonged intracellular acidification (45 minutes) with lactic acid or hypercapnia did not produce significant irreversible damage to the ventricular trabeculae, as judged by developed isometric contractile tension or intracellular pH.

METHODS

Preparation

Mongrel dogs of either sex were anaesthetized using 60 mg/kg pentobarbital and their hearts were removed via a lateral intercostal incision. The hearts were placed in cold Krebs-Henseleit solution and free-running trabeculae (1 mm thick) were removed from the free wall of the right ventricle. The muscle was isometrically suspended in a horizontal bath of a Jasco model TI02 tissue fluorescence measuring assembly with the light beam from a Jasco model CAF100 intracellular ion analyzer (Jasco Inc., Easton, Maryland) focused on its undersurface. One end of the trabeculae was tied to a Grass FT03 force transducer (Grass Instrument Co., Quincy, Mass.), and the other end was anchored to the horizontal bath. Stainless steel stimulation electrodes were placed on either side within 5 mm of the muscle to provide field stimulation at a site away from the light beam. The muscles were stimulated at 0.3 Hz with a voltage 50% above threshold. The 3 × 5 cm horizontal bath had an aeration inlet and re-circulation inlet and outlet; it was kept at 37°C with a Peltier effect thermoelectric device. The solutions were re-circulated at 25 ml/min to external water-jacketed reservoirs at 37°C. A temperature probe and miniature pH probe (MI 404, M440, Microelectrode Inc., Londonderry, NH) were placed in the bath adjacent to the muscle.

Equipment

Light from the 50 Watt lamp of the Jasco CAF100 fluorimeter passed through a rotating filter assembly which placed either a 440 nm or 500 nm filter into the light path. The incident light was reflected from a mirror on the Jasco TI02 and focused on the muscle surface through a quartz glass window. Epifluorescence returned along the same path and was reflected from a mirror through a 540 nm filter, and focused on the photomultiplier tube (PMT) of the Jasco CAF100 for measurement. Fluorescence emission at each excitation wavelength (FI_{500}, FI_{440}), fluorescence ratio (FI_{500}/FI_{440}), tension, temperature, and pH were measured simultaneously and recorded on a six-channel chart recorder (model SE-400, BBC-Metrawatt/Goerz, Broomfield, Colorado).

The fluorescence spectra (Figure 1 only) were scanned with a Perkin-Elmer Fluorescence Spectrophotometer (model LS-5, Oak Brook, Illinois) using a 5 mm slit width. The sample chamber of the Spectrophotometer was adapted to accommodate an aeration tube and tissue suspension hooks inside a 1 cm square cuvette.

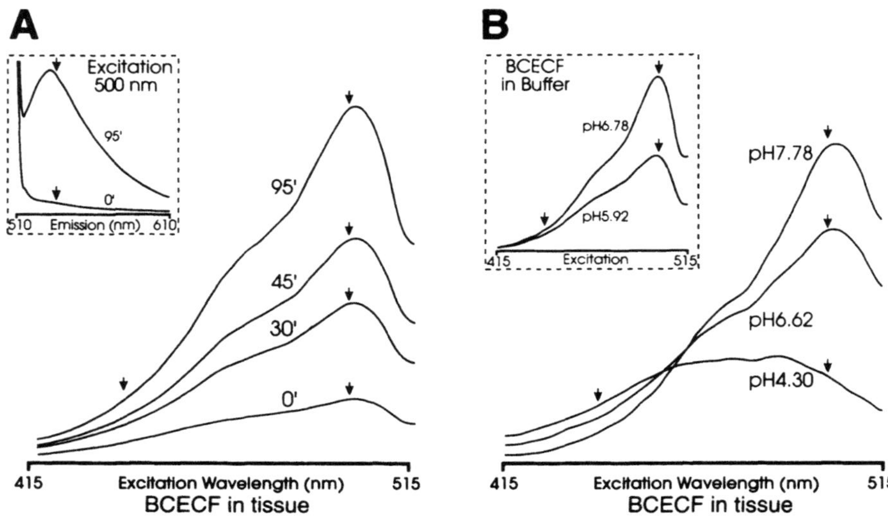

Figure 1. Fluorescence scans of BCECF were done in a Perkin Elmer Spectrofluorimeter. Excitation spectra between 415 nm and 515 nm are shown with emission measured at 540 nm. Scans were done with a 5 mm slit width and light intensity was plotted in arbitrary units as a function of wavelength (λ). Arrows indicate FI_{440} at 440 nm and FI_{500} at 500 nm. HEPES buffer solution was maintained at 37°C. **A**. The fluorescence excitation spectrum of a ventricular trabeculum at pH 7.45 before (**0′**) and after the addition of 8 μM BCECF-AM for 30 minutes (**30′**), 45 minutes (**45′**), and 95 minutes (**95′**). The excitation peak in the tissue was at 507 nm, and in buffer at 502 nm. ***Inset:*** The emission scan from 510 nm to 610 nm with excitation at 500 nm. Spectra are shown from tissue alone (**0′**) and with BCECF-AM (**95′**) at pH 7.45. Arrows indicate FI_{500} at 540 nm. The emission peak in the tissue was at 532 nm, and in the buffer at 527 nm. **B**. Fluorescence excitation spectrum of a trabeculum loaded with BCECF at pH 7.78, 6.62, and 4.30. The isosbestic point occurs at 463 nm. ***Inset:*** 0.58 μM BCECF free acid in a buffer of cytosolic electrolyte concentrations at pH 6.78 and pH 5.92. The isosbestic point occurs at 415 nm.

Dye loading

The tissues were bathed in re-circulating Krebs-Henseleit (Bicarb) solution or Krebs-HEPES (HEPES) solution (5 ml bath volume, 20 ml total re-circulating volume). After equilibration at optimal preload tension in either Bicarb or HEPES solution, bubbled with either 95% O_2—5% CO_2 or 100% O_2 respectively, at 37°C, the muscles were exposed to a final concentration of 8 μM 2′,7′-bis(carboxyethyl)-5,6-carboxyfluorescein acetoxy methyl ester (BCECF-AM). Intracellular esterases cleave cytosolic BCECF-AM and trap BCECF free acid inside the cells. BCECF in buffer solution has a peak emission at 540 nm when excited at 500 nm (FI_{500}). This fluorescence is pH sensitive, exhibiting higher intensity at higher pH values. Fluorescence at 540 nm emission with 440 nm excitation (FI_{440}) is not pH sensitive in the buffer and may be used as an indication of dye concentration. BCECF-AM does not contribute to the fluorescence at FI_{500} or FI_{440}, hence liberation of free

BCECF acid inside the cells results in an increase in FI_{440}. In one to two hours, the cells in the trabeculae were loaded with BCECF after which the tissues were washed with five changes of fresh solution over 30 minutes to remove any BCECF-AM and BCECF from outside the cells before any experiments were performed. Autofluorescence accounted for 12% ± 1% of the total signal at 500 nm (n = 9) and 75% ± 3% at 440 nm. Autofluorescence was subtracted from the total fluorescence measurements.

Experimental protocol

BCECF free acid, untreated trabecular tissue, or BCECF-loaded tissue was added to, or mounted in a cuvette containing oxygenated HEPES-buffered medium and their excitation and emission fluorescence spectra were obtained using the Perkin-Elmer Spectrophotometer. The unstimulated muscles were stretched to 0.5 to 0.7 grams preload tension. The progression of loading after the addition of 8 μM final concentration of BCECF-AM was monitored in two tissues by performing multiple scans over a 95 minute period. Spectral scans were also obtained in HEPES-buffer acidified with acetic acid or alkalinized with NaOH medium to different pH values. A similar protocol was performed in the absence of trabeculae: five different concentrations of BCECF free acid in an intracellular electrolyte solution (120 mM KCl, 1 mM $MgCl_2$, 25 mM HEPES, pH 7.00) were exposed to different pH values and fluorescent spectra were scanned.

In separate experiments, different concentrations of BCECF free acid were added to a solution of intracellular electrolytes (contents above) in the Jasco CAF-100. Fluorescence measurements were made (540 nm emission with 440 nm or 500 nm excitation) at different medium pH values (with NaOH/HCl). During calibrations, the muscles were pH-clamped using 4.5 μM nigericin and 80 mM potassium (12,18). The dye fluorescence was measured at steady state after changing the pH of the bathing solution. Calibrations were performed at various times between 1 and 16 hours after loading with BCECF. Four tissues were subjected to repeated calibrations at various times over a 12 hour period. One tissue was tested 24 hours after loading with BCECF. Calibrations were also performed using 4.5 μM nigericin and different KCl concentrations for each pH calibration. In two trabeculae the medium pH and 4.5 μM nigericin were maintained constant and KCl was added in 40 mM concentration steps.

Solutions

Bicarbonate-buffered solutions contained (in mM): NaCl (118), KCl (4.7), $MgSO_4$ (1.2), KH_2PO_4 (1.4), Glucose (11), $NaHCO_3$ (26), $CaCl_2$ (2.5). HEPES-buffered solutions had the same composition except that $NaHCO_3$ was replaced with 25 mM HEPES. The pH of the solutions was adjusted to 7.40 with NaOH. Acidification was achieved by addition of (+) lactic acid or acetic acid from 10 mM stock solutions, or by bubbling 100% CO_2 in the medium. Chemicals were purchased from Sigma Chemical Company (St. Louis, MO).

Statistics

Students paired **t**-test was used to compare the effects of any treatment to control. In experiments testing for reperfusion damage, a blocked ANOVA was performed, followed by individual analysis using Duncan's test to find differences between tissues in control conditions and those subjected to acidifications, or after the removal of the acidification (washout). Slopes were compared using a **t**-test. Only one tissue from each animal was used for this study. Results are expressed as the mean ± S.E., n = number of animals in sample.

RESULTS

The addition of BCECF-AM to the bathing medium results in incorporation of BCECF into the cells of the trabeculae (Figure 1A), producing maximal fluorescence emission at 532 nm (Figure 1A, inset) with excitation at 507 nm (Figure 1B). Maximal emission of fluorescence of BCECF free acid in a high potassium intracellular electrolyte buffer solution occurs at 527 nm when excited at 500 nm (Figure 1B, inset), therefore there was a shift in the peaks to 532 nm emission with 500 nm excitation inside the muscle. There was also a shift of the isosbestic point from 440 nm to 463 nm in the excitation spectra (540 nm emission) (Figure 1B). Since the isosbestic point is shifted when the dye is intracellular, calibrations must be done with BCECF in situ. BCECF free acid was added to the high potassium electrolyte buffer solution until the change in FI_{440} was equal to the average change in FI_{440} seen at the completion of tissue loading with BCECF-AM. Cytosolic BCECF was estimated to be 0.58 μM. BCECF-AM in solution or solution alone did not exhibit pH-dependent changes in fluorescence in the range measured. Before dye-loading, tissue exhibited less than 3% of the pH-dependent fluorescence of tissue loaded with BCECF.

Dissociated $BCECF^-$ exhibits maximal fluorescence ratio (R_{Max}) whereas BCECF H has minimal fluorescence ratio (R_{Min}). The proton activity can be calculated by equation [1] (19).

$$[H^+] = K_a \times \frac{(R_{Max} - R)}{(R - R_{Min})} \times \frac{F_{440_{BASE}}}{F_{440_{ACID}}} \qquad [1]$$

$F_{440_{BASE}}/F_{440_{ACID}}$ ratio (the ß factor) was calculated at 0.681 for BCECF in situ and was not affected by the amount of dye loaded. Since the ß factor was a constant, it was combined with the dissociation constant, K_a, to produce the apparent dissociation constant, K_a'. A Henderson-Hasselbalch logarithmic transformation of equation [1] yields equation [2] which is used to fit calibration curves.

$$pH_i = pK_a' - \log\frac{(R_{Max} - R)}{(R - R_{Min})} \qquad [2]$$

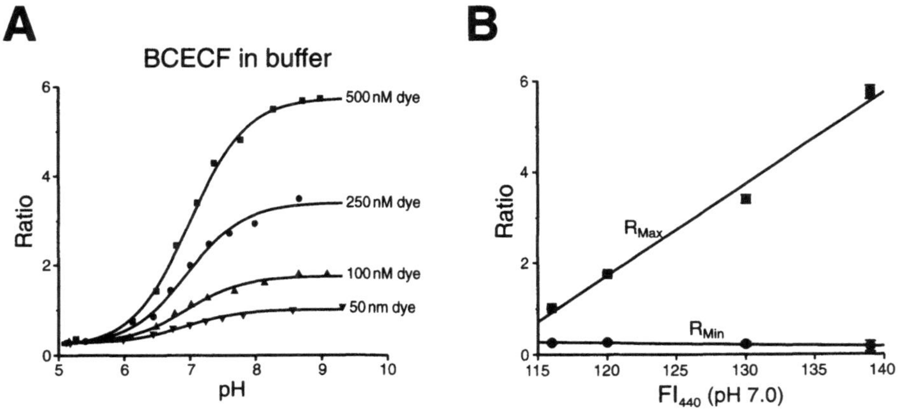

Figure 2. Multiple calibrations were done with different concentrations of BCECF free acid in high potassium intracellular electrolyte solution. **A.** The steady-state fluorescence ratio (Jasco CAF100 fluorimeter) of FI_{500}/FI_{440} is plotted against pH. The lines are described by a best fit to equation [2] using a χ^2 test (see text). Concentration of free acid is indicated to the right of each plot. **B.** The parameters of the best fit lines in A, $\mathbf{R_{Max}}$ and $\mathbf{R_{Min}}$, are plotted against FI_{440}, representing the amount of dye, measured at pH 7.0 during the calibration. Lines were fit by linear regression (■, R_{Max}, $r^2 = 0.99$; ●, R_{Min}, $r^2 = 0.96$).

Figure 2A shows the pH-dependence of the ratio of fluorescence of various concentrations of BCECF free acid in a high potassium intracellular electrolyte medium. Calibration curves were constructed by plotting the fluorescence ratios against the pH of the bathing medium, and the data to equation [2] were fitted using a χ^2 method of least squares. The value of pKa′ was found to be 6.968 ± 0.027 ($n = 5$). Since FI_{440} may be linearly correlated with dye concentration ($r^2 = 0.98$, data not shown), we used FI_{440} to represent dye concentration. Different concentrations of BCECF in solution produce linearly correlated R_{Max} values (Figure 2B; intercept = −22.5, slope = 0.202, $r^2 = 0.988$). There is little change in R_{Min} with the different concentrations of BCECF (slope = −0.003, $p > 0.05$).

During the course of experiments with BCECF loaded in the trabeculae, FI_{440} showed a slow, progressive decline of about 30% per hour. This decline was also observed in the dark after the tissues were left in the recirculating bath for over eight hours. The drop in FI_{440} was more prominent during acidic to alkaline transitions in the bathing solutions. Thus this decrease in FI_{440}, or in BCECF, was considered to be largely due to leakage of the dye, and not due to photobleaching. Since there is a decrease in intracellular BCECF with time, equation [2] needs to be corrected to account for dye leakage.

At the end of each experiment the tissues were pH-clamped with nigericin, an ionophore selective for hydrogen and potassium ions, and 80 mM KCl (12,18). The calibration curves for three different tissues are shown in Figure 3A. The parameters from the best-fit curves in Figure 3A were plotted against the amount

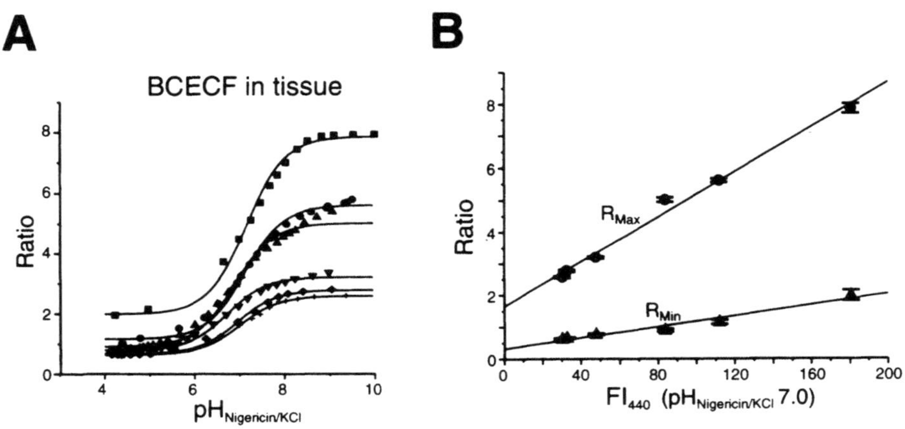

Figure 3. Multiple calibrations were done in BCECF-loaded canine ventricular trabeculae pH-clamped using 4.5 μM nigericin and 80 mM KCl. Six calibrations from 3 preparations are shown: 2 from each tissue. **A.** The steady-state fluorescence ratio (Jasco CAF100 fluorimeter) of FI_{500}/FI_{440} is plotted against pH. The lines are described by a best fit to equation [2] using a χ^2 test. **B.** The parameters of the best fit lines in A are plotted against the amount of dye loading (FI_{440}) measured at pH 7.0 during the calibration. Lines were fit by linear regression (●, R_{Max}, $r^2 = 0.99$; ▲, R_{Min}, $r^2 = 0.96$). Measurements were corrected for autofluorescence.

of BCECF loaded in the tissue as indicated by FI_{440} at pH 7.0 (Figure 3B) showing a similar correlation to that seen with BCECF free acid. For each experiment the curve had different R_{Max} and R_{min} values. The apparent dissociation constant (pK_a') did not change significantly in different experiments, hence it was taken as a constant. In our experimental conditions, the mean pK_a' from 23 tissues was 7.135 ± 0.027. As was found with BCECF free acid, R_{Max} and R_{Min} of the tissue-loaded dye were both dependent on the amount of dye loading (Figure 3B). A linear regression of R_{Max} and of R_{Min} with FI_{440} (plots shown in Figure 3B) yields the correction equations [3] and [4].

$$R_{Max} = 1.652 + .0351 \times FI_{440} \quad (r^2 = 0.989) \qquad [3]$$

$$R_{Min} = 0.3196 + .00859 \times FI_{440} \quad (r^2 = 0.957) \qquad [4]$$

R_{Max}, is more sensitive to changes in BCECF concentration than is R_{Min}, however since the values of both R_{Max} and R_{Min} are important in equation [2], both correction equations were incorporated into the calculation of pH_i. The corrected formula for the calculation of pH_i incorporates equations [3] and [4] into equation [2] to yield equation [5].

$$pH_i = 7.135 - \log \frac{(1.652 + 0.0351 \times FI_{440} - R)}{(R - 0.3196 - 0.00859 \times FI_{440})} \qquad [5]$$

Due to the shift in the isosbestic point, the value of FI_{440} was also pH-dependent. The slopes and intercepts of the correction equations [3] and [4] are pH-dependent. Since the magnitude of the changes in FI_{440} with pH is small we assumed this variation to be insignificant. To calculate the amount of error produced by this assumption, we re-plotted the slope and intercept data in Figure 3B using FI_{440} values measured at different medium pH values ranging from 6.5 to 8.5. The slopes and intercepts in equations [3] and [4] were themselves linearly correlated with pH (data not shown). The assumption that FI_{440} is not pH-dependent results in an error of 0.027 per calculated pH unit using equation [5] in the range of pH 6.5 to 8.5. The pH_i predictions using this equation would hence have a variability of 0.055 per unit of pH_i (contributed to by the pH-dependence of FI_{440} and by the variability in pK_a'), i.e.: 94.5% confidence. In the absence of any corrections for dye leak (using equation [2]), the calculated pH_i values were off by as much as 4 to 5 pH units if the measurement was made a few hours before the calibration.

Figure 4 shows the effect of changing potassium concentration on the calibration curves of tissue with 4.5 μM nigericin. In the same tissue, three consecutive calibrations were performed with different concentrations of KCl. Although there was a decrease in R_{Max} related to dye leakage, there was no change in the calculated pK_a' with 40 to 120 mM KCl. In a separate experiment on the same preparation but with pH was held constant, KCl concentration was increased from 40 to 140 mM in 20 mM steps (Figure 4, inset). Fluorescence of BCECF-loaded tissue was not appreciably affected by different concentrations of KCl between 80 and 120 mM. Similar results were observed in two tissues. The use of a high-potassium solution of intracellular electrolytes, instead of adding KCl to HEPES-buffered solution, did not produce a calibration curve different from that done with the muscle in HEPES-buffered solution with KCl added.

To verify that we were in fact measuring cytosolic pH, NH_4Cl experiments were carried out. In six experiments using five different dogs, ammonium chloride changed the intracellular pH without affecting the extracellular pH (Figure 5, pH_o, second trace from top). In bicarbonate-buffered solutions, it was difficult to change solutions without a small change in pH of the bathing solution, hence there was a small shift in extracellular pH (pH_o), but pH_i did not follow the changes in pH_o. Upon addition of 10 mM NH_4Cl there was an immediate alkalization of the cytosol, followed by positive inotropy (Figure 5). After washout of NH_4Cl there was an immediate acidification of the cytosol, and a return of pH_i and twitch tension to pre-exposure level. Similar observations were made in HEPES (Figure 5A) and bicarbonate-buffered (Figure 5B) Krebs solutions.

The dependence of pH_i on pH_o was found by changing pH_o either with CO_2 or acetic acid (Figure 6). In Bicarbonate-buffered solutions, with each addition of acetic acid there was a rapid transient drop in pH_o followed by a return to steady state showing a relatively small acidification of the medium (Figure 6B). With the same tissue in HEPES-buffered solution each addition of acid had a less pronounced effect on pH_o and pH_i (Figure 6A). Although pH_i appears to decrease more in bicarbonate-containing buffer, with each transient acidification a further drop in FI_{440}

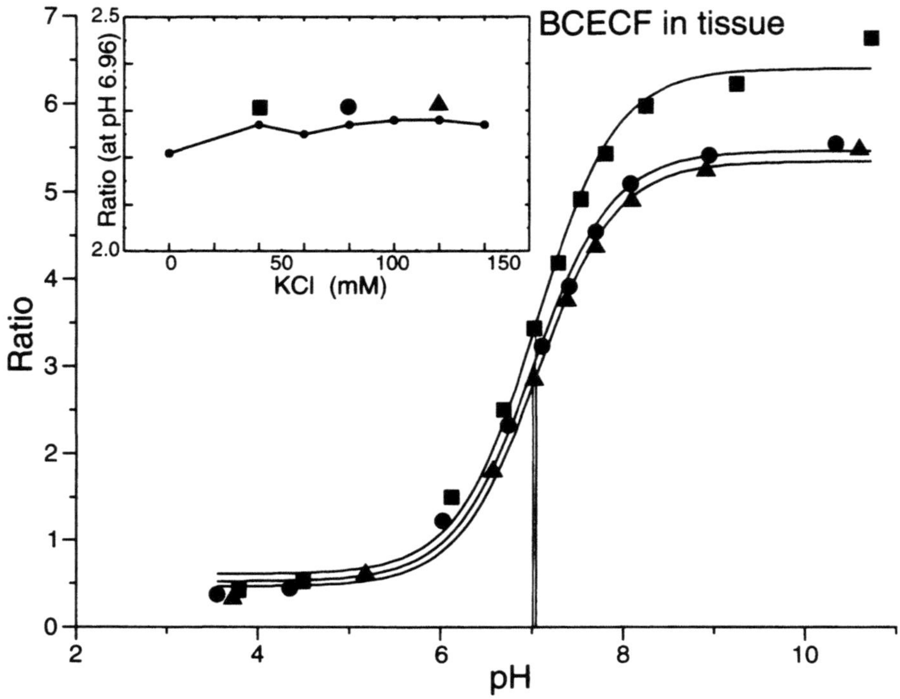

Figure 4. Multiple calibrations were done in BCECF-loaded canine ventricular trabeculae pH-clamped using 4.5 μM nigericin. Three calibrations from the same tissue are shown. The steady-state fluorescence ratio (Jasco CAF100 fluorimeter) of FI_{500}/FI_{440} is plotted against pH. Different concentrations of KCl were used in each curve (■, 40 mM; ●, 80 mM; ▲, 120 mM). The lines are described by a best fit to equation [2] using a χ^2 test. Lines dropped to the pH axis indicate pKa of each curve. R_{Max}, and R_{Min} show dye leak-related changes. **Inset** shows the dependence of the steady-state fluorescence ratio on KCl concentrations (pH was kept constant at 6.96) stepped up to 140 mM.

was observed, suggesting a leak of the dye. The dye leak is also indicated by the relative difference between the ratio trace and the tension trace since cytosolic acidification would be reflected in the twitch tension.

The slopes of the dependence of pH_i on pH_o were collected from the six preparations acidified with CO_2 or five tissues acidified with acetic acid (Table 1). There was no difference between the slopes of pH_i as a function of pH_o in bicarbonate buffer or in HEPES buffer, but the slopes of pH_i as a function of pH_o in tissues acidified with CO_2 were less than in those acidified with acetic acid ($p < 0.003$ in bicarbonate buffer; $p < 0.09$ in HEPES buffer; $n = 5$, steady-state pH from paired data was used for analysis). The slope of tension as a function of pH_o was not significantly changed by the different modes of acidification (with acetic acid or with CO_2) but was less pH-dependent in the bicarbonate buffer (slope of 0.80 ± 0.094, n = 5, with CO_2; 0.857 ± 0.128, n = 4, with acetic acid; tension normalized at pH 7.4) as compared to HEPES buffer (1.465 ± 0.196 with CO_2; 1.051 ± 0.125,

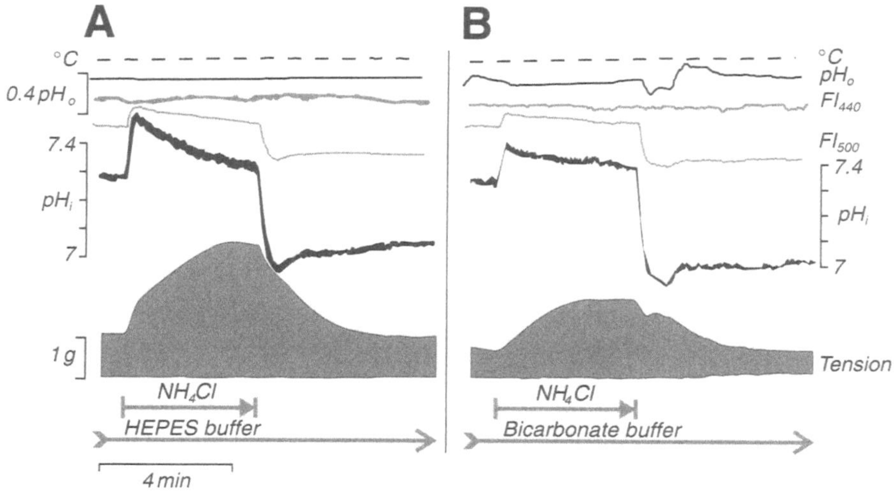

Figure 5. Sample tracings show the effects of a short pulse of NH_4Cl (10 mM) on the measured parameters of the function of canine ventricular trabeculae in **A**. HEPES-buffered medium and **B**. bicarbonate-buffered medium. Lines from top to bottom represent temperature (Temp), pH of the bathing solution (pH_o), FI_{440}, FI_{500}, cytosolic pH (pH_i), and tension. Calibration bars from top to bottom describe pH_o, pH_i, and tension. Similar results were found in 6 experiments with 5 tissues from different animals.

n = 4, with acetic acid). Acidification with hydrochloric acid produced results similar to those with acetic acid. pH was easier to titrate with acetic acid.

The effect of acidification on tissue function is shown in Figures 7 and 8. The muscle strips in medium buffered at pH 7.442 ± 0.023 had pH_i values of 7.240 ± 0.054. The tissues were then subjected to acidification by CO_2 (n = 4 in bicarbonate buffer, n = 5 in HEPES buffer) or by lactic acid (n = 5 in bicarbonate buffer, n = 6 in HEPES buffer) for 45 minutes. Figure 7 shows data from a typical experiment during CO_2-acidification in HEPES- and Bicarbonate-buffered media (Figure 7A and B respectively). pH_i and tension both decreased rapidly following acidification. The twitch tension recovered slightly (22% ± 10% of the lowest tension immediately after acidification, n = 13) during the 45 minutes acidification even though the pH_i did not recover during the same period. pH_i was more difficult to control in bicarbonate-buffered medium when CO_2 was used for acidification. pH_o returned to control levels upon removal of the excess CO_2. In both bicarbonate buffer and HEPES buffer, the replacement of the acidic medium with medium at control pH resulted in an initial transient positive inotropy as compared to control followed by a steady state level not different from control (n = 11). The results from similar experiments in bicarbonate-buffered or HEPES-buffered solution are shown in Figure 8A and B respectively. Re-alkalization of the bathing medium to control pH resulted in a return of all parameters to pre-test values (p = 1).

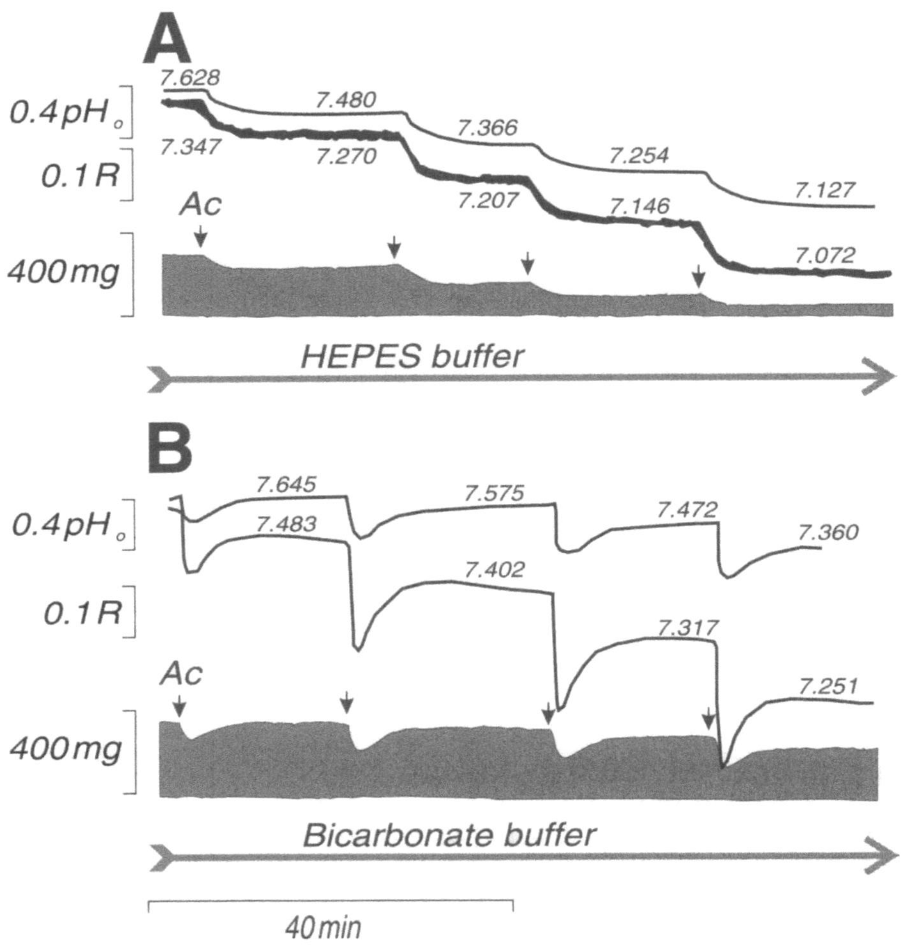

Figure 6. Sample tracings show the effects of a progressively increasing concentration of acetic acid on the pH_i and tension from one canine ventricular trabeculum bathed in **A**. HEPES-buffered medium and **B**. bicarbonate-buffered medium. Acetic acid was added to the bathing solution in small steps (↓, Ac) to acidify the medium. From top to bottom, the tracings and calibration bars represent pH of the bathing solution (pH_o), ratio of fluorescence, and tension. Steady-state values of pH_o and pH_i are shown over the trace for each step.

DISCUSSION

Fluorescent dyes such as BCECF are used to measure pH_i in cardiac trabeculae simultaneously with tension. BCECF leaks out of the cells of the tissue, making long experimental protocols difficult to perform. Previous studies have shown a slow time-dependent decrease in FI_{440} (10,11,12), but few have investigated the reasons

Table 1. Change in pH*i* per change in pH*o*

Medium	Acidification	Mean ± SE†	(n)	p*
Bicarb.	CO_2	0.375 ± 0.045	(6)	<0.003
Bicarb.	Acetic Acid	0.727 ± 0.074	(5)	
HEPES	CO_2	0.541 ± 0.089	(6)	<0.09
HEPES	Acetic Acid	0.593 ± 0.104	(5)	

*Statistics were performed on paired data, n = 5.
†Values from each experiment were calculated as the slope of the plot of pH*i* against pH*o*.

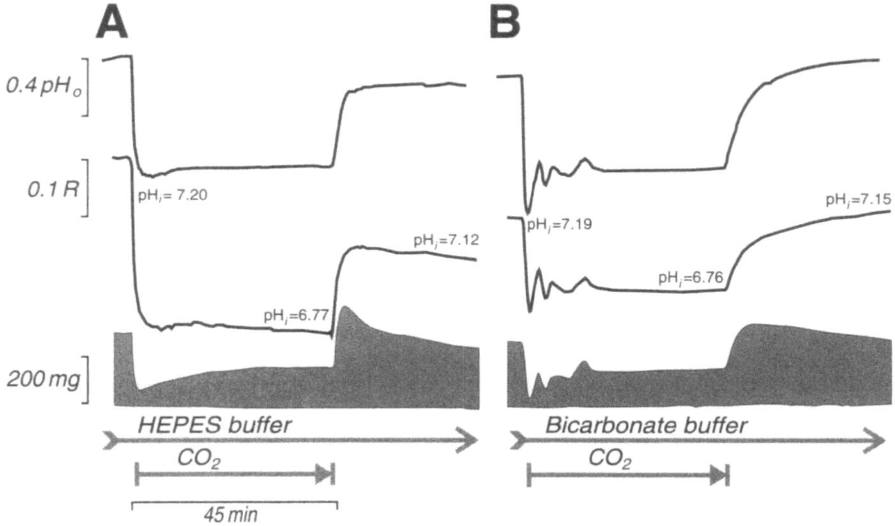

Figure 7. Sample tracings from one canine ventricular trabeculum exposed to a period of acidosis. In the presence of **A**. HEPES-buffered medium and of **B**. bicarbonate-buffered medium, CO_2 was bubbled in the medium to maintain pH_o at pH 6.65 for 45 minutes. Traces from top to bottom represent medium pH (pH_o), ratio, and twitch tension. Steady-state values of pH_o and pH_i are shown for control pH, acidic medium, and after a wash with medium at control pH.

for this phenomenon. We also observed a similar slow decrease in fluorescence with time, which was accelerated with acid to base transitions. We describe a method for correcting the calibrations to account for this dye leakage. This technique estimates pH_i more closely during long experimental protocols using cardiac trabeculae. We investigated the recovery of pH_i simultaneously with tension after 45 minutes of acidification with a high concentration of CO_2 and with the addition of lactic acid in bicarbonate- and HEPES-buffered media. We could not observe tissue damage

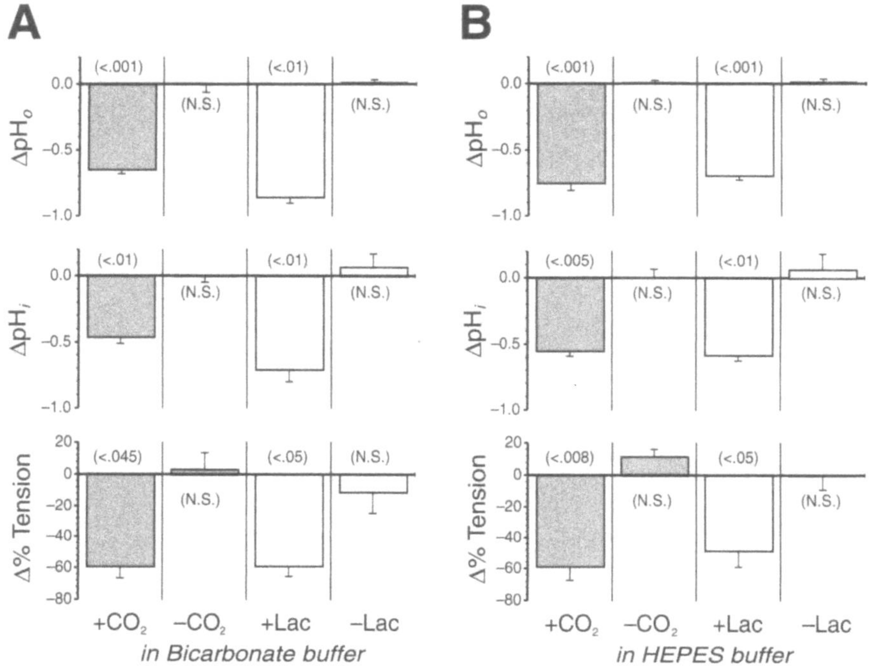

Figure 8. The canine ventricular trabeculae were acidified with CO_2 (▧, **+CO₂**) (see Figure 7) or with the addition of lactic acid (△, **+Lac**) in the presence of either bicarbonate buffer (**A**) or HEPES-buffer (**B**). Data reported as change from control (**Δ[pH]ₒ**, **Δ[pH]ᵢ**, and **Δ%Tension**) in the presence of acid and after its removal (**−CO₂**, **−Lac**). Measurements were performed at steady-state. In all conditions, acidification of pH_o produced significant acidification of pH_i and significant reduction in tension (p values in parentheses). Removal of the acidic conditions after 45 minutes resulted in a return of all parameters to pre-acidic values. In control conditions, the bathing medium pH was 7.442 ± 0.023, cytosolic pH was 7.240 ± 0.054 and tension was taken as 100% (n = 20). (n = 4 in bicarbonate buffer acidified with CO_2, n = 5 in HEPES buffer acidified with CO_2, n = 5 in bicarbonate buffer acidified with lactic acid, n = 6 in HEPES buffer acidified with lactic acid)

caused by the acidification and realkalinization as judged by the recovery of generated tension and of pH_i, which was complete.

The calibration of BCECF in situ was previously found not to be different from that of BCECF in buffer (20). We observed a shift in the excitation and emission peaks, and in the isosbestic point, a phenomenon that has been observed by others (21). Unpredictable cytosolic ionic strengths and interaction of BCECF with intracellular proteins may change the in situ fluorescence of the dye (12,22), hence calibrations need to be performed with BCECF in situ. In our preparations there was leakage of the dye from the cells over time, which caused a slow reduction in the ratio. A calibration of BCECF done in a tissue at the end of a long experiment is therefore invalid for calculating pH_i from a part of the experiment done long before the calibration. The leakage of the dye from the cells was dependent on the treatment performed. Transitions from acid to base result in faster dye leakage than transitions from base to acid or steady pH maintenance. This suggests the presence of an acid transporter on the membrane. Although such a transporter has not been described in cardiac tissue, the use of probenecid has been advocated for the prevention of dye leak from cardiac tissue (7) as well as from other cell types (13). FI_{440} decreases with time even in the absence of any incident light, hence the phenomenon is not due to photo-bleaching of the dye, or if any photo-bleaching does occur, then the products do not fluoresce at 540 nm when excited at 440 nm or 500 nm. This is concluded from the fact that the formula parameters from calibrations done at different times in the same muscle strip (time-dependent decrease in cytosolic dye concentration) match the parameters from calibrations done with different initial cytosolic dye concentrations. This is also evidence that compartmentalization of the dye does not arise slowly during experiments. We did not attempt to further characterize the leakage phenomenon. It has been assumed that the only change in time during an experiment with BCECF is a downward shift of the calibration curve (10,11), implying that relative changes intracellular pH should be reliable regardless of the amount of leakage of the dye. However, since R_{Max} changes more than R_{Min}, the slope of the linear portion also changes, decreasing with dye leakage. If one is to use the calibrations at the end of an experiment where the amount of dye loaded is low, and extrapolate those values to the beginning of the experiment where the amount of dye loaded is high, the extrapolated results are not only offset from the real pH_i, but measured changes are exaggerated. As a result of the leakage of the dye, in order to obtain more reliable results one must limit the experiment to a short duration, and only relative changes can be described unless the calibration is done immediately after the experimental protocol. This is limiting to the kind of experiments that could be performed using BCECF. Only a few studies have incorporated measurements corrected for the dye leakage in smooth muscle (12) or renal tubule cells (16). In our preparation, uncorrected calibration curves consistently gave us an overestimation of both absolute pH_i values and of relative pH_i changes (data not shown).

The calculated pH_i values are most reliable in the range of pH 6 to 8 since beyond these values the ratio approaches the asymptotes at R_{Min} and R_{Max}. A linear correction for the data in this range may also be used (12) but at the expense of greater variability in the results (data not shown). Contribution to the pH_i calculation by the pH-dependence of FI_{440} is minimal (0.055 units per pH unit) and was left out of the equation by necessity. If anything, the shift in the isosbestic point actually serves to make the ratio more sensitive to pH changes since FI_{440} will change in the opposite direction from FI_{500}, producing a larger change in the ratio per change in pH unit. The ß factor in equation [1] was found to be constant, therefore the apparent pK_a' did not change with pH. In order to minimize any error in calculated pH_i, the FI_{440} values at pH 7.0 were used to find the correction factors. The calculations of pH*i* will be slightly underestimated for pH_i values greater than pH 7.0 and slightly overestimated for pH_i values less than pH 7.0 (up to 0.027 per pH unit).

Our medium-to-tissue volume ratio was 4000:1 and the muscle surface was continuously superfused by rapid flow; hence the contribution of extracellular dye to fluorescence was negligible. The ammonium chloride pulse has been used to induce transient cytosolic pH changes without affecting extracellular pH. In solution NH_4Cl is in equilibrium with NH_3 which freely diffuses through the cell membranes. Inside the cells NH_3 and NH_4^+ establish an equilibrium, thereby taking up H^+ and resulting in alkalization of the cytosol. The removal of the ammonium chloride produces the same series of events in the reverse direction, resulting in acidification of the cytosol. Experiments with the ammonium chloride pulse demonstrated that pH_i was being measured independently of pH_o.

pH_i directly affects the force of contraction by interaction of H^+ with binding sites for Ca^{2+} on troponin C (1,24) and by more indirect effects such as on calcium release sites (25), potassium channels (26), calcium channels (27), and other mechanisms (24,28,29). The relationship between pH_i and cell or muscle contraction has been described in detail elsewhere (30,31,32). The buffering capacity of the cells is affected by not only membrane events such as the sodium-hydrogen exchanger, anion exchangers, and the sodium-dependent anion exchangers, but also intracellular enzymes and metabolism. We found that the dependence of pH_i on pH_o was not contingent on the presence of bicarbonate in the medium. The observation that tension is less dependent on pH_i when bicarbonate is present in the medium, implies some additional role of bicarbonate on tension development which is independent of the changes in pH_i.

Ischemia involves many simultaneous events. Thirty minutes of in vitro ischemia suffices to produce damage during reperfusion (16,33). Simulated metabolic acidosis, with lactic acid, has been shown histologically to damage endocardium and cardiac cells in multicellular preparations (15). Simulated respiratory acidosis, with CO_2, did not result in myocardial damage upon restoration of pH_o, whereas ischemia, by stopped flow, produced reperfusion-induced damage (16). Lazdunski et al. (17) proposed that ischemia-reperfusion damage was due to calcium overload resulting from enhanced sodium-calcium exchange activity in response to a cytosolic sodium overload. The sodium overload was proposed to be due to the activity of sodium-hydrogen exchange (17,33,34). Although sodium levels increase in cardiac cells during acidosis (35,36), the sodium-hydrogen exchanger was found not to contribute to increased cytosolic calcium concentrations in single cells (37) nor to play a role in low-flow ischemic conditions. Furthermore, in our experimental conditions (decreased pH_o), it is unlikely that the H^+ gradient was sufficient to produce an increase in Na^+ in the cytosol, as has been shown by Vaughan-Jones and Wu (38). During ischemic events it is also expected that the Na^+-H^+ exchanger does not produce enough of an increase in cytosolic Na^+ to result in Ca^{2+} overload damage (39). Perhaps a summation effect is required during reperfusion, combining the simultaneous attenuation of sodium pump activity or an increase in Na^+ influx through other channels.

Previous studies have found mild acidosis to be cardioprotective in hypoxia (40). Severe acidosis could initiate tissue necrosis (28) whereas mild cytosolic acidosis in hypoxic conditions did not produce tissue damage (5). We show that merely changing pH_o by 0.7 units is not sufficient to produce tissue damage. When only acidification is imposed on the tissue, the Na^+-H^+ exchanger is rarely implicated as producing any damage (39,6,7). It has been found that amiloride protects the heart from post-ischemic reperfusion damage, but the Na^+-H^+ exchanger specific blocker EIPA does not (41). Amiloride also inhibits Na^+ channels, Na^+-Ca^{2+} exchange, and other Na^+-Ca^{2+} exchange blockers also prevent reperfusion-induced damage (41). These data point toward some mechanism other than the Na^+-H^+ exchanger and it is difficult to implicate the Na^+-H^+ exchanger for explaining the increase in cytosolic calcium during acidosis (24). Hypoxia induces free radical formation on reperfusion that can directly produce tissue damage or cause additional sodium influx (14,42). It is therefore more likely that the reintroduction of oxygen may produce tissue damage by the liberation of free radicals (42).

The slight increase in tension during the acidification reflects extra calcium available for release from the sarcoplasmic reticulum (SR), but not an increase in cytosolic resting calcium *per se* since there was no increase in resting tension. Acidification could increase the amount of cytosolic free calcium by displacing it from cytosolic calcium-buffering proteins. The free calcium may be sequestered into the sarcoplasmic reticulum and increase the amount of calcium available for release (30). Another possible reason for the delayed increase in tension during acidification is that during cytosolic acidification there is less calcium released from the sarcoplasmic reticulum (25), allowing calcium to accumulate inside the SR. With less calcium released from the SR, more calcium will be taken up from the medium, and will result in increased calcium storage. The calcium in the intracellular storage sites is increased, and this may eventually produce the small increase in twitch tension seen during the acidic treatment and may also explain the reactive inotropy upon removal of the acid medium when the calcium release channel is disinhibited.

In summary, the calibration curves for the fluorescence of BCECF must be adjusted to compensate for dye leakage from the cells throughout each experiment; hence the pH-dependence of the dye fluorescence is dependent on the concentration of the dye. These changes are linear such that measurement of FI_{440} provides the parameters required to correct for the calibrations. Once the correction parameters are determined for a particular tissue type and equipment they could be employed in all subsequent experiments requiring pH measurements with the same equipment and in the same tissue type. We found that the acidification of the cytosol is not by itself responsible for the tissue damage caused by post-ischemic reperfusion. The presence or absence of bicarbonate in the buffer medium does not change the cellular control of pH_i during or after acidification. There is an immediate and complete recovery of pH_i and tension upon removal of the acid load imposed by either hypercapnia or lactic acid.

FOOTNOTES

List of abbreviations

BCECF	2′,7′-bis(carboxyethyl)-5,6-carboxyfluorescein
BCECF-AM	2′,7′-bis(carboxyethyl)-5,6-carboxyfluorescein acetoxy methyl ester
FI_{440}	Fluorescent Intensity with 440 nm excitation, 540 nm emission

FI_{500}	Fluorescent Intensity with 500nm excitation, 540nm emission
pH_i	cytosolic pH
pH_o	pH of bathing medium

ACKNOWLEDGEMENTS

Supported by grants from Heart and Stroke Foundation and the Medical Research Council of Canada. A. Guia was a recipient of Heart and Stroke Foundation of Canada Traineeship Award.

REFERENCES

1. Fabiato A, Fabiato F. 1978. Effects of pH on the myofilaments and the sarcoplasmic reticulum of skinned cells from cardiac and skeletal muscles. J Physiol 276:233–255.
2. Moody W, Hagiwara S. 1982. Block of inward rectification by intracellular H^+ in immature oocytes of the starfish *Mediaster aequalis*. J Gen Physiol 79:115–130.
3. Roos A, Boron WF. 1981. Intracellular pH. Physiol Rev 61:296–435.
4. Bountra C, Vaughan-Jones RD. 1989. Effect of intracellular and extracellular pH on contraction in isolated, mammalian cardiac muscle. J Physiol 418:163–187.
5. Bright CM, Ellis D. 1992. Intracellular pH changes induced by hypoxia and anoxia in isolated sheep heart Purkinje fibers. Exp Physiol 77:165–175.
6. Kohmoto O, Spitzer KW, Movsesian MA, Barry WH. 1990. Effects of intracellular acidosis on $[Ca^{2+}]_i$ transients, transsarcolemmal Ca^{2+} fluxes, and contraction in ventricular myocytes. Circ Res 66:622–632.
7. Mohabir R, Lee H-C, Kurz RW, Clusin WT. 1991. Effects of ischemia and hypercarbic acidosis on myocyte calcium transients, contraction, and pH_i in perfused rabbit hearts. Circ Res 69: 1525–1537.
8. Borzak S, Kelly RA, Kramer BK, Matoba Y, Marsh JD, Reers M. 1990. In situ calibration of fura-2 and BCECF fluorescence in adult rat ventricular myocytes. Am J Physiol 259:H973–H981.
9. Bright GR, Fisher GW, Rogowska J, Taylor DL. 1984. Fluorescence ratio imaging microscopy: temporal and spatial measurements of cytoplasmic pH. J Cell Biol 98:717–724.
10. Aalkjær C, Cragoe EJ. 1988. Intracellular pH regulation in resting and contracting segments of rat mesenteric resistance vessels. J Physiol 402:391–410.
11. Putnam RW, Grubbs RD. 1990. Steady-state pH_i, buffering power, and effect of CO_2 in a smooth muscle-like cell line. Am J Physiol 258:C461–C469.
12. Yu J, Zheng JJ, Ong BY, Bose R. 1991. Intracellular pH measurement with fluorescent dye in canine basilar arteries. Blood Vessels 28:464–474.
13. Gérard C, Boudier JA, Mauchamp J, Verrier B. 1990. Evidence for probenecid-sensitive organic anion transporters on polarized thyroid cells in culture. J Cell Physiol 144:354–364.
14. Allen DG, Orchard CH. 1987. Myocardial contractile function during ischemia and hypoxia. Circ Res 60:153–168.
15. Carter G, Gavin JB. 1989. Endocardial damage induced by lactate, lowered pH and lactic acid in non-ischemic beating hearts. Pathol 21:125–130.
16. Cobbe SM, Poole-Wilson PA. 1980. The time of onset and severity of acidosis in myocardial ischaemia. J Mol Cell Cardiol 12:745–760.
17. Lazdunski M, Frelin C, Vigne P. 1985. The sodium/hydrogen exchange system in cardiac cells: its biochemical and pharmacological properties and its role in regulating internal concentrations of sodium and internal pH. J Mol Cell Cardiol 17:1029–1042.
18. Eisner DA, Nichols CG, O'Neill SC, Smith GL, Valdeolmillos M. 1989. The effects of metabolic inhibition on intracellular calcium and pH in isolated rat ventricular cells. J Physiol 411:393–418.
19. James-Kracke MR. 1992. Quick and accurate method to convert BCECF fluorescence to pH_i: calibration in three different types of cell preparations. J Cell Physiol 151:596–603.
20. Rink TJ, Tsien RY, Pozzan T. 1982. Cytoplasmic pH and free Mg^{2+} in lymphocytes. J Cell Biol 95:189–196.
21. Thomas JA, Buchsbaum RN, Zimniak A, Racker E. 1979. Intracellular pH measurements in Ehrlich Ascites tumor cells using spectroscopic probes generated in situ. Biochemistry 18:2210–2218.

22. Reers M, Kelly RA, Smith TW. 1989. Calcium and proton activities in rat cardiac mitochondria: effect of matrix environment on behavior of fluorescent probes. Biochem J 257:131–142.
23. Noël J, Tejedor A, Vinay P, Laprade R. 1989. Fluorescence measurement of intracellular pH on proximal tubule suspensions, the need for a BCECF sink. Renal Physiol Biochem 12:371–387.
24. Orchard CH, Kentish JC. 1990. Effects of changes of pH on the contractile function of cardiac muscle. Am J Physiol 258:C967–C981.
25. Rousseau E, Pinkos J. 1990. pH modulates conducting and gating behavior of single calcium release channels. Pflügers Arch 415:645–647.
26. Harvey RD, Ten Eick RE. 1989. On the role of sodium ions in the regulation of the inward rectifying potassium conductance in cat ventricular myocytes. J Gen Physiol 94:329–348.
27. Irisawa H, Sato R. 1986. Intra- and extracellular actions of proton on the calcium current of isolated guinea pig ventricular cells. Circ Res 59:348–355.
28. Hoxworth JM, Xu K, Zhou Y, Lust WD, LaManna JC. 1999. Cerebral metabolic profile, selective neuron loss and survival of acute and chronic hyperglycemic rats following cardiac arrest and resuscitation. Brain Res 821:467–479.
29. Kaila K, Vaughan-Jones RD, Bountra C. 1987. Regulation of intracellular pH in sheep cardiac Purkinje fiber: interactions among Na^+, H^+, and Ca^{2+}. Can J Physiol Pharmacol 65:963–969.
30. Balnave CD, Vaughan-Jones RD. 2000. Effect of intracellular pH on spontaneous Ca^{2+} sparks in ventricular myocytes. J Physiol 528:125–137.
31. MacLeod KT, Harding SE. 1991. Effects of phorbol ester on contraction, intracellular pH and intracellular Ca^{2+} in isolated mammalian ventricular myocytes. J Physiol 444:481–498.
32. Vaughan-Jones RD, Eisner DA, Lederer WJ. 1987. Effects of changes of intracellular pH on contraction in sheep cardiac Purkinje fibers. J Gen Physiol 89:1015–1032.
33. Karmazyn M. 1988. Amiloride enhances postischemic ventricular recovery: possible role of Na^+-H^+ exchange. Am J Physiol 255:H608–H615.
34. Meng, H-P, Pierce GN. 1990. Protective effects of 5-(N,N-dimethyl)amiloride on ischemia-reperfusion injury in hearts. Am J Physiol 258:H1615–H1619.
35. Bielen FV, Bosteels S, Verdonck F. 1990. Consequences of CO_2 acidosis for transmembrane Na^+ transport and membrane current in rabbit cardiac Purkinje fibers. J Physiol 427:325–345.
36. Kaila K, Vaughan-Jones RD. 1987. Influence of sodium-hydrogen exchange on intracellular pH, sodium and tension in sheep cardiac Purkinje fibers. J Physiol 390:93–118.
37. Hayashi H, Miyata H, Noda N, Kobayashi A, Hirano M, Kawai T, Yamazaki N. 1992. Intracellular Ca^{2+} concentration and pH_i during metabolic inhibition. Am J Physiol 262:C628–C634.
38. Vaughan-Jones RD, Wu M-L. 1990. Extracellular H^+ inactivation of Na^+-H^+ exchange in the sheep cardiac Purkinje fiber. J Physiol 428:441–466.
39. Imai S, Shi A-Y, Ishibashi T, Nakazawa M. 1991. Na^+/H^+ exchange is not operative under low-flow ischemic conditions. J Mol Cell Cardiol 23:505–517.
40. Greene HL, Weisfeldt ML. 1977. Determinants of hypoxic and posthypoxic contracture. Am J Physiol 232:H526–H533.
41. Harrison SM, Lancaster MK. 1994. The effects of NiCl on intracellular Ca^{2+}, Na^+, and H^+, during metabolic inhibition in rat ventricular cells. Biophys J 66:A95.
42. McCord JM. 1985. Oxygen-derived free radicals in postischemic tissue injury. N Eng J Med 312:159–163.

II. Pathogenesis of Ischemic Heart Disease

Pathophysiology of Cardiovascular Disease,
edited by N.S. Dhalla, H. Rupp,
A. Angel & G.N. Pierce
Kluwer Academic Publishers, Boston, 2004

Redox Regulation of Ischemic Heart Disease

Dipak K. Das

Cardiovascular Research Center, University of Connecticut School of Medicine, Farmington, Connecticut

Summary. In this review, attempts are made to establish the role of reactive oxygen species as signaling molecules which regulate cellular and molecular basis of ischemic heart disease. Ischemia/reperfusion is a classical example for free radical signaling, because partial or mild ischemia can lead to the generation of reactive oxygen species (ROS) which simultaneously perform execution and repair of the cardiomyocytes. Execution occurs with the death signaling in which ROS play a crucial role while repair process is mediated by adaptive response in which oxygen free radicals function as typical signaling molecules through the activation of receptor tyrosine kinases, protein kinase C and MAP kinases as well as induction of redox-sensitive transcription factors and genes.

Key words: Redox signaling, Ischemic heart disease, Ischemia/Reperfusion Injury, Apoptosis, Gene expression, Survival and Death Signal, Ischemic Preconditioning, Stress Adaptation.

INTRODUCTION

Evidence is rapidly accumulating to support the role of reactive oxygen species as intracellular second messengers. The long-held view of oxygen free radicals being detrimental to the biological tissues was challenged after the recent discovery that these reactive species can function as signaling molecules (1). Involvement of oxygen free radicals in mitogenic stimulation of cell as well as growth factor- and cytokine-induced signal transduction strongly suggest that the reactive oxygen species function as second messengers (2). Moreover, the mitogenic signals mediated through

Correspondence: Dipak K. Das, Ph.D., Cardiovascular Research Center, University of Connecticut, School of Medicine, 263 Farmington Avenue, Farmington, CT 06030-1110. Tel: 860 679-3687, Fax: 860 679 4606, E-mail: DDAS@NEURON.UCHC.EDU.

the generation of reactive oxygen species, activate many transcription factors including nuclear transcription factor NFκB and genes including antioxidant enzymes and bcl-2 (3).

A large number of degenerative diseases, including coronary heart disease, have been linked with the overproduction of oxygen-derived free radicals. The results of many studies, including our own, have demonstrated that excessive production of reactive oxygen species in concert with drastic reduction of antioxidant reserve play a crucial role in the pathophysiology of ischemic heart disease (4). Myocardial ischemia and reperfusion cause the cardiomyocytes to face conditions that shift their redox status to undergo a drastic chage subjecting them to oxidative stress (5). Interventions with oxygen free radical scavengers or antioxidant therapy have been found to be cardioprotective against ischemic reperfusion injury (6).

Recent studies have documented that coronary heart diseases cause cardiomycyte death not only by necrosis but also by apoptosis (7). Reperfusion of ischemic myocardium results in apoptotic cell death and DNA fragmentation (8). In concert, ischemia/reperfusion is associated with the induction of a number of both pro- and anti-apoptotic genes and transcription factors (9). Ebselen, a gluthathione peroxidase mimic, was found to reduce cardiomyocyte apoptosis (10). The hearts from the transgenic mice overexpressing GSHPx-1 gene were resistant to ischemia/reperfusion injury while those from the GSHPx-1 knockout animals, devoid of any copy of GSHPx-1, was extremely vulnerable to the cellular injury as compared to wild-type controls (11).

The results mentioned above clearly indicate that cardiomyocyte death induced by ischemia/reperfusion is redox-regulated and that reactive oxygen species not only trigger the cell death but also function as intracellular signaling molecules. This review will discuss the intracellular signaling pathways potentiated by the reactive oxygen species that lead to the induction of gene expression and apoptotic cell death.

REDOX SIGNALING: ROLE OF MITOCHONDRIA

A number of sources have been identified for the generation of oxygen-derived free radicals in the heart during ischemia and reperfusion which include catecholamine oxidation, xanthine dehydrogenase-xanthine oxidase conversion, microsomal respiratory chain via cytochrome P-450 monooxygenase system and mitochondrial uncoupling (12,13). It appears that all of these sources actively take part in the process of free radical production, because inhibition of any of these sources can reduce the cellular injury to some extent. Mitochondria seems to play a crucial role in the process of production and propagation of reactive oxygen species as described below.

Life needs a continuous supply of energy, and cells depend on mitochondriala for its production and supply. Hypoxia or ischemia causes rapid energy consumption by depleting ATP which cannot be readily regenerated because of the absence of both glucose and oxygen. The mitochondrial respiratory chain is also a major source of reactive oxygen species. Upon reperfusion, ischemic heart mitochondria overproduce hydroxyl radicals ($OH^{\cdot}$) which can be reduced by pretreating the hearts with $OH^{\cdot}$ scavengers (14). Reduction of $OH^{\cdot}$ is associated with the amilioration of ischemic reperfusion injury suggesting that mitochondrial free radical production plays a role in the pathophysiology of cellular injury. We now know that mitochondria play a major role in both necrotic and apoptotic cell death. While partial ischemia or hypoxia causes both necrotic and apoptotic cell death, complete ischemia can only cause cell necrosis, because a certain amount of ATP is required to execute apoptosis. During reperfusion, the permeability transition pore in the inner mitochondrial membrane that is normally closed, becomes opened, presumably due to the influx of Ca^{2+} and/or oxidants (15). Mitochondria rapidly loses their proton-motive force due to the leakage of protons through the mitochondrial permeability transition pore and rapidly hydrolyze cellular ATP resulting in swelling, bursting the outer mitochondrial membrane, and releasing cytochrome C and other intermembrane proteins, triggering apoptosis.

It should be clear from the above discussion that mitochondria serve as judge to decide whether a cell would live or die, and if die, whether the cell should die of necrosis or apoptosis. During partial or reversible ischemia, mitochondria of the cardiomyocytes can maintain a certain level of ATP, enough to execute apoptosis and simultaneously activate a cascade of enzymes responsible for the repair and prevention of cell death. Thus, there are two opposing forces existing in the ischemic mitochondria, one to execute that include proteases, nucleases, and reactive oxygen species, and another to extricate that include the repairing enzymes. During apoptosis, cytochrome c is released, activating caspase 3 which serves as the final executioner for apoptotic cell death. On the other hand, the anti-death gene, bcl-2, located on the outer mitochondrial membrane, tries to oppose the apoptotic process. In a recent study, our laboratory demonstrated that bcl-2 activity is downregulated in the ischemic-reperfused myocardium when apoptotic cardiomyocytes appear (Figure 1) (16). Myocardial adaptation to ischemia upregulated bcl-2 and prevented apoptosis. Bcl-2 gene appears to play a crucial role in sending the survival signal to the ischemic myocardium because inhibition of bcl-2 gene with antisense ODN completely blocked cardioprotective effects of preconditioning (Figure 2 and Figure 3).

INTRACELLULAR PATHWAYS FOR REDOX SIGNALING

Recent studies implicate that reactive oxygen species are not only the destructive elements for the cells, but also they are essential for the biological and physiological function of the cells. Biological cells including cardiomyocytes contain enzymes that can simultaneously generate reactive oxygen species and intracellular redox buffer in response to a specific stress. Depending on the amounts of antioxidant reserve and oxygen free radicals, the reactive oxygen species are either destroyed or persist. Thus, the oxygen free radicals fulfil the defination of a second messenger which are either up-regulated or down-regulated after a physiologic stimuli like ischemia. A number of growth factors and cytokines have been found to induce oxidative stress and overproduce specific antioxidant enzymes (17,18). Environ-

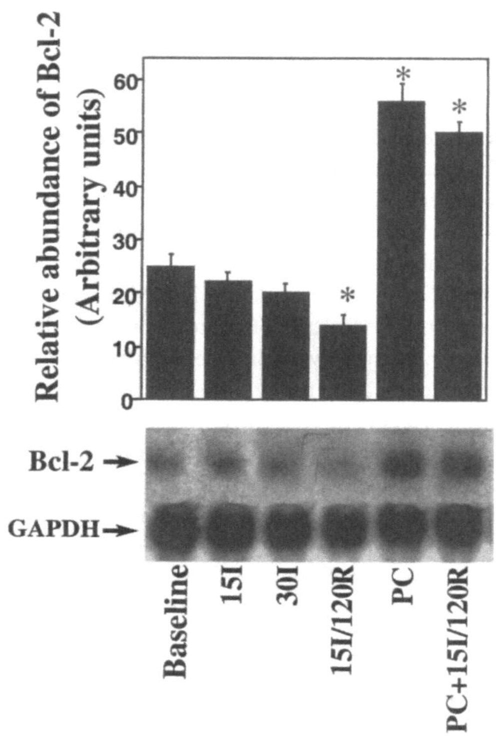

Figure 1. *Effects of ischemia/reperfusion and preconditioning (adaptation) on the Bcl-2 mRNA expression.* As revealed by the Northern blot analysis, the induction of Bcl-2 gene expression was reduced significantly in the ischemic reperfused myocardium while the Bcl-2 gene expression was increased in the preconditioned heart (Top). The housekeeping gene, GAPDH, is shown at the bottom.

mental stresses including heat stress, oxidative stress and ischemia/reperfusion also produce oxidative stress which is then translated into the induction of antioxidant enzymes (19).

Perhaps the finding that the production of reactive oxygen species during the agonist-induced activation of the nuclear transcription factor, NFκB, provided the first concrete evidence for the role of reactive oxygen species as a second messenger. NFκB regulates the inducible expression of a number of genes involved in cell survival and execution. For example, NFκB has been found to control anti-apoptotic gene, bcl-2, and pro-apoptotic factors, bax and p53, in the ischemic/reperfused myocardium (20). Diverse extracellular signals from IL-1, TNFα, H_2O_2, etc. converge into development of oxidative stress which leads to activation of NFκB (21). Such activation of NFκB can be blocked by antioxidants such as vitamin E (22) or α-lipoic acid (23). Antioxidants such as N-acetyl cystein can prevent NFκB activation by diverse stimuli, including H_2O_2, by supressing the generation of reactive oxygen species (24).

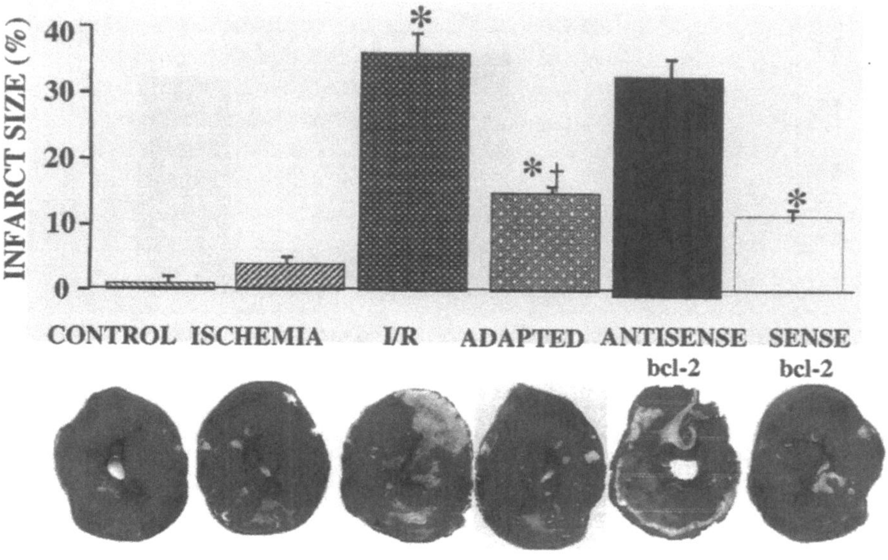

Figure 2. *Effects of antisense bcl-2 ODN on myocardial infarct size of ischemia/reperfused and preconditioned (adapted) heart.* Hearts were perfused for 150 min and served as control, or hearts were subjected to 30 min ischemia or 30 min ischemia followed by 120 min reperfusion. The preconditioned group was subjected to four episodes of, 5 min ischemia followed by 10 min of reperfusion. (Bottom) Representative hearts revealing various degrees of myocardial infarction. (Top) The percent infarct size of control ([ ]), ischemia ([ ]), ischemia and reperfusion ([ ]), preconditioned ([ ]), antisense bcl-2 ([ ]), and sense bcl-2 ([ ]) transfected rat myocardium. Results are expressed as Means ± SEM of six different rats per group. *$p < 0.05$ compared with the control group of sample.

Receptor tyrosine kinase (RTK) plays an important role in redox signaling by autophosphorylation of several tyrosines along their own intracellular tails. Recent studies from several laboratories including our own have implicated that myocardial adaptation to ischemic stress occurs through the activation of several tyrosine kinases (25). Phosphorylation of tyrosine kinases has been shown to be linked with the activation of both phosapholipase C and phospholipase D leading to the activation of multiple kinases including protein kinase C (PKC) and MAP kinases. Phospholipase D appears to play an important role in the intracellular signaling process. Cohen et al. also demonstrated that PLD is critical in ischemic adaptation (26). Activation of PLD was documented in ischemic reperfused hearts (27). Phospholipase D catalyzes the terminal diester bond of phosphatidylcholine which results in the formation of choline and phosphatidic acid. Phosphatidic acid then serves as the substrate for diacylglycerol synthesis by the action of phosphatidic acid on phosphohydrolase. Diacyl glycerol may itself serve as a second messenger by activating PKC. Work done by Eskildsen-Helmond et al. (28) also suggests a link between ischemic adaptation and activation of PLD, eventually resulting in potentiation of PKC isoenzymes. In a more recent study, Fryer et al. (29) demonstrated that

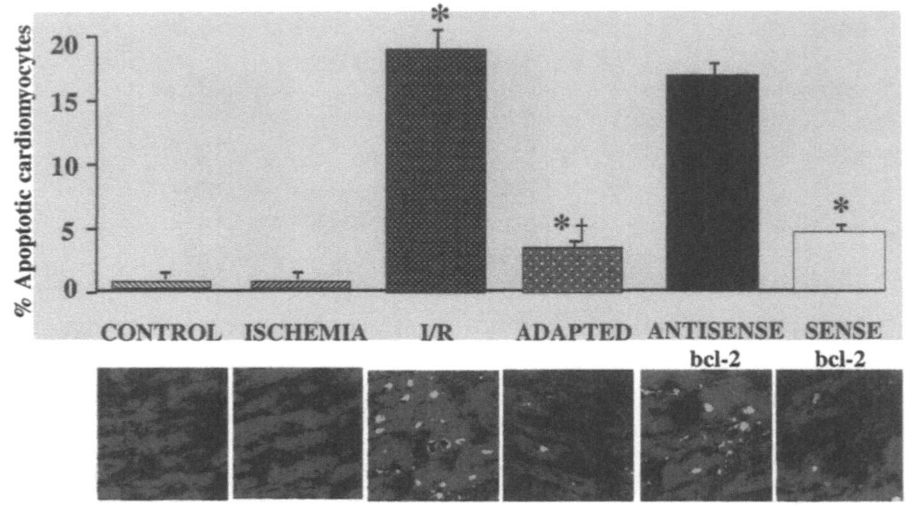

Figure 3. *Effects of antisense bcl-2 ODN on cardiomyocyte apoptosis of ischemia/reperfused and preconditioned (adapted) heart.* Hearts were perfused for 150 min and served as control, or hearts were subjected to 30 min ischemia or 30 min ischemia followed by 120 min reperfusion. The preconditioned group was subjected to four episodes of, 5 min ischemia followed by 10 min of reperfusion. (Bottom) Representative hearts revealing various degrees of cardiomyocyte apoptosis (spots are apoptotic cardiomyocytes revealed after TUNEL staining in conjunction with antibody against α-myosin heavy chain antibody). (Top) The percent infarct size of control ([ ]), ischemia ([ ]), ischemia and reperfusion ([ ]), preconditioned ([ ]), antisense bcl-2 ([ ]), and sense bcl-2 ([ ]) transfected rat myocardium. Results are expressed as means ± SEM of six different rats per group. *p < 0.05 compared with the control group of sample.

pretreatment with tyrosine kinase inhibitors partially attenuated ischemic adaptation in the rat heart.

The mitogen-activated protein (MAP) kinases, a serine/threonine protein kinase family, play an essential role in mediating intracellular signal transduction events. In response to extracellular stimulation, MAP kinases are rapidly activated and in turn regulate cellular functions by inducing the phosphorylation of proteins, such as an oncogene product c-jun, S6 ribosomal protein kinase, and MAP kinase activated protein kinase 2 (30,31). MAPKAP kinase 2 has been implicated in a novel mammalian stress activated signal transduction pathway initiated by a variety of mitogens, pro-inflammatory cytokines, or environmental stresses, where it regulates its substrate molecules by serine/threonine phosphorylation (32). In the case of rat heart, a mitogen-activated protein kinase cascade has already been identified (33). These authors have demonstrated that MAPK isoforms p^{42} MAPK and P^{44} MAPK and two peaks of MEK were activated by more than 10-fold in perfused hearts or ventricular myocytes exposed to PMA for 5 min. In our own study, we identified the participation of MAP kinase cascades in the ischemically adapted rat hearts (34). The results of our study demonstrated that a kinase cascade involving tyrosine kinase-phospholipase D-MAP kinases-MAPKAP kinase 2 is triggering after ischemic stress.

The intracellular signaling mechanisms that lead to adaptation require one or more members of MAP kinase cascades. Among the three distinct MAP kinase families, stress-activated protein kinase (SAPK), also known as c-Jun NH_2-terminal kinases (JNK), and p38 MAP kinase are known to be regulated by extracellular stresses including environmental stress, oxidative stress, heat shock and UV radiation (35). JNKs and p38 MAP kinase appear to be involved in distinct cellular functions, because they possess different cellular targets and are located on different signaling pathways. Thus, JNK kinases activate c-Jun while p38 MAP kinase stimulates MAPKAP kinase. A recent study demonstrated that ischemic adaptation triggered a tyrosine kinase-regulated signaling pathway leading to the translocation and activation of p38 MAP kinase and MAPKAP kinase 2 (36).

The results of a recent study documented that 30 min of ischemia followed by 2 hr of reperfusion increased the induction of JNK1, c-JUN and p38 MAP kinase proteins (37). Ischemic adaptation also enhanced these kinases compared to control. However, subsequent ischemia/reperfusion-mediated increase in JNK1, p38 and c-Jun was blocked by this adaptation. Fifteen minutes of perfusion with anisomycin increased the amount of both JNK1 and p38 MAP kinases as well as c-Jun in the heart, all of which were decreased in amount after subsequent ischemia/reperfusion. Curcumin, a JNK and c-Jun inhibitor, blocked the ischemia/reperfusion- and adaptation-mediated increase in JNK1 and c-Jun while it had no effect on p38 MAP kinase. SB 203580, an inhibitor for P38 MAPK phosphorylation, on the other hand, was equally effective in reducing the amount of p38 MAP kinase, but exterted no effects on JNK1 and c-Jun. Ischemia/reperfusion-mediated increased myocardial infarction was reduced by treating the hearts with anisomycin, a dual activator of P38 MAPK and JNK. The cardioprotective effects of preconditioning were abolished by either curcumin or SB 203580. Thus, it appears that activation of SAPKs are obligatory for myocardial adaptation to ischemia, but the activation is only transient, and the activities rapidly come down to near baseline levels after subsequent ischemia and reperfusion (37).

Evidence suggests that MAPKAP kinase 2 is a crucial step leading to gene expression and myocyte adaptation resulting in adaptive cardioprotection. This unique protein kinase is highly expressed in heart muscle suggesting that it also may be expressed and functioning in the myocardium in response to stress. MAP KAP kinase 2 also has been shown to have increased activity when subjected to oxidative stress as well as heat shock (33,36). This increased activity of MAPKAP kinase 2 in association with heat shock protein gives rise to the hypothesis that this kinase may be one of the critical factors involved with ultimate transcription of proteins leading to adaptive protection of the heart. In cultured myocytes, the activity of MAPKAP kinase 2 was found to increase when the myocytes were subjected to oxidative stress as well as heat shock (33). Heat shock proteins (HSP) are early targets of phosphorylation by a variety of stress conditions. HSP-27, HSP-32 and HSP-70 can be induced by oxidative stress and ischemic preconditioning (38). Heat shock proteins have been found to be not only cardioprotective through reduction of infarct size but evidence also suggests that they are important in delayed protection against infarction, the so called "second window" of protection.

As mentioned earlier, it has been demonstrated that cellular PKC activation is an important step in the mechanism of adaptive protection of heart (39). The PKC hypothesis received further support from the observations that any agent that can activate PKC can also precondition the heart. For example, phenylephrine, an α_1 agonist, angiotensin AT_1 and bradykinin B_2 receptors can activate PKC (40), and they can also precondition the hearts when infused prior to ischemia (39,41). A variety of stress signals can also translocate and activate PKC. For example, mechanical stress induced by stretching can activate PKC in cultured myocytes (42). Immediately after stretching, activation of phosphatidyl inositol turnover was observed suggesting a role of phospholipase C in PKC activation. Even a short-term ischemia or ischemia followed by reperfusion were previously shown to translocate and activate PKC. Furthermore, both α_1-receptor stimulation and Ca^{2+} ion can translocate and activate PKC (43).

A recent study from our laboratory demonstrated the inhibition of the enhanced tyrosine kinase phosphorylation during ischemic adaptation by DMTU (44). DMTU also inhibited adaptation-mediated increased phosphorylation of p38 MAP kinase and MAPKAP kinase 2 activity. However, DMTU had no effect on the translocation and activation of protein kinase C (PKC) resulting from adaptation. The cardioprotective effect of adaptation was abolished by both DMTU and SN 50. Ischemic adaptation resulted in the nuclear translocation and activation of NFκB. Increased NFκB binding was blocked by both DMTU and SN 50. The results of this study demonstrate that reactive oxygen species play a crucial role in signal transduction mediated by adaptation. This signaling process appears to be potentiated by tyrosine kinase phosphorylation resulting in the activation of p38 MAP kinase and MAPKAP kinase 2 leading to the activation of NFκB, suggesting a role of oxygen free radicals as second messenger. Free radical signaling seems to be independent of PKC, although PKC is activated during adaptation process suggesting the role of two separate signaling pathways in the adapted heart.

REGULATION OF EXECUTION AND REPAIR BY REDOX SIGNALING

Although ischemia/reperfusion has been found to cause cardiomyocyte death, precise physiological role of redox signaling is not clear. Ischemia/reperfusion, especially intermittant ischemia commonly known as ischemic preconditioning, lead to the activation of both G-proteins and receptor tyrosine kinases potentiating a signaling cascade resulting in the activation of multiple kinases that leads to the induction of the activation of several redox-sensing transcription factors and genes. Such intracellular events ultimalely dictate the cells to survive or die. The changes in gene expression are likely to influence the physiologic function of the cardiomyocytes during post-ischemic survival.

A number of recent studies clearly demonstrated that redox signaling play a physiological role in myocardial survival during post-ischemic period. As mentioned

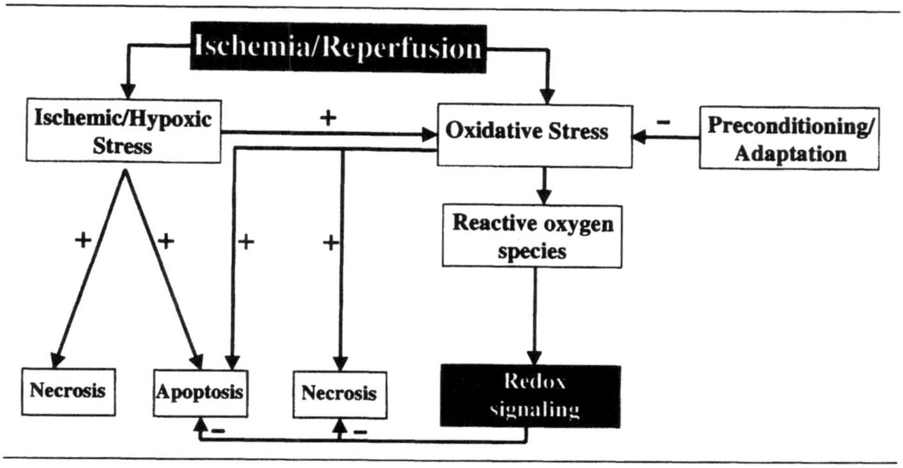

Figure 4. Redox Signaling by Ischemia/Reperfusion.

earlier, oxygen-free radicals are generated during ischemia/reperfusion. When heart is adapted to ischemic stress by repeated short-term ischemia and reperfusion, the generation of the reactive oxygen species is rapidly increased, but does not increase at the same rate during subsequent ischemia and reperfusion. The same pattern of the development of oxidative stress is observed when hearts are treated with endotoxin, IL-1 ot TNFα. Interestingly, these interventions lead to the development of oxidative stress within a very short period, but reduce/inhibit subsequent oxidative stress development when hearts are subjected to ischemia/reperfusion. Additionally, such adaptive response is associated with the induction of the expression of a number of stress proteins, including HSP 27, HSP 70 and heme oxygenase and antioxidant enzymes that include Mn-SOD and glutathione peroxidase (36). Such adaptive response is also associated with the increased binding activity of NFκB. Increased activity of NFκB and induction of the protective proteins can be blocked by pretreating the hearts with a oxygen free radical scavenger such as DMTU (44). More interestingly, an endotoxin-derived compound, monophosphoryl lipid A (MLA), was found to induce iNOS mRNA through a tyrosine kinase-dependent mechanism and protect the heart from ischemic reperfusion injury.

Thus, it seems likely that one of the major functions of redox signaling in the ischemic myocardium is to synthesize stress-inducible proteins through the activation of transcription factors such as NFκB or by potentiating induction of other inducible proteins such as iNOS (Figure 4). However, more study is necessary to completely unveil the physiologic function of redox cycling.

REDOX-SENSING TRANSCRIPTION FACTORS AND GENES

NFκB and AP-1 are two well-known redox-sensitive transcription factors. NFκB is a critical regulator for gene expression induced by diverse stress signals including mutagenic, oxidative and ischemic stresses. Activation of NFκB is likely to be involved in the induction of gene expression associated with the ischemic adaptation. This nuclear transcription factor is a member of Rel transcription factor family which is involved in the regulation of stress defense mechanisms. Survey of literature suggests that NFκB is an oxidative stress responsive transcription factor, and that reactive oxygen intermediates play a crucial role in the activation of the factor. Recent studies from our laboratory demonstrated ischemic adaptation translocated and increased the binding of NFκB in heart. NFκB binding activity is very low in non-ischemic control hearts (45). Reperfusion of ischemic myocardium significantly increases the translocation of NFκB from cytosol to nucleus. Perfusion of the heart with DMTU inhibits NFκB translocation from cytosol to nucleus (44). NFκB binding activity is increased in the ischemically adapted hearts.

AP-1 is another redox-sensitive signaling molecule which also plays an important regulatory role in cellular responses to stress induced by external factors including UV radiation, phorbol esters, and TNFα (46). The binding site of AP-1 is recognized by Jun family member homodimers and Jun/Fos family member heterodimers. The balance between Jun and Fos is very critical for gene expression. Stress induced by ischemia/reperfusion was previously shown to induce the activation of c-Jun (47). AP-1 is a redox-sensitive signaling molecule which plays an important regulatory role in cellular responses to stress induced by external factors. The AP-1 transcription factor complex is composed of a group of proteins encoded by the Jun and Fos families which bind to the AP-1 consensus sequences. Regulation of AP-1 in response to external stimuli is mediated by members of the MAPK family. AP-1 regulates the activation of transcription of a variety of genes. Electrophoretic mobility shift assay indicated increased AP-1 binding activity in the ischemic/reperfused rat heart compared to control perfused group (48). DMTU inhibited this binding activity significantly. Similar effects were also observed for ischemically adapted group. Myocardial adaptation induced by cyclic episodes of short term ischemia and reperfusion results in the activation of NFκB but not AP-1.

In various cell lines, wild-type-P53 (transcription factor, pro-apoptotic gene) induced by DNA damage has been shown to induce programmed cell death or apoptosis. It is reported that P53 protein functions as an active transcription factor in lesioned brain (49). How activation of p53 promotes apoptosis is unclear, but it might involve Bax, a series of p53-inducible genes or signaling through Fas-related pathways (50). There are other p53 effectors which include caspases that execute apoptotic cell death (51). Several studies on the master controller gene of apoptosis, p53, reveals conflicting results (52). Our recent study showed an increase in p53 activity in the ischemic/reperfused myocardium (48). Such an increase in p53 activity was prevented by adapting the heart to ischemia.

Bcl-2 is a well-known anti-death gene which function as an intracellular antioxidant. Recent analysis of the bcl-2 gene family reveals complex network regulating apoptosis. Within this bcl-2 gene family, some of the candidates can supress apoptosis, while the others can induce apoptosis (53). Among the proteins coded by the genes of this family, Bcl-2 and Bcl-x_L act as cell death repressors (54), whereas Bax and an alternatively sliced Bcl-x product, Bcl-xs, promote cell death (7). When in excess over Bcl-2, Bax counterracts the repressive action of Bcl-2 on apoptosis. Similarly, excess Bcl-xs antagonize the function of Bcl-xL. Thus, a critical balance between the Bcl-2, Bax and Bcl-$x_{L/S}$ molecules may determine the fate of cells in response to cytotoxic agents or environmental stress. It has been reported that p53 activate the transcription of the Bax gene via p53-response elements while down regulating Bcl-2 expression at the same time (50).

Apoptosis initiated by various different stimuli can be blocked by overexpressing Bcl-2. For example, the activation of Bcl-2 was associated with the inhibition of apoptosis in the adapted myocardium (48). A downregulation of this anti-death gene occurred in concert with the significant amount of apoptosis in the ischemic/reperfused myocardium.

DEATH SIGNAL BY ROS

Programmed cell death or apoptosis is recognized as a physiological counter part of cell replication and is the contributing cause of cardiomyocyte cell death during ischemia/reperfusion, myocardial infarction, and heart failure (55). Apoptosis is an energy requiring process, needs *de novo* pro apoptotic gene expression (p53, bax etc.) and is directed by an in born genetic program. The terminal result of this program is the fragmentation of nuclear DNA which is associated with ultrastuctural changes in cellular morphology, while the functional integrity of the cell membrane still remains intact. Two different mechanisms of cell death have been evaluated so far during myocardial infarction, necrosis and apoptosis. The important distinguishing features of cell necrosis are rupture of cell membrane, cell swelling, plasma membrane breakdown, clumping of nuclear chromatin, swelling and disruption of sarcoplasmic reticulum and mitochondria, and appearance of granular densities in the matrix of mitochondria. Due to rupture of the sarcoplasmic reticulum in the necrotic cardiomyocytes, calcium overloading takes place which causes disturbances in other electrolytes. On the other hand, apoptosis occurs in the absence of membrane rupture and is characterized by internucleosomal cleavage of DNA by a Ca^{2+} and Mg^{2+} dependent endonuclease. Ultrastructural features of apoptosis include segregation of nuclear chromatin and condensation of the cytoplasm. The apoptotic cell surface develops protuberances known as apoptotic bodies which are generally engulfed or digested by adjacent cells. Again apoptosis is an active, strongly regulated, energy-requiring process whereas necrosis is a passive process and occurs in response to lethal injury. Untimely initiation of apoptotic cell death in the myocardium might also play an important role in the pathogenesis of various other myocardial diseases such as heart failure, posttransplantation rejection, chemotherapy-induced cardiac dysfunction, viral infections etc.

The common inducers of apoptosis include oxygen free radicals/oxidative stress and Ca^{2+}, which are also implicated in the pathogenesis of myocardial ischemic reperfusion injury. Cardiomyocytes exposed to hypoxia revealed apoptotic cell death as evidenced by DNA fragmentation in conjunction with the expression of Fas mRNA (56). In a recent study, apoptotic and necrotic myocyte cell deaths associated with ischemia/reperfusion were shown to be independent contributing variables of infarct size in rats (31). Another study has shown apoptosis to be a feature of human vascular pathology; including restenotic lesions and, to a lesser extent, atherosclerotic lesions, suggesting that apoptosis may modulate the cellularity of lesions that produce human vascular obstruction (57).

Apoptotic cell death is a function of the duration of reperfusion and that even up to 1 hr of ischemia does not induce apoptosis (48). Apoptotic cells become apparent only in the 90 and 120 min reperfused hearts. None of the ischemic hearts showed any evidence of apoptosis. These results were corroborated with the findings of DNA fragmentation which showed increased ladders of DNA bands only in the 120 min reperfused hearts. The presence of apoptotic cells and DNA fragmentation in the myocardium were completely abolished by pre-treating the myocardium with ebselen, a glutathione peroxidase mimic which also reduced the ischemic reperfusion injury (8). In another related study, SOD plus catalase were found to amiliorate the apoptotoic cell death (58).

Although cardiomyocyte apoptosis occurs only after prolonged reperfusion following an ischemic insult, the signal for apoptosis is initiated during ischemia. Translocation of phosphatidyl serine and phosphatidyl ethanolamine, a hallmark for apoptosis, was found to occur during ischemia, but apoptosis did not become apparent until hearts were reperfused following an ischemic insult (59). In this study, isolated rat hearts subjected to 30 min global ischemia followed by 2 hr of reperfusion, showed apoptotic cardiomyocytes as expected. Assay of phosphatidylethanolamine (PE) and phosphatidylserine (PS) topography in cardiomyocytes using 2,4,6-trinitrobenzenesulfonate (TNBS) demonstrated that about 30% of total phosphatidylethanolamine and 5% of total phosphatidylserine were available for the TNBS derivative in the control hearts. The availability of both PE and PS was significantly increased in cardiomyocytes from the hearts after 30 min of ischemia, and this level was maintained up to 2 hr of reperfusion following ischemia. This suggests that ischemia results in a significant loss of normal asymmetric sarcolemmal phospholipid distribution, possibly with an outward migration of PE and PS.

Caspases enzymes are a family of cysteine proteases that play a central role in apoptosis. In humans at least seven of the ten currently known family members participate in one of two distinct signaling pathways: activation of proinflammatory cytokines and promotion of apoptotic cell death. As mentioned previously, during apoptosis, a loss of electrochemical gradient across the inner membrane occurs in mitochondria resulting in uncoupling of oxidative phosphorylation, generation of free radical and Ca^{2+} overload into the cytosol. In addition, cytochrome C and possibly other proteins are released from mitochondria into the cytosol. Mitochondria play a very important role in apoptosis. The mechanism by which caspases are activated is still unknown. However, recent studies have established that activation of proximal caspases leads to the activation of the distal caspases in the process of apoptosis (24). Caspase 3 (ced-3) activation begins when caspase-9 binds to Apaf-1 (ced-4 is homologous to the recently identified human protein, Apaf-1), this reaction is initiated by the release of cytochrome C in the cytosol. Caspase-9 is directly activated by Apaf-1 and cytochrome-c (60). Significant accumulation of cytochrome c in the cytosol, over myofibrils, and near intercalated discs of cardiomyocytes in failing hearts is reported in a heart transplantation study. The cytochrome c was also associated with the activation of caspase-3 and cleavage of its substrate protein kinase-c delta, but not poly (ADP-ribose) polymerase. Whereas, there was no accumulation of cytosolic cytochrome c or caspase-3 in the control hearts (61). In another study, rabbits subjected to 30 min of coronary artery occlusion followed by 3 h of reperfusion demonstrated significant activation of protein expression of caspases-2, -3 and -7. In addition, this study also successfully demonstrated selective cleavage of poly(ADP-ribose) polymerase (PARP) into apoptotic fragments in ischemic/reperfused myocardium. Administration of broad-spectrum caspase inhibitor YVAD-cmk (4.8 mg/kg) partially blocked caspase activation and reduced apoptotic cells. This study indicated that caspases are critical mediators of myocardial injury induced by ischemia/reperfusion and that inhibition of caspases may be beneficial in myocardial infarction (62). Bialik S et al. (63) reported that deprivation of serum and glucose, components of ischemia in vivo, induces apoptosis. This manifestation of apoptosis was blocked by zVAD-fmk, a peptide caspase inhibitor. In contrast to control cells, apoptotic cardiomyocytes revealed cytoplasmic accumulation of cytochrome C from mitochondria. Caspase processing was inhibited by the zVAD-fmk. During myocardial ischemia cytochrome C is released from mitochondria, and this is one of the earliest events in ischemic myocardium. Therefore, the possible scenerio of ischemia-induced myocardial apoptosis suggests: elevation of cytosolic Ca^{2+} concentration, release of cytochrome C from mitochondria which can associate with Apaf-1 factor and pro-caspase-9, triggering the activation of caspase-3 and apoptosis.

SURVIVAL SIGNAL BY REDOX SIGNALING

There is increasing evidence that a variety of antioxidants and antioxidative enzymes play a major role in myocardial protection against acute stress such as ischemia or hypoxia. Reactive oxygen species including superoxide anions (O_2^-), hydroxyl radicals ($OH^\cdot$) and singlet oxygen ($^1O_2$) are believed to be the key factors causing reperfusion injury. To defend itself against free radical attack, the heart like many other organs is equipped with its own defense system (64). The first line of defense comprises antioxidants such as α-tocopherol, ascorbic acid, glutathione and several antioxidant enzymes like SOD, catalase and glutathione peroxidase. SOD catalyzes the dismutation reaction whereas catalase detoxifies H_2O_2 efficiently. A large number of studies exist in the literature to support a direct role of oxygen free radicals in apoptosis. An expression vector containing SOD was found to delay the event of apoptosis in cultured sympathetic neurons. This study also proved that if SOD was injected after the oxidative stress, it had no effect on apoptosis at all. As mentioned earlier, a seleno peroxide mimic, ebselen, could reduce the apoptotic cell death and DNA fragmentation in concert with the reduction of myocardial ischemic reperfusion injury (8). During the apoptotic changes, the activity of Cu/Zn-type superoxide dismutase in the tadpole tail markedly increased with a concomitant inhibition in the catalase activity. The apoptotic process was found to be increased by the addition of H_2O_2 and aminotriazole, a potent catalase inhibitor (65). In another study, it was found that Cu/Zn-SOD, Mn-SOD and catalase in cultured neutrophils were significantly effective in delaying the event of apoptosis, suggesting that reactive oxygen species plays a role in the neutrophil apoptosis (66). Transgenic mice overexpressing GSHPx-1 gene were resistant to while knockout mice devoid of any copy of GSHPx-1 gene were susceptible to myocardial ischemia reperfusion injury (11). In concert, overexpression of GHSPx-1 gene was associated with significantly reduced number of apoptotic cardiomyocytes, and GSHPx-1 gene knockout mice mouse hearts had significantly higher number of apoptotic cells compared to those found in wild-type control.

Redox regulation of cardiomyocytte apoptosis in the ischemic reperfusion myocardium is further supported from the observation that ischemic adaptation-induced inhibition of cardiomyocyte apoptosis was blocked by a $OH^{\cdot}$ radical scavenger, DMTU as well as a NFκB blocker, SN-50 peptide (44). In concert, NFκB and DMTU prevented the cardioprotective effects of myocardial adaptation to ischemia.

ANTIOXIDANTS AND REDOX SIGNALING

Mammalian hearts are protected from the cellular injury by their own defence system which includes various intracellular antioxidants; such as glutathione, α-tocopherol, ascorbic acid, β-carotene; and antioxidant enzymes that include SOD, catalase and glutathione peroxidase. These cellular compounds reduce/eliminate the oxidative stress by directly quenching the reactive oxygen species before they damage vital cellular components, and therefore, they can be considered as part of the first line of defense against the external stress. Often, because of the inadequancy of the intracellular antioxidants or due to the presence of increased amount of oxidative stress, the reactive oxygen species may reach their targets that include nucleic acids, proteins, and lipids. This results in the injury to the cellular components causing DNA strand breaks, protein degradation, and lipid peroxidation. Mammalian cells are also protected by a second line defense system consisting of several lipolytic and proteolytic enzymes, proteases, phospholipases, etc. which are involved in the systematic recognition and removal of the injured cellular components (67). Recent studies from our laboratory indicated that myocardial cells possess an inducible pathway for the antioxidant defense. The signal transduction pathway by which various stress signals are translated into oxidative stress leading to the modulation of antioxidants/antioxidant enzymes are different, but the patterns of the inducible enzymes are strikingly similar. The expression of the inducible genes in response to environmental stress has been shown to be the reflection of the ultimate adaptive responses and viewed as the third line of defense (64).

Although cells such as cardiomyocytes need sufficient antioxidant reserve to maintain a reducing environment for their own protection, these cells also need a certain redox state (defined as the ratio of the concentrations of oxidizing equivalents to the concentrations of reducing equivalents) to maintain normal hormonal and ionic homeostasis. GSH by functioning as the principle redox buffer protects the cells from external injury through redox signaling. A good example of redox cycling is the change in GSH/GSSG ratio in the cardiomyocytes as a consequence of ischemic or oxidative stress and restoration of the ratio after antioxidant therapy. Ischemia/reperfusion is also associated with a reduction of antioxidant reserve which include lowering of certain antioxidant enzymes. On the other hand, ischemic adaptation prior to a lethal ischemic episode prevents the ischemia/reperfusion-mediated lowering of antioxidant activities. For example, the activities of Mn-SOD and GSHPx are lowered after a prolonged reperfusion. Such lowering of the enzyme activities is prevented by subjecting the hearts to ischemic adaptation prior to prolonged ischemia/reperfusion. Cytokines such as IL-1 or TNFα as well as endotoxin also upregulate these enzyme activities through the oxidative stress adaptation (68). Another antioxidant enzyme heme oxygenase (also known as HSP 32) is also adaptively increased in response to ischemia/reperfusion (69). As mentioned earlier, antioxidant and anti-death protein bcl-2 is reduced in the ischemic myocardium which is prevented by adapting the hearts to ischemia.

SUMMARY AND CONCLUSION

It appears from the above discussion, survival and death of the cardiomyocytes depend critically on their redox state (Figure 5). While sufficient antioxidant reserve warrants a reducing environment for the survival of the myocytes, the reserve is

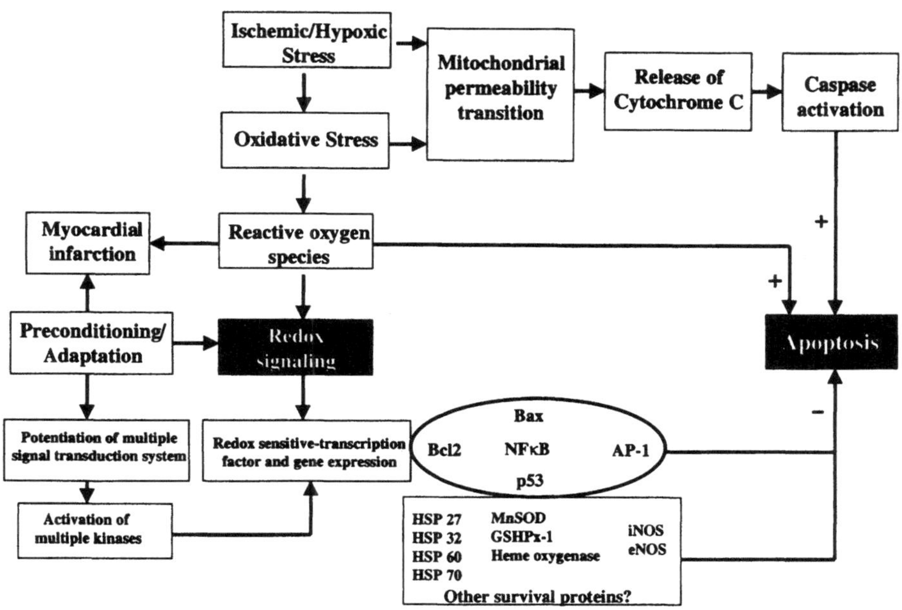

Figure 5. Redox Regulation of Anti-Death Signal.

exhausted during ischemic heart disease leading to a change in redox state. Such changes in the redox state determine whether the cardiomyocytes should die, and if they die, whether they die of necrosis or apoptosis. Ultimate survival or death depends on the redox signaling and transcription regulation of several redox sensing-genes and DNA repair processes.

LIST OF ABBREVIATIONS

NfκB: Nuclear transcription factor κB
AP-1: Activator protein-1
MAP: Mitogen-activated protein kinases
PKC: Protein kinase C
JNK: c-JunNH_2-terminal kinases
SAPK: Stress-activated protein kinases
HSP: Heat shock protein
PLD: Phospholipase D
PMA: Phorbol ester
ATP: Adenosine tri-phosphate
GSHPx: Glutathione peroxidase
SOD: Superoxide dismutase
Ca^{2+}: Calcium
TNF: Tumor necrosis factor
DMTU: Dimethyl thiourea
MLA: Monophosphoryl lipid A
IL-1: Interleukin-1
H_2O_2: Hydrogen peroxide
ESR: Electron Spin Resonance
HPLC: High performance liquid chromatography

ACKNOWLEDGEMENTS

This study was supported by NIH HL 34360, HL 22559 and HL 33889.

REFERENCES

1. Herrlich P, Bohmer FD. 2000. Redox regulation of signal transduction in mammalian cells. Biochem Pharmacol 59:35–41.
2. Rosette C, Karin M. 1996. Ultraviolet light and osmotic stress: Activation of the JNK cascade through multiple growth factor and cytokine receptors. Science 274:1194–1197.
3. Dalton TP, Shertzer HG, Puga A. 1999. Regulation of gene expression by reactive oxygen. Annu Rev Pharmacol Toxicol 39:67–101.
4. Das DK, Maulik N. 1994. Evaluation of antioxidant effectiveness in ischemia reperfusion tissue injury methods. Methods Enzymol 233:601–610.
5. Meerson FZ, Kagan VE, Kozlov P, Beekina LM, Khipenko YV. 1982. The role of lipid peroxidation in pathogenesis of ischemic damage and the antioxidant protection of the heart. Basic Res Cardiol 77:465–472.
6. Tosaki A, Droy-Lefaix MT, Pali T, Das DK. 1993. Effects of SOD, catalase and a novel anti-arrhythmic drug, EGB 671, on reperfusion-induced arrhythmias in isolated rat hearts. Free Rad Biol Med 14:361–370.
7. Olivetti G, Abbi R, Quaini F, Kajstura J, Cheng W, Nitahara JA, Quaini E, Loreto C Di, Beltrami

CA, Krajewski S, Reed JC, Anversa P. 1997. Apoptosis in the failing heart, New Eng J Med 336:1131–1141.
8. Maulik N, Yoshida T, Das DK. 1998. Oxidative stress developed during the reperfusion of ischemic myocardium induces apoptosis. Free Rad Biol Med 24:869–875.
9. Nishio Y, Kashiwagi A, Taki H, Shinozaki K, Maeno Y, Kojima H, Maegawa H, Haneda M, Hidaka H, Yasuda H, Horiike K, Kikkawa R. 1998. Altered activities of transcription factors and their related gene expression in cardiac tissues of diabetic rats. Diabetes 47:1318–1325.
10. Das DK, Maulik N. 1998. Apoptosis in ischemia reperfusion injury. In: *Biological Oxidants and Antioxidants* edited by L Packer, ASH Ong AOCS Press, IL. pp. 165–177.
11. Maulik N, Yoshida T, Das DK. 1999. Regulation of cardiomyocyte apoptosis in ischemic reperfused mouse heart by glutathione peroxidase. Mol Cell Biochem 196:13–21.
12. Das DK, Maulik N. 1995. Protection against free radical injury in the heart and cardiac performance. In: *Exercise and Oxygen Toxicity* edited by CK Sen, L Packer, O Hanninen. Elsevier Science, Amsterdam.
13. Kukreja RC, Hess ML. Oxygen radicals, neutrophil-derived oxidants, amyocardial reperfusion injury. In: *Pathophysiology of reperfusion injury* edited by Das DK, CRC Press, pp. 221–242, Boca Raton, FL.
14. Das DK, George A, Liu X, Rao PS. 1989. Detection of hydroxyl radicals in the mitochondria of ischemic-reperfused myocardium by trapping with salicylate. Biochem Biophys Res Commun 165:1004–1009.
15. Halestrap AP, Kerr PM, Javadov S, Woodfield KY. 1998. Elucidating the molecular mechanism of the permeability transition pore and its role in reperfusion injury of the heart. Biochim Biophys Acta 1366:79–94.
16. Maulik N, Engelman RM, Rousou JA, Flack JE, Deaton D, Das DK. 1999. Ischemic preconditioning reduces apoptosis by upregulating anti-death gene bcl-2. Circulation 100(suppl II):369–375.
17. Maulik N, Watanabe M, Engelman D, Engelman RM, Kagan VE, Kisin E, Tyurin V, Cordis GA, Das DK. 1995. Am J Physiol (Cell): 269:C907–C916.
18. Sharma HS, Das DK. 1997. Role of cytokines in myocardial ischemia and reperfusion. Mediators Inflammation 6:175–184.
19. Maulik N, Watanabe M, Engelman D, Engelman RM, Das DK. 1995. Oxidative stress adaptation improves postischemic ventricular recovery. Mol Cell Biochem 144:67–74.
20. Bromme HJ, Holz J. 1996. Apoptosis in the heart; when and why. Mol Cell Biochem 163/164:261–275.
21. Beg AA, Baltimore D. 1996. An essential role for NFκB in preventing TNF-alpha-induced cell death. Science 274:787–789.
22. Suzuki YJ, Packer L. 1993. Inhibition of NFκB DNA binding activity by α-tocopheryl succinate. Biochem Mol Biol Int 31:693–700.
23. Suzuki YJ, Aggarwal BB, Packer L. 1992. α-lipoic acid is a potent inhibitor of NFκB activation in human T cells. Biochem Biophys Res Commun 189:1709–1715.
24. Screck R, Rieber P, Baeuerle PA. 1991. Reactive oxygen intermediates as apparently widely used messengers in the activation of the NFκB transcription factor and HIV-1. EMBO J 10:2247–2258.
25. Das DK. 1998. Ischemic preconditioning: Role of multiple kinases in signal amplification and modulation. In: Adv Organ Biol, edited by Das DK, Jay Press, CT.
26. Cohen MV, Liu Y, Liu GS, Wang P, Cordis GA, Das DK, Downey JM. 1996. Phospholipase D plays a major role in ischemic preconditioning in rabbit heart. Circulation 94:1713–1718.
27. Moraru II, Popescu L, Maulik N, Liu X, Das DK. 1992. Phospholipase D signaling in ischemic heart. Biochim Biophys Acta 1139:148–154.
28. Eskildensen K, Helmond YEG, Gho BCG, Bezstarosti K, Dekkers DHW, Loe KS, Van Heusten H. 1998. Protein Kinase C in the Mechanism of Ischemic Preconditioning in the Myocardium. Ann NY Acad Sci 793:210–225.
29. Fryer RM, Schultz Hsu JE, Gross GJ. 1998. Pretreatment with tyrosine kinase inhibitors partially attenuates ischemic preconditioning in rat hearts. Am J Phisiol 275:H2009–H2015.
30. Anderson NG, Maller JI, Tonks NK, Sturgill TW. 1990. Requirement for integration of signals from two distinct phosphorylation pathways for activation of Mapkinase. Nature (Lond) 343:651–653.
31. Seger R, Krebs EG. 1995. The MAPK signaling cascade. FASEB J 9:726–735.
32. Cobb MH, Goldsmith EJ. 1995. How MAP kinases are regulated? J Biol Chem 270:14843–14846.
33. Zu YL, Ai Y, Gilchrist A, Maulik N, Watras J, Sha'afi RI, Das DK, Huang CK. 1997. High expression and activation of MAP kinase-activated protein kinase 2 in myocardium. J Mol Cell Cardiol 29:2150–2168.

34. Maulik N, Watanabe M, Zu YL, Huang CK, Cordis GA, Schley JA, Das DK. 1996. Ischemic preconditioning triggers the activation of MAP kinases and MAPKAP kinase 2 in rat hearts. FEBS Lett 396:233–237.
35. Kyriakis JM, Banerjee P, Nikolakaki E, Dai T, Rubie EA, Ahmad MF, Avruch J, Woodgett JD. 1994. The stress-activated protein kinase subfamily of c-Jun kinases. Nature 369:156–160.
36. Maulik N, Yoshida T, Zu YL, Sato M, Banerjee A, Das DK. 1998. Ischemic preconditioning triggers tyrosine kinase signaling: potential role for MAPKAP kinase 2. Am J Physiol 275:H1857–H1864.
37. Sato M, Cordis GA, Maulik N, Das DK. SAPKs regulation of ischemic preconditioning. Am J Physiol (in press).
38. Benjamin IJ, McMillan R, Scholich K, Mullenix JB, Wittpoth C, Poppleton HM, Pierre SC, Lindorfer MA, Garrison JC. 1999. Facilitation of signal onset and termination by adenylyl cyclase. Science 283:1328–1331.
39. Mitchell MB, Meng X, Brown J, Harken AH, Banerjee A. 1995. Preconditioning of isolated rat heart is mediated by protein kinase C. Circ Res 76:73–81.
40. Dixon BS, Sharma RV, Dickerson T, Fortune J. 1994. Bradykinin and angiotensin II: activation of protein kinase C in arterial muscle. Am J Physiol 266:C1406–C1420.
41. Sato M, Engelman RM, Otani H, Maulik N, Rousou JA, Flack JE 3rd, Deaton DW, Das DK. 2000. Myocardial protection by preconditioning of heart with losartan, an angiotensin type II type-1 receptor blocker: implication of bradykinin-dependent and bradykinin-independent mechanisms. Circulation 102(suppl. III):346–351.
42. Yazaki Y, Komuro I, Yamazaki T, Tobe K, Maemura K, Kadowaki T, Nagai R. 1993. Role of protein kinase system in the signal transduction of stretch-mediated protooncogene expression and hypertrophy of cardiac myocytes. Mol Cell Biochem 119:11–16.
43. Henrich CJ, Simpson PC. 1988. Differential acute and chronic response of protein kinase C in cultured neonatal rat heart myocytes to $alpha_1$-adrenergic and phosbol ester stimulation. J Mol Cell Cardiol 20:1081–1085.
44. Das DK, Maulik N, Sato M, Ray P. 1999. Reactive oxygen species function as second messenger during ischemic preconditioning of heart. Mol Cell Biol 196:59–67.
45. Maulik N, Goswami S, Galang N, Das DK. 1999. Differential regulation of bcl-2, AP-1 and NFκB on cardiomyocyte apoptosis during myocardial ischemic stress adaptation. Febs Lett 443:331–336.
46. McMahon SB, Monroe JG. 1992. Role of primary response genes in generating cellular responses to growth factors. FASEB J 62:2707–2715.
47. Artuc M, Karman D, Jurgovsky K, Schadendorf D. 1997. Humanmelonoma cell lines selected in vitro displaying various levels of drug resistance against cisplatin, fotemustine, vindestine or etoposide: modulation of protooncogene expression. Anticancer 17:4359–4370.
48. Maulik N, Goswami S, Galang N, Das DK. 1999. Differential regulation of Bcl-2, AP-1 and NF-kappaβ on cardiomyocyte apoptosis during myocardial ischemic stress adaptation. FEBS Lett 443:331–336.
49. Hughes PE, Alexi T, Yoshida T, Schreiber SS, Knusel B. 1996. Excitotoxic lesion of rat brain with quinolinic acid induces expression of p53 messenger RNA and protein and p53 inducible genes BAX and GADD-45 brain areas showing DNA fragmentation. Neuroscience 74:1143–1160.
50. Miyashita T, Krajewski S, Krajewska, Wang MH, Lin K, Liebermann D, Hoffmann B, Reed JC. 1994. Tumor suppressor p53 is a regulator of bcl-2 and bax gene expression in vitro and in vivo. Oncogene 9:1799–1805.
51. Thornberry NA, Lazebnik Y. 1998. Caspases: enemies within. Science 281:1312–1316.
52. Clark AR, Purdie CA, Harrision DJ, Morris RG, Bird CC, Hooper ML, Wyllie AH. 1993. Thymocyte apoptosis induced by p53-dependent and independent pathways. Nature 362:849–852.
53. Williams GT, Smith CA. 1993. Molecular regulation of apoptosis: genetic controls on cell death. Cell 74:777–779.
54. Nunez G, London L, Hockenbery D, Alexander M, Mckearn JP, Korsmeyer SJ. 1999. Deregulated Bcl-2 gene expression selectively prolongs survival of growth factor-deprived hemopoietic cell lines. J Immunol 144:3602–3610.
55. Maclellan WR, Schneider MD. 1997. Death by design. Programmed cell death in cardiovascular biology and disease. Circ Res 81:137–144.
56. Tanaka M, Inada T, Fujiwara H, Ohtani S, Yamasaki K, Fujiwara T, Yokota R, Sasayama S, Doyama K. 1998. Expression of heat shock protein after ischemic preconditioning in rabbit hearts. Jap Circ J 62:512–516.
57. Kajstura J, Cheng W, Rreiss K, Clark WA, Sonnenblick EH, Krajewski S, Reed JC, Olivetti G, Anversa

P. 1996. Apoptotic and necrotic myocyte cell deaths are independent contributing variables of infarct size in rats. Lab Invest 74:86–107.
58. Galang N, Maulik N. Apoptotic cell death during ischemia/reperfusion and its attenuation by antioxidant therapy. Nutritional Environ Interact (Toxicology) (in press).
59. Maulik N, Kagan VE, Tyrin VA, Das DK. 1998. Redistribution of phosphatidyl-ethanolamine and phosphatidyl-serine precedes reperfusion-induced apoptosis. Am J Physiol 274:H242–H248.
60. Pang G, O'Rourke K, Dixit VM. 1997. Caspase-9, Bcl-XL, and Apaf-1 form a ternary complex. J Biol Chem 273:584–590.
61. Narula J, Pandey P, Arbustini E, Haider N, Narula N, Kolodgie FD, Dal Bello B, Semigran MJ, Bielsa-Masdeu A, Dec GW, Israels S, Ballester M, Virmani R, Saxena S, Kharbanda S. 1999. Apoptosis in heart failure: release of cytochrome c from mitochondria and activation of caspase-3 in human cardiomyopathy. Proc Natl Acad Sci, USA 96:8144–8149.
62. Holly TA, Drincic A, Byun Y, Nakamura S, Harris K, Klocke FJ, Cryns VL. 1999. Caspase inhibition reduces myocyte cell death induced by myocardial ischemia and reperfusion in vivo. J Mol Cell Cardiol 31:1709–1715.
63. Bialik S, Cryns VL, Drincic A, Miytata S, Wollowick AL, Srinivasan A, Kitsis RN. 1999. The mitochondrial apoptotic pathway is activated by serum and glucose deprivation in cardiac myocytes. Circ Res 85:403–414.
64. Das DK, Maulik N, Moraru II. 1995. Gene expression in acute myocardial stress. Induction by hypoxia, ischemia/reperfusion, hyperthermia and oxidative stress. J Mol Cell Cardiol 27:181–193.
65. Hanada H, Kashiwagi A, Takehara Y, Kanno T, Yabuki M, Sasaki J, Inoue M, Utsumi K. 1997. Do reactive oxygen speciec underlie the mechanism of apoptosis in the tadpole tail. Free Rad Biol Med 23:294–301.
66. Oishi K, Machida K. 1997. Inhibition of neutrophil apoptosis by antioxidants in culture medium. Scandinavian J Immunol 45:21–27.
67. Davies KJ. 1986. A intracellular proteolytic systems may function as secondary antioxidant defenses: a hypothesis. J Free Rad Biol Med 2:155–173.
68. Maulik N, Wei ZJ, Engelman RM, Lu D, Moraru II, Rousou JA, Das DK. 1993. Interleukin-1a preconditioning reduces myocardial ischemic reperfusion injury. Circulation 88(suppl II):387–394.
69. Tosaki A, Das DK. 2002. The role of heme oxygenase signaling in various disorders. Mol Cell Biochem 232:149–157.

Pathophysiology of Cardiovascular Disease,
edited by N.S. Dhalla, H. Rupp,
A. Angel & G.N. Pierce
Kluwer Academic Publishers, Boston, 2004

Cell Biology of Acute Reperfusion Injury

Claudia Schäfer and Hans-Michael Piper

Physiological Institute, Justus Liebig University, Aulweg 129, D-35392 Giessen, Germany

Summary. In ischemic-reperfused myocardium, necrosis of cardiomyocytes may develop not only due to the ischemic conditions but also due to the specific circumstances of reperfusion. Reperfusion-induced hypercontracture is a major cause of lethal reperfusion-injury of myocardium. Hypercontracture of myofibrils is caused by reenergetisation of Ca^{2+}-overloaded cardiomyocytes. Ca^{2+}-overload is due to the preceding ischemic period. Upon reperfusion, Ca^{2+}-overload leads to rapid oscillations of cytosolic Ca^{2+}-concentration. Rapid normalisation of pH favors hypercontracture, prolonged acidosis protects against it. Inhbition of reperfusion-induced Ca^{2+}-oscillations (inhibition of Ca^{2+}-uptake or Ca^{2+}-release by the sarcoplasmic reticulum; inhibition of the reverse mode of the sarcolemmal Na^+/Ca^{2+}-exchanger) or of pH_i recovery (simultaneous inhibition of the Na^+/H^+-exchanger and the Na^+/HCO_3^- symporter of sarcolemma) during the vulnerable phase of reperfusion can protect the ischemic-reperfused cardiomyocyte against reperfusion injury.

Key words: Reperfusion, Protection, Cardiomyocytes.

INTRODUCTION

The development of cell injury in ischemic tissue starts with a deficit in the cellular balance of energy. The hydrolysis of ATP and the production of lactate acidify the cytosol. Acidosis inhibits glycolysis and anaerobic ATP production. The cellular reserves of energy are depleted and cation pumps regulating the normal intracellular ionic milieu fail due to a lack of energy. The Na^+-K^+-ATPase is inhibited and the sarcolemmal Na^+ gradient is diminished. Na^+, Ca^{2+} and H^+ accumulate in the

Corresponding author: Dr. C. Schäfer, Physiological Institute, Justus Liebig University, Aulweg 129, D-35392 Giessen. Tel: ++49-641-9947241, Fax: ++49-641-9947239, E-mail: Claudia.Schaefer@physiologie.med.uni-giessen.de.

cytosol. The accumulation of Ca^{2+} in the cytosol during ischemia can contribute ischemic and reperfusion injury of myocardial cells.

After ischemia, the disturbance of cation homeostasis can be rapidly reversed when the cardiomyocytes are reoxygenated at a normal extracellular pH (1,2). The recovery to a preischemic cytosolic Ca^{2+} concentration ($[Ca^{2+}]_i$) can be divided in three phases: 1) a initial rapid fall of $[Ca^{2+}]_i$ due to sequestration of Ca^{2+} into the sarcoplasmatic reticulum (SR) (phase 1); 2) a period of Ca^{2+} oscillations due to cycles of transient release and re-uptake of Ca^{2+} by the SR (phase 2); 3) reestablishment of a normal resting level of $[Ca^{2+}]_i$ (phase 3) due to extrusion of excess Ca^{2+} across the sarcolemma (3, 4). Transsarcolemmal extrusion of Ca^{2+} from the reoxygenated cells is mediated by a tandem action of the Na^+ pump, creating a normal transsarcolemmal Na^+ gradient, with the Na^+/Ca^{2+} exchange mechanism, using the Na^+ gradient for the extrusion of Ca^{2+} (5). During reoxygenation, Ca^{2+}-oscillations occurring simultaneously with the recovery of intracellular pH elicit uncontrolled activation of the myofibrils and thereby development of hypercontracture (1,2,4). Hypercontracture is caused by strong contractile activation which causes irreversible cell injury by disruption of cytoskeletal structures (6–8). After prolonged ischemic conditions cardiomyocytes become increasingly susceptible to hypercontracture and this seems due to weakening of the cytoskeletal architecture (9,10). Reoxygenation-induced hypercontracture has been shown to represent an important cause for severe tissue injury in reperfused myocardium. In tissue hypercontracture causes mutual disruption of adjacent cells and thereby necrosis of cardiomyocytes (11).

PROTECTION AGAINST CELLULAR REPERFUSION INJURY BY PROLONGATION OF ACIDOSIS

Reperfusion protocols that prevent a rapid renormalisation of ischemic intracellular pH have been found to protect myocardial structure and function (12–16). It has been shown on isolated cardiomyocytes that reperfusion conditions preserving a low intracellular pH (pH_i) prevent reoxygenation-induced hypercontracture of cardiomyocytes (1,2). The Na^+/H^+ exchanger (NHE) and the HCO_3^--dependent H^+-transport, commonly called the Na^+/HCO_3^- symporter (NBS) (17,18), play major roles in reperfusion-induced H^+-extrusion (Figure 1). These transporters are therefore the main targets for pH_i control in the reperfused cardiomyocyte.

During recovery from ischemic conditions, the full extent of reoxygenation-induced hypercontracture develops gradually in cardiomyocytes. At the beginning of phase 1 of reoxygenation, cytosolic acidosis inhibits activation of the myofibrils due to acidotic desensitisation to Ca^{2+} (19). Hypercontracture develops progressively during the second phase of reoxygenation when the spontaneous recovery of pH_i occurs. The time course of pH_i recovery is determined by the sarcolemmal H^+ transport mechanisms. Both, inhibitors of NBS and of NHE slow down the rapidity of pH_i recovery (1). In rat ventricular cardiomyocytes NBS accounts for the minor part of H^+ extrusion during recovery from intracellular acidosis. Only the simultaneous inhibition of NHE and NBS prevents the rapid pH_i recovery

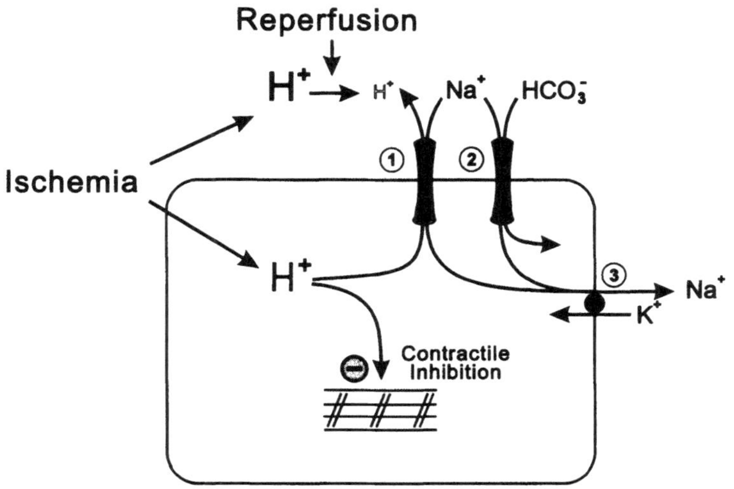

Figure 1. Schematic representation of pH-regulation during reperfusion. Protons accumulated during ischemia. During reperfusion Na^+-K^+-ATPase is activated (3), Na^+-gradient is restored and protons can be extruded via the Na^+/H^+-exchanger (1) and the Na^+/HCO_3^--symporter (2). Accumulation of protons during reperfusion inhibits contractile activation of myofibrils.

(1,14,17,18) and only with continued acidosis cardiomyocytes are protected against reoxygenation-induced hypercontracture (1).

PROTECTION AGAINST CELLULAR REPERFUSION INJURY BY INHIBITION OF REOXYGENATION-INDUCED CA^{2+}-OSCILLATIONS

The slowdown of pH_i recovery during phase 2 has also a significant impact on the generation of SR-dependent Ca^{2+}oscillations (1,2). Reoxygenation-induced spontaneous Ca^{2+} oscillations are less frequent if the H^+-extrusion systems are inhibited. When both H^+-extruding mechanisms are inhibited simultaneously, oscillations are effectively suppressed; apparently due to the acidotic inhibition of the SR Ca^{2+}-release channel. Reoxygenation-induced Ca^{2+}-oscillations can also be suppressed directly by inhibition of the Ca^{2+}-release channel with ryanodine or inhibition of the Ca^{2+}-ATPase with thapsigargin or CPA (Figure 2). All these maneuvers reduce the oscillation frequency and therefore the development of hypercontracture during phase 2 (4). It was found that the volatile anaesthetic halothane protects against reperfusion—induced hypercontracture and cell death also by this mechanism (4).

PROTECTION AGAINST CELLULAR REPERFUSION INJURY BY INHIBITION OF THE REVERSE MODE OF THE NA^+/CA^{2+}-EXCHANGER

Under ischemic conditions the Na^+/Ca^{2+} exchanger (NCE) represents an important route by which Ca^{2+} enters the ischemic cardiomyocyte (20–24). Decrease of the

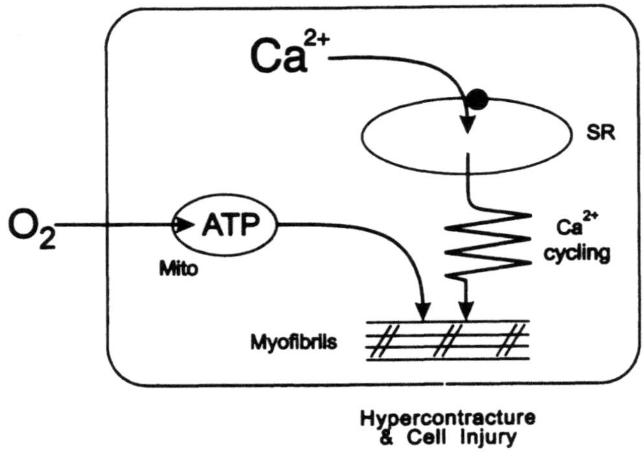

Figure 2. Schematic representation of Ca^{2+}-oscillations and hypercontracture development during reperfusion. Ca^{2+}-uptake and -release by the sarcoplasmic reticulum induce oscillations of cytosolic Ca^{2+} and, triggered by oscillations, excessive contractile activation of myofibrils leads to hypercontrature development in the re-energised cell.

transmembrane Na^+ gradient and depolarisation of the sarcolemma leads to a reduction in the net-outward transport of Ca^{2+} via the forward mode of NCE and favors instead a net Ca^{2+} influx via the reverse mode of NCE (Figure 3). During the early phase of reoxygenation, the NCE changes again its directional mode. During the very first few minutes of reoxygenation, however, the equilibrium conditions at the cell membrane still favor the reverse mode of NCE operation. This influences significantly the outcome of reoxygenation. When the reverse mode of NCE was inhibited during the early phase of reoxygenation, the reoxygenation-induced oscillations of cytosolic Ca^{2+} were markedly reduced in cardiomyocytes. Application of an inhibitor of the reverse mode of the NCE, KB-R 7943, or reperfusion under Ca^{2+}-free conditions reduced the oscillatory activity and protected against hypercontracture (25). The data indicated that during the early phase of reoxygenation Ca^{2+} influx through NCE in reverse mode operation is responsible for a part of the SR-dependent Ca^{2+}-oscillations. Consistent with the protection against hypercontracture of cardiomyocytes is the finding that the inhibitor of NCE reverse mode (KB-R 7943) protects ischemic- reperfused hearts from severe injury even if applied during reperfusion (25).

CONCLUSION

Severe cellular injury during the acute phase of reperfusion is manifested by the development of hypercontracture. Hypercontracture represents true reperfusion injury, since it can be prevented by interventions during reperfusion. Protection

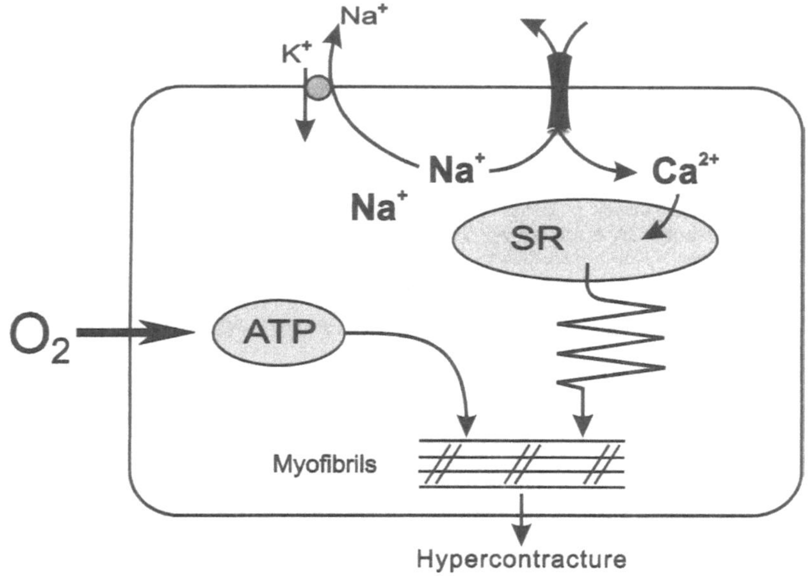

Figure 3. Schematic representation of functional coupling of Na^+/K^+-ATPase with Na^+/Ca^{2+}-exchanger in reverse mode operation. During the early phase of reperfusion the reverse mode of Na^+/Ca^{2+} exchanger is activated, extracellular Ca^{2+} enters the cell and triggers the development of Ca^{2+}-oscillations. These oscillations elicit together with reenergetisation the development of hypercontracture.

against hypercontracture can be achieved, e.g., by prolongation of ischemic cytosolic acidosis and inhibition of reoxygenation-induced Ca^{2+}-oscillations during the early phase of reoxygenation. The pH-regulating systems (Na^+/H^+-exchanger and Na^+/HCO_3^- symporter) and the Na^+/Ca^{2+}-exchanger represent important cellular targets for a causal therapy directed against lethal reperfusion injury in the heart.

REFERENCES

1. Schäfer C, Ladilov YV, Siegmund B, Piper HM. 2000. Importance of bicarbonate transport for protection of cardiomyocytes against reoxygenation injury. Am J Physiol Heart Circ Physiol 278: H1457–1463.
2. Ladilov YV, Siegmund B, Piper HM. 1995. Protection of reoxygenated cardiomyocytes against hypercontracture by inhibition of Na^+/H^+ exchange. Am J Physiol 268:H1531–H1539.
3. Siegmund B, Zude R, Piper HM. 1992. Recovery of anoxic-reoxygenated cardiomyocytes from severe Ca^{2+}-overload. Am J Physiol 263:H1262–H1269.
4. Siegmund B, Schlack W, Ladilov YV, Balser C, Piper HM. 1996. Halothane protects cardiomyocytes against reoxygenation-induced hypercontracture. Circulation 96:4372–4379.
5. Siegmund B, Ladilov YV, Piper HM. 1994. Importance of sodium for recovery of calcium control in reoxygenated cardiomyocytes. Am J Physiol 267:H506–H513.
6. Siegmund B, Koop A, Klietz T, Schwartz P, Piper HM. 1990. Sarcolemmal integrity and metabolic competence of cardiomyocytes under anoxia-reoxygenation. Am J Physiol 258:H285–H291.
7. Ganote CE. 1983. Contraction band necrosis and irreversible myocardial injury. J Mol Cell Cardiol 15:67–73.

8. Hohl C, Ansel A, Altschuld R, Brierly GP. 1982. Contracture of isolated rat heart cells on anaerobic to aerobic transition. Am J Physiol 242:H1022–1030.
9. Ladilov YV, Siegmund B, Balser C, Piper HM. 1997. Simulated ischemia increases the susceptibility of rat cardiomyocytes to hypercontracture. Circ Res 80:69–75.
10. Ganote C, Armstrong S. 1993. Ischaemia and the myocyte cytoskeleton: review and speculation. Cardiovasc Res. 27:1387–1403.
11. Barrabes JA, Garcia-Dorado D, Ruiz-Meana M, Piper HM, Solares J, Gonzales MA, Oliveras J, Pilar Herrejon M, Soler Soler J. 1996. Myocardial segment shrinkage during coronary reperfusion in situ. Pflügers Arch Eur J Physiol 431:519–526.
12. Bugge E, Munch-Ellingsen J, Ytrehus K. 1996. Reduced infarct size in the rabbit heart in vivo by ethylisopropyl-amiloride. A role for Na^+/H^+ exchange. Bas Res Cardiol 91:203–209.
13. Bugge E, Ytrehus K. 1995. Inhibition of sodium-hydrogen exchange reduces infarct sice in the isolated rat heart—a protective additive to ischaemic preconditioning. Cardiovasc. Res. 29:269–274.
14. Grace AA, Kirschenlohr HL, Metcalfe JC, Smith GA, Weissberg PL, Cragoe EJ, Vandenberg VL. 1993. Regulation of intracellular pH in the perfused heart by external HCO_3^- and Na^+-H^+ exchange. Am J Physiol 265:H289-H298.
15. Kitakaze M, Weisfeldt ML, Marban E. 1988. Acidosis during early reperfusion prevents myocardial stunning in reperfused ferret hearts. J Clin Invest 82:920–927.
16. Maddaford TG, Pierce GN. 1997. Myocardial dysfunction is associated with activation of Na^+/H^+ exchange immediately during reperfusion. Am J Physiol 273:H2232–H2239.
17. Vandenberg JI, Metcalfe JC, Grace AA. 1993. Mechanisms of pH_i recovery after global ischemia in the perfused heart. Circ Res 72:993–1003.
18. Lagadic-Gossmann D, Buckler J, Vaughan-Jones KJ. 1992. Role of bicarbonate in pH recovery from intracellular acidosis in the Guinea-pig ventricular myocyte. J Physiol Lond 458:361–384.
19. Marban E, Kusuoka H. 1987. Maximal Ca^{2+} activated force and myofilament Ca^{2+} sensitivity in intact mammalian hearts. J Gen Physiol 90:609–623.
20. Ladilov Y, Haffner S, Balser-Schafer C, Maxeiner H, Piper HM. 1999 Jun. Cardioprotective effects of KB-R7943: a novel inhibitor of the reverse mode of Na^+/Ca^{2+}exchanger. Am J Physiol. 276: H1868–1876.
21. Haigney M, Lakatta EG, Stern MD, Silverman HS. 1994. Sodium channel blockade reduces sodium loading and sodium-dependent calcium loading. Circulation 90:391–399.
22. Murphy E, Perlmann M, London RE, Steenbergen C. 1991. Amiloride delays the ischemia-induced rise in cytosolic free calcium. Circ Res 68:1250–1258.
23. Tani M, Neely JR. 1990. Na^+ accumulation increases Ca^{2+} overload and impairs function in anoxic rat hearts. J Mol Cell Cardiol 22:57–72.
24. Ziegelstein RC, Zweier JL, Mellits ED, Younes A, Lakatta EG, Stern MD, Silverman HS. 1992. Dimethylthiourea, an oxygen radical scavenger, protects isolated cardiac myocytes from hypoxic injury by inhibition of Na^+-Ca^{2+} exchange and not by its antioxidant effects. Circ Res 70:804–811.
25. Schäfer C, Ladilov Y, Inserte J, Schäfer M, Haffner S, Garcia-Dorado D, Piper HM. 2001. Role of the reverse mode of the Na^+/Ca^{2+} exchanger in reoxygenation-induced cardiomyocyte injury. Cardiovasc Res. 51:241–50.

Pathophysiology of Cardiovascular Disease,
edited by N.S. Dhalla, H. Rupp,
A. Angel & G.N. Pierce
Kluwer Academic Publishers, Boston, 2004

Reactive Oxygen Species, Na^+/H^+ Exchange and Ca^{2+} Overload During Myocardial Ischemia/Reperfusion Injury: The Paradox Explained?

Emily C. Rothstein and Pamela A. Lucchesi

Department of Physiology & Biophysics, University of Alabama at Birmingham, Birmingham, AL 35294 USA

Summary. Reperfusion of the myocardium following an ischemic episode is associated with profound contractile and metabolic dysfunction. Reactive oxygen species (ROS), such as hydrogen peroxide (H_2O_2) and hydroxyl (•OH) radicals, are produced during reperfusion after a transient ischemic attack of the myocardium. Reperfusion is also associated with the activation of the cardiac Na^+/H^+ exchanger isoform-1 (NHE-1), which restores intracellular pH (pH_i) towards normal following ischemia-induced acidosis. Activation of NHE-1 also produces undesirable secondary effects leading to the exacerbation of tissue injury, a phenomenon termed the "pH paradox". Increased generation of ROS plays an important role in reperfusion-induced myocardial stunning and NHE-1 activation. However, the exact relationship between ROS, NHE-1 activity and intracellular Ca^{2+} overload has not been defined. Recent development of new, more specific NHE-1 inhibitors has enhanced our understanding of the link between ROS burst, the pH paradox, and Ca^{2+} overload during I/R injury.

Key words: pH paradox, Ca^{2+}, kinases, Na^+/H^+ exchanger, H_2O_2.

INTRODUCTION

Cardiovascular disease resulting from myocardial ischemia and infarction is a major cause of morbidity and mortality in industrialized nations. Myocardial ischemia is

Correspondence To: Pamela A. Lucchesi, Ph.D., University of Alabama at Birmingham, Department of Physiology & Biophysics, 1918 University Blvd., MCLM-986, Birmingham, AL 35294-0005. Phone: (205) 934-5742, Fax: (205) 975-9028, E-mail: lucchesi@physiology.uab.edu.

defined as an imbalance between the oxygen demands of cardiac tissue and the supply of oxygenated blood. Hallmarks of ischemia include hypoxia, reduced supply of metabolic substrates and reduced removal of metabolic byproducts (1). The degree of imbalance between oxygen supply and demand is variable and dictates the severity of the ischemic episode (2). When ischemia is of sufficient severity and persists long enough, myocytes become irreversibly injured and undergo cellular necrosis. Brief periods of ischemia are associated with reversible myocardial injury. With reversible ischemia, the intensity of this damage is not sufficient to cause cell death but is sufficient to produce dysfunction of key cellular organelles. Reperfusion of cardiac tissue following a reversible ischemic episode contributes additively to the effects of ischemia, resulting in cardiac dysfunction. These reperfusion-induced injuries include reduced tension development, arrhythmias, mitochondrial swelling and Ca^{2+} overload, and are referred to as myocardial stunning (1). Myocardial stunning may occur in clinical situations such as during unstable angina or exercise- or stress-induced ischemia and after cardiopulmonary bypass or myocardial infarction (for review, see (3)).

Ischemia/reperfusion (I/R) injury is intimately linked to the generation of reactive oxygen species (ROS), alterations in pH, and Ca^{2+} overload in the myocardium. These changes should exert cardiovascular protection but instead actually exacerbate tissue injury and are referred to as the oxygen paradox, pH paradox, and Ca^{2+} paradox, respectively. Many recent advances have been made to elucidate the relationship of these events to myocardial injury.

ROS INVOLVEMENT IN ISCHEMIA/REPERFUSION INJURY

ROS have been considered to be a major factor in reperfusion injury of the myocardium. ROS include both free radicals, which are extremely reactive due to unpaired electrons in their outer orbit, and other oxidant species such as hydrogen peroxide (H_2O_2). The highly reactive hydroxyl radicals (•OH) are formed by the reaction of O_2^- with H_2O_2 or by the reaction of iron (II) with H_2O_2 (4). High levels of ROS cause lipid peroxidation and cytotoxicity (5). Lower levels of ROS lead to protein modification affecting both inter-molecular and intra-molecular interactions. For example, oxidation of cysteine residues and/or formation of intra-molecular disulfide bridges result in protein conformational changes. ROS can also lead to protein dimerization or crosslinking formation of inter-molecular disulfide bridges or dityrosine formation (6,7). This ROS-derived protein peroxidation ultimately leads to altered enzyme and ion channel function. The combination of these ROS-mediated cellular changes could provide the rationale for coincident myocardial dysfunction amidst ROS increases (5).

ROS are produced during reperfusion after a transient ischemic attack of the myocardium. Multiple studies have suggested a link between ROS and myocardial injury following I/R. Bolli and coworkers, using spin resonance to quantitate increases in ROS, were the first to show that the greatest increase in ROS generation occurs during the initial moments of reperfusion following an ischemic episode (8). Hydroxyl radicals have been demonstrated to cause lipid peroxidation and myocardial injury, and is thought to trigger the contractile dysfunction observed during reperfusion (1). Bolli's group concluded that at least 50% of myocardial stunning is due to ROS (9). In addition, whole animal studies have shown reduced myocardial injury with antioxidant intervention. Specifically, amelioration of ROS activity, using scavengers of free radicals such as catalase and superoxide dismutase, reduces myocardial stunning and reperfusion arrhythmias (Zwier, 1987). In addition, results of the CHAOS (Cambridge Heart Antioxidant Study) trial showed decreased incidence of cardiovascular events in patients with coronary disease following treatment with the antioxidant vitamin E (10).

A number of mechanisms are capable of generating ROS within the ischemic heart. Within the myocardium and the endothelium, the oxidation of xanthine and hypoxanthine by xanthine oxidase generates superoxide (O_2^-) that is further reduced to H_2O_2 and •OH. During I/R there is infiltration of the myocardium by neutrophils that produce several types ROS, including O_2^-, H_2O_2 and OCL^- (1). Exposure of non-ischemic myocardium or cultured myocytes to ROS can produce cell injury similar to that seen in I/R. In a cell culture model of I/R injury using both neonatal and adult rat ventricular myocytes, we showed decreased cardiac myocyte contractility following brief exposure to low (50 µM) concentrations of H_2O_2, similar to the free radical burst that occurs during I/R, that emulates the cellular dysfunction seen with I/R injury (11,12).

ROS have numerous intracellular targets including second messenger pathways, L-type Ca^{2+} and K^+ channels, ion transporters and contractile proteins (4). In cardiac myocytes O_2^- and H_2O_2 have been shown to inhibit Na^+ and Ca^{2+} pumps, accelerate rundown of L-type Ca^{2+} currents, activate Na^+/Ca^{2+} exchange (13), and deplete internal caffeine-sensitive stores by inhibiting the sarcoplasmic reticulum Ca^{2+}-ATPase (1,14). These effects are independent of metabolic inhibition since FCCP or 2-deoxyglucose does not mimic the effects of ROS (14). ROS-generating systems have also been shown to reduce myofibrillar ATPase activity, which could attenuate myofilament responsiveness to Ca^{2+} and force development within the myocardium (15).

NHE-1 AND THE PH PARADOX

The Na^+/H^+ exchanger isoform-1 (NHE-1) plays a major role in the control of intracellular pH (pH_i) in cardiac myocytes (for review, see (16)). Under normal conditions, the basal activity of this antiporter is low but can be dramatically increased by intracellular acidosis, and in response to neurohormonal activation (17). In response to I/R injury, NHE-1 activation corrects the intracellular acidosis, but has secondary effects, due to Na^+ influx, that lead to Ca^{2+} loading of the myocardium and exacerbation of tissue injury (16,18,19). This increase in intracellular Na^+ alters the activity of the Na^+/Ca^{2+} exchanger (NCX), leading to increased intracellular Ca^{2+}. Data from Stowe's laboratory showed that NHE-1 inhibition reduced Na^+ and Ca^{2+} loading during I/R leading to an improvement in cardiovascular function and reduced infarct size in isolated guinea pig hearts (20). Their work indicates that cytoplasmic Na^+ loading and Ca^{2+} accumulation in response to NHE-1 inhibition clearly implicate NHE-1 and slowed forward-mode and/or enhanced reverse-mode NCX largely responsible for mediating the large increase in Ca^{2+} due to an increase in Na^{+} on reperfusion with cardiac injury. This implies that NHE-1 activity is at its peak during initial reperfusion (when ROS burst is known to occur and the cytosolic pH is decreased) since inhibiting NHE-1 just before reperfusion was equally effective as inhibiting NHE-1 prior to ischemia.

Considerable evidence indicates that NHE-1 inhibition exerts beneficial effects on I/R injury. Karmazyn et al., using isolated rat hearts, were among the first show inhibition of NHE-1 enhancing recovery of the myocardium following low-flow ischemia (21). Recently, clinical studies with new, more specific, NHE inhibitors have exhibited mixed results. A study by Hendrikx et al. showed a significant improvement in cardiovascular function when the NHE-1 inhibitor HOE694 was given prior to ischemia and with only partial protection when administered prior to reperfusion (22). In addition, a study using isolated working rat hearts demonstrated a dose-dependent reduction of the incidence and the duration of reperfusion arrhythmias with the NHE-1 specific inhibitor HOE642 (23). The GUARDIAN (GUARD During Ischemia Against Necrosis) trial suggested a possible cardioprotective effect of HOE642 for specific clinical settings, such as in patients undergoing coronary artery bypass graft surgery, when administration of the inhibitor was possible prior to the onset of myocardial ischemia (for review, see (24)).

The regulation of NHE-1 involves both changes in expression as well as in the activation of existing transporter proteins. In cardiac myocytes, the type of regulation may be dependent on the length of the ischemic period. Protein expression appears to play a major role in exchanger activity during chronic ischemia, whereas phosphorylation may be more important in response to acute I/R (25). Phosphorylation of the exchanger directly is associated with an increase in maximum velocity, whereas phosphorylation of a NHE regulatory factor results in a shift in pH_i dependence towards more alkaline pH_i values and an increase in maximum velocity (17).

REGULATION OF NHE-1 BY ROS

NHE-1 activation has been shown to be an important catalyst for I/R injury, we were interested in studying the effects of ROS on NHE-1 activation. Using pH_i measurements and in-gel kinase assays, our group has shown H_2O_2-stimulated NHE-1 activity in neonatal rat ventricular myocytes (11,26). Also, Sanbaitis et al. used a change in sarcolemmal H^{+} efflux rate to confirm H_2O_2-induced activation of NHE-1 in adult rat ventricular myocytes (27). These studies agree with work from Karmazyn's laboratory using NHE-1 inhibition to improve myocardial function following ROS stress alone (28) and ROS stress with I/R (29).

The mechanisms by which ROS activate NHE involve activation of multiple signal transduction cascades. We and others have shown that ROS and hypoxia activate members of the mitogen activated protein kinase (MAPK) family, including p38, c-jun NH_2-terminal kinase (JNK) and extracellular signal-regulated kinases

(ERK1/2) (11,12,27,30). Interestingly, all three MAPK pathways are activated in the myocardium by hypoxia and ischemia reperfusion (31,32). Anti-oxidants such as N-acetyl cysteine prevent activation of p38 MAPK during ischemia and a scavenger •OH diminishes the activation of p38, JNK, and ERK1/2 in reperfused hearts (33,34).

There is strong evidence that oxidative stress regulates NHE-1 by phosphorylation Recent studies from Fliegel's laboratory demonstrated that MAPK-dependent pathways including p90 ribosomal S6 kinase and ERK1/2 and are important in regulating NHE-1 during ischemia and reperfusion of the rat myocardium (35). We have shown that H_2O_2-induced activation of NHE-1 in neonatal and rat ventricular myocytes involves ERK1/2 MAPK (11,12). On the other hand, Snabiatis et al. demonstrated that src tyrosine kinases and protein kinase C are also required from the activation of NHE-1 by H_2O_2 (27).

THE ROLE OF ROS IN CA^{2+} HOMEOSTASIS

Intracellular Ca^{2+} overload plays a key role in the pathogenesis of I/R injury. Increased intracellular Ca^{2+} can alter excitation-contraction (E-C) coupling and intracellular signaling events. The myocardium has been suggested to undergo arrhythmias, contractile dysfunction, and cell death in the setting of Ca^{2+} overload (36).

Although several studies have attempted to correlate the three phenomena of increased ROS production, acidosis, and Ca^{2+} overload, most used very high doses of ROS. High concentrations of H_2O_2 (1–10 mM) can produce Ca^{2+} overload via inhibition of Na^+ and Ca^{2+} pumps, increased Ca^{2+} influx through voltage gated Ca^{2+} channels and inhibition the sarcoplasmic reticulum Ca^{2+}-ATPase (1,37,38). However, these concentrations of H_2O_2 are ~10–100 fold higher than those observed for the burst in OFR production during I/R (39,40). Little has been studied on the effect of the lower doses of ROS on myocyte contractility and $[Ca^{2+}]_i$.

NHE-1 IS A LINK BETWEEN ROS AND THE CA^{2+} PARADOX

Our lab has shown that low concentrations of H_2O_2 (50 μM) lead to a diastolic Ca^{2+} overload, an increased activation of NHE-1, and a decreased contractility through a MAP kinase-dependent pathway (11,12,26).

We recently determined the involvement of the ERK1/2 MAPK pathway and the activation of NHE-1 in the diastolic Ca^{2+} overload seen following H_2O_2 treatment of neonatal rat ventricular myocytes. Treatment with the MEK inhibitor U0126 (5 μM) or the NHE-1 inhibitor HOE642 (5 μM) attenuated this diastolic Ca^{2+} overload. These data indicate that H_2O_2-induced Ca^{2+} overload is partially mediated by NHE-1 activation, secondary to phosphorylation of NHE-1 by the ERK1/2 MAPK pathway, suggesting a correlation between the increase in ROS, the NHE, and Ca^{2+} overload.

NCX is likely to be involved in NHE-dependent increases in intracellular Ca^{2+} since it is known to be a major factor in the regulation of Ca^{2+} in the cardiomyocyte. This exchanger is a secondary active transporter that is responsive to changes in both Na^+ and Ca^{2+} concentration gradients and to changes in membrane potential. NCX can also be regulated by other factors such as phosphatidylinositol-4,5-bisphosphate (PIP_2) and phosphorylation (41). Iwamoto et al. demonstrated a modulation of exchanger phosphorylation and activity via protein kinase C (42). Also shown was a rise in $[Ca^{2+}]_i$ after H_2O_2 exposure that was reduced by PKC inhibitors (43). NCX is known to be sensitive to changes in overall cellular Na^+ or changes in sub-sarcolemmal Na^+. Therefore, even small changes in ion concentrations close to the membrane could affect exchanger activity. Hilgemann et al. have suggested that Na^+ binding to the intracellular surface of the exchanger can lead to an inactivated state (44). The extent of inactivation depends on available Na^+ and Ca^{2+} concentrations. In response to increased intracellular Na^+, NCX can reverse to exchange intracellular Na^+ for extracellular Ca^{2+}. With reverse exchange, increased cytoplasmic Na^+ would induce an influx of Ca^{2+}. The role of reversal of the exchanger in cardiomyocytes is controversial. Both inactivation and reversal of the exchanger would lead to an increase in intracellular Ca^{2+} levels. NCX is important during the removal of intracellular Ca^{2+} in the cardiomyocyte. An early study by Tani et al. suggest that reperfusion Ca^{2+} uptake occurs at least partially by Na^+/Ca^{2+} exchange based on a correlation between increased intracellular Na^+ and Ca^{2+} uptake. In addition, this study showed decreased ventricular function recovery related to the extent of Ca^{2+} uptake during reperfusion (45). Interestingly, work by Goldhaber directly relates ROS to Na^+/Ca^{2+} exchange showing enhancement of NCX activity in cardiomyocytes by high levels (1 mM) of H_2O_2 (13).

Based upon our own studies and available data in the literature, we propose the following model that links alterations in ROS with activation of NHE-1 and intracellular Ca^{2+}overload in cardiac myocytes (Figure 1). According to this model, activation of NHE-1 by H_2O_2 requires activation of the ERK1/2 MAPK (11,12,35,46), which may be facilitated by Src and PKC (47). ERK1/2 phosphorylates the carboxy cytoplasmic tail of NHE-1 leading to the increase in NHE-1 activity. The resultant increase in intracellular Na^+, in turn, leads to modification of NCX and the resultant intracellular Ca^{2+} overload.

ALTERNATIVE MECHANISMS OF ROS-MEDIATED I/R INJURY

Other elements involved in E-C coupling such as myofilament contractile and regulatory proteins may also play a role in the H_2O_2-induced alterations in cardiac myocyte $[Ca^{2+}]_i$ and function. The persistent contractile dysfunction in reperfusion injury, despite the presence of normal or elevated intracellular Ca^{2+} has led to the hypothesis that the primary mechanism responsible for the observed contractile dysfunction is decreased sensitivity of myofilaments to Ca^{2+}. Ca^{2+} overload may activate a Ca^{2+}-dependent protease (calpain) that causes the proteolysis of myofilament regulatory proteins including Troponin I (TnI) and α-actinin, and that Ca^{2+} overload is associated with myofilament disassembly. However, the role of TnI degradation in myocardial stunning is controversial. While some studies have shown degradation of TnI during I/R (48–51), others have shown no change in TnI levels (52–54). Also, alterations in sarcomeric organization could lead to myofilament con-

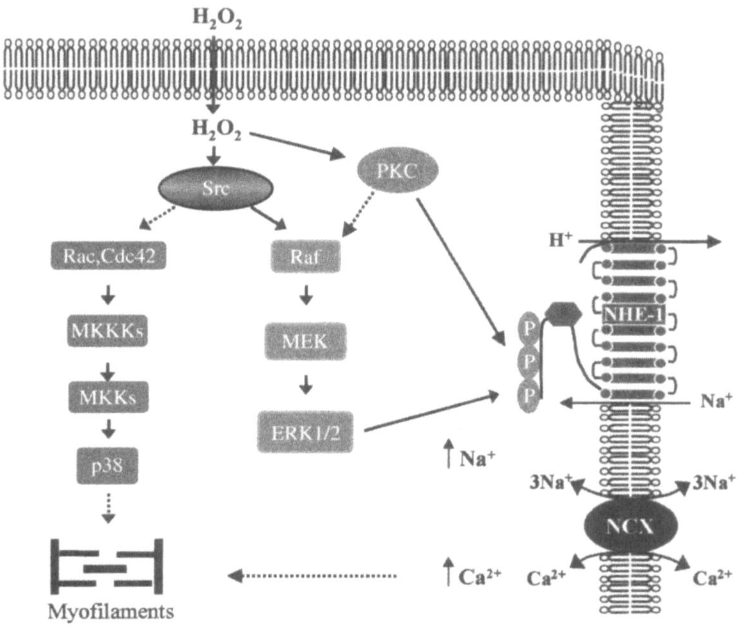

Figure 1. Proposed link between alterations in ROS, intracellular Ca^{2+} overload and contractile dysfunction in cardiac myocytes. Exposure of cardiac myocytes to H_2O_2 results in the activation of src and PKC. 1. Src activation leads to the phosphorylation and activation of ERK1/2 MAPK, which phosphorylate the COOH tail of NHE-1 thereby increasing exchanger activity. PKC may also directly phosphorylate NHE-1. The net rise in intracellular Na^+ alters NCX activity leading to intracellular Ca^{2+} overload. 2. Activation of p38 MAPK and Ca^{2+} overload may contribute to contractile dysfunction by altering myofilament organization.

tractile dysfunction associated with oxidant-induced I/R injury. MAPKs potentially play a major role in regulating this contractile dysfunction. JNK or p38 MAPK can regulate both sarcomeric assembly and stress fiber formation in cardiomyocytes (55,56). Interestingly, simultaneous over-expression of both JNK and p38 leads to myofibrillar disassembly (57). Other studies have shown that different isoforms of p38 MAPK are involved in sarcomeric assembly and apoptosis (56–58). Even in the absence of Ca^{2+} overload, ROS may decrease contractility by altering myofibrillar ATPase, decreasing creatine kinase activity, stimulating myosin phosphatase, or by metabolic depletion of ATP (59,60). Thus, it seems clear that the principal initiator of myocardial stunning is oxidant stress while the principal mediator is a decrease in the maximal Ca^{2+}-induced activation of the myofilaments.

CONCLUSIONS

These recent advances have contributed to the understanding of the chain of events that link the ROS burst with Ca^{2+} overload and the subsequent contractile dysfunction. It is likely that ROS-induced stimulation of sarcolemmal NHE-1 activity pays an important role in myocardial to the myocardium during I/R injury. Further clinical trials are necessary to determine the protective effects of NHE-1 inhibitor therapy against ischemia related myocardial injury.

ACKNOWLEDGMENTS

HOE642 was a generous gift of Hoechst Marion Roussel. Studies cited by this review were supported by NIH grants RO1-HL63318 and 5RO1-HL56046 (PAL).

REFERENCES

1. Goldhaber JI, Weiss JN. 1992. Oxygen free radicals and cardiac reperfusion abnormalities. Hypertension 20:118–127.
2. Reimer K, Jennings R. 1991. Myocardial ischemia, hypoxia, and infarction. In: The Heart and Cardiovascular System. Eds: HA Fozzard, E Haber, RB Jennings, AM Katz and HE Morgan, 1875–1973. New York: Raven Press.
3. Kloner RA, Arimie RB, Kay GL, Cannom D, Matthews R, Bhandari A, Shook T, Pollick C, Burstein S. 2001. Evidence for stunned myocardium in humans: a 2001 update. Coron Artery Dis 12:349–356.
4. Park JL, Lucchesi BR. 1999. Mechanisms of myocardial reperfusion injury. Ann Thorac Surg 68:1905–1912.
5. Lefer DJ, Granger DN. 2000. Oxidative stress and cardiac disease. Am J Med 109:315–323.
6. Droge W. 2002. Free radicals in the physiological control of cell function. Physiol Rev 82:47–95.
7. Finkel T. 2000. Redox-dependent signal transduction. FEBS Lett 476:52–54.
8. Bolli R, Patel BS, Jeroudi MO, Lai EK, McCay PB. 1988. Demonstration of free radical generation in "stunned" myocardium of intact dogs with the use of the spin trap alpha-phenyl N-tert-butyl nitrone. J Clin Invest 82:476–485.
9. Bolli R, Jeroudi MO, Patel BS, DuBose CM, Lai EK, Roberts R, McCay PB. 1989. Direct evidence that oxygen-derived free radicals contribute to postischemic myocardial dysfunction in the intact dog. Proc Natl Acad Sci U S A 86:4695–4699.
10. Stephens NG, Parsons A, Schofield PM, Kelly F, Cheeseman K, Mitchinson MJ. 1996. Randomised controlled trial of vitamin E in patients with coronary disease: Cambridge Heart Antioxidant Study (CHAOS). Lancet 347:781–786.
11. Sabri A, Byron KL, Samarel AM, Bell J, Lucchesi PA. 1998. Hydrogen peroxide activates mitogen-activated protein kinases and Na^+-H^+ exchange in neonatal rat cardiac myocytes. Circ Res 82: 1053–1062.
12. Wei S, Rothstein EC, Fliegel L, Dell'Italia LJ, Lucchesi PA. 2001. Differential MAP kinase activation and Na^+/H^+ exchanger phosphorylation by H(2)O(2) in rat cardiac myocytes. Am J Physiol Cell Physiol 281:C1542–C1550.
13. Goldhaber JI. 1996. Free radicals enhance Na^+/Ca^{2+} exchange in ventricular myocytes. Am J Physiol 271:H823–H833.
14. Goldhaber JI, Liu E. 1994. Excitation-contraction coupling in single guinea-pig ventricular myocytes exposed to hydrogen peroxide. J Physiol 477:135–147.
15. Ventura C, Guarnieri C, Caldarera CM. 1985. Inhibitory effect of superoxide radicals on cardiac myofibrillar ATPase activity. Ital J Biochem 34:267–274.
16. Karmazyn M, Gan XT, Humphreys RA, Yoshida H, Kusumoto K. 1999. The myocardial Na^+/H^+ exchange: structure, regulation, and its role in heart disease. Circ Res 85:777–786.
17. Bianchini L, Pouyssegur J. 1996. Regulation of the Na^+/H^+ exchanger isoform NHE1: role of phosphorylation. Kidney Int 49:1038–1041.
18. Fliegel L, Frohlich O. 1993. The Na^+/H^+ exchanger: an update on structure, regulation and cardiac physiology. Biochem J 296:273–285.
19. Tani M. 1990. Mechanisms of Ca^{2+} overload in reperfused ischemic myocardium. Annu Rev Physiol 52:543–559.
20. An J, Varadarajan SG, Camara A, Chen Q, Novalija E, Gross GJ, Stowe DF. 2001. Blocking Na^+/H^+ exchange reduces $[Na^+]_i$ and $[Ca^{2+}]_i$ load after ischemia and improves function in intact hearts. Am J Physiol Heart Circ Physiol 281:H2398–H2409.
21. Karmazyn M. 1988. Amiloride enhances postischemic ventricular recovery: possible role of Na+-H+ exchange. Am J Physiol 255:H608–H615.

22. Hendrikx M, Mubagwa K, Verdonck F, Overloop K, Van Hecke P, Vanstapel F, Van Lommel A, Verbeken E, Lauweryns J, Flameng W. 1994. New Na^+/H^+ exchange inhibitor HOE 694 improves postischemic function and high-energy phosphate resynthesis and reduces Ca2+ overload in isolated perfused rabbit heart. Circulation 89:2787–2798.
23. Scholz W, Albus U, Counillon L, Gogelein H, Lang HJ, Linz W, Weichert A, Scholkens BA. 1995. Protective effects of HOE642, a selective sodium-hydrogen exchange subtype 1 inhibitor, on cardiac ischaemia and reperfusion. Cardiovasc Res 29:260–268.
24. Avkiran M, Marber MS. 2002. Na^+/H^+ exchange inhibitors for cardioprotective therapy: progress, problems and prospects. J Am Coll Cardiol 39:747–753.
25. Dyck JRB, Maddaford TG, Pierce GN, Fliegel L. 1995. Induction of expression of the sodium-hydrogen exchanger in rat myocardium. Cardiovas Res 29:203–208.
26. Rothstein EC, Byron KL, Reed RE, Fliegel L, Lucchesi PA. 2002. H_2O_2-Induced Ca^{2+} Overload in Ventricular Myocytes Involves ERK1/2 MAP Kinases: Role for NHE-1 Dependent and Independent Pathways. Am J Physiol published April 11, 2002, 10.1152/ajpheart.00198.2002 83.
27. Snabaitis AK, Hearse DJ, Avkiran M. 2002. Regulation of sarcolemmal Na(+)/H(+) exchange by hydrogen peroxide in adult rat ventricular myocytes. Cardiovasc Res 53:470–480.
28. Hoque AN, Karmazyn M. 1997. Effect of sodium-hydrogen exchange inhibition on functional and metabolic impairment produced by oxidative stress in the isolated rat heart. Can J Physiol Pharmacol 75:326–334.
29. Myers ML, Farhangkhoee P, Karmazyn M. 1998. Hydrogen peroxide induced impairment of post-ischemic ventricular function is prevented by the sodium-hydrogen exchange inhibitor HOE 642 (cariporide). Cardiovasc Res 40:290–296.
30. Pimentel DR, Amin JK, Xiao L, Miller T, Viereck J, Oliver-Krasinski J, Baliga R, Wang J, Siwik DA, Singh K, Pagano P, Colucci WS, Sawyer DB. 2001. Reactive oxygen species mediate amplitude-dependent hypertrophic and apoptotic responses to mechanical stretch in cardiac myocytes. Circ Res 89:453–460.
31. Bogoyevitch MA, Gillespie-Brown J, Ketterman AJ, Fuller SJ, Ben-Levy R, Ashworth A, Marshall CJ, Sugden PH. 1996. Stimulation of the stress-activated mitogen-activated protein kinase subfamilies in perfused heart. p38/ERK mitogen-activated protein kinases and c-Jun N-terminal kinases are activated by ischemia/reperfusion. Circ Res 79:162–173.
32. Knight RJ, Buxton DB. 1996. Stimulation of c-Jun kinase and mitogen-activated protein kinase by ischemia and reperfusion in the perfused rat heart. Biochem Biophys Res Commun 218:83–88.
33. Sugden PH, Clerk A. 1998. "Stress-responsive" mitogen-activated protein kinases (c-Jun N-terminal kinases and p38 mitogen-activated protein kinases) in the myocardium. Circ Res 83:345–352.
34. Clerk A, Fuller SJ, Michael A, Sugden PH. 1998. Stimulation of "stress-regulated" mitogen-activated protein kinases (stress-activated protein kinases/c-Jun N-terminal kinases and p38- mitogen-activated protein kinases) in perfused rat hearts by oxidative and other stresses. J Biol Chem 273:7228–7234.
35. Moor AN, Gan XT, Karmazyn M, Fliegel L. 2001. Activation of Na^+/H^+ exchanger-directed protein kinases in the ischemic and ischemic-reperfused rat myocardium. J Biol Chem 276:16113–16122.
36. Avkiran M. 1999. Rational basis for use of sodium-hydrogen exchange inhibitors in myocardial ischemia. Am J Cardiol 83:10G–17G; discussion 17G–18G.
37. Josephson RA, Silverman HS, Lakatta EG, Stern MD, Zweier JL. 1991. Study of the mechanisms of hydrogen peroxide and hydroxyl free radical- induced cellular injury and calcium overload in cardiac myocytes. J Biol Chem 266:2354–2361.
38. Xu KY, Zweier JL, Becker LC. 1997. Hydroxyl radical inhibits sarcoplasmic reticulum Ca(2+)-ATPase function by direct attack on the ATP binding site. Circ Res 80:76–81.
39. Banerjee A, Grosso MA, Brown JM, Rogers KB, Whitman GJ. 1991. Oxygen metabolite effects on creatine kinase and cardiac energetics after reperfusion. Am J Physiol 261:H590–H597.
40. Zweier JL, Flaherty JT, Weisfeldt ML. 1987. Direct measurement of free radical generation following reperfusion of ischemic myocardium. Proc Natl Acad Sci U S A 84:1404–1407.
41. Philipson KD, Nicoll DA. 2000. Sodium-calcium exchange: a molecular perspective. Annu Rev Physiol 62:111–133.
42. Iwamoto T, Pan Y, Nakamura TY, Wakabayashi S, Shigekawa M. 1998. Protein kinase C-dependent regulation of Na^+/Ca^{2+} exchanger isoforms NCX1 and NCX3 does not require their direct phosphorylation. Biochemistry 37:17230–17238.
43. Ward CA, Moffat MP. 1995. Role of protein kinase C in mediating effects of hydrogen peroxide in guinea-pig ventricular myocytes. J Mol Cell Cardiol 27:1089–1097.

44. Hilgemann DW, Collins A, Matsuoka S. 1992. Steady-state and dynamic properties of cardiac sodium-calcium exchange. Secondary modulation by cytoplasmic calcium and ATP. J Gen Physiol 100:933–961.
45. Tani M, Neely JR. 1989. Role of intracellular Na^+ in Ca^{2+} overload and depressed recovery of ventricular function of reperfused ischemic rat hearts. Possible involvement of Na^+-H^+ and Na^+-Ca^{2+} exchange. Circ Res 65:1045–1056.
46. Moor AN, Fliegel L. 1999. Protein kinase-mediated regulation of the Na^+/H^+ exchanger in the rat myocardium by mitogen-activated protein kinase-dependent pathways. J Biol Chem 274:22985–22992.
47. Snabaitis AK, Hiroyuki Y, Avkiran M. 2000. Roles of mitogen-activated protein kinases and protein kinase C in alpha-1A adrenoceptor-mediated stimulation of the sarcolemmal Na^+-H^+ exchanger. Circ Res 86:214–220.
48. Gao WD, Liu Y, Mellgren R, Marban E. 1996. Intrinsic myofilament alterations underlying the decreased contractility of stunned myocardium. A consequence of Ca^{2+}-dependent proteolysis? Circ Res 78:455–465.
49. Gao WD, Atar D, Liu Y, Perez NG, Murphy AM, Marban E. 1997. Role of troponin I proteolysis in the pathogenesis of stunned myocardium. Circ Res 80:393–399.
50. McDonough JL, Arrell DK, Van Eyk JE. 1999. Troponin I degradation and covalent complex formation accompanies myocardial ischemia/reperfusion injury. Circ Res 84:9–20.
51. Van Eyk JE, Powers F, Law W, Larue C, Hodges RS, Solaro RJ. 1998. Breakdown and release of myofilament proteins during ischemia and ischemia/reperfusion in rat hearts: identification of degradation products and effects on the pCa-force relation. Circ Res 82:261–271.
52. Feng J, Schaus BJ, Fallavollita JA, Lee TC, Canty JM, Jr. 2001. Preload Induces Troponin I Degradation Independently of Myocardial Ischemia. Circulation 103:2035–2037.
53. Kim SJ, Kudej RK, Yatani A, Kim YK, Takagi G, Honda R, Colantonio DA, Van Eyk JE, Vatner DE, Rasmusson RL, Vatner SF. 2001. A novel mechanism for myocardial stunning involving impaired Ca^{2+} handling. Circ Res 89:831–837.
54. Thomas SA, Fallavollita JA, Lee TC, Feng J, Canty JM, Jr. 1999. Absence of troponin I degradation or altered sarcoplasmic reticulum uptake protein expression after reversible ischemia in swine. Circ Res 85:446–456.
55. Thorburn J, Xu S, Thorburn A. 1997. MAP kinase- and Rho-dependent signals interact to regulate gene expression but not actin morphology in cardiac muscle cells. Embo J 16:1888–1900.
56. Zechner D, Thuerauf DJ, Hanford DS, McDonough PM, Glembotski CC. 1997. A role for the p38 mitogen-activated protein kinase pathway in myocardial cell growth, sarcomeric organization, and cardiac-specific gene expression. J Cell Biol 139:115–127.
57. Wang Y, Su B, Sah VP, Brown JH, Han J, Chien KR. 1998. Cardiac hypertrophy induced by mitogen-activated protein kinase kinase 7, a specific activator for c-Jun NH2-terminal kinase in ventricular muscle cells. J Biol Chem 273:5423–5426.
58. Wang Y, Huang S, Sah VP, Ross J, Jr., Brown JH, Han J, Chien KR. 1998. Cardiac muscle cell hypertrophy and apoptosis induced by distinct members of the p38 mitogen-activated protein kinase family. J Biol Chem 273:2161–2168.
59. Shattock MJ. 1997. Myocardial stunning: do we know the mechanism? Basic Res Cardiol 92:18–22.
60. Hein S, Scheffold T, Schaper J. 1995. Ischemia induces early changes to cytoskeletal and contractile proteins in diseased human myocardium. J Thorac Cardiovasc Surg 110:89–98.

Pathophysiology of Cardiovascular Disease,
edited by N.S. Dhalla, H. Rupp,
A. Angel & G.N. Pierce
Kluwer Academic Publishers, Boston, 2004

Calcium Signalling During Acute Myocardial Ischemia: New Insights From Optical Mapping Techniques

William T. Clusin

Cardiovascular Divison
Stanford University School of Medicine
Stanford, Ca 94305, USA
650-723-7395
wclusin@stanford.edu

Summary. Acute myocardial ischemia leads to an increase in the systolic and diastolic levels of the intracellular calcium transient during the first few minutes. This overall increase is associated with beat-to-beat fluctations in the peak and trough levels of the transients—calcium transient alternans. New advances in fluorescence imaging techniques have allowed the spatial distribution of calcium transient alternans to be studied during ischemia, and to be correlated with similar imaging of the cardiac action potential. These studies provide further evidence that alternans phenomena are responsible for the genesis of ventricular fibrillation during acute ischemia, and suggest specific molecular mechanisms that could be involved.

Key words: Calcium, Ischemia, Fluorescence, Action Potentials, Fibrillation.

INTRODUCTION

Substantial advances have occurred in the last few decades in our understanding of how cardiac ischemia affects cytosolic calcium levels, and the role that calcium may play in the arrhythmias of acute ischemia. Principal advances that have occurred since 1985 include the use of fluorescent calcium indicators to record cytosolic calcium transients from intact hearts, and the use of high resolution optical imaging to characterize the spatial distribution of abnormalities in the calcium transient under conditions where ventricular fibrillation occurs.

There are two types of arrhythmias in humans where the activation of abnormal ion currents by cytosolic calcium is clearly the cause of the arrhythmia. These are the triggered arrhythmias associated with digitalis toxicity, and the hereditary arrhythmias associated with mutations in the calcium release channel (also called the ryanodine receptor). The latter group of arrhythmias are frequently precipitated by exercise or catecholamine release (1–3). There is considerable evidence that the more common lethal arrhythmias that occur in the pre-hospital phase of myocardial infarction—chiefly ventricular fibrillation—also involve calcium activated membrane currents. A key piece of evidence is the fact that drugs which prevent calcium overload, such as beta blockers and the non-di-hydropyridine calcium channel blockers, can prevent ventricular fibrillation when given prior to coronary artery occlusion (4,5). This observation has held up consistently in animals despite the inability of calcium channel blockers to favorably impact survivial in prospective studies of patients with coronary artery disease. Fibrillation during ischemia can also be prevented by ryanodine, which blocks the calcium release channel (6). However, additional research is needed to determine with certainty whether abnormalities in cytosolic calcium directly mediate the initiation of lethal arrhythmias during ischemia.

Role of alternans in acute ischemic arrhythmias

Extensive research has shown that alternans phenomena, including alternans of the action potential duration, force of contraction and the electrocardiographic T-wave immediately precede the onset of ventricular fibrillation during coronary artery occlusion in large mammalian hearts. There is general agreement that T-wave alternans during the early phase of ischemia is caused by fluctuations in action potential duration (APD) and that these fluctuations in APD can be out of phase in different regions of the heart (7). This spatial heterogeneity of action potential duration leads to dispersion of refractoriness in which the gradient of repolarization reverses with every beat, and repetitive beats, such as those produced by re-entry or a rotor, encounter a constantly changing pattern of refractoriness that degenerates into chaos.

There is also general agreement about the following points: First, it is generally agreed that mechanical alternans occurs in the first few minutes of ischemia, which is the same time period when APD alternans develops. Second, there is clear evidence that mechanical alternans during ischemia is due to alternation in the amplitude of the calcium transient (calcium transient alternans) and not to fluctuations in myofilament length (8). Third, there is clear evidence that mechanical alternans and APD alternans both occur in conditions other than ischemia, such as rate jumps and premature stimuli, and that under these conditions, the relationship between the two phenomena is sufficiently reproducible to imply causation—i.e. causation of one phenomenon by the other, or of both phenomena by some common factor. Fourth, it is clearly established that intracellular calcium can effect several different membrane currents, and that these calcium-activated or calcium modified currents could explain coupling between electrical alternans and mechanical alternans. A key point, here is whether the relationship between the calcium transient and the action potential is concordant (longer action potential with large calcium transient) or discordant (longer action potential with small calcium transient). Modulation of ion currents by cytosolic calcium could explain either of these phenomena, but it is important to establish which phenomenon occurs with ischemia, and whether the relation between the calcium transient and action potential is constant from one experiment to another, and in all animal species.

Previous studies establishing a possible relationship between APD alternans, calcium transient alternans and ventricular fibrillation during ischemia fall into seven categories:

A. *Floating Microelectrode studies.* A study by Downar (9) first showed that APD alternans occurs in the first 4 min. of ischemia in the pig heart and this is associated with T-wave alternans. APD alternans occurred with little change in action potential amplitude.
B. *Studies of t-wave alternans in open chest animals.* Computer techniques for mapping t-wave alternans on the surface of the ischemic pig heart were developed by Carson et al. (10). T-wave alternans was shown to eminate from hot-spots a few mm across. Similar maps were constructed by Konta (11) who showed that the severity of t-wave alternans, as shown by the maps, predicted which hearts would develop ventricular fibrillation.
C. *Computer models of APD alternans.* APD alternans is known to develop if the slope of the APD restitution curve exceeds 1, and the tissue is paced above a rate threshold. Two laboratories have used computer modeling to show that in sheets of simulated cardiac cells, spatially out of the phase alternans could develop on this basis alone, with flow of electric current through gap junctions organizing the cells into independent "domains. (12,13). These models made no predictions about the behavior of calcium transients, and seem not to have considered the effects of calcium activated currents, or flow of calcium ions across gap junctions. These studies leave open the possibility that ischemia simply changes the restitution curve so that APD alternans develops, and that this has nothing to do with calcium or with the processes that produce calcium transient alternans. In view of these studies, it is important to determine if alternating calcium transients are usually or always present in specific regions of the ischemic heart having APD alternans.
D. *Recordings of action potentials with potentiometric dyes and MAP electrodes during ischemia.* APD alternans during ischemia has been studied with the Franz monophasic action potential electrode in the ischemic rabbit heart by Lee et al. (14) and by Kurz et al. (15). In these studies it was possible to record APD alternans simultaneously at two sites that were out of phase, by using two MAP electrodes. It was also possible to record APD and calcium transient alternans from the same heart by placing the MAP electrode near an 11-mm diameter fiberoptic probe. Much more detailed spatial resolution of action potentials in

intact hearts can be obtained using optical mapping with the potentiometric dye di-4-ANEPPS, together with high resolution imaging cameras (see below).

E. *Physiologic recordings of calcium transient alternans in cardiac muscle.* There have been a total of 10 peer-reviewed papers published which contain actual recordings of calcium transient alternans in cardiac tissue. Such papers first appeared in 1988, when it became possible to record calcium transients from multicellular cardiac preparations, including intact hearts (14). Most of the papers published before 1999 are mentioned in a review by Euler (16). Five of the 10 papers include attempts to record the action potential concurrently. Three of the papers pertain to calcium transient alternans induced by ischemia, and an additional paper involves simulated ischemia. The studies describing alternans in conditions other than ischemia involve rapid pacing, or premature beats, or alternans occurring spontaneously, sometimes in hypertrophied or failing hearts. This is a remarkably small amount of research given the strong evidence that electromechanical alternans is a major cause of death in humans.

F. *Voltage clamp experiments.* Chudin et al. (17) used voltage clamp to impose action potentials of fixed duration on rabbit ventricular myocytes which had shown concordant APD and $[Ca^{++}]_i$ transient alternans when stimulated rapidly without voltage clamp (Figure 1). Their recordings showed the same degree of $[Ca^{++}]_i$ transient alternans when APD was fixed as when it varied. They conclude that the $[Ca^{++}]_i$ alternans was not being caused by variations in APD and that the reverse may be true. A similar result was obtained by Orchard et al. (18) who

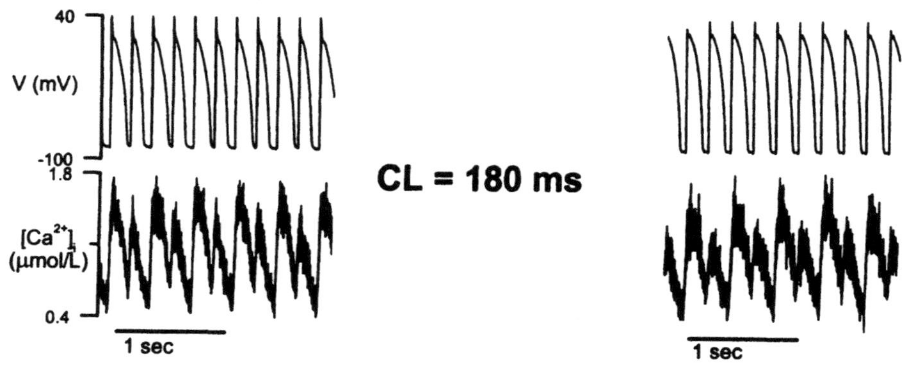

Figure 1. Action potential duration and $[Ca^{++}]_i$ transient alternans in an isolated rabbit ventricular myocyte during rapid pacing. In the left panel, APD alternans is evident with no change in action potential amplitude. There is associated calcium transient alternans, with the tall $[Ca^{++}]_i$ transients accompanying the broader action potentials. The diastolic $[Ca^{++}]_i$ level also fluctuates, so that the shorter transients arise from a higher baseline. In the right hand panel, a voltage clamp circuit has been used to impose action potential waveforms of constant duration at the same frequency as the left hand panel. This maneuver has no effect on the $[Ca^{++}]_i$ transients, which show the same degree of alternans as in the unclamped condition. From Chudin et al. (17).

used square voltage pulses to study mechanical alternans induced by acidosis in ferret heart muscle. They observed concordant alternating tail currents, (larger tail with stronger contraction) which they ascribed to calcium-activated inward current. The above two studies are consistent with older results by Morad and Trautwein (19) showing that moderate changes in APD do not affect contraction strength during single beats in the mammalian heart.

G. *Pharmacologic Studies.* Drugs that prevent calcium cycling by the sarcoplasmic reticulum, such as caffeine and ryanodine can prevent both electrical and mechanical alternans (20). This would be consistent with the hypothesis that electrical alternans resulted from variations in a calcium activated ion current. Ryanodine can also suppress arrhythmias during ischemia (6). Pretreatment of pig hearts with verapamil prevents both t-wave alternans and ventricular fibrillation (21). This suggests a role of calcium overload in electrical alternans, although this observation is less specific than observations involving ryanodine.

Calcium transient alternans during ischemia in the rabbit heart

Experimental studies of calcium transient alternans during ischemia have been most extensive in the rabbit heart. The ability to record cytosolic calcium transients from multicellular preparations was first established in rabbit hearts where it was found that arterial perfusion of the heart with the fluorescent indicator, indo-l, allowed acquisition of beat-to-beat transients from the epicardial surface (22). This observation has been repeated in many laboratories, and a number of groups have studied the effects of ischemia. These studies showed that, in contrast to what was previously believed, brief periods of ischemia (2–10 minutes) could produce an increase in cytosolic calcium that was not accompanied by cell necrosis, and was reversible on cessation of ischemia. Regular action potentials and calcium transients can occur under these conditions, with an increase in both the peak systolic and end-diastolic level of the calcium transients. This result has been obtained in three types of experiments: (a.) Total ischemia in rabbit hearts loaded with indo-l by arterial perfusion (14,23); (b.) Total ischemia in ferret hearts where the epicardial cells have been loaded with aequorin (24); and (c.) Low flow ischemia in rat hearts loaded with indo-l by arterial perfusion (25). The rise in diastolic calcium is thought to be due to failure of the sarcoplasmic reticulum to fully sequester calcium. The rise in calcium levels is associated with broadening of the calcium transient. There is a decrease in cellular pH under these conditions, and a fall in pH can increase systolic and diastolic calcium levels in the absence of ischemia (23).

After 2–4 minutes of ischemia, calcium transient alternans is frequently seen (Figure 2). There are beat-to-beat fluctuations in the peak systolic and end diastolic level of the transient, with the larger transients arising from a lower end-diastolic level and the smaller transients arising from a higher level. The presumed mechanism is that the SR has taken up more calcium prior to the large transients, and therefore releases more. Concurrent recordings of mechanical force obtained with

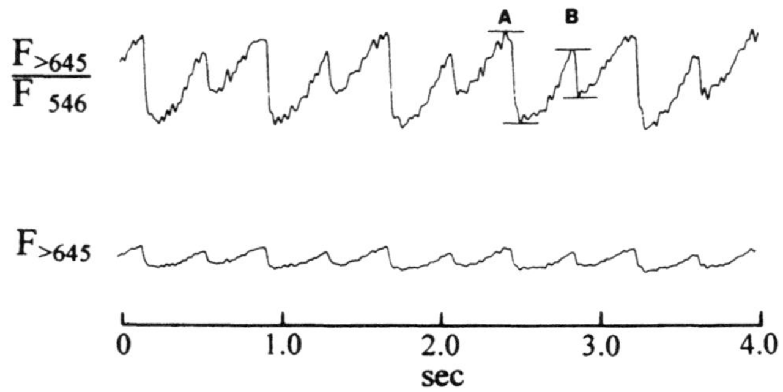

Figure 2. Calcium transient alternans induced by 2.5 min. of ischemia in a blood perfused rabbit heart. Recordings from the same heart prior to ischemia did not show alternans. The heart is loaded with the long wave length $[Ca^{++}]_i$ indicator, Fura Red, which gives a fluorescence decrease in response to a rise in cytosolic calcium. The heart is illuminated at 546 nm and emissions are collected with a 645 nm long pass filter. A ratio of fluorescence emissions to reflected excitation light is used to reduce certain artifacts. For each pair of beats, an alternans ratio can be calculated as 1 − (B/A) were B is the net amplitude of the smaller transients and A is the net amplitude of the larger transients. From Wu and Clusin (29).

epicardial strain gauge transducers or an intracavitary balloon show mechanical alternans in which relaxation is more complete prior to the strong beat. The extent of diastolic mechanical alternans is greater with the strain gauge transducer than the intracavitary balloon.

Calcium transient alternans during ischemia can occur under conditions in which the action potentials are full sized, and fluctuate in duration only, as in Figure 1. Recordings showing this have been obtained in rabbit hearts by simultaneous use of an optical probe and a monophasic action potential electrode (14). However, in these experiments, the two recordings were obtained from regions of myocardium several mm apart, so the relationship between the action potential and the calcium transient during ischemia remains to be fully defined.

In the intact animal, or human, ischemia occurs in the presence of blood. Wu and Clusin (26) compared the effects of ischemia in saline-perfused versus blood-perfused rabbit hearts. They developed an alternans ratio, defined as 1—B/A where A is the net amplitude (end-diastolic value to peak systolic value) of the taller transient and B is the net amplitude of the smaller transients (Figure 2). Alternans occurred in 9 of 11 blood-perfused ischemic hearts (82%) and in 6 of 14 saline perfused hearts (43%). The mean alternans ratio was 22.8% in the blood perfused hearts and 7.3% in the saline-perfused hearts ($P < 0.005$). These results indicate that occurrence of alternans is greatly potentiated by the presence of blood. Soluable factors present in blood including thrombin (27), and platelet release products (28) might contribute to this effect.

Recordings of calcium transient alternans in the presence of blood can only be obtained if a long wave length indicator is used. Wu and Clusin used the indicator, Fura Red, where fluorescence decreases on binding calcium (Figure 2). Qian et al. (29) used rhod-2 in blood perfused hearts with equally good results. Rhod-

2 had been used by others in saline perfused hearts. Both Fura Red and rhod-2 give larger fluorescence changes (ΔF/F) during the calcium transients than indo-1. With both indicators, the calcium-dependent fluorescence change greatly exceeds the motion artifact, which allows single wave-length calcium transients to be used without the need for a fluorescence ratio.

Spatial heterogeneity of calcium transient alternans

Qian et al. (29) have recently studied the uniformity of calcium transient alternans in the rabbit heart by optical mapping with a high-resolution digital camera. This camera could acquire images of 100 × 100 pixels every 3.3 msec. Based on earlier recordings of mechanical alternans with epicardial strain gauges, it was expected that alternans could be out of phase in discrete regions of the epicardial surface during ischemia.

Figure 3 shows an example of calcium transient alternans that is out of phase in two pixel regions from the same ischemic heart. In A, the first and odd beats show large calcium transients, while in B, the second and even beats show large calcium

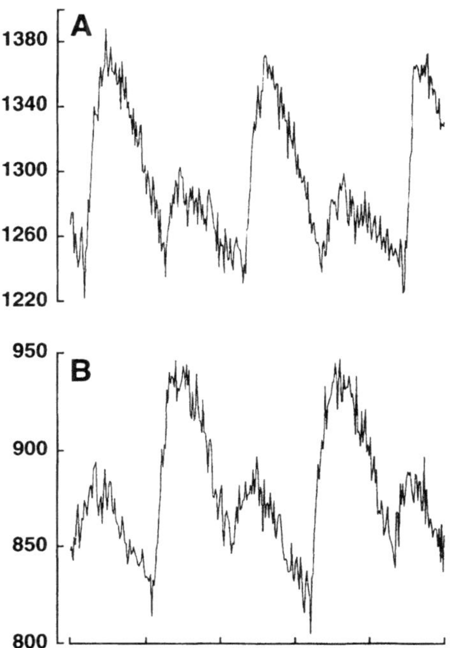

Figure 3. $[Ca^{2+}]_i$ transient alternans from 2 different pixels in a blood perfused ischemic rabbit heart loaded with the fluorescent indicator rhod-2 AM. The heart is illuminated by a laser at 532 nm, and emissions are collected by a high speed imaging camera with a 600 nm long pass filter. The heart did not show alternans prior to ischemia. The location of these pixels is shown in Figure 5 ([50,71] for A; [20,40] for B). From Qian et al. (29).

transients. The physical distance between these pixels is 11 mm, which means that the cells within these pixel regions would not have any direct electrotonic coupling.

Similar recordings from 8 pixels in a different heart are shown in Figure 4. Here two pixels are displayed in each of four panels. In the top panel, the pixel shown in green (1) has uniform calcium transients from beat-to-beat, while the pixel shown in white shows marked calcium transient alternans, with the first and odd beats taller and the second and even beats smaller. The physical distance beteen these pixels is 3.6 mm. When an alternans map was constructed for this heart (see below) pixel 1 turned out to lie at the boundary between two regions where alternans was out of phase. The second and third panels (pixels 3–6) show pairs of pixels with alternans that is out phase at the two sites. The physical distance between the pixels is 8 mm for 3 and 4 and 6 mm for 5 and 6. The bottom pair of pixels (7 and 8) shows alternans that is of the same phase, but the extent of alternans (alternans ratio) is markedly different. The physical distance between these pixels is 2.6 mm.

Qian et al. (29) found that the spatial distribution of alternans could be analyzed by construction of alternans maps. The principal feature of the alternans maps was

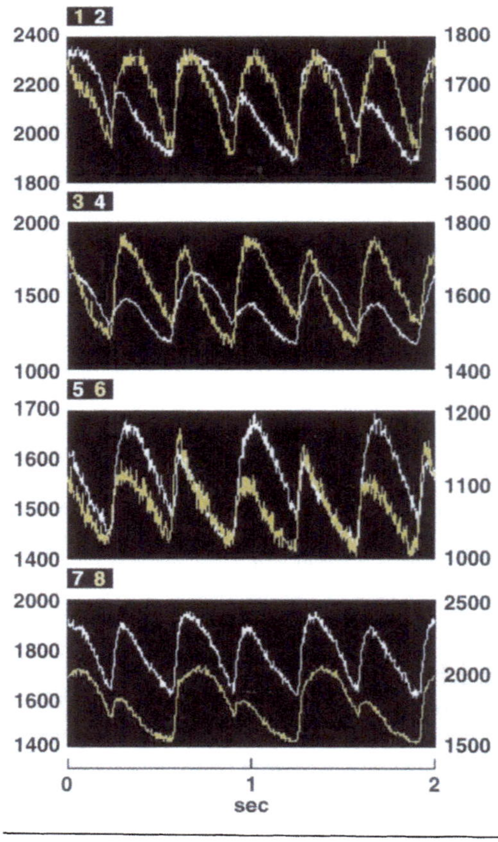

Figure 4. Calcium transients from selected pixels in a different blood-perfused rabbit heart. The duration of ischemia is 4 min. No alternans was present prior to ischemia. The middle two panels show pairs of pixels where alternans is out of phase. In the top panel, pixel one shows uniform calcium transients. In the bottom panel, both pixels show alternans of the same phase, but the alternans ratio is much larger for pixel 8 than 7. From Qian et al. (29).

that alternans occurs in discrete regions, and that regions where alternans is out of phase are separated by a boundary along which pixels showing uniform calcium transients can be found. Two types of alternans maps are possible, qualitative and quantitative. For a qualitative map, two different symbols are used for alternans that is out of phase, and a third symbol is used for uniform calcium transients. In the case of a quantitative map, the same is done, but the alternans ratio values (Figure 2) are displayed for each pixel.

Figure 5 shows a quantitative alternans map for the same experiment as Figure 3. Pixels with a low alternans ratio (less than 0.25; yellow highlight), including those with no alternans (alternans ratio = 0), form a boundary between two regions with high alternans ratios where alternans is out of phase. Inspection of the ratio values

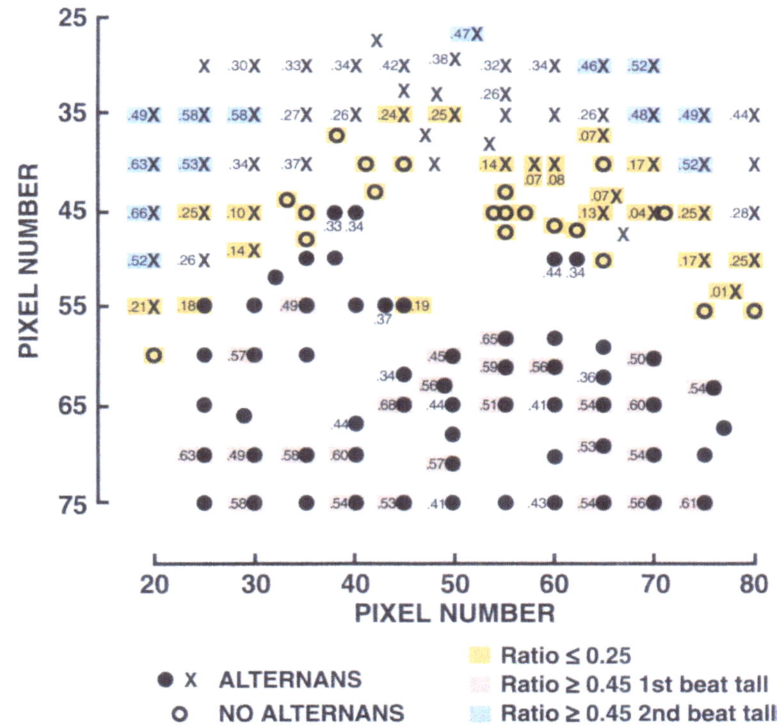

Figure 5. Quantitative alternans map for the experiment in Figure 3. One hundred thirty six pixels are plotted. Pixels showing alternans with the odd transient taller are plotted as filled circles (•). All of these pixels form a contiguous region located at the bottom of the figure. Pixels having the opposite pattern with even transients taller are shown as X's. These pixels form a contiguous region at the top of the figure. Twenty pixels have uniform calcium transients and are plotted as open circles (o's). These are located along the boundary between the two regions that are out of phase. Alternans ratios have been calculated for most of the pixels. Alternans ratios less than or equal to 0.25 are highlighted in yellow. These form a boundary between the two regions with marked alternans (blue and tan highlights). Thirty six pixels that were interrogated showed atypical fluctuations in $[Ca^{2+}]_i$ and were not plotted. Adapted from Qian et al. (29).

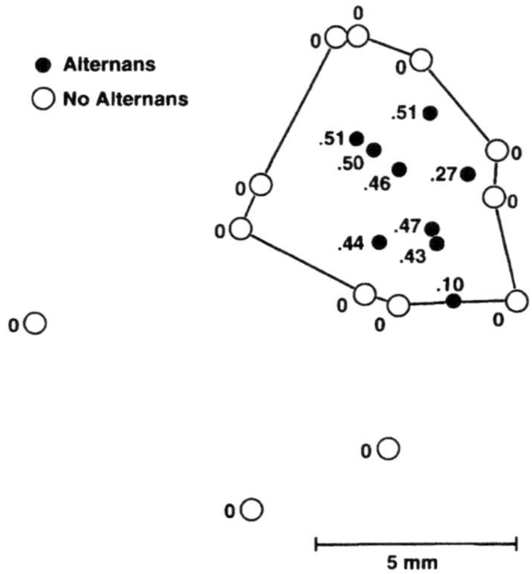

Figure 6. Quantitative alternans map for a different heart, in which only one region with alternans is present. Alternans ratio values are shown for pixels with alternans. Pixels not showing alternans have a ratio value of zero. Pixels were interrogated quasi-systematically using a mouse. From Qian et al. (29).

shows that the ratio behaves as a continuous variable, transitioning gradually from high values (blue and tan highlights), where there is marked alternans, to zero at the boundary, to high values again where there is marked alternans of opposite phase.

Figure 6 shows a quantitative alternans map from another heart. In this map there is only one region with alternans, which is entirely surrounded by pixels showing no alternans. All pixels with alternans had the same phase relationship, and the pixels with relatively low alternans ratios (0.10 and 0.27) are closer to the boundary than the other pixels.

Qian et al. (29) discussed several possible explanations for the spatial patterns of calcium transient alternans. Based on the depth of penetration of the excitation and emission light, it was estimated that each pixel would include about 360 myocytes. It is clear that if the phase of alternans within these cells were completely random, there should be no significant variation in the amplitude of calcium transients from odd to even beats. Clearly, some physical process acts to synchronize the phase relationship of alternans within each pixel and among nearby pixels. One possibility is that the duration of the cardiac action potential determines the amplitude of each calcium transient, and varies from beat to beat. In that case, electrotonic interactions could tend to synchronize alternans in nearby cells. The experiment in Figure 1 argues against this possibility and there is other evidence that the duration of the cardiac action potential does not control the amplitude of the calcium transient on a beat-to-beat basis.

An alternative way in which the synchronization might occur would involve movement of cytosolic calcium across gap junctions so that a large transient in one cell would favor the occurrence of a large transient in the adjoining cells to which it was coupled. The fact that calcium can cross gap junctions and modulate calcium release in adjoining cells has been proven by the study of triggered propagated contractions (30,31) in which waves of calcium-induced calcium release can propagate from cell to cell in the absence of action potentials. In the case of alternans, wave propagation would not be involved, since the cells do have action potentials, with nearly simultaneous upstrokes. However, movement of calcium across gap junctions could synchronize the alternation of the calcium transients during ischemia so that the large transients would occur together in adjoining cells, followed by small transients on the subsequent beat.

Studies of APD alternans during ischemia using potentiometric dyes

Three different experimental techniques have been used to study APD alternans, each of which has different drawbacks. Use of floating microelectrodes is painstaking, but persuasive, and a few recordings have been obtained with this technique which show APD alternans that is out of phase at different sites in the same ischemic heart (32). A technique that is somewhat more manageable is the contact monophasic electrode invented by Franz. Recordings with this device have confirmed that APD alternans can be out of phase at two different sites on the same ischemic heart. The principal drawback of this technique is the low spatial resolution, which is based on the fact that recordings are obtained from a circular area approximately 3 mm across. As noted above, significant variations in the calcium transient alternans ratio can occur within this distance (29). A modified version of this device allows optical recordings of calcium transients to be obtained from the same region of cells that produces the MAP (33). However, spatial resolution is still not be as good as that obtained by high resolution optical imaging.

High resolution optical imaging can be used to record simultaneous action potentials from many points on the heart surface using potentiometric dyes, which change their fluorescence in a linear fashion with changes in membrane potential. In a particularly elegant study, Choi and Salama (34), showed that calcium transients and action potentials could be recorded concurrently from small regions of saline-perfused, non-ischemic guinea pig hearts that were co-loaded with rhod-2 AM and with the potentiometric dye RH-237.

Potentiometric dyes have two important limitations. First, the change in fluorescence with these dyes is smaller than for the long wave length calcium indicators or indo-l, and no ratio signal can be obtained. Consequently motion artifact is a more serious problem. Some laboratories utilize contraction blockers such as D-600, which would be inappropriate for studies involving ischemia. Among the available potentiometric dyes, di-4-ANEPPS gives one of the largest fluorescence changes, but this dye cannot be used in hearts co-loaded with rhod-2 or Fura Red because all three dyes have approximately the same emission spectrum. The second limitation of potentiometric dyes is that the fluorescence signal deteriorates somewhat during ischemia.

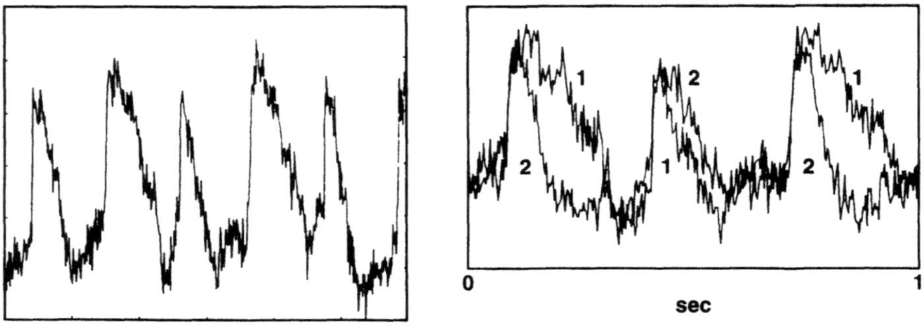

Figure 7. Optically recorded action potentials from a blood-perfused rabbit heart stained with the potentiometric dye di-4-ANEPPS, and rendered globally ischemic for three min. by stopping flow. There is marked APD alternans with a variation in APD_{50} of about 60%. A control recording showed uniform action potentials prior to ischemia. The right hand panel shows superimposition of recordings from two pixels in the same heart. For the first and odd numbered beats, the action potential from pixel 1 is much longer than the action potential from pixel 2. For the middle beat, the reverse is true. This fluctuating repolarization sequence would strongly favor the development of fibrillation. The physical distance between these pixels was 4 mm. A map of contiguous pixels shows two regions of APD alternans that is out of phase, and separated by a boundary.

In spite of these limitations, Qian et al. (35) were able to record APD alternans in a series of blood perfused hearts loaded with di-4-ANEPPS. A total of seven hearts were studied, of which six showed calcium transient alternans in at least some pixels during ischemia. Action potentials were uniform prior to ischemia and never developed alternans until after the second minute of ischemia. Figure 7 shows action potentials recorded from individual pixels in one of the hearts that developed alternans. These recordings were obtained at three minutes of ischemia. The left hand portion of the figure shows five consecutive action potentials from one pixel. The right hand portion of the figure shows superimposition of action potentials from two pixels where alternans is out of phase. The physical distance between these pixels was 4 mm. During the first and last beat, the action potential from the pixel labelled 1 is much longer than the action potential from the pixel labelled 2. During the middle beat, action potential 2 is longer. For this heart it was possible to construct a map of APD alternans similar to the maps of calcium transient alternans shown in Figure 5 and 6. As with calcium transient alternans, there were two discrete regions showing alternans that was out of phase. These regions were separated by a roughly linear boundary. It is presumed that if calcium transients could have been recorded from the same two regions, they would also show alternans that was out of phase. However, this assumption requires further experimental proof, which may require additional refinement of the recording techniques.

Feasibility of recording calcium transients in vivo

One important drawback of the isolated rabbit or rat heart is that it does not develop ventricular fibrillation during brief periods of ischemia. The most reliable way to observe ventricular fibrillation during ischemia, is to perform experiments in open chest dog or pig hearts. A model of ischemia-induced ventricular fibrillation has been developed in which VF can be induced in the canine heart during a predictable period of time by simultaneous occlusion of the LAD and circumflex arteries during pacing at 3 per second. In this model VF occurs after 2–3 min. of ischemia but is prevented or greatly delayed by pre-treatment with the calcium channel blocker diltiazem (5). Onset of ventricular fibrillation also occurs predictably during occlusion of the left anterior descending artery in pigs, typically in the first 5–20 min.

In order to correlate changes in the calcium transient with onset of ventricular fibrillation, it would be necessary to record calcium transients in a large animal heart in vivo during occlusion of a coronary artery. This poses several problems. First, blood would have to be present in the heart. Second, motion artifact and immobilization would present much greater problems; and third, loading of the heart would be complicated by the tendency of blood cells to take up and consume the cell permeant AM form of the dye.

Figure 8 shows a recording of calcium transients that was obtained from the left ventricular surface of the heart of an open chest pig using the cell-permeant indicator rhod-2 and a special fiber-optic probe mount that was designed to stabilize the optical apparatus and and eliminate motion. Calcium transients have now been recorded from open chest pigs with both rhod-2 and Fura Red (36). Two methods of loading have been developed. One method is to use an angioplasty catheter with a perfusion port to occlude the left anterior descending artery and perfuse distal to the occlusion with oxygenated saline solution containing rhod-2 AM. The angioplasty catheter allows red cells to be washed out of the vasculature while the loading proceeds. The second method is to infuse saline solution containing the AM indicator into the extracellular space using a very small needle and a syringe pump. The hand-held fiberoptic probe is then placed directly over the stained region of subepicardial tissue. Based on the above methods, it is clear that recordings of calcium transient alternans at the onset of ventricular fibrillation are practical to obtain.

Relationship between calcium transient and APD alternans

As noted above, there are a number of mechanisms by which calcium transient and APD alternans could be interrelated. In some of these schemes a calcium-activated ion current would be involved, so that the abnormality in intracellular calcium handling would be the primary cause of the arrhythmias. Not all possible schemes require this, however, and since intracellular calcium can produce either an inward current or an outward current, the relationship between the calcium transient and the action potential could be either concordant or discordant. Four specific possibilities need to be considered:

1. It is possible that APD alternans is purely due to changes in the restitution curve, and that this does not involve cytosolic calcium. Calcium transient alternans could occur by a different mechanism, and there could be cells which exhibited APD alternans without calcium transient alternans, and vice versa. This needs to

A

B

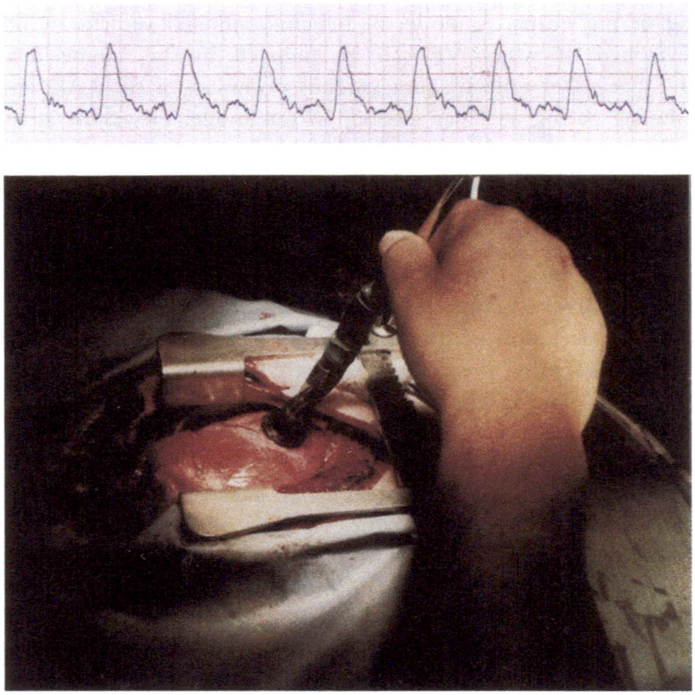

Figure 8. Recording of calcium transients obtained from the left ventricular surface of a pig heart loaded with rhod-2 AM. Rhod-2 was loaded into the left anterior descending artery using an angioplasty catheter. A hand held fiber-optic probe is used to illuminate the heart surface, and capture fluorescence emissions. The probe is immobilized by a tight fit between the plastic receptacle that contains the optic fibers (touching heart) and a C-shaped probe immobilizing device sutured to the heart (not visible). Three of the eight fiber optic bundles used to collect fluorescence emissions are visible as they insert into the tip of the probe. Fura Red has also been used successfully to record calcium-dependent fluorescence transients from open chest pig hearts.

be tested experimentally by making simultaneous recordings of both signals from the same cells in a large number of ischemic hearts.

2. If AP alternans involved beat-to-beat variation in the amplitude as well as the duration of the action potential, there could be variations in the amount of trigger calcium, which could lead to fluctuations in calcium transient amplitude. In this case, action potential amplitude alternans would be the primary event and calcium transient alternans a secondary phenomenon. While there are some published recordings in which AP amplitude does vary from beat to beat, there are many examples like Figure 1 in which only APD varies. Morad and Trautwein (19) previously showed that action potential duration does not contol contraction strength on a beat to beat basis, except when the action potential is very short.

3. APD alternans could be a consequence of calcium transient alternans, with the long action potentials coupled to the taller transients (concordant alternans; Figure 1). This would reflect the operation of a calcium-activated inward current, such as electrogenic sodium/calcium exchange, which is believed to contribute to the maintenance of the action potential plateau.
4. APD alternans could be a consequence of calcium transient alternans in which the tall calcium transient accompanied the briefer action potential (discordant alternans). There are two known mechanisms that could explain this. First, a rise in cytosolic calcium is known to favor more rapid inactivation of the L-type calcium current, which maintains the action potential plateau. Faster inactivation of the calcium channel presumably explains the reduction in action potential duration with flash photolysis of caged calcium (Nitr-5, ref. 37) or direct injection of calcium into cardiac Purkinje fibers (38). A second mechanism for discordant alternans would involve calcium activated chloride channels (39,40), which help mediate phase I repolarization of the action potential in some, but not all cardiac cells.

With schemes one, three and four above, the calcium transient alternans would be a primary phenomenon and not a simple consequence of variation in the action potential. If this is the case, a molecular explanation needs to be found whereby ischemia would produce variations in the amount of calcium released from the sarcoplasmic reticulum. One possibility is that effects of ischemia on cytosolic pH are responsible for fluctuation in calcium release from the SR. Mohabir et al. (23) measured cytosolic pH in the rabbit heart using the fluorescent pH indicator BCECF and found that 3 min. of ischemia produced a fall in cytosolic pH to 6.65. Acidosis was shown by Orchard et al. (18) to produce mechanical alternans and alternating inward tail currents in voltage-clamped papillary muscles of the ferret heart. They induced intracellular acidosis using 15% and 30% CO_2 solutions, which had pH values of 6.7 and 6.85, respectively. Thus, it seems quantitatively plausible that ischemia-induced acidity is the entire explanation for calcium transient alternans. If that were the case, the next question would be whether acidity is titrating a single residue on a single protein, such as the calcium release channel, or whether more than one titration site is involved. Orchard et al. did specifically postulate that acidity was producing calcium transient and tail current alternans through an effect on the SR.

A second possibly relevant result is the fact that tissue hypoxia plus nitric oxide produces abnormal opening of the sarcoplasmic reticulum calcium release channel (41). It is possible that elevated nitric oxide levels occur with ischemia that would lead to abnormal opening of the calcium release channel.

It is clear that further studies of calcium transient and action potential alternans during ischemia with simultaneous recordings of the two phenomena in the same cells is the key to understanding which of the above schemes is correct. Although there are considerable technical challenges in obtaining these recordings, methodologic improvements have proceeded to the point where it should be to resolve these issues definitively.

REFERENCES

1. Priori SG, Napolitano C, Tiso N, Memmi M, Vignati G, Boise R, Sorrentino V, Daniele G. 2001. Mutations in the cardiac ryanodine receptor gene (hRyR2) underlie catecholaminergic polymorphic ventricular tachycardia. Circulation 103:196–200.
2. Laitinen PJ, Brown KM, Pippo K, Swan H, Devaney JM, Brahmbhatt B, Donarum EA, Marimo M, Tiso N, Viitasalo M, Toivonen L, Stephan DA, Kontula K. 2001. Mutations of the cardiac ryanodine receptor (RyR2) gene in familial polymorphic ventricular tachycardia. Circulation 103:485–490.
3. Tiso N, Stephan DA, Nava A, Bagattin A, Devaney JM, Stanchi F, Larderet G, Brahmbhatt B, Brown K, Bauce B, Muriago M, Basso C, Thiene G, Danieli GA, Rampazzo A. 2001. Identification of mutations in the cardiac ryanodine receptor gene in families affected with arrhythmogenic right ventricular cardiomyopathy type 2 (ARVD2) Hum Mol Genet 10:189–194.
4. Kaumann AJ, Aramendia P. 1968. Prevention of ventricular fibrillation induced by coronary ligation. J Pharm Exp Ther 164:326–332.
5. Clusin WT, Bristow MR, Baim DS, Schroeder JS, Jaillon P, Brett P, Harrison DC. 1982. The effects of diltiazem and reduced serum ionized calcium on ischemic ventricular fibrillation in the dog. Circ Res 50:518–526.
6. Thandroyen FT, McCarthy J, Burton KP, Opie LH. 1988. Ryanodine and caffeine prevent ventricular arrhythmias during acute myocardial ischemia and reperfusion in rat heart. Circ Res 62:306–314
7. Pastore JM, Girouard SD, Laurita KR, Akar FG, Rosenbaum DS. 1999. Mechanism linking T-wave alternans to the genesis of cardiac fibrillation. Circulation 99:1385–1394.
8. Kihara Y, Morgan JP. 1991. Abnormal calcium handling is the primary cause of mechanical alternans: Study in ferret ventricular muscles. Am J Physiol 261:H1746–H1755.
9. Downar E, Janse MJ, Durrer D. 1977. The effects of acute coronary artery occlusion on subepicardial transmembrane potentials in the intact porcine heart. Circulation 56:217–224.
10. Carson DL, Cardinal R, Savard P, Vermeulen M. 1986. Characterization of unipolar waveform alternation in acutely ischaemic porcine myocardium. Cardiovascular Res 20:521–527.
11. Konta T, Ikeda K, Yamaki M, Nakamura K, Honma K, Kubota I, Yasui S. 1990. Significance of discordant ST alternans in ventricular fibrillation. Circulation 78:1047–1059.
12. Watanabe MA, Fenton FH, Evans SJ, Hastings HM, Karma A. 2001. Mechanism of discordant alternans. J Cardiovasc Electrophys 12:196–209.
13. Qu Z, Garfinkel A, Chen P-S, Weiss JN. 2000. Mechanisms of discordant alternans and induction of re-entry in simulated cardiac tissue. Circulation 102:1664–1670.
14. Lee HC, Mohabir R, Smith N, Franz MR, Clusin WT. 1988. Effect of ischemia on calcium-dependent fluorescence transients in rabbit hearts containing indo-1. Correlation with monophasic action potentials and contraction. Circulation 78:1047–1059.
15. Kurz RW, Mohabir R, Ren XL, Franz MR. 1993. Ischaemia induced alternans of action potential duration in the intact heart: Dependence on coronary flow, preload and cycle length. Eur Heart J 14:1410–1420.
16. Euler DE. 1999. Cardiac alternans: mechanisms and pathophysiological significance. Cardiovasc Res 42:583–590.
17. Chudin E, Goldhaber J, Garfinkel A, Weiss J, Kogan B. 1999. Intracellular Ca^{++} dynamics and stability of ventricular tachycardia. Biophys J 77:2930–2941.
18. Orchard CH, McCall E, Kirby MS, Boyett MR. 1991. Mechanical alternans during acidosis in ferret ventricular muscle. Circ Res 68:69–76.
19. Morad M, Trautwein W. 1968. The effect of the duration of the action potential on contraction in the mammalian heart muscle. Pfluger's Archiv 299:66–82.
20. Saitoh H, Bailey JC, Surawicz B. 1989. Action potential duration alternans in dog Purkinje and ventricjular muscle fibers. Further evidence in support of two different mechanisms. Circulation 80:1421–1431.
21. Hashimoto H, Suzuki K, Miyake S, Nakashima M. 1983. Effects of calcium antagonists on the electrical alternans of the ST segment and on associated mechanical alternans during acute coronary occlusion in dogs. Circulation 68:667–672.
22. Lee HC, Smith N, Mohabir R, Clusin WT. 1987. Cytosolic calcium transients from the beating mammalian heart. Proc Natl Acad Sci 84:7793–7797.
23. Mohabir R, Lee HC, Kurz RW, Clusin WT. 1991. Effects of ischemia and hypercarbic acidosis on myocyte calcium transients, contraction and pH in perfused rabbit hearts. Circ Res 69:1525–1537.

24. Kihara Y, Grossman W, Morgan JP. 1989. Direct measurement of changes in intracellular calcium transients during hypoxia, ischemia and reperfusion of the intact mammalian heart. Circ Res 65:1029–1044.
25. Figueredo VM, Brandes R, Weiner MW, Massie BM, Camacho SA. 1993. Endocardial versus epicardial differences of intracellular free calcium under normal and ischemic conditions. Circ Res 72:1082–1090.
26. Wu YM, Clusin WT. 1997. Calcium transient alternans in blood-perfused ischemic hearts: observations with fluorescent indicator Fura Red Am J Physiol 273:H2161–H2169.
27. Chien WW, Mohabir R, Clusin WT. 1990. Effect of thrombin on calcium homeostasis in chick embryonic heart cells: Receptor operated calcium entry with IP3 and a pertussin toxin-sensitive G protein as second messengers. J Clin Invest 86:1436–1443.
28. Chien WW, Mohabir R, Newman D, Leung LLK, Clusin WT. 1990. Effect of platelet release products on cytosolic calcium in cardiac myocytes. Biochem Biophys Res Com 170:1121–1127.
29. Qian YW, Clusin WT, Lin SF, Han J, Sung RJ. 2001. Spatial heterogeneity of calcium transient alternans during the early phase of myocardial ischemia in the blood perfused rabbit heart. Circ 104:2082–2087.
30. Miura M, Boyden PA, Ter Keurs HEDJ. 1998. $[Ca^{2+}]_i$ waves during triggered propagated contractions in intact trabeculae. Am J Physiol 274:H266–H276.
31. Lamont C, Luther PW, Balke CW, Wier WG. 1998. Intercellular Ca^{2+} waves in rat heart muscle. J Physiol 512:669–676.
32. Janse MJ, Kleber AF, Downar E, Durrer D. 1977. Changements electrophysiologiques pendant l'ischemie myocardique et mechanisme possible des troubles du rhythme ventriculaire. Ann Cardiol Angeiol 26 (suppl)551–554.
33. Clusin WT, Han J, Qian Y-W. 1999. Simultaneous recordings of calcium transients and action potentials from small regions of the perfused rabbit heart. Pace 22:834a.
34. Choi BR, Salama G. 2000. Simultaneous maps of optical action potentials and calcium transients in guinea-pig hearts: mechanisms underlying concordant alternans. J Physiol 529:171–188.
35. Qian Y-W, Lin S-F, Sung RJ, Clusin WT. 2001 Spatial heterogeneity of action potential alternans in blood perfused ischemic rabbit hearts. Biophys J 80:643a.
36. Clusin WT, Vriens P, Qian Y-W. 1998. Cytosolic calcium transients from in vivo porcine hearts. J Investigative Med 46:193a.
37. Clusin WT. 2000. The role of cytosolic calcium in electrical and mechanical alternans during ischemia. In: *Monophasic Action Potentials: Bridging Cell and Bedside*. Ed. M.R. Franz, 209–225. Armonk, NY. Futura Press.
38. Isenberg G. 1975. Is potassium conductance of cardiac Purkinje fibers controlled by $[Ca^{2+}]_i$? Nature 253:273–274.
39. Zygmunt AC, Gibbons WR. 1991. Calcium-activated chloride current in rabbit ventricular myocytes. Circ Res 68:424–437.
40. Zygmunt AC. 1994. Intracellular calcium activates a chloride current in canine ventricular myocytes. Am J Physiol 267:H1984–H1995.
41. Eu JP, Sun J, Xu L, Stamler JS, Meissner G. 2000. The skeletal muscle calcium release channel: Coupled O_2 sensor and NO signaling functions. Cell 102:499–509.

Pathophysiology of Cardiovascular Disease,
edited by N.S. Dhalla, H. Rupp,
A. Angel & G.N. Pierce
Kluwer Academic Publishers, Boston, 2004

Effects of Ischemia on Cardiomyocyte Connexin-43 Distribution and Phosphorylation Studied in in vivo and in vitro Models

Stéphane Tanguy, Madhumathy Jeyaraman, Bradley W. Doble, Zhisheng Jiang, Robert R. Fandrich, and Elissavet Kardami

Institute of Cardiovascular Sciences, Departments of Human Anatomy and Cell Sciences and Physiology, University of Manitoba, Winnipeg, Canada

Summary. The gap junction protein connexin-43 (Cx43) exists mainly in the phosphorylated state in the normal heart. We have investigated short-term effects of ischemia on cardiac Cx43 phosphorylation and distribution, in four models: global ischemia of the ex vivo perfused heart, left ventricular ischemia induced by irreversible coronary ligation in vivo, simulated ischemia on isolated adult myocyte pellets, and neonatal cardiomyocytes incubated in a hypoxia chamber. Antibody AB.13-800 that recognizes specifically the 41 kDa nonphosphorylated form of cardiac Cx43 labeled intercalated discs (ICDs) in myocytes from perfused rat hearts subjected to 30 min global ischemia; also in myocytes at the infarct border 6 hours post-infarction. Ischemia induced a sharp increase in the 41 kDa Cx43 from perfused hearts, isolated adult myocyte pellets and neonatal myocyte cultures subjected to hypoxia. The protein phosphatase type 1/2A inhibitors okadaic acid and calyculin A, tested in the in vitro models, decreased ischemia-induced Cx43 dephosphorylation. The 41 kDa Cx43 was present in both Triton-soluble as well as Triton-insoluble (enriched in ICDs) cardiac membrane fractions, assessed by western blotting. We conclude that ischemia causes dephosphorylation of cardiomyocyte Cx43 in vivo as well as in vitro, and that this phenomenon occurs irrespectively of stage (neonatal or adult) or presence of cell contact. Cx43 dephosphorylation occurs at

Corresponding Author: Elissavet Kardami, Institute of Cardiovascular Sciences, SBRC 3008, 351 Taché Avenue, Winnipeg, Manitoba, Canada R2H 2A6, FAX: 204-233 6723, TEL: 204 235 3519, E-mail: ekardami@sbrc.ca.

ICDs, is mediated at least in part by PP1/2-type phosphatases, and would be expected to affect GJ function and contribute to ischemia-induced conductance and contractile changes.

Key words: Ischemia models, gap junctions, phosphorylation.

INTRODUCTION

Gap junctions (GJ) are specialized regions of intercellular contact containing membrane channels through which cells exchange ions as well as metabolites and second messengers of less than 1 kDa. The GJ hemi-channels in each cell consist of hexameric aggregates of connexins; members of a multigene family (1). Each connexin has four conserved transmembrane domains, two highly conserved extracellular regions and two unique cytoplasmic regions (2). Individual tissues, and even cells, may express multiple connexin isoforms, contributing to distinct conductance and permeability properties. Three major connexin isoforms are expressed in cardiomyocytes (3): Cx43, the major channel protein of working ventricular cardiomyocytes; Cx45, found in the atrioventricular node and adjoining His bundles; and Cx40, expressed in the atria, and the fast conducting tissue of the His-Purkinjee system.

In the adult heart, GJ localized predominantly at the ICDs between cardiomyocytes, provide cell-to-cell electrical continuity, support coordinated contraction, allowing the heart to perform as an effective pump (3). Changes therefore in Cx43 levels and properties would be expected to affect cardiac function.

Regulation of Cx43 can occur at multiple levels such as in transcription, translation, post-translational modification, assembly and degradation (1). Changes in the biochemical environment (pH, membrane composition) can alter GJ properties dramatically (increase or decrease permeability). Changes can occur rapidly, within milliseconds, or slowly, over several hours. Regulation of connexins at the level of gene expression is effected by hormones, cAMP, growth factors and oncogenes (4). Phosphorylation of connexins is an important regulatory mechanism, having profound effects on trafficking, assembly, turnover, conductance and overall permeability (5). It is important to note that Cx43 in the normal heart is predominantly phosphorylated (6) pointing to a link between Cx43 phosphorylation and normal cardiac function.

Cardiac pathology is accompanied by GJ changes, acutely and chronically. GJ remodeling occurs after myocardial infarction at the border zone between scar and viable myocardium: GJ are no longer restricted to the ICDs but distribute along the whole sarcolemmal membrane (7). Decreased Cx43 levels are detected in hibernating, hypofunctional human myocardium (8). In general, overall Cx43 expression may increase in the early stages of compensatory hypertrophic growth while chronic ventricular dysfunction and heart failure are associated with diminished Cx43 expression and reduced GJ (9). Changes in GJ levels and distribution have been proposed to underlie the development of arrhythmias, and exacerbate ventricular dysfunction (3). The critical role of GJ in arrhythmogenesis and contraction was demonstrated recently, by conditional inactivation of the Cx43 gene in the heart; this resulted in loss/decrease of Cx43, spontaneous ventricular arrhythmias, and contractile abnormalities (10,11). Preservation of GJ function is therefore an important therapeutic target for the diseased heart.

Ischemia induces dephosphorylation of Cx43 in the heart (7) (13). Since phosphorylation has been demonstrated to affect coupling and/or membrane association in many systems, Cx43 dephosphorylation would be expected to impact on its properties in situ and affect cardiac function. Manipulations therefore aimed at controlling Cx43 phosphorylation, or preventing its dephosphorylation, may also have therapeutic implications. In vitro models of ischemia, assuming they reflect pertinent in vivo responses, can be used to explore ways of manipulating Cx43 phosphorylation and offer many advantages since they allow tighter control of conditions as well as the testing of a large number of physiologically important molecules and "drugs". For this purpose we have examined and compared the effects of ischemia on the accumulation and subcellular localization of the 41 kDa d-Cx43 in four different rat models: the perfused heart subjected to global ischemia, the in vivo heart after myocardial infarction, isolated adult cardiomyocytes subjected to simulated ischemia, and neonatal cardiomyocytes subjected to hypoxia.

METHODS

Animals and Models of Ischemia: Spague-Dawley rats (250 g) were used throughout, following the guidelines of the Canadian Council on Animal Care. For the in vivo experiments, left ventricular myocardial infarction was induced surgically by irreversible coronary ligation, as described (14). For the ex vivo experiments, hearts, obtained rapidly after the animal was killed by decapitation, were perfused through the aorta at a constant pressure of 80 cm H_2O using oxygenated Krebs-Henseleit buffer (pH 7.4, 37.4°C) solution. Hearts were paced at 4 Hz (240 bpm) and function monitored via a water-filled balloon introduced in the left ventricle and connected to a pressure transducer (12,15). Hearts from "control" or "ischemic" groups were submitted to 20 min of stabilization in normoxic conditions. The "ischemic" group was subjected to complete interruption of flow for 30 min under normothermic conditions.

In vitro ischemia: Adult cardiomyocytes were isolated as described (15) and were used in an in vitro model of simulated ischemia (mineral oil-overlay of myocyte pellets) as described (16). Neonatal myocyte cultures were prepared and maintained in culture for one week as described (17,18). They were incubated in a hypoxia chamber for 4 hours, in the presence of hypoxic medium, following procedures described previously by others (19).

Immunofluorescence for Cx43 was performed as described (17). Hearts were frozen in an OCT-dry ice bath after 30 min ischemia or after stabilization in perfusion buffer. Sections (7 μm) were post-fixed for 15 min with freshly made 1% paraformaldheyde in cold phosphate buffered saline. Sections were observed and photographed using a Nikon Diaphot microscope equipped with epifluorescence optics.

Western blotting for Cx43 was performed as described (17). Total heart extracts were obtained from 50 mg of minced ventricular tissue homogenized and sonicated in SDS/PAGE sample buffer. Lysates were centrifuged at 13,000 rpm for 20 min, to remove residual tissue. Subcellular fractions were obtained from 100 mg of minced ventricular tissue (15). Cytosolic fractions (=C) were recovered as supernatant after homogenization in buffer containing Hepes (20 mmol/L, pH = 7.4) and supplements (beta-glycerophosphate; 60 mmol/L, EGTA; 5 mmol/L, EDTA; 5 mmol/L, PMSF; 1 mmol/L, sodium fluoride; 5 mmol/L, sodium orthovanadate; 1 mmol/L, leupeptin; 2 μg/mL, pepstatin; 2 μg/mL, aprotinin 2 μg/mL, E64 2 μg/mL) and centrifugation for 1 hour at 100,000xrpm at 4°C. Pellets were resuspended with a glass homogenizer in 0.5% Triton X-100 in PBS containing supplements as above, sonicated and incubated on ice for 30 minutes. Supernatants, obtained after 1 hour of centrifugation (100,000xrpm) constituted the Triton-X-soluble membrane fraction (=T). The final fraction (=S) was obtained by resuspending insoluble pellets in double strength SDS/PAGE buffer containing supplements as above with a glass homogenizer, and centrifugation at 13,000xrpm for 20 minutes to remove any insoluble residue. All procedures were conducted at 4°C. Fractions were stored at −70°C. Protein concentration was determined using the BCA method (Sigma) with bovine serum albumin as a standard. Antigen-antibody complexes were visualized by enhanced chemiluminescence (Pierce; Rockford, IL).

Antibodies: Rabbit polyclonal antibodies (r.AB), recognizing Cx43.res.367-382, have been described and characterized (18) and mouse monoclonal antibodies (m.AB) recognizing Cx43.res.250-270 were purchased from Transduction Laboratories (Lexington, KY). Monoclonal antibodies raised against Cx43.res.360-376, AB.13-800, were characterized previously (20) and purchased from Zymed Laboratories (San Fransisco, CA). For immunofluorescence, r.AB, m.AB and AB.13-800 were used at 1 : 1000, 1 : 200, and 1 : 200 dilutions, respectively. For western blotting, they were used at 1 : 20,000, 1 : 10,000, and 1 : 2000, respectively. Rabbit polyclonal antibodies against N-cadherin were purchased from Santa Cruz laboratories (Santa Cruz, CA), and used at 1 : 100 dilution.

Statistics: An unpaired two-tailed Student t-test was used to assess the statistical significance ($p < 0.05$) of differences between values from "non-ischemic" and "ischemic" samples for the densitometric measurements from Western blots.

RESULTS AND DISCUSSION

Cx43 distribution after global ischemia of the ex vivo perfused heart is shown in Figure 1. Antibodies r.AB and m.AB produced the well known punctate staining at the ICDs of cardiomyocytes of non-ischemic hearts (Figure 1,a,b). In contrast, AB.13-800 did not label cardiac myocyte ICDs in these hearts (Figure 1,c), in agreement with our previous observations (20); simultaneous staining for N-cadherin points to the location of ICDs in the same field (Figure 1,d,h). After 30 min ischemia, r.AB, and m.AB continued to label ICDs (Figure 1,e,f). In contrast to the lack of staining of non-ischemic tissue, AB.13-800 now clearly labeled the ICDs of myocytes (Figure 1,g). Qualitatively similar observations were made in all cardiomyocytes in all fields observed (taken from several transverse sections per heart, from at least 6 different ischemic or non-ischemic hearts). Global ischemia there-

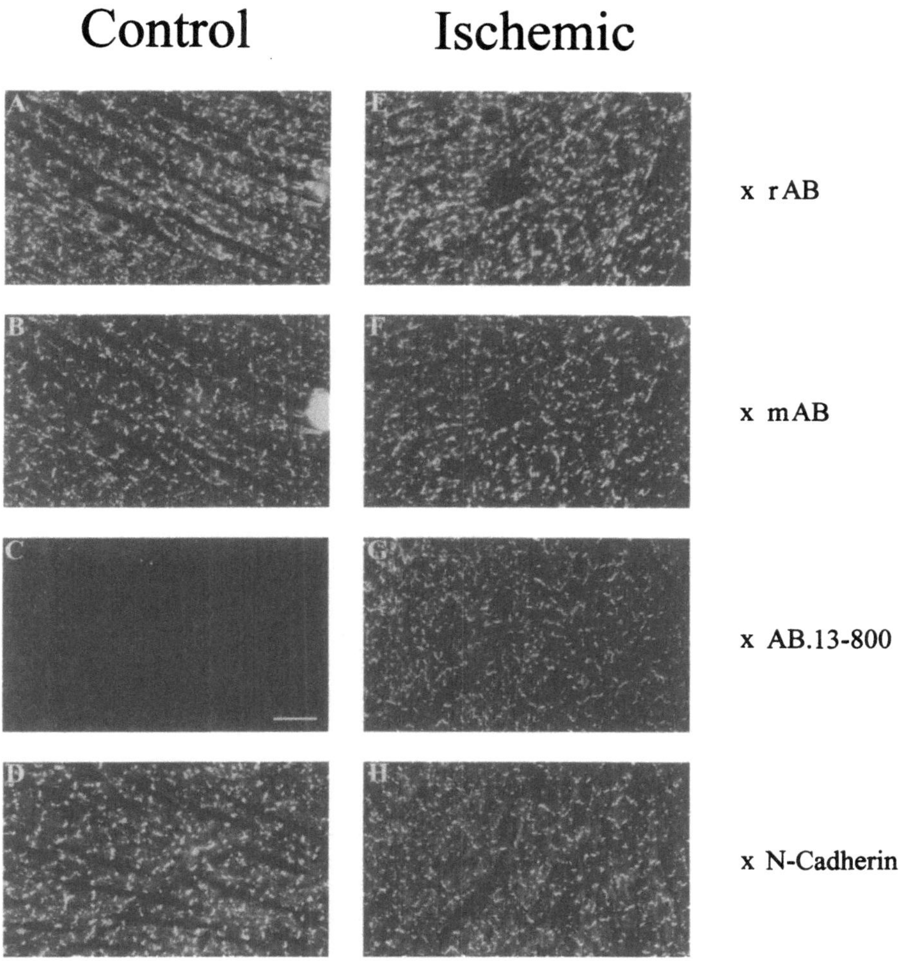

Figure 1. *Localization of d-Cx43 to intercalated discs of ischemic hearts.* A, B, C and D, sections from non-ischemic hearts. E, F, G, H, sections from ischemic hearts: A, B: Double immunofluorescence staining for Cx43, using r.AB and m.AB, respectively (see Methods for antibody description). E, F: Double immunofluorescence staining for Cx43, using r.AB and m.AB. C, D and G, H: Double immunofluorescence using AB.13-800 (C, G) and an anti-N-cadherin antibody (D, H), respectively. Bar = 100 μm.

fore resulted in accumulation of dephosphorylated (d-) Cx43 to the ICDs of all cardiomyocytes, in agreement with another report (13).

To determine the effect of ischemia in vivo, we looked at the distribution of Cx43 6 hours after myocardial infarction induced by permanent coronary ligation. As reported by others (7), anti-Cx43 staining was lost from the main bulk of the infarct containing necrotic myocytes. The border zone between viable myocytes and

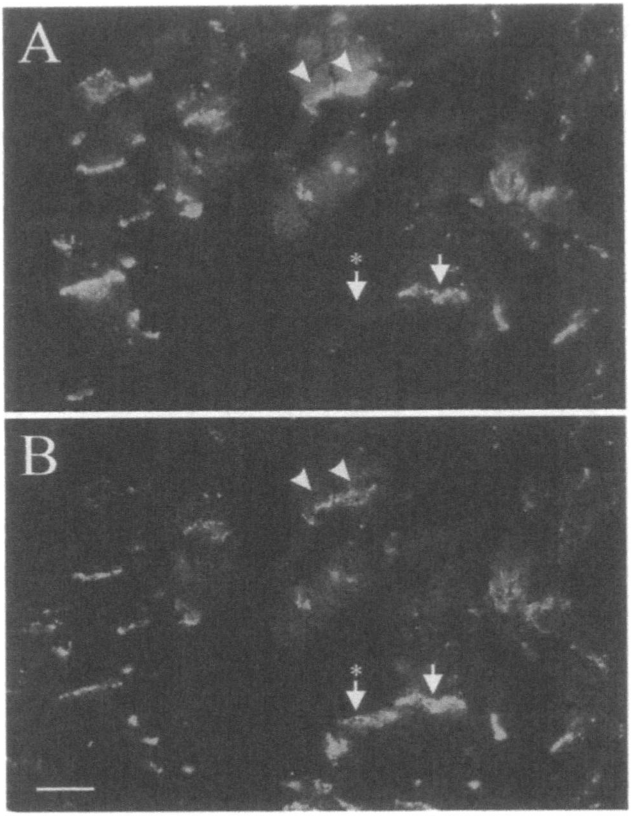

Figure 2. *Localization of d-Cx43 to intercalated disks of myocytes at the infarct border.* A, B: Double immunofluorescencestaining for d-Cx43 or total Cx43, respectively. Arrowheads point to ICDs staining intensely for d-Cx43; arrows point to ICDs that stain weakly or negatively (asterisk) for d-Cx43 but strongly for total (presumably predominantly phosphorylated) Cx43. Bar = 50 μM.

necrotic tissue displayed variable intensity of anti-Cx43 distribution at ICDs (Figure 2A), probed with r.AB. This antibody detects total Cx43, decreases in relative intensity of staining therefore point to decreased levels of Cx43. The ICDs of myocytes closest to the infarct stained weakly (arrowheads, Figure 2A), implying loss of GJ in these cells. The AB.13-800 produced strong staining of ICDs in myocytes at the infarct border, demonstrating presence of d-Cx43 at these sites (Figure 2B); occasionally, ICDs could be identified staining only for total and not for d-Cx43 (arrow with asterisk, Figure 2). Interestingly, staining for d-Cx43 was strongest in those border myocytes that presented weaker staining for total Cx43 (arrowheads, compare A and B pattern in Figure 2). Thus ICD heterogeneity is present at the infarct border: Cx43 can exist as mostly phosphorylated, or dephosphorylated, or a mixture

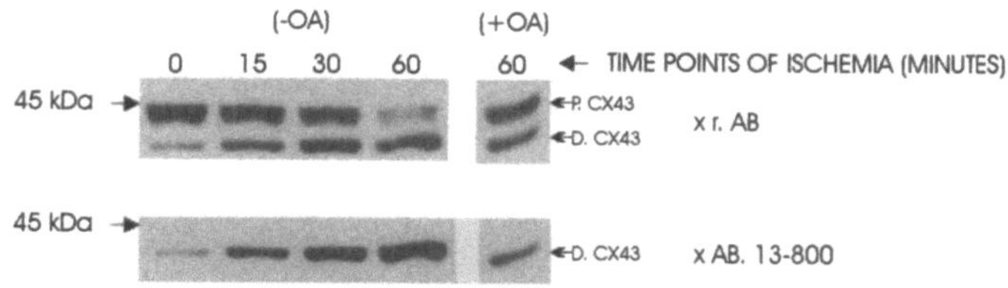

Figure 3. *Ischemia induces time-dependent Cx43 dephosphorylation in isolated adult myocytes.* Western blot of myocyte lysates probed for total (r.AB) or d-Cx43 (AB.13-800), at different time points (0, 15, 30 and 60 min), in the absence of okadaic acid (OA) as indicated. OA pretreatment decreases levels of d-CX43. Migration corresponding to 45 kDa is indicated.

of phosphorylated and dephosphorylated species. Also, it would appear that in myocytes at the infarct border d-Cx43 increases, while total, and phosphorylated, Cx43 decreases. It is possible that p-Cx43 is more vulnerable to ischemia-associated proteolysis and that dephosphorylation at the ICDs may prevent or delay its degradation. Inhibition of the proteasomal degradation pathway leads to preferential accumulation of d-Cx43 (21) and is cardioprotective.

We used two widely used in vitro models to examine the effects of simulated ischemia on Cx43 dephosphorylation. In the first, isolated adult myocytes were pelleted and subjected to simulated ischemia by the oil-overlay method, for various time points (0–60 min). As shown in Figure 3, simulated ischemia caused time dependent increases in d-Cx43, followed by corresponding decreases in p-Cx43. Cx43 dephosphorylation upon ischemia therefore does not depend on intercellular communication and contact since isolated adult myocytes are not connected at ICDs. The second in vitro model consisted of confluent cultures of differentiated neonatal cardiomyocytes subjected to hypoxic medium and a hypoxia-chamber inclubation for 4 h. Hypoxia in these cultures was also accompanied by Cx43 dephosphorylation, as shown in Figure 4. Cx43 dephosphorylation therefore occurs irrespectively of stage (neonatal vs adult myocytes), and method of oxygen deprivation.

We have used in vitro systems to test the role of PP1/2A type phosphatases in Cx43 dephosphorylation. As we reported recently (22), Cx43 dephosphorylation was partially prevented by okadaic acid (Figure 3) or calyculin A (Figure 4); both compounds are inhibitors of PP1/2A-type phosphatases. In subsequent experiments

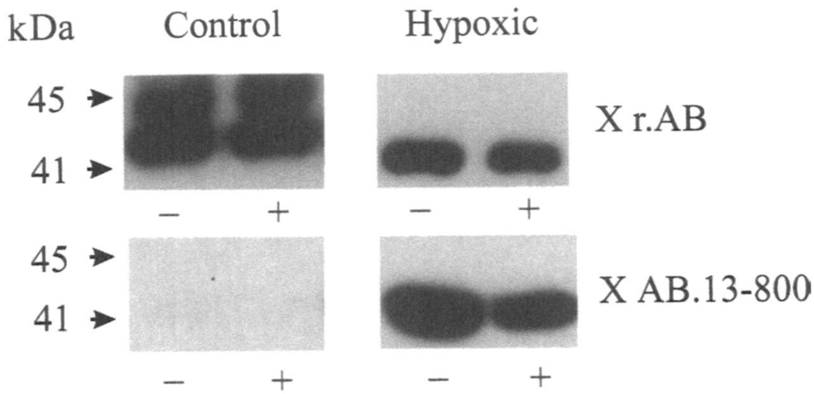

Figure 4. *Hypoxia induces dephosphorylation of Cx43 in neonatal cardiomyocyte cultures; Calyculin A decreases Cx43 dephosphorylation.* Western blot of myocyte lysates probed for total (r.AB) or d-Cx43 (AB.13-800), as indicated. (+) and (−) denote presence or absence of calyculin A respectively.

we were able to demonstrate that complete prevention of ischemia-induced Cx43 dephosphorylation could be achieved by combined inhibition of PP1 and PP2B-type (calcineurin) phosphatases (Jeyaraman and Kardami, unpublished observations).

While we established that Cx43 dephosphorylation occurs in all models of ischemia, an important issue that needs to be explored further is the localization and state of assembly/aggregation of d-Cx43, since phosphorylation is known to affect these properties. As shown in Figure 1, d-Cx43 localizes to the ICDs, however this does not necessarily show that d-Cx43 is a component of functional GJ. To approach this issue one may consider certain properties of the channels: Mature and functional intercellular GJ are highly aggregated structures that are not soluble in membrane detergents such as Triton X-100 (23), and are deemed to be composed of p-Cx43 (24). To determine whether d-Cx43 was distributed in the aggregated, Triton-insoluble cell fraction, we used the ex vivo heart global ischemia model. Hearts were subjected to subcellular fractionation and Western blotting analysis of cardiac cytosolic- (=Cyt), Triton-soluble membrane- (=T), and Triton-insoluble (but SDS soluble)- (=S) protein fractions. The S fraction is enriched in aggregated GJ proteins. Figure 5 shows the migration pattern as well as distribution of Cx43 before and after ischemia. It is important to remember that increased levels of Cx43 phosphorylation display correspondingly decreased electrophoretic mobility (24). Before ischemia, Cx43 is present mostly in the S fraction, as predominantly p-Cx43; bands P1–4, in order of decreasing mobility and increasing phosphorylation. A small amount of d-Cx43 (P0) is also detectable in this preparation. The T fraction contains both d-CX43 and some p-Cx43, the less phosphorylated species (P1, P2). After 30 min ischemia, the S-fraction consists mainly of d-Cx43 (the P0 species), while relative levels of p-Cx43 are reduced. The T-fraction consists entirely of d-Cx43 after ischemia. In the experiment shown in Figure 5, Cx43 was detected using only

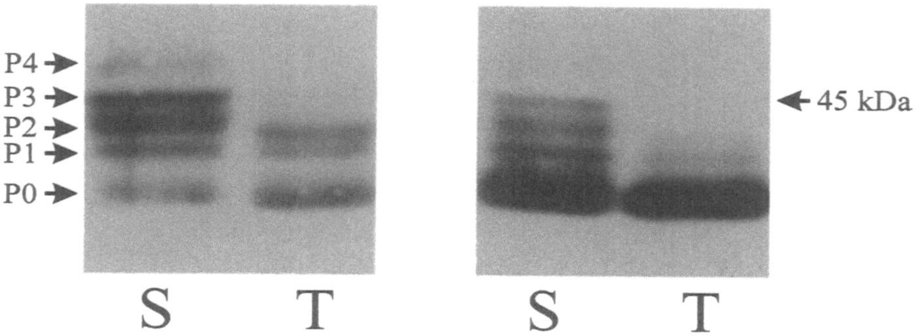

Figure 5. Subcellular distribution of the dephosphorylated and phosphorylated Cx43 species, before and after ischemia, as indicated. Western blot of myocardial membrane fractions, probed for total Cx43: T denotes the Triton-soluble, while S denotes the Triton-insoluble fractions. P1–4 point to the migration of Cx43 species possessing increasing levels of phosphorylation. P0 represent d-Cx43. Samples were analyzed in large format 9% polyacrylamide gels. Equal amounts of protein (20 µg) were loaded per lane.

the p.AB. In another experimental series we used all three different anti-Cx43 antibodies (r.AB, m.AB and AB.13-800) to assess the status and distribution of p- and d-Cx43. As shown in Figure 6, Cx43 was absent in the cytosolic fraction, as expected for a membrane protein. In non-ischemic hearts, the bulk of Cx43 was phosphorylated and was present in the S fraction (Figure 6). Ischemia caused a sharp increase in the 41 kDa Cx43, detected by all antibody preparations, accompanied by a nearly complete disappearance of the more extensively p-Cx43 species detectable by both r.AB and m.AB. The remaining phosphorylated forms were composed mostly of the less phosphorylated ~43 kDa Cx43 species and were recovered in the S fraction, a finding consistent with their continuous presence at ICDs. Most of the increased 41 kDa Cx43 was also recovered in the S fraction, confirming that d-Cx43 can exist in a highly aggregated state, and may form functional channels. A slight increase in 41 kDa Cx43 seen in the T fraction may reflect some redistribution of Cx43 from the S to the T fraction upon dephosphorylation, and/or prevention of Cx43 phosphorylation during ischemia. Presence of 41 kDa Cx43 in both the T and S fractions (Figures 5 and 6) suggested that it can exist in a "less" or "more" soluble form at the membrane, and that factors other than, or in addition to, its state of phosphorylation may affect its solubility properties. As a component of the connexon between myocytes Cx43 would be subjected to extensive thiol group cross-linking resulting in decreased solubility (23).

To quantitate the relative proportions of d-Cx43 and p-Cx43 in non-ischemic and ischemic hearts we analyzed total cardiac lysates by western blotting and densitometry. This method was chosen as to limit dephosphorylation of Cx43 during handling; phosphatase inhibitors may not completely block Cx43 dephosphorylation during more prolonged handling procedures such as the subcellular fractionation used for the above (unpublished observations). We thus calculated that d-Cx43

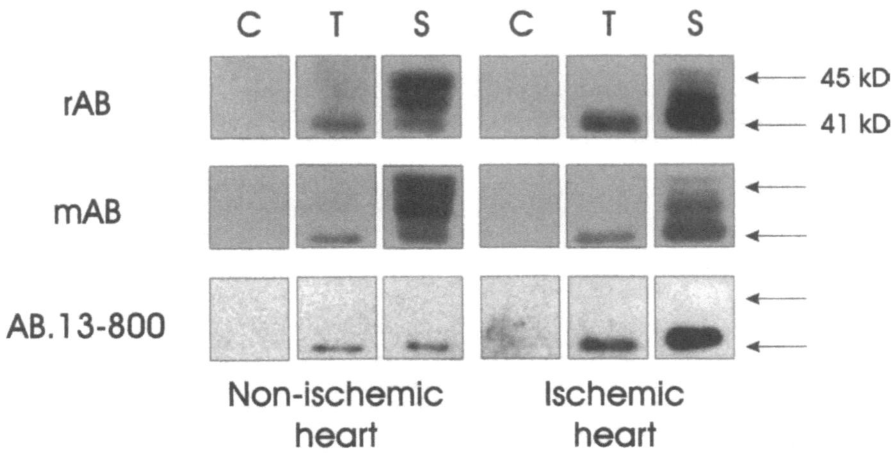

Figure 6. Subcellular distribution of d-Cx43 in cardiac fractions as a function of ischemia, assessed by western blotting and three different anti-Cx43 antibody preparations (r.AB, m.AB and AB.13-800), as indicated. C, T and S denote cytosolic, Triton X-100 soluble membrane, and Triton X-100 insoluble (SDS soluble) particulate fractions, respectively. Equal amounts of protein per fraction (20 μg) were loaded per lane. Arrows point to the migration of the 45 and 41 kDa Cx43 bands.

constituted 2.34% of total Cx43 in non-ischemic, compared to 72.5% in ischemic hearts (r.AB, and m.AB) ($P < 0.05$). Total levels of Cx43 did not seem to decrease during this period.

The functional consequences of Cx43 dephosphorylation during ischemia are a matter of some debate, since studies have suggested increased or decreased GJ permeability in ischemia. In oocytes (25), or in membrane artificial layers (26) d-Cx43 channels show increased conductance properties. According to a recent report, GJ-permeable fluorescent dyes can still migrate between ischemic myocytes of the ex vivo heart. GJ channels furthermore have been shown to allow the spread of contracture that occurs after ischemia (27). Because these conditions cause dephosphorylation of Cx43 (13), they would imply that cardiac GJ composed of d-Cx43 are permeable to ions and small metabolites, and are thus are capable of conduction. Similar findings have been reported for ischemic astrocytes (28). In contrast, another group reported that Cx43 dephosphorylation correlates with loss of electrical coupling (13) and development of arrhythmias. Calcium overload causes irreversible cardiomyocyte injury during ischemia and this would result in closure as well as loss of GJ (13). ATP depletion (rapidly occuring in ischemia, and associated with Cx43 dephosphorylation) decreases GJ conductance (29). In these studies, inhibition of PP1 type phosphatases (that would prevent Cx43 dephosphorylation) prevented the loss in gap junctional communication, providing a link between Cx43 dephosphorylation and decreased conductance.

The reason for these apparent discrepancies is not yet known. Because Cx43 can be phosphorylated at multiple sites it may exist in multiple species, each presenting a different level/states of phosphorylation (Figure 5). As we have shown (22) early ischemia is associated with conversion of Cx43 to less phosphorylated species, but not complete dephosphorylation; complete dephosphorylation is observed at later time points (30–60 min ischemia; data not shown). It is likely that different phosphorylation states of Cx43 have different properties; it is also likely that electrical conductance and metabolic coupling are regulated in a different manner (30).

Our own recently published data (22) have suggested an association between prevention of Cx43 dephosphorylation by okadaic acid and improvement in cardiac structural integrity that becomes compromised during ischemia. Manipulations therefore that target the phosphorylation state of Cx43, by affecting the phosphatase(s) and kinase(s) involved deserve further consideration as potential cardioprotective treatments.

ACKNOWLEDGENTS

This work was supported by the Canadian Institutes for Health Research and the Heart and Stroke Foundation of Canada (EK). ST and ZJ were supported by postdoctoral fellowships from the Manitoba Health Research Council.

REFERENCES

1. Kumar NM, Gilula NB. 1996. Gap junction communication channel. Cell 84:381–388.
2. Beyer EC. 1993. Gap junctions. Int Rev Cytol: 1–37.
3. Saffitz JE, Yamada KA. 2000. Closing the gap in understanding the regulation of intercellular communication. Cardiovasc Res 45:807–809.
4. Hossain MZ, Boynton AL. 2000. Regulation of Cx43 gap junctions: the gatekeeper and the password. Sci STKE 54:E1.
5. Lampe PD, Lau AF. 2000. Regulation of gap junctions by phosphorylation of connexins. Arch Biochem Biophys 384:205–215.
6. Lau AF, Hatch-Pigott V, Crow DS. 1991. Evidence that heart connexin43 is a phosphoprotein. J Mol Cell Cardiol 23(6):659–663.
7. Huang XD, Sandusky GE, Zipes DP. 1999. Heterogeneous loss of connexin43 protein in ischemic dog hearts. J Cardiovasc Electrophysiol 10(1):79–91.
8. Kaprielian RR, Gunning M, Dupont E, Sheppard MN, Rothery SM, Underwood R, Pennell DJ, Fox K, Pepper J, Poole-Wilson PA, Severs NJ. 1998. Downregulation of immunodetectable connexin43 and decreased gap junction size in the pathogenesis of chronic hibernation in the human left ventricle. Circulation 97:651–660.
9. Dupont E, Matsushita T, Kaba RA, Vozzi C, Coppen SR, Khan N, Kaprielian R, Yacoub MH, Severs NJ. 2001. Altered connexin expression in human congestive heart failure. J Mol Cell Cardiol 33:359–371.
10. Gutstein DE, Morley GE, Vaidya D, Liu F, Chen FL, Stuhlmann H, Fishman GI. 2001. Heterogeneous expression of Gap junction channels in the heart leads to conduction defects and ventricular dysfunction. Circulation 104:1194–1199.
11. Gutstein DE, Morley GE, Tamaddon H, Vaidya D, Schneider MD, Chen J, Chien, KR Stuhlmann H, Fishman GI. 2001. Conduction slowing and sudden arrhythmic death in mice with cardiac-restricted inactivation of connexin43. Circ Res 88:333–339.
12. Curtis MJ, Macleod BA, Tabrizchi R, Walker MJ. 1986. An improved perfusion apparatus for small animal hearts. J Pharmacol Methods 15:87–94.
13. Beardslee MA, Lerner DL, Tadros PN, Laing JG, Beyer EC, Yamada KA, Kleber AG, Schuessler RB, Saffitz JE. 2000. Dephosphorylation and intracellular redistribution of ventricular connexin43 during electrical uncoupling induced by ischemia. Circ Res 87:656–662.
14. Jiang ZS, Padua RR, Ju H, Doble BW, Jin Y, Hao J, Cattini PA, Dixon IM, Kardami E. 2002. Acute protection of ischemic heart by FGF-2: involvement of FGF-2 receptors and protein kinase C. Am J Physiol Heart Circ Physiol 282:H1071–H1080.

15. Padua RR, Merle PL, Doble BW, Yu CH, Zahradka P, Pierce GN, Panagia V, Kardami E. 1998. FGF-2-induced negative inotropism and cardioprotection are inhibited by chelerythrine: involvement of sarcolemmal calcium-independent protein kinase C. J Mol Cell Cardiol 30:2695–2709.
16. Armstrong SC, Ganote CE. 1992. Effects of the protein phosphatase inhibitors okadaic acid and calyculin A on metabolically inhibited and ischaemic isolated myocytes. J Mol Cell Cardiol 24:869–884.
17. Doble BW, Chen Y, Bosc DG, Litchfield DW, Kardami E. 1996. Fibroblast growth factor-2 decreases metabolic coupling and stimulates phosphorylation as well as masking of connexin43 epitopes in cardiac myocytes. Circ Res 79:647–658.
18. Doble BW, Ping P, Kardami E. 2000. The epsilon subtype of protein kinase C is required for cardiomyocyte connexin-43 phosphorylation. Circ Res 86(3):293–301.
19. Zhao J, Renner O, Wightman L, Sugden PH, Stewart L, Miller AD, Latchman DS, Marber MS. 1998. The expression of constitutively active isotypes of protein kinase C to investigate preconditioning. J Biol Chem 273:23072–23079.
20. Nagy JI, Li WE, Roy C, Doble BW, Gilchrist JS, Kardami E, Hertzberg EL. 1997. Selective monoclonal antibody recognition and cellular localization of an unphosphorylated form of connexin43. Exp Cell Res 236:127–136.
21. Beardslee MA, Laing JG, Beyer EC, Saffitz JE. 1998. Rapid turnover of connexin43 in the adult rat heart. Circ Res 83:629–635.
22. Jeyaraman M, Tanguy S, Fandrich RR, Lukas A, Kardami E. 2002. Ischemia-induced dephosphorylation of cardiomyocyte connexin-43 is reduced by okadaic acid and caliculin A but not fostriecin. Mol Cell Biochem In press.
23. Manjunath CK, Page E. 1986. Rat heart gap junctions as disulfide-bonded connexon multimers: their depolymerization and solubilization in deoxycholate. J Membr Biol 90:43–57.
24. Musil LS, Goodenough DA. 1991. Biochemical analysis of connexin43 intracellular transport, phosphorylation, and assembly into gap junctional plaques. J Cell Biol 115:1357–1374.
25. Moreno AP, Saez JC, Fishman GI, Spray DC. 1994. Human connexin43 gap junction channels. Regulation of unitary conductances by phosphorylation. Circ Res 74:1050–1057.
26. Kim DY, Kam Y, Koo SK, Joe CO. 1999. Gating connexin43 channels reconstituted in lipid vesicles by mitogen-activated protein kinase phosphorylation. J Biol Chem 274:5581–5587.
27. Ruiz-Meana M, Garcia-Dorado D, Lane S, Pina P, Inserte J, Mirabet M, Soler-Soler J. 2001. Persistence of gap junction communication during myocardial ischemia. Am J Physiol Heart Circ Physiol 280:H2563–H2571.
28. Li WE, Ochalski PA, Hertzberg EL, Nagy JI. 1998. Immunorecognition, ultrastructure and phosphorylation status of astrocytic gap junctions and connexin43 in rat brain after cerebral focal ischaemia. Eur J Neurosci 10:2444–2463.
29. Duthe F, Plaisance I, Sarrouilhe D, Herve JC. 2001. Endogenous protein phosphatase 1 runs down gap junctional communication of rat ventricular myocytes. Am J Physiol Cell Physiol 281: C1648–C1656.
30. Kwak BR, Jongsma HJ. 1996. Regulation of cardiac gap junction channel permeability and conductance by several phosphorylating conditions. Mol Cell Biochem 157:93–99.

Pathophysiology of Cardiovascular Disease,
edited by N.S. Dhalla, H. Rupp,
A. Angel & G.N. Pierce
Kluwer Academic Publishers, Boston, 2004

Temperature Dependence of Ischemic Preconditioning

J.S. Juggi, E.A.R. Ghadhanfar, M. Barac-Nieto, and S. Joseph

Department of Physiology
Faculty of Medicine, Kuwait University, Kuwait

Summary. Temperature dependence of ischemic preconditioning was evaluated in the globally ischemic model of Langendorff perfused rat heart. The results of these studies indicate that hyperthermic (42°) preconditioning offered optimal preservation of myocardial and coronary vascular functions during prolonged periods of ischemia. The cardioprotection was significantly better than at normothermic or hypothermic preconditioning temperatures. The exact mechanism(s) of this protection is not clear, but the results of this study do indicate a role of HSP70, particularly the nuclear translocation of this protein during hyperthermic preconditioning.

Key words: hyperthermic ischemic preconditioning, cardiac contractile function, heat shock proteins, coronary vascular dynamics, cardioprotection.

INTRODUCTION

The cardioprotective role of ischemic preconditioning (PC) to prolonged periods of regional or global myocardial ischemia is well established (1). Brief periods of myocardial ischemia-reperfusion before a prolonged ischemic insult offer a significant protection to myocardial structural and functional integrity. Beneficial effects of PC have been extrapolated to the clinical settings and the additive protective role of PC combined with cardioplegic interventions has been evaluated during open-heart surgery procedures (2,3).

Address for Correspondence: Prof. J.S. Juggi, Department of Physiology, Faculty of Medicine, Kuwait University, P.O. Box 24923 Safat 13110, Kuwait. Telephone: (965) 5319593, Fax: (965) 533-8937, e-mail:<oommen@hsc.kuniv.edu.kw>.

Several studies have been carried out to identify the cardioprotective mechanism of PC (4,5,6,7). Some of the studies have reported that cardioprotection by PC is mediated through the induction of heat-shock and other stress-induced proteins particularly during the delayed phase of cardioprotection (8). The studies reported in the literature have applied PC at normothermic temperature.

Recent studies confirm the expression of heat stress related antioxidative genes and proteins in the heart following exposure to hyperthermia. The ischemic tolerance of the isolated perfused hearts harvested several hours after whole body hyperthermia, was found to be significantly improved and this improvement was linked to the expression of various heat shock proteins (9). Synthesis of stress-induced heat shock proteins has also been demonstrated in isolated perfused non-preconditioned hearts subjected to ischemia at different elevated temperatures. The degree of expression of HSP71 was found to be dependent on the ischemic temperature (10). We describe here the temperature dependent cardioprotective effects of ischemic preconditioning on the post-ischemic recovery of myocardial and coronary vascular functions as related to the expression of HSP 70.

EFFECT OF PRECONDITIONING TEMPERATURE ON POST-ISCHEMIC MYOCARDIAL FUNCTION

The temperature-dependent effects of preconditioning were studied in isolated rat hearts perfused with standard Krebs-Henseleit buffer by the Langendorff technique. Pressure-controlled perfusion was conducted for all the experimental protocols. Control perfusion temperature was kept constant at 34°C and the preconditioning temperature varied from 10°C (hypothermic) to 34°C (control), 37°C (normothermic) and 42°C (hyperthermic). The hearts were subjected to 2 episodes of ischemic preconditioning at the desired temperature, each episode followed by 10 min. of reperfusion at 34°C. Thereafter all the heart were subjected to sustained global ischemia for 60 min. at 34°C followed by reperfusion for 30 min. at 34°C. Under control conditions left ventricular end-diastolic pressure (LVEDP) was kept constant at 5 mm Hg. Left ventricular maximum developed pressure (Pmax), coronary flow (CF) and coronary vascular resistance (CVR) were monitored under control and experimental conditions by an online monitoring and haemodynamic data acquisition system (Hugo-Sachs Electronik, Germany). Myocardial temperature was continuously monitored by a needle thermistor probe inserted into the intraventricular system. (Thermalert. TH-5 Physitemp, USA).

The effect of different preconditioning temperatures on the post-ischemic recovery in the left ventricular (LV) global contractility is shown in Table 1. Post-ischemic recovery of Pmax was significantly reduced after unprotected ischemia-reperfusion accompanied by a 7-fold increase in LVEDP. Hypothermic preconditioning at 10°C failed to significantly improve the post-ischemic recovery in LV contractility when compared with unprotected ischemia-reperfusion. However, preconditioning at the control temperature of 34°C, at the normothermic temperature of 37°C and at the hyperthermic temperature of 42°C significantly improved the post-ischemic recovery in LV contractility. Optimum protection was achieved at the hyperthermic pre-

Table 1. Effect of preconditioning temperature on the post-ischemic recovery of myocardial contractility and coronary vascular dynamics

	P_{max} (mm Hg)		LVEDP (mm Hg)		CF $ml \cdot min^{-1}$		CVR $mm Hg \cdot ml^{-1} \cdot min^{-2}$	
	Con	Rep	Con	Rep	Con	Rep	Con	Rep
Isch. 34° (n = 5)	92 ± 4	41 ± 4*	4.8 ± 0.3	33.8 ± 1.7*	10.2 ± 0.6	4.5 ± 0.5*	5.2 ± 0.3	12.1 ± 1.3*
PC 10°C + Isch. 34°C (n = 5)	107 ± 6	57 ± 4*†	4.9 ± 0.1	23.5 ± 2.7*†	11.1 ± 0.5	5.6 ± 0.6*	4.6 ± 0.2	10.0 ± 1.3*
PC 34°C + Isch. 34°C (n = 5)	108 ± 6	68 ± 6*†	4.9 ± 0.2	16.3 ± 3.3*†	11.3 ± 0.7	6.3 ± 1*	4.6 ± 0.3	9.3 ± 1.5*
PC 37°C + Isch. 34°C (n = 5)	111 ± 6	74 ± 8*†	4.9 ± 0.2	17.6 ± 3.4*†	13.3 ± 0.5	7.7 ± 0.5*†‡	3.9 ± 0.2	7.1 ± 0.5*†
PC 42°C + Isch. 34°C (n = 5)	106 ± 5	74 ± 6*†‡	5 ± 0.3	17.3 ± 2.9*†	12.9 ± 0.4	7.7 ± 0.5*†‡	4.0 ± 0.1	7.0 ± 0.5*†

The data was computed at 30 min reperfusion and expressed as Mean ± SEM.
Con, Control; Rep, Reperfusion; P_{max}, left ventricular developed pressure; LVEDP, left ventricular end-diastolic pressure; CF, coronary flow; CVR, coronary vascular resistance.
*, $p < 0.05$ compared with respective controls; †, $p < 0.05$ compared to Isch. 34°C; ‡, $p < 0.05$ compared to PC 10°C + Isch. 34°C.

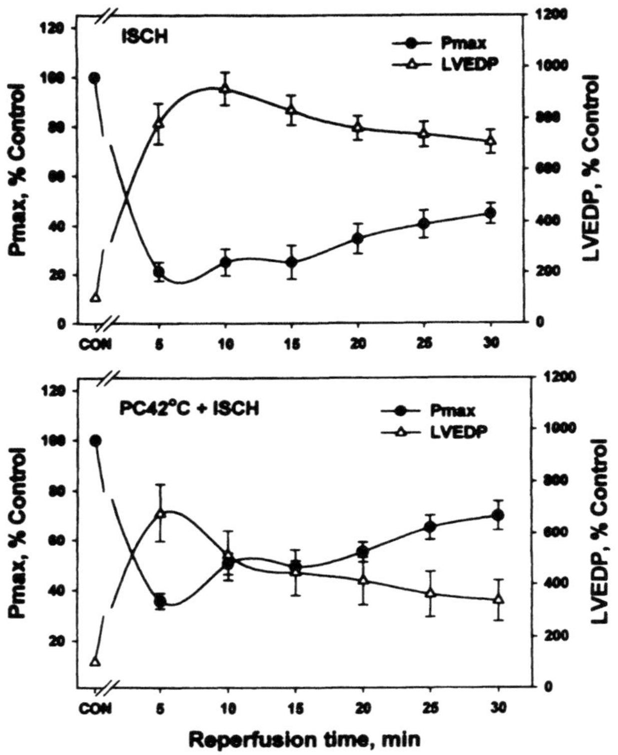

Figure 1. Time-course of reperfusion recovery in the left ventricular contractility after unprotected ischemia at 34°C and after hyperthermic (42°C) preconditioned ischemia.
Normalized Pmax and LVEDP values (mean ± SEM) were compared during the 30 min reperfusion. ISCH, unprotected ischemia at 34°C, PC-42°C + ISCH, hyperthermic (42°C) preconditioned ischemia.

conditioning temperature (Table 1). The time course of reperfusion recovery in Pmax and LVEDP after unprotected and protected (PC, 42°C) ischemia-reperfusion is shown in Figure 1. During the early reperfusion period there was a considerable increase in the LVEDP accompanied by a large fall in Pmax after unprotected global ischemia-reperfusion indicating an early reperfusion-induced LV contracture. With continuing reperfusion some recovery in Pmax accompanied by some improvement in LVEDP were evident. Hyperthermic preconditioning prevented the development of early ischemic contracture, significantly improved recovery in Pmax and was followed by a significant further reduction in LVEDP.

Coronary vascular dynamics also showed significant improvement with hyperthermic preconditioning. Post-ischemic recovery in coronary flow was 44% after unprotected ischemia-reperfusion as compared to 64% after hyperthermic preconditioning at 42°C (Table 1). Coronary flow changes were accompanied by a sig-

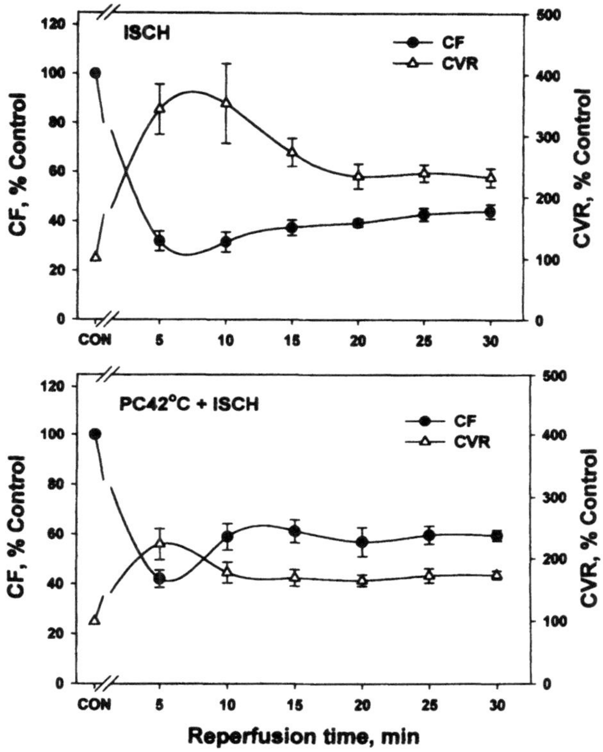

Figure 2. Time-course of reperfusion recovery in the left ventricular coronary flow as influenced by the recovery of coronary vascular resistance after unprotected ischemia at 34°C and after hyperthermic (42°C) preconditioned ischemia.
Normalized CF and CVR values (mean ± SEM) were compared during the 30 min reperfusion.
ISCH, unprotected ischemia at 34°C; PC-42°C + ISCH, hyperthermic (42°C) preconditioned ischemia.

nificant increase in coronary vascular resistance. Coronary vascular resistance increased more than 2-fold after unprotected ischemia-reperfusion. However, with hyperthermic preconditioning this doubling in coronary vascular resistance was curtailed to around 1.75 fold the control value (Table 1).

The time-course of reperfusion recovery in coronary flow and the recovery in coronary vascular resistance after unprotected and protected (PC, 42°C) ischemia-reperfusion are shown in Figure 2. During the early reperfusion period there was a significant dissociation in the recovery of coronary vascular resistance and coronary flow. Coronary vascular resistance considerably increased immediately after reperfusion accompanied by a significant depression in coronary flow. However, after hyperthermic preconditioning (42°C), the early increase in coronary vascular resistance was attenuated with improved recovery in coronary flow and a further decrease in coronary vascular resistance (Figure 2).

In summary hyperthermic preconditioning (42°C) produced optimal preservation resulting in 25% increase in the myocardial contractile tolerance and 16% increase in the coronary vascular tolerance to ischemia.

The cellular mechanisms of heat-stress mediated cardioprotection are still not fully understood. Recent studies however, indicate the expression of heat stress related and antioxidative genes and proteins in the heart following exposure to whole body hyperthermia (9). A direct correlation between the amount of heat shock protein and the degree of myocardial protection following whole body hyperthermia has also been demonstrated by (11).

CARDIOPROTECTIVE ROLE OF HEAT SHOCK PROTEINS

In order to elucidate the mechanism of enhanced cardioprotection by hyperthermic preconditioning, the induction of heat shock proteins was studied in the hyperthermic (42°C) preconditioned ischemia-reperfused hearts. The hearts were freeze-clamped at the end of the 30 min. reperfusion period.

Specific antibodies against HSC70 and HSP70 (Stressgen) were used in Western blots of left ventricular homogenates and nuclear fractions derived from the freeze-clamped hearts. The primary antibodies were detected using HRP conjugated goat-anti-rabbit IGg, followed by chemiluminescent ECL detection, autoradigraphy and densitometry. Ischemic (60 min, 34°C) reperfused (30 min, 34°C) rat heart (n = 3); preconditioned (2 cycles of 5 min ischemic episodes at 42°C, separated by a 10 min perfusion interval at 34°C) ischemic (60 min, 34°C) reperfused (30 min, 34°C) rat hearts (n = 3); control perfused rat hearts (34°C, n = 3) and intact freeze-clamped rat hearts (n = 3) were used. The results are shown in Figures 3, 4. Expression of

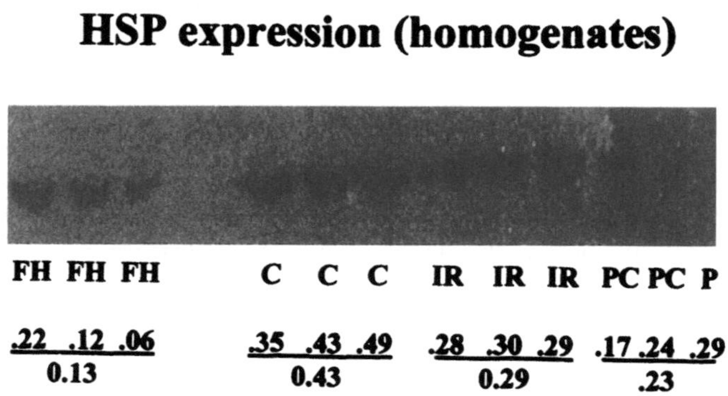

Figure 3. Expression of HSP70 in left ventricular homogenates of freeze-clamped fresh (FH), control perfused (C), ischemia-reperfused (IR) and hyperthermic preconditioned ischemia-reperfused (PC) hearts. Data from 12 hearts (3 per group) are shown. For each, 20 micrograms protein were loaded in each lane of the gel. The relative densities of the ECL imune-detected bands to that of a protein standard are given (underlined numbers). The average relative density for each group is shown below the lines. Note the significantly higher expression of HSP70 in C compared to F ($P < 0.03$) and to PC ($P < 0.01$) hearts.

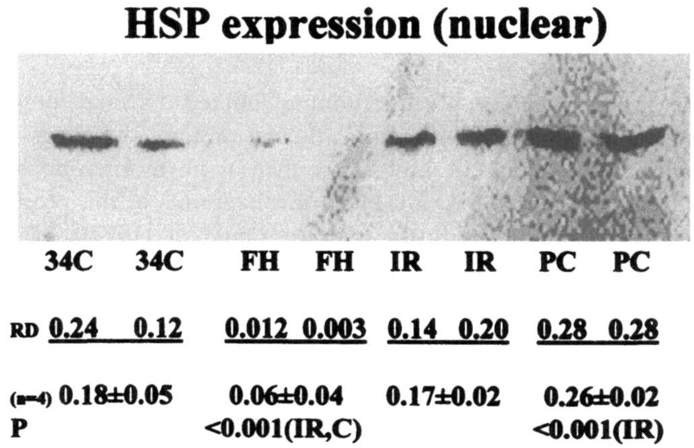

Figure 4. Expression of HSP70 in nuclear fractions isolated from the left ventricles of freeze-clamped fresh hearts (FH), control hearts perfused at 34C (34C), ischemia reperfused hearts (IR) and hyperthermic-preconditioned ischemia-reperfused hearts (PC). Data from 6 hearts, (2 per group) are shown. For each, 20 micrograms of nuclear proteins were loaded in each lane of the gel. The relative densities (RD) of the ECL-immuno-detected bands to that of a protein standard is given (underlined numbers) for each lane. For each group, the average relative density of 4 experiments similar to the two illustrated is given below the line. Note the significantly lower nuclear expression of HSP70 in intact fresh hearts (FH) than in control perfused (34C) or ischemia reperfused (IR) hearts, and the significantly higher nuclear expression of HSP70 in preconditioned (PC) than in IR, 34C or FH hearts.

HSC (not shown) was higher ($P < 0.05$) in left ventricle homogenates of control perfused hearts (0.64 ± 0.05) compared to that in ischemia-reperfused hearts (0.47 ± 0.02) or preconditioned (0.23 ± 0.03) hearts. Expression of HSP in left ventricular homogenates (Figure 3) was higher in control perfused hearts (0.35 ± 0.03) than in ischemic-reperfused (0.29 ± 0.03 $P < 0.03$), preconditioned (0.23 ± 0.02 $P < 0.03$) or intact (0.13 ± 0.01 $P < 0.01$) hearts. By contrast nuclear expression of HSP was significantly higher in preconditioned hearts (0.26 ± 0.03) compared to that in ischemia-reperfused (0.17 ± 0.01 $P < 0.01$), control perfused (0.18 ± 0.01) and intact (0.06 ± 0.01) hearts. From these results it may be speculated that nuclear translocation of HSP participates in the protective effects of hyperthermic preconditioned ischemia-reperfused rat hearts. These results are complimentary to earlier reports which indicate an improvement in the ischemic tolerance of isolated perfused hearts harvested after several hours of whole body hyperthermia. The fundamental improvement was linked to increased expression of various heat-shock proteins (12). Increased synthesis of heat shock proteins has been demonstrated in isolated perfused hearts subjected to ischemia (non-preconditioned) at increased temperatures (10). This study documents a specific increase in the nuclear but not in total expression of heat shock proteins in the hyperthermic preconditioned hearts.

Ischemic preconditioning, particularly at hyperthermic temperatures provided a significant protection to the coronary vasculature. Although the mechanism of this protection is not clear from the results of this study, recent reports indicate the induction of heat shock proteins in coronary endothelial cells by ischemic preconditioning (13).

In summary, hyperthermic preconditioning offered optimal preservation of myocardial and coronary vascular functions during prolonged periods of ischemia. The cardioprotection was significantly better than at normothermic or hypothermic preconditioning temperatures. The exact mechanism(s) of this protection is not clear, but the results of this study do indicate a role of HSP70, particularly the nuclear translocation of this protein during hyperthermic preconditioning.

REFERENCES

1. Murry CE, Jennings RB, Reimer KA. 1986. Preconditioning with ischemia: a delay of lethal cell injury in ischemic myocardium. Circulation 74:1124–1136.
2. Yellon DM, Alkhulaifi AM, Pugsley WB. 1993. Preconditioning the human myocardium. Lancet 342:276–277.
3. Juggi JS, Al-Awadi F, Joseph S. Telahoun G, Prahash A. 1997. Ischemic preconditioning is not additive to preservation with hypothermia or crystalloid cardioplegia in the globally ischemic rat heart. Mol Cell Biochem 176:303–313.
4. Downey JM, Liu GS, Thornton JD. 1993. Adenosine and the anti-infarct effects of preconditioning. Cardiovascular Res 27(1):3–8.
5. Das DK, Engelman RM, Kimura Y. 1993. Molecular adaptation of cellular defenses following preconditioning of the heart by repeated ischemia. Cardiovasc Res 27:578–584.
6. Schulz JE, Hsu AK, Groww GJ. 1996. Morphine mimics the cardioprotective effect of ischemic preconditioning via a glibenclamide-sensitive mechanism in the rat heart. Circ Res 78:1100–1104.
7. Baines CP, Goto M, Downey JM. 1997. Oxygen radicals released during ischemic preconditioning contribute to cardioprotection in the rabbit myocardium. J Moll Cell Cardiol 29:207–216.
8. Marber MS, Latchman DS, Walker JM, Yellon JM. 1993. Cardiac stress protein elevation 24 hours after brief ischemia or heat stress is associated with resistance to myocardial infarction. Circulation 88:1264–1272.
9. Qian YZ, Shipley JB, Levasseur JE, Kukreja RC. 1998. Dissociation of heat shock proteins expression with ischemic tolerance by whole body hyperthermia in rat heart. J Mol Cell Cardiol 30(6): 1163–1172.
10. Currie RW. 1987. Effects of ischemia and perfusion temperature on the synthesis of stress-induced (heat shock) proteins in iscolated and perfused rat heart. J Mol Cell Cardiol 19(8):795–808.
11. Hutter MM, Sievers Re, Barbosa V, Wolfe CL. 1994. Heat shock proteins induction in rat hearts. A direct correlation between the amount of heat shock proteins induced and the degree of myocardial protection. Circulation 89:355–362.
12. Knowlton AA. 1995. The role of heat shock proteins in the heart. J Mol Cell Cardiol 27:121–131.
13. Amrani M, Latif N, Morrison K, Gray CC, Jayakumar J, Corbett J, Goddwin AT, Dunn MJ, Yacoub MH. 1988. Relative induction of heat shock protein in coronary endothelial cells and cardiomyocytes: implications for myocardial protection. J Thorac Cardiovasc Surg 115(1):200–209.

Pathophysiology of Cardiovascular Disease,
edited by N.S. Dhalla, H. Rupp,
A. Angel & G.N. Pierce
Kluwer Academic Publishers, Boston, 2004

Mechanism of Cardioprotective Effect of Remote Aortic Preconditioning

Manjeet Singh and Ajay Sharma

Department of Pharmaceutical Sciences and Drug Research Punjabi University
Patiala-147002
Punjab, India

Summary. The present study is designed to investigate the mechanism of cardioprotective effect of remote aortic preconditioning. Four episodes, each episode comprising of 5 min aortic occlusion followed by reperfusion for 5 min were employed to produce remote preconditioning. Heart was isolated, perfused on Langendorff's apparatus and subjected to 30 min of sustained global ischaemia followed by reperfusion for 120 min. Coronary effluent was analysed for LDH and CK release to assess the degree of cardiac injury. Myocardial infarct size was estimated macroscopically using TTC staining. Remote aortic preconditioning reduced release of LDH and CK in coronary effluent and decreased myocardial infarct size. Reserpine, a biogenic amine depletor, ruthenium red, a sarcoplasmic reticulum Ca^{2+} channel blocker, ethylisopropylamiloride, (EIPA) a Na^{+}/H^{+} exchange inhibitor and gadolinium, a mechanosensitive channel blocker, did not modulate the cardioprotective effect of remote aortic preconditioning. On the other hand, verapamil, a L type Ca^{2+} channel blocker, and 2′,4′-dichlorobenzamil, a Na^{+}/Ca^{2+} exchange inhibitor, attenuated the cardioprotective effect of remote aortic preconditioning. On the basis of these results, it may be suggested that the cardioprotective effect of remote aortic preconditioning may be mediated through opening of L-type Ca^{2+} channels or activation of Na^{+}/Ca^{2+} exchange.

Key words: Remote preconditioning, Ischaemia-reperfusion, Calcium channels, Na^{+}/H^{+} exchange, Na^{+}/Ca^{2+} exchange.

Correspondence: Dr. Manjeet Singh, Professor and Head, Department of Pharmaceutical Sciences and Drug Research, Punjabi University Patiala-147002, Punjab, India. Phone no. 0175-2282461-6, Ext: 6184, Fax: 91-(0175)-2283073 2282437, E-mail: manjeet@pbi.ernet.in

INTRODUCTION

The cardioprotective effect of ischaemic preconditioning has been initially demonstrated against sustained ischaemia and reperfusion induced myocardial necrosis (1). It has been documented that transient occlusion of circumflex artery has induced protection of the myocardium supplied by left anterior descending coronary artery (2). Moreover short occlusions of renal or mesenteric artery also precondition the myocardium against ischaemia and reperfusion injury (3–5). These studies suggest that it is possible to precondition the heart by giving ischaemic stress in remote regions and this phenomenon has been termed as "remote preconditioning". The activation of angiotensin receptors are not noted to be involved in remote renal preconditioning (6). Thus mechanism of the cardioprotective effect of remote preconditioning is far from clear. Therefore, the present study has been designed to investigate the role of (i) endogenous biogenic amines, (ii) gadolinium sensitive mechanosensitive channels, (iii) L-type calcium channels, (iv) Na^+/H^+ exchange and (v) Na^+/Ca^{2+} exchange in the cardioprotective effect of remote aortic preconditioning.

MATERIALS AND METHODS

Wistar rats (120–280 g) of either sex having free access to food and water were employed in the present study.

Remote aortic preconditioning

Rats were anaesthetized with thiopental sodium (40 $mg\,kg^{-1}$ i.p.). A 2 cm long incision was given on left side of the abdomen. Abdominal aorta was isolated below the point of origin of renal artery and a silken suture was passed beneath it. The aorta was occluded by tying a shoe lace knot around it and knot was untied for reperfusion. Four episodes, each episode comprising of 5 min occlusion and 5 min reperfusion, were used to produce remote aortic preconditioning.

Sustained global ischaemia and reperfusion in isolated perfused rat heart

Heart was removed immediately after remote aortic preconditioning and mounted on Langendorff's apparatus. The heart was retrogradely perfused at a constant pressure of 80 mmHg, with Krebs Henseleit (K-H) buffer (NaCl 118 mM; KCl 4.7 mM; $CaCl_2$ 2.5 mM; $MgSO_4.7H_2O$ 1.2 mM; $NaHCO_3$ 25 mM; KH_2PO_4 1.2 mM; $C_6H_{12}O_6$ 11 mM), pH 7.4, maintained at 37°C bubbled with 95% O_2 and 5% CO_2. Flow rate was maintained between 6–8 ml per min. The preparation was allowed to stabilize for 10 min. Global ischaemia was produced by blocking the inflow of K-H buffer for 30 min followed by reperfusion for 120 min. Coronary effluent was collected before sustained global ischaemia, immediately, 5 min and 30 min after reperfusion for estimation of LDH and CK.

Enzymatic estimations

LDH and CK were estimated spectrophotometrically by 2,4-DNPH method (7) and Hughes method (8), respectively.

LDH catalyses the following reaction:

Lactate + NAD → Pyruvate + NADH

The pyruvate so formed couples with 2,4-dinitrophenylhydrazine (2,4-DNPH) to give corresponding hydrazone which gives brown colour in alkaline medium. The intensity of this colour is proportional to the amount of LDH activity and is measured spectrophotometrically at 440 nm.

CK catalyses the following reaction:

Creatine Phosphate + ADP → Creatine + ATP

At pH 7.4, CK catalyses the forward reaction. The creatine so formed, reacts with diacetyl and ∝-naphthol in alkaline medium to give pink colour. The intensity of this colour is proportional to the enzyme activity which is measured spectrophotometrically at 520 nm .

Infarct size measurement

Heart was removed from Langendorff's apparatus and washed with K-H solution. Left ventricle was excised and made free from right ventricle and both auricles. It was refrigerated overnight. Frozen ventricle was cut into transverse slices of about 3 mm thickness. These slices were stained by incubating for 20 min in 1% solution of triphenyltetrazolium chloride (TTC) in tris buffer pH 7.4 (9). TTC is converted to red formazone pigment by NADH and dehydrogenase enzyme, therefore TTC stained the viable cells deep red (10). The infarcted cells have lost the enzyme and cofactor (10) and thus remained unstained. Percentage infarction was measured by volume and weight method (11).

Experimental protocol

Sixteen groups of animals were employed in the present study.

Group 1 (Sham; n = 5): Rats were subjected to surgical procedure for isolation of aorta but aorta was not occluded. Hearts were excised 40 min after the isolation of aorta and isolated hearts were perfused on Langendorff's apparatus and was subjected to global ischaemia for 30 min followed by reperfusion for 120 min.

Group II (Remote Aortic Preconditioning; n = 5): Rats were subjected to remote aortic preconditioning. Hearts were excised immediately after last episode of preconditioning, perfused on Langendorff's apparatus and were subjected to global ischaemia for 30 min followed by reperfusion for 120 min.

Group III (Reserpine Treatment and Remote Aortic Preconditioning; n = 5): Reserpine (1 mg/kg, i.p.), a biogenic amine depletor, was administered 24 h before subjecting the rats to remote aortic preconditioning. Rest of protocol was same as described in Group II.

Group IV (Verapamil Treatment and Remote Aortic Preconditioning; n = 5): Verapamil (25 mg/kg, i.p.), a L-type Ca^{2+} channel blocker, was administered 30 min before subjecting the rats to remote aortic preconditioning and rest of protocol was same as described in Group II.

Group V (Ruthenium Red Treatment and Remote Aortic Preconditioning; n = 5): Ruthenium red (165 µg/kg i.v.), ryanodine sensitive sarcoplasmic reticulum Ca^{2+} channels blocker, was administered 5 min before subjecting the rats to remote aortic preconditioning and rest of protocol was same as described in Group II.

Group VI (2′,4′-dichlorobenzamil Treatment and Remote Aortic Preconditioning; n = 5): 2′,4′-dichlorobenzamil (225 µg/kg, i.v.), Na^{+}/Ca^{2+} exchange inhibitor, was administered 5 min before subjecting the rats to remote aortic preconditioning and rest of protocol was same as described in Group II.

Group VII (Ethylisopropylamiloride i.e. EIPA Treatment and Remote Aortic Preconditioning; n = 5): EIPA (12 µg/kg, i.v.), Na^{+}/H^{+} exchange inhibitor, was administered 5 min before subjecting the rats to remote aortic preconditioning and rest of protocol was same as described in Group II.

Group VIII (Gadolinium Treatment and Remote Aortic Preconditioning; n = 5): Gadolinium (0.5 mg/kg, i.p.), a mechanosensitive stretch channel blocker, was administered 30 min before subjecting the rats to remote aortic preconditioning and rest of protocol was same as described in Group II.

In separate groups (group IX–XVI), same dose levels of the above mentioned pharmacological interventions were also administered to rats without subjecting them to remote aortic preconditioning in order to study their *per se* effect.

Statistical analysis

Values were expressed as mean ± SEM. One way ANOVA was employed to calculate the statistical significance. Studentised range test and Dunnett's test were used as post-hoc tests for multiple comparisons between groups and comparison with control group, respectively. $p < 0.05$ was considered to be statistically significant.

RESULTS

Effect of remote aortic preconditioning (RAPC) and pharmacological interventions on LDH and CK release

Sustained ischaemia followed by reperfusion produced a marked increase in release of LDH in coronary effluent (Figure 1 and 2). Remote aortic preconditioning (RAPC) markedly attenuated ischaemia and reperfusion-induced increase in release of LDH and CK (Figure 1 and 2). The administration of reserpine (1 mg/kg, i.p.), verapamil (25 mg/kg, i.p.), ruthenium red (165 µg/kg, i.v.), gadolinium (0.5 mg/kg, i.p.) and 2′,4′-dichlorobenzamil (225 µg/kg, i.v.) *per se* produced no marked effect on ischaemia and reperfusion-induced increase in release of LDH and CK in coronary effluent. However, EIPA (12 µg/kg, i.v.) *per se* attenuated ischaemia and reperfusion-induced increase in release of LDH and CK. Moreover, administration of

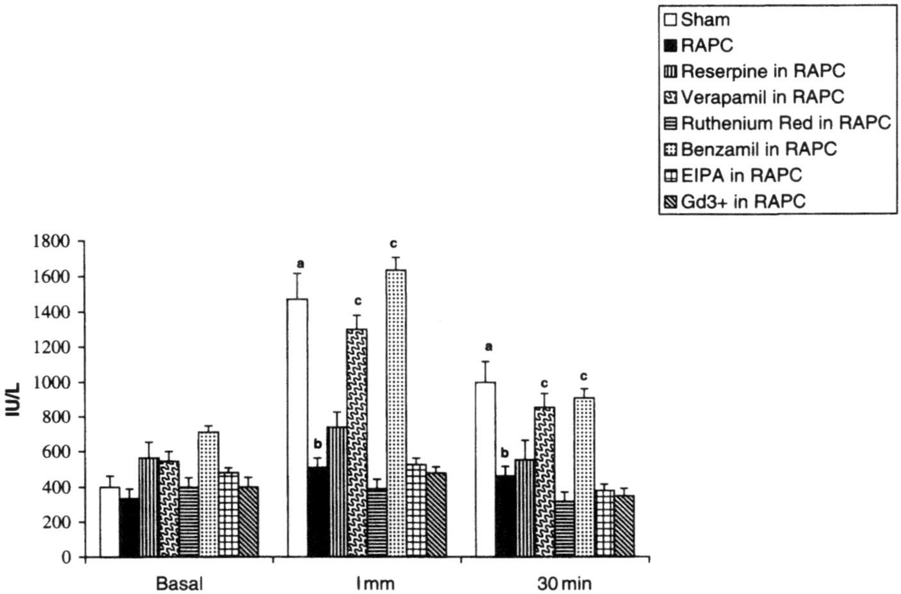

Figure 1. Effect of remote aortic preconditioning (RAPC) and pharmacological interventions on LDH release.

Isolated perfused rat heart subjected to global ischaemia for 30 min and 120 min of reperfusion. LDH was estimated in coronary effluent after stabilisation (basal), immediately (Imm) and after 30 min of reperfusion. Values are mean ± S.E.M. for five animals. a = $p < 0.05$ Vs Basal value; b = $p < 0.05$ Vs Sham; c = $p < 0.05$ Vs RAPC.

verapamil and 2′,4′ dichlorobenzamil attenuated RAPC induced decrease in release of LDH and CK (Figure 1 and 2). However, administration of reserpine, ruthenium red, EIPA and gadolinium did not modulate the effect of RAPC on release of LDH and CK (Figure 1 and 2).

Effect of remote aortic preconditioning (RAPC) and pharmacological interventions on myocardial infarct size

Remote aortic preconditioning (RAPC) significantly reduced myocardial infarct size measured by volume and weight method (Figure 3). The administration of reserpine (1 mg/kg, i.p.), verapamil (25 mg/kg, i.p.), ruthenium red (165 μg/kg, i.v.), gadolinium (0.5 mg/kg, i.p.) and 2′,4′-dichlorobenzamil (225 μg/kg, i.v.) *per se* produced no notable effect on infarct size whereas EIPA (12 μg/kg, i.v.) treatment markedly reduced myocardial infarct size. Administration of Verapamil and 2′,4′-dichlorobenzamil attenuated RAPC induced decrease in infarct size (Figure 3). No such attenuation of cardioprotective effect of RAPC was noted with reserpine, ruthenium red, EIPA and gadolinium (Figure 3).

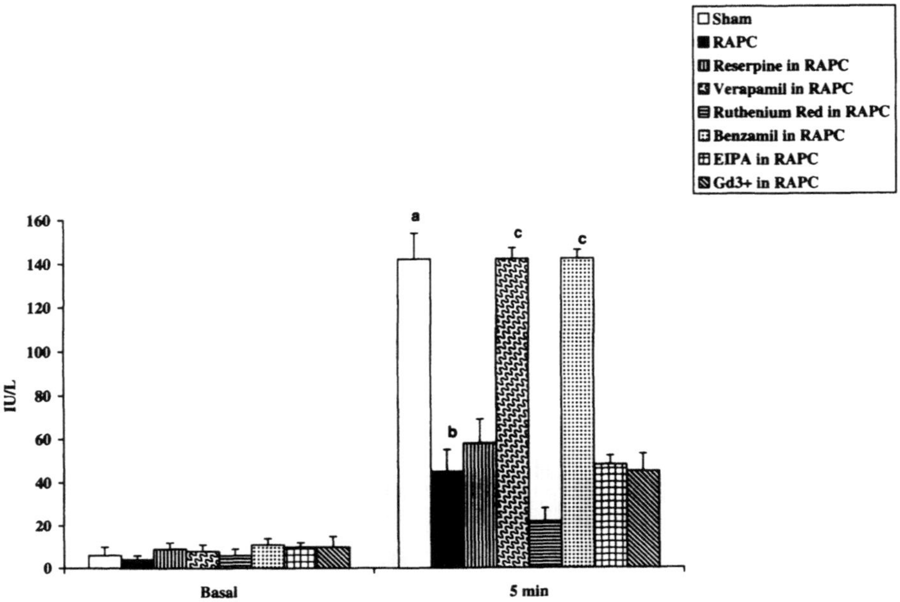

Figure 2. Effect of remote aortic preconditioning (RAPC) and pharmacological interventions on CK release.
Isolated perfused rat heart subjected to global ischaemia for 30 min and 120 min of reperfusion. CK was estimated in coronary effluent after stabilisation (basal) and after 5 min of reperfusion. Values are mean ± S.E.M. for five animals. a = $p < 0.05$ Vs Basal value; b = $p < 0.05$ Vs Sham; c = $p < 0.05$ Vs RAPC.

DISCUSSION

It is noted in our laboratory that peak release of LDH occurs either immediately or 30 min after reperfusion and peak CK release is observed to occur after 5 min of reperfusion (12,13). Therefore, samples of coronary effluent were collected at these time intervals, in the present study, to estimate the release of LDH and CK.

The cardioprotective effect of norepinephrine preconditioning (13) and attenuation of ischaemic preconditioning induced cardioprotection with α-adrenoceptor blockers and reserpine (14–16) suggest that release of endogenous biogenic amines may be involved in observed cardioprotective effect of RAPC. However, this contention is not supported by the present results because depletion of biogenic amines with reserpine has not prevented the noted cardioprotective effect of RAPC.

Increase in cytosolic Ca^{2+} is reported to trigger preconditioning like cardioprotection against Ca^{2+} paradox-induced cardiac injury (17). Stretch is reported to activate L- type Ca^{2+} channels and consequently increase the Ca^{2+} influx (18). Therefore, it is probable that the aortic occlusion during RAPC may open L-type Ca^{2+} channels of the myocardium as a result of occlusion induced-overload or stretch. This

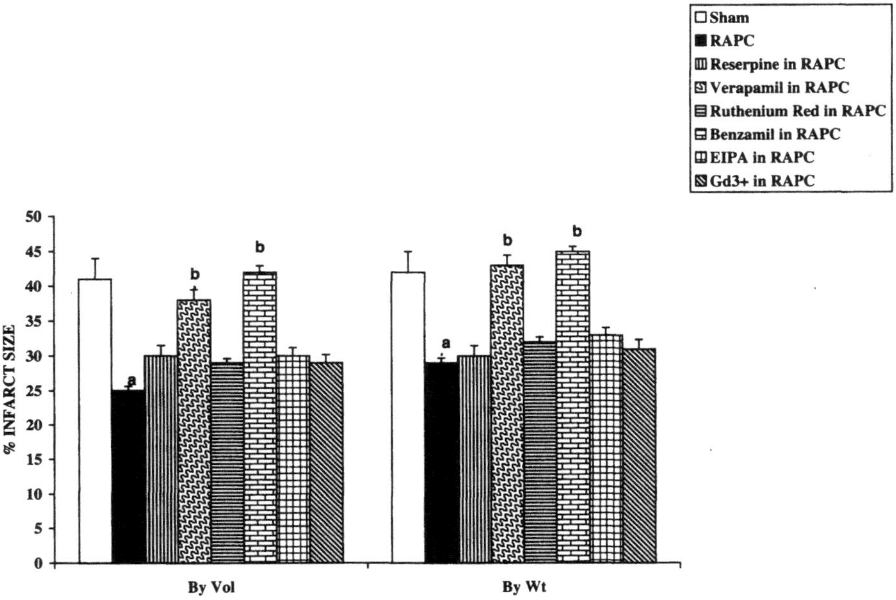

Figure 3. Effect of remote aortic preconditioning (RAPC) and pharmacological interventions on myocardial infarct size.
Myocardial infarct size was estimated by volume and weight method in isolated perfused rat heart subjected to global ischaemia for 30 min and 120 min of reperfusion. Values are mean ± S.E.M. for five animals. a = $p < 0.05$ Vs Sham; b = $p < 0.05$ Vs RAPC.

contention is supported by the results of the present study because verapamil, a L-type Ca^{2+} channel blocker, has attenuated the cardioprotective effect of RAPC. Moreover, stretch is reported to release Ca^{2+} from sarcoplasmic reticulum (19). However, ruthenium red, a blocker of ryanodine sensitive sarcoplasmic reticulum Ca^{2+} channels, has not modulated the noted cardioprotective effect of RAPC. This observation suggest that intracellular release of Ca^{2+} from sarcoplasmic reticulum may not be involved in the cardioprotective effect of RAPC.

Mechanosensitive (MS) channels are a family of plasma membrane bound channels that are gated by mechanical deformation (20). MS channels have been demonstrated to be present in myocardial tissue (21). Therefore, opening MS channels in response to aortic occlusion-induced stretch may account for the cardioprotective effect of remote aortic preconditioning. However, this contention is not supported by the results of the present study because gadolinium, a stretch sensitive MS channel blocker, did not attenuate the cardioprotective effect of remote aortic preconditioning.

Activation of Na^{+}/H^{+} exchange as a result of ischaemia-reperfusion may lead to intracellular Ca^{2+} overload as a result of activation of Na^{+}/Ca^{2+} exchange to handle the increase in intracellular sodium (22,23). EIPA is known to inhibit Na^+/H^+ exchange in the myocardium (24,25). Therefore, the cardioprotective effect of *per se* treatment of EIPA against sustained ischaemia and reperfusion injury, observed in the present study, may be due to inhibition of Na^+/H^+ exchange and consequent prevention of intracellular Ca^{2+} overload. On the other hand, it is interesting to note that administration of EIPA before RAPC has not modulated the cardioprotective effect of RAPC. It suggests that Na^+/H^+ exchange is neither activated nor it may be involved in the cardioprotective effect of RAPC. Na^+/Ca^{2+} exchange in myocardium is activated with ischaemia and reperfusion (26). 2′,4′-dichlorobenzamil is a Na^+/Ca^{2+} exchange inhibitor (27). Therefore, attenuation of cardioprotective effect of RAPC with administration of 2′,4′-dichlorobenzamil implicate the activation of Na^+/Ca^{2+} exchange in RAPC which may be responsible for its cardioprotective effect.

On the basis of the above discussion, it may be concluded that the cardioprotective effect of remote aortic preconditioning may be mediated through opening of L-type Ca^{2+} channels or activation of Na^+/Ca^{2+} exchange.

REFERENCES

1. Murry CE, Jennings RB, Reimer KA. 1986. Preconditioning with ischaemia: a delay of lethal cell injury in ischaemic myocardium. Circulation 74:1124–1136.
2. Przyklenk K, Bauer B, Ovize M, Kloner RA, Whittaker P. 1993. Regional ischaemic "preconditioning" protects remote virgin myocardium from subsequent sustained coronary occlusion. Circulation 87:893–899.
3. Gho BCG, Eskildsen-helmond YEG, de Zeeuw S, Lamers JMJ, Berdouw BD. 1976. Does protein kinase C plays a pivotal role in the mechanism of ischaemic preconditioning. Cardiovasc Drugs Ther 10:775–786.
4. Takaoka K, Yabe T, Morikawa S, Inubushi T, Kinoshita M. 1999. Renal ischaemia/reperfusion remotely improves myocardial energy metabolism during myocardial ischaemia via adenosine receptors in rabbits: effects of "remote preconditioning". J Am Coll Cardiol 33:556–564.
5. Schoemaker RG, van Heijningen CL. 2000. Bradykinin mediates cardiac preconditioning at a distance. Am J Physiol 278:H1571–H1576.
6. Kaur K. 1998. Acute and delyed cardioprotective effects of remote renal preconditioning in rat. M. Pharm Thesis, Punjabi University, Patiala.
7. King J. 1959. A routine method for the estimation of lactic dehydrogenase activity. J Med Lab Tech 16:265–272.
8. Hughes BP. 1961. A method for the estimation of serum creatine kinase and aldose activity in the normal and pathological sera. Clin Chim Acta 7:597–603.
9. Fishbein MC, Meerbaum S, Rit J, Lando U, Kanmatsuse K, Merair JC, Corday E, Ganz W. 1981. Early phase acute myocardial infarct size quantification: validation of the triphenyltetrazolium chloride tissue enzyme staining technique. Am Heart J 101:593–600.
10. Klein HH, Puschmann S, Schaper J, Schaper W. 1981. The mechanism of the tetrazolium reaction in identifying experimental myocardial infarction. Virchows Arch 393:287–297.
11. Chopra K, Singh M, Kaul N, Andrabi KI, Ganguly NK. 1992. Decrease of myocardial size with desferrioxamine. Possible role of oxygen free radicals in its ameliorative effects. Mol Cell Biochem 113:71–76.
12. Parikh V, Singh M. 1999. Possible role of adrenergic component and cardiac mast cell degranulation in preconditioning-induced cardioprotection. Pharmacol Res 40:129–131.
13. Sharma A, Singh M. 2001. Protein kinase C activation and cardioprotective effect of preconditioning with oxidative stress in isolated rat heart. Mol Cell Biochem 219:1–6.
14. Sharma A, Singh M. 1997. The possible role of adrenergic component in ischaemic preconditioning. Meth Find Exp Clin Pharmacol 19:493–499.

15. Sharma A, Singh M. 2000. Possible mechanism of cardioprotective effect of ischaemic preconditioning in isolated rat heart. Pharmacol Res 41:635–640.
16. Sharma A, Singh M. 2000. Possible mechanism of cardioprotective effect of angiotensin preconditioning in isolated rat heart. Eur J Pharmacol 406:85–92.
17. Miyawaki H, Ashraf M. 1997. Calcium as a mediator of ischaemic preconditioning. Circ Res 80: 790–799.
18. Gannier F, White E, Garnier D, Le Guennec JY. 1996. A possible mechanism for large stretch-induced increase in intracellular calcium in isolated guinea-pig ventricular myocytes. Cardiovasc Res 32: 158–167.
19. Tanaka Y, Hata S, Ishiro H, Ishii K, Nakayama K. 1994. Stretch release calcium from intracellular storage site in canine cerebral arteries. Can J Physiol 72:19–24.
20. Morris CE. 1990. Mechanosensitive ion channels. J Membr Biol 113:93–107.
21. Bustamante JO, Ruknudin A, Sachs F. 1991. Stretch-activated channels in heart cells relevance to cardic hypertrophy J. Cardiovasc Pharmacol 17:S110–S113.
22. Ohara F, Sugimoto T, Yamamoto N, Okhubo K, Maeda K, Ozaki T, Seki J Gato T. 1999. Preischaemic and postischaemic treatment with a new Na^+/H^+ exchange inhibitor, FR 183998, shows cardioprotective effects in rat with cardiac ischaemia and reperfusion. J Cardiovasc Pharmacol 34:848–856.
23. Yamamoto S, Matsui K, Kitano M, Ohashi N. 2000. SM-20550, a new Na^+/H^+ exchange inhibitor and its cardioprotective effect in ischaemic/reperfused isolated rat hearts by preventing Ca^{2+}-overload. J Cardiovasc Pharmacol 35:855–862.
24. Meng HP, Maddaford TG, Pierce GN. 1993. Effect of amiloride and selected analogues on post ischaemic recovery of cardic contractile function. Am J Physiol 264:H1831–H1835.
25. Bugge E, Munch-Ellingsen J, Ytrehus K. 1996. Reduced infarct size in the rabbit heart in vivo by ethylisopropyl-amiloride. A role for Na^+/H^+ exchange. Basic Res Cardiol 91:203–209.
26. Tani M, Neely JR. 1989. Role of intracellular Na^+ in Ca^{2+} overload and depressed recovery of ventricular function of reperfused ischaemic rat hearts: possible involvement of Na^+/H^+ and Na^+/Ca^{2+} exchange. Circ Res 65:1045–1056.
27. Deshpande SB, Fukuda A, Nishino H. 1997. 3-Nitropropionic acid increases intracellular calcium in cultured astrocytes by reverse operation of $Na^+/Ca2^+$ exchanger. Exp Neurol 145:38–45.

Pathophysiology of Cardiovascular Disease,
edited by N.S. Dhalla, H. Rupp,
A. Angel & G.N. Pierce
Kluwer Academic Publishers, Boston, 2004

Inhibition of Degranulation of Resident Cardiac Mast Cells May Prevent Ischaemia and Reperfusion-Induced Myocardial Injury

Manjeet Singh and Harjot K. Saini

Department of Pharmaceutical Sciences and Drug Research, Punjabi University, Patiala-147002, Punjab, India

Summary. The present study is designed to investigate the role of resident cardiac mast cells in ischaemia and reperfusion-induced apoptotic and necrotic death of cardiomyocytes. Isolated perfused rat heart was subjected to 30 min of global ischaemia followed by reperfusion for 120 min. The extent of mast cell degranulation was determined by the release of mast cell peroxidase (MPO) in the coronary effluent. Release of LDH, CK in the coronary effluent and estimation of infarct size are employed to assess the extent of myocardial injury. DNA fragmentation analysis using terminal deoxynucleotidyl transferese-mediated UTP nick end-labeling (TUNEL) and agarose gel electrophoresis are employed to estimate apoptotic death of cardiomyocytes. Disodium cromoglycate (DSCG, 100 μM), theophylline (100 μM) and FR 167653 (2.2 μM) treatment before global ischaemia and reperfusion are noted to reduce degranulation of resident cardiac mast cells, measured in terms of MPO release and has produced cardioprotective effect assessed as infarct size and release of LDH and CK. DSCG, theophylline and FR 167653 treatment has also markedly reduced the number of TUNEL positive cardiomyocytes and ventricular DNA smearing. These results tentatively suggest that DSCG, theophylline and FR 167653 inhibit degranulation of resident cardiac mast cells and consequently prevent ischaemia and reperfusion induced apoptosis and apoptosis induced necrosis in cardiomyocytes.

Key words: Ischaemia-Reperfusion, Cardioprotection, MPO, Cardiac Mast Cells, Degranulation.

Correspondence Address: Dr. Manjeet Singh, Department of Pharmaceutical Sciences & Drug Research, Punjabi University, Patiala-147002, Punjab, India. Telephone: 91-(0175)-2284002, Fax: 91-(0175)-2283073/2282437, E-mail: manjeet@pbi.ernet.in

INTRODUCTION

Mast cells are reported to be present in the heart mainly in atria adjacent to small capillaries and venules (1). Degranulation of resident cardiac mast cells and consequent release of cytotoxic mediators from them have been tentatively implicated in ischaemia and reperfusion-induced myocardial injury (2–8). Mast cell degranulation is noted to produce apoptotic death of cardiomyocytes (9). Mast cell stabilizers have been reported to prevent ischaemia and reperfusion-induced myocardial injury (10–12). Proinflammatory cytokines such as TNFα and IL-1β are noted to depress human myocardium and they have been implicated in ischaemia and reperfusion-induced myocardial dysfunction (13–15). TNFα is released during ischaemic conditions by cardiomyocytes (16,17) coronary endothelial cells (18), macrophages (19) and mast cells (20). Thus the present study has been designed to investigate the role of resident cardiac mast cells and release of cytokines from them in ischaemia and reperfusion-induced apoptotic and necrotic death of cardiomyocytes.

MATERIALS AND METHODS

Wistar albino rats of either sex weighing 200–300 gm were used in the present study. The animal experiments were conducted in accordance with guidlines of US National Institute of Health for Care and use of Laboratory animals. FR-167653 (Fujisawa Pharmaceutical Co. Ltd., Osaka, Japan), theophylline (German Remedies Ltd., Goa, India) and disodium cromoglycate (DSCG) (Rallis India Ltd., Bombay, India) were dissolved in normal saline, in minimum amount of 0.1N HCl and in minimum amount of distilled water, respectively. The drug solutions were diluted with Krebs Henseleit (K-H) physiological solution.

Isolated rat heart preparation

Rats were heparinised (500 I.U, i.p.) and sacrificed by stunning. Hearts were rapidly excised and immediately mounted on Langendorff's apparatus (21). The heart was enclosed in a double walled jacket and the temperature of which was maintained at 37°C by circulating hot water. The preparation was perfused with K-H solution (NaCl 118 mM; KCl 4.7 mM; $CaCl_2$ 2.5 mM; $MgSO_4.7H_2O$ 1.2 mM; $NaHCO_3$ 25 mM; KH_2PO_4 1.2 mM and $C_6H_{12}O_6$ 11 mM) pH 7.4, maintained at 37°C and bubbled with 95% O_2 and 5% CO_2. The coronary flow rate was maintained 6–9 ml/min and perfusion pressure was kept constant at 70 mmHg. Global ischaemia was produced for 30 min by closing the inflow of physiological solution and it was followed by reperfusion for 120 min. Two thin silver electrodes fixed at the ventricular apex and origin of aorta were employed to record ECG (BPL, MK 801, Bangalore, India) for monitoring heart rate.

Isolation of cardiomyocytes

Cardiomyocytes were isolated using modified method of Tytgat (22). Hearts were perfused for 5 min with calcium free Tyrode solution containing 3 mM EGTA. It was followed by 30 min perfusion with 50 ml calcium free Tyrode solution containing 0.12 U/ml collagenase type III fraction A (Sigma-Aldrich, St Louis, USA), 0.12 mg/ml protease type XIV (Sigma-Aldrich, St Louis, USA), 1 mg/ml BSA (SISCO Research Lab. Pvt. Ltd., Bombay, India) and 0.515 g/50 ml MEM, Eagle (Hi Media, Bombay, India) in recirculating manner and heart was perfused for further 5 min with low calcium solution. The heart was removed from the Langendorff's apparatus. Auricles, root of aorta and right ventricle were removed and left ventricle was cut into three to four slices. These were placed in low calcium solution containing 1 mg/ml BSA and 0.01 gm/ml MEM, Eagle maintained at 37°C, pH 7.4 and continuously bubbled with 95% O_2 and 5% CO_2. The tissue slices were incubated till cell suspension was achieved by gentle agitation. The cell suspension was filtered through 250 μm nylon mesh to eliminate large superfluous cellular material and was centrifuged (1200 rpm) for 5 min. The obtained cell pellet was washed with 10 ml of Tyrode solution containing BSA and MEM at 37°C and recentrifuged at the same speed. The cellular viability was examined using 0.4% trypan blue in 0.9% NaCl using improved Naubauer hemocytometer.

Assessment of myocardial injury

To determine the extent of myocardial injury, release of lactate dehydrogenase (LDH) and creatine kinase (CK) were measured in coronary effluent using 2,4-DNPH method (23) and Hughes method (24), respectively. Values were expressed as International Units (IU) per litre of coronary effluent.

Myocardial infarct size

Heart was removed from the Langendorff's apparatus and auricles, root of aorta and right ventricle were removed. Left ventricle was kept overnight at 4°C and was cut into uniform slices of about 2–3 mm thickness. The slices were incubated in 1% triphenyl tetrazolium chloride (TTC) at 37°C in 0.2 M Tris buffer (pH 7.4) for 20 min (25,26). The viable myocardium was stained brick red while the infarcted portion remained unstained. Infarct size was measured macroscopically by volume and weight method (27,28).

Assessment of mast cell degranulation

Mast cell peroxidase (MPO), a marker of mast cell degranulation was estimated quantitatively in coronary effluent using the method of Keller et al. (2). 2.0 ml of coronary effluent was mixed with 0.4 ml of O-dianisidine (0.8 mM), 0.4 ml of NaH_2PO_4 (50 mM) and 0.8 ml of distilled water. The reaction was initiated by adding 0.4 ml of 0.0003% H_2O_2 at room temperature in dark. After 10 min, the reaction was stopped by adding 0.4 ml of 0.2% sodium azide. The absorbance was measured at 492 nm using spectrophotometer (Beckman DU 640B, Nyon, Switzerland). The results were expressed as IU/min.

TUNEL staining

Apoptotic cell death in isolated rat heart subjected to 30 min ischaemia followed by 120 min reperfusion was evaluated using terminal deoxy nucleotidyl transferase mediated—UTP nick end labeling (TUNEL) method (29) using APO-DIRECT™ Kit (Pharmingen, San Diego, CA, USA). TUNEL—positive cardiomyocytes were counted under fluorescent microscope and apoptotic index was calculated by the following formula.

$$\text{Apoptotic Index} = \frac{\text{Number of TUNEL positive cell nuclei}}{\text{Number of total cell nuclei}} \times 100$$

DNA fragmentation analysis using agarose gel electrophoresis

DNA was isolated from cardiomyocytes and DNA fragmentation was identified using method described by Maulik et al. (30). DNA samples were electrophoresed on 1.8% agarose gel with 0.5 µg/ml ethidium bromide in 0.5 × TBE buffer (89 mM Tris buffer; 89 mM boric acid and 2 mM EDTA) for 1.50 hr at 50 V, 400 mA and 3 W in a submarine electrophoresis unit (Pharmacia Biotech, Freiburg, Germany).

Experimental protocol

The present study is comprised of five groups and each group consists of 10 experiments. Heart obtained from rat sacrified by stunning was perfused with K-H solution and allowed to stabilize for 10 min. Cardiomyocytes DNA was extracted from left ventricle. Moreover, left ventricle was employed to perform TUNEL staining. A diagrammatic representation of experimental protocol is shown in Figure 1.

Group I (Sham control group, n = 10)

Isolated rat heart was perfused with K-H solution for 160 min after stabilization for 10 min.

Group II (Control group, n = 10)

After stabilization, isolated rat heart was perfused with K-H solution for 10 min and then subjected to 30 min global ischaemia followed by reperfusion for 120 min.

Group III (Theophylline treated group, n = 10)

After stabilization, isolated rat heart was perfused with K-H solution containing theophylline (100 µM) for 10 min and then subjected to 30 min global ischaemia followed by reperfusion for 120 min.

Group IV (FR 167653 treated group, n = 10)

After stabilization, isolated rat heart was perfused with K-H solution containing FR 167653 (2.2 µM) for 10 min and then subjected to 30 min global ischaemia followed by reperfusion for 120 min.

Group V (DSCG treated group, n = 10)

After stabilization, isolated rat heart was perfused with K-H solution containing DSCG (100 µM) for 10 min and then subjected to 30 min global ischaemia followed by reperfusion for 120 min.

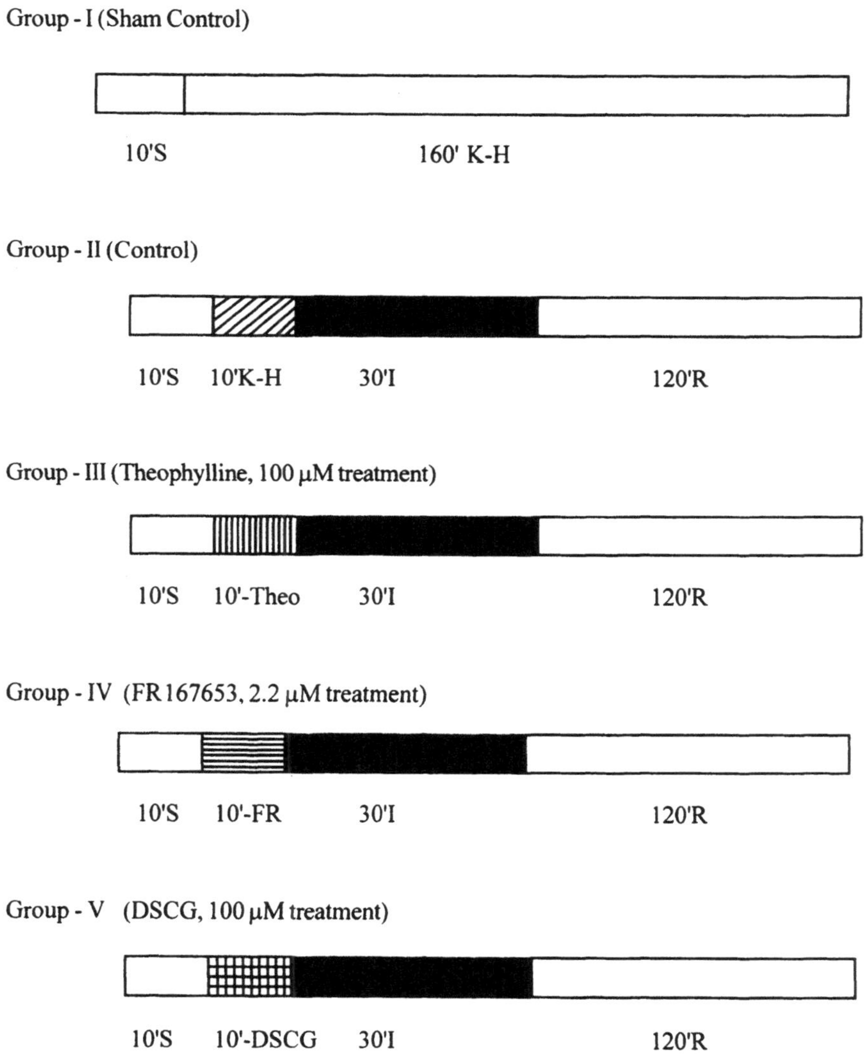

Figure 1. Diagrammatic representation of experimental protocol. **S**-Stabilization; **K-H**-perfusion with K-H solution; **I**-global ischaemia; **R**-reperfusion with K-H solution; **10′-Theo**—perfusion with K-H solution containing 100 μM theophylline; **10′ FR**—perfusion with K-H solution containing 2.2 μM FR 167653; **10′DSCG**—perfusion with K-H solution containing 100 μM disodium cromoglycate.

Coronary effluent was collected immediately after stabilization, before global ischaemia and immediately, 5 min and 30 min after reperfusion. Apoptotic index, infarct size and analysis of DNA fragmentation using agarose gel electrophoresis was carried out at the end of 120 min reperfusion preceded by 30 min of ischaemia. ECG was recorded after stabilization, 5, 15 and 30 min of global ischaemia and 5, 15, 30, 60 and 120 min of reperfusion in all groups.

Statistical analysis

Values for enzymatic data and infarct size were expressed as mean ± S.E.M. Statistical significance was calculated using one way analysis of variance. Studentized range test and Dunnett's test were used as post-hoc tests for multiple comparison between groups and comparison with control group, respectively. $p < 0.05$ was considered to be statistically significant.

RESULTS

Global ischaemia followed by reperfusion significantly reduced the heart rate (222 ± 14.5 to 84 ± 11.8) and coronary flow rate (7.93 ± 0.34 to 2.37 ± 0.5). Theophylline (100 μM), FR 167653 (2.2 μM) and DSCG (100 μM) treatment produced no marked effect on heart rate and coronary flow rate.

Effect of pharmacological interventions on ischaemia and reperfusion-induced LDH and CK release

Global ischaemia for 30 min followed by reperfusion for 120 min increased the release of LDH and CK noted immediately, 30 min and 5 min after reperfusion, respectively. Theophylline (100 μM), FR 167653 (2.2 μM) and DSCG (100 μM) treatment before global ischaemia markedly decreased the elevated release of LDH and CK in coronary effluent (Figure 2 and Figure 3).

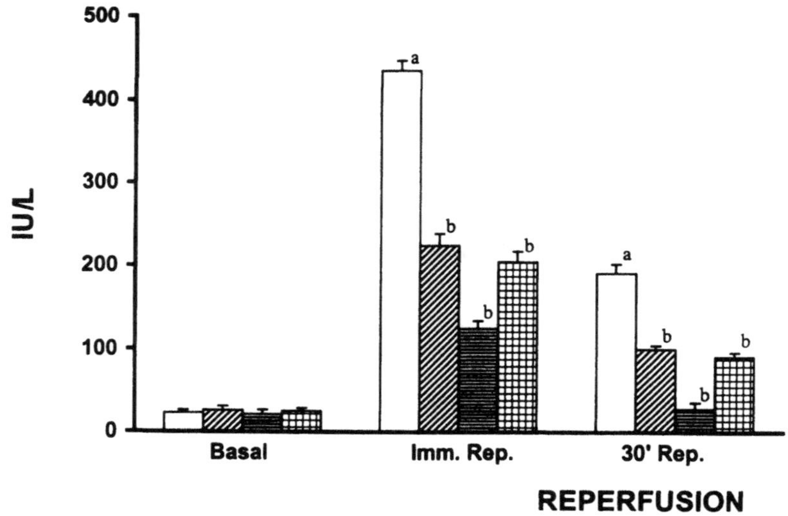

Figure 2. Effect of 100 μM theophylline (▨), 2.2 μM FR 167653 (▤) and 100 μM DSCG (▦) treatment on LDH release as compared to control (□). LDH was estimated in coronary effluent after stabilization (basal), immediately after reperfusion (Imm. Rep.) and 30 min after reperfusion (30' Rep.). Values are expressed as mean ± SEM. a = $p < 0.05$ vs basal value; b = $p < 0.05$ vs control.

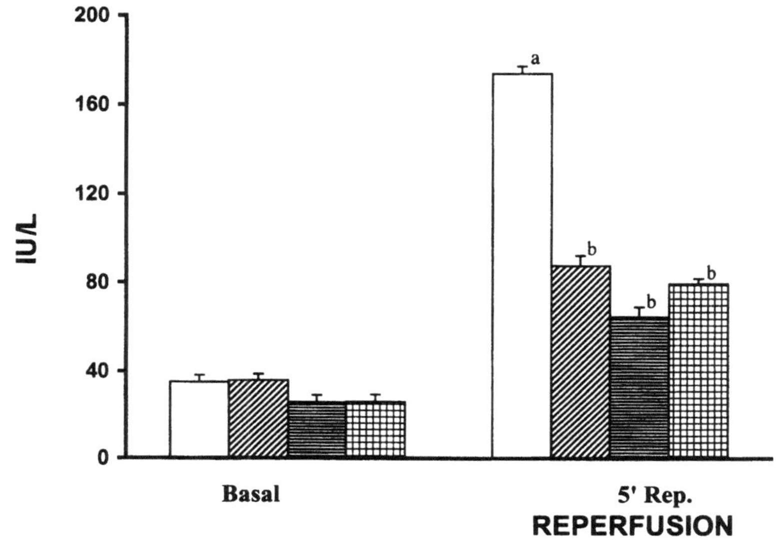

Figure 3. Effect of 100 µM theophylline (▨), 2.2 µM FR 167653 (▤) and 100 µM DSCG (▦) on CK release as compared to control (▭). CK was estimated in coronary effluent after stabilization (basal) and 5 min after reperfusion (5′ Rep.). Values are expressed as mean ± SEM. a = $p < 0.05$ vs basal value; b = $p < 0.05$ vs control.

Effect of pharmacological interventions on myocardial infarct size

The extent of myocardial infarct size in control experiments as a result of global ischaemia for 30 min and reperfusion for 120 min was recorded as percentage infarct size of left ventricle i.e. 35.81 ± 1.33 and 37.15 ± 1.37 calculated by volume and weight method, respectively (Figure 4). Theophylline (100 µM), FR 167653 (2.2 µM) and DSCG (100 µM) treatment before global ischaemia significantly reduced the myocardial infarct size.

Effect of pharmacological interventions on ischaemia and reperfusion-induced MPO release

Global ischaemia for 30 min followed by reperfusion for 120 min produced marked increase in MPO release in coronary effluent recorded immediately after reperfusion. Theophylline (100 µM), FR 167653 (2.2 µM) and DSCG (100 µM) treatment before global ischaemia markedly attenuated the ischaemia, and reperfusion-induced increase in release of MPO (Figure 5).

Effect of pharmacological interventions on apoptotic cell death

Global ischaemia for 30 min followed by reperfusion for 120 min produced significant increase in apoptotic index. Theophylline (100 µM), FR 167653 (2.2 µM) and

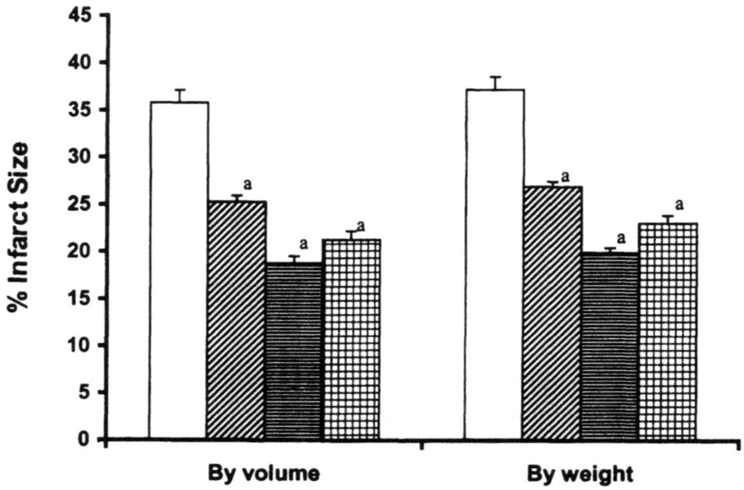

Figure 4. Effect of 100 μM theophylline (▨), 2.2 μM FR 167653 (▤) and 100 μM DSCG (▦) on myocardial infarct size as compared to control (▭) measured by volume and weight method. Values are expressed as mean ± SEM. a = $p < 0.05$ vs control.

DSCG (100 μM) treatment before global ischaemia significantly decreased the apoptotic index (Figure 6, 7 and 8). FR 167653 was more effective than theophylline and DSCG to reduce the number of TUNEL positive cardiomyocytes.

Effect of pharmacological interventions on ventricular DNA smearing

No ventricular cardiomyocyte DNA smearing was noted in sham control hearts. Global ischaemia for 30 min followed by reperfusion for 120 min produced marked ventricular cardiomyocyte DNA smearing. Theophylline (100 μM), FR 167653 (2.2 μM) and DSCG (100 μM) treatment before global ischaemia markedly attenuated ischaemia and reperfusion-induced ventricular cardiomyocyte DNA smearing. FR 167653 was more effective than theophylline and DSCG to reduce ventricular DNA smearing (Figure 9).

DISCUSSION

Mast cells are reported to be present in the heart (31). MPO is identified histochemically to be present in the mast cells and release of this enzyme is correlated to degranulation of mast cells (2). The peak release of MPO occurs immediately after reperfusion (7) and within 3–6 min of reoxygenation after ischaemia (2). There-

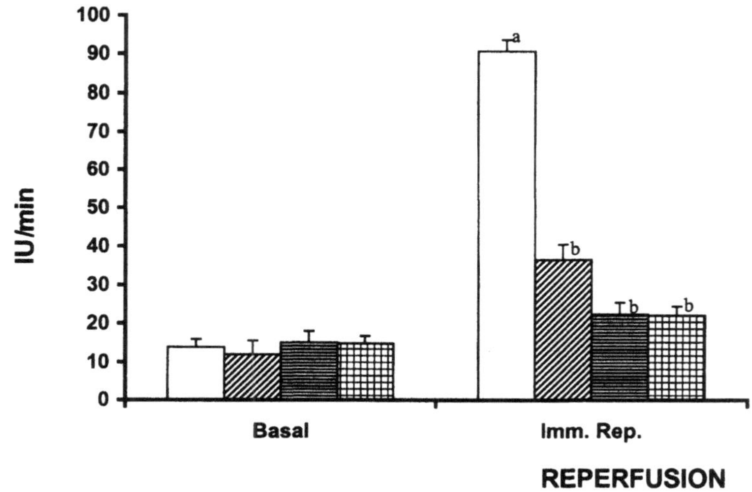

Figure 5. Effect of 100 µM theophylline (▨), 2.2 µM FR 167653 (▤) and 100 µM DSCG (▦) on MPO release as compared to control (□). MPO was estimated in coronary effluent after stabilization (basal), immediately after reperfusion (Imm. Rep.). Values are expressed as mean ± SEM. a = $p < 0.05$ vs basal value; b = $p < 0.05$ vs control.

fore, in the present study the observed increase in MPO release in coronary effluent immediately after reperfusion has demonstrated degranulation of mast cells in isolated rat heart preparation as a result of ischaemia followed by reperfusion. This contention is also supported by our earlier histological data (5).

DSCG is reported to inhibit accummulation of intracellular calcium (32) and consequently stabilize mast cells (5). Therefore, DSCG may have attenuated the increased release of MPO as a result of ischaemia and reperfusion employed in the present study by preventing the degranulation of resident cardiac mast cells. IL-1β and TNFα are reported to modulate mast cell secretory functions (33,34) and theophylline is noted to attenuate the release of IL-1β and TNFα and interferon gamma from mast cell (35). Resident cardiac mast cells are noted to be the main source of TNFα released during ischaemia and reperfusion (20). Moreover, theophylline is a nonselective phosphodiesterase inhibitor and consequently increases cAMP concentration (36). Increase in cAMP concentration is documented to decrease the release of mediators from mast cells (37). Therefore, it may be suggested that theophylline induced decrease in release of MPO immediately after reperfusion observed in the present study may be due to its action on resident cardiac mast cells mediated through inhibition of phosphodiesterase or prevention of cytokine evoked degranulation of mast cells. Increase in expression of p38 MAPK is noted during ischaemia

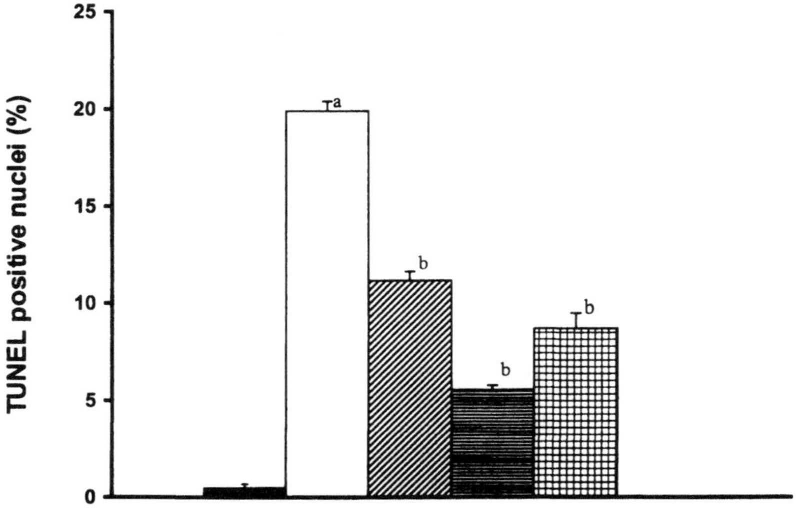

Figure 6. Effect of 100 μM theophylline (▨), 2.2 μM FR 167653 (▤) and 100 μM DSCG (▦) treatment on apoptotic index as compared to control (□) and sham control (■). Values are expressed as mean ± SEM. a = $p < 0.05$ vs sham control value; b = $p < 0.05$ vs control.

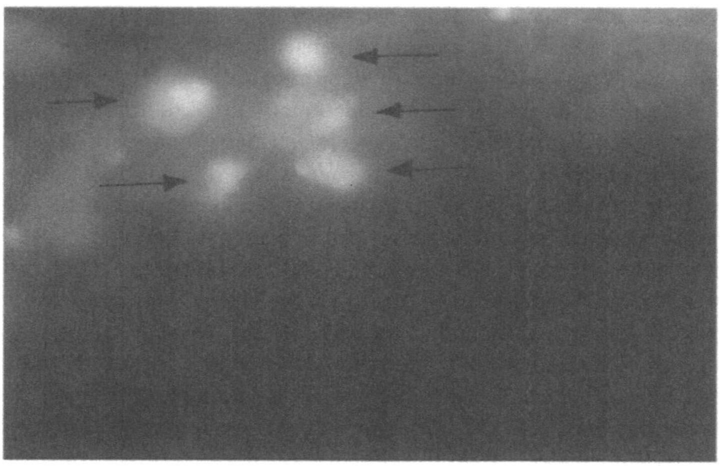

Figure 7. TUNEL positive nuclei (indicated by arrow) seen under fluorescent microscope in a section of left ventricle obtained from isolated rat heart subjected to 30 min of global ischaemia followed by reperfusion for 120 min.

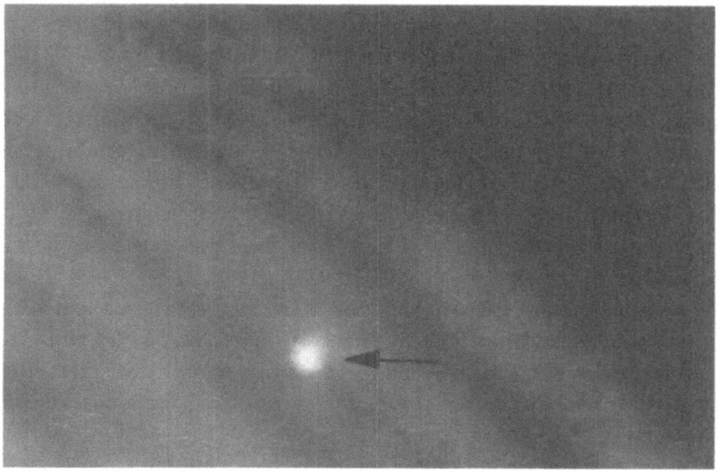

Figure 8. TUNEL positive nuclei (indicated by arrow) seen under fluorescent microscope in a section of left ventricle obtained from isolated rat heart treated with FR 167653 (2.2 μM) before subjected to 30 min of global ischaemia followed by reperfusion for 120 min.

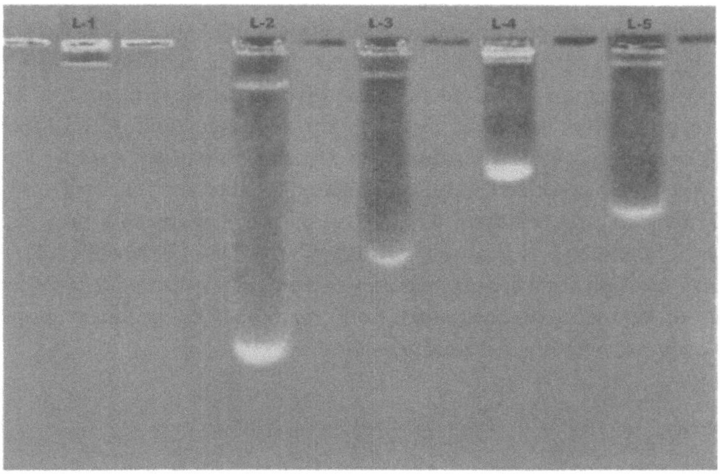

Figure 9. Electrophoretic pattern of DNA extracted from rat cardiomyocytes. **L-1, L-2, L-3, L-4** and **L-5** represent DNA extracted from **sham control** rat heart, DNA extracted from **control** isolated rat heart subjected to 30 min global ischaemia followed by reperfusion for 120 min, DNA extracted from **theophylline** (100 μM) treated isolated rat heart and subjected to global ischaemia for 30 min followed by reperfusion for 120 min, DNA extracted from **FR 167653** (2.2 μM) treated isolated rat heart subjected to global ischaemia for 30 min followed by reperfusion for 120 min and DNA extracted from **DSCG** (100 μM) treated isolated rat heart subjected to global ischaemia for 30 min followed by reperfusion for 120 min respectively. **L** represents **Lane.**

and reperfusion (38,39) and it regulates the release of mediators from human cultured mast cells (40). The release of monocyte chemoattractant protein-1 (MCP-1) from bone marrow derived mast cells (BMMC) is inhibited by SB 202190, a p38 MAPK inhibitor (41). FR 167653 is a potent inhibitor of p38 MAPK and its activity is comparable to SB 203580 (42). Thus, the FR 167653 induced decrease in MPO release, observed in the present study may be due to its inhibitory effect on p38 MAPK and prevention of degranulation of resident cardiac mast cells. Therefore, it may be tentatively concluded that DSCG, theophylline and FR 167653, perhaps by different mechanisms of action, may have prevented the degranulation of resident cardiac mast cells.

DSCG, theophylline and FR 167653 have prevented ischaemia and reperfusion-induced myocardial injury measured in terms of infarct size, LDH and CK release. It suggests that the cardioprotective effect of DSCG, theophylline and FR 167653 in study at hand may be mediated through prevention of degranulation of resident cardiac mast cells.

Theophylline is reported to cause apoptosis in eosinophils by decreasing the expression of bcl-2 protein (43), which is antiapoptotic in nature. This observation is not supported by the results of the present study because theophylline contrary to the above report has reduced the apoptotic index and ventricular DNA smearing. It is interesting to note that all the three pharmacological interventions employed in the present study have markedly reduced ischaemia and reperfusion-induced increase in apoptotic index and ventricular DNA smearing. Apoptotic index is a marker of apoptotic process initiated in cells (44,45). Apoptosis is an active gene directed process of cell death (46), characterized by fragmentation of chromosomal DNA (47). Myocardium subjected to ischaemia and reperfusion do not demostrate clear cut or pure DNA laddering and appear to represent mixtures of DNA smearing (48). Ischaemia without any subsequent reperfusion has not produced any DNA smearing but it has produced significant infarct or necrotic cell death (49). Apoptosis and necrosis are reported to coexist in acute ischaemia and reperfusion (48,50–53). Therefore, it is probable to suggest that DSCG, theophylline, and FR-167653 may have prevented degranulation of resident cardiac mast cells and consequent release of cytotoxic mediators from mast cells to produce apoptosis and apoptosis induced necrosis in cardiomyocytes.

REFERENCES

1. Sperr WR, Bankl HC, Mundigler G, Klappacher G, Globschmidt K, Agis H, Simon P, Laufer P, Imhof M, Radaskiewicz T, Glogar D, Lechner K, Valent P. 1994. The human cardiac mast cells: localization, isolation, phenotype and functional characterization. Blood 84:3876–3884.
2. Keller AM, Clancy RM, Barr ML, Marboe CC, Cannon PJ. 1988. Acute reoxygenation injury in the isolated rat heart: role of resident cardiac mast cells. Circ Res 63:1044–1052.
3. Kaur H, Parikh V, Sharma A, Singh M. 1997. Effect of amiloride a Na^+/H^+ exchange inhibitor on cardioprotective effect of ischaemic preconditioning: possible involvement of resident cardiac mast cells. Pharmacol Res 36:95–102.
4. Parikh V, Singh M. 1997. Resident cardiac mast cells and cardioprotective effect of ischaemic preconditioning in isolated rat heart. J Cardiovasc Pharmacol 30:149–156.

5. Parikh V, Singh M. 1998. Cardiac mast cell stabilization and cardioprotective effect of ischaemic preconditioning in isolated rat heart. J Cardiovasc Pharmacol 31:779–785.
6. Parikh V, Singh M. 1999. Possible role of cardiac mast cells in norepinephrine-induced myocardial preconditioning. Methods Find Exp Clin Pharmacol 21:269–274.
7. Parikh V, Singh M. 1999. Possible role of adrenergic component and cardiac mast cell degranulation in preconditioning-induced cardioprotection. Pharmacol Res 40:129–137.
8. Parikh V, Singh M. 1999. Possible role of cardiac mast cell degranulation and preservation of nitric oxide release in isolated rat heart subjected to ischaemic preconditioning. Mol cell Biochem 199:1–6.
9. Hara M, Matsumori A, Ono K, Kido H, Hwang MW, Miyamoto T, Iwasaki A, Okada M, Nakatani K, Sasayama S. 1999. Mast cells cause apoptosis of cardiomyocytes and proliferation of other intramyocardial cells in vitro. Circulation 100:1443–1449.
10. Parenteau GL, Clark RE. 1991. Prevention of ischaemia-reperfusion injury by the allergy drug lodoxamide tromethamine. Ann Thorac Surg 52:832–838.
11. Wang P, Downey JM, Cohen MV. 1996. Mast cell degranulation does not contribute to ischaemic preconditioning in isolated rabbit hearts. Basic Res Cardiol 91:458–467.
12. Humphreys RA, Kane KA, Parratt JR. 1998. Lack of involvement of mast cell degranulation in the antiarrhythmic effect of preconditioning in rats. J Cardiovasc Pharmacol 31:418–423.
13. Cain BS, Meldrum DR, Dinarello CA, Meng X, Joo KS, Banerjee A, Harken AH. 1999. Tumour necrosis factor-alpha and interleukin-1 beta synergistically depress human myocardial function. Crit Care Med 27:1309–1318.
14. Squadrito F, Altavilla D, Squadrito G, Saitta A, Campo GM, Arlotta M, Quartarone C, Ferlito M, Caputi AP. 1999. Cyclosporin-A reduces leukocyte accumulation and protects against myocardial ischaemia-reperfusion injury in rats. Eur J Pharmacol 365:159–168.
15. Pomerantz BJ, Reznikov LL, Harken AH, Dinarello CA. 2001. Inhibition of caspase 1 reduces human myocardial ischaemic dysfunction via inhibition of IL-8 and IL-1β. Proc Natl Acad Sci USA 98: 2871–2876.
16. Gurevitch J, Frolkis I, Yuhas Y, Paz Y, Matsa M, Mohr R, Yakirevich V. 1996. Tumor necrosis factor-alpha is released from the isolated rat heart undergoing ischemia and reperfusion. J Am Coll Cardiol 28:247–252.
17. Meldrum DR, Cleveland JC JR, Cain BS, Meng X, Harken AH. 1998. Increased myocardial tumor necrosis factor-alpha in a crystalloid-perfused model of cardiac ischemia-reperfusion injury. Ann Thorac Surg 65:439–443.
18. Herskowitz A, Choi S, Ansari AA, Wesselingh S. 1995. Cytokine mRNA expression in post ischaemic/reperfusion myocardium. Am J Pathol 146:419–426.
19. Kapadia S, Lee J, Torre-Aminone G, Birdsall HH, Ma TS, Mann DL. 1995. Tumor necrosis factor-α gene and protein expression in adult feline myocardium after endotoxin administration. J Clin Invest 96:1042–1052.
20. Frangogiannis NG, Lindsey ML, Michael LH, Youker KA, Bressler RB, Mendozo LH, Spengler RN, Smith CW, Entman ML. 1998. Residual cardiac mast cells degranulate and release preformed TNFα initiating the cytokine cascade in experimental canine myocardial ischaemia/reperfusion. Circulation 98:699–710.
21. Langendorff O. 1895. Untersuchungen am uberlebender saugethierherzen. Pflungers Archiv fur die Gesmate Physiologie 61:291–332.
22. Tytgat J. 1994. How to isolate cardiac myocytes? Cardiovasc Res 28:280–283.
23. King J. 1959. A routine method for estimation of lactate dehydrogenase activity. J Med Lab Tech 16:265–272.
24. Hughes B. 1961. A method for the estimation of serum creatine kinase and aldose activity in normal and pathological sera. Clin Chim Acta 7:597–603.
25. Lie JT, Pairdero PC, Holley KE, Tilus JL. 1975. Macroscopic enzyme mapping verification of large homogenous experimental myocardial infarcts of predictable size and location in dogs. J Thorac Cardiovasc Surg 69:599–606.
26. Fishbein MC, Meerbaum S, Rit J, Lando U, Kanmatsube K, Mercier JC, Corday E, Ganz W. 1981. Early phase acute myocardial infarct size quantification: Validation of triphenyltetrazolium chloride tissue enzyme staining technique. Am Heart J 101:563–600.
27. Banka N, Anand IS, Chakravarti RN, Sharma Pl, Wahi PL. 1981. Macroscopic measurements of experimental myocardial infarct size in rhesus monkey—A comparison of two methods. Bull PGI 15:147–150.

28. Chopra K, Singh M, Kaul N, Andrabi KI, Ganguly NK. 1992. Decrease of myocardial infarct size with desferroxamine. Possible role of of oxygen free radicals in its ameliorative effect. Mol Cell Biochem 113:71–76.
29. Allen RT, Hunter III WJ, Agarwal DK. 1997. Morphological and biochemical characterization and analysis of apoptosis. J Pharmacol Toxol Meth 37:215–228.
30. Maulik N, Kagan EV, Tyurin AV, Das DK. 1998. Redistribution of phosphatidyl ethanolamine and phosphatidylserine preceds reperfusion-induced apoptosis. Am J Physiol 274:H242–248.
31. Marone G, de Crescenzo G, Florio G, Granata F, Dente V, Genovese A. 1999. Immunological modulation of human cardiac mast cells. Neurochem Res 24:1195–1202.
32. White JR, Ishizaka T, Ishizaka K, Shaafi R. 1984. Direct demonstration of increase intracellular concentratin of free calcium as measured by quin-2 in stimulated rat peritoneal mast cells. Proc Natl Acad Sci USA 81:3978–3982.
33. Hogaboam CM, Befus AD, Wallace JL. 1993. Modulation of rat mast cell reactivity by IL-1 beta. Divergent effects on nitric oxide and platelet activating factor release. J Immunol 151:3767–3774.
34. Brzezinska-Blaszczyk E, Forczmanski M, Pletrzak A. 2000. The action of tumour necrosis factor-alpha on rat mast cells. J Interferon cytokine Res 20:377–381.
35. Paleman RA, Kips JC, Pauwels RA. 1998. Therapeutic activities of theophylline in chronic obstructive pulmonary disease. Clin Exp Allergy 281: Suppl 3:53–56.
36. Dent G, Giembycz MA, Evans PM, Rabe KF, Barnes PJ. 1994. Supression of human eosinophil respiratory burst and cyclic AMP hydrolysis by inhibitors of type IV phosphodiesterase: Interaction with the beta adrenoceptor agonist albuterol. J Pharmacol Exp Ther 271:1167–1174.
37. Shichijo M, Inagaki N, Kimata M, Serizawa I, Saito H, Nagai H. 1999. Role of 3′,5′-adenosine monophosphate in the regulation of chemical mediator release and cytokine production from cultured human mast cells. J Allergy Clin Immunol 103:S 421–428.
38. Ma XL, Kumar S, Gao F, Louden CS, Lopez BL, Christopher TA, Wang C, Lee JC, Feuerstein GZ, Yue TL. 1999. Inhibition of p38 mitogen-activated protein kinase decreases cardiomyocytes apoptosis and improves cardiac function after myocardial ischaemia and reperfusion. Circulation 99:1685–1691.
39. Barancik M, Htun P, Strohm C, Kilian S, Schaper W. 2000. Inhibition of the cardiac p38-MAPK pathway by SB 203580 delays ischaemic cell death. J Cardiovasc Pharmacol 35:474–483.
40. Kimata M, Inagaki N, Kato T, Miura T, Serizawa I, Nagai H. 2000. Roles of mitogen-activated protein kinase pathways for mediator release from human cultured mast cells. Biochem Pharmacol 60:589–594.
41. Teshima R, Onose J, Okunuki H, Sawada J. 2000. Effect of Ca (2+) ATPase inhibitors on MCP-1 release from bone marrow-derived mast cells and the involvement of p38 MAPK activation. Int Arch Allergy Immunol 121:34–43.
42. Takahashi S, Keto Y, Fujita T, Uchiyama T, Yamamoto A. 2001. FR 167653, a p38 mitogen-activated protein kinase inhibitor, prevents *Helicobacter pyroli*-induced gastritis in Mongolian gerbils. J Pharmacol Exp Ther 296:48–56.
43. Chung IY, Nam-Kung EK, Lee NM, Chang HS, Kim DJ, Kim YH, Park CS. 2000. The down-regulation of Bcl-2 expression is necessary for theophylline-induced apoptosis of eosinophil. Cell Immunol 203:95–102.
44. Zhao Z-Q, Nakamura M, Wang NP, Wilcox JN, Shearer S, Ronson RS, Guyton RA, Vinten-Johansen J. 1999. Reperfusion induces myocardial apoptotic cell death. Circulation (Suppl) 99:I–705 (Abstract).
45. Gavrieli Y, Sherman Y, Ben-Sasson SA. 1992. Identification of programmed cell death in situ via specific labeling of nuclear DNA fragmentation. J Cell Biol 119:493–501.
46. Steller H. 1995. Mechanisms and genes of cellular suicide. Science 267:1445–1449.
47. Wyllie AH. 1980. Glucocorticoid-induced thymocyte apoptosis is associated with endogenous endonuclease activitation. Nature 284:555–556.
48. Buja LM. 1998. Modulation of myocardial response to ischaemia. Lab Invest 78:1345–1373.
49. Kaur H, Sharma A, Singh M. 2002. Pharmacological intervention and modulation of ischaemia-reperfusion-induced apoptotic and necrotic myocardial cell death. Thesis Punjabi University 56–57.
50. Kajstura J, Cheng W, Reiss K, Clark WA, Sonnenblick EH, Krajewski S, Reed JC, Oliverti G, Anversa P. 1996. Apoptotic and necrotic myocyte cell deaths are independent contributing variables of infarct size in rats. Lab Invest 74:86–107.

51. Saraste A, Pulkki K, Kallajoki M, Henriksen K, Parvinen M, Voipo-Pulkki LM. 1997. Apoptosis in human acute myocardial infarction. Circulation 95:320–323.
52. Veinot JP, Gattinger DA, Fliss H. 1997. Early apoptosis in human myocardial infarcts. Human Path 28:485–492.
53. Ohno M, Takemura G, Ohno A, Misao J, Hayakawa Y, Shinya M, Fujiwara T, Fujiwara H. 1998. Apoptotic myocytes in infarct area in rabbit heart may be oncotic myocytes with DNA fragmentation: Analysis by immunogold electron microscopy combined with in situ nick end-labeling. Circulation 98:1422–1430.

Pathophysiology of Cardiovascular Disease,
edited by N.S. Dhalla, H. Rupp,
A. Angel & G.N. Pierce
Kluwer Academic Publishers, Boston, 2004

Mechanisms of Cardioprotection against Ischemia Reperfusion Injury

Raja B. Singh and Naranjan S. Dhalla

Institute of Cardiovascular Sciences, St. Boniface General Hospital Research Centre, and Department of Physiology, Faculty of Medicine, University of Manitoba, Winnipeg, Canada R2H 2A6

Summary. The high incidence of morbidity and mortality associated with acute myocardial ischemia (AMI) has stimulated substantial research to salvage the dying myocardium. Early restoration of flow to the ischemic myocardium by different interventions has significantly reduced the mortality associated with the acute coronary syndrome but could not improve the morbidity. Accordingly, the possibility of reperfusion injury, and not ischemic injury, as the cause of post ischemic myocardial dysfunction is being considered seriously. Ventricular arrhythmias, myocardial stunning and cell death are the prominent sequels to post ischemic reperfusion. Although thrombolytic and other therapies like percutaneous transluminal angioplasty and coronary artery bypass grafting have significantly increased the rate of survival of AMI patients, these procedures have also spurred interest in the phenomena of ischemia reperfusion injury. In view of the immense clinical consequences of ischemia reperfusion in humans, a large number of animal studies have been carried out to understand the pathophysiology and therapy. This article elaborates different mechanisms which have been suggested to cause ischemia reperfusion injury as well as therapeutic modalities that have shown promise in reducing the associated cardiac abnormalities.

***Key words*:** Ischemia-reperfusion, oxidative stress, calcium overload, ischemic preconditioning, cardioprotection.

INTRODUCTION

Although restoration of flow is essential for the salvage of the ischemic myocardium, reperfusion has been shown to depress cardiac function that recovers gradually over

Address for Correspondence: Dr. Naranjan S. Dhalla, Professor and Director, Institute of Cardiovascular Sciences, St. Boniface General Hospital Research Centre, 351 Tache Avenue, Winnipeg, Manitoba, Canada R2H2A6. Tel: (204) 235-3417, Fax: (204) 233-6723, E-mail: cvso@sbrc.ca.

a period of time (1). This phenomenon of post ischemic mechanical cardiac dysfunction upon restoration of flow has been coined as myocardial stunning (2) and is considered to be a complex syndrome rather than a single entity. The post ischemic dysfunction upon reperfusion was first demonstrated by Tennant and Wiggers (3) and then by Jennings et al. (4). Myocardial stunning can be defined as the cardiac abnormality that persists in a previously ischemic myocardium on reperfusion despite the restoration of normal or near normal blood flow and when no irreversible damage has been done to the myocardium (5). It is noteworthy that myocardial stunning is clearly differentiated from myocardial hibernation which is reflected by a persistent contractile dysfunction in a viable myocardial tissue with reduced coronary flow. Furthermore, hibernation is an adaptive condition of the myocardium as has been shown by Diamond et al. (6) and popularized by Rahimtoola (7). Nevertheless, both, myocardial hibernation and stunning share a common denominator of being reversible in nature. In fact both, hibernation and stunning are milder forms of myocardial infarction and it remains to be seen if they share any common mechanisms with infarction.

It should be pointed out that ventricular arrhythmias, due to ischemia reperfusion, occur within seconds of the reflow and are seen in all the species studied. They vary from ventricular premature beats to severe ventricular fibrillation and many mechanisms have been proposed to explain their origin. The oxidant stress as a consequence of excess production of free radicals causes damage to the lipid membranes and alters the membrane permeability to different substances especially cations. Some studies have indicated that mechanisms responsible for free radical production increase the susceptibility of heart to arrhythmias (8) but this would vary from one animal model to the other depending upon many different factors such as host defense (9). Stimulation of the adrenergic receptors and increase in the cyclic adenosine monophosphate (cAMP) levels as well as abnormalities in cationic homeostasis are some of the other causes of ischemia reperfusion induced arrhythmias forwarded.

Different experimental models have been used to study ischemia reperfusion injury under in vivo and in vitro conditions. Such studies are designed to elaborate the mechanism of action of different factors involved in the disease pathology and are considered to closely mimic the disease in humans. Though many findings can be reproduced in different animal models, variation in the results within these studies exists and thus it is not prudent to extrapolate these results to humans. Unlike animal models, the study of reperfusion injury in humans has some inherent problems that include assessment of regional function, formation of collaterals, duration and severity of ischemia, and absence of baseline values in most cases. It is difficult to assess the degree of restoration of flow to an ischemic area as well as to differentiate between silent ischemia, stunning and hibernation. Since different parameters like preload, afterload, heart rate change as myocardial stunning in man evolves; the status of cardiac function cannot be assessed at different time points (5). Most animal studies involve abrupt reperfusion, whereas it does not seem to be the case in humans. Nonetheless, the results of ischemia reperfusion (IR) injury in animal models are reliable mainly because the coronary artery is completely occluded, the time frame of ischemia can be established, collateral formation can be ascertained using angiographic methods and most importantly, baseline values can be obtained for comparison between pre and post ischemic parameters.

(I) PATHOPHYSIOLOGY OF MOCARDIAL ISCHEMIA REPERFUSION INJURY

(a) Oxidative stress

Two decades of research aimed at solving the problems associated with IR injury has resulted in the emergence of the oxyradical and calcium hypotheses as two significant pathogenic mechanisms that explain most (if not all) of the cardiac abnormalities seen in IR injury. Exhaustive research in this area is making it obvious that the two are not mutually exclusive. Zweier et al. (10) and Bolli et al. (11) have separately shown the production of reactive oxygen species (ROS) during IR and its association with contractile dysfunction. It has been shown that IR injury can be attenuated by administering mercaptopropionyl glycine (MPG) a minute before reperfusion, indicating that the ROS are produced during the first minute of reperfusion and peak between four to ten minutes (11); this finding was later confirmed directly by using the spin trap method (12). Under physiological conditions different scavenging mechanisms effectively clear the free radicals but under conditions of stress, the free radical production far outweighs the scavenging capabilities of the cardiomyocyte. ROS have been shown to attack the cellular membranes and contractile proteins, which could explain in part the contractile dysfunction seen with IR. Several sources for the generation of ROS due to IR include:

(1) Mitochondria

Mitochondria are the main source of these free radicals that are constantly being produced even under normal physiologic conditions (13). Mitochondria produce majority (80–90%) of the high-energy phosphates (adenosine triphosphate; ATP) in mammalian tissues by the process of oxidative phosphorylation involving the respiratory chain. The mitochondrial respiratory chain (or the electron transport chain) consists of enzymes and coenzymes that convey high-energy electrons from the reduced nicotinamide adenine dinucleotide ($NADH^+$)—reduced flavin adenine dinucleotide ($FADH^+$) pathway to molecular oxygen. Cytochrome oxidase reduces molecular oxygen (in complex IV) to form water thus coupling ATP production with the generation of water. The high-energy of electrons is converted to ATP and about 98 % of the electrons are tightly coupled to the ATP production while 1–2% results in the superoxide anion (O_2^-) formation (14). Some investigators have indicated this to happen at the level of the NADH-ubiquinone-complex or complex I and at complex II (14), while others have indicated this to happen mainly at complex III (15,16,17). But these free radicals are neutralized by the natural defenses of the cell that include the mitochondrial (manganese), cytosolic (copper and zinc) superoxide dismutases (Mn SOD and CuZn SOD), glutathione peroxidase (GSH-PO) and phospholipid hydroperoxide glutathione peroxidase

(13,18). Various SOD enzymes scavenge O_2^- to form H_2O_2, which though not a free radical, acts as a second messenger and can regulate the activity of enzymes essential for calcium release, cell growth or apoptosis; these enzymes include phospholipase A_2, C and D, Src Kinase, p38 mitogen activated protein kinases (MAPK) c-Jun kinase (JNK) and survival kinase (Akt/PKB) (19). H_2O_2 forms water and molecular oxygen under the enzymatic action of catalase and or glutathione peroxidases. It forms the highly reactive hydroxyl radical under conditions of stress via the Fenton (20) and or the Haber Weiss (21) pathways.

(2) NAD(P)H oxidases

The neutrophil NAD(P)H oxidases are commonly known as the phagocytic oxidases while those found in the vasculature or the cardiac cells are called the non-phagocytic types. Virtually every cell type in the vasculature has been shown to produce ROS that regulates key cellular events like cell growth, acting as a signaling molecule (22). In the vascular cells and cardiomyocytes, NAD(P)H oxidases are the major source of superoxide production in response to various stimuli like cytokines and growth factors and are believed to be essential for normal physiological responses like growth, migration and matrix remodeling. NAD(P)H oxidases in the cardiovascular system are poorly understood though it is believed that non-phagocytic cell oxidases are similar to the nuetrophilic NAD(P)H oxidases. The phagocytic NAD(P)H oxidase is a membrane bound electron transport chain that mediates the flow of electrons from the substrates to molecular oxygen forming superoxide. For NADH (23,24) and NAD(P)H (25), there is controversy regarding the preferred substrate for superoxide generation. Though similar, the non-phagocytic oxidases produce superoxide chronically and are always in an active state compared to the phagocytic oxidases that are activated by the association of cytosolic agonists with cytochrome. The latter produce 10–100 times more superoxide than the former but in short bursts (19). The exact downstream effectors for the oxidase produced ROS are poorly understood but p38 MAPK, JNK and members of the stress activated kinase family have been postulated (23).

(3) Superoxide and peroxynitrite

It has been suggested that about 0.4–2% of all the oxygen consumed during oxidative phosphorylation is converted to ROS (14,26). Though small, this is enough to cause severe pathologies in the living organisms if not detoxified by the antioxidant mechanisms. Melov et al. (27,28) have showed that transgenic mice deficient in mitochondrial SOD have a maximum life span of 8 days only. Apart from the mitochondrial electron transport chain complexes, O_2^- can be formed by the phagocytic and nonphagocytic NAD(P)H oxidases (23,24,25), xanthine oxidase (XO) (29), cyclooxygenase (COX), lipooxygenase (LOX) and cytochrome P_{450} mono-oxygenase; these enzymatic reactions occur in cardiac myocytes, fibroblasts and endothelial cells during normoxia, anoxia and upon reoxygenation (30). Cardiac myocytes and fibroblasts differ in their response to superoxide anion. Fibroblasts are more resilient and proliferate secreting the transforming growth factor (TGF-β) (an important stimulus for fibrosis) as compared to the apoptosis and the activation of various kinases seen in cardiac myocytes (31). In an attempt to unify the oxyradical and calcium overload hypotheses, many authors have implicated O_2^- in damage to the sarcoplasmic reticulum (SR) by modifying the uptake and release of Ca^{2+} causing cytosolic Ca^{2+} overload (32,33). To this end, it was shown that ROS from the xanthine-XO system depressed the SR Ca^{2+} uptake in homogenates and isolated SR vesicles, respectively (34,35). This would imply an abnormal regulation of Ca^{2+} by SR Ca^{2+} ATPase (SERCA) or the SR ryanodine sensitive Ca^{2+} release channel (RyR). Okabe et al. (36,37) have reported that ROS depress the SR Ca^{2+} uptake and enhances the release from the Ca^{2+} release channel. More recently, they have shown that O_2^- modulates the gating characteristics of the SR Ca^{2+} release channel or RyR, and this is brought about by a degradation of calmodulin (the calcium binding protein) by the O_2^-. This modulation of the RyR causes enhanced release and subsequent accumulation of Ca^{2+} in the cytosol (38). Recent work from our laboratory has shown preservation of SR function and gene expression for SR Ca^{2+} pump (SERCA2a) and ryanodine sensitive Ca^{2+} release channel (RyR) (39) as well as SR protein phosphorylation (calcium\calmodulin dependent kinase, CaMK, and exogenous cAMP dependent protein kinase, PKA) (40) by treatment of IR hearts with SOD and catalase indicating a role for ROS in causing SR dysfunction.

Superoxide readily forms highly toxic peroxynitrite ($ONOO^-$) via a biradical reaction with nitric oxide (NO) (41). Peroxynitrite is not a free radical itself as it shares two unpaired electrons on superoxide and nitric oxide that form a new bond (42,43). NO is produced by the same cell types that produce the superoxide anion under conditions of stress (vascular endothelium, nutrophils and myocytes). Peroxynitrite has been suggested to have both cardiotoxic and cardioprotective effects. It induces cellular injury by causing lipid peroxidation (44), DNA fragmentation as seen in apoptosis (45), damage to proteins and lipids in the plasma (46), depletion of antioxidants like glutathione and cysteine as well as nitration of proteins causing organ dysfunction (47). Most of its effects are mediated by its intermediate $ONOOH^-$. It reacts with carbon dioxide (CO_2) to form CO_3^- and NO_2^- to eventually form the nitronium ion (NO_3^-) that facilitates protein nitration and alters organ function (48). Absence of CO_2 would explain the inability of peroxynitrite to cause profound damage under in vitro conditions. In IR, activated neutrophils and endothelium produce NO via the endothelial NO synthase (eNOS) and the inducible NOS (iNOS) (49). The release of NO coincides with the superoxide production during the early phase of reperfusion and thus facilitates peroxynitrite formation (50). Yasmin et al. (51) have shown that peroxynitrite formation peaks in the early seconds of reperfusion and this coincides with the formation of both NO and superoxide. Some of the other toxic effects of peroxynitrite include membrane sulfhydryl group oxidation of the red blood cells without hemolysis at concentrations of 0.1–2 mM (52) while Kondo et al. (53) have shown hemolysis to occur at lower concentrations of 100–500 μM suggesting a possible dose dependent effect.

Peroxynitrite aids white blood cells to kill infectious and cancerous cells due to its cytocidal oxidant properties (54) and has also been shown to cause apoptosis in white cells themself (45,54). Thiols (glutathione) neutralize the effects of peroxynitrites and the protection is believed to be due to depression in the degradation of the myocardial enzyme aconitase, which is a part of the mitochondrial respiratory chain (55).

(b) Calcium Hypothesis

Disturbance in Ca^{2+} homeostasis is postulated to cause contractile abnormalities due to IR injury. Three possible mechanisms that could explain these abnormalities due to disturbed Ca^{2+} regulation are: (1) decreased responsiveness of the myofilaments to Ca^{2+}, (2) Ca^{2+}overload, and (3) excitation contraction uncoupling due to SR dysfunction. Two of these plausible mechanisms of IR injury are discussed here.

(1) Decreased responsiveness to Ca^{2+}

Though the exact underlying pathology responsible for decreased responsiveness of the myofilaments to Ca^{2+} is not known, evidence available indicates some structural abnormalities at the level of myofilaments that would alter their response to Ca^{2+} during excitation contraction. Previous studies from our laboratory have shown that ROS cause oxidation of the critical thiol groups in the myofilaments (56). Such structural modifications of the myofilaments could explain the decreased responsiveness to Ca^{2+} as well as the contractile abnormalities due to IR injury. ROS cause a reduction in the content of reduced glutathione and increase in oxidized glutathione (57). Taken together, changes in the content of these have been shown to decrease the Ca^{2+} sensitivity of the myofilaments (58). It would be interesting to point out that this effect of ROS on the myofilaments may possibly provide for the integration of the calcium and oxyradical hypotheses. Another important factor that could alter not only the contractility of myofilaments but also their response to Ca^{2+} during excitation contraction coupling is their proteolytic degradation by the Ca^{2+} activated neutral protease.

(2) Calcium overload

Oxygen deficiency during ischemia causes a shift in myocardial metabolism resulting in decreased levels of ATP and accumulation of hydrogen ions and lactate. This acidosis causes activation of the sodium-hydrogen exchanger (NHX) resulting in sodium (Na^+) overload. Until reperfusion begins, Na^+ accumulates and the acidosis prevents the activity of the sodium-calcium exchanger (NCX). Reperfusion of the myocardium causes reversal of the acidosis and activates the NCX. This pushes Ca^{2+} into the cardiomyocytes in exchange for Na^+ resulting in the Ca^{2+} overload. This was a simple hypothesis forwarded by Grinwald (59) in 1982 and still holds good. Subsequent work by different investigators has shown that $[Na^+]_i$ rises rapidly during ischemia and remains so for about 8–10 minutes after reperfusion whereas the acidosis takes about 30 seconds to resolve. Grinwald (60) in a subsequent work showed that the functioning of sodium-potassium pump failed in IR, which could possibly explain the Ca^{2+} overload seen on reperfusion. This could be due to the effect of ROS on the sarcolemmal Na^+-K^+ ATPase. Some studies have shown that cytosolic Ca^{2+} concentration rises even during ischemia but none of the effects of calcium overload are seen during ischemia alone. Impaired myofilament responsiveness to Ca^{2+} (62) and the proteolytic degradation of the various proteins (63) occur during reperfusion. Thus, reperfusion would seem essential for occurrence of Ca^{2+} overload and the different cardiac abnormalities associated with IR injury. Some of the mechanisms responsible for Ca^{2+} overload are discussed below.

(i) Sodium-hydrogen exchanger. The NHX or antiporter is one of the membrane transporters that regulate the cardiomyocyte pH, the others being the Na^+-HCO_3^- symport, Cl^--HCO_3^- and the Cl^--OH^- exchangers. The first two are activated under conditions of acidosis while the latter two respond to alkaline conditions (64). The NHX is the most important of these all and has five different isoforms of which NHX1 is the primary isoform present in cardiomyocytes (65). NHX1 maintains the intracellular pH and cell volume by exchanging H^+ for Na^+ in a stoichiometric ratio of 1 : 1 (66). The glycosylated NHX1 protein has a molecular weight of 110 kDa. It has 2 principal domains, a 500-amino acid transmembrane domain and a 315-amino acid hydrophilic carboxy C-terminus cytoplasmic domain that regulates the activity of the NHX1 via phosphorylation dependent reactions (67). Acidosis is the main stimulus for the activation of the NHX1 (68) and most of the cell H^+ is extruded through it. The antiporter has an H^+ sensor that is receptive to changes in the pH where proton binding to a NHX1 oligomer causes conformational changes causing its activation. The NHX can be activated by other means too apart from acidosis like ischemic metabolites such as H_2O_2 (69) and lysophosphatidylcholine (70). Most of these act by stimulation of the phosphoinositide hydrolysis and activation of PKC resulting in phosphorylation and activation of NHX1 (65) and MAPKs (69). Under normal conditions the rate of Na^+ entry via NHX and efflux by Na^+-K^+ ATPase would balance each other (71). During ischemia ATP depletion causes the inhibition of Na^+-K^+ ATPase (72,73) or Na^+-K^+ pump failure (60) resulting in accumulation of intracellular Na^+ and eventually Ca^{2+} overload by exchanging Na^+ for Ca^{2+} via the NCX (reverse mode) (74). Na^+-K^+ ATPase has been suggested to be able to remove the acidosis induced Na^+ load all by itself in the absence of Na^+ entry (75). NHX inhibitors like amiloride attenuate the ouabain induced elevation in $[Na^+]$ and the entry of $Ca2^+$ through the NCX (71) suggesting Na^+ overload to be the primary defect. But for the concomitant Na^+ and eventual Ca^{2+} overload, NHX activation on reperfusion would have been beneficial, as it would restore the cellular pH to normal. Some recent studies have strongly implicated the NCX in the genesis of Ca^{2+} overload by demonstrating its overexpression in animals that showed enhanced IR injury possibly by increasing the Ca^{2+} entry (76).

(ii) Sodium-calcium exchanger. The NCX extrudes calcium for Na^+ under normal conditions but has a reverse mode that gets activated during IR causing efflux of

Ca^{2+}. The NCX is normally driven by an electrochemical gradient and the Na^+ to Ca^{2+}ratio is 3:1 (77). A reversal in the ionic concentrations or membrane potential can put the NCX in the reverse mode. In IR, the reverse mode of the NCX plays a substantial role in the development of Ca^{2+} overload and IR injury (78). Sarcolemmal depolarization and increased Na^+ concentration are instrumental in this role of the reverse mode. Failure of the Na^+-K^+ ATPase due to ATP depletion during ischemia (73) causes depolarization of the sarcolemma and Na^+ overload, which is further accentuated by the intracellular acidosis. Hearts reperfused with low Na^+ concentration buffers have shown a depression in the development of Ca^{2+} overload implicating Na^+overload and the NCX in the genesis of Ca^{2+} overload and IR injury (79). Similar results have been obtained with NCX (80) and NHX inhibitors (81). This is consistent with the observation that over expression of NCX is associated with an increased susceptibility to IR injury (76).

From the above discussion, it would emerge that genesis of Ca^{2+} overload during IR depends upon the status of the NHX and NCX. An important corollary to this observation is that for Na^+ overload to develop, the failure of the Na^+-K^+ pump is needed (60,72,73). ATP hydrolysis/depletion during ischemia would inhibit the Na^+-K^+ ATPase and this would lead to depolarization of the sarcolemmal membrane especially during phase 2 of the action potential leading to a reversal of the NCX. Na^+-K^+ ATPase depression would also result in reversal of the ionic concentration gradient that normally exists between Na^+ and Ca^{2+}causing an extrusion of Na^+ and not Ca^{2+} by the reverse mode of the NCX.

(II) ISCHEMIC PRECONDITIONING AND CARDIOPROTECTION

Ischemic preconditioning (IPC) of the myocardium is probably one of the most powerful tools for cardioprotection available that can be reproduced easily and all animal models experimented with, have shown IPC including humans. Understanding the precise mechanisms involved in IPC would help exploit its therapeutic potential in treating various cardiac abnormalities. Murray et al. (82) showed that anesthetized dogs exposed to four 5-minute ischemic episodes (with 5 minutes of reperfusion in between) prior to a 40-minute occlusion of the circumflex coronary artery and subsequent reperfusion showed a reduction in the infarct size. This group had hypothesized that repeated episodes of ischemia would deplete the myocardium of all its ATP whereas the intervening reperfusion would help flush the ischemic metabolites and reduce the osmotic load of the tissue. But no cumulative loss of ATP was found compared to the first episode of ischemia. The protection afforded was lost when the sustained period of ischemia was prolonged to 3 hours indicating that IPC was actually delaying the imminent cell death only (82,83). Subsequent investigators have broadened the perspective of IPC by adding other correlates to the initial reduction in infarct size like changes in ST-segment elevation, alterations in cardiac function and effects on arrhythmias (84). It was suggested that attenuated ST- segment elevation provided immediate evidence of the presence of IPC. The occurrence of IPC in animals (rabbits) that do not form significant collaterals indicates that IPC is independent of collateral formation.

ST-segment attenuation was lost on prolonged reperfusion proving indirectly that early IPC is a short acting phenomenon (85).

IPC occurs in two distinct phases. The one described by Murray et al. (82) is the early or classical phase of preconditioning while two groups subsequently described a delayed phase of cardioprotection or the late phase (86,87). The early phase begins within minutes of the first ischemic stimulus [coronary occlusion for 60 to 90 seconds during percutaneous transluminal coronary angioplasty (PTCA) is associated with less ST-segment elevation and reduced pain with subsequent procedures], affords potent protection which, lasts for only 120 to 180 minutes (82,85). The late phase begins 12 to 24 hours after the ischemic stimulus but lasts for 72 to 96 hours (88,89). The two phases differ in duration and mechanism of development but both employ a *trigger* and a *mediator* of various processes leading to end *effector* eventually. Little interest has been shown in the early or classical preconditioning primarily because of the short duration and doubtful clinical significance. This phase is characterized by a ligand (adenosine, noradrenaline, bradykinins, opioids and/or ROS) interaction with the G-protein linked receptors and activation of the intracellular targets. These trigger intracellular signaling pathways involving the protein kinase C (PKC), tyrosine kinases (TrK), mitogen activated protein kinases (MAPK) and opening of ATP- sensitive potassium ATP (K_{ATP}) channels both in the sarcolemma and mitochondria. It does not protect the heart against the effects of stunning or infarction. But during any subsequent ischemic insult the high energy phosphate (HEP) depletion, ATP utilization, lactate and H^+ accumulation are much slower, which would best describe a reduced energy demand of the PC myocardium (90,91).

Most work on IPC has been done in the area of late or delayed PC also called the second window of protection (SWOP), which involves complex intracellular signaling pathway activation and protein synthesis. NO, PKC, TrK, adenosine receptor, K_{ATP} channels, opiod receptors, cyclooxygenase-2 and aldose reductase are some of the sites implicated in the mechanisms of IPC. It has been hypothesized that the different triggers result in the translocation of the PKC and subsequent synthesis of various cardioprotective molecules like heat shock proteins (92) and NOS (93). Adenosine, NO, ROS and opioid receptor agonists are probably the triggers of late IPC.

(a) Adenosine receptors

Adenosine is an autacoid released by myocytes, neutrophils and endothelium into the extracellular fluid during early ischemia and mediates it effects through specific A_1, A_{2a}, A_{2b} and A_3 receptors. It exerts its effects on all the three phases of cardioprotection i.e. preconditioning, ischemia and reperfusion. Drury et al. (94) in 1929 showed that tissue extracts containing adenosine produced bradycardia, hypotension and coronary vasodilation. Later, Berne (95) demonstrated a relation among coronary flow, oxygen consumption by the myocardium and the adenosine levels whereas Baxter et al. (96) suggested the role of adenosine in IPC, which is mediated through metabolic changes and the K_{ATP} channels probably through the A_1 and A_3

receptors. These effects can be blocked by the K_{ATP} channel blockers glibenclamide (97,98), 5-hydroxydecanoate (5-HD) (104) and glyburide (105) making adenosine receptors the upstream and the K_{ATP} channels downstream of the same pathway.

(b) ATP sensitive potassium channels (K_{ATP})

The role of K_{ATP} channels is becoming an essential part of IPC as more and more studies agree on the common downstream role of K_{ATP}. Different studies have suggested that the K_{ATP} channel opening is integral to the effects seen with ischemia (99), adensoine receptor agonists (both A_1 and A_3) (100), δ_1- opioid receptor agonist (101) and heat stress (102). ATP was the first known modulator of the function of potassium channels and came to be known as ATP-sensitive potassium channels. High concentrations of ATP close the K_{ATP} channel under normal physiologic conditions while ADP activates the opening of these channels. Much of the work on their role in IPC comes from the use of pharmacological blockers; however in the recent past focus has shifted from the surface channels to the ones located on the inner side of the mitochondrial membrane.

(i) Mitochondrial ATP sensitive potassium channels (mitoK_{ATP}). Garlid et al. (103) have shown a role for the mito K_{ATP} channels downstream of the mediators mentioned above. Though the exact mechanism remains elusive, it is known that K_{ATP} blockers like 5-HD (104), glibenclamide (98) and glyburide (105) abolish the protection afforded by IPC. This blockade has been confirmed in human tissue as well (106). So adenosine receptor activation stimulates the mitoK_{ATP} channel opening through an unknown G-protein coupled pathway (107), an effect that can be blocked by the K_{ATP} channel blockers suggesting adenosine to be the upstream and mitoK_{ATP} channels to be the downstream of the same pathway. What links these two players of this pathway to each other was ascertained, at least in part, by two studies that indicated a role for PKC. Adenosine (A1) receptor stimulation activates PKC and inhibits PKA, both of which are known to effect mitoK_{ATP} gating. PKC in particular has been shown to activate the K_{ATP} (sarcolemmal) channel opening (108,109). Probably it phosphorylates the K_{ATP} channel protein in a way so as to cause its activation. Most of the triggers mentioned above increase the cyclic-guanosine monophosphate (c-GMP) levels either directly by acting on the myocytes or via the endothelium; c-GMP levels of the preconditioned hearts has been shown to be higher than the virgin hearts (110,111). A recent study has shown that NO-donors stimulate c-GMP dependent protein kinase (PKG) that facilitates K_{ATP} channel opening, an effect that was blocked by selective PKG inhibitors (112).

(ii) Sarcolemmal ATP sensitive potassium channels. The sarcolemmal K_{ATP} channels were first described by Noma (113) in ventricular myocytes. These have a repolarizing effect on the potassium currents thereby shortening the action potential duration (APD) (114) and causing the early repolarization ("injury current of ischemia") thought to be the basis of the ST-segment elevation due to ischemic insult (115). The study of these channels was based upon two hypotheses: (i) K_{ATP} channel blockers would protect against IR injury and arrhythmias in particular, (ii) K_{ATP}

channel openers would be protective by shortening the APD thereby reducing the Ca^{2+} entry into the cell. Studies with K_{ATP} channel openers revealed shortening of the APD and recovery of post ischemic contractile function suggesting a possible association between cardioprotection and APD (116). Some of the studies however showed that there was conservation of ATP during ischemia and mitochondrial protection with K_{ATP} channel openers (117,118). APD shortening should cause reduced contractile function, which did not augment well with ATP conservation suggesting another mechanism for the effects mediated by K_{ATP} channel openers. Studies with aprikalim, a K_{ATP} channel opener, showed protection that was more than what would be seen with cardioplegia alone (119). These results indicated the APD shortening effect not to be as important as believed and some other studies with K_{ATP} channel openers like bimakalim, cromokalim reduced the infarct size in dogs without affecting the APD (120,121,122). This separated the cardioprotection from APD shortening and the surface K_{ATP} channels. Nevertheless, the effect on the APD is critical to the mitochondrial ability to regulate the calcium concentration and protects against calcium overload. Kir6.2, the inward rectifying potassium channel and SUR (sulfonyl urea receptor), the ATP binding cassette protein together form the K_{ATP} channels. SUR provides for binding of the sulphonylurea drugs, K_{ATP} channel openers and ATP and ADP. Jovanovic et al. (123) have shown that Kir6.2/SUR2A gene delivery along with pharmacologic activation of the K_{ATP} channel resulted in a phenotype that could effectively handle intracellular Ca^{2+} overload irrespective of the degree of metabolic insult. Transgenic mice deficient in K_{ATP} channel have been shown to lose their ability to maintain intracellular Ca^{2+} homeostasis (124). This provides an exciting new potential for gene therapy for K_{ATP} channel in IR injury.

(c) Protein kinase C

Baxter et al. (125) proposed a role for PKC in IPC. It has been implicated in the preconditioning seen with adenosine and heat stress (92) mostly by facilitating the K_{ATP} channel opening. Most of the evidence is indirect and comes from the use of pharmacological inhibitors like chelerythrine. Baxter and Yellon hypothesized that IPC is a result of PKC translocation from the cytosol to the nucleus. Ping et al. (126) showed that the epsilon (ε) (and the η) isoforms of PKC are translocated (and activated) during IPC without affecting the total PKC activity. Ventura et al. (170) have shown κ-opioid receptor stimulation to cause translocation of PKC to the nucleus resulting in an increased transcription of the opioid peptide gene. It was further shown that PKC isoform ε, but not η is necessary for late IPC (127) and that the activation of PKC was blocked by L-NAME suggesting the role of NO in PKC activation (128). Studies with NO donors in the absence of ischemia resulted in translocation and activation of PKC similar to that seen in IPC (128). Though speculative, this would suggest a role for the NO derived ROS like peroxynitrite in PKC mediated IPC. This viewpoint was supported by the fact that MPG blocked the activation and translocation of PKC ε (129).

(d) Tyrosine kinases and mitogen activated protein kinases

The TrKs are a group of receptor and non-receptor kinases involved in downstream intracellular signaling of different pathways like PKC mediated IPC. Dawn et al. (130) have shown that inhibition of the non-receptor TrKs of Src family abolished IPC seen with PKC activation. Some studies have indicated that chelerythrine blocked the IPC induced activation of these Src family members (Src and Lck) suggesting their role downstream of PKC ε (131). Gustein et al. (22) have shown that opioids can cause activation of the extracellular signal regulated kinase (ERK) and p38 MAPK. These observations would suggest a role for TrK and MAPK in the PKC signaling mediated IPC.

(e) Opioids

Opioid receptor agonists have been shown to be protective against cardiovascular abnormalities (132,133) and the IR injury induced arrhythmias (134). In particular the delta1 (δ1)-opioid receptor has been reported to provide adaptation to stress and protect against arrhythmias (135). In 1995, Schultz et al. (136) showed that the opioid antagonist naloxone blocked the IPC induced effect on the infarct size. Later it was reported that the preconditioning seen with opioids is mediated by the peripheral receptors (137). Subsequent studies with morphine have revealed that the preconditioning effect of opioids is manifest at the level of the cardiomyocytes in the intact hearts and a role for the mitoK_{ATP} channels has been suggested (138). Fryer et al. (171) have shown the involvement of the opioids in the SWOP. They reported TAN 67, a δ1-opioid receptor agonist, to reduce the infarct size after 24–48 hours of administration and not after 12 hours. This effect could be blocked by the administration of BNTX (a selective δ1-antagonist), glibenclamide and 5-hydroxydecanoate indicating a role for mitochondrial K_{ATP} channels as well.

(f) Nitric oxide and reactive oxygen species in delayed IPC

The essential role of ROS for the development of late IPC was demonstrated by the abolition of late IPC by administration of MPG (139) alone and in combination with catalase and SOD (140). MPG has also been shown to block the late IPC developed against arrhythmias and infarction (141). This has been confirmed by the intracoronary administration of ROS generating solutions in rabbits producing late IPC response (142). Development of late IPC in the presence of adenosine blockers (143) and the loss of IPC against stunning (144) and infarction (145) by the administration of L-NAME suggested a role for NO as a trigger in late IPC. These studies further showed iNOS to be the isoform involved in the development of late PC. Guo et al. (146) showed that iNOS, but not eNOS, gene is upregulated in late PC and that selective targeting of the iNOS gene resulted in loss of cardioprotection from IPC, while pretreatment with NO donors mimiced cardioprotection similar to the one afforded by IPC against stunning and infarction even in the absence of ischemia (129). Leesar et al. (147) have shown that pretreatment with nitroglycerin (NTG) 24 hours before coronary angioplasty affords cardioprotection evidenced by an attenuated ST-segment elevation, less chest pain and reduced regional ventricular wall abnormalities. Recent studies have implicated NO as both the trigger and mediator of late PC. NO and its second messenger, cGMP, have been shown to produce multiple effects on the myocardium like antagonism of β-adrenergic stimulation, reduced myocardial oxygen consumption and contractility as well as opening of K_{ATP} channel (148,149). Another interesting study to distinguish between the isoforms of NOS involved in IPC employed tetrahydrobiopterin (BH_4), the cofactor needed for NO production by both iNOS and eNOS. NO is produced continuously by iNOS but in a sudden spurt by eNOS and both need BH4 for this. Takano et al. (150) in this study showed that BH4 is needed for the cardioprotection seen with late IPC. Blocking BH4 formation on day 1 prior to the first ischemic stimulus did not abolish the protection seen by late PC whereas by blocking the formation of BH4 on day 2, the cardioprotection of late PC was abolished. It was hypothesized that there was enough BH4 to support the early formation of NO by eNOS but the subsequent NO originated from iNOS.

(III) PHARMACOLOGICAL MEDIATORS OF CARDIOPROTECTION

(a) Antioxidants

Different studies in the past two decades have provided substantial evidence implicating ROS in the pathology of IR injury. The free radicals generated during the first few minutes of reperfusion include superoxide and hydroxyl radical, as well as oxidants such as peroxynitrite and H_2O_2. Major antioxidant systems include: (i) scavengers like **SOD**, catalase, dimethylthiourea, MPG, vitamin E and ascorbic acid and β-carotene, (ii) glutathione (GSH) and GSH peroxidase and (iii) chelators like desferroxamine. SOD is considered the first line of defense and consists of copper/zinc, iron and manganese SOD (Cu, Zn-SOD, Fe-SOD and Mn-SOD). Mn-SOD is generated on the inner side of the mitochondrial membrane and is critical to the ROS neutralization. Melov et al. (27,28) have showed that transgenic mice deficient in mitochondrial SOD have a maximum life span of 8 days only. Chen et al. (151) have shown that overexpression of the Mn-SOD in transgenic mice protects against IR injury in mice and expression of human Mn-SOD in transgenic mice has been shown to protect against the eproxidation of the membrane lipids (152). SOD confers protection by dismutating the superoxide anion to molecular oxygen and H_2O_2. **Catalase** is another free radical scavenger that acts in tandem with SOD and further reduces the H_2O_2 to water and molecular oxygen (153,154,155). Previous work from our laboratory has shown preservation of SR function, gene expression and SR protein phosphorylation on treatment of IR hearts with SOD and Catalase (39,40). SOD administration has also been shown to reduce arrhythmias associated with repefusion (156) and thrombolysis (157). Studies with **Dimethylthiourea and MPG** have shown significant improvement in cardiac function in IR hearts by scavenging the highly reactive hydroxyl radical when administered before reperfusion (158,159,160). Effects of oxidative stress have also been attenuated by the use of antioxidants like **Vitamin E** (α-Tocopherol) analogues, ascorbic acid and a novel lipid peroxidation inhibitor. Vitamin E gets incorporated into cells rather slowly due to its lipophilicity and hence does not find much application in acute ischemic or reperfusion insult. IRF1 042, a Vitamin E analogue has been shown to reduce infarct size in a dose dependent manner in IR injury (161). It has been reported to act by inhibiting the nuclear factor κB. Shimizu et al. (162) have shown **H290/51** (an indole derivative) to reduce the infarct size in pigs by inhibiting the lipid peroxidation of the membranes.

(b) Opioids

Cardiac tissue, both atria (163) and ventricles (163,164), has been shown to synthesize all three types of opioid peptides (enkephalin, endorphin and dynorphin) and it is well established that they are released into the circulation under conditions of stress (132). Of the three opioid receptor types characterized so far (δ, μ, κ) (165,166), functional studies have indicated the presence of δ and κ receptors in the adult cardiac tissue (167,168). Opioids are involved in different cardiovascular pathophysiologies like congestive heart failure (133), IR induced arrhythmias (134) and the effects of circulatory shock on the baroreceptors (132). Maslov et al. (135) have shown that hypoxia or stress adaptation of the heart against arrhythmias activates the δ-opioid receptor in providing protection. But the most significant evidence for the role of opioids in cardioprotection came from the work of Schultz et al. (136) who showed that opioid receptor antagonist nalaxone blocked the cardioprotection afforded by IPC on the infarct size. It was later shown that the cardioprotective effect of opioids was mediated through peripheral opioid receptors in rabbit hearts (137). As discussed earlier, the cardiac tissue releases opioids endogenously in response to stimuli like ischemia and the left ventricle has even been suggested to behave like an endocrine organ producing enkephlins (164). Takasaki et al. (169) have further suggested that proenkephlin products are the likely mediators of IPC in the rabbit heart. The cardioprotective effect of the opioids is exerted at the levels of the cardiomyocyte and mediated via the activation of the K_{ATP} channels. This was indicated by blocking the concentration dependent effect of morphine in isolated cardiomyocytes by nalaxone, glibenclamide and 5-hydroxydecanoate (138). Significantly, Ventura et al. (170) showed that κ-opioid receptor stimulation causes translocation of PKC to the nucleus resulting in increased transcription of the opioid peptide gene. Fryer et al. (171) have shown opioids to have role in the SWOP. Bolling et al. (172) could protect hearts exposed to cold cardioplegia of 18 hours or global ischemia for 2 hours by using a δ-opioid receptor agonist, DADLE, showing promise for organ preservation and during bypass surgeries.

(c) Beta Blockers

β-blockers have been used for a long time mostly to treat hypertension. But they have also shown promise in angina pectoris, myocardial infarction and even congestive heart failure. Timolol, propranolol, atenolol, metoprolol, carvedilol are some of the β-blockers that have proven to be effective. Most effects seen can be safely attributed to as class effects; these would include correction of hypertension, chest pain, improvement in supraventricular tachyarrhythmias and ventricular arrhythmias and prevention of reinfarction (173). The antiischemic effects seen with β-blockers probably result from a reduction in the heart rate and contractility, which means reduced oxygen consumption. It is not fully understood as to how β-blockers reduce mortality and the incidence of sudden cardiac death. Beta-1-blockade seems to be the major mechanism responsible for the similar actions of selective and non-selective β-blockers. This doesnot seem to be the mechanism of reducing the incidence of sudden cardiac death. Carvedilol is a non-selective β-blockers that has shown much promise in the recent past due to its ability to block both beta 1 and 2 and also alpha 1 receptors (174). It shares all the other properties of β-blockers and in addition decreasing the afterload on the heart due to its ability to block the α1 receptors. Different studies have shown the ability of carvedilol to reduce the infarct size in IR animal models (175). It has been shown to reduce the neutrophil accumulation (176) as well as incidence of ventricular arrhythmias (177) and inhibit lipid peroxidation of the myocardial cell membranes (176).

(d) Angiotensin converting enzyme inhibitors

Angiotensin converting enzyme (ACE) converts angiotensin I (AI) to angiotensin II (AII) and degrades bradykinin. A II is cardiotoxic at pathophysiological levels and causes cardiomyocyte necrosis by stimulating the formation of ROS from non-phagocytic NAD(P)H oxidases (179). The cardiovascular actions of A II are mainly mediated through the AT1 receptor as seen by reduced infarct size in studies using AT1 receptor blockers (180). ACE inhibitors have been shown to limit the infarct size, prevent reperfusion arrhythmias and reduce reinfarction (181). ACE inhibitor not only prevents the conversion of AI to AII but also attenuates bradykinin degradation. ACE inhibitors (182) and AT1 receptor blockers (183) have both shown protection against IR injury. Even pretreatment with ACE inhibitors has been shown to benefit against myocardial infarction (184). Angiotensin receptor 1 (AT1) blockade causes an increase in the levels of AII. Increased AII has been hypothesized to activate the AT2 receptors in the endothelium causing increased formation of bradykinin leading to cardioprotection (185). The reduction in infarct size seen with candesartan (AT1 receptor blocker) was abolished by the concomitant use of bradykinin receptor blocker and cyclooxygenase inhibitor indicating a role for bradykinin and prostaglandins in mediating the effects of candesartan (186).

CONCLUSION

Innumerable studies conducted to elaborate the mechanism of IR injury narrow down to the Ca^{2+} and the oxyradical hypotheses. The two are not mutually exclusive and many points of convergence have been speculated but a careful assessment would be needed before they could be integrated and the results extrapolated to humans. Of the different cardioprotective mechanisms studied, IPC is the most powerful method and if adequately exploited could prove to be a powerful therapeutic tool. Elucidation of the roles of different molecules like adenosine receptors, NO, bradykinin, SOD, Na^{+}-K^{+} ATPase, NCX, PKC and the K_{ATP} channels has given a new direction to research in the area of IR injury whereas genetic manipulation of these targets could provide new powerful tools for cardioprotection.

ACKNOWLEDGMENTS

The work reported here was supported by a grant from the Canadian Institute for Health Research (CIHR Group in Experimental Cardiology). NSD hold the Pharmaceutical Research and Development chair supported by Merck Frost, Canada.

REFERENCES

1. Heyndrickx GR, Millard RW, McRitchie RJ, Maroko PR, Vatner SF. 1975. Regional myocardial functional and electrophysiological alterations after brief coronary artery occlusion in conscious dogs. J Clin Invest 56:978–985.
2. Braunwald E, Kloner RA. 1982. The stunned myocardium: prolonged, postischemic ventricular dysfunction. Circulation 66:1146–1149.
3. Tennant R, Wiggers C. 1931. The effect of coronary occlusion on myocardial contraction. Am J Physiol 112:351–361.
4. Jennings RB, Sommers HM, Smyth GA, Flack HA, Linn H. 1960. Myocardial necrosis induced by temporary occlusion of a coronary artery in the dog. Arch Pathol 70:68–78.
5. Bolli R. 1992. Myocardial "stunning" in man. Circulation 86:1671–1691.
6. Diamond GA, Forrester JS, deLuz PL, Wyatt HL, Swan HJ. 1978. Post-extrasystolic potentiation of ischemic myocardium by atrial stimulation. Am Heart J 95:204–209.
7. Rahimtoola SH. 1989. The hibernating myocardium. Am Heart J 117:211–221.
8. Bernier M, Hearse DJ, Manning AS. 1986. Reperfusion-induced arrhythmias and oxygen-derived free radicals. Studies with "anti-free radical" interventions and a free radical-generating system in the isolated perfused rat heart. Circ Res 58:331–340.
9. Connaughton M, Kelly FJ, Haddock PS, Hearse DJ, Shattock MJ. 1996. Ventricular arrhythmias induced by ischaemia-reperfusion are unaffected by myocardial glutathione depletion. J Mol Cell Cardiol 28:679–688.
10. Zweier JL, Kuppusamy P, Williams R, Rayburn BK, Smith D, Weisfeldt ML, Flaherty JT. 1989. Measurement and characterization of postischemic free radical generation in the isolated perfused heart. J Biol Chem 264:18890–18895.
11. Bolli R, Jeroudi MO, Patel BS, Aruoma OI, Halliwell B, Lai EK, McCay PB. 1989. Marked reduction of free radical generation and contractile dysfunction by antioxidant therapy begun at the time of reperfusion. Evidence that myocardial "stunning" is a manifestation of reperfusion injury. Circ Res 65:607–622.
12. Bolli R, Patel BS, Jeroudi MO, Lai EK, McCay PB. 1988. Demonstration of free radical generation in "stunned" myocardium of intact dogs with the use of the spin trap alpha-phenyl N-tert-butyl-nitrone. J Clin Invest 82:476–485.
13. Chance B, Sies H, Boveris A. 1979. Hydroperoxide metabolism in mammalian organs. Physiol Rev 59:527–605.
14. Boveris A, Cadenas E, Stoppani AO. 1976. Role of ubiquinone in the mitochondrial generation of hydrogen peroxide. Biochem J 156:435–444.
15. Cadenas E, Davies KJ. 2000. Mitochondrial free radical generation, oxidative stress and aging. Free Radical Biol Med 29:222–230.
16. Raha S, Robinson BH. 2000. Mitochondria, oxygen free radicals, disease and aging. Trends Biochem Sci 25:502–508.
17. Sugioka K, Nakano M, Totsune-Nakano H, Minakami H, Tero-Kubota S, Ikegami Y. 1988. Mechanism of O2-generation in reduction and oxidation cycle of ubiquinones in a model of mitochondrial electron transport system. Biochem Biophys Acta 936:377–385.

18. Augustin W, Wiswedel I, Noack H, Reinheckel T, Reichelt O. 1997. Role of endogenous and exogenous antioxidants in the defence against functional damage and lipid peroxidation in rat liver mitochondria. Mol Cell Biochem 174:199–205.
19. Griendling KK, Sorescu D, Ushio-Fukai M. 2000. NAD(P)H oxidase: role in cardiovascularbiology and disease. Circ Res 86:494–501.
20. Fenton H. 1894. Oxidation of tartaric acid in the presence of iron. J Chem Soc:899.
21. Haber F, Weiss J. 1934. The catalytic decomposition of hydrogen peroxide by iron salts. Proc R Soc A:332.
22. Suzuki YJ, Ford GD. 1999. Redox regulation of signal transduction in cardiac and smooth muscle. J Mol Cell Cardiol 31:345–353.
23. Griendling KK, Minieri CA, Ollerenshaw JD, Alexander RW. 1994. Angiotensin II stimulates NADH and NADPH oxidase activity in cultured vascular smooth muscle cells. Circ Res 74: 1141–1148.
24. Mohazzab KM, Kaminski PM, Wolin MS. 1994. NADH oxidoreductase is a major source of superoxide anion in bovine coronary artery endothelium. Am J Physiol 266:2568–2572.
25. Pagano PJ, Ito Y, Tornheim K, Gallop PM, Tauber AI,.Cohen RA. 1995. An NADPH oxidase superoxide-generating system in the rabbit aorta. Am J Physiol 268:2274–2280.
26. Boveris A. 1984. Determination of the production of the superoxide radicals and hydrogen peroxide in mitochondria. Methods Enzymol 105:429–435.
27. Melov S, Coskun P, Patel M, Tuinstra R, Cottrell B, Jun AS, Zastawny TH, Dizdaroglu M, Goodman SI, Huang TT, Miziorko H, Epstein CJ, Wallace DC. 1999. Mitochondrial disease in superoxide dismutase2 mutant mice. Proc Natl Acad Sci USA 96:846–851.
28. Melov S, Schneider JA, Day BJ, Hinerfeld D, Coskun P, Mirra SS, Crapo JD, Wallace DC. 1998. A novel neurological phenotype in mice lacking mitochondrial manganese sueroxide dismutase. Natl Genet 18:159–163.
29. Kuppusamy P, Zweier JL. 1989. Characterization of free radical generation by xanthine oxidase. Evidence for hydroxyl radical generation. J Biol Chem 264:9880–9884.
30. Kim KS, Takeda K, Sethi R, Pracyk JB, Tanaka K, Zhou YF, Yu ZX, Ferrans VJ, Bruder JT, Kovesdi I, Irani K, Goldschmidt-Clermont P, Finkel T. 1998. Protection from reoxygenation injury by inhibition of rac1. J Clin Invest 101:1821–1826.
31. Li PF, Dietz R, von Harsdorf R. 1999. Superoxide induces apoptosis in cardiomyocytes, but proliferation and expression of transforming growth factor-beta1 in cardiac fibroblasts. FEBS Lett 448:206–210.
32. Kukreja RC, Hess ML. 1992. The oxygen free radical system: from equations through membrane protein interactions to cardiovascular injury and protection. Cardiovasc Res 26:641–655.
33. Opie L. 1992. Cardiac metabolism: emergence, decline and resurgence. Part II. Cardiovasc Res 26:817–830.
34. Hess ML, Okabe E, Ash P, Kontos HA. 1984. Free radical mediation of the effects of acidosis on the calcium transport by cardiac sarcoplasmic reticulum in whole heart homogenates. Cardiovasc Res 18:149–157.
35. Okabe E, Hess ML, Oyama M, Ito H. 1983. Characterization of the free radical mediated damage to canine cardiac sarcoplasmic reticulum. Arch Biochem Biophys 225:164–177.
36. Okabe E, Kuse K, Sekishita T, Suyama N, Tanaka K, Ito H. 1991. The effect of ryanodine on oxygen free radical induced dysfunction of cardiac sarcoplasmic reticulum. J Pharmacol Exp Ther 256:868–875.
37. Okabe E, Odajima C, Taga R, Kukreja RC, Hess ML, Ito H. 1988. The effect of oxygen free radical on calcium permeability and calcium loading at steady state in cardiac sarcoplasmic reticulum. Mol Pharmacol 34:388–394.
38. Midori K, Okabe E. 1998. Superoxide anion radical-triggered Ca^{2+} release from the cardiac sarcoplasmic reticulum through ryanodine receptor Ca^{2+} channel. Mol Pharmacol 53:497–503.
39. Temsah RM, Netticadan T, Chapman D, Takeda S, Mochizuki S, Dhalla NS. 1999. Alterations in sarcoplasmic reticulum function and gene expression in ischemic-reperfused rat heart. Am J Physiol 277:584–594.
40. Netticadan T, Temsah R, Osada M, Dhalla NS. 1999. Status of Ca2+/calmodulin protein kinase phosphorylation of cardiac SR proteins in ischemia-reperfusion. Am J Physiol 277:384–391.
41. Beckman JS, Beckman TW, Chen J, Marshall PA, Freeman BA. 1990. Apparent hydroxyl radical production by peroxynitrite: implications for endothelial injury from nitric oxide and superoxide. Proc Natl Acad Sci USA 87:1620–1624.

42. Beckman JS, Koppenol WH. 1996. Nitric oxide, superoxide, and peroxynitrite: the good, the bad, and ugly. Am J Physiol 271:1424–1437.
43. Pryor WA, Squadrito GL. 1995. The chemistry of peroxynitrite: a product from the reaction of nitric oxide with superoxide. Am J Physiol 268:699–722.
44. Radi R, Beckman JS, Bush KM, Freeman BA. 1991. Peroxynitrite-induced membrane lipid peroxidation: the cytotoxic potential of superoxide and nitric oxide. Arch Biochem Biophys 288: 481–487.
45. Lin KT, Xue JY, Sun FF, Wong PY. 1997. Reactive oxygen species participate in peroxynitrite-induced apoptosis in HL-60 cells. Biochem Biophys Res Commun 230:115–119.
46. Van der Vliet A, Smith D, O'Neill CA, Kaur H, Darley-Usmar V, Cross CE, Halliwell B. 1994. Interactions of peroxynitrite with human plasma and its constituents: oxidative damage and antioxidant depletion. Biochem J 303:295–301.
47. Ma XL, Lopez BL, Liu GL, Christopher TA, Ischiropoulos H. 1997. Peroxynitrite aggravates myocardial reperfusion injury in the isolated perfused rat heart. Cardiovasc Res 36:195–204.
48. Squadrito GL, Pryor WA. 1998. The nature of reactive species in systems that produce peroxynitrite. Chem Res Toxicol 11:718–719.
49. Dinerman JL, Lowenstein CJ, Snyder SH. 1993. Molecular mechanisms of nitric oxide regulation. Potential relevance to cardiovascular disease. Circ Res 73:217–222.
50. Liu P, Hock CE, Nagele R, Wong PY. 1997. Formation of nitric oxide, superoxide, and peroxynitrite in myocardial ischemia-reperfusion injury in rats. Am J Physiol 272:2327–2336.
51. Yasmin W, Strynadka KD, Schulz R. 1997. Generation of peroxynitrite contributes to ischemia-reperfusion injury in isolated rat hearts. Cardiovasc Res 33:422–432.
52. Soszynski M, Bartosz G. 1996. Effect of peroxynitrite on erythrocytes. Biochim Biophys Acta 1291:107–114.
53. Kondo H, Takahashi M, Niki E. 1997. Peroxynitrite-induced hemolysis of human erythrocytes and its inhibition by antioxidants. FEBS Lett 413:236–238.
54. Lin KT, Xue JY, Wong PY. 1997. Peroxynitrite. An apoptotic agent in HL-60 cells. Adv Exp Med Biol 407:413–419.
55. Cheung PY, Danial H, Jong J, Schulz R. 1998. Thiols protect the inhibition of myocardial aconitase by peroxynitrite. Arch Biochem Biophys 350:104–108.
56. Suzuki S, Kaneko M, Chapman DC, Dhalla NS. 1991. Alterations in cardiac contractile proteins due to oxygen free radicals. Biochim Biophys Acta 1074:95–100.
57. Ferrari R, Ceconi C, Curello S, Guarnieri C, Caldarera CM, Albertini A, Visioli O. 1985. Oxygen-mediated myocardial damage during ischemia and reperfusion: role of the cellular defenses against oxygen toxicity. J Mol Cell Cardiol 17:937–945.
58. Bauer SF, Schwarz K, Ruegg JC. 1989. Glutathione alters calcium responsiveness of cardiac skinned fibers. Basic Res Cardiol 84:591–596.
59. Grinwald PM. 1982. Calcium uptake during post-ischemic reperfusion in the isolated rat heart: influence of extracellular sodium. J Mol Cell Cardiol 14:359–365.
60. Grinwald PM. 1992. Sodium pump failure in hypoxia and reoxygenation.J Mol Cell Cardiol 24:1393–1398.
61. Kato K, Shao Q, Elimban V, Lukas A, Dhalla NS. 1998. Mechanism of depression in cardiac sarcolemmal Na^+-K^+-ATPase by hypochlorous acid. Am J Physiol 275:826–831.
62. Carrozza JP Jr, Bentivegna LA, Williams CP, Kuntz RE, Grossman W, Morgan JP. 1992. Decreased myofilament responsiveness in myocardial stunning follows transient calcium overload during ischemia and reperfusion. Circ Res 71:1334–1340.
63. Gao WD, Atar D, Liu Y, Perez NG, Murphy AM, Marban E. 1997. Role of troponin I proteolysis in the pathogenesis of stunned myocardium. Circ Res 80:393–399.
64. Leem CH, Lagadic-Gossmann D, Vaughan-Jones RD. 1999. Characterization of intracellular pH regulation in the guinea-pig ventricular myocyte. J Physiol 517:159–180.
65. Wakabayashi S, Shigekawa M, Pouyssegur J. 1997. Molecular physiology of vertebrate Na+/H+ exchangers. Physiol Rev 77:51–74.
66. Aronson PS. 1985. Kinetic properties of the plasma membrane Na+-H+ exchanger. Annu Rev Physiol 47:545–560.
67. Counillon L, Pouyssegur J. 1995. Structure-function studies and molecular regulation of the growth factor activatable sodium-hydrogen exchanger (NHE-1). Cardiovasc Res 29:147–154.
68. Wu ML, Vaughan-Jones RD. 1997. Interaction between Na+ and H+ ions on Na-H exchange in sheep cardiac Purkinje fibers. J Mol Cell Cardiol 29:1131–1140.

69. Sabri A, Byron KL, Samarel AM, Bell J, Lucchesi PA. 1998. Hydrogen peroxide activates mitogen-activated protein kinases and Na+-H+ exchange in neonatal rat cardiac myocytes. Circ Res 82: 1053–1062.
70. Hoque AN, Haist JV, Karmazyn M. 1997. Na(+)-H+ exchange inhibition protects against mechanical, ultrastructural, and biochemical impairment induced by low concentrations of lysophosphatidylcholine in isolated rat hearts. Circ Res 80:95–102.
71. Frelin C, Vigne P, Lazdunski M. 1984. The role of the Na+/H+ exchange system in cardiac cells in relation to the control of the internal Na+ concentration. A molecular basis for the antagonistic effect of ouabain and amiloride on the heart. J Biol Chem 259:8880–8885.
72. Karmazyn M, Gan XT, Humphreys RA, Yoshida H, Kusomoto K. 1999. The myocardial Na(+)-H(+) exchange: structure, regulation, and its role in heart disease. Circ Res 85:777–786.
73. Griese M, Perlitz V, Jungling E, Kammermeier H. 1988. Myocardial performance and free energy of ATP-hydrolysis in isolated rat hearts during graded hypoxia, reoxygenation and high K^+-perfusion. J Mol Cell Cardiol 20:1189–1201.
74. Kim D, Cragoe EJ Jr, Smith TW. 1987. Relations among sodium pump inhibition, Na-Ca and Na-H exchange activities, and Ca-H interaction in cultured chick heart cells. Circ Res 60: 185–193.
75. Duan J, Moffat MP. 1992. Contractile and electrophysiologic effects of realkalization in cardiac tissues: role of Na/H exchange and increased [Ca]i. Adv Exp Med Biol 311:435–436.
76. Cross HR, Lu L, Steenbergen C, Philipson KD, Murphy E. 1998. Overexpression of the cardiac Na+/Ca2+ exchanger increases susceptibility to ischemia/reperfusion injury in male, but not female, transgenic mice. Circ Res 83:1215–1223.
77. Egger M, Niggli E. 1999. Regulatory function of Na-Ca exchange in the heart: milestones and outlook. J Membr Biol 168:107–130.
78. Pierce GN, Meng H. 1992. The role of sodium-proton exchange in ischemic/reperfusion injury in the heart. Na(+)-H+ exchange and ischemic heart disease. Am J Cardiovasc Pathol 4:91–102.
79. Kusuoka H, Camilion de Hurtado MC, Marban E. 1993. Role of sodium/calcium exchange in the mechanism of myocardial stunning: protective effect of reperfusion with high sodium solution. J Am Coll Cardiol 21:240–248.
80. Ladilov Y, Haffner S, Balser-Schafer C, Maxeiner H, Piper HM. 1999. Cardioprotective effects of KB-R7943: a novel inhibitor of the reverse mode of Na+/Ca2+ exchanger. Am J Physiol 276: 68–76.
81. Lochner A, Genade S, Tromp E, Theron S, Trollip G. 1998. Postcardioplegic myocardial recovery: effects of halothane, nifedipine, HOE 694, and quinacrine. Cardiovasc Drugs Ther 12:267–277.
82. Murry CE, Jennings RB, Reimer KA. 1986. Preconditioning with ischemia: a delay of lethal cell injury in ischemic myocardium. Circulation 74:1124–1136.
83. Murry CE, Richard VJ, Jennings RB, Reimer KA. 1991. Myocardial protection is lost before contractile function recovers from ischemic preconditioning. Am J Physiol 260:796–804.
84. Birnbaum Y, Hale SL, Kloner RA. 1996. Progressive decrease in the ST segment elevation during ischemic preconditioning: is it related to recruitment of collateral vessels? J Mol Cell Cardiol 28:1493–1499.
85. Sebbag L, Brooks SE, Jennings RB. 1997. Cardioprotection, lost 3 hours after ischemic preconditioning (PC) can be fully restored by repeated preconditioning in dogs: significance of ST-segment changes for detecting cardioprotection. Eur Heart J 18:162.
86. Kuzuya T, Hoshida S, Yamashita N, Fuji H, Oe H, Hori M, Kamada T, Tada M. 1993. Delayed effects of sublethal ischemia on the acquisition of tolerance to ischemia. Circ Res 72:1293–1299.
87. Marber MS, Latchman DS, Walker JM, Yellon DM. 1993. Cardiac stress protein elevation 24 hours after brief ischemia or heat stress is associated with resistance to myocardial infarction. Circulation 88:1264–1272.
88. Tang XL, Qiu Y, Park SW, Sun JZ, Kalya A, Bolli R. 1996. Time course of late preconditioning against myocardial stunning in conscious pigs. Circ Res 79:424–434.
89. Bolli R. 2000. The late phase of preconditioning. Circ Res 87:972–983.
90. Jennings RB, Sebbag L, Schwartz LM, Crago MS, Reimer KA. 2001. Metabolism of preconditioned myocardium: effect of loss and reinstatement of cardioprotection. J Mol Cell Cardiol 33: 1571–1588.
91. Murry CE, Richard VJ, Reimer KA, Jennings RB. 1990. Ischemic preconditioning slows energy metabolism and delays ultrastructural damage during a sustained ischemic episode. Circ Res 66: 913–931.

92. Qian YZ, Shipley JB, Levasseur JE, Kukreja RC. 1998. Dissociation of heat shock proteins expression with ischemic tolerance by whole body hyperthermia in rat heart. J Mol Cell Cardiol 30: 1163–1172.
93. Bolli R, Bhatti ZA, Tang XL, Qiu Y, Zhang Q, Guo Y, Jadoon AK. 1997. Evidence that late preconditioning against myocardial stunning in conscious rabbits is triggered by the generation of nitric oxide. Circ Res 81:42–52.
94. Drury AN, Szent-Gyorgyi A. 1929. The physiological activity of adenine compound with special reference to their action upon the mammalian heart. J Physiol (Lond) 68:213–237.
95. Berne RM. 1963. Cardiac nucleotides in hypoxia: Possible role in regulation of coronary blood flow. Am J Physiol 204:317–332.
96. Baxter GF, Marber MS, Patel VC, Yellon DM. 1994. Adenosine receptor involvement in a delayed phase of myocardial protection 24 hours after ischemic preconditioning. Circulation 90:2993–3000.
97. Liu GS, Thornton J, Van Winkle DM, Stanley AW, Olsson RA, Downey JM. 1991. Protection against infarction afforded by preconditioning is mediated by A1 adenosine receptors in rabbit heart. Circulation 84:350–356.
98. Grover GJ, Sleph PG, Dzwonczyk S. 1992. Role of myocardial ATPsensitive potassium channels in mediating preconditioning in dog heart and their possible interaction with adenosine A1 receptors. Circulation 86:1310–1316.
99. Bernardo NL, D'Angelo M, Okubo S, Joy A, Kukreja RC. 1999. Delayed ischemic preconditioning is mediated by opening of ATP-sensitive potassium channels in the rabbit heart. Am J Physiol 276:1323–1330.
100. Takano H, Bolli R, Black RG Jr, Kodani E, Tang XL, Yang Z, Bhattacharya S, Auchampach JA. 2001. A(1) or A(3) adenosine receptors induce late preconditioning against infarction in conscious rabbits by different mechanisms. Circ Res 88:520–528.
101. Fryer RM, Hsu AK, Eells JT, Nagase H, Gross GJ. 1999. Opioid-induced second window of cardioprotection: potential role of mitochondrial KATP channels. Circ Res 84:846–851.
102. Hoag JB, Qian YZ, Nayeem MA, D'Angelo M, Kukreja RC. 1997. ATP-sensitive potassium channel mediates delayed ischemic protection by heat stress in rabbit heart. Am J Physiol 273: 2458–2464.
103. Garlid KD, Paucek P, Yarov-Yarovoy V, Murray HN, Darbenzio RB, D'Alonzo AJ, Lodge NJ, Smith MA, Grover GJ. 1997. Cardioprotective effect of diazoxide and its interaction with mitochondrial ATP-sensitive K+ channels. Possible mechanism of cardioprotection. Circ Res 81:1072–1082.
104. Auchampach JA, Grover GJ, Gross GJ. 1992. Blockade of ischaemic preconditioning in dogs by the novel ATP dependent potassium channel antagonist sodium 5-hydroxydecanoate. Cardiovasc Res 26:1054–1062.
105. Gross GJ, Auchampach JA. 1992. Blockade of ATP-sensitive potassium channels prevents myocardial preconditioning in dogs. Circ Res 70:223–233.
106. Tomai F, Crea F, Gaspardone A, Versaci F, De Paulis R, Penta de Peppo A, Chiariello L, Gioffre PA. 1994. Ischemic preconditioning during coronary angioplasty is prevented by glibenclamide, a selective ATP-sensitive K+ channel blocker. Circulation 90:700–705.
107. Kirsch GE, Codina J, Birnbaumer L, Brown AM. 1990. Coupling of ATP-sensitive K+ channels to A1 receptors by G proteins in rat ventricular myocytes. Am J Physiol 259:820–826.
108. Hu K, Duan D, Li GR, Nattel S. 1996. Protein kinase C activates ATP-sensitive K+ current in human and rabbit ventricular myocytes. Circ Res 78:492–498.
109. Speechly-Dick ME, Grover GJ, Yellon DM. 1995. Does ischemic preconditioning in the human involve protein kinase C and the ATP-dependent K+ channel? Studies of contractile function after simulated ischemia in an atrial in vitro model. Circ Res 77:1030–1035.
110. Yamaguchi F, Nasa Y, Yabe K, Ohba S, Hashizume Y, Ohaku H, Furuhama K, Takeo S. 1997. Activation of cardiac muscarinic receptor and ischemic preconditioning effects in situ rat heart. Heart Vessels 12:74–83.
111. Iliodromitis EK, Papadopoulos CC, Markianos M, Paraskevaidis IA, Kyriakides ZS, Kremastinos DT. 1996. Alterations in circulating cyclic guanosine monophosphate (c-GMP) during short and long ischemia in preconditioning. Basic Res Cardiol 91:234–239.
112. Han J, Kim N, Kim E, Ho WK, Earm YE. 2001. Modulation of ATP-sensitive potassium channels by cGMP-dependent protein kinase in rabbit ventricular myocytes. J Biol Chem 276:22140–22147.
113. Noma A. 1983. ATP-regulated K+ channels in cardiac muscle. Nature 305:147–148.
114. D'Alonzo AJ, Darbenzio RB, Parham CS, Grover GJ. 1992. Effects of intracoronary cromakalim on postischaemic contractile function and action potential duration. Cardiovasc Res 26:1046–1053.

115. Shaw RM, Rudy Y. 1997. Electrophysiologic effects of acute myocardial ischemia: a theoretical study of altered cell excitability and action potential duration. Cardiovasc Res 35:256–272.
116. Cole WC, McPherson CD, Sontag D. 1991. ATP-regulated K+ channels protect the myocardium against ischemia/reperfusion damage. Circ Res 69:571–581.
117. Grover GJ, Newburger J, Sleph PG, Dzwonczyk S, Taylor SC, Ahmed SZ, Atwal KS. 1991. Cardioprotective effects of the potassium channel opener cromakalim: stereoselectivity and effects on myocardial adenine nucleotides. J Pharmacol Exp Ther 257:156–162.
118. McPherson CD, Pierce GN, Cole WC. 1993. Ischemic cardioprotection by ATP-sensitive K+ channels involves high-energy phosphate preservation. Am J Physiol 265:1809–1818.
119. Pignac J, Bourgouin J, Dumont L. 1994. Cold cardioplegia and the K+ channel modulator aprikalim (RP 52891): improved cardioprotection in isolated ischemic rabbit hearts. Can J Physiol Pharmacol 72:126–132.
120. Yao Z, Gross GJ. 1994. Effects of the KATP channel opener bimakalim on coronary blood flow, monophasic action potential duration, and infarct size in dogs. Circulation 89:1769–1775.
121. Grover GJ, D'Alonzo AJ, Parham CS, Darbenzio RB. 1995. Cardioprotection with the KATP opener cromakalim is not correlated with ischemic myocardial action potential duration. J Cardiovasc Pharmacol 26:145–152.
122. Hamada K, Yamazaki J, Nagao T. 1998. Shortening of action potential duration is not prerequisite for cardiac protection by ischemic preconditioning or a KATP channel opener. J Mol Cell Cardiol 30:1369–1379.
123. Jovanovic N, Jovanovic S, Jovanovic A, Terzic A. 1999. Gene delivery of Kir6.2/SUR2A in conjunction with pinacidil handles intracellular Ca2+ homeostasis under metabolic stress. FASEB J 13:923–929.
124. Miki T, Tashiro F, Iwanaga T, Nagashima K, Yoshitomi H, Aihara H, Nitta Y, Gonoi T, Inagaki N, Miyazaki J, Seino S. 1997. Abnormalities of pancreatic islets by targeted expression of a dominant-negative KATP channel. Proc Natl Acad Sci USA 94:11969–11973.
125. Baxter GF, Goma FM, Yellon DM. 1995. Involvement of protein kinase C in the delayed cytoprotection following sublethal ischaemia in rabbit myocardium. Br J Pharmacol 115:222–224.
126. Ping P, Zhang J, Qiu Y, Tang XL, Manchikalapudi S, Cao X, Bolli R. 1997. Ischemic preconditioning induces selective translocation of protein kinase C isoforms epsilon and eta in the heart of conscious rabbits without subcellular redistribution of total protein kinase C activity. Circ Res 81: 404–414.
127. Qiu Y, Ping P, Tang XL, Manchikalapudi S, Rizvi A, Zhang J, Takano H, Wu WJ, Teschner S, Bolli R. 1998. Direct evidence that protein kinase C plays an essential role in the development of late preconditioning against myocardial stunning in conscious rabbits and that epsilon is the isoform involved. J Clin Invest 101:2182–2198.
128. Ping P, Takano H, Zhang J, Tang XL, Qiu Y, Li RC, Banerjee S, Dawn B, Balafonova Z, Bolli R. 1999. Isoform-selective activation of protein kinase C by nitric oxide in the heart of conscious rabbits: a signaling mechanism for both nitric oxide-induced and ischemia-induced preconditioning. Circ Res 84:587–604.
129. Takano H, Tang XL, Qiu Y, Guo Y, French BA, Bolli R. 1998. Nitric oxide donors induce late preconditioning against myocardial stunning and infarction in conscious rabbits via an antioxidant-sensitive mechanism. Circ Res 83:73–84.
130. Dawn B, Xuan YT, Qiu Y, Takano H, Tang XL, Ping P, Banerjee S, Hill M, Bolli R. 1999. Bifunctional role of protein tyrosine kinases in late preconditioning against myocardial stunning in conscious rabbits. Circ Res 85:1154–1163.
131. Ping P, Zhang J, Zheng YT, Li RC, Dawn B, Tang XL, Takano H, Balafanova Z, Bolli R. 1999. Demonstration of selective protein kinase C-dependent activation of Src and Lck tyrosine kinases during ischemic preconditioning in conscious rabbits. Circ Res 85:542–550.
132. Holaday JW. 1983. Cardiovascular effects of endogenous opiate systems. Annu Rev Pharmacol Toxicol 23:541–594.
133. Liang CS, Imai N, Stone CK, Woolf PD, Kawashima S, Tuttle RR. 1987. The role of endogenous opioids in congestive heart failure: effects of nalmefene on systemic and regional hemodynamics in dogs. Circulation 75:443–451.
134. Lee AY, Chen YT, Kan MN, P'eng FK, Chai CY, Kuo JS. 1992. Consequences of opiate agonist and antagonist in myocardial ischaemia suggest a role of endogenous opioid peptides in ischaemic heart disease. Cardiovasc Res 26:392–395.

135. Maslov LN, Krylatov AV, Lishmanov YB. 1996. On the involvement of endogenous opiate receptor agonists in the antiarrhythmic effects of adaptation. Bull Exp Biol Med 121:20–21.
136. Schultz JE, Rose E, Yao Z, Gross GJ. 1995. Evidence for involvement of opioid receptors in ischemic preconditioning in rat hearts. Am J Physiol 268:2157–2161.
137. Schultz JJ, Hsu AK, Gross GJ. 1997. Ischemic preconditioning is mediated by a peripheral opioid receptor mechanism in the intact rat heart. J Mol Cell Cardiol 29:1355–1362.
138. Liang BT, Gross GJ. 1999. Direct preconditioning of cardiac myocytes via opioid receptors and KATP channels. Circ Res 84:1396–1400.
139. Tang XL, Takano H, Rizvi A, Turrens JF, Qiu Y, Wu WJ, Zhang Q, Bolli R. 2002. Oxidant species trigger late preconditioning against myocardial stunning in conscious rabbits. Am J Physiol Heart Circ Physiol 282:281–291.
140. Sun JZ, Tang XL, Park SW, Qiu Y, Turrens JF, Bolli R. 1996. Evidence for an essential role of reactive oxygen species in the genesis of late preconditioning against myocardial stunning in conscious pigs. J Clin Invest 97:562–576.
141. Yamashita N, Hoshida S, Taniguchi N, Kuzuya T, Hori M. 1998. A "second window of protection" occurs 24h after ischemic preconditioning in the rat heart. J Mol Cell Cardiol 30:1181–1189.
142. Tang XL, Takano H, Rizvi A, Turrens JF, Qiu Y, Wu WJ, Zhang Q, Bolli R 2002. Oxidant species trigger late preconditioning against myocardial stunning in conscious rabbits. Am J Physiol Heart Circ Physiol 282:281–291.
143. Maldonado C, Qiu Y, Tang XL, Cohen MV, Auchampach J, Bolli R. 1997. Role of adenosine receptors in late preconditioning against myocardial stunning in conscious rabbits. Am J Physiol 273: 1324–1332.
144. Bolli R, Manchikalapudi S, Tang XL, Takano H, Qiu Y, Guo Y, Zhang Q, Jadoon AK. 1997. The protective effect of late preconditioning against myocardial stunning in conscious rabbits is mediated by nitric oxide synthase. Evidence that nitric oxide acts both as a trigger and as a mediator of the late phase of ischemic preconditioning. Circ Res 81:1094–1107.
145. Takano H, Manchikalapudi S, Tang XL, Qiu Y, Rizvi A, Jadoon AK, Zhang Q, Bolli R. 1998. Nitric oxide synthase is the mediator of late preconditioning against myocardial infarction in conscious rabbits. Circulation 98:441–449.
146. Guo Y, Jones WK, Xuan YT, Tang XL, Bao W, Wu WJ, Han H, Laubach VE, Ping P, Yang Z, Qiu Y, Bolli R. 1999. The late phase of ischemic preconditioning is abrogated by targeted disruption of the inducible NO synthase gene. Proc Natl Acad Sci USA 96:11507–11512.
147. Leesar MA, Stoddard MF, Dawn B, Jasti VG, Masden R, Bolli R. 2001. Delayed preconditioning-mimetic action of nitroglycerin in patients undergoing coronary angioplasty. Circulation 103: 2935–2941.
148. Gross SS, Wolin MS. 1995. Nitric oxide: pathophysiological mechanisms. Annu Rev Physiol 57: 737–769.
149. Kelly RA, Balligand JL, Smith TW. 1996. Nitric oxide and cardiac function. Circ Res 79:363–380.
150. Takano H, Tang XL, Qui Y, Bolli R. 1998. The tetrahydrobiopterin synthesis inhibitor N-acetylsertotonin abrogates late preconditioning against myocardial stunning in conscious rabbits. Circulation (suppl.I): I-586. Abstract.
151. Chen Z, Siu B, Ho YS, Vincent R, Chua CC, Hamdy RC, Chua BH. 1998. Overexpression of MnSOD protects against myocardial ischemia/reperfusion injury in transgenic mice. J Mol Cell Cardiol 30:2281–2289.
152. Ibrahim W, Lee US, Yen HC, St Clair DK, Chow CK. 2000. Antioxidant and oxidative status in tissues of manganese superoxide dismutase transgenic mice. Free Radic Biol Med 28:397–402.
153. Myers ML, Bolli R, Lekich RF, Hartley CJ, Roberts R. 1985. Enhancement of recovery of myocardial function by oxygen free-radical scavengers after reversible regional ischemia. Circulation 72:915–921.
154. Przyklenk K, Kloner RA. 1986. Superoxide dismutase plus catalase improve contractile function in the canine model of the "stunned myocardium". Circ Res 58:148–156.
155. Gross GJ, Farber NE, Hardman HF, Warltier DC. 1986. Beneficial actions of superoxide dismutase and catalase in stunned myocardium of dogs. Am J Physiol 250:372–377.
156. Nejima J, Knight DR, Fallon JT, Uemura N, Manders WT, Canfield DR, Cohen MV, Vatner SF. 1989. Superoxide dismutase reduces reperfusion arrhythmias but fails to salvage regional function or myocardium at risk in conscious dogs. Circulation 79:143–153.

157. Mehta JL, Nichols WW, Saldeen TG, Chandna VK, Nicolini FA, Lawson DL, ter Riet MF. 1990. Superoxide dismutase decreases reperfusion arrhythmias and preserves myocardial function during thrombolysis with tissue plasminogen activator. J Cardiovasc Pharmacol 16:112–120.
158. Bolli R, Jeroudi MO, Patel BS, Aruoma OI, Halliwell B, Lai EK, McCay PB. 1989. Marked reduction of free radical generation and contractile dysfunction by antioxidant therapy begun at the time of reperfusion. Evidence that myocardial "stunning" is a manifestation of reperfusion injury. Circ Res 65:607–622.
159. Bolli R, Zhu WX, Hartley CJ, Michael LH, Repine JE, Hess ML, Kukreja RC, Roberts R. 1987. Attenuation of dysfunction in the postischemic 'stunned' myocardium by dimethylthiourea. Circulation 76:458–468.
160. Myers ML, Bolli R, Lekich RF, Hartley CJ, Roberts R. N-2-mercaptopropionylglycine improves recovery of myocardial function after reversible regional ischemia. 1986. J Am Coll Cardiol 8: 1161–1168.
161. Altavilla D, Deodato B, Campo GM, Arlotta M, Miano M, Squadrito G, Saitta A, Cucinotta D, Ceccarelli S, Ferlito M, Tringali M, Minutoli L, Caputi AP, Squadrito F. 2000. IRFI 042, a novel dual vitamin E-like antioxidant, inhibits activation of nuclear factor-kappaB and reduces the inflammatory response in myocardial ischemia-reperfusion injury. Cardiovasc Res 47:515–528.
162. Shimizu M, Wang QD, Sjoquist PO, Ryden L. 1998. The lipid peroxidation inhibitor indenoindole H290/51 protects myocardium at risk of injury induced by ischemia-reperfusion. Free Radic Biol Med 24:726–731.
163. Pintus G, Pinna GG, Ventura C. 1994. Evaluation of opioid peptide gene expression by solution hybridization RNase protection. Boll Soc Ital Biol Sper 70:213–219.
164. Weil J, Eschenhagen T, Fleige G, Mittmann C, Orthey E, Scholz H. 1998. Localization of preproenkephalin mRNA in rat heart: selective gene expression in left ventricular myocardium. Am J Physiol 275:378–384.
165. Paterson SJ, Robson LE, Kosterlitz HW. 1983. Classification of opioid receptors. Br Med Bull 39: 31–36.
166. Dhawan BN, Cesselin F, Raghubir R, Reisine T, Bradley PB, Portoghese PS, Hamon M. 1996. International Union of Pharmacology. XII. Classification of opioid receptors. Pharmacol Rev 48: 567–592.
167. Ventura C, Spurgeon H, Lakatta EG, Guarnieri C, Capogrossi MC. 1992. Kappa and delta opioid receptor stimulation affects cardiac myocyte function and Ca2+ release from an intracellular pool in myocytes and neurons. Circ Res 70:66–81.
168. Wittert G, Hope P, Pyle D. 1996. Tissue distribution of opioid receptor gene expression in the rat. Biochem Biophys Res Commun 218:877–881.
169. Takasaki Y, Wolff RA, Chien GL, van Winkle DM. 1999. Met5-enkephalin protects isolated adult rabbit cardiomyocytes via delta-opioid receptors. Am J Physiol 277:2442–2450.
170. Ventura C, Maioli M, Pintus G, Posadino AM, Tadolini B. 1998. Nuclear opioid receptors activate opioid peptide gene transcription in isolated myocardial nuclei. J Biol Chem 273:13383–13386.
171. Fryer RM, Hsu AK, Eells JT, Nagase H, Gross GJ. 1999. Opioid-induced second window of cardioprotection: potential role of mitochondrial KATP channels. Circ Res 84:846–851.
172. Bolling SF, Su TP, Childs KF, Ning XH, Horton N, Kilgore K, Oeltgen PR. 1997. The use of hibernation induction triggers for cardiac transplant preservation. Transplantation 63:326–329.
173. Hjalmarson A. 1994. Empiric therapy with beta-blockers. Pacing Clin Electrophysiol 17:460–466.
174. Feuerstein GZ, Ruffolo RR Jr. 1996. Carvedilol, a novel vasodilating beta-blocker with the potential for cardiovascular organ protection. Eur Heart J 17 Suppl B:24–29.
175. Hamburger SA, Barone FC, Feuerstein GZ, Ruffolo RR Jr. 1991. Carvedilol (Kredex) reduces infarct size in a canine model of acute myocardial infarction. Pharmacology 43:113–120.
176. Bril A, Slivjak M, DiMartino MJ, Feuerstein GZ, Linee P, Poyser RH, Ruffolo RR Jr, Smith EF 3rd. 1992. Cardioprotective effects of carvedilol, a novel beta adrenoceptor antagonist with vasodilating properties, in anaesthetised minipigs: comparison with propranolol. Cardiovasc Res 26: 518–525.
177. Brunvand H, Kvitting PM, Rynning SE, Berge RK, Grong K. 1996. Carvedilol protects against lethal reperfusion injury through antiadrenergic mechanisms. J Cardiovasc Pharmacol 28:409–417.
178. Ruffolo RR Jr, Feuerstein GZ. 1997. Pharmacology of carvedilol: rationale for use in hypertension, coronary artery disease, and congestive heart failure. Cardiovasc Drugs Ther 11 Supp:247–256.
179. Sorescu D, Griendling KK. 2002. Reactive oxygen species, mitochondria, and NAD(P)H oxidases in the development and progression of heart failure. Congest Heart Fail 8:132–140.

180. Shimizu M, Wang QD, Sjoquist PO, Ryden L. 1998. Angiotensin II type 1 receptor blockade with candesartan protects the porcine myocardium from reperfusion-induced injury. J Cardiovasc Pharmacol 32:231–238.
181. Smith WH, Ball SG. 2000. ACE inhibitors in heart failure: an update.Basic Res Cardiol 95 Suppl 1:18–14.
182. Pfeffer MA. 2000. Enhancing cardiac protection after myocardial infarction: rationale for newer clinical trials of angiotensin receptor blockers. Am Heart J 139:23–28.
183. Yoshiyama M, Kim S, Yamagishi H, Omura T, Tani T, Yanagi S, Toda I, Teragaki M, Akioka K, Takeuchi K, et al. 1994. Cardioprotective effects of the angiotensin II type I receptor antagonist TCV-116 on ischemia-reperfusion injury. Am Heart J 23:919–923.
184. Lazar HL, Bao Y, Rivers S, Bernard SA. 2002. Pretreatment with angiotensin-converting enzyme inhibitors attenuates ischemia-reperfusion injury. Ann Thorac Surg 73:1522–1527.
185. Liu YH, Yang XP, Sharov VG, Nass O, Sabbah HN, Peterson E, Carretero OA. 1997. Effects of angiotensin-converting enzyme inhibitors and angiotensin II type 1 receptor antagonists in rats with heart failure. Role of kinins and angiotensin II type 2 receptors. J Clin Invest 99:1926–1935.
186. Jalowy A, Schulz R, Dorge H, Behrends M, Heusch G. 1998. Infarct size reduction by AT1-receptor blockade through a signal cascade of AT2-receptor activation, bradykinin and prostaglandins in pigs. J Am Coll Cardiol 32:1787–1796.

III. The Cellular Basis for Vascular Disease

Pathophysiology of Cardiovascular Disease,
edited by N.S. Dhalla, H. Rupp,
A. Angel & G.N. Pierce
Kluwer Academic Publishers, Boston, 2004

Modulation of Arachidonic Acid Metabolism by Phorbol 12-Myristate 13-Acetate in Monocytes and Myocytes

Biao Lu and Patrick C. Choy

Department of Biochemistry and Medical Genetics, Faculty of Medicine, University of Manitoba, Winnipeg, Manitoba, Canada. R3E 0W3

Summary. The regulation of arachidonate metabolism and eicosanoid production is controlled by the cytosolic phospholipase A_2 ($cPLA_2$) and the cyclooxygenases (COX). In this study, the regulation of the arachidonate-eicosanoid cascade in premoncytic cells (U937) and cardiac myocytes (H9c2) was investigated. Cells were treated with PMA, and the release in arachidonate and the production of PGE_2 were monitored. PMA caused an immediate enhancement of arachidonate release and an elevated PGE_2 production at a later time point. Under identical experimental conditions, the $cPLA_2$, COX-1 and COX-2 enzyme protein levels were analyzed by immunoblotting assay. In U937 cells, PMA caused a small increase in $cPLA_2$ protein, whereas an induction of COX-2 expression and an up-regulation of COX-1 expression were observed. In H9c2 cells, PMA treatment caused the up-regulation of all three enzymes, but the up-regulation of COX-1 was marginal when compared with COX-2. Our study clearly shows that PMA caused an increase in arachidonate release and the subsequent production of PGE_2 in both U937 and H9c2 cells. The action of PMA is mediated via the modulation of $cPLA_2$ and COX-2 expression in both cell types. Our results confirm the important roles of these two enzymes in the regulation of the arachidonate-eicosanoid cascade in both monocytes and cardiac myocytes.

Key words**:** arachidonate, eicosanoid, phospholipase A_2, cyclooxygenase, monocyte.

INTRODUCTION

Arachidonic acid (AA) is an essential fatty acid and an important precursor for the biosynthesis of bioactive lipid mediators. The vast majority of cellular AA is stored

Address for Correspondence: Dr. P.C. Choy, Department of Biochemistry and Medical Genetics, Faculty of Medicine, University of Manitoba, Winnipeg, Manitoba, Canada. R3E 0W3. Tel: (204)789-3375, Fax: (204)789-3942, E-mail: pchoy@ms.umanitoba.ca.

in acylated form in membrane phospholipids. The initial step for the synthesis of the AA metabolites is the release of the free fatty acid from membrane phospholipids by a group of hydrolytic enzymes (reviewed in 1–3). The principal hydrolytic enzyme for the release of AA is the cytosolic phospholipase A_2 ($cPLA_2$), which preferably hydrolyzes AA from the sn-2 position of phospholipids (1–3). The free AA by itself is an important second message in a variety of cellular processes (4–8). In addition, it can be converted to the eicosanoids (1–3) by the action of cyclooxygenases (COX) to produce prostaglandins and thromboxanes. Hence, both PLA_2 and COX appear to be key enzymes for the production of inflammatory mediators (reviewed in 9,10).

Mammalian organs contain structurally diverse forms of PLA_2. PLA_2 are classified into several major types including secretary PLA_2 ($sPLA_2$), calcium-independent PLA_2, and the 97 kDa cytosolic PLA_2 ($cPLA_2$). The $cPLA_2$ plays an important role in mediating AA release, because cells lacking $cPLA_2$ are generally devoid of eicosanoid synthesis, and $cPLA_2$ preferentially hydrolyzes sn-2 arachidonic acid (1–3). The COX has two isoforms: COX-1 and COX-2. The two enzymes catalyze the committed step in prostaglandin synthesis (11,12). COX-1 is constitutively expressed in most tissues, whereas COX-2 is inducible and subjected to transcriptional regulation upon varieties stimuli. Of interest, the human $cPLA_2$ and COX-2 genes are mapped to the same chromosome 1q25, and the promoters share some similarities in both 5′ and 3′ untranslated regions suggesting the possibility of coordinated regulation (12–14).

The regulation of AA metabolism is controlled by multiple interconnected pathways at multiple stages via different mechanisms, however, the detailed mechanisms and regulations have not been fully delineated. We hypothesized that the regulation of arachidonate-eicosanoid cascade is mediated through the modulation of the key enzymes. In this study, we investigated the modulation of AA metabolism by phorbol 12-mysristate 13-acetate (PMA) in two cultured cell models, the human premonocytic U937 cells and the rat heart myoblastic H9c2 cells. The purposes of this study are two-folds: (1) to study the effects of PMA on the production of AA and PGE_2; and (2) to study the tissue-specific expression and the coordinated regulation of the three key enzymes: $cPLA_2$, COX-1, and COX-2 by PMA.

MATERIALS AND METHODS

Materials

Dulbecco's modified Eagle's medium, PMRI-1640 medium, phosphate-buffered saline, phorbol 12-myristate 13-acetate (PMA), bovine serum albumin (BSA), and other chemicals were obtained from Sigma Chemical Company (St. Louis, MO). Polyclonal anti-COX-2 antibodies, monoclonal anti-$cPLA_2$ antibody, and polyclonal anti-Actin antibodies were purchased from Santa Cruz Biotechnology, Inc. (Santa Cruz, CA). Monoclonal anti-COX-1 antibody and PGE_2 EIA assay kit were purchased form Cayman Chemical Co. (Ann Arbor, MI). Ionomycin (a calcium ionophore) was obtained from Biomol Inc. (Plymouth Meeting, PA).

[5,6,8,9,11,12,14,15-^{3}H]-arachidonic acid (209 mCi/mmol) was obtained from Amersham Pharmacia Biotech (Buckinghamshire, England). All lipid standards were obtained from Serdary Research Laboratory (London, Ontario, Canada). Thin layer chromatograph plates (SIL G-25) were the products of Mecherey-Nagel (Postfach, Germany).

Cell culture and labeling

H9c2 cells, a rat heart myoblast cell line was obtained from the American Type Culture Collection. They were cultured in culture dishes with Dulbecco's modified Eagles medium containing 10% fetal bovine serum, 100 units/mL of penicillin G, 10 μg/mL of streptomycin, and 0.25 μg/mL amphotericin B. The cells were incubated at 37°C in an atmosphere of 95% humidified air and 5% carbon dioxide until 70% confluence was achieved. In some experiments, cells were switched to serum-free medium containing 0.1% BSA and incubated for 12 hours before the appropriate amount of PMA in vehicle or vehicle alone (0.01% dimethylsulfoxide) were added for the prescribed periods of time.

The human premonocytic cell line U-937 was obtained from the American Type Culture Collection. These cells were grown in cell culture dishes with RPMI-1640 medium containing 10% fetal bovine serum, 100 units/mL of penicillin G, 10 μg/mL of streptomycin, and 0.25 μg/mL amphotericin B. Cells were seeded in flasks at a density of 3×10^5/mL for 12 hours and radio-labeled with 1.0 μCi/mL [^{3}H] AA for 12 h in serum-free RPMI-1640 medium containing 0.1% bovine serum albumin. At the end of the labeling, cells were washed 3 times with the same medium to remove any nonspecific binding of AA to the cell surface. Subsequently, cells were incubated in the serum-free medium containing 0.1% BSA and the appropriate amount of PMA for the prescribed amount of time.

Measurement of AA-release

The AA released from the cells was determined as described previously (15,16). Briefly, cells treated with PMA or vehicle were stimulated with 5 μM calcium ionophore (Ionomycin, Biomol Research Labs., Inc.) for 15 min, and the labeled AA released into the medium was determined. The recovered medium was acidified with 50 μL of glacial acetic acid, an a 0.8-mL aliquot was resolved by thin layer chromatography in a solvent system consisting of hexane/diethyl ether/acetic acid (70:30:1, v/v). The fatty acid fraction on the thin-layer chromatographic plate was visualized by iodine vapor, and its radioactivity was determined by liquid scintillation counting.

Measurement of PGE_2 release

The effect of PMA on the production of PGE_2 in U937 and H9c2 cells were determined. Following a quiescent period in serum-free medium for 12 h, cells were treated with PMA for 0–72 h. After the treatment, the medium was collected to determine the endogenous production of PGE_2 by an enzyme immunoassay (EIA) using a commercial prostaglandin E_2 EIA Kit (Cayman Chemical). Briefly, 50 µL of the standard PGE_2 or samples was added to the pre-coated mouse anti-rabbit IgG macrotitre plates. Subsequently, 50 µL PGE_2 tracer and 50 µL monoclonal antiserum of PGE_2 were added into each well and the mixture was incubated for 18 h at 4°C. Following the incubation, the content in each well was removed and the wells were washed with PBS buffer containing 0.05% Tween-20 for 12 times. After wash, wells were incubated with 200 µL Ellman's reagent for 90 min at room temperature on a shaker, or until a yellow color developed. The absorbance of solution was measured by a mocroplate reader at $A_{412\,nm}$. PGE_2 concentrations were calculated from a standard curve.

Electrophoresis and Western blot analyses

Cells were collected and washed in ice-cooled PBS buffer and the cell pellet was obtained by a centrifugation at 1000 g for 10 min. at 4°C. The cell lysate was prepared by treating cells on ice for 20 min with the lysis buffer containing 15 nM Tris, pH 8.0, 0.5% Tween X-100, 1 nM DTT, 0.5 nM EDTA, and 1 mM phenylmethysulfonyl fluoride (PMSF). Subsequently, the supernatant was obtained by centrifugation at 4000 g for 8 min. The protein concentration of the supernatant was determined by the Bio-Rad protein assay kit (Bio-Rad Laboratories), using bovine serum albumin as standard.

Samples containing 30 to 50 µg of protein were subjected to sodium dodecylsulfate/8.0% polyacrylamide gel electrophoresis (SDS-PAGE) with pre-stained protein markers as references. Protein fractions in the gel were transferred to the Immobilon-P transfer membrane (Millipore). The membrane was incubated with the monoclonal anti-$cPLA_2$ (1:200–1:500), anti-COX-1 (1:1000), polyclonal anti-COX-2 (1:500), or polyclonal anti-Actin (1:2000) antibodies. The membranes were then exposed to secondary antibodies that were coupled to horse-radish peroxidase. The protein bands were detected on film using a Western blotting detection reagent kit (Amersham). The relative intensities of protein bands were acquired using Bio-Rad MAZ MultiImaging System (Bio-Rad Laboratories), and data were digitized and analyzed by the Quantity One software supplied by Bio-Rad Laboratories.

Statistical analysis

The data were analyzed by a two-tail independent Student's *t* test. In all cases, the level of significance was defined as $P < 0.05$. Results are presented as the mean ± standard deviation.

RESULTS

PMA induced arachidonic acid release in U937 cells

We reported earlier that PMA caused an enhancement of AA release in H9c2 cells (15). In this study, the effect of PMA on AA release in human premonocytic U937 cells was investigated. Labeled arachidonic acid was incorporated into U937

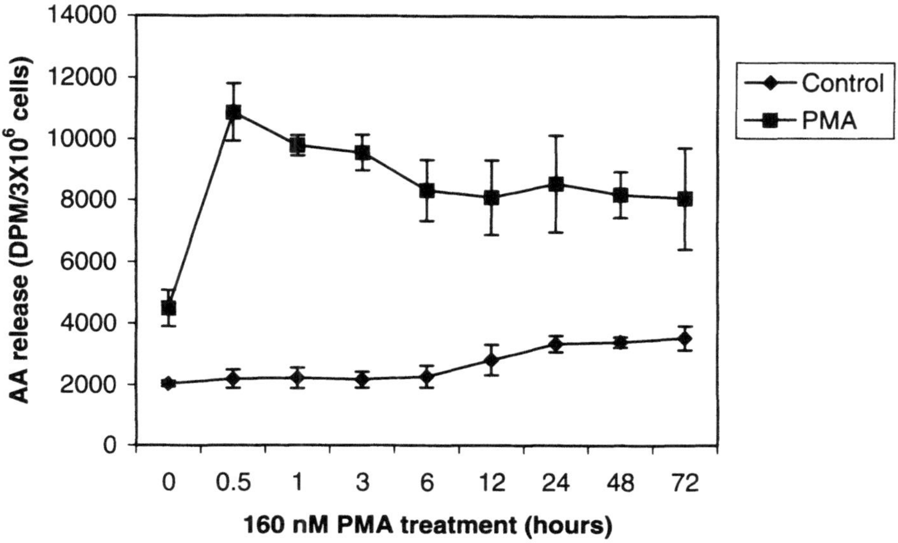

Figure 1. PMA-induced arachidonic acid release in U937 cells. U937 cells were metabolically labeled with [^{3}H] arachidonic acid, and incubated in the absence or presence of 160 nM PMA for 0–72 h. After PMA treatment, the cells were stimulated with 5 μM Ionomycin for 15 min. An aliquot of the medium was taken and labeled arachidonic acid was isolated by thin-layer chromatographic (TLC) and quantified by scintillation counting. Data are expressed as means ± S.D. of three separate sets of experiments.

cells by incubating the cells in a serum-free RPMI-1640 medium containing 0.1% BSA, and [^{3}H] arachidonic acid (1.0 uCi/mL) was added into the incubating mixture. The cells were incubated for 12 h at 37°C, and then washed with the unlabeled medium. For each study, the labeled cells were incubated with the serum-free medium (with 1% BSA) and 160 nM PMA for various time periods. Subsequent to the prescribed period of incubation, the cells were stimulated with calcium ionophore (Ionomycin, 5 μM) for 15 min, and the amount of labeled AA released to the medium was determined. As shown in Figure 1 an increase in AA release was sustained at all time points (up to 72 h) of PMA treatment, but the maximum increase in AA release was obtained at 0.5 h of incubation. Our results indicate that the enhancement of AA release by PMA treatment in U937 cells was similar to that obtained in H9c2 cells (15).

PMA induced PGE_2 release in U937 and H9c2 cells

AA is the precursor for eicosanoid biosynthesis. In this study, the release of PGE_2 into the medium following PMA stimulation in U937 and H9c2 cells were determined. Cells were stimulated by PMA as described in the previous section, and the amount of PGE_2 released into the medium was determined by enzyme immunoas-

Table 1. The effect of PMA on PGE_2 release in U937 cells

PMA treatment (nM)	PGE_2 release (pg/ml)
0	9.54 ± 2.08
1	193.56 ± 28.14[a]
10	249.90 ± 27.12[a]
100	256.37 ± 28.60[a]
200	238.00 ± 4.51[a]
400	217.68 ± 75.31[a]

U937 cells were incubated in serum-free PMRI-1640 medium containing 0.1% BSA with increasing concentration of PMA for 48h. The cell-free medium was recovered and PGE_2 concentration was determined by EIA. Results are means ± standard deviation of two separate sets of experiments.
[a] $P < 0.05$ when compared with the control (untreated).

say (EIA). As shown in Figure 2a, the PMA treatment did not cause a detectable increase in PGE_2 production during the first 12-h treatment. Interestingly, the amount of PGE_2 in the medium was marginally elevated at 24h of treatment, and a substantial increase was observed at 48 and 72h of incubation. In a dose-response study, a 19-fold increase in PGE_2 production was observed when cells were treated with 1nM PMA (Table 1). Maximum increase in PGE_2 release (26-fold increase) was observed by the treatment with 100nM PMA under identical experimental conditions.

PMA treatment also caused a moderate increase of PGE_2 release in H9c2 cells. In a time course study, an increase in PGE_2 release was observed at 3h of incubation. A higher degree of increase in PGE_2 release was observed during 24 and 48h of incubation. Further increase in the incubation time did not cause a high degree of increase in PGE_2 release in the H9c2 cells (Figure 2b). Our results indicated that PMA treatment caused a time dependent enhancement of PGE_2 release in both U937 and H9c2 cells. A more rapid response to PMA treatment was observed in H9c2 cells than U937 cells. In both cases, the enhancement of AA release by PMA precedes the increase in PGE_2 release in both cell types.

Effect of PMA on $cPLA_2$, COX-1, and COX-2 in U937 and H9c2 cells

The PMA-induced enhancement of AA release and PGE_2 production could be caused by the increase in the respective enzymes. Hence, quantitative determination of enzymes obtained from the control and PMA treated cells were analyzed by SDS-PAGE. As shown in Figure 3a $cPLA_2$ level was marginally higher in cells treated with 160nM PMA during the first 0.5h (116% ± 9%) and reached a maximum at 24h (153% ± 12%). After 24h of incubation, the level of $cPLA_2$ was drastically reduced (62% ± 10% at 48h, 38% ± 1% at 72h). It appears that PMA caused a biphasic response in the modulation of $cPLA_2$ enzyme levels, with an early up-regulation followed by a down-regulation. The biphasic change of $cPLA_2$ levels may in part contribute to the observed increase in AA release in the

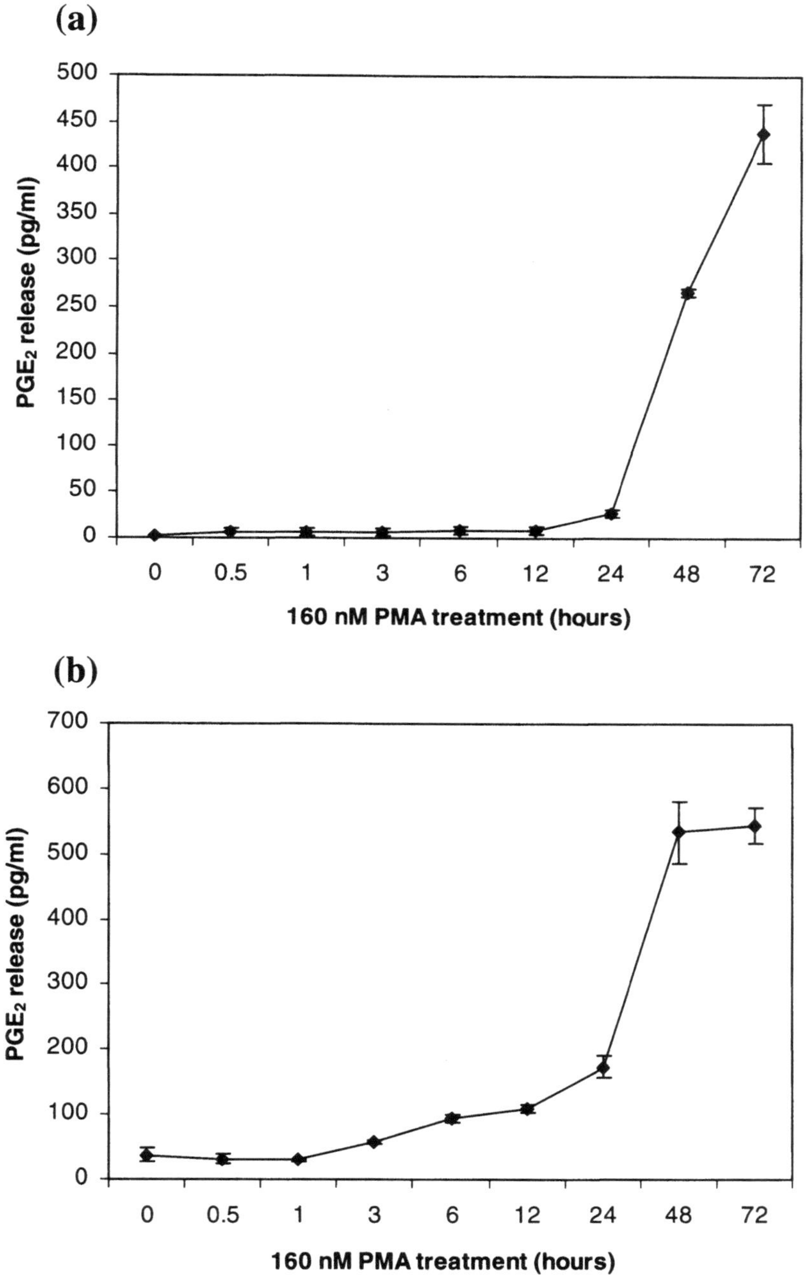

Figure 2. PMA-induced PGE_2 release in U937 and H9c2 cells. U937 (Figure a) and H9c2 (Figure b) cells were incubated in serum-free PMRI-1640 or DMEM-medium containing 0.1% BSA in the presence of 160 nM PMA for 0–72 h. After PMA treatment, a aliquot of the medium was taken for the determination of PGE_2 production by EIA. Data are expressed as means ± S.D. of four experiments.

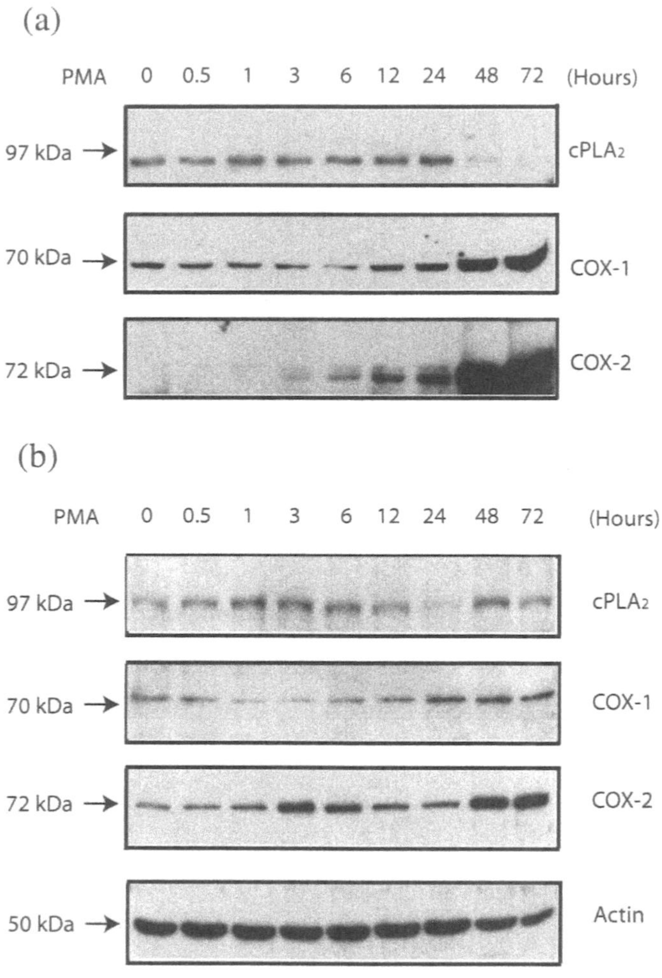

Figure 3. Effect of PMA on $cPLA_2$, COX-1, and COX-2 in U937 and H9c2 cells. U937 (Figure a) or H9c2 (Figure b) cells were incubated with 160nM PMA for 0–72h. The cell lysates containing 50mg of protein were subjected to SDS-PAGE (8.0%). The protein bands in the gel were transferred to PVDF sheet and treated with respective primary antibodies. Protein bands corresponding to $cPLA_2$, COX-1, and COX-2 were detected by a coupled peroxidase color development system.

early phase of PMA stimulation, followed by a moderate reduction after prolonged treatment.

A similar pattern of response on the level of $cPLA_2$ by PMA treatment was also obtained in H9c2 cells. Treatment with PMA caused an immediate increase of the $cPLA_2$ level, and the stimulation was maintained at the first 6h of incubation. The effect was diminishing at 12h, and totally eliminated at 24h of incubation. This observation is consistent with the observed biphasic response of PGE_2 release by PMA stimulation in H9c2 cells (Figure 2b).

The effect of PMA on the levels of COX-1 and COX-2 enzyme proteins was also studied under the same experimental conditions. In U937 cells, PMA treatment caused a slight decrease in COX-1 protein level during the first 6h of incubation, and returned to the control level after 12h of incubation. The COX-1 level was subsequently up-regulated between 24–72h, with a 6-fold increase in enzyme level at the 72h time point. In contrast, COX-2 protein was not detected in untreated U937 cells. Upon PMA treatment, COX-2 protein was induced and became detectable 3 hours after PMA-treatment. The enzyme level was found to increase linearly between 3–24h, and exponentially between 24–72h of incubation. The change in COX-2 enzyme levels is consistent with the observed increase in PGE_2 release in U937 cells (Figure 2a), suggesting that it may play a significant role in PGE_2 production.

Treatment of H9c2 cells with PMA also caused a biphasic response in COX-1 level during a time course study. The enzyme level was reduced during the first 6 hours of treatment, and subsequently returned to normal level after 24 hours of treatment. The level of COX-2 was slightly elevated between 3–24h of incubation, and greatly stimulated at 48h and 72h of incubation (Figure 3b). The immunostaining of actin in the same blot was employed as a control to exclude the possibility of unequal loading of the protein sample into the gel. The change in COX-2 enzyme level is consistent with the observed increase in PGE_2 release in H9c2 cells (Figure 2b).

The efficacy of PMA in the modulation of COX-1 and COX-2

The efficacy of PMA in the modulation of COX-1 and COX-2 proteins in U937 and H9c2 cells were conducted in a does-response study. U937 cells were incubated with various concentrations of PMA (1–400nM) in serum-free RPMI-1640 medium containing 0.1% BSA for 48h. An aliquot of the cell lysate containing 50mg of protein was subjected to SDS-PAGE (8.0%). As shown in Figure 4a, COX-1 levels were effectively up-regulated at low PMA levels (1nm PMA with 5-fold increase) and were not significantly affected by increasing PMA concentrations between 10–400nM. In contrast, COX-2 was not detectable in U937 cells in the absence of PMA (Figure 4a). However, COX-2 was induced upon PMA treatment, and the levels increased in a does-dependent manner with an optimal concentration of PMA between 50–200nM. A similar set of experiment was conducted with H9c2 cells (Figure 4b). The COX-1 levels in these cells were not significantly affect by increasing PMA concentrations between 1–100nM. A marginal increases were noticed with high does of PMA treatment (>100nM). However, COX-2 was significantly up-regulated by increasing PMA concentrations between 1–200nM. The result obtained from this study indicates that an optimal concentration of PMA (160nM) was used in our study. It also illustrates the diversity of response by different cell types to low concentrations of PMA in the modulation of enzyme levels.

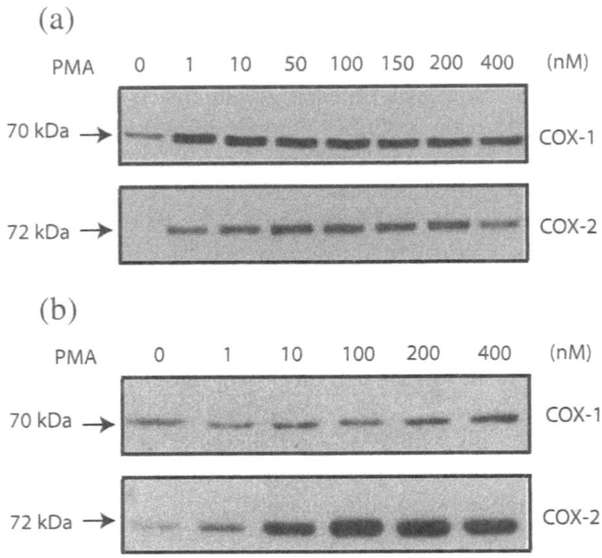

Figure 4. Does response for the effects of PMA on enzyme protein synthesis of COX-1 and COX-2 in U937 and H9c2 cells. U937 (Figure a) or H9c2 (Figure b) cells were incubated in serum-free RPMI-1640 medium containing 0.1% BSA with increasing concentration of PMA for 48h. The cell lysates containing 50mg of protein were subjected to SDS-PAGE (8.0%). The protein bands in the gel were transferred to PVDF sheet and treated with respective primary antibodies. Protein bands corresponding to COX-1, and COX-2 were detected by a coupled peroxidase color development system.

DISCUSSION

The central role of arachidonic acid metabolism in the production of eicosanoids in mammalian cells is well documented (12,13). The production of prostanoids is regulated by the arachidonate-eicosanoid cascade. In this cascade, the $cPLA_2$ is responsible for the release of the free AA from the membrane phospholipids, whereas COX is responsible for the eicosanoid formation (reviewed in 2,17–19). Our results demonstrated that PMA caused an immediate increase of AA release followed by a subsequent enhancement of PGE_2 production in monocytes and myocytes.

The effect of PMA on the activation of $cPLA_2$ has been well-documented. PMA mediates its effect at both the transcriptional and post-translational levels (19–22). It is clear that the increase in $cPLA_2$ level observed in this study was stimulated at the protein level. The increase in enzyme level does not seem to account for the increase in AA release after PMA treatment. Hence, PMA might modulate $cPLA_2$ at the post-translational level. We have shown in an earlier study that PMA causes the activation of $cPLA_2$, possibly via the indirect action of protein kinase C (16). Although protein kinase C has the ability to phosphorylate $cPLA_2$ in vitro, $cPLA_2$ is not activated by the phosphorylation process. In addition, protein kinase C does not have the ability to directly phosphorylate $cPLA_2$ under in vivo conditions (22,23). An alternate mechanism is that protein kinase C causes the activation of the p42/p44 MAPK cascade, which may result in the phosphorylation of $cPLA_2$.

In this study, the relationship between the up-regulation of COX and production of eicosanoid has been amply demonstrated. PMA elicited an immediate increase in AA release in both U937 and H9c2 cells, but the increase did not result in a comparable PGE_2 production in the first 24 h. The enhancement of PGE_2 production appears to result from the elevated levels of COX-1 and COX-2 in both cell types under sustained stimulation of PMA.

There are two isoforms of cyclooxygenase, COX-1 and COX-2, which are coded by separate and distinct genes (11,12). COX-1 has been shown to be constitutively expressed in most tissues (24). COX-2 is an inducible enzyme, which is subjected to transcriptional regulation (25). It is clear from this study that the effect of PMA was more pronounced with COX-2 than COX-1. The fact that the temporal change in COX-2 level corresponded closely with the change in PGE_2 release in both cell types is indicative of the importance of COX-2 in the production of prostanoids under external stimuli.

The increase in COX-2 expression by PMA treatment was caused by an enhancement at the transcriptional level. Since PMA has been shown to cause the transformation of the premoncytic cells into monocytic cells, the increase in COX-2 expression in the U937 cells might be a consequence of cellular transformation. Alternatively, an increase in COX-2 expression was also observed in H9c2 cells, indicating that the enhancement of transcription might be independent of the transformation process.

In summary, our study clearly shows that PMA caused an increase in arachidonate release and the subsequent production of PGE_2 in both U937 and H9c2 cells. The action of PMA is mediated via the modulation of $cPLA_2$ and COX-2 expression in both cell types. Our results confirm the important roles of these two enzymes in the regulation of the arachidonate-eicosanoid cascade in both monocytes and cardiac myocytes.

ACKNOWLEDGEMENTS

We thank Drs. T.R. Xu, B.M. Sun, and D. Lee for excellent technical support and useful discussion. This work was supported by Canadian Institute for Health Research.

REFERENCES

1. Vance DE. 1991. Phospholipid metabolism and cell signaling in eukaryotes. *In* Biochemistry of Lipids, Lipoproteins and Membranes. D.E. Vance and J.E. Vance, editors. 205–240. Elsevier Science Publishers, Amsterdam.
2. Leslie CC. 1997. Properties and regulation of cytosolic phospholipase A2. J Biol Chem 272: 16709–16712.
3. Dennis EA. 1994. Diversity of group types, regulation, and function of phospholipase A2. J Biol Chem 269:13057–13060.

4. Lambeau G, Lazdunski M. 1999. Receptors for a growing family of secreted phospholipases A2. TiPS 20:162–170.
5. Dennis EA, Rhee SG, Billah MM, Hannun YA. 1991. Role of phospholipase in generating lipid second messengers in signal transduction. FASEB J 5:2068–2077.
6. Cao Y, Pearman AT, Zimmerman GA, McIntyre TM, Prescott SM. 2000. Intracellular unesterified arachidonic acid signals apoptosis. PNAS 97:11280–11285.
7. Scorrano L, Penzo D, Petronilli V, Pagano F, Bernardi P. 2001. Arachidonic acid causes cell death through the mitochondrial permeability transition. J Biol Chem 276:12035–12040.
8. Rao GN, Baas AS, Glasgow WC, Eling TE, Runge MS, Alexander RW. 1994. Activation of mitogen-activated protein kinases by arachidonic acid and its metabolites in vascular smooth muscle cells. J Biol Chem 269:32586–32591.
9. Smith WL. 1989. The eicosanoids and their biochemical mechanisms of action. Biochem J 259: 315–324.
10. Smith WL, DeWitt DL, Garavito RM. 2000. Cyclooxygenases: structural, cellular, and molecular biology. Annu Rev Biochem 69:145–182.
11. Yoloyama C, Tanabe T. 1989. Cloning of human gene encoding prostaglandin endoperoxide synthase and primary structure of the enzyme. Biochem Biophys Res Commun 165:888–894.
12. Hla T, Neilson D. 1992. Human cyclooxygnase-2 cDNA. PNAS 89:7384–7388.
13. Wu T, Ikezono T, Angus CW, Shelhamer JH. 1994. Characterization of the promoter for the human 85 kDa cytosolic phospholipase A2 gene. Nucleic Acids Research 23:5093–5098.
14. Clark JD, Schievella AR, Nalefski EA, Lin LL. 1995. Cytosolic Phospholipase A_2. J Lipid Mediators Cell Signalling 12:83–117.
15. Tran K, Wong JT, Lee E, Chan AC, Choy PC. 1996. Vitamin E potentiates arachidonate release and phospholipase A_2 activity in rat heart myoblastic cells. Biochem J 319:385–391.
16. Wong JT, Tran K, Pierce GN, Chan AC, O K, Choy PC. 1998. Lysophosphatidylcholine stimulates the release of arachidonic acid in human endothelial cells. J Biol Chem 272:16709–16712.
17. Needleman P, Turk J, Jakschik BA, Morrison AR, Lefkowith JB. 1986. Arachidonic acid metabolism. Ann Rev Biochem 55:69–102.
18. Herschman HR. 1996. Prostaglandin synthase 2. Biochim Biophys Acta 1299:125–140.
19. Funk CD. 2001. Prostaglandins and leukotrienes: Advances in eicosanoid biology. Science 294:1871–1875.
20. Kuroda A, Sugiyama E, Taki H, Mino T, Kobayashi M. 1997. Interleukin-4 inhibits the gene expression and biosynthesis of cytosolic phospholipase A2 in lipopolysaccharide stimulated U937 macrophage cell line and freshly prepared adherent rheumatoid synovial cells. Biochem Biophys Res Commun 230:40–43.
21. Nakamura T, Lin LL, Kharbanda S, Knopf J, Kufe D. 1992. Macrophage colony stimulating factor activates phosphatidylcholine hydrolysis by cytoplasmic phospholipase A2. EMBO Journal 11: 4917–4922.
22. Lin LL, Wartmann M, Lin AY, Knopf JL, Seth A, Davis RJ. 1993. cPLA2 is phosphorylated and activated by MAP kinase. Cell 72:269–278.
23. Seger R, Krebs EG. 1995. The MAPK signaling cascade. FASEB J 9:726–735.
24. Xu XM, Tang JL, Chen XM, Wang LH, Wu KK. 1997. Involvement of two sp1 elements in basal endothelial prostaglandin H synthase-1 promoter activity. J Biol Chem 272:6943–6950.
25. Habib A, Creminon C, Frobert Y, Grassi J, Pradelles P, Maclouf J. 1993. Demonstration of an inducible cyclooxygenase in human endothelial cells using antibodies raised against the carboxyl-terminal region of the cyclooxygnase-2. J Biol Chem 268:23448–23454.

Pathophysiology of Cardiovascular Disease,
edited by N.S. Dhalla, H. Rupp,
A. Angel & G.N. Pierce
Kluwer Academic Publishers, Boston, 2004

Endothelium-Derived Hyperpolarizing Factor(s). Does it Exist and What Role Does it Play in the Regulation of Blood Flow?

Chris R. Triggle,[1] Malarvannan Pannirselvam,[1]
Todd J. Anderson,[2] and Hong Ding[3]

Departments of Pharmacology & Therapeutics[1] and Medicine[2] and Smooth Muscle Research Group, Faculty of Medicine, University of Calgary, Calgary, Alberta, T2N 4N1 and Department of Pharmacology[3], Yale University School of Medicine, Boyer Center, New Haven, CT. 06536

Summary. The endothelium is a source of many substances that play important roles in the short- and long-term regulation of the cardiovascular system. In this review we focus on endothelium-derived hyperpolarizing factor, or EDHF. EDHF is often referred to as the "third pathway" as, in addition to nitric oxide (NO) and prostacyclin (PGI_2) EDHF seems to play an important role as the "third" endothelium-derived relaxing factor. Considerable debate is on going concerning the nature of EDHF and, indeed, whether a unique molecule even exists. EDHF, by definition, mediates its action by directly, or indirectly, opening K-channels. In most instances the action of EDHF is abolished by the combination of two K-channel toxins, apamin and charybdotoxin; however, the channels that these inhibitors interact with would seem to be located on endothelial rather than vascular smooth muscle cells. The cellular mechanisms whereby EDHF mediates vascular smooth muscle hyperpolarization may involve myo-endothelial gap junctions thus negating the role for a true chemical mediator. Endothelial dysfunction is a common feature of cardiovascular disease, including the cardiovascular dysfunction associated with diabetes, and the role that changes in the nature/function of EDHF play in this process is currently an area of considerable interest. Therapeutic and dietary interventions that restore endothelial function may prove to be of tremendous benefit in the treatment of cardiovascular disease.

Key words: endothelium-derived relaxing factors, endothelium-derived hyperpolarizing factors, nitric oxide, potassium channels, diabetes.

Author for correspondence: Chris R. Triggle, phone 403 220 6853, fax 403 270 0497, E-mail: triggle@ucalgary.ca.

INTRODUCTION

The endothelium plays an essential role in both the short-term and long-term regulation of vascular function and, over the last few years, the increasing knowledge of the endothelium has highlighted its integral role in a number of pathophysiological states including cardiovascular diseases, cancer and inflammatory diseases. Vascular endothelial cells regulate arterial blood flow and blood pressure by releasing various vasoactive substances and notably prostacylin (1), endothelium-derived relaxing factor, now known to be nitric oxide or NO (2,3). Endothelial cells are not only the source of a number of factors that acutely regulate blood flow as well as angiogenic growth factors, but they are also pivotally involved in the recruitment and adhesion of leukocytes and platelets, and they express adhesion molecules and growth factors. In this chapter we discuss the evidence supporting the existence of another family of vasoactive substances—the endothelium-derived hyperpolarizing factors, or EDHF, and recent advances made in the understanding of the possible identity of EDHFs. The chemical nature and physiological and pathophysiological significance of EDHF is currently an area of intense investigation and the endothelium, and EDHF, is a potential target in a variety of diseases and also are targets for new therapeutic approaches to treat these diseases (4).

Endothelium-derived hyperpolarizing factor is defined as the non-nitric oxide and non-prostanoid and endothelium-derived factor that is assumed to be released by either mechanical activation, such as shear stress, or chemical activation, via, for instance, acetylcholine, and mediates vascular smooth muscle relaxation in association with hyperpolarization of the smooth muscle cell (see (4–7)).

NATURE OF EDHF

There are a number of candidate molecules that, to varying extents, fulfill some of the criteria that would be expected of an EDHF and some key aspects are summarized below and also in Figure 1.

NO, PGI_2 can hyperpolarize smooth muscle although this does appear to be highly tissue dependent and, in the case of NO, also linked to the presence of mediators of active tone. Thus, Bolotina et al., 1994 (8) and Mistry and Garland (9) have reported that NO directly, via a soluble guanylyl cyclase (sGC)-independent mechanism, stimulates charybdotoxin-sensitive K^+ channels in the rabbit aorta and rat small mesenteric arteries respectively. NO activates K_{ATP} channels in rat mesenteric arteries (10) and, by a process dependent on passive stretch, in guinea-pig coronary arteries (11), an apamin-sensitive K^+ channel (12) and BK_{Ca} in rabbit middle cerebral arteries (13).

Epoxyeicosatrienoic acids, EETs, generated via an epoxygenase from arachidonic acid, are attractive candidates for EDHFs in a number of vascular beds, notably in the coronary vascular bed (14,15). Cytochrome P450 (CYP) enzymes can metabolize arachidonic acid to vasoactive substances and the CYP 2J2 isoform, that is expressed in both human endothelial and vascular smooth muscle, can generate

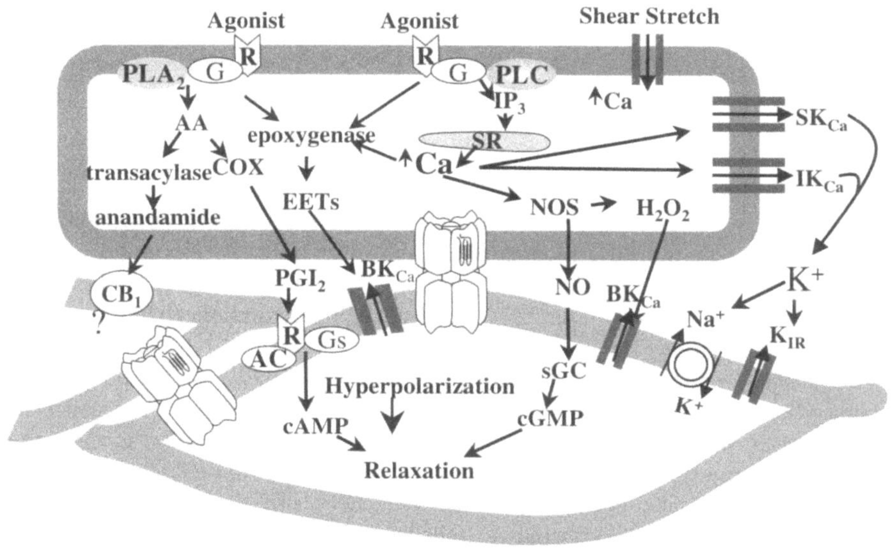

Figure 1. A single endothelial cell is depicted overlying two vascular smooth muscle cells.

Endothelial cells can be activated by both chemical (i.e. acetylcholine and bradykinin which both activate cell signalling events via G-protein coupled receptors) and mechanical (shear stress) activation. G-protein coupled receptor activation can lead to the activation of a number of endothelial cell pathways that include:

a/ Phospholipase A_2 (PLA_2) to release arachidonic acid that:

i/ via cyclooxygenase (COX), can produce prostacyclin (PGI_2) that, in turn, activates adenylate cyclase (AC) in smooth muscle cells and raises cyclic AMP (cAMP) levels thus leading to vascular smooth muscle relaxation.

ii/ via epoxygenase, can produce epoxyeicosatrienoic acids (EETs) which can hyperpolarize vascular smooth muscle by activating large conductance calcium-activated K-channels (BK_{Ca}).

iii/ via transacylase to produce anandamide that can activate cannabinoid 1 (CB_1) receptors and also lead to vascular smooth muscle relaxation.

B/ Phospholipase C (PLC) to produce inositol trisphosphate that can release intracellular Ca^{2+} from sarcoplasmic reticulum (SR) which in turn activates nitric oxide synthase (NOS) to produce nitric oxide (NO) and, in situations of low substrate (L-arginine) availability, also hydrogen peroxide (H_2O_2). H_2O_2 may hyperpolarize vascular smooth muscle by activating large conductance calcium-activated K-channels (BK_{Ca}).

NO activates vascular smooth muscle soluble guanylyl cyclase (sGC) to produce cyclic GMP (cGMP) and may also activate K-channels directly or indirectly via cGMP and G-kinase (not shown).

Small conductance calcium-activated and intermediate calcium-activated K-channels (SK_{Ca} and IK_{Ca} respectively) associated with endothelial cells are activated by the increase in endothelial cell calcium and this will lead to the efflux of K^+ from the cell.

An increase in K^+ in the milieu between endothelial and vascular smooth muscle cells may activate Na,K-ATPase as well as inward rectifier K-channels (K_{IR}) leading to the hyperpolarization of smooth muscle.

Gap junctions made up of connexins may also provide both electrical and chemical coupling between endothelial cells and vascular smooth muscle cells as well as between endothelial and endothelial cells.

EETs and an accumulation of evidence suggests that, at the very least, these products have important physiological and pathophysiological effects in the cardiovascular system (15). Furthermore, EETs may not only play a role in the regulation of vascular tone and homeostasis, but also possess anti-atherogenic activity that may be comparable to that of NO (16).

Isoprostanes are generated from polyunsaturated fatty acids, such as arachidonic acid, by PLA_2 under circumstances when oxygen free radicals are elevated. Isoprostanes may serve as both endothelium-derived contracting factors (EDCFs), as thromboxanes are powerful constrictors, as well as EDHFs, as, for instance, 8-iso PGE_2 is a vasodilator. EDHF-mediated responses have been reported to be sensitive to PLA_2 inhibitors and Janssen (17) has advanced the hypothesis that isoprostanes may be candidate molecules for EDHF.

Anandamide, an endogenous cannabinoid, is another arachidonic acid product via a transacylase and also has been advanced as a possible EDHF in some vascular beds (18). However, the current view is that anandamide may be an important regulator of vascular tone, but is unlikely to be an EDHF (see review (7)).

Carbon monoxide, like NO, can also activate sGC and is also synthesized in endothelial and vascular smooth muscle cells by haeme oxygenase 2 (19), however, it is considerably less potent than NO (as an activator of sGC) and seems unlikely that it fulfills the role of being an EDHF (5).

Hydrogen peroxide, H_2O_2, may be produced in the endothelium from superoxide anions via superoxide dismutase. H_2O_2 activates Ca^{2+}-activated K-channels in VSMC (20–21), endothelial cells (22) and also relaxes and hyperpolarizes endothelium-denuded porcine coronary arteries (23) and thus H_2O_2 fulfills some of the criteria expected of an EDHF (23).

In situations where the eNOS substrate, L-arginine, or cofactor, tetrahydrobiopterin, has been depleted, for instance, in disease states where there is high oxidative stress, eNOS can generate superoxide anions (24) as can CYP and NAPH.

Matoba et al., 2000 (25) studied ACh-mediated relaxation of small mesenteric arteries from eNOS -/- mice and demonstrated that catalase, an enzyme that dismutates H_2O_2 to water and oxygen, inhibited EDHF-mediated hyperpolarization and relaxation. Catalase has, however, been shown to inhibit COX-dependent 8-iso prostaglandin $F_{2\alpha}$ (26), thus suggesting that H_2O_2 may not be the final mediator of EDH.

Potassium ion as an EDHF was first suggested by Edwards et al. (1998) and they reported that in the rat hepatic artery an ACh-mediated an increase in K_o (from 4.6 to 11.6 mM), as measured with a K^+-sensitive microelectrode (27). They also recorded membrane potential with sharp glass microelectrodes and illustrated that ACh can hyperpolarize both vascular and endothelial cells with endothelial cell hyperpolarization inhibited by a combination of apamin and charybdotoxin and vascular hyperpolarization by a combination of ouabain and barium (30 μM). Thus, the hypothesis was advanced that it is the apamin-sensitive small conductance calcium-activated K^+ channels (SK_{Ca}) and, the charybdotoxin-sensitive intermediate conductance calcium-activated K^+ channels (IK_{Ca}) on endothelial cells that regulate EDHF release and the ouabain-sensitive Na^+, K^+-ATPase and inward rectifying K^+

channel (K_{ir}) on the vascular smooth muscle that mediate the vascular actions of EDHF. An increase in extracellular K^+ activates Na^+, K^+-ATPase and also opens K_{ir} on VSMCs. An increase in K^+_o by 5 mM mimicked the effects of ACh, and comparable data was reported for the rat mesenteric artery preparation. Although the data from Edwards et al. (1998) provided very strong support that small changes in extracellular potassium, which were already known to cause vascular smooth muscle relaxation, may represent the putative EDHF the conclusion that EDHF is K^+, exiting from endothelial cells following the ACh-mediated opening of pamin/charybdotoxin-sensitive K^+ channels, has not been universally accepted (for instance see (6)).

That K^+ may be an EDHF is an attractive hypothesis that would place potassium ions along side NO as a cell-signalling mediator that likely evolved as an early regulator of vascular function (see also (28)). Because of the similarity of K^+-induced vasodilatation to that mediated by EDHF in other vessels it was concluded that K^+ might be a universal EDHF (27); however, this now seems unlikely.

Myoendothelial gap junctions can, via intercellular channels, provide a pathway for the passage of small water-soluble molecules (<1000 Da), including cAMP, cGMP, inositol trisphosphate as well as inorganic ions, including Ca^{2+}, but not proteins, to pass between adjacent connecting cells. Myoendothelial gap junctions are established by the docking of the two connexon hemichannels supplied by the endothelial and smooth muscle cells. Thirteen rodent connexins have been identified to date (see review (29)) and connexin 43 has been reported to be the dominant gap junction protein present in both VSMC and EC (30–31). Van Kempen and Jongsma used immunohistochemical techniques to study the distribution of connexins 37, 40 and 43 in bovine, micropig and rat aorta and coronary vessels and concluded that connexin 40 is the constitutive connexin that was found between VSMC and EC with connexin 43 only between VSMC and connexin 37 between EC; connexin 45 has also been located in smooth muscle (32). The co-localization of connexin 40 and 43 has also been reported in both EC and VSMC (33) and the conductance properties of the heterocellular gap junction channel differ for connexons formed from connexins 40 and 43 (34). Since expression of these connexins varies between vascular beds there is clearly the potential for specialization of function within the circulation and it is thus possible that, rather than multiple chemical EDHFs, EDH may be explained by the heterogeneity in the distribution of the connexins that make up the myoendothelial gap junctions.

ENDOTHELIUM DYSFUNCTION

Despite the fact that vascular complications are the leading causes of mortality in type I and type II diabetes there is a paucity of data specifically on the pathogenesis, prevention and the treatment of the vascular complications in diabetes. This is clearly illustrated by the finding that type II diabetic patients suffer from a mortality rate 3–4 times that of the general population (35). Thus a better understanding of the role played by the endothelium in contributing to the progression of the cardiovascular complications of diabetes should lead to a decrease in both the morbidity and mortality associated with this disease.

NO is a key endothelium-derived factor that regulates cardiovascular function by promoting vasodilatation, inhibiting platelet aggregation, white blood cell adhesion and smooth muscle cell proliferation (36). Endothelial dysfunction is defined as an impaired endothelium-dependent relaxation (for instance, to acetylcholine) resulting from a reduction in the bioavailability/bioactivity of NO in the vessel wall (37). A dysfunctional endothelium contributes to several pathological conditions, including altered anticoagulant and anti-inflammatory properties of the endothelium, impaired modulation of vascular growth, and dysregulation of vascular remodeling. It is of interest that the EET family, as potential EDHFs, also possess, in addition to their actions as direct modulators of vascular tone, anti-inflammatory actions (16), anti-migratory actions on SMCs (38), and a protective role against hypoxia-reoxygenation injury in SMCs (39). In addition, epoxyeicosatrienoic acid-mediated renal vasodilation to arachidonic acid is enhanced in spontaneously hypertensive rats (40) suggesting that EET synthesis may be up-regulated in partial compensation for the elevated blood pressure. EETs are thus not simply vasodilator substances.

Studies with the db/db -/- mouse, a mouse model of type II diabetes which has a defective leptin receptor and exhibits hyperglycemia, hyperinsulinemia, insulin resistance and obesity, have demonstrated endothelial dysfunction that is characterized by a reduced vascular relaxation-response to acetylcholine (41). Of interest was that the same study also reported that the endothelial dysfunction was corrected by the acute administration of the co-factor for nitric oxide synthase, tetrahydrobiopterin, and also improved by the cell permeable polyethylene analogue of superoxide dismutase, PEG-SOD.

CONCLUSIONS

The endothelium plays an important role in maintaining a healthy circulation. Reductions in the bio-availabilty of NO may be the basis of the endothelial dysfunction that is characteristically observed at the early stages of cardiovascular disease. The contribution of EDHF in the maintenance of cardiovascular function in situations when NO production/bioavailability is an important area for further investigation.

ACKNOWLEDGEMENTS

Research data referenced in this chapter from the laboratories of the authors was supported by: AHFMR (T.D., M.P.); CDA (T.D. and C.R.T.); CIHR (C.R.T.); HSFC (H.D.).

REFERENCES

1. Moncada S, Gryglewski R, Bunting S, and Vane JR. 1976. An enzyme isolated from arteries transforms prostaglandin endoperoxides to an unstable substance that inhibits platelet aggregation. Nature 263:663–665.

2. Furchgott RF, Zawadzki JV. 1980. The obligatory role of endothelial cells in the relaxation of arterial smooth muscle by acetylcholine. Nature 288:373–376.
3. Palmer RM, Ferrige AG, Moncada S. 1987. Nitric oxide release accounts for the biological activity of endothelium-derived relaxing factor. Nature 327:524–526.
4. Ding H, Triggle CR. 2001. Relaxing blood Vessels. Pharmaceutical News 8:42–49.
5. Feletou M, Vanhoutte PM. 1999. The third pathway: endothelium-dependent hyperpolarization. J Physiol Pharmacol 50:525–534.
6. Ding H, Triggle CR. 2000. Novel endothelium-derived relaxing factors. Identification of factors and cellular targets. J Pharmacol Toxicol Methods 44:441–452.
7. McGuire JJ, Ding H, Triggle CR. 2001. Endothelium-derived relaxing factors: a focus on endothelium-derived hyperpolarizing factor(s). Can J Physiol Pharmacol 79:443–470.
8. Bolotina VM, Najibi S, Palacino JJ, Pagano PJ, Cohen RA. 1994. Nitric oxide directly activates calcium-dependent potassium channels in vascular smooth muscle. Nature 368:850–853.
9. Mistry DK, Garland CJ. 1998. Nitric oxide (NO)-induced activation of large conductance Ca2+-dependent K+ channels (BK(Ca)) in smooth muscle cells isolated from the rat mesenteric artery. Br J Pharmacol 124:1131–1140.
10. Garland JG, McPherson GA. 1992. Evidence that nitric oxide does not mediate the hyperpolarization and relaxation to acetylcholine in the rat small mesenteric artery. Br J Pharmacol 105:429–435.
11. Parkington HC, Tonta MA, Coleman HA, Tare M. 1995. Role of membrane potential in endothelium-dependent relaxation of guinea-pig coronary arterial smooth muscle. J Physiol 484(Pt 2): 469–480.
12. Murphy ME, Brayden JE. 1995. Nitric oxide hyperpolarizes rabbit mesenteric arteries via ATP-sensitive potassium channels. J Physiol 486(Pt 1):47–58.
13. Dong H, Waldron GJ, Cole WC, Triggle CR. 1998. Roles of calcium-activated and voltage-gated delayed rectifier potassium channels in endothelium-dependent vasorelaxation of the rabbit middle cerebral artery. Br J Pharmacol 123:821–832.
14. Campbell WB, Harder DR. 1999. Endothelium-derived hyperpolarizing factors and vascular cytochrome P450 metabolites of arachidonic acid in the regulation of tone. Circ Res 84:484–488.
15. Fleming I. 2001. Cytochrome p450 and vascular homeostasis. Circ Res 89:753–762.
16. Node K, Huo Y, Ruan X, Yang B, Spiecker M, Ley K, Zeldin DC, Liao JK. 1999. Anti-inflammatory properties of cytochrome P450 epoxygenase-derived eicosanoids. Science 285: 1276–1279.
17. Janssen LJ. 2002. Are endothelium-derived hyperpolarizing and contracting factors isoprostanes? Trends Pharmacol Sci 23:59–62.
18. Randall MD, Alexander SP, Bennett T, Boyd EA, Fry JR, Gardiner SM, Kemp PA, McCulloch AI, Kendall DA. 1996. An endogenous cannabinoid as an endothelium-derived vasorelaxant. Biochem Biophys Res Commun 229:114–120.
19. Zakhary R, Gaine SP, Dinerman JL, Ruat M, Flavahan NA, Snyder SH. 1996. Heme oxygenase 2: endothelial and neuronal localization and role in endothelium-dependent relaxation. Proc Natl Acad Sci USA 93:795–798.
20. Barlow RS, White RE. 1998. Hydrogen peroxide relaxes porcine coronary arteries by stimulating BKCa channel activity. Am J Physiol 275:H1283–H1289.
21. Hayabuchi Y, Nakaya Y, Matsuoka S, Kuroda Y. 1998. Hydrogen peroxide-induced vascular relaxation in porcine coronary arteries is mediated by Ca2+-activated K+ channels. Heart Vessels 13: 9–17.
22. Bychkov R, Pieper K, Ried C, Milosheva M, Bychkov E, Luft FC, Haller H. 1999. Hydrogen peroxide, potassium currents, and membrane potential in human endothelial cells. Circulation 99:1719–1725.
23. Beny JL, der Weid PY. 1991. Hydrogen peroxide: an endogenous smooth muscle cell hyperpolarizing factor. Biochem Biophys Res Commun 176:378–384.
24. Stroes E, Hijmering M, van Zandvoort M, Wever R, Rabelink TJ, van Faassen EE. 1998. Origin of superoxide production by endothelial nitric oxide synthase. FEBS Lett 438:161–164.
25. Matoba T, Shimokawa H, Nakashima M, Hirakawa Y, Mukai Y, Hirano K, Kanaide H, Takeshita A. 2000. Hydrogen peroxide is an endothelium-derived hyperpolarizing factor in mice. J Clin Invest 106:1521–1530.
26. Watkins MT, Patton GM, Soler HM, Albadawi H, Humphries DE, Evans JE, Kadowaki H. 1999. Synthesis of 8-epi-prostaglandin F2alpha by human endothelial cells: role of prostaglandin H2 synthase. Biochem J 344 Pt 3:747–754.

27. Edwards G, Dora KA, Gardener MJ, Garland CJ, Weston AH. 1998. K+ is an endothelium-derived hyperpolarizing factor in rat arteries. Nature 396:269–272.
28. Vanhoutte PM. 1998. Vascular biology. Old-timer makes a comeback. Nature 396:213, 215–213, 216.
29. Kumar NM, Gilula NB. 1996. The gap junction communication channel. Cell 84:381–388.
30. Christ GJ, Spray DC, el Sabban M, Moore LK, Brink PR. 1996. Gap junctions in vascular tissues. Evaluating the role of intercellular communication in the modulation of vasomotor tone. Circ Res 79:631–646.
31. Christ GJ, Brink PR. 1999. Analysis of the presence and physiological relevance of subconducting states of Connexin43-derived gap junction channels in cultured human corporal vascular smooth muscle cells. Circ Res 84:797–803.
32. van Kempen MJ, Jongsma HJ. 1999. Distribution of connexin37, connexin40 and connexin43 in the aorta and coronary artery of several mammals. Histochem Cell Biol 112:479–486.
33. Little TL, Beyer EC, Duling BR. 1995. Connexin 43 and connexin 40 gap junctional proteins are present in arteriolar smooth muscle and endothelium in vivo. Am J Physiol 268:H729–H739.
34. Brink P. 2000. Gap junction voltage dependence. A clear picture emerges. J Gen Physiol 116:11–12.
35. Haffner SM, Lehto S, Ronnemaa T, Pyorala K, Laakso M. 1998. Mortality from coronary heart disease in subjects with type 2 diabetes and in nondiabetic subjects with and without prior myocardial infarction. N Engl J Med 339:229–234.
36. Rubanyi GM. 1993. The role of endothelium in cardiovascular homeostasis and diseases. J Cardiovasc Pharmacol 22 Suppl 4:S1–14.
37. De Vriese AS, Verbeuren TJ, Van d V, Lameire NH, Vanhoutte PM. 2000. Endothelial dysfunction in diabetes. Br J Pharmacol 130:963–974.
38. Sun J, Sui X, Bradbury JA, Zeldin DC, Conte MS, Liao JK. 2002. Inhibition of vascular smooth muscle cell migration by cytochrome p450 epoxygenase-derived eicosanoids. Circ Res 90: 1020–1027.
39. Yang B, Graham L, Dikalov S, Mason RP, Falck JR, Liao JK, Zeldin DC. 2001. Overexpression of cytochrome P450 CYP2J2 protects against hypoxia-reoxygenation injury in cultured bovine aortic endothelial cells. Mol Pharmacol 60:310–320.
40. Pomposiello SI, Carroll MA, Falck JR, McGiff JC. 2001. Epoxyeicosatrienoic acid-mediated renal vasodilation to arachidonic acid is enhanced in SHR. Hypertension 37:887–893.
41. Pannirselvam M, Verma S, Anderson TJ, Triggle CR. 2002. Cellular basis of endothelial dysfunction in small mesenteric arteries from spontaneously diabetic (db/db -/-) mice: role of decreased tetrahydrobiopterin bioavailability. Br J Pharmacol 136:255–263.

Pathophysiology of Cardiovascular Disease,
edited by N.S. Dhalla, H. Rupp,
A. Angel & G.N. Pierce
Kluwer Academic Publishers, Boston, 2004

Cytokinesis in Smooth Muscle

Newman L. Stephens

Departments of Physiology Faculty of Medicine
University of Manitoba, Manitoba, Winnipeg, Canada R3E 3J7

Summary. Cytokinesis plays important roles in smooth muscle cell hyperplasia and hypertrophy. Since in proliferating cells the contractile machinery is of the non-muscle type we tested the hypothesis that one of the regulatory enzymes for cytokinesis, the 210 kDa molecular mass non-muscle type myosin light chain kinase (nmMLCK-210), is present at the cell's contraction ring. Immunocytochemistry showed the presence of the enzyme in the interphase nucleus and suggested it may be a transcription factor; however cis-platin cross-linking studies of DNA to the enzyme did not substantiate this. At anaphase the enzyme migrated to the contraction ring zone of the dividing cell where we had demonstrated the presence of non-muscle myosin heavy chain, α-sm-actin and α-tubulin. Thus a complete contractile machinery existed at this site. As cytokinesis proceeds a mid-body structure develops which is made of a centromeric protein, CENP-F, that is initially bound to the kinetochore, but dissociates from it and migrates to the midbody or telophase disc zone. It is involved in the final stage of cytokinesis. We have obtained preliminary evidence to show that CENP-F is localized to the nucleus in smooth muscle. We conclude that the location of non-muscle myosin light chain kinase at the contraction ring suggests it plays a role in cytokinesis.

Key words**:** cytokinesis, non-muscle myosin light chain kinase, smooth muscle myosin light chain, contraction ring.

INTRODUCTION

A discussion of smooth muscle cytokinesis might seem out of place in a book devoted to "Signal Transduction and Cardiac Hypertrophy" but is perhaps justified,

Address for Mailing proof: Newman L Stephens, University of Manitoba, Dept of Physiology, 425BMSB; 730 William Avenue, Winnipeg, Manitoba, CANADA, R3E 3J7. Tel: 204-789-3778, Fax: 204-789-3941, E-mail: nstephe@ms.umanitoba.ca.

if one can stretch a point, by the realization that muscle cell hypertrophy can result from delayed cytokinesis. Furthermore since it is such a highly conserved process in the cell cycle it is likely that what one learns from its study in the airway smooth muscle cell (ASMC) could be used to elucidate the pathogenesis of myocardial cell hypertrophy. Certainly what we have observed in regulation of ASMC we have also seen in the saphenous venous smooth muscle cell and in breast cancer cells (unpublished observation).

Cytokinesis is the final process in cell division that immediately follows upon karyokinesis. It is of brief duration, occupying about 1 hour in a 29 hour cycle in proliferating canine AMSCs. Its function is to bring about equal partitioning of cytoplasmic contents between daughter cells. It has to be very tightly regulated to ensure this, so both spatial and temporal requirements must be met (1–5).

Cytokinesis has been described as being comprised of anywhere from 4 to 5 phases (4). These are: 1). Specification of the cleavage furrow site which occur in early anaphase. This requires the actions of astral spindles (microtubules) and actin filaments (microfilaments). The role of polar spindles has not been determined but in cultured cells they may specify the position of the contraction furrow which is the ingressing structure developing at the contraction ring (CR). The ring itself is made up of actin filaments which are synthesized under the control of Rho-kinase. 2). Assembly of the contractile ring. Myosin II assembly develops adjacent to the actin filaments; the two constitute the contractile machinery needed for cytokinesis. A number of regulatory proteins are also found in this location. Whether myosin is obligatory has been disputed on the basis of data emanating from Dictyostelium cells (6). It was shown that Dictyostelium cells were capable of dividing despite the absence of myosin. The ability of such cells to divide was said to be due to "traction-mediated fission" of the myocyte in which different parts of the cell pull in different directions because of adhesive forces between the cell and the extracellular matrix. It thus appears that cytokinesis can occur via one mechanism, that is myosin-dependent and another which is based on cell adhesion. 3). Actomyosin contraction leads to development of a cleavage furrow. 4). The fourth stage involves development of membrane fusion to physically separate the 2 daughter cells. 5). The fifth stage is the development of a three-dimensional disc which encompasses the entire midzone and whose distribution survives disassembly of interpolar microtubules. The disc has been called the telophase disc (7). As shown by immunocytochemistry, the disc appears to contain myosin but not actin. The myosin in the disc likely interacts with actin in the cell cortex to bring about cytokinesis.

We now consider the proteins that regulate cytokinesis:

1. Small G proteins, their modulators and target molecules

The G proteins regulate the synthesis of the actin network which develops at the contraction ring. Chief among these are Rho GTPases which exist in an active GTP-bound form and an inactive GDP bound form; the two provide a switching mechanism. These bound nucleotides are regulated by GTPase-activating proteins (GAPs) (8), GDP dissociation inhibitors (GDs) and exchange factors (GEFs) (9). The importance of Rho is shown by the observation that clostridium botulinus C3 transferase, a potent Rho inhibitor, prevents cleavage furrow formation (10).

There is a wide array of Rho downstream targets such as P1–3 kinase, PLD, phosphatidyl 4-phosphate 5-kinase, PKN, rhophilin, rhotekin, multiple tyrosine kinases and the Rho-associated coiled-coil protein kinases ROCK-1 and ROCK-II (11). Another question is what targets Rho to the contraction ring. The answer is that a mammalian Rho homologue, Rho 1, localizes to clustered cortical actin patches and is involved in the translocation of Rho to the cleavage furrows (12).

Very recently a new enzyme, citron kinase which is a splice variant of Rho and contains PDZ (PSD-95, discs large, Z01) and SH3 binding domains at its C terminus, and a zinc finger and a Rho-binding sequence within the C-terminal half of the molecule colocalizes with Rho to the cleavage furrow (13). It appeared in distinct spots associated with tubulin bundles at the midbody when transfected with Hela cells. It is interesting also that citron kinase mutants with C-terminal truncations were capable of delaying cytokinesis and of producing multinucleated cells.

We have recently observed the presence of nonmuscle myosin light chain kinase (MLCK-210) at the contraction ring where it may be serving to phosphorylate the 20 kDa myosin light chain (MLC20) of myosin heavy chain (MHC) and bring about furrow ingression and cytokinesis (14).

To bring about actin assembly not only is Rho needed but so also is CDC 42. How it regulates cytokinesis is still not known but the action of a CDC 42 downstream effector molecular (p21-activated kinases—PAKS) that contains sequences known as formin homology regions or FH domains (15) may be responsible.

2. MLC 20 phosphorylation and cytokinesis

Actomyosin is not only important as a structural component of the contractile ring but is also responsible for the force and shortening needed to produce furrow ingression in the so-called "purse-string model" of cytokinesis.

Myosin appears at the cleavage furrow shortly after the appearance of actin filaments.

The orientation of actin is not only as per the "purse string" model i.e. it is not only a simple equatorial band at the furrow. Many filaments are aligned perpendicular to the equatorial filaments. This has led to the development of the so-called "isotropic contraction" model which states that the less-ordered equatorial network of actin and myosin would produce contraction forces spreading out in diverse orientations (1,16). The purse-string model, on the other hand, would produce forces perpendicular to the long axis of the dividing cell.

The role of MLC20 is key to contraction; phosphorylation of certain residues is a *sine qua non*. The phosphorylation is brought about by myosin light chain kinase (MLCK) which phosphorylates serine-19 and threonine-18. The contribution of the latter is about 25% of the total. The regulatory role of MLC20 phosphorylation is controversial. In Dictyostelium it is said not to be required (6) but in Drosophila (17) is required. The MLCK is of the non-muscle variety (nmMLCK).

Quite apart from MLC20 phosphorylation at serine-19 and threonine-18, phosphorylation of the carboxyl terminus (the tail) is also important. Impairment of this process leads to myosin filaments that cannot assemble or that disassemble poorly. Such filaments fail to bring about cytokinesis. The localization of phosphorylated MLC20 has been seen in the midzone of the cell in anaphase (18).

Another interesting observation is that Cdc2 kinase phosphorylation of MLC20 acts as a regulator and prevents premature cytokinesis.

3. The role of intracellular Ca^{2+}

Since actin and myosin filaments are present at the contraction ring, their activation must require Ca^{2+}. In support of this a transient increase in $[Ca^{2+}]_I$ has been reported during anaphase which regulates cytokinesis (19). The Ca^{2+} changes are driven by cyclic changes in IP3 levels. Several genes may be involved in this process. These include those encoding phosphotidyl 4-kinase, Protein kinase C, GTP-binding proteins (12) and protein containing a Src homology region which has 3 domains (20). Mutations in genes controlling proteins that regulate cytoskeleton function (21) and phosphatase activity (22) affect cytokinesis.

4. The role of protein phosphorylation

Since a fairly complete contractile machinery is present at the contraction furrow and in many ways resembles that of smooth muscle and other non-muscle contractile systems, it seems reasonable to expect that regulation is brought about by phosphorylation at Ser 19 and Thr 18 residues in MLC20, by myosin light chain kinase. However no consensus exists regarding the precise role of phosphorylation. In Dictyostelium MLC20 phosphorylation is not needed for contraction (6) at the furrow, but in germ line cells of Drosophila it was required (17).

Another fact, recently recognized is that regulation of the assembly state of myosin via phosphorylation of residues in the carboxy terminus, is also important for cytokinesis (23).

Work published by Satterwhite et al. (24) indicated that $p34^{CDC2}$ phosphorylation of MLC 20 at Ser-1 and Ser-2 could inhibit myosin activity and consequently myosin-dependent contraction of the cleavage furrow. Subsequent dephosphorylation by phosphatases followed by phosphorylation at serine-19 and threonin 18 would allow for activation of actomyosin ATPase and cytokinesis.

The kinase activity of purified $p34^{cdc2}$ kinase is low and perhaps other kinases are involved for full phosphorylation e.g. protein kinase C and citron kinase.

This review provides a framework for discussion of our own work on airway smooth muscle cell cytokinesis.

Using immunocytochemistry and confocal microscopy we have noted that during interphase of proliferating canine airway smooth muscle cells (ASMC) non-muscle myosin light chain kinase (nmMLCK-210) is present in the nucleus. See Figure 1. Cross-linking studies with cis-platin and Western blotting showed no binding of the enzyme to genomic DNA and suggest nmMLCK-210 is not involved

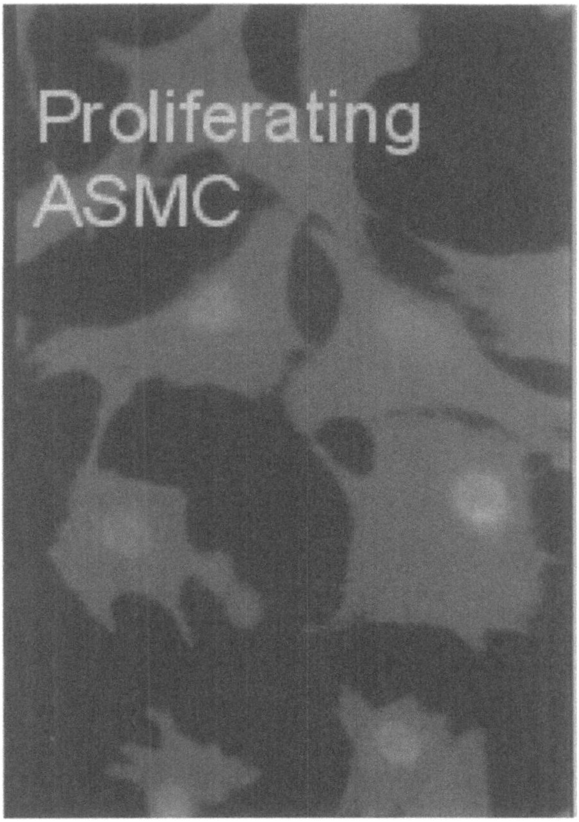

Figure 1. Cultured proliferating tracheal smooth muscle cells at 75% confluence. Immunohistochemistry was carried out using primary antibody against nonmuscle myosin light chain kinase of 210 molecular mass (nmMLCK-210). The secondary antibody was conjugated with Cy3. Cells which are in interphase show intranuclear localization of nmMLCK-210.

in gene transcription. At telophase the enzyme appeared in the cytoplasm at the contraction ring and disappeared from the nucleus. See Figure 2. Smooth muscle α-actin and myosin heavy chain were also demonstrated by Western blots, at the contraction ring. Thus most of the proteins required for contraction are assembled at the contraction ring and are sufficient for cytokinesis. Figure 3 shows a cell in cytokinesis phase. Two daughter nuclei are seen. The picture is an immunocytochemical one stained for nmMLCK-210 and shows there is no enzyme in the nuclei and that most of it is in the cytoplasm in the region of the contraction ring presumably biding its time to regulate cytokinesis.

As far as we know, this is the first report of the presence of nmMLCK-210 at the contraction ring of the dividing airway smooth muscle cell.

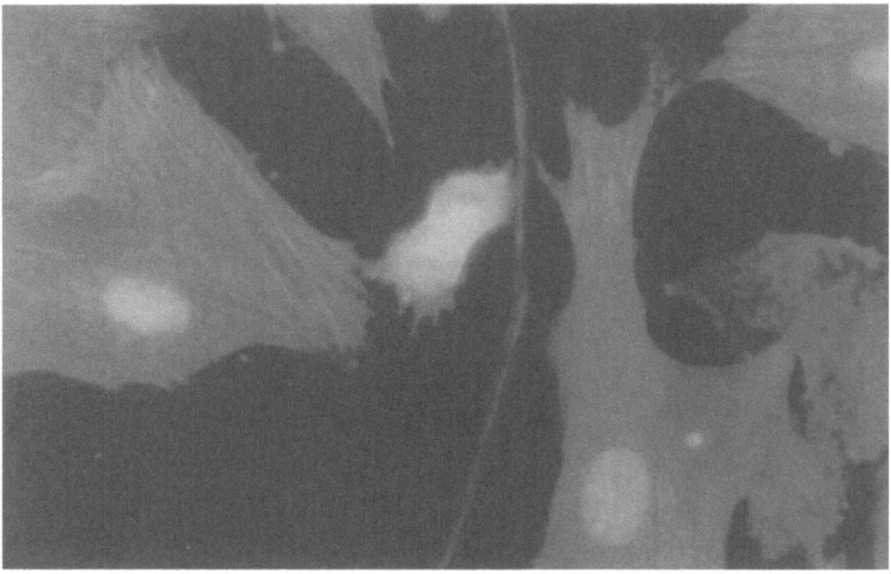

Figure 2. Tracheal smooth muscle cells which have been arrested (withdrawal of 10% FBS from the medium) for 7 days. The preparation has been treated in the same way as cells in Figure 1. While MLCK is present in the cytoplasm almost none is seen in the dividing nucleus.

One other unique structure develops in the final stages of cytokinesis. This is the so-called mid-body structure (25) also termed the telophase disc. It is made up of centromeric protein F (CENP-F) which develops at the kinetochore that overlies the centromeric gene of the chromosome (25). By late telophase it transfers to the midbody zone where it takes part in the completion of cytokinesis. Using autoimmune serum (26) obtained from rheumatoid arthritis patients and which contains CENP-F antibodies we demonstrated binding to an antigen located in the daughter nuclei presumably at the chromosomes. See Figure 4. Migration of CENP-F to the very middle of the mid-body structure, as reported by others, has not been seen, thus far, by us.

The function of the telophase disc has not yet been established, but its position at the midbody and the presence of myosin in it suggests that cytokinesis may involve the interaction of the disc organelle with actin in the cell cortex to complete cytokinesis in mammalian cells.

5. Microtubule and microfilament interaction during cytokinesis

From what has been already stated above it is clear that temporal and spatial interaction between the two must be occurring. The microtubules (astral spindles) are involved in specification of the contraction ring site which microfilaments develop

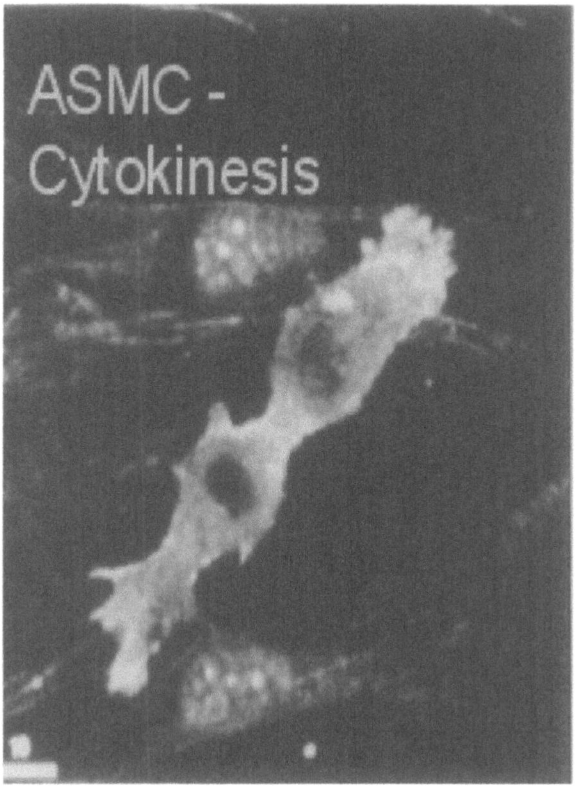

Figure 3. A single tracheal smooth muscle cell is seen in cytokinesis. Immunocytochemistry was carried out for nmMLCK-210 as in Figure 1. The daughter nuclei are "empty" and the enzyme is seen at the contraction ring and at peripheral focal adhesions.

adjacent to this site and are responsible for furrow ingression (27–29). Experimentally nocodazole treatment to reduce the number of midzone microtubules leads to inhibition of cytokinesis. Several lines of investigation suggest that, in Dictyostelium microtubules are important, at least for starting furrow ingress but whether they are needed for further evolution of the furrow is not clear (30,31).

The ability of astral microtubules to specify development of the contraction furrow seems to depend upon distance such that there is a critical location at which the ring forms (32).

The development of multi-nucleate smooth muscle cells has evoked interest, for example, in vascular smooth muscle cells from hypertensive animals. We ourselves have seen development of a new supercontractile airway smooth muscle cell which is multi-nucleated (unpublished observations). Such multi-nucleation has been

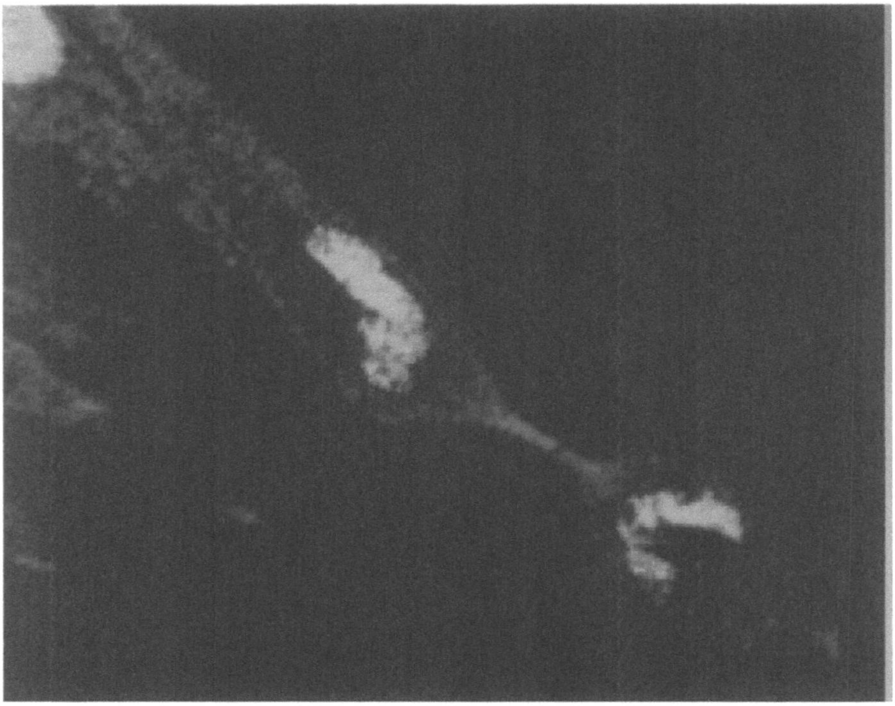

Figure 4. Immunocytochemistry microphotograph of a single tracheal smooth muscle cell at telophase/cytokinesis stained with Cy3 to show α-tubulin, and with FITC to show CENP-F; the latter was visualized with auto-immune serum.

reported in *Carenorhabiditis elegans* embryos where null mutants for a ZEN-4 protein (a motor protein) result in multi-nucleation (33).

In addition to the role of astral spindles in specifying the site-determination of the contraction ring it seems that the central spindle may be important in cytokinesis.

CYTOKINESIS IN PROLIFERATING AIRWAY SMOOTH MUSCLE CELLS

Cytokinesis, which is undoubtedly the most dramatic morphological event in the cell cycle, plays a role in airway smooth muscle cell (ASMC) hyperplasia by allowing cell division to proceed and, in hypertrophy, by delaying it. It has been investigated in a variety of cells by several workers (34,35). In view of the recent, considerable interest in cellular hyperplasia and hypertrophy in chronic asthma (36–43) its consideration is important. We describe some of the proteins involved in cytokinesis and demonstrate that non muscle myosin light chain kinase of molecular mass 210 kDa (nmMLCK-210) is present at the dividing cell's contraction ring along with other components of the contractile machinery, suggesting it plays an important part in cytokinesis. The rationale is that karyokinesis and cytokinesis occur in proliferating cells in which the mature, differentiated smooth muscle type of contractile and regulatory proteins change from contractile to non-contractile types. Thus smooth muscle type myosin heavy chain isoforms (SM-1, SM-2) are replaced by those of non-muscle type (nmMHC-A, nmMHC-B) and so it is likely that smMLCK-108 will be replaced by nmMLCK-210. For cytokinesis it is likely that for furrow ingress nmMLCK-210 must be the enzyme responsible.

Citron kinase (13) has also been described in the contraction ring zone of the dividing cell and is said to be responsible for phosphorylation of MLC20 also. Thus two phosphorylating enzymes, at least, have been shown to be present at the contraction ring. Activation by Ca^{2+} and calmodulin leads to phosphorylation of MLC20 which activates actomysin ATPase and initiates cross-bridge cycling (1). Just what are the amino acid residues that are phosphorylated has not yet been identified. For activation of MLC20, phosphorylation of serine-19 and threonine-18 is required (44). Phosphorylation of serine-1 and 2 and threonine-9 inhibits the ATPase (45).

Using immunocytochemistry and confocal microscopy we have noted that during interphase of proliferating ASMC, nmMLCK-210 is present in the nucleus. See Figure 1 and Figure 2. Cross-linking studies with cis-platin and Western blotting showed no binding of the enzyme to genomic DNA which suggests that nmMLCK-210 is not a transcription factor.

With progress to the telophase stage the enzyme appeared in the cytoplasm at the contraction ring (See Figure 3) and disappeared from the nucleus. It is possible that migration from the nucleus to the contraction ring indicates activation of the enzyme.

One other unique structure develops in the final stages of cytokinesis. This is the mid-body structure (7), a 3-dimensional body situated at the middle of the contraction ring; it is also knows as the telophase disc. It is made up of centromeric protein-F (CENP-F) which develops at the kinetochore that overlies the centromeric gene of the chromosome (25). By late telophase it transfers to the mid-body zone where it participates in the completion of cytokinesis. Using auto-immune serum (26) obtained from rheumatoid arthritis patients and CENP-F antibodies kindly provided by Dr. J B Rattner of the University of Alberta, at Calgary, we demonstrated binding of both antibodies to an antigen located at the chromosome. Migration of CENP-F to the very middle of the telophase disc as reported by others (25) has not been seen by us thus far.

In addition to localizing nmMLCK-210 immunoreactivity at the contraction ring we also located it at the periphery of prospective daughter cells where it was at the regions of focal adhesions. See Figure 4.

CONCLUSION

While considerable information exists relating to cytokinesis in a variety of different cell lines nothing, thus far, has been reported for airway smooth muscle. In view of the fact that hypertrophy of this cell likely plays an important role in the narrowing of airways, and that is a hallmark of chronic asthma, this is surprising.

We believe that our report of the presence of non-muscle myosin light chain kiase in the nucleus during interphase and its translocation to the contraction ring in telophase/cytokinesis—and presumably its activation—is the first in the field for any cell type. We feel it is probably the most important enzyme participating in activating the contractile machinery at the contraction and thus regulating the final phase of cell division.

The presence of Rho kinase and its activation is probably not directly involved in contraction. It's most likely function is in actin filamentogenesis at the contraction ring. Citron kinase has been reported at the contraction ring and very likely regulates furrow ingression. Just how it interacts with nonmuscle myosin light chain kinase to turn on the contractile machinery is unknown. Experiments in which the kinases are blocked and the effects on cytokinesis studied, need to be conducted.

ACKNOWLEDGEMENTS

The work described was made possible by operating grants from the Canadian Institute of Health Research and the Manitoba Institute of Child Health Research. Autoimmune serum was kindly provided by Dr. Frixos Paraskevas of the University of Manitoba. Antibodies against CENP-F were the gift of Dr. Rattner of the University of Saskatchewan.

REFERENCES

1. Fishkind DJ, Wang Y-C. 1995. New horizons for cytokinesis. Current Opinion in Cell Biology 7:23–31.
2. Satterwhite LL, Pollard TD. 1992. Cytokinesis. Current Opinion in Cell Biology 4(4):43–52.
3. Wolf WA, Chero T-L, Chisholm RL. 1999. Regulation of cytokinesis. Cellular and Molecular Life Sciences 55:108–120.
4. Glotzer M. 1997. The mechanism and control of cytokinesis. Current Opinion in Cell Biology 9:815–823.
5. Hales KG, Erfei B, Wu J-Q, Adam JC, Yu I-C, Pringle JR. 1999. Cytokinesis: An emerging unified theory for eukaryoties. Current Opinion in Cell Biology 11:717–725.
6. Ostrow B, Chen P, Chisholm RL. 1994. Expression of a myosin RLC phosphorylation site-mutant complements the cytokinesis and developmental defects of dictyostelium RMLC null cells. J Cell Biol 127:1945–1955.
7. Andreassen PR, Palmer DK, Wener MH, Margolis RL. 1991. Telophase disc: A new mammalian mitotic organello that bisects telophase cells with a possible function in cytokinesis. J Cell Sci 99: 523–534.
8. Larmarche N, Hall A. 1994. Gaps for rho-related GTPases. Trends Genet 10:436–440.
9. Narumiya S. The small GTPase rho: cellular functions and signal transduction. J Biochem (Tokyo), 228. 120:215–228.
10. Kishi K, Sasaki T, Kuroda S, Itoh T, Takai Y. 1993. Regulation of cytoplasmic division of Xenopus embryo by rhop21 and its inhibitory GDP/GTP exchange protein (rho GD1). J Cell Biol 120:1187–1195.
11. Amanio M, Ito M, Kumura K, Fukata Y, Chihara K, Nakano T. 1996. Phosphorylation and activation of myosin by rho-associated kinase. J Biol Chem 271:20246–20249.
12. Yamochi W, Tanaka K, Nonaka H, Maeda A, Musha T, Takai Y. 1994. Growth site localization of Rho I small GTP-binding protein and its involvement in bud formation in S cervisiae. J Cell Biol 125:1077–1093.

13. Madaule P, Eda M, Watanabe N, Fryisawa K, Matsuoka T, HB, et al. 1998. Role of citron kinase as a target of the small GTPase rho in cytokinesis. Nature 394:491–494.
14. Kong H, Cheng Z-Q, Stephens DL, Stephens NL. 2001. Study of canine airway smooth muscle muscle cell (CASMC) growth: The possible role of non-muscle myosin light chain kinase (NM-MLCK) in cytokinesis. Am J Resp and Cell Mol Biol 163(5):A973.
15. Swan KA, Severson AF, Cartero JC, Martin PR, HS, et al. 1998. A c.Elegans FH gene required for a late step in embryonic cytokinesis. J Cell Sci 111:2017–2027.
16. Fishkind DJ, Wang Y-L. 1993. Orientation and 3-D organization of actin filaments in dividing cultured cells. J Cell Biol 123:837–848.
17. Jordan P, Karess P. 1997. Myosin light chain-activating phosphorylation sites are required for oogenesis in drosophila. J Cell Biol 139:1805–1819.
18. DeBiasio RL, Lohka MJ, Wilson KL, Post PL, Taylor DL. 1996. Myosin II transport organization and phosphorylation evidence for cortical flow/solation-contraction coupling during cytokinesis and cell locomotion. Mol Biol Cell 7:1259–1282.
19. Ciapa B, Pesando D, Wilding M, Whittaker M. 1994. Cell cycle Ca2+ transients driven by cyclic changes in IP3 levels. Nature 368:875–878.
20. Bauer F, Urdaci M, Aigle N, Crouzel M. 1993. Alteration of a yeast SH3 protein leads to conditional viability with defects in cytokinesis. Mol Cell Biol 13:5070–5084.
21. Leo DJ, Reed SI. 1993. Morphogenesis in the yeast cell cycle: regulation by CDC38 and cyclins. J Cell Biol 120:1305–1320.
22. Konshute N, Yamano H, Niva H, Yoshida T, Yanagida M. 1993. Negative regulation of mitosis by the fission yeast proteins phosphatase. Genes Dev 7:1059–1071.
23. Egelhoff TT, Lec RJ, Spudich JA. 1993. Dictostelium myosin heavy chain phosphorylation sites regulate myosin filament assembly and localization in vivo. Celll 75:363–371.
24. Satterwhite LL, Lohka MJ, Wilson KL, Corden JL. 1992. Phosphorylation of myosin-II RLC by cyclin p34cdc2i a mechansism for the timing of cytokinesis. J Cell Biol 118:595–605.
25. Liao H, Winkfein RJ, Mack G, Rattner JB, Yen TJ. 1995. CENP-F is a protein of the nuclear matrix that assembles onto kineto-chores at late G2 and is rapidly degreaded after mitosis. J Cell Biol 130(3):507–518.
26. Earnshaw WC, Rothfield N. 1985. Identification of a family of human centromeric proteins using autoimmune sera from patients with scleroderma. Chormosoma 19:313–321.
27. Mabuchi I. 1986. Biochemical aspects of cytokinesis. Int Rev Cytol 101:105–245.
28. Rappaport R. 1986. Establishment of the mechanism of cytokinesis in animal cells. Int Rev Cytol 105:245.
29. Bray D, White JG. 1988. Cortical flow in animal cells. Science 239:883–888.
30. Neujahr R, Albrecht R, Kohler J, Matzner M, Schwartz J, Westphal M. 1998. Microtubule-mediated centrosome mobility and the positioning of cleavage furrows in multinmucleate myosin II-null cells. J Cell Sci 111:1227–1240.
31. Wheatley SP, Wang W. 1996. Midzone microtubule bundles are continuously required for cytokinesis in cultured epithelial cells. J Cell Biol 135:981–989.
32. Harris AK, Grewalt SL. 1989. Simulation testing of mechanisms for inducing the formation of the contractile ring in cytokinesis. J Cell Biol 109:2215–2223.
33. Raich WB, Moran AN, Rothamn JH, Hardin J. 1998. Cytokinesis and midzone microtubule organization in C.Elegans require the kinesion-like portein ZEN-4. Mol Biol Cell 8:2605–2615.
34. Wolfert L. 1960. The mechanics and mechanism of cleavage. Int Rev Cytol 10:163–216.
35. Sabry JH, Moores SL, Ryan S, Zhang J-H, Spudich JA. 1997. Myosin heavy chain phosphorylation sites regulate myosin localization during cytokinesis in live cells. Mol Biol Cell 8:2605–2615.
36. Huber HL, Koessler KK. 1922. The pathology of bronchial asthma. Arch Int Med 30:689.
37. Heard BE, Hossain S. 1973. Hyperplasia of bornchial muscle in asthma. J Pathol 110:319–331.
38. Ebina M, Takahashi T, Chuba T, Motomiya M. 1993. Cellular hypertrophy and hyperplasia of aireay smooth muscle underlying bronchial asthma. Am Rev Resp Dis 130:720–726.
39. Sabonya R. 1984. Quantitative structural alterations in long-standing allergic asthma. Am Rev Respir Dis 130:2289–2922.
40. Dunnill DMS, Massarella GR, Anderson JA. 1969. Anatomy of the bronchi in normal subjects, in status asthmaticus, in chronic bornchitis, and in emphysema. Thorax 24:176–179.
41. Takizawa T, Wurtbeck WMT. 1971. Mucus and mucous gland size in the major bronchi of patients with chronic bronchitis, asthma and asthmatic bornchitis. Am Rev Resp Dis 104:331–336.

42. Carroll NG, Elliott J, Morton AR, James AL. 1993. The structure of large and small airways in non-fatal and fatal asthma. Am Rev Resp Dis 147:405–410.
43. Kuzoano K, Bosken CH, Pare PD, Bai TR. 1993. Small airways dimensions in asthma and in COPD. Am Rev Resp Dis 148:1220–1228.
44. Colburn JC, Muchnoff CH, Hsu L-C, Slaughter CA, Kamm KE. 1977. Sites phosphorylated in myosin light chain in contracting smooth muscle. J Biol Chem 263(5):19166–19173.
45. Morrison DL, Sanghera JE, Stewart J, Sutherland C, Walsh MP, Pelek SL. 1996. Phosphorylation and activation of smMLCK by MAPkinase and CDK-1. Biochem Cell Biol 74:549–557.

Pathophysiology of Cardiovascular Disease,
edited by N.S. Dhalla, H. Rupp,
A. Angel & G.N. Pierce
Kluwer Academic Publishers, Boston, 2004

ADP-Ribosylation and the Cardiovascular System

Lorraine Yau and Peter Zahradka

Institute of Cardiovascular Sciences
St. Boniface General Hospital Research Centre
and
Department of Physiology
University of Manitoba

Summary. Post-translational modification of proteins with ADP-ribose alters their physical and functional characteristics. The relationship between ADP-ribosylation and several bacterial toxins (e.g. cholera toxin, pertussis toxin, diphtheria toxin) is well established. In contrast, the function of the endogenous ADP-ribosyl transferase enzyme is poorly understood. The latter statement also applies to poly(ADP-ribose) polymerase, although PARP has become familiar due to its association with apoptosis. This article summarizes the enzymology of ADP-ribosylation, reviews the various cellular processes in which it may participate and examines the possible functions of ADP-ribosylation in cardiovascular tissues. Poly(ADP-ribosyl)ation, which has been linked to ischemia-reperfusion injury, is known to participate in DNA repair. In contrast, mono(ADP-ribosyl)ation is best known for modulating G protein function. In the cardiovascular system, mono(ADP-ribosyl)ation may exert effects by modifying growth factors or by transducing the intracellular effects of nitric oxide. Alternatively, mono(ADP-ribosyl)ation could influence cell migration by modifying cytoskeletal proteins. By also examining the role of ADP-ribosylation in other systems, considerable insight into the possible contributions of ADP-ribosylation to cardiovascular function can be extrapolated to both normal and pathological conditions.

Key words: ADP-ribosylation, PARP, ADP-ribosyl transferase, cardiac, vascular.

OVERVIEW

ADP-ribosylation is a post-translational modification of proteins that involves the transfer of an ADP-ribose moiety from NAD^+ to an acceptor amino acid. ADP-ribosylation reactions can be divided into 2 distinct classes: mono(ADP-ribosyl)ation

Corresponding author: Peter Zahradka, Institute of Cardiovascular Sciences, St. Boniface Research Centre, 351 Tache Avenue, Winnipeg, MB, Canada R2H 2A6. Tel: 204-235-3507, Fax: 204-233-6732, E-mail: peterz@sbrc.ca.

and poly(ADP-ribosyl)ation. Both are considered important for a variety of cellular functions including cell signaling, cell differentiation, cell growth, cell proliferation, cell death and DNA repair. However, mono(ADP-ribosyl)ation controls enzymatic activity directly while poly(ADP-ribosyl)ation modulates the binding of proteins to nucleic acids.

POLY(ADP-RIBOSYL)ATION

Poly(ADP-ribosyl)ation is catalyzed by the enzyme poly(ADP-ribose) polymerase (PARP, EC 2.4.2.30), which transfers ADP-ribose to nuclear acceptor proteins. This reaction is thought to be ubiquitous in eukaryotic cells, with the possible exception being yeast. Poly(ADP-ribosyl)ation is associated with a variety of DNA-related processes including chromatin condensation, replication, recombination, DNA repair, gene expression, cellular differentiation, growth and transformation (1–4) and, more recently, the phenomena of apoptosis (5–8). A correlation also exists between PARP activity in mononuclear leukocytes of mammals and species-specific life span (3,9,10). ADP-ribosylation is a mechanism for the alteration of chromatin structure and this property may be the foundation of its effects on several cellular functions (11–13). A shuttle mechanism that links the synthesis and degradation (i.e. ADP-ribosylation cycle) of poly(ADP-ribose) polymers with changes in PARP activity and DNA binding (14) also accounts for the contribution of PARP to various nuclear processes (15). Furthermore, the ADP-ribosylation cycle (Figure 1) has been described in a variety of cell types including lymphocytes (16), endothelial cells (17,18), fibroblasts (19), smooth muscle cells (SMCs) (20) and hepatocytes (21,22). Degradation of poly(ADP-ribose) polymers is carried out by the enzyme poly(ADP-ribose) glycohydrolase (3). The apparently constitutive activation of this enzyme accounts for the short half-life of poly(ADP-ribose) polymers within the cell nucleus. Poly(ADP-ribose) glycohydrolase has been reviewed elsewhere (3,23) and will not be discussed here.

POLY(ADP-RIBOSE) POLYMERASE (PARP)

PARP is a nuclear enzyme of MW 113,000 that consists of 1014 amino acids. The gene encoding PARP is located on chromosome 1q, the same as that for the proto-oncogene *trk* and for the cytokine TGF-β (24). From biochemical and genetic studies, three functional domains have been identified: i) an N-terminal 46-kDa DNA binding domain that contains two Zn^{2+} finger structures, a nuclear localization signal and the newly described 24-kDa apoptotic domain (which is generated after cleavage of PARP by caspase-3 (25)); ii) a 22-kDa automodification domain, and iii) a 54-kDa NAD^{+} binding domain at the C-terminus (26). Within this protein, two functional modules have been identified: the N-terminus, which acts as the sensor for DNA nicks, and the C-terminus which synthesizes the poly(ADP-ribose) polymers. Within each of these domains, there are strongly conserved regions (e.g. Zn^{2+} fingers, NAD^{+} binding domain, DNA binding domain, β-sheet structures, Rossman fold structure), and homology of this enzyme is greater than 90% among

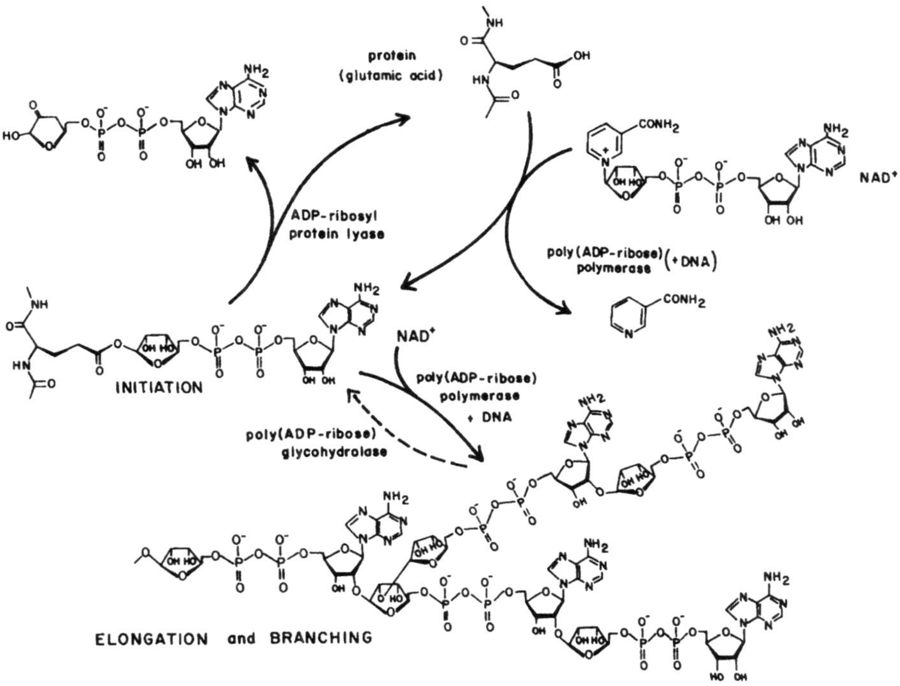

Figure 1. Poly(ADP-ribosyl)ation.

mammals (with 61% similarity and 43% identity to Drosophila) (26). It is interesting that PARP is so highly conserved in eucaryotes despite the high energy cost associated with the ADP-ribosylation cycle, which points to an important role for this enzyme in cellular function.

MECHANISM OF ACTION

PARP is a multifunctional enzyme having the capacity to catalyze 3 separate reactions: i) initiation (1 reaction per acceptor site), ii) elongation (greater than 200 reactions per site), and iii) branching (5–7 reactions per site) (26–28). Polymers of greater than 100 units in length have been detected. Using NAD^+ as a substrate, PARP catalyzes a reaction that involves the transfer and subsequent synthesis of ADP-ribose polymers, either linear or branched, on nuclear proteins. Furthermore, PARP is a catalytic dimer with an intermolecular automodification reaction (29,30). Experiments utilizing purine-base substituted NAD^+ analogs have conclusively demonstrated that the presence of an amino group within the purine base at a specific position is important for efficient catalysis (31). Furthermore, PARP is dependent on the presence of DNA strand breaks for its full enzymatic activity (the Zn^{2+} fingers serve as sensors (14)), and PARP is an important component of efficient DNA excision repair (4,32).

PARP SUBSTRATES

Over the years, more than 30 nuclear proteins have been proposed as targets for poly(ADP-ribosyl)ation, including histones (H1, H2B) (33–35), DNA polymerase-α (36,37), endonucleases (38), topoisomerase I (39,40), DNA ligase II (41), HMG proteins (42,43), RNA polymerases (44,45), Fos (46), p53 (47–49) and PARP itself (i.e. automodification) (37,50–53). However, whether these proteins are true endogenous substrates of PARP should be considered carefully since many of the experimental procedures employed for the identification of these acceptor proteins involved the use of permeabilized cells and nuclear extracts from cells. These procedures in and of themselves would have resulted in cellular, nuclear and DNA damage, and thus PARP activation, with the consequence of having artifactually induced ADP-ribosylation.

CELL BIOLOGY

PARP is a critical regulatory component of the cellular response to stress, including DNA damage. PARP is expressed constitutively and usually found bound to chromatin and requires the presence of DNA nicks or strand breaks (single or double stranded) for activation (4). It participates in DNA repair by associating with strand breaks, and the subsequent elaboration of poly(ADP-ribose) polymers then signals the cell to switch off DNA synthesis temporarily (32). This serves as a mechanism to ensure that lesions are not replicated before DNA repair can be completed and may indicate that PARP is a survival factor that plays an essential and positive role during DNA damage recovery (54). Interestingly, PARP has also been shown to play a role in apoptosis or programmed cell death (PCD) (8). The mechanism behind this action is postulated to involve a decrease in a cell's energy pools (i.e. NAD^+) upon massive activation of PARP (by DNA damage/strand breaks) (55–58). Since the degree of PARP activation is dependent on the amount of DNA damage, this may be the signal that determines whether the cell either repairs the damage or, if the damage is too great, induces PCD. Activation of the apoptotic program involves cleavage of PARP by proteases such as protease resembling ICE (interleukin-β converting enzyme) (prICE) or caspase-3 (25,59,60) and, ultimately, PARP cleavage may represent a critical juncture at which point a cell must decide whether it is able to recover or undergo PCD or necrotic cell death. The role of PARP in apoptosis was confirmed by observations that 3-aminobenzamide (3AB), an inhibitor of PARP, could rescue cells from undergoing PCD (8,19,61,62). Furthermore, PARP has been reported to modify p53, a tumour suppressor protein that is associated with activation of apoptosis (49,63,64).

Increased PARP activity has also been found in proliferating cells (37,65) and the transcription of PARP has been detected before the onset of DNA synthesis early in the cell cycle (66). Moreover, activation of PARP may contribute to regulation of gene activity as has been observed with phosphoenolpyruvate carboxykinase (PEPCK) and procollagen gene expression (67–69). PARP may also participate in the transcription of ribosomal RNA, although it has no comparable role in ribosomal protein expression (70). Furthermore, PARP has been reported to exist associated with the nuclear matrix in rat testis (71). Interestingly, activation of PARP in this system involves both automodification and heteromodification reactions at the nuclear matrix level with both histones and nuclear matrix acting as acceptors (72). Finally, NF-κB-dependent gene expression may be regulated by PARP (73–75), with possible repercussions in the sequelae of both apoptosis and inflammatory processes (76).

A possible role for PARP in cell proliferation and differentiation has been identified through the application of known PARP inhibitors. For example, 3AB inhibits phytohemagglutinin-induced lymphocyte proliferation (77), cytokine-dependent proliferation of hemopoietic cells (78), and the growth of both smooth muscle and hepatic cells in response to mitogen treatment (70,79). In addition, the differentiation of mononuclear (80) and Friend erytholeukemia (81) is inhibited by 3AB. A more direct indication that PARP is essential for cell differentiation has been obtained in studies of HL-60 and NB4 cells (82,83) where over-expression of PARP was shown to block neutrophilic differentiation. The investigations by Bhattia et al. (83) are unique in that they did not rely on PARP inhibitors to provide the correlation between poly(ADP-ribosyl)ation and a specific cellular process. The limitations of solely relying on inhibitors is directly related to the questionable specificity of agents employed as PARP inhibitors, and several studies have raised doubts about this strategy (69,84). Future studies will therefore be required to confirm or refute many of the associations that have already been reported.

PARP IN THE CARDIOVASCULAR SYSTEM

In the cardiovascular system, inhibition of PARP by 3AB has been shown to attenuate cell death associated with oxidant stress in isolated cardiac myoblast preparations and in experimental models of cardiac ischemia (85,86). As a result, inhibition of PARP may be therapeutic for myocardial infarction due to a prevention in the decline of high energy phosphates during ischemia/reperfusion (21,85). Conversely, activation of PARP may contribute to the pathophysiology of reperfusion injury via the PARP suicide hypothesis (86). It is suspected that ischemia/reperfusion results in DNA damage and therefore activates PARP which then depletes energy pools (both NAD^+ and ATP) and results in activation of the cell death program (i.e. PARP suicide hypothesis) (55–58,86). It is also plausible that an inhibition of mononuclear cell activation (80) could limit development of oxygen radicals in the tissues, and by this mechanism reduce the damage sustained during reperfusion.

In the vasculature, PARP may play a role in the pathogenesis of endothelial dysfunction, having a role in peroxynitrite-induced cytotoxicity (20). Development of endothelial dysfunction in this model has been demonstrated to improve with 3AB treatment (20). As well, inhibitors of PARP such as 3AB and HMBA (hexamethylenebisacetamide) have been shown to suppress SMC proliferation (87,88).

The study by Thyberg et al. (87) also revealed significant changes in cytoskeletal organization and gene expression. Furthermore, based on the structural properties of the cells, they concluded HMBA prevented phenotypic modulation in response to culture conditions. Interestingly, Weissberg et al. (89) had previously proposed that the effects of HMBA on the differentiation status of SMCs are mediated through TGF-β and not poly(ADP-ribosyl)ation. Nevertheless, it was observed that inhibitors of poly(ADP-ribosyl)ation instead prevented SMC proliferation, thus dissociating cell growth processes from cell differentiation.

LESSONS LEARNED FROM PARP KNOCKOUT MICE

The preparation of $PARP^{-/-}$ knockout mice (KO-1: (90), KO-2: (91), KO-3: (92)) has permitted the detection of additional PARP isoforms within the cell, since $PARP^{-/-}$ is not a developmentally lethal phenotype (90), and poly(ADP-ribose) polymers are present in $PARP^{-/-}$ mice. Furthermore, cells and cell lines derived from neuronal cells and fibroblast cells of $PARP^{-/-}$ mice are able to undergo DNA repair after exposure to reactive oxygen species (93). Nevertheless, cells derived from the KO-1 mouse had a marginally slower rate of proliferation and skin hyperplasia was observed in about 1/3 of older mice (90). Interestingly, $PARP^{-/-}$ mice are resistant to drug-induced inflammation (90) and this may be related to a requirement for PARP in the activation of NF-κB-dependent target genes (73). Recently, both PARP-2 and PARP-3 have been identified and cloned (94,95), and PARP-2, a 62-kDa protein that is also DNA damage-dependent, may account for the residual poly(ADP-ribose) synthesis observed in PARP-1 deficient cells exposed to DNA damaging agents (96). Taken together, it may be concluded that PARP is dispensible during embryogenesis and that its DNA repair function is sufficiently important for genome protection to warrant a redundant enzyme activity. Alternatively, PARP-2 may play a more significant role in DNA repair, since, interestingly, the generation of knockout mice has indicated that PARP (or PARP-1 as it should now be identified) may play a critical role in cell death and in various inflammatory processes (96). The presence of a PARP gene family may suggest that specific PARP enzymes have specific cellular functions and provide the cell with an additional regulatory mechanism for distinct cell processes. This premise is supported by the inclusion of both tankyrase and VPARP in this family (97). Tankyrase, which is a telomere-associated protein, may regulate telomere length and consequently contribute to cellular longevity. In contrast, VPARP is localized to the cytosolic vault complex (98). Although the function of the vault complex is not completely understood, a role in nucleo-cytoplasmic trafficking has been proposed (99). VPARP may therefore explain the presence of a cytoplasmic poly(ADP-ribosyl)ating activity in association with mRNA (100,101).

MONO(ADP-RIBOSYL)ATION REACTIONS

Mono(ADP-ribosyl)ation reactions are catalyzed by a family of mono(ADP-ribosyl)transferases (mART) found in viruses, bacteria and eucaryotic cells

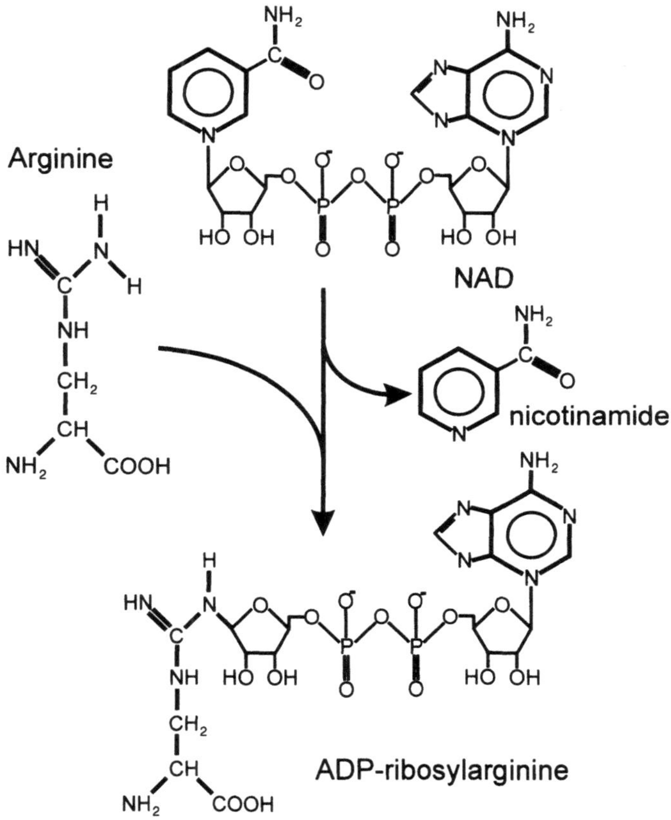

Figure 2. Mono(ADP-ribosyl)ation.

(102–104). This is a reversible post-translational modification of proteins in eucaryotic and prokaryotic cells that involves the transfer of an ADP-ribose moiety from NAD^+ to an acceptor amino acid (Figure 2). The reaction catalyzed by mART's is distinct from that catalyzed by PARP in several ways. Most notably, the chain length of ADP-ribose units is shorter (one compared to 100–200), but the chemical nature of the glycosidic linkage, the character of the enzymes involved and the site of reaction are also distinct (104,105). The role of mono(ADP-ribosyl)ation in cell function is related to cellular metabolism and has been associated with cellular differentiation (106), cellular proliferation (87), cell migration (107–110) and both intra- and extra-cellular communication (111–114).

Catabolism of mono(ADP-ribose) is catalyzed by ADP-ribosyl(amino acid) transferase hydrolases (EC 3.2.2.19). The end result is the regeneration of free, unmodified amino acid (arginine, cysteine, diphthamide or asparagine). Hydrolases have been identified in bacterial, avian and mammalian systems, appearing to be ubiquitous in eukaryotic tissues, and thus supporting the premise that ADP-ribosylation reactions are important mediators of cellular regulation. The best characterized of the hydrolases is the ADP-ribosylarginine hydrolase in eucaryotic cells (102,115,116). This hydrolase activity was detected in the soluble fraction of turkey erythrocytes (117), cultured mouse cells (118), rat skeletal muscle (119) and rat brain (120). Thus a eucaryotic ADP-ribosylation cycle is hypothesized based on the reversible modification of proteins by transferases and hydrolases (102,116), much like the reversible phosphorylation modification of proteins by kinases and phosphatases (121). Additionally, ADP-ribose can be processed by phosphodiesterases and phosphatases as seen with the GPI-anchored skeletal muscle mART, generating a phosphoribosyl-integrin and 5′AMP, although the physiological consequences of this structure have yet to be examined (122,123).

MONO(ADP-RIBOSYL)TRANSFERASE (mART)

The discovery of eucaryotic protein ADP-ribosylation (124–126) resulted in the identification of mARTs. Bacterial toxins such as cholera toxin, pertussis toxin, diphtheria toxin and botulinum C3 endotoxin ADP-ribosylate a variety of eucaryotic and prokaryotic acceptor proteins, and modulate their associated cellular functions (103,127–132). A variety of animal tissues and organs are also now known to contain mART and mART activity (105). Vertebrate mART activity was first detected in turkey erythrocytes (133,134), rat liver homogenates (135) and Xenopus tissues (122,136,137). Presently, 6 avian mARTs (4 turkey erythrocyte ARTs: A, B, C and A′, and chicken ARTs from hen liver and heterophil granules) and more than 13 mammalian mARTs (e.g. ART 1–5, ART I–IV from rat brain, and 4 mARTs from rat adrenal medulla) have now been purified and/or cloned and characterized (122,138,139). This subject has been reviewed recently by Okazaki and Moss (122).

Of the mammalian mARTs, expression was initially identified in cardiac and skeletal muscle, leukocytes/lymphocytes, brain and testis (105,106,140–149). Subsequently, mART expression was demonstrated in hepatoma cells (150,151), SMCs (152) and endothelial cells where these enzymes participated in signal transduction and prostacyclin production (153).

Vertebrate mARTs (e.g. NAD:arginine(ADP-ribosyl)transferases (EC 2.4.2.31)) are a family of enzymes ranging in molecular weight from approximately 25- to 36-kDa and 66- and 69-kDa (4 distinct ART's from rat brain—I to IV and rat adrenal medulla) (122). Despite a lack of overall similarity in the amino acid sequences (i.e. lack of conservation among sequences), there appear to be 3 regions (regions I, II, III) of sequence homology that make up the catalytic site. A precise configuration of these regions is required for the ADP-ribose transfer reaction (122,154,155). Conserved regions house an NAD-binding cleft composed of an α-helix bent over a β-strand, along with an arginine or histidine residue and an active-site glutamate which are critical for enzymatic activity (122,154). Furthermore, 2 separate groups of enzymes can be defined by the presence of either arginine (cholera and pertussis toxin-like) or histidine (diphtheria toxin-like) in region I. Region II contains a specific amino acid motif (aromatic-hydrophobic-Ser-X-Ser-hydrophobic) and region III contains the active-site glutamate. For more details, please refer to the review by Okazaki and Moss (122). Interestingly, several of the mammalian mARTs have regions of homology to glycosylphosphatidylinositol (GPI)-anchored proteins, and thus may be membrane bound with an extracellular orientation (104,156,157). Secretory and intracellular counterparts also exist and there may therefore be 3 discreet families of transferases (158).

ENZYMOLOGY OF mART

A common mechanism of substrate binding and catalysis apparently exists between and among both bacterial and eucaryotic mARTs. Mammalian and bacterial mARTs are believed to act via an SN2-like mechanism in which nucleophilic attack on NAD^+ is followed by displacement of the nicotinamide and inversion of configuration (155). In this enzymatic reaction, a single ADP-ribose moiety is transferred to the acceptor substrate. Amino acid-specific acceptors for mARTs include arginine, cysteine, diphthamide, histidine and asparagine of which arginine-mARTs are the best characterized thus far (104). Mechanisms for regulating mARTs have not been described since little research has been done in this area, however, Weng et al. (159) did recently report on the auto(ADP-ribosyl)ation of ART5 as a mechanism for the regulation of mART activity.

mART SUBSTRATES

Vertebrate mARTs are typically species- and tissue-specific with respect to acceptor proteins (160). Although a large variety of proteins are ADP-ribosylated, the majority of mARTs are arginine-specific (and can also modify other simple guanidino derivatives) and a variety of potential substrates have been described (113,122,161,162). Acceptors of ADP-ribose in these mammalian (in vitro) systems include casein, ovalbumin and simple guanidino compounds such as arginine and agmatine (104). Additional target proteins include Ha-ras-p21, transducin, protein kinases (with subsequent reduction in activity), and skeletal Ca^{2+}-ATPase (with subsequent decrease in activity) (137,163). Finally, endogenous arginine-dependent ADP-ribosylation of G proteins (i.e. G_s, G_i and G_o) is well documented in mammalian cells and tissues (140,143,164–166).

Endogenous vertebrate ADP-ribosylation of cysteine residues is less common, but has been observed in human and bovine erythrocytes via the actions of a cysteine-specific mART purified from erythrocyte and platelet membranes. This cys-mART is also reported to modify $G_i\alpha$ (122,167,168). Interestingly, ADP-ribosyl-cysteine linkages can occur as the result of a nonenzymatic reaction involving ADP-ribose and NAD^+ glycohydrolases. In this case, an ADP-ribosyl-thiazolidine linkage is generated in place of the thioglycoside ADP-ribosyl-cysteine linkage, the type normally observed with pertussis toxin catalyzed reactions (122,169).

CELL BIOLOGY OF mART

Bacterial mARTs have a variety of biological consequences and functions within eucaryotic cells and organisms. For example, cholera toxin ADP-ribosylates an arginine residue of $G_s\alpha$, the stimulatory α-subunit of the heterotrimeric GTP-binding (G) proteins, and irreversibly inhibits its GTPase activity. The result is activation of adenylyl cyclase and a subsequent increase in cellular cAMP levels (124,126,170). On the other hand, pertussis toxin ADP-ribosylates a cysteine residue in the α-subunit of G_i and G_o which blocks the inhibitory effect of the α_i-subunit on adenylyl cyclase and results in activation of the cAMP cascade and/or uncoupling of the G protein from its receptor (115,171). Nevertheless, both modifications result in abnormalities in cellular electrolyte flux and fluid secretion in the small intestines (172–174). Diphtheria toxin (and *Pseudomonas aeruginosa* exotoxin A) ADP-ribosylates a diphthamide (i.e. a modified histidine) residue in elongation factor-2 (EF-2) resulting in the inactivation of EF-2. This disrupts the regulatory mechanism for translocation of elongating proteins across the endoplasmic reticulum membrane and results in inhibition of protein synthesis (175–178). Interestingly, a fusion protein of IL-2 and diphtheria toxin has been successfully applied for the treatment of neoplastic diseases. This fusion protein targets cells with IL-2 receptors and the resultant ADP-ribosylation of EF-2 inhibits cell proliferation (179). Finally, exotoxin C2 and C3 both modify components of the cytoskeletal network. Botulinum C2 exotoxin ADP-ribosylation of an arginine causes inhibition of (non-muscle) actin polymerization and subsequent degradation of the microfilament/cytoskeletal network in cells (127). Botulinum C3 exotoxin ADP-ribosylation of an asparagine residue on the small GTP-binding protein Rho results in the inactivation and attenuation of Rho-associated signaling pathways (180), resulting in decreased tyrosine phosphorylation and phosphatidylinositol 3-kinase (PI3-kinase) activation. Consequently, an altered GTP-binding protein interaction with the cytoskeletal elements has the potential for modulating cell proliferation and differentiation via disturbances in the organization of the actin cytoskeleton (127,181–186).

As mentioned previously, eucaryotic mARTs are involved in a variety of cell functions, including cell differentiation, proliferation, migration and cellular communication (87,106–114). Specific modification of certain cytoskeletal/intermediate filament proteins would result in the alteration of the above-mentioned cell functions. For example, desmin is ADP-ribosylated by an endogenous arg-mART in skeletal muscle (187–189). This modification results in desmin disassembly (189), with consequences related to cell differentiation, migration, proliferation and intracellular communication (161). ADP-ribosylation of desmin is an important regulatory mechanism in differentiating muscle cells, with mART activity increasing dramatically after myoblast fusion (106). In addition, desmin from SMCs has been shown to react with an antibody specific for ADP-ribosylated-arginine (121). A second example would be the modification of actin filaments. ADP-ribosylation of actin in human neutrophils is associated with inhibition of neutrophil chemotaxis (110). Additionally, non-muscle β/γ-actin, skeletal muscle α-actin and smooth muscle γ-actin are ADP-ribosylated in vitro with resultant inhibition of actin polymerization (190). Furthermore, endogenous ADP-ribosylation of heterophil non-muscle actin was proposed as a mechanism for regulating heterophil phagocytosis, secretion and chemotaxis (functions dependent on actin de/polymerization) (190). The final example is the ADP-ribosylation of integrin α7, the receptor for the extracellular protein laminin, on skeletal muscle cells by mART. This modification may modulate cell-cell or cell-matrix interactions, with effects on cellular differentiation and/or proliferation (113,123,191).

Other eucaryotic cell functions that may be altered by mART include neuronal transmission (in the brain, ADP-ribosylation of G proteins can modify neuronal signal transmission) (139,140), and regulation of T-cell proliferation and cytolytic activity (ADP-ribosylation of a $p56^{lck}$-associated protein) (162). More recently, Lupi et al. (192) have reported the modification of the β-subunit of $G_{\beta\gamma}$ by arg-mART as a novel mechanism of regulating G protein-mediated signal transduction. This modification prevented the inhibition of calmodulin-stimulated type I adenylyl cyclase. Finally, modification of MARCKS and tubulin by mART has been demonstrated to lead to changes in cytoskeletal and microtubular assembly/organization (193–195), as well as alterations in the ability of MARCKS to bind calmodulin and interact with the plasma membrane (194).

From a more clinical point of view, in experimental diabetic neuropathy, mART activity has been associated with a reduction of substance P axonal transport and subsequent onset of diabetic peripheral neuropathy (196). It was observed that exogenous ADP-ribosylation reactions were reduced in extracts from diabetic retinas, suggesting increased cytosolic endogenous ADP-ribosylation activity. Treatment with an inhibitor of ADP-ribosylation (silybin) increased the extent of exogenous ADP-ribosylation (196).

MONO(ADP-RIBOSYL)ATION IN THE CARDIOVASCULAR SYSTEM

In the cardiovascular system, arg-mART has been found in relatively high levels in neonatal and adult rat hearts (145). In cardiac muscle, ADP-ribosylation reactions have been associated with developmental regulation (145). Furthermore, Jones and Baird (197) have shown that bFGF, a growth factor intimately associated with the cardiovascular system, angiogenesis and the musculoskeletal system, is ADP-ribosylated on the surface of adult bovine arch endothelial cells (and in human hepatoma cells). This modification may represent an additional regulatory mechanism, in conjunction with phosphorylation, for this potent mitogen. In the past, bFGF has been found to be both phosphorylated and ADP-ribosylated in vitro (198–201). Similarly, ADP-ribosylation of PDGF has been shown to reduce its mitogenic activity (202). These represent new and unique ways by which ADP-ribosylation can control the response of cardiovascular tissues to mitogenic factors.

Nitric oxide (NO) is a second messenger produced by NO synthase (NOS) in response to a variety of signals that is able to stimulate endogenous ADP-ribosylation in vitro. Moreover, NO, acting as a free radical molecule, can activate ADP-ribosylation of several proteins in vivo (203–205) and result in cellular damage (206).

In the vasculature, NO-dependent ADP-ribosylation of G proteins by an endogenous ART in smooth muscle increases the activity of adenylyl cyclase and results in vasodilatation via a decrease in the activation of phospholipase C (207). As a result, it is postulated that hypertension-associated decreases in eNOS expression and activity would decrease the amount of NO produced, thereby leading to decreased ADP-ribosylation of G proteins and, consequently, increased vessel constriction since activation of G proteins by agonists is unopposed (207). Moreover, ADP-ribosylation was discovered in SMCs derived from bovine coronary artery (152). Here, this modification was found to modulate $G_s\alpha$ and thereby play a role in activating K^+ channels as induced by 11,12-EET. It is also of note that metabolism of ADP-ribose to adenosine can result in dilation of coronary arteries (208).

Stimulation of NO production in vascular tissues has been shown to cause growth inhibition. This observation has led to trials in which the use of NO generators as therapy for preventing the development of arterial lesions following revascularization procedures (e.g. balloon angioplasty, bypass graft surgery) was evaluated. Among the NO generators that have been tested is L-arginine (209,210). However, arginine and its derivatives (e.g. arginine methylester) are also known to inhibit mono(ADP-ribosyl)ation by operating as a pseudosubstrate and thus preventing modification of the target protein (135). Since there is evidence that inhibition of arginine-dependent mART blocks cell proliferation (87,211), L-arginine could influence vascular lesion formation via this mechanism rather than through NO. The report by Uemura et al. (212) has shed some light on this interesting question. In this study, both L-arginine and D-arginine were compared for their ability to block vascular lesion development, and it was observed that both enantiomers were equally effective. Since D-arginine is not a substrate of nitric oxide synthase, these data support the premise that arginine functions through its action as an inhibitor of mono(ADP-ribosyl)ation.

As was described earlier, inhibitors of poly(ADP-ribosyl)ation have been shown to prevent the differentiation and proliferation of SMCs (87). These investigators also tested a specific inhibitor of mono(ADP-ribosyl)ation in their study. A reduction in PDGF- and serum-induced DNA synthesis, as determined by thymidine autoradiography, was reported with the inhibitor MIBG (*meta*-iodobenzylguanidine). Interestingly, little effect was seen with MIBG in a separate experiment measuring radiolabelled thymidine incorporation in response to PDGF and serum. This inconsistency is puzzling in the absence of additional information. However, it may be significant that MIBG was effective only at concentrations above 100 μM, whereas it has been previously reported that inhibition of cell proliferation by MIBG is maximal at 50 μM (213). It is also known that MIBG interferes with mitochondrial respiration (214) and that this effect is not coupled to inhibition of cell proliferation (213). Since blocking respiration leads to the production of free radicals, thymidine autoradiography may be sufficiently sensitive to detect free radical-induced DNA repair processes. However, since there have been no additional studies examining this issue, further speculation into the role of mono(ADP-ribosyl)ation in relation to SMC growth is not possible.

CONCLUSIONS

It is becoming evident that ADP-ribosylation events have the potential to control multiple cellular processes. The definitive role of PARP in mediating DNA repair (15) and its connection to apoptosis (215) appear to have direct relevance to the cardioprotective effects of PARP inhibitors (216). Similarly, mono(ADP-ribosyl)ation has been shown to influence cytoskeletal organization, migration and intracellular signal transduction, all of which are essential for vascular function. Taken together, the studies described in this review suggest that ADP-ribosylation may constitute a novel participant in the cardiovascular system and vasculature, by regulating key cellular processes such as development, differentiation, proliferation and SMC contractility (i.e. vasoactivity). Further studies will nevertheless be required to clarify how ADP-ribosylation can contribute to so many different systems and whether they could eventually be used as targets for therapeutic intervention.

REFERENCES

1. D'Amours D, Desnoyers S, D'Silva I, Poirier GG. 1999. Poly(ADP-ribosyl)ation reactions in the regulation of nuclear functions. Biochem J 342:249–268.
2. de Murcia G, Schreiber V, Molinete M, Saulier B, Poch O, Masson M, Niedergang C, Menissier de Murcia J. 1994. Structure and function of poly(ADP-ribose) polymerase. Mol Cell Biochem 138: 15–24.
3. Lagueux J, Shah GM, Menard L, Thomassin H, Duchaine C, Hengartner C, Poirier GG. 1994. Poly(ADP-ribose) catabolism in mammalian cells. Mol Cell Biochem 138:45–52.
4. Shall S. 1994. The function of poly (ADP-ribosylation) in DNA breakage and rejoining. Mol Cell Biochem 138:71–75.
5. Negri C, Donzelli M, Bernardi R, Rossi L, Burkle A, Scovassi AI. 1997. Multiparametric staining to identify apoptotic human cells. Exp Cell Res 234:174–177.
6. Pieper AA, Verma A, Zhang J, Snyder SH. 1999. Poly (ADP-ribose) polymerase, nitric oxide and cell death. Trends Pharmacol Sci 20:171–181.
7. Rosenthal DS, Ding R, Simbulan-Rosenthal CM, Vaillancourt JP, Nicholson DW, Smulson M. 1997. Intact cell evidence for the early synthesis, subsequent late apopain-mediated suppression, of poly(ADP-ribose) during apoptosis. Exp Cell Res 232:313–321.
8. Scovassi AI, Poirier GG. 1999. Poly(ADP-ribosylation) and apoptosis. Mol Cell Biochem 199: 125–137.
9. Burkle A, Muller M, Wolf I, Kupper JH. 1994. Poly(ADP-ribose) polymerase activity in intact or permeabilized leukocytes from mammalian species of different longevity. Mol Cell Biochem 138:85–90.
10. Grube K, Burkle A. 1992. Poly(ADP-ribose) polymerase activity in mononuclear leukocytes of 13 mammalian species correlates with species-specific life span. Proc Natl Acad Sci USA 89:11759–11763.
11. Hayaishi O, Ueda K. 1977. Poly(ADP-ribose) and ADP-ribosylation of proteins. Annu Rev Biochem 46:95–116.
12. Kleczkowska HE, Althaus FR. 1996. The role of poly(ADP-ribosyl)ation in the adaptive response. Mutat Res 358:215–221.
13. Purnell MR, Stone PR, Whish WJ. 1980. ADP-ribosylation of nuclear proteins. Biochem Soc Trans 8:215–227.
14. Zahradka P, Ebisuzaki K. 1982. A shuttle mechanism for DNA-protein interactions. The regulation of poly(ADP-ribose) polymerase. Eur J Biochem 127:579–585.
15. Lindahl T, Satoh MS, Poirier GG, Klungland A. 1995. Post-translational modification of poly(ADP-ribose) polymerase induced by DNA strand breaks. Trends Biochem Sci 20:405–411.
16. Schraufstatter IU, Hyslop PA, Hinshaw DB, Spragg RG, Sklar LA, Cochrane CG. 1986. Hydrogen peroxide-induced injury of cells and its prevention by inhibitors of poly(ADP-ribose) polymerase. Proc Natl Acad Sci USA 83:4908–4912.

17. Kirkland JB. 1991. Lipid peroxidation, protein thiol oxidation and DNA damage in hydrogen peroxide-induced injury to endothelial cells:role of activation of poly(ADP-ribose)polymerase. Biochim Biophys Acta 1092:319–325.
18. Thies RL, Autor AP. 1991. Reactive oxygen injury to cultured pulmonary artery endothelial cells:mediation by poly(ADP-ribose) polymerase activation causing NAD depletion and altered energy balance. Arch Biochem Biophys 286:353–363.
19. Yamamoto K, Tsukidate K, Farber JL. 1993. Differing effects of the inhibition of poly(ADP-ribose) polymerase on the course of oxidative cell injury in hepatocytes and fibroblasts. Biochem Pharmacol 46:483–491.
20. Szabo C, Cuzzocrea S, Zingarelli B, O'Connor M, Salzman AL. 1997. Endothelial dysfunction in a rat model of endotoxic shock. Importance of the activation of poly (ADP-ribose) synthetase by peroxynitrite. J Clin Invest 100:723–735.
21. Bowes J, Ruetten H, Martorana PA, Stockhausen H, Thiemermann C. 1998. Reduction of myocardial reperfusion injury by an inhibitor of poly (ADP-ribose) synthetase in the pig. Eur J Pharmacol 359:143–150.
22. Stubberfield CR, Cohen GM. 1988. NAD^+ depletion and cytotoxicity in isolated hepatocytes. Biochem Pharmacol 37:3967–3974.
23. Desnoyers S, Shah GM, Brochu G, Hoflack JC, Verreault A, Poirier GG. 1995. Biochemical properties and function of poly(ADP-ribose) glycohydrolase. Biochimie 77:433–438.
24. Smulson ME. 1994. Poly(ADP-ribose) polymerase gene on chromosome 1q: early role in differentiation linked replication; gene on human chromosome 13q: marker of carcinogenesis. Mol Cell Biochem 138:77–84.
25. Smulson ME, Pang D, Jung M, Dimtchev A, Chasovskikh S, Spoonde A, Simbulan-Rosenthal C, Rosenthal D, Yakovlev A, Dritschilo A. 1998. Irreversible binding of poly(ADP)ribose polymerase cleavage product to DNA ends revealed by atomic force microscopy: possible role in apoptosis. Cancer Res 58:3495–3498.
26. Alvarez-Gonzalez R, Pacheco-Rodriguez G, Mendoza-Alvarez H. 1994. Enzymology of ADP-ribose polymer synthesis. Mol Cell Biochem 138:33–37.
27. Alvarez-Gonzalez R, Jacobson MK. 1987. Characterization of polymers of adenosine diphosphate ribose generated in vitro and in vivo. Biochemistry 26:3218–3224.
28. Kiehlbauch CC, Aboul-Ela N, Jacobson EL, Ringer DP, Jacobson MK. 1993. High resolution fractionation and characterization of ADP-ribose polymers. Anal Biochem 208:26–34.
29. Bauer PI, Buki KG, Hakam A, Kun E. 1990. Macromolecular association of ADP-ribosyltransferase and its correlation with enzymic activity. Biochem J 270:17–26.
30. Mendoza-Alvarez H, Alvarez-Gonzalez R. 1993. Poly(ADP-ribose) polymerase is a catalytic dimer and the automodification reaction is intermolecular. J Biol Chem 268:22575–22580.
31. Oei SL, Griesenbeck J, Buchlow G, Jorcke D, Mayer-Kuckuk P, Wons T, Ziegler M. 1996. NAD^+ analogs substituted in the purine base as substrates for poly(ADP-ribosyl) transferase. FEBS Lett 397:17–21.
32. Satoh MS, Lindahl T. 1992. Role of poly(ADP-ribose) formation in DNA repair. Nature 356: 356–358.
33. Boulikas T. 1988. At least 60 ADP-ribosylated variant histones are present in nuclei from dimethylsulfate-treated and untreated cells. EMBO J 7:57–67.
34. Boulikas T. 1990. Poly(ADP-ribosylated) histones in chromatin replication. J Biol Chem 265: 14638–14647.
35. Huletsky A, Niedergang C, Frechette A, Aubin R, Gaudreau A, Poirier GG. 1985. Sequential ADP-ribosylation pattern of nucleosomal histones. ADP- ribosylation of nucleosomal histones. Eur J Biochem 146:277–285.
36. Simbulan CM, Suzuki M, Izuta S, Sakurai T, Savoysky E, Kojima K, Miyahara K, Shizuta Y, Yoshida S. 1993. Poly(ADP-ribose) polymerase stimulates DNA polymerase alpha by physical association. J Biol Chem 268:93–99.
37. Yoshida S, Simbulan CM. 1994. Interaction of poly(ADP-ribose)polymerase with DNA polymerase alpha. Mol Cell Biochem 138:39–44.
38. Tanaka Y, Yoshihara K, Itaya A, Kamiya T, Koide SS. 1984. Mechanism of the inhibition of Ca^{2+}, Mg^{2+}-dependent endonuclease of bull seminal plasma induced by ADP-ribosylation. J Biol Chem 259:6579–6585.
39. Ferro AM, Olivera BM. 1984. Poly(ADP-ribosylation) of DNA topoisomerase I from calf thymus. J Biol Chem 259:547–554.

40. Kasid UN, Halligan B, Liu LF, Dritschilo A, Smulson M. 1989. Poly(ADP-ribose)-mediated post-translational modification of chromatin-associated human topoisomerase I. Inhibitory effects on catalytic activity. J Biol Chem 264:18687–18692.
41. Creissen D, Shall S. 1982. Regulation of DNA ligase activity by poly(ADP-ribose). Nature 296: 271–272.
42. Tanuma S, Johnson GS. 1983. ADP-ribosylation of nonhistone high mobility group proteins in intact cells. J Biol Chem 258:4067–4070.
43. Tanuma S, Yagi T, Johnson GS. 1985. Endogenous ADP ribosylation of high mobility group proteins 1 and 2 and histone H1 following DNA damage in intact cells. Arch Biochem Biophys 237:38–42.
44. Furneaux HM, Pearson CK. 1980. Intracellular NAD^+ content and ADP-ribose polymerase activity of serum-stimulated baby hamster kidney fibroblasts. J Cell Physiol 105:401–407.
45. Taniguchi T, Suzuki S, Shizuta Y. 1985. Poly (ADP-ribosyl)ation of RNA polymerase II from wheat germ. Biochem Biophys Res Commun 127:526–532.
46. Amstad PA, Krupitza G, Cerutti PA. 1992. Mechanism of c-fos induction by active oxygen. Cancer Res 52:3952–3960.
47. Malanga M, Pleschke JM, Kleczkowska HE, Althaus FR. 1998. Poly(ADP-ribose) binds to specific domains of p53 and alters its DNA binding functions. J Biol Chem 273:11839–11843.
48. Pleschke JM, Kleczkowska HE, Strohm M, Althaus FR. 2000. Poly(ADP-ribose) binds to specific domains in DNA checkpoint proteins. J Biol Chem 275:40974–40980.
49. Simbulan-Rosenthal CM, Rosenthal DS, Luo R, Smulson ME. 1999. Poly(ADP-ribosyl)ation of p53 during apoptosis in human osteosarcoma cells. Cancer Res 59:2190–2194.
50. Adamietz P. 1987. Poly(ADP-ribose) synthase is the major endogenous nonhistone acceptor for poly(ADP-ribose) in alkylated rat hepatoma cells. Eur J Biochem 169:365–372.
51. Kawaichi M, Ueda K, Hayaishi O. 1981. Multiple autopoly(ADP-ribosyl)ation of rat liver poly(ADP-ribose) synthetase. Mode of modification and properties of automodified synthetase. J Biol Chem 256:9483–9489.
52. Kreimeyer A, Wielckens K, Adamietz P, Hilz H. 1984. DNA repair-associated ADP-ribosylation in vivo. Modification of histone H1 differs from that of the principal acceptor proteins. J Biol Chem 259:890–896.
53. Yoshihara K, Hashida T, Yoshihara H, Tanaka Y, Ohgushi H. 1977. Enzyme-bound early product of purified poly(ADP-ribose) polymerase. Biochem Biophys Res Commun 78:1281–1288.
54. de Murcia JM, Niedergang C, Trucco C, Ricoul M, Dutrillaux B, Mark M, Oliver FJ, Masson M, Dierich A, LeMeur M, Walztinger C, Chambon P, de Murcia G. 1997. Requirement of poly(ADP-ribose) polymerase in recovery from DNA damage in mice and in cells. Proc Natl Acad Sci USA 94:7303–7307.
55. Berger NA. 1985. Poly(ADP-ribose) in the cellular response to DNA damage. Radiat Res 101:4–15.
56. Carson DA, Seto S, Wasson DB, Carrera CJ. 1986. DNA strand breaks, NAD metabolism, and programmed cell death. Exp Cell Res 164:273–281.
57. Tanizawa A, Kubota M, Hashimoto H, Shimizu T, Takimoto T, Kitoh T, Akiyama Y, Mikawa H. 1989. VP-16-induced nucleotide pool changes and poly(ADP-ribose) synthesis: the role of VP-16 in interphase death. Exp Cell Res 185:237–246.
58. Wintersberger U, Wintersberger E. 1985. Poly ADP-ribosylation–a cellular emergency reaction? FEBS Lett 188:189–191.
59. Lazebnik YA, Kaufmann SH, Desnoyers S, Poirier GG, Earnshaw WC. 1994. Cleavage of poly(ADP-ribose) polymerase by a proteinase with properties like ICE. Nature 371:346–347.
60. Scovassi AI, Denegri M, Donzelli M, Rossi L, Bernardi R, Mandarino A, Frouin I, Negri C. 1998. Poly(ADP-ribose) synthesis in cells undergoing apoptosis: an attempt to face death before PARP degradation. Eur J Histochem 42:251–258.
61. Kuo ML, Chau YP, Wang JH, Shiah SG. 1996. Inhibitors of poly(ADP-ribose) polymerase block nitric oxide-induced apoptosis but not differentiation in human leukemia HL-60 cells. Biochem Biophys Res Commun 219:502–508.
62. Malorni W, Rivabene R, Straface E, Rainaldi G, Monti D, Salvioli S, Cossarizza A, Franceschi C. 1995. 3-Aminobenzamide protects cells from UV-B-induced apoptosis by acting on cytoskeleton and substrate adhesion. Biochem Biophys Res Commun 207:715–724.
63. Beneke R, Geisen C, Zevnik B, Bauch T, Muller WU, Kupper JH, Moroy T. 2000. DNA excision repair and DNA damage-induced apoptosis are linked to Poly(ADP-ribosyl)ation but have different requirements for p53. Mol Cell Biol 20:6695–6703.

64. Kapasi AA, Singhal PC. 1999. Aging splenocyte and thymocyte apoptosis is associated with enhanced expression of p53, bax, and caspase-3. Mol Cell Biol Res Commun 1:78–81.
65. Simbulan-Rosenthal CM, Rosenthal DS, Boulares AH, Hickey RJ, Malkas LH, Coll JM, Smulson ME. 1998. Regulation of the expression or recruitment of components of the DNA synthesome by poly(ADP-ribose) polymerase. Biochemistry 37:9363–9370.
66. Wein KH, Netzker R, Brand K. 1993. Cell cycle-related expression of poly(ADP-ribosyl) transferase in proliferating rat thymocytes. Biochim Biophys Acta 1176:69–76.
67. Ghani QP, Hussain MZ, Zhang J, Hunt KH. 1992. Control of procollagen gene transcription and prolyl hydroxylase activity by poly(ADP-ribose) In: ADP-ribosylation reactions. Ed. GG Poirier and P Moreau, 211–217. New York: Springer-Verlag.
68. Leverence RR, Beale EG, Granner DK. 1988. 3-Aminobenzamide inhibits poly(ADP ribose) synthetase activity and induces phosphoenolpyruvate carboxykinase (GTP) in H4IIE hepatoma cells. Arch Biochem Biophys 260:667–673.
69. Yau L, Elliot T, Lalonde C, Zahradka P. 1998. Repression of phosphoenolpyruvate carboxykinase gene activity by insulin is blocked by 3-aminobenzamide but not by PD128763, a selective inhibitor of poly(ADP-ribose) polymerase. Eur J Biochem 253:91–100.
70. Zahradka P, Yau L. 1994. ADP-ribosylation and gene expression. Mol Cell Biochem 138:91–98.
71. Quesada P, d'Erme M, Parise G, Faraone-Mennella MR, Caiafa P, Farina B. 1994. Nuclear matrix-associated poly(ADPribosyl)ation system in rat testis chromatin. Exp Cell Res 214:351–357.
72. Quesada P, Tramontano F, Faraone-Mennella MR, Farina B. 2000. The analysis of the poly(ADPR) polymerase mode of action in rat testis nuclear fractions defines a specific poly(ADP-ribosyl)ation system associated with the nuclear matrix. Mol Cell Biochem 205:91–99.
73. Hassa PO, Hottiger MO. 1999. A role of poly (ADP-ribose) polymerase in NF-kappaB transcriptional activation. Biol Chem 380:953–959.
74. Kameoka M, Ota K, Tetsuka T, Tanaka Y, Itaya A, Okamoto T, Yoshihara K. 2000. Evidence for regulation of NF-kappaB by poly(ADP-ribose) polymerase. Biochem J 346:641–649.
75. Oliver FJ, Menissier-de Murcia J, Nacci C, Decker P, Andriantsitohaina R, Muller S, de la Rubia G, Stoclet JC, de Murcia G. 1999. Resistance to endotoxic shock as a consequence of defective NF-kappaB activation in poly (ADP-ribose) polymerase-1 deficient mice. EMBO J 18:4446–4454.
76. Pero RW, Axelsson B, Siemann D, Chaplin D, Dougherty G. 1999. Newly discovered anti-inflammatory properties of the benzamides and nicotinamides. Mol Cell Biochem 193:119–125.
77. Ittel ME, Jongstra-Bilen J, Rochette-Egly C, Mandel P. 1983. Involvement of polyADP-ribose polymerase in the initiation of phytohemagglutinin induced human lymphocyte proliferation. Biochem Biophys Res Commun 116:428–434.
78. Colon-Otero G, Sando JJ, Sims JL, McGrath E, Jensen DE, Quesenberry PJ. 1987. Inhibition of hemopoietic growth factor-induced proliferation by adenosine diphosphate-ribosylation inhibitors. Blood 70:686–693.
79. Cui Y, An S, Jabr S, Maturana JA, Wu JM, Gutstein WH. 1993. The anti-proliferative effects of nicotinamide and 3-aminobenzamide on human smooth muscle cells in vitro. Biochem Mol Biol Int 31:935–944.
80. Francis GE, Gray DA, Berney JJ, Wing MA, Guimaraes JE, Hoffbrand AV. 1983. Role of ADP-ribosyl transferase in differentiation of human granulocyte-macrophage progenitors to the macrophage lineage. Blood 62:1055–1062.
81. Ebisuzaki K, Casley WL, Griffiths A, Wheaton L. 1991. Temporal mapping of the differentiation pathway of the murine erythroleukemia cell. Cancer Res 51:1668–1673.
82. Bhatia M, Kirkland JB, Meckling-Gill KA. 1995. Modulation of poly(ADP-ribose) polymerase during neutrophilic and monocytic differentiation of promyelocytic (NB4) and myelocytic (HL-60) leukaemia cells. Biochem J 308:131–137.
83. Bhatia M, Kirkland JB, Meckling-Gill KA. 1996. Overexpression of poly(ADP-ribose) polymerase promotes cell cycle arrest and inhibits neutrophilic differentiation of NB4 acute promyelocytic leukemia cells. Cell Growth Differ 7:91–100.
84. Lea MA, Barra R, Randolph V, Kuhr WG. 1984. Effects of nicotinamide and structural analogs on DNA synthesis and cellular replication of rat hepatoma cells. Cancer Biochem Biophys 7:195–202.
85. Bowes J, McDonald MC, Piper J, Thiemermann C. 1999. Inhibitors of poly (ADP-ribose) synthetase protect rat cardiomyocytes against oxidant stress. Cardiovasc Res 41:126–134.
86. Thiemermann C, Bowes J, Myint FP, Vane JR. 1997. Inhibition of the activity of poly(ADP ribose) synthetase reduces ischemia-reperfusion injury in the heart and skeletal muscle. Proc Natl Acad Sci USA 94:679–683.

87. Thyberg J, Hultgardh-Nilsson A, Kallin B. 1995. Inhibitors of ADP-ribosylation suppress phenotypic modulation and proliferation of smooth muscle cells cultured from rat aorta. Differentiation 59:243–252.
88. Grainger DJ, Hesketh TR, Weissberg PL, Metcalfe JC. 1992. Hexamethylenebisacetamide selectively inhibits the proliferation of human and rat vascular smooth-muscle cells. Biochem J 283:403–408.
89. Weissberg PL, Grainger DJ, Shanahan CM, Metcalfe JC. 1993. Approaches to the development of selective inhibitors of vascular smooth muscle cell proliferation. Cardiovasc Res 27:1191–1198.
90. Wang ZQ, Auer B, Stingl L, Berghammer H, Haidacher D, Schweiger M, Wagner EF. 1995. Mice lacking ADPRT and poly(ADP-ribosyl)ation develop normally but are susceptible to skin disease. Genes Dev 9:509–520.
91. Trucco C, Oliver FJ, de Murcia G, Menissier-de Murcia J. 1998. DNA repair defect in poly(ADP-ribose) polymerase-deficient cell lines. Nucleic Acids Res 26:2644–2649.
92. Masutani M, Suzuki H, Kamada N, Watanabe M, Ueda O, Nozaki T, Jishage K, Watanabe T, Sugimoto T, Nakagama H, Ochiya T, Sugimura T. 1999. Poly(ADP-ribose) polymerase gene disruption conferred mice resistant to streptozotocin-induced diabetes. Proc Natl Acad Sci USA 96:2301–2304.
93. Shieh WM, Ame JC, Wilson MV, Wang ZQ, Koh DW, Jacobson MK, Jacobson EL. 1998. Poly(ADP-ribose) polymerase null mouse cells synthesize ADP-ribose polymers. J Biol Chem 273: 30069–30072.
94. Ame JC, Rolli V, Schreiber V, Niedergang C, Apiou F, Decker P, Muller S, Hoger T, Menissier-de Murcia J, de Murcia G. 1999. PARP-2, A novel mammalian DNA damage-dependent poly(ADP-ribose) polymerase. J Biol Chem 274:17860–17868.
95. Johansson M. 1999. A human poly(ADP-ribose) polymerase gene family (ADPRTL):cDNA cloning of two novel poly(ADP-ribose) polymerase homologues. Genomics 57:442–445.
96. Shall S, de Murcia G. 2000. Poly(ADP-ribose) polymerase-1: what have we learned from the deficient mouse model? Mutat Res 460:1–15.
97. Smith S. 2001. The world according to PARP. Trends Biochem Sci 26:174–179.
98. Kickhoefer VA, Siva AC, Kedersha NL, Inman EM, Ruland C, Streuli M, Rome LH. 1999. The 193-kD vault protein, VPARP, is a novel poly(ADP-ribose) polymerase. J Cell Biol 146:917–928.
99. Chugani DC, Rome LH, Kedersha NL. 1993. Evidence that vault ribonucleoprotein particles localize to the nuclear pore complex. J Cell Sci 106:23–29.
100. Chypre C, Maniez C, Mandel P. 1988. Cytoplasmic poly(ADP-ribose) polymerase associated with free messenger ribonucleoprotein particles in rat brain. J Neurochem 51:561–565.
101. Jesser M, Chypre C, Hog F, Mandel P. 1993. Cytoplasmic poly(ADP-ribose)polymerase from mouse plasmacytoma free messenger ribonucleoprotein particles:purification and characterization. Biochem Biophys Res Commun 195:558–564.
102. Moss J, Zolkiewska A, Okazaki I. 1997. ADP-ribosylarginine hydrolases and ADP-ribosyltransferases. Partners in ADP-ribosylation cycles. Adv Exp Med Biol 419:25–33.
103. Okazaki IJ, Moss J. 1996. Structure and function of eukaryotic mono-ADP-ribosyltransferases. Rev Physiol Biochem Pharmacol 129:51–104.
104. Zolkiewska A, Okazaki IJ, Moss J. 1994. Vertebrate mono-ADP-ribosyltransferases. Mol Cell Biochem 138:107–112.
105. Soman G, Graves DJ. 1988. Endogenous ADP-ribosylation in skeletal muscle membranes. Arch Biochem Biophys 260:56–66.
106. Kharadia SV, Huiatt TW, Huang HY, Peterson JE, Graves DJ. 1992. Effect of an arginine-specific ADP-ribosyltransferase inhibitor on differentiation of embryonic chick skeletal muscle cells in culture. Exp Cell Res 201:33–42.
107. Norgauer J, Kownatzki E, Seifert R, Aktories K. 1988. Botulinum C2 toxin ADP-ribosylates actin and enhances O_2- production and secretion but inhibits migration of activated human neutrophils. J Clin Invest 82:1376–1382.
108. Saxty BA, Yadollahi-Farsani M, Kefalas P, Paul S, MacDermot J. 1998. Inhibition of chemotaxis in A7r5 rat smooth muscle cells by a novel panel of inhibitors. Br J Pharmacol 125:152–158.
109. Saxty B, Yadollahi-Farsani M, Kefalas P, Paul S, MacDermot J. 1998. ADP-ribosyltransferase inhibitors and pseudo-substrates inhibit rat smooth muscle cell migration in vitro. Biochem Soc Trans 26:S130
110. Stasia MJ, Jouan A, Bourmeyster N, Boquet P, Vignais PV. 1991. ADP-ribosylation of a small size GTP-binding protein in bovine neutrophils by the C3 exoenzyme of Clostridium botulinum and effect on the cell motility. Biochem Biophys Res Commun 180:615–622.

111. Brune B, Dimmeler S, Molina y Vedia L, Lapetina EG. 1994. Nitric oxide:a signal for ADP-ribosylation of proteins. Life Sci 54:61–70.
112. Hilz H. 1997. ADP-ribose. A historical overview. Adv Exp Med Biol 419:15–24.
113. Zolkiewska A, Moss J. 1993. Integrin alpha 7 as substrate for a glycosylphosphatidylinositol-anchored ADP-ribosyltransferase on the surface of skeletal muscle cells. J Biol Chem 268:25273–25276.
114. Zolkiewska A, Nightingale MS, Moss J. 1992. Molecular characterization of NAD: arginine ADP-ribosyltransferase from rabbit skeletal muscle. Proc Natl Acad Sci USA 89:11352–11356.
115. Takada T, Okazaki IJ, Moss J. 1994. ADP-ribosylarginine hydrolases. Mol Cell Biochem 138: 119–122.
116. Williamson KC, Moss J. 1990. Mono-ADP-ribosyltransferases and ADP-ribosylarginine hydrolases: a Mono-ADP-ribosylation cycle in animal cells. In: ADP-ribosylating toxins and G proteins: Insights into signal. Ed. J Moss, M Vaughan, 493–510. Washington: American Society for Microbiology.
117. Moss J, Tsai SC, Adamik R, Chen HC, Stanley SJ. 1988. Purification and characterization of ADP-ribosylarginine hydrolase from turkey erythrocytes. Biochemistry 27:5819–5823.
118. Smith KP, Benjamin RC, Moss J, Jacobson MK. 1985. Identification of enzymatic activities which process protein bound mono(ADP-ribose). Biochem Biophys Res Commun 126:136–142.
119. Chang YC, Soman G, Graves DJ. 1986. Identification of an enzymatic activity that hydrolyzes protein-bound ADP-ribose in skeletal muscle. Biochem Biophys Res Commun 139:932–939.
120. Moss J, Stanley SJ, Nightingale MS, Murtagh JJ, Jr., Monaco L, Mishima K, Chen HC, Williamson KC, Tsai SC. 1992. Molecular and immunological characterization of ADP-ribosylarginine hydrolases. J Biol Chem 267:10481–10488.
121. Graves JD, Krebs EG. 1999. Protein phosphorylation and signal transduction. Pharmacol Ther 82: 111–121.
122. Okazaki IJ, Moss J. 1999. Characterization of glycosylphosphatidylinositiol-anchored, secreted, and intracellular vertebrate mono-ADP-ribosyltransferases. Annu Rev Nutr 19:485–509.
123. Zolkiewska A, Moss J. 1997. The alpha 7 integrin as a target protein for cell surface mono-ADP-ribosylation in muscle cells. Adv Exp Med Biol 419:297–303.
124. Gill DM, Meren R. 1978. ADP-ribosylation of membrane proteins catalyzed by cholera toxin: basis of the activation of adenylate cyclase. Proc Natl Acad Sci USA 75:3050–3054.
125. Honjo T, Nishizuka Y, Hayaishi O. 1968. Diphtheria toxin-dependent adenosine diphosphate ribosylation of aminoacyl transferase II and inhibition of protein synthesis. J Biol Chem 243: 3553–3555.
126. Moss J, Vaughan M. 1977. Mechanism of action of choleragen. Evidence for ADP-ribosyltransferase activity with arginine as an acceptor. J Biol Chem 252:2455–2457.
127. Aktories K. 1994. Clostridial ADP-ribosylating toxins: effects on ATP and GTP-binding proteins. Mol Cell Biochem 138:167–176.
128. Aktories K. 1997. Rho proteins: targets for bacterial toxins. Trends Microbiol 5:282–288.
129. Gierschik P. 1992. ADP-ribosylation of signal-transducing guanine nucleotide-binding proteins by pertussis toxin. Curr Top Microbiol Immunol 175:69–96.
130. Lerm M, Schmidt G, Aktories K. 2000. Bacterial protein toxins targeting rho GTPases. FEMS Microbiol Lett 188:1–6.
131. Moss J. 1987. Signal transduction by receptor-responsive guanyl nucleotide-binding proteins: modulation by bacterial toxin-catalyzed ADP-ribosylation. Clin Res 35:451–458.
132. Popoff MR. 1998. Interactions between bacterial toxins and intestinal cells. Toxicon 36:665–685.
133. Moss J, Stanley SJ. 1981. Histone-dependent and histone-independent forms of an ADP-ribosyltransferase from human and turkey erythrocytes. Proc Natl Acad Sci USA 78:4809–4812.
134. Moss J, Stanley SJ, Watkins PA. 1980. Isolation and properties of an NAD- and guanidine-dependent ADP- ribosyltransferase from turkey erythrocytes. J Biol Chem 255:5838–5840.
135. Moss J, Stanley SJ. 1981. Amino acid-specific ADP-ribosylation. Identification of an arginine-dependent ADP-ribosyltransferase in rat liver. J Biol Chem 256:7830–7833.
136. Godeau F, Belin D, Koide SS. 1984. Mono(adenosine diphosphate ribosyl) transferase in Xenopus tissues. Direct demonstration by a zymographic localization in sodium dodecyl sulfate—polyacrylamide gels. Anal Biochem 137:287–296.
137. Tsuchiya M, Shimoyama M. 1994. Target protein for eucaryotic arginine-specific ADP-ribosyltransferase. Mol Cell Biochem 138:113–118.

138. Fujita H, Okamoto H, Tsuyama S. 1995. ADP-ribosylation in adrenal glands: purification and characterization of mono-ADP-ribosyltransferases and ADP-ribosylhydrolase affecting cytoskeletal actin. Int J Biochem Cell Biol 27:1065–1078.
139. Matsuyama S, Tsuyama S. 1991. Mono-ADP-ribosylation in brain: purification and characterization of ADP-ribosyltransferases affecting actin from rat brain. J Neurochem 57:1380–1387.
140. Duman RS, Terwilliger RZ, Nestler EJ. 1991. Endogenous ADP-ribosylation in brain: initial characterization of substrate proteins. J Neurochem 57:2124–2132.
141. Hara N, Mishima K, Tsuchiya M, Tanigawa Y, Shimoyama M. 1987. Mono(ADP-ribosyl)ation of $Ca2^{+}$-dependent ATPase in rabbit skeletal muscle sarcoplasmic reticulum and the effect of poly L-lysine. Biochem Biophys Res Commun 144:856–862.
142. Obara S, Mishima K, Yamada K, Taniguchi M, Shimoyama M. 1989. DNA-regulated arginine-specific mono(ADP-ribosyl)ation and de-ADP-ribosylation of endogenous acceptor proteins in human neutrophils. Biochem Biophys Res Commun 163:452–457.
143. Obara S, Yamada K, Yoshimura Y, Shimoyama M. 1991. Evidence for the endogenous GTP-dependent ADP-ribosylation of the alpha-subunit of the stimulatory guanyl-nucleotide-binding protein concomitant with an increase in basal adenylyl cyclase activity in chicken spleen cell membrane. Eur J Biochem 200:75–80.
144. Peterson JE, Larew JS, Graves DJ. 1990. Purification and partial characterization of arginine-specific ADP-ribosyltransferase from skeletal muscle microsomal membranes. J Biol Chem 265: 17062–17069.
145. Piron KJ, McMahon KK. 1990. Localization and partial characterization of ADP-ribosylation products in hearts from adult and neonatal rats. Biochem J 270:591–597.
146. Soman G, Haregewoin A, Hom RC, Finberg RW. 1991. Guanidine group specific ADP-ribosyltransferase in murine cells. Biochem Biophys Res Commun 176:301–308.
147. Soman G, Mickelson JR, Louis CF, Graves DJ. 1984. NAD: guanidino group specific mono ADP-ribosyltransferase activity in skeletal muscle. Biochem Biophys Res Commun 120:973–980.
148. Soman G, Tomer KB, Graves DJ. 1983. Assay of mono ADP-ribosyltransferase activity by using guanylhydrazones. Anal Biochem 134:101–110.
149. Yamashita A, Sato E, Yasuda H, Kurokawa T, Ishibashi S. 1991. Reduction of mono(ADP-ribosyl)ation of 20kDa protein with maturation in rat testis: involvement of guanine nucleotides. Biochim Biophys Acta 1091:46–50.
150. Ledford BE, Leno GH. 1994. ADP-ribosylation of the molecular chaperone GRP78/BiP. Mol Cell Biochem 138:141–148.
151. Leno GH, Ledford BE. 1989. ADP-ribosylation of the 78-kDa glucose-regulated protein during nutritional stress. Eur J Biochem 186:205–211.
152. Li PL, Chen CL, Bortell R, Campbell WB. 1999. 11,12-Epoxyeicosatrienoic acid stimulates endogenous mono-ADP-ribosylation in bovine coronary arterial smooth muscle. Circ Res 85:349–356.
153. Halldorsson H, Bodvarsdottir T, Kjeld M, Thorgeirsson G. 1992. Role of ADP-ribosylation in endothelial signal transduction and prostacyclin production. FEBS Lett 314:322–326.
154. Domenighini M, Rappuoli R. 1996. Three conserved consensus sequences identify the NAD-binding site of ADP-ribosylating enzymes, expressed by eukaryotes, bacteria and T-even bacteriophages. Mol Microbiol 21:667–674.
155. Takada T, Iida K, Moss J. 1995. Conservation of a common motif in enzymes catalyzing ADP-ribose transfer. Identification of domains in mammalian transferases. J Biol Chem 270:541–544.
156. Ferguson MA, Williams AF. 1988. Cell-surface anchoring of proteins via glycosyl-phosphatidylinositol structures. Annu Rev Biochem 57:285–320.
157. Low MG. 1989. Glycosyl-phosphatidylinositol: a versatile anchor for cell surface proteins. FASEB J 3:1600–1608.
158. Okazaki IJ, Moss J. 1998. Glycosylphosphatidylinositol-anchored and secretory isoforms of mono-ADP-ribosyltransferases. J Biol Chem 273:23617–23620.
159. Weng B, Thompson WC, Kim HJ, Levine RL, Moss J. 1999. Modification of the ADP-ribosyltransferase and NAD glycohydrolase activities of a mammalian transferase (ADP-ribosyltransferase 5) by auto-ADP-ribosylation. J Biol Chem 274:31797–31803.
160. Taniguchi M, Tsuchiya M, Shimoyama M. 1993. Comparison of acceptor protein specificities on the formation of ADP-ribose.acceptor adducts by arginine-specific ADP-ribosyltransferase from rabbit skeletal muscle sarcoplasmic reticulum with those of the enzyme from chicken peripheral polymorphonuclear cells. Biochim Biophys Acta 1161:265–271.

161. Huang HY, Graves DJ, Robson RM, Huiatt TW. 1993. ADP-ribosylation of the intermediate filament protein desmin and inhibition of desmin assembly in vitro by muscle ADP-ribosyltransferase. Biochem Biophys Res Commun 197:570–577.
162. Wang J, Nemoto E, Dennert G. 1996. Regulation of CTL by ecto-nictinamide adenine dinucleotide (NAD) involves ADP-ribosylation of a $p56^{lck}$-associated protein. J Immunol 156:2819–2827.
163. Okazaki IJ, Moss J. 1996. Mono-ADP-ribosylation: a reversible posttranslational modification of proteins. Adv Pharmacol 35:247–280.
164. Donnelly LE, Boyd RS, MacDermot J. 1992. Gs alpha is a substrate for mono(ADP-ribosyl) transferase of NG108-15 cells. ADP-ribosylation regulates Gs alpha activity and abundance. Biochem J 288:331–336.
165. Jacquemin C, Thibout H, Lambert B, Correze C. 1986. Endogenous ADP-ribosylation of Gs subunit and autonomous regulation of adenylate cyclase. Nature 323:182–184.
166. Molina y Vedia L, Nolan RD, Lapetina EG. 1989. The effect of iloprost on the ADP-ribosylation of Gs alpha (the alpha-subunit of Gs). Biochem J 261:841–845.
167. Saxty BA, van Heyningen S. 1995. The purification of a cysteine-dependent NAD^+ glycohydrolase activity from bovine erythrocytes and evidence that it exhibits a novel ADP- ribosyltransferase activity. Biochem J 310:931–937.
168. Tanuma S, Kawashima K, Endo H. 1988. Eukaryotic mono(ADP-ribosyl)transferase that ADP-ribosylates GTP-binding regulatory Gi protein. J Biol Chem 263:5485–5489.
169. McDonald LJ, Wainschel LA, Oppenheimer NJ, Moss J. 1992. Amino acid-specific ADP-ribosylation: structural characterization and chemical differentiation of ADP-ribose-cysteine adducts formed nonenzymatically and in a pertussis toxin-catalyzed reaction. Biochemistry 31:11881–11887.
170. Cassel D, Pfeuffer T. 1978. Mechanism of cholera toxin action: covalent modification of the guanyl nucleotide-binding protein of the adenylate cyclase system. Proc Natl Acad Sci USA 75:2669–2673.
171. Ui M, Katada T. 1990. Bacterial toxins as probe for receptor-Gi coupling. Adv Second Messenger Phosphoprotein Res 24:63–69.
172. Field M. 1976. Regulation of active ion transport in the small intestine. Ciba Found Symp 42: 109–127.
173. Moss J, Vaughan M. 1988. ADP-ribosylation of guanyl nucleotide-binding regulatory proteins by bacterial toxins. Adv Enzymol Relat Areas Mol Biol 61:303–379.
174. Sharp GW. 1973. Action of cholera toxin on fluid and electrolyte movement in the small intestine. Annu Rev Med 24:19–23.
175. Collier RJ. 1967. Effect of diphtheria toxin on protein synthesis: inactivation of one of the transfer factors. J Mol Biol 25:83–98.
176. Collier RJ. 1975. Diphtheria toxin: mode of action and structure. Bacteriol Rev 39:54–85.
177. Iglewski WJ. 1994. Cellular ADP-ribosylation of elongation factor 2. Mol Cell Biochem 138: 131–133.
178. Wick MJ, Frank DW, Storey DG, Iglewski BH. 1990. Structure, function, and regulation of Pseudomonas aeruginosa exotoxin A. Annu Rev Microbiol 44:335–363.
179. vanderSpek J, Cosenza L, Woodworth T, Nichols JC, Murphy JR. 1994. Diphtheria toxin-related cytokine fusion proteins: elongation factor 2 as a target for the treatment of neoplastic disease. Mol Cell Biochem 138:151–156.
180. Kumagai N, Morii N, Fujisawa K, Nemoto Y, Narumiya S. 1993. ADP-ribosylation of rho p21 inhibits lysophosphatidic acid-induced protein tyrosine phosphorylation phosphatidylinositol 3-kinase activation in cultured Swiss 3T3 cells. J Biol Chem 268:24535–24538.
181. Kishi K, Sasaki T, Kuroda S, Itoh T, Takai Y. 1993. Regulation of cytoplasmic division of Xenopus embryo by rho p21 and its inhibitory GDP/GTP exchange protein (rho GDI). J Cell Biol 120: 1187–1195.
182. Maehama T, Sekine N, Nishina H, Takahashi K, Katada T. 1994. Characterization of botulinum C3-catalyzed ADP-ribosylation of rho proteins and identification of mammalian C3-like ADP-ribosyltransferase. Mol Cell Biochem 138:135–140.
183. Nishiki T, Narumiya S, Morii N, Yamamoto M, Fujiwara M, Kamata Y, Sakaguchi G, Kozaki S. 1990. ADP-ribosylation of the rho/rac proteins induces growth inhibition, neurite outgrowth and acetylcholine esterase in cultured PC-12 cells. Biochem Biophys Res Commun 167:265–272.
184. Paterson HF, Self AJ, Garrett MD, Just I, Aktories K, and Hall A. 1990. Microinjection of recombinant p21rho induces rapid changes in cell morphology. J Cell Biol 111:1001–1007.

185. Ridley AJ, Hall A. 1992. The small GTP-binding protein rho regulates the assembly of focal adhesions and actin stress fibers in response to growth factors. Cell 70:389–399.
186. Ridley AJ, Paterson HF, Johnston CL, Diekmann D, Hall A. 1992. The small GTP-binding protein rac regulates growth factor-induced membrane ruffling. Cell 70:401–410.
187. Graves DJ, Huiatt TW, Zhou H, Huang HY, Sernett SW, Robson RM, McMahon KK. 1997. Regulatory role of arginine-specific mono(ADP-ribosyl)transferase in muscle cells. Adv Exp Med Biol 419:305–313.
188. Huang HY, Zhou H, Huiatt TW, Graves DJ. 1996. Target proteins for arginine-specific mono (ADP-ribosyl) transferase in membrane fractions from chick skeletal muscle cells. Exp Cell Res 226:147–153.
189. Yuan J, Huiatt TW, Liao CX, Robson RM, Graves DJ. 1999. The effects of mono-ADP-ribosylation on desmin assembly-disassembly. Arch Biochem Biophys 363:314–322.
190. Terashima M, Mishima K, Yamada K, Tsuchiya M, Wakutani T, Shimoyama M. 1992. ADP-ribosylation of actins by arginine-specific ADP-ribosyltransferase purified from chicken heterophils. Eur J Biochem 204:305–311.
191. Zolkiewska A, Thompson WC, Moss J. 1998. Interaction of integrin alpha 7 beta 1 in C2C12 myotubes and in solution with laminin. Exp Cell Res 240:86–94.
192. Lupi R, Corda D, Di Girolamo M. 2000. Endogenous ADP-ribosylation of the G protein beta subunit prevents the inhibition of type 1 adenylyl cyclase. J Biol Chem 275:9418–9424.
193. Aktories K, Wegner A. 1992. Mechanisms of the cytopathic action of actin-ADP-ribosylating toxins. Mol Microbiol 6:2905–2908.
194. Chao D, Severson DL, Zwiers H, Hollenberg MD. 1994. Radiolabelling of bovine myristoylated alanine-rich protein kinase C substrate (MARCKS) in an ADP-ribosylation reaction. Biochem Cell Biol 72:391–396.
195. Scaife RM, Wilson L, Purich DL. 1992. Microtubule protein ADP-ribosylation in vitro leads to assembly inhibition and rapid depolymerization. Biochemistry 31:310–316.
196. Donadoni ML, Gavezzotti R, Borella F, Di Giulio AM, Gorio A. 1995. Experimental diabetic neuropathy. Inhibition of protein mono-ADP-ribosylation prevents reduction of substance P axonal transport. J Pharmacol Exp Ther 274:570–576.
197. Jones EM, Baird A. 1997. Cell-surface ADP-ribosylation of fibroblast growth factor-2 by an arginine-specific ADP-ribosyltransferase. Biochem J 323:173–177.
198. Boulle N, Jones EM, Auguste P, Baird A. 1995. Adenosine diphosphate ribosylation of fibroblast growth factor-2. Mol Endocrinol 9:767–775.
199. Feige JJ, Baird A. 1989. Basic fibroblast growth factor is a substrate for protein phosphorylation and is phosphorylated by capillary endothelial cells in culture. Proc Natl Acad Sci USA 86: 3174–3178.
200. Vilgrain I, Baird A. 1991. Phosphorylation of basic fibroblast growth factor by a protein kinase associated with the outer surface of a target cell. Mol Endocrinol 5:1003–1012.
201. Vilgrain I, Gonzalez AM, Baird A. 1993. Phosphorylation of basic fibroblast growth factor (FGF-2) in the nuclei of SK-Hep-1 cells. FEBS Lett 331:228–232.
202. Saxty BA, Yadollahi-Farsani M, Upton PD, Johnstone SR, MacDermot J. 2001. Inactivation of platelet-derived growth factor-BB following modification by ADP-ribosyltransferase. Br J Pharmacol 133:1219–1226.
203. Brune B, Lapetina EG. 1989. Activation of a cytosolic ADP-ribosyltransferase by nitric oxide-generating agents. J Biol Chem 264:8455–8458.
204. Schuman EM, Meffert MK, Schulman H, Madison DV. 1994. An ADP-ribosyltransferase as a potential target for nitric oxide action in hippocampal long-term potentiation. Proc Natl Acad Sci USA 91:11958–11962.
205. Zhang J, Snyder SH. 1992. Nitric oxide stimulates auto-ADP-ribosylation of glyceraldehyde-3-phosphate dehydrogenase. Proc Natl Acad Sci USA 89:9382–9385.
206. Parrado J, Bougria M, Ayala A, Machado A. 1999. Induced mono-(ADP)-ribosylation of rat liver cytosolic proteins by lipid peroxidant agents. Free Radic Biol Med 26:1079–1084.
207. Kanagy NL, Charpie JR, Webb RC. 1995. Nitric oxide regulation of ADP-ribosylation of G proteins in hypertension. Med Hypotheses 44:159–164.
208. Zhang DX, Zou AP, Li PL. 2001. Adenosine diphosphate ribose dilates bovine coronary small arteries through apyrase- and 5′-nucleotidase-mediated metabolism. J Vasc Res 38:64–72.
209. Taguchi J, Abe J, Okazaki H, Takuwa Y, Kurokawa K. 1993. L-arginine inhibits neointimal formation following balloon injury. Life Sci 53:387–392.

210. Suzuki T, Hayase M, Hibi K, Hosokawa H, Yokoya K, Fitzgerald PJ, Yock PG, Cooke JP, Suzuki T, Yeung AC. 2002. Effect of local delivery of L-arginine on in-stent restenosis in humans. Am J Cardiol 89:363–367.
211. Loesberg C, van Rooij H, Smets LA. 1990. Meta-iodobenzylguanidine (MIBG), a novel high-affinity substrate for cholera toxin that interferes with cellular mono(ADP-ribosylation). Biochim Biophys Acta 1037:92–99.
212. Uemura S, Fathman CG, Rothbard JB, Cooke JP. 2000. Rapid and efficient vascular transport of arginine polymers inhibits myointimal hyperplasia. Circulation 102:2629–2635.
213. Cornelissen J, Wanders RJ, Van den Bogert C, Van Kuilenburg AB, Elzinga L, Voute PA, Van Gennip AH. 1995. Meta-iodobenzylguanidine (MIBG) inhibits malate and succinate driven mitochondrial ATP synthesis in the human neuroblastoma cell line SK-N-BE(2c). Eur J Cancer 4:582–586.
214. Loesberg C, van Rooij H, Nooijen WJ, Meijer AJ, Smets LA. 1990. Impaired mitochondrial respiration and stimulated glycolysis by m-iodobenzylguanidine (MIBG). Int J Cancer 46:276–281.
215. Duriez PJ, Shah GM. 1997. Cleavage of poly(ADP-ribose) polymerase: a sensitive parameter to study cell death. Biochem Cell Biol 75:337–349.
216. Gilad E, Zingarelli B, Salzman AL, Szabo C. 1997. Protection by inhibition of poly (ADP-ribose) synthetase against oxidant injury in cardiac myoblasts In vitro. J Mol Cell Cardiol 29:2585–2597.

Pathophysiology of Cardiovascular Disease,
edited by N.S. Dhalla, H. Rupp,
A. Angel & G.N. Pierce
Kluwer Academic Publishers, Boston, 2004

Influence of the Myocardium Contractile State in the Coronary Venous Flow Deviation Phenomenon

Otoni Moreira Gomes, Alfredo Inácio Fiorelli, Eros Silva Gomes, and Nelson Luiz Teixeira Barros-Moraes

Fundação Cardiovascular São Francisco de Assis / ServCor
Rua José do Patrocínio, 522
Santa Mônica—31525-160
Belo Horizonte—MG / Brasil
Tel./Fax.: 31-55-3442-7488
E-mail: servicor@servicor.com.br

Summary. Twenty-Four mongrel dogs were anesthetized and submitted to normothermic extracorporeal circulation and to selective perfusion of the coronary arteries with flows of 10.0, 5.0 and 2.5 ml/kg body weight/min. A catheter was introduced inside the coronary sinus, in a 2.0 to 2.5 cm extension, for the flow measurements. The following conditions where investigated: Group I—with hearts beating spontaneously; Group II—under ventricular fibrilation; Group III—under stimulation by isoproterenol (200 mcg); Group IV—under artificial estimulation by pace-maker. Results in Group I hearts demonstrated an inverse relashionship between the coronary arterial flow and its proportional drainage in coronary sinus. This phenomenon was not altered ($p < 0.05$) in Group II (fibrilating) and IV (pace-maker, 80–180 bpm). In Group II hearts, under isoproterenol stimulation an inversion of the above pattern occured with linear increasing of the coronary sinus proportional drainage with the coronary arterial flow augmentation. It is concluded that the coronary venous flow distribution pattern is altered by the improvement in myocardial contractility induced by isoproterenol.

Key words: Coronary Circulation, Cardiovascular Physiology, Myocardium Ischemia, Coronary Venous Drainage.

Correspondence: Otoni Moreira Gomes, MD. Rua Manoel Lopes Coelho, 365—Itapoã—31710-530—Belo Horizonte—MG-Brasil. Tel.: 31-55-3452-7143, E-mail: gomes@servicor.com.br.

INTRODUCTION

The concept of coronary circulation reserve has been established as the maximum capacity for coronary flow increase triggered by the myocardium oxygen demand, with normal range between 400 and 500% rising in the basal flow, depending this variation of the coronary arterial system vasodilation (1). Nevertheless, previous researches from our Institution (2) showed that coronary flow distribution changes in accordance with its magnitude; as higher the coronary arterial flow, lower its proportional drainage in the coronary sinus. This condition was named coronary deviation phenomenon and understood as of great functional importance pointing that the amount of coronary flow increase do not means proportional amount improvement in the left ventricle myocardium perfusion.

The present investigation analyses the influence of different cardiac contractility states in the coronary flow deviation phenomenon.

MATERIAL AND METHOD

Twenty-four mongrel dogs, weighing 8–30 kg, where anesthetized with sodium thiopental (33 mg/kg dosis) maintained with controled respiration and submitted to a median longitudinal thoracotomy for exposition of the heart and great vessels.

The following groups were considered: Group I (Control)—six animals studied with spontaneous cardiac activity.

Group II—six animals studied during electrically induced ventricular fibrillation.

Group III—six animals studied during isoproterenol administration (200 mcg).

Group IV—six animals studied during cardiac frequency variation by peacemaker stimulation (Medtronic 5880-A) with electrodes implanted in the right ventricle.

It was established normothermic extracorporeal circulation with total drainage of the venous return and arterial flow perfusion (110–120 ml/kg of body weight) through the femoral artery. The prime volume was composed by homologous blood, ringer solution, manitol, heparin and sodium bicardonate (according to laboratorial control results).

The central venous pressure, myocardium temperature, mean arterial pressure (femoral artery) and coronary artery perfusion pressure were controled.

For the coronary artery perfusion an isolated roller pump was employed and an arterial cannula fixed in the ascending aorta proximally to the aortic occlusion clamp.

After the beginning of the extracorporeal circulation the right atrium was openned and a 6 mm diameter PVC catheter introduced 2.0–2.5 cm in the coronary sinus (Figure 1), which was tied over the catheter by a 3-0 polyester stitch.

After the aortic clamping the left chambers were drained and aspirated through the interatrial septum.

All coronary flow drainage measures were done after eight minutes of coronary arterial perfusion with flows of 10.0, 5.0 and 2.5 ml/kg of body weight.

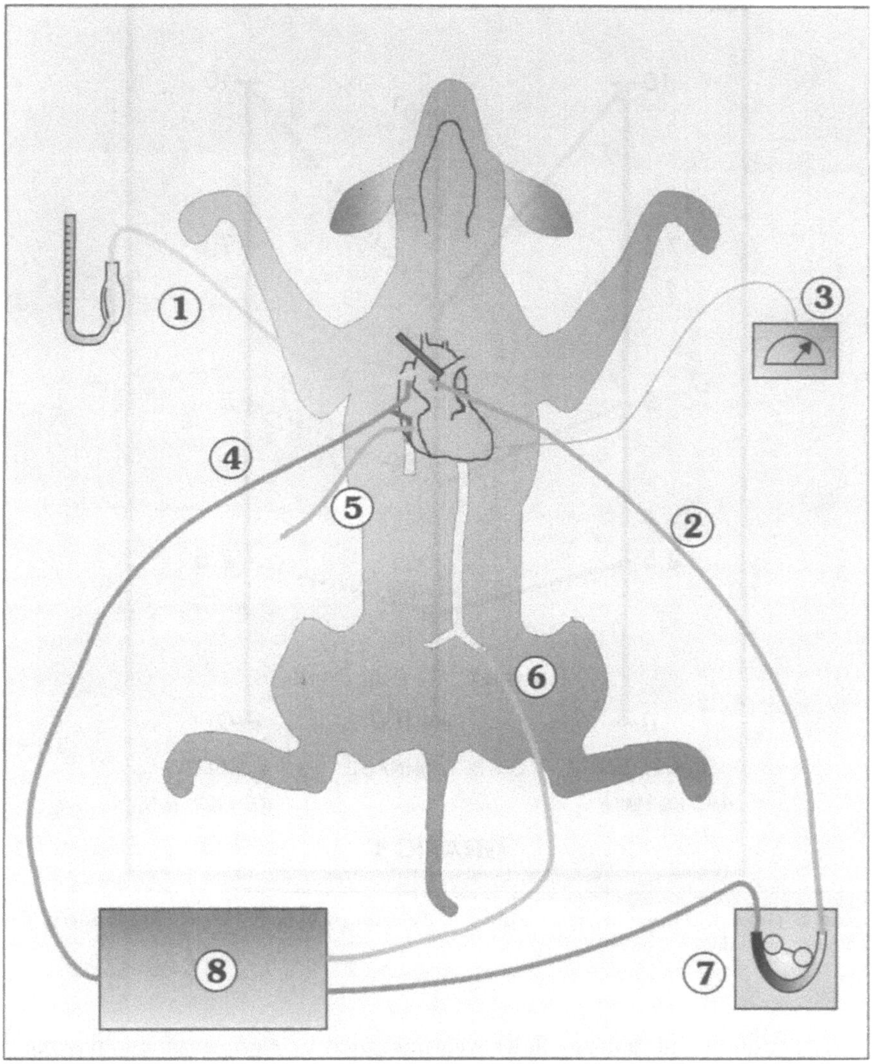

Figure 1. Escheme of the extracorporeal perfusion and drainage lines. 1—Coronary Perfusion, 2—Venous Line, 3—Arterial Line, 4—Coronary Sinus Drainage.

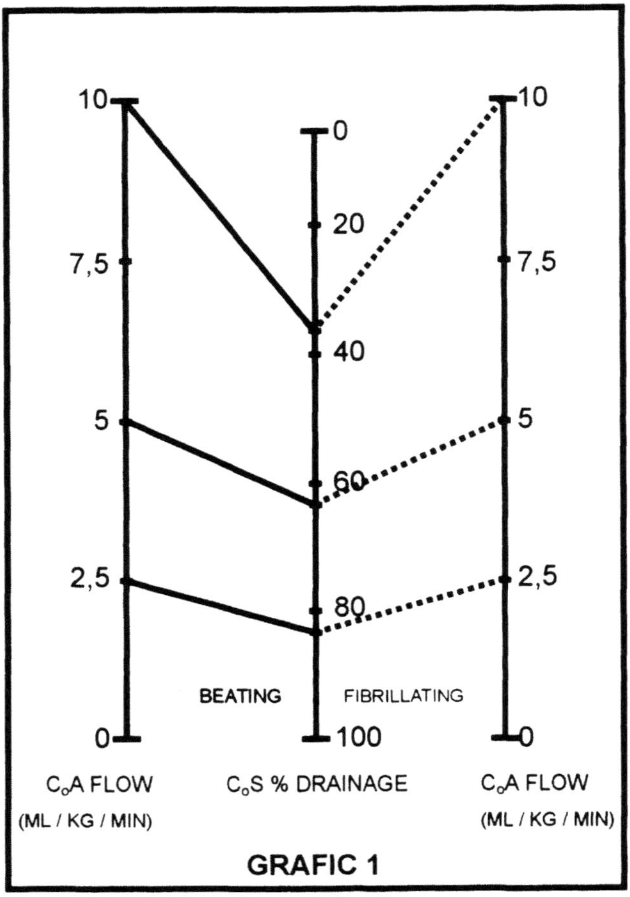

Figure 2. Group II. Percent drainage variation in the coronary sinus. CoAF—Coronary arterial flow; CoS—Coronary sinus.

The perfusion and drainage flows were measured by electromagnetic flowmeters (Statham).

Statistical analysis was performed (ANOVA and t student) with significance level of 0.05.

COMMENTS

Vieussens (3), in the 1706 century by the first time described the existence of direct communications between the coronary arteries and cardiac chambers. Thebesius (4), in 1708, described the coronary venous drainage to the right chambers. Hoffman

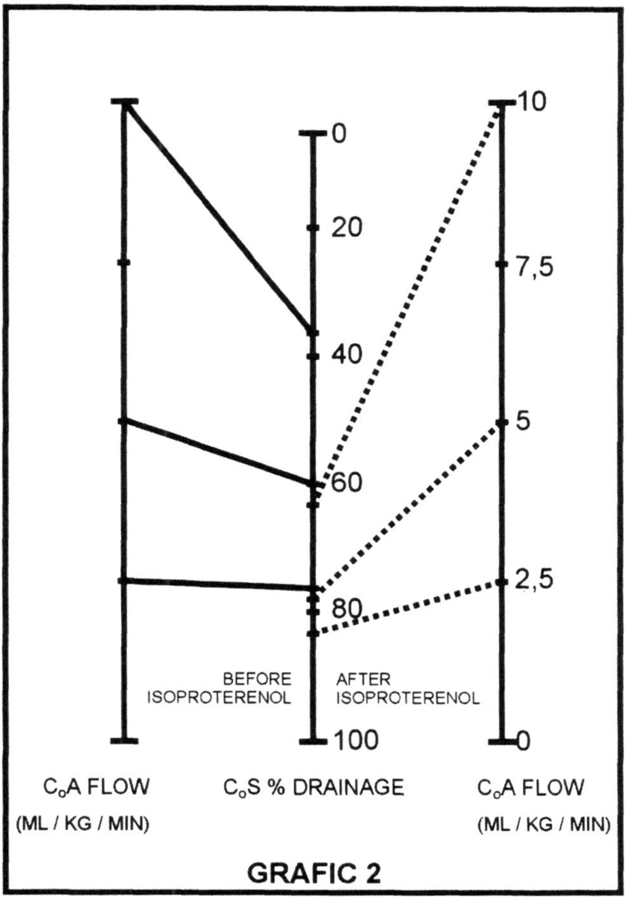

Figure 3. Group III. Isoproterenol action; CoAF—Coronary arterial flow; CoS—Coronary sinus.

et al. (5) in 1971 confirmed the existence of the coronary arterial and venous connections with the cardiac cavities and Robertson (1941) (6), admitted like Thebesius that such intercommunicating system would serve for fast ventricular wall damping, during sudden heart hyperfunction.

In the present study the inhotropic activity enhancement by isoproterenol directed the coronary flow for the coronary sinus pathway reducing the flow to the cardiac chambers.

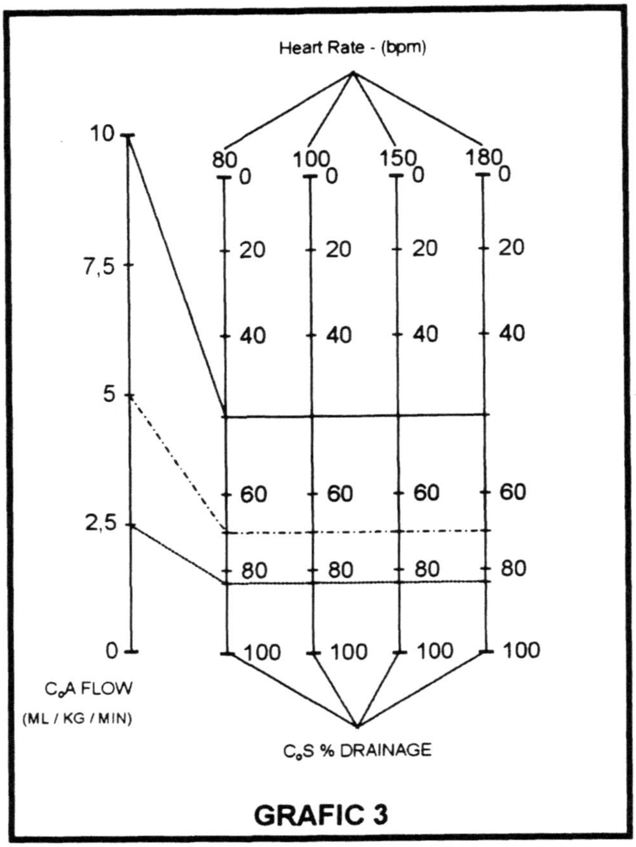

Figure 4. Heart rate and coronary sinus drainage behaviour. CoAF—Coronary arterial flow; CoS—Coronary sinus.

Other concerns arised regarding the classic understanding of the coronary circulation physiology is related with the myocardium adaptation to the coronary flow variation having Sabinston and Blalock (7), in 1958, described the following physiological mechanisms as responsible for the coronary flow adaptation: mean aortic arterial pressure, coronary arteries diameter, arteries tonus, arterial pO_2, blood viscosity and right atrium venous pressure. Nowadays it is admitted that the pO_2 is the main determinant trigger of the coronary flow regulatory changes (8).

Table 1. Results from group I animals

OBS. N°	Coronary Arterial Flow—ml/kg/min 10.0		5.0		2.5	
1	85.0*	70.8%	55.6	91.6%	29.0	96.6%
2	40.0	26.6%	25.0	33.3%	16.0	42.1%
3	33.0	25.3%	24.0	36.9%	24.0	75.0%
4	20.0	16.6%	30.0	50.0%	18.0	60.0%
5	40.0	33.3%	46.0	76.6%	24.0	80.0%
6	55.0	55.5%	40.00	80.0%	20.0	80.0%
Mean	45.5	38.0%	36.8	61.4%	21.8	72.3%

*Coronary Sinus Drainage Flow—ml/min.

Table 2. Ventricular fibrillation and percent drainage variation in the coronary sinus

OBS.	Coronary Arterial Flow—ml/kg/min 10.0		5.0		2.5	
N°	B	F	B	F	B	F
1	51.0*	51.0	89.0	88.5	99.0	99.0
2	36.5	36.2	66.0	66.0	87.0	87.5
3	38.6	38.2	71.5	72.0	86.0	86.0
4	27.9	30.0	80.0	80.0	94.0	94.0
5	43.0	43.0	52.0	52.0	76.0	78.0
6	22.3	22.0	48.0	48.3	64.5	64.0
Mean	36.5	36.7	67.8	67.8	84.4	84.7

B—Beating.
F—Fibrillating.
*—%.

Table 3. Isoproterenol and percent drainage variation in the coronary sinus

OBS.	Coronary Arterial Flow—ml/kg/min 10.0		5.0		2.5	
N°	C	I	C	I	C	I
1	33.0*	68.0	52.0	67.0	76.0	83.0
2	42.0	73.0	78.0	92.0	84.0	92.5
3	28.0	51.0	64.0	82.0	92.5	95.0
4	53.0	82.0	69.0	89.0	82.5	93.0
5	36.0	49.0	53.0	69.0	66.0	75.0
6	31.0	52.0	49.0	67.0	58.0	69.0
Mean	37.1	62.5	60.8	77.6	76.5	84.6

C—Control.
I—During Isoproterenol Effect.
*—%.

Table 4. Heart rate and coronary sinus drainage

	Coronary Arterial Flow—ml/kg/min											
OBS.	10.0				5.0				2.5			
N°	80*	120	150	180	80	120	150	180	80	120	150	180
1	80**	80	82	81	98	98	98	98	99	99	99	99
2	58	59	58	58	92	90	92	91	98	99	98	99
3	52	52	52	52	71	71	71	71	78	79	77	78
4	27	27	27	27	52	52	52	52	61	60	61	61
5	39	39	39	40	64	64	64	64	74	74	74	74
6	43	42	43	41	68	68	68	68	81	81	82	81
Mean	49.8	49.8	50.1	49.8	74.1	73.8	74.1	74.0	81.8	82.0	81.8	82.0

*—Heart Rate—bpm.
**—% Drainage in the Coronary Sinus.

COMMENTS

Vieussens (3), in the 1706 century by the first time described the existence of direct communications between the coronary arteries and cardiac chambers. Thebesius (4), in 1708, described the coronary venous drainage to the right chambers. Hoffman et al. (5) in 1971 confirmed the existence of the coronary arterial and venous connections with the cardiac cavities and Robertson (1941) (6), admitted like Thebesius that such intercommunicating system would serve for fast ventricular wall damping, during sudden heart hyperfunction.

In the present study the inhotropic activity enhancement by isoproterenol directed the coronary flow for the coronary sinus pathway reducing the flow to the cardiac chambers.

Other concerns arised regarding the classic understanding of the coronary circulation physiology is related with the myocardium adaptation to the coronary flow variation having Sabinston and Blalock (7), in 1958, described the following physiological mechanisms as responsible for the coronary flow adaptation: mean aortic arterial pressure, coronary arteries diameter, arteries tonus, arterial pO2, blood vis-cosity and right atrium venous pressure. Nowadays it is admitted that the pO2 is the main determinant trigger of the coronary flow regulatory changes (8).

The results of the present investigation do suggest also important participation of the communicating system between coronary vessels and cardiac cavities in the regulation of the nourishing flow, considering that increasing in the arterial flow do not mean proportionally, more perfusion in the myocardial region draining to the coronary sinus.

This understanding open another possible interpretation to the coronary circulation reserve concept including now a proeminent role of the coronary-cavity communications through Thebesius and Vieussens systems. This understanding maker easier the interpretation of weird clinicall conditions like the absence of sintomatology in patients with severe coronary artery obstructions in opposition to the presence of ischemic syndromes in people with normal coronary angiographic patterns.

The analysis of the present results seems to expose a new frame of understanding for some adaptative myocardial alterations in pathologic and special pharmacodynamic conditions.

For example, the rate increasing by pacemaker stimulation has not changed the blood distribution at coronary microcirculation level, oppositely to the expected assumption that a shortening of the dyastolic period would promote a mechanical blockage of the shunt to cardiac cavities. The present findings probably may explain the absence of coronary ischemia in many patients with tachicardia crisis.

The isoproterenol administration resulted always in higher flow drainage to the coronary sinus venous system, may be with better utilization by the left ventricle myocardium. It is noticeable that this drug effect in the coronary flow deviation phenomenon occurred even with heart rate between 80–120 bpm outstanding the participation of the drug inotropic action to explain the observed changes in the microcirculation behaviour. Newer investigation should be done to explain whether the above described circulatory changes are exclusively related to isoproterenol effects or may be reproduced by others inotropic agents.

REFERENCES

1. Rushmer RF. 1970. Cardiovascular dynamics, 3rd ed., W.B. Saunders, Philadephia.
2. Gomes OM, Moraes NLTB, Fiorelli AI, Armelin E, Bittencourt D, Pileggi F, Macruz R, Décourt LV, Zerbini EJ. 1979. Fenômeno de derivação do fluxo coronário: primeira demonstração experimental. Arq Bras Cardiol, 32:141.
3. Vieussen R.—**Apud** Rushmer RF. 1970. Cardiovascular dynamics, 3rd. ed. W.B. Saunders, Philadelphia.
4. Thebesius AC. 1708. Dissertatio medica de circulo sanguinis in corde. Lugduni Batavorum, Elzevier.
5. Hoffmann E, Gebhardt CH, Kisseler B. 1971. Morphologic der ventrikulocoronaren Anastomosen bei Herzen mit intaktem und pathologisch verandertem Coronarge—Fa., System. Langenbecks. Arch Chir 328:180.
6. Robertson HF. 1941. The physiology, pathology and clinical significance of experimental coronary sinus obstruction. Its relation to cardiac surgery, coronary thrombosis and nutrition of the heart by Thebesians vessel or coronary sinus backflow. Surgery, 9:1.
7. Sabinston DC, Blalock A. 1958. Physiologic and anatomic determinants of coronary blood flow and their relationship to myocardial revascularization. Surgery, 44:406.
8. Nasser MG. 1970. Regulation of coronary flow—In Rushmer, R.F., ed.—Cardiovascular dynamics 3rd. ed. W.B. Saunders, Philadelphia.

Pathophysiology of Cardiovascular Disease,
edited by N.S. Dhalla, H. Rupp,
A. Angel & G.N. Pierce
Kluwer Academic Publishers, Boston, 2004

Natriuretic Peptides and Endothelin in Aortic Stenosis and Heart Transplantation

Igor Riecansky,[1] Martin Vrsansky,[1]
Katarina Danova,[2] and Ivan Pechan[2]

[1] *Department of Cardiology,* [2] *Department of Clinical Biochemistry*
Slovak Institute of Cardiovascular Diseases
Pod Krasnou Horkou 1, 833 48 Bratislava, Slovakia
e-mail: vrsansky@pobox.sk, tel.: +421-903-106328

Summary. In the first study we have tried to test usefulness of ANP and BNP in the timing of aortic valve surgery. 23 patients were examined and compared to 20 controls. Echocardiography and 6-minute walk test were performed and pre-, peak- and post-exercise samples were taken from each patient. ANP and BNP plasma concentrations were measured by immuno-radiometric assay.

All natriuretic peptide plasma levels exhibited wide interpatient variability. All ANP and BNP plasma levels were found to be significantly higher than in controls. Both ANP and BNP levels had a non-significant, but rising trend with the NYHA classification. Only ANP showed statistically significant changes with exercise, which, however, did not increase the test validity. There was a correlation between AVG and BNP1,2, but none was found with AVAI and LVMI. Sensitivity of the test to detect a critical aortic stenosis with AVG >100 mmHg was 77% for ANP and 100% for BNP, specificity was 57% and 29%, respectively. However, it was not sensitive enough to detect borderline gradients. Wide inter-patient variability makes the test an insufficient screening tool, however, it could be used to exclude significant aortic stenosis.

In the second group we have studied the development of ANP, BNP and endothelin-1 (ET-1) levels in 17 patients after heart transplantation. The blood samples for were taken before the heart transplantation and on days 1, 3, 5, 7, 14, 21 and 28. ANP and BNP levels decreased after the heart transplantation, with a subsequent rise with a peak on days 5 and 3, respectively. ET-1 levels increased steeply in the 1st day, with a subsequent steady fall to pre-transplant.

Key words: aortic stenosis, heart transplant, natriuretic peptides, endothelin, ANP, BNP, ET-1.

INTRODUCTION

Neuroendocrine activation plays an important role in the pathophysiology of heart disease. Natriuretic peptides and endothelins have been object of intensive research in the past decade and their diagnostic, prognostic and therapeutic potentials in heart failure and ischemic heart disease are widely recognized.

We have focused on two areas where the research has been less intensive—valvular heart disease and heart transplantation. The objective of our first study was to find a relationship between hemodynamic and clinical characteristics and natriuretic peptide plasma levels in aortic stenosis and to investigate their possible use in timing of aortic valve surgery. In the second study we have studied natriuretic peptide and endothelin plasma concentrations in patients in their first month after heart transplantation.

Natriuretic peptides

Natriuretic peptides are a family of structurally similar peptides with diverse effects on cardiovascular, renal and endocrine homeostasis. They include four hormones.

Atrial natriuretic peptide (ANP) was identified in 1984 by Kangawa and Matsuo (1). It is a 28-amino acid polypeptide molecule containing a 17-amino acid ring. ANP is synthesized and secreted predominantly in the atria. The gene is located on the short arm of the first chromosome. Its expression yields pre-pro-ANP, a 150-amino acid precursor. Cleavage of a 5′-terminal signal peptide gives pro-ANP_{1-126}, which is stored in atrial granules. Before secretion it is cleaved by a serine protease to produce active C-terminal ANP_{99-126} (ANP) and inactive N-terminal ANP_{1-98} (N-ANP) (2). ANP has a short serum half-life of 2.5 minutes. On the contrary, the more stable N-ANP has a half-life of 1-2 hours, which results in its 50-fold higher plasma concentrations, making it a useful diagnostic tool.

Brain natriuretic peptide (BNP), a 32-amino acid polypeptide with a 17-amino acid ring, was first isolated in porcine brain in 1988 (3), however, it was soon discovered to be synthesized and secreted predominantly in the ventricular myocardium (2). Its 108-amino acid precursor, pro-BNP_{1-108} is cleaved to active C-terminal BNP_{77-108} and inactive N-terminal BNP_{1-76}. Plasma concentrations of BNP are only 20% of those of ANP, however, its half-life is much longer—20 minutes.

C-type natriuretic peptide (CNP), a 22-amino acid peptide is produced in vascular endothelium, brain and kidney tissue. It is a potent paracrine vasodilator and vascular cell growth inhibitor, with limited natriuretic effects (2). Dendroaspis natriuretic peptide (DNP), a 38-amino acid peptide, was isolated in green mamba snake venom (*Dendroaspis angusticeps*). It has marked natriuretic and diuretic effects.

Natriuretic peptide plasma concentrations are increased in diseases with increased atrial or ventricular pressures (4). However, precise mechanisms are yet unclear. Some authors argue that wall stretch and not increased intracavitary pressure is responsible for natriuretic peptide release (5). It has also been documented that angiotensin II and endothelin increase secretion of natriuretic peptides. Leskinen (6) proved that release of ANP in response to increased atrial pressure is mediated by endothelin, however, endothelin receptor antagonists did not have any effect on ANP concentrations in the absence of increased atrial stretch.

Natriuretic peptides act through natriuretic peptide receptors (NPR). There are three types of NPRs. NPR-A is a receptor for ANP and BNP, NPR-B binds CNP. Both are transmembrane proteins consisting of an extracellular domain with a hormone binding site and intracellular domains with protein kinase and guanylate cyclase activities. Binding of a natriuretic peptide to the extracellular domain results in release of inhibition that the protein kinase exerts on guanylate cyclase and subsequent production of cyclic guanosine monophosphate (cGMP).

NPR-C is a clearance receptor without protein kinase or guanylate cyclase domains. Binding of a natriuretic peptide, predominantly ANP, to NPR-C results in its endocytosis and lysosomal degradation. Another pathway of natriuretic peptide elimination, more important in BNP, is enzymatic cleavage of the ring between amino acids 105 and 106 by neutral endopeptidase (NEP) (2).

Both ANP and BNP exert diverse effects on kidneys and cardiovascular system. They stimulate diuresis and natriuresis by inhibition of sodium transport in the collecting duct, dilate of both arteries and veins, inhibit renin and aldosterone secretion, stimulate vagal tone and decrease sympathetic activity (2).

Plasma ANP and BNP concentrations are considerably increased in heart failure. BNP, due to its secretion in the ventricles and inactivation by NEP, seems to be a better marker of systolic left ventricular dysfunction expressed by ejection fraction than both ANP and N-ANP (7). BNP also seems to be the best marker of diastolic left ventricular (LV) dysfunction and LV hypertrophy.

ANP and BNP plasma concentrations are markedly increased in patients with myocardial infarction. Subsequently, ANP levels fall with a smaller peak on the 3rd day. BNP levels continue to rise for 16 hours and fall thereafter. Patients with LV dysfunction have a second BNP peak on the 5th day, resulting from remodelation of the ventricular myocardium. BNP and N-ANP are highly sensitive independent predictors of mortality in patients after myocardial infarction (8).

BNP may be useful in differential diagnosis of breathlessness. Normal plasma levels exclude systolic LV dysfunction, but increased levels need to be confirmed by echocardiography (9). BNP and N-ANP can also be used in screening for asymptomatic LV dysfunction (10). N-ANP is an independent indicator of mortality, therefore it can be used to evaluate prognosis in patients with congestive heart failure and normal ejection fraction (11). BNP plasma concentrations are better markers of prognosis in chronic heart failure than LV ejection fraction alone (12). Treatment of heart failure with ACE inhibitors carries better prognosis if their dosages are titrated against BNP levels instead of empirical treatment (13).

Natriuretic peptides may be administered intravenously. Infusion of carperitide, a recombinant ANP, has strong diuretic and natriuretic properties, vasodilatation is less pronounced in patients treated with ACE inhibitors. Infusion of nesiritide, a recombinant BNP, decreases systemic and peripheral vascular resistance, increases cardiac output and diuresis. Another approach in influencing concentrations of natriuretic peptides is neutral endopeptidase (NEP) inhibition. Candoxatril can be administered orally. It has both diuretic and vasodilatating effects. The potential of these drugs to improve prognosis of heart failure has yet to be evaluated in randomized studies.

Endothelins

Endothelin (ET) was isolated by Yanagisawa in 1988 from a porcine cellular culture. Later it was discovered to be a family of three peptides—ET-1, ET-2 and ET-3. They consist of a hydrophobic C-terminal part and an N-terminal part with two disulphidic bonds between cysteine residues (14).

ET-1 is synthesized mainly in endothelial cells, but it was also detected in vascular smooth muscle, hepatocytes, astrocytes, neurons, macrophages, endometrium, Sertoli cells and breast glandular cells. ET-2 is produced predominantly in kidneys, but also in myocardium, brain and placenta. Unlike the ET-1 and ET-3, it does not circulate in plasma. ET-3 was detected in brain, hypophysis, lungs, pancreas and spleen. ET-2 and ET-3 have no specific effects on cardiovascular system (15).

The gene for ET-1 is located on the 6th chromosome. Its expression in the endothelial cells is stimulated by growth factors, interleukins, insulin, vasopressin, bradykinin and angiotensin II, noradrenaline; it is inhibited by NO, prostacycline and ANP. Expression of the gene generates a 201-amino acid precursor prepro-ET-1. This is cleaved to a 38-amino acid pro-ET-1, also called big ET-1. Transformation of the big ET-1 is mediated by endothelin converting enzyme (ECE) (8). 75% of endothelin production on the basal cell surface, which results in its very low plasma concentrations and predominant paracrine and autocrine function. However, plasma ET-1 levels could be a sufficient marker of its local production (16).

ET receptors are transmembrane proteins associated with G-protein that activates phospholipase C. Production of inositol triphosphate results in increase of intracellular calcium concentration and subsequent contraction of vascular smooth muscle. There are two types of ET receptors. ET_A is situated mainly on vascular smooth muscle cells and mediates vasoconstricting effects of ET-1. ET_B is located predominantly on endothelial cells and its activation leads to release of vasodilating substances, whereas on vascular smooth muscle cells it causes vasoconstriction, moreover it also seems be a clearance receptor (8).

Endothelins are hormones with strong vasoconstricting activity. They cause coronary vasoconstriction, bronchoconstriction, have positive inotropic and chronotropic effects, increase sodium and water excretion, renin, aldosterone, adrenaline and ANP concentrations. ET-1 also have strong mitogenic properties and stimulate cytokine production (8).

Several experimental and clinical studies have proved that endothelins are increased in heart failure and correlate with mortality (17). They rise in acute myocardial infarction, intravenous infusion of ET-1 reduces coronary flow by 90% (18). Endothelins may be important in a number of conditions, such as arterial hypertension, myocardial hypertrophy, neointimal proliferation after coronary intervention, Raynaud syndrome, as well as bronchial asthma and renal failure. Treatment with bosentan, a non-selective endothelin antagonist, leads to reduction of blood pressure, decrease of systemic and peripheral resistances and increase of cardiac output (19).

MATERIALS AND METHODS

Aortic stenosis

The first group included 23 patients (12 men, 11 women, age 63 ± 7 years) admitted for cardiological evaluation during a period of one month. 2 patients had a mild aortic regurgitation, 3 of them had significant coronary artery disease. Patients with other heart disorders, hepatic or renal dysfunction were excluded. None of the subjects had had congestive heart failure clinically or on a chest X-ray. They were compared to 20 age and sex matched healthy controls.

The patients were divided into the New York Heart Association (NYHA) classes and presence of syncopes or angina was noted. Echocardiography was performed in each patient by the same examiner on Hewlett Packard Sonos 2000. Parameters measured included left ventricular (LV) and left atrial (LA) systolic and end-diastolic diameters, posterior wall and interventricular septum thickness, LV ejection fraction. As a degree of severity of the aortic stenosis, peak aortic valve pressure gradient (AVG—101 ± 23 mm Hg) was calculated from modified Bernoulli equation, and aortic valve area index (AVAI—0.29 ± 0.05 cm^2/m^2) was calculated from the continuity equation. Left ventricular mass index (LVMI—237 ± 91 g/m^2), as a measure of left ventricular hypertrophy, was calculated using Devereux formula. All patients underwent coronary angiography which discovered significant stenoses in 3 patients, and in 4 patients peak-to-peak gradients were measured invasively.

Three samples were taken from each patient. The first sample (ANP1, BNP1) was taken in the morning between 8–9:00 before breakfast, after 20 minutes' rest in supine position. The second specimen (ANP2, BNP2) was taken immediately after the patient had completed a 6 minute walk test. The performance in meters was recorded. The third one (ANP3, BNP3) was taken again after 20 minutes of rest.

Heart transplantation

The second group included 17 patients after heart transplantation—16 men, 1 woman, 12 with a diagnosis of dilated cardiomyopathy and 5 with ischemic chronic heart failure, with average age of 42 ± 8 years. The blood samples were taken before the heart transplantation and on days 1, 3, 5, 7, 14, 21 and 28 after a minimum of 20 minutes of rest in supine position.

Assays and statistics

Edetic acid—aprotinine test tubes were used for natriuretic peptides, plain edetic acid tubes were used for endothelin. The samples were transported on ice, immediately centrifuged, and plasma was stored at −20°C within 30 minutes. ANP and BNP plasma concentrations were measured with commercially available immunoradiometric assay (IRMA) by Shionogi, Japan, with upper reference limits of 50 pg/ml for ANP and 18.4 pg/ml for BNP. The measured natriuretic peptide levels were corrected for the standard packed cell volume of 0.40. ET-1 levels were determined with enzyme-linked immunosorbent assay (ELISA) by R&D Systems, USA, with reference intervals of 0.3–0.9 pg/ml.

The measured values were expressed as mean ± standard error of the mean. The data were tested for statistically significant differences with Student t-test and paired t-test using Microsoft Excel software. Pearson and Spearman correlation were used to test the relationship between natriuretic peptide levels and echocardiographic parameters and NYHA classification, respectively.

RESULTS

Aortic stenosis

All natriuretic peptide plasma levels exhibited wide inter-patient variability (ANP1,2,3 69.4 ± 27.3, 86.1 ± 47.7, 46.9 ± 43.0 pg/ml (Figure 1) and BNP1,2,3 169.0 ± 151.2, 184.6 ± 166.4, 170.7 ± 150.6 pg/ml (Figure 2), respectively. All plasma levels were found to be significantly higher than in controls (ANP1,2,3—$p < 0.001$; BNP1,2,3—$p = 0.001$). Surprisingly, only ANP showed statistically

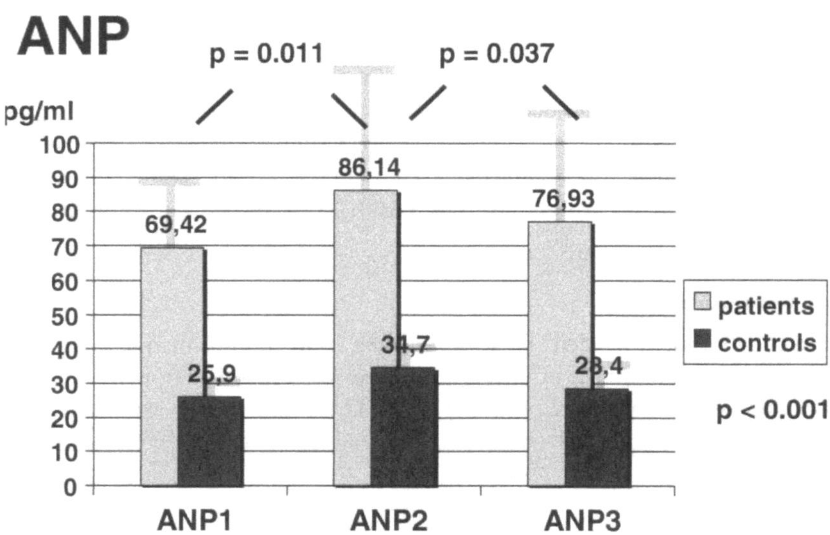

ANP1 – pre-exercise, ANP2 – peak-exercise, ANP3 – post-exercise plasma ANP concentrations

Figure 1. pre-, peak- and post-exercise mean ANP plasma concentrations.

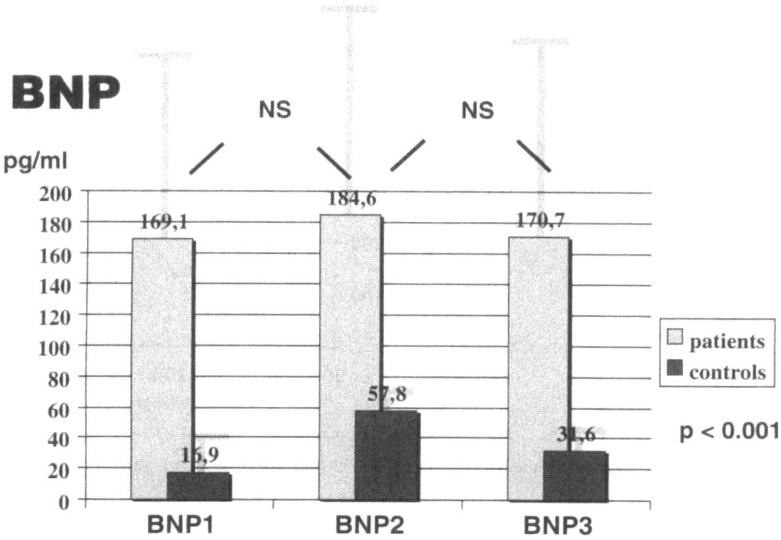

BNP1 – pre-exercise, BNP2 – peak-exercise, BNP3 – post-exercise plasma BNP concentrations

Figure 2. Pre-, peak- and post-exercise BNP mean plasma concentrations.

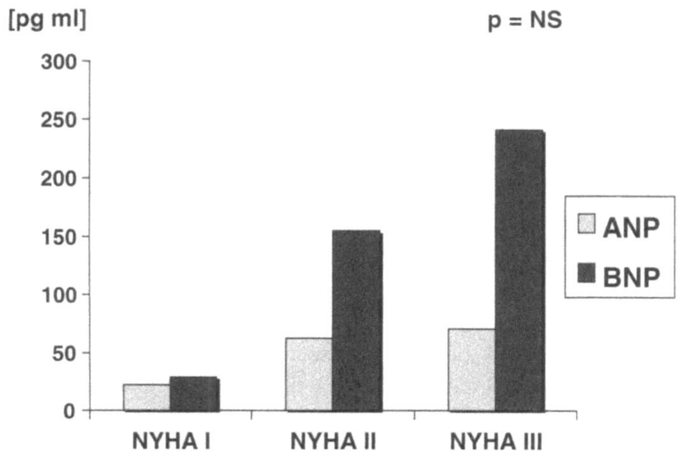

Figure 3. Resting mean ANP and BNP versus NYHA classification.

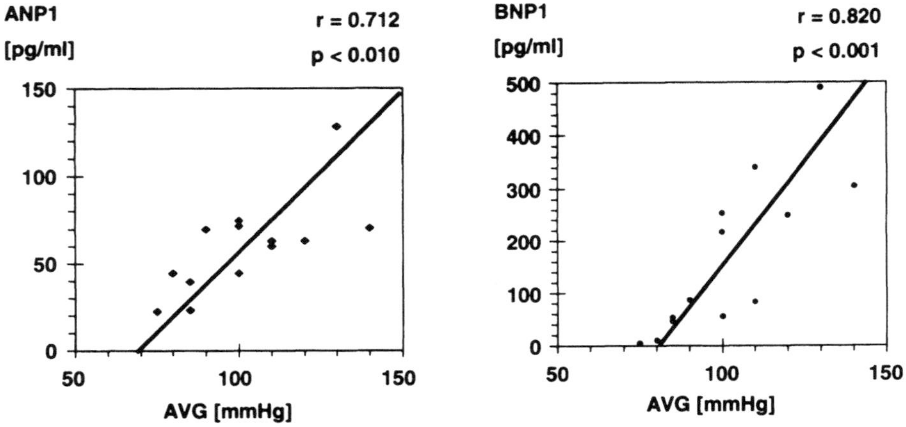

Figure 4. Correlation of resting ANP and BNP plasma concentrations with aortic valve gradient (AVG).

significant changes between pre-, peak- and post-exercise levels (ANP 1–2—$p = 0.011$, ANP 2–3—$p = 0.037$). Exercise testing did not improve test validity, probably due to inconsistent performance at the 6 minute walk test.

Both ANP1 and BNP1 had a non-significant rising trend with NYHA classification (Figure 3) and ANP1 correlated weakly with performance at the 6 minute walk test. We found a strong statistically significant correlation between AVG and BNP1 ($r = 0.820$, $p < 0.001$) and BNP2 levels (0.821, $p < 0.001$) (Figure 4). There was no correlation between ANP or BNP and AVAI, LVMI or other indices. Logarithmic transformation of the plasma concentrations did not improve the correlation results. Considering upper limits as 2 standard deviations above mean of the control group, we have found that sensitivity of the test for the patients with AVG over 100 mm Hg was 77.7% for ANP1 and 100% for BNP1; and specificity was 57.1% and 28.6%, respectively. However, the test was not sensitive enough to detect borderline gradients under 100 mm Hg.

Heart transplantation

On the first day after the heart transplantation the plasma concentrations of ANP fell significantly ($p < 0.05$) from 285.3 ± 71.3 pg/ml pre-transplant to 53.6 ± 12.1 pg/ml. In the next days the values increased to 117.8 ± 32.4 pg/ml on day 5, followed by a steady decline to 78.1 ± 18.8 pg/ml on day 28 (Figure 5). On the contrary, BNP levels did not change significantly in the first three days, with 434.2 ± 108.6 pg/ml pre-transplant and 403.4 ± 43.2 pg/ml on day 1. Afterwards the BNP values rose non-significantly to 715.7 ± 142.1 pg/ml on day 14, with subsequent gradual decline to 340.0 ± 87.8 pg/ml, which is lower than the pre-transplant values. ET-1 levels increased steeply from 1.60 ± 0.27 pg/ml pre-transplant to 2.47

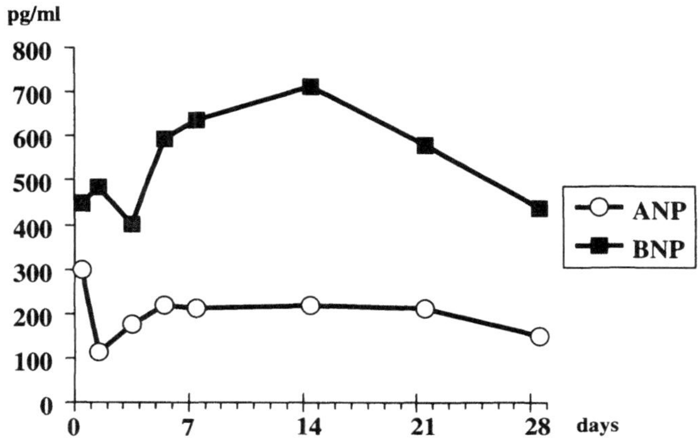

Figure 5. Mean ANP and BNP plasma concentrations after heart transplantation.

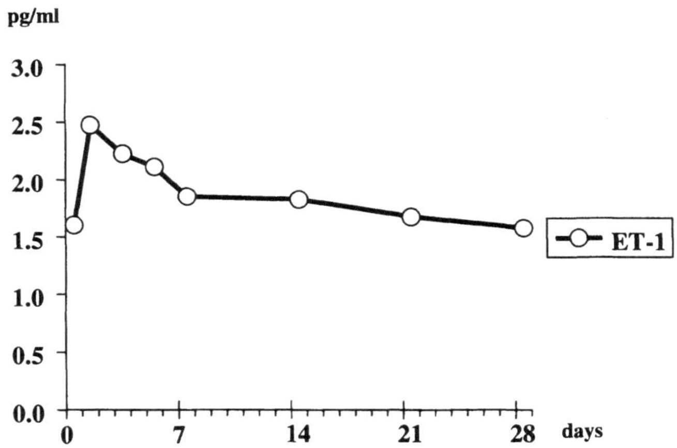

Figure 6. Mean ET-1 plasma concentrations after heart transplantation.

± 1.46 pg/ml on day 1, with a subsequent steady fall to pre-transplant levels on day 28, 1.57 ± 0.13 pg/ml (Figure 6).

DISCUSSION

Aortic stenosis

There are only a few studies regarding natriuretic peptides in valvular disease, aortic stenosis in particular. Consistently with our findings, there was a wide inter-patient variability (20). Correlation was found between ANP or BNP levels and different characteristics such as end-systolic wall stress (21), left ventricular mass index and left atrial pressure (22,23), and aortic valve gradient (24). However, the studies used small numbers of patients and the results of different studies are inconsistent.

We have found both ANP and BNP plasma concentrations to be significantly higher in patients with severe aortic stenosis than in control subjects. Moreover, we have found that the natriuretic peptides correlate with the severity of aortic stenosis expressed by the aortic valve peak pressure gradient, BNP being a better marker of severity. There was also a non-significant trend of correlation with symptoms of aortic stenosis expressed by the NYHA classification and performance at the 6 minute walk test. We have not found exercise testing to improve the validity of the test. The BNP plasma concentration had a high sensitivity and negative predictive value to detect critical aortic stenosis, although its specificity and positive predictive value was low. However, it was insensitive in lower gradients where it could be helpful in decision-making concerning the timing of surgery. Due to the wide inter-patient variability, the plasma natriuretic peptides do not seem to be suitable as a screening tool for detection of severe aortic stenosis indicated for aortic valve replacement and can by no means replace echocardiography, but it could still be used to exclude it. However, larger numbers of patients, particularly with borderline gradients, would be needed to confirm these conclusions.

Heart transplantation

Heart transplantation leads to normalization of hemodynamics, which results in decreased renin-angiotensin-aldosterone and catecholamine levels. Similar changes were expected to occur in natriuretic peptides, but the ANP and BNP levels remain increased, though lower than before the heart transplantation. No specific mechanism of this finding has yet been postulated. Geny et al. (25) hypothesized that extensive atrial mass remaining after transplantation may sustain high natriuretic peptide levels, but these are increased irrespective of surgical technique, i.e. standard versus bicaval orthotopic transplantation with small atrial mass. Denervation of the heart could also be responsible for ANP hypersecretion. Corticosteroids and cyclosporin may influence natriuretic peptide gene expression either by direct stimulation, or indirectly through increased preload and afterload. Cyclosporin may decrease renal clearance of ANP. Geny et al (26) proved association between increased concentrations of ET-1 and ANP. Another theory blames inflammatory processes. In animal experiments there was a correlation between ANP and allograft rejection, however, this is still controversial in humans.

Endothelin levels also rise after the transplantation and remain increased for 3–12 months (27). There probably is a connection with systemic hypertension that usually develops in these patients. The heart transplantation as a surgical procedure gives rise to a combination of factors, such as acute physiological stress, tissue damage and endothelial dysfunction that may together with chronic endothelial impairment and immunosupressive therapy contribute to increased ET-1 concentrations after the operation.

CONCLUSION

Neuroendocrine activation is an important process in pathophysiology of heart disease. Natriuretic peptides have been studied most extensively in heart failure, where they seem to have important diagnostic and prognostic implications. Their possible therapeutic usage seems promising, too. However, they may also contribute to diagnosis and management of other heart disorders associated with pressure or volume overload. Although nonspecific, their high negative predictive value enables exclusion of a serious heart disease.

REFERENCES

1. Kangawa K, Matsuo H. Purification and complete amino acid sequence of alpha-human atrial natriuretic peptide. Biochem Biophys Res Commun 1984; 118:131–139.
2. Stein BC, Levin RI. Natriuretic peptides: physiology, therapeutic potential, and risk stratification in ischemic heart disease. Am Heart J 1998; 135:914–923.
3. Sudoh T, Minamino N, Kangawa K, Matsuo H. A new natriuretic peptide in human brain. Nature 1988; 332:78–81.
4. Raine AE, Erne P, Burgisser E, Muller FB, Bolli P, Burkart F, Buhler FR. Atrial natriuretic peptide and atrial pressure in patients with congestive heart failure. N Engl J Med 1986; 315:533–537.
5. Edwards BS, Zimmerman RS, Schwab TR, Heublein DM, Burnett JC. Atrial stretch, not pressure, is the principal determinant controlling the acute release of atrial natriuretic factor. Circ Res 1988; 62:191–195.
6. Leskinen H, Vuolteenaho O, Ruskoaho H. Combined inhibition of endothelin and angiotensin II receptors blocks volume load induced cardiac hormone release. Circ Res 1997; 80:114–123.
7. Mc Donagh TA, Robb SD, Murdoch DR, Morton JJ, Ford I, Morrison CE, Tunstall-Pedoe H, McMurray JJV, Dargie HJ. Biochemical detection of left ventricular systolic dysfunction. Lancet 1998; 351:9–13.
8. Hall C, Rouleau JL, Moye L, de Champlain J, Bichet D, Klein M, Sussex B, Packer M, Rouleau J, Arnold MO. N-ANP, an independent predictor of long term prognosis after myocardial infarction. Circulation 1994; 89:1934–1942.
9. Cowie MR, Struthers AD, Wood DA, Coats AJS, Thompson SG, Poole-Wilson PA, Sutton GC. Value of natriuretic peptides in assessment of patients with possible new heart failure in primary cary. Lancet 1997; 350:1349–1353.
10. Lerman A, Gibbons RJ, Rodeheffer RJ, Bailey KR, McKinley LJ, Heublein DM, Burnett JC Jr. Circulating N-terminal atrial natriuretic peptide as a marker for symptomless left ventricular dysfunction. Lancet 1993; 341:1105–1109.
11. Francis GS, Benedict C, Johnstone DE, Kirlin PC, Nicklas J, Liang CS, Kubo SH, Rudin-Toretsky E, Yusuf S. for the SOLVD investigators. Comparison of neuroendocrine activation in patients with left ventricular dysfunction with and without congestive heart failure. A substudy of the studies of left ventricular dysfunction (SOLVD). Circulation 1990; 82:1724–1729.
12. McDonagh TA, Cunningham AD, Morrison CE, McMurray JJ, Ford I, Morton JJ, Dargie HJ. Left ventricular dysfunction, natriuretic peptides, and mortality in an urban population. Heart. 2001 Jul; 86(1):21–26.
13. Troughton RW, Framptom CM, Yandle TG, Espiner EA, Nicholls MG, Richards AM. Treatment of heart failure guided by plasma N-BNP concentrations. Lancet 2000; 355:1126–1130
14. Yanagisawa M, Kurihana H, Kimura S. A novel potent vasoconstrictor peptide produced by vascular endothelial cells. Nature 1988; 332:411–415.
15. Lazurova I, Kramer HJ. Endothelins—physiology, pathophysiology and impact on arterial hypertension. Vnitr Lek 2000; 46:719–723.
16. Frelin CH, Guedin H. Why are circulating concentrations of endothelin-1 so low? Cardiovasc Res 1994; 28:1613–1622.
17. Levin ER. Endothelins. New Engl J Med 1995; 333:356–363.
18. Watanabe T, Suzuki N, Shimamoto N, Fujino M, Imada A. Endothelin in myocardial infarction. Nature 1990; 27:1613–1618.
19. Schiffrin EL. Endothelin and endothelin antagonists in hypertension. J Hypertens 1998; 16:1891–1895.

20. Prasad N, Bridges AB, Lang CC. Brain natriuretic peptide concentrations in patients with aortic stenosis. Am Heart J 1997; 133:477–479.
21. Ikeda T, Matsuda K, Itoh H, Shirakami G, Miyamoto Y, Yoshimasa T, Nakao K, Ban T. Plasma levels of brain and atrial natriuretic peptides elevate in proportion to left ventricular end-systolic wall stress in patients with aortic stenosis. Am Heart J 1997; 133:307–314.
22. Qi W, Mathisen P, Kjekshus J, Simonsen S, Bjornerheim R, Endresen K, Hall C. Natriuretic peptides in patients with aortic stenosis. Am Heart J 2001; 142:725–732.
23. Hrudova J, Linhart A, Vondracek V, Humhal J, Holm F, Jachymova M, Ort J, Mareckova Z, Horky K. Humoral profile of patients with severe aortic stenosis. Cas Lek Cesk 2000; 139:74–78.
24. Talwar S, Downie PF, Squire IB, Davies JE, Barnett DB, Ng LL. Plasma N-terminal pro BNP and cardiotrophin-1 are elevated in aortic stenosis. Eur J Heart Fail 2001; 3:15–19.
25. Geny B, Charloux A, Lampert E, Lonsdorfer J, Haberey P, Piquard F. Enhanced brain natriuretic response to peak exercise in heart transplant recipients. J Appl Physiol 1998; 85:2270–2276.
26. Geny B, Piquard F, Follenius M, Thiranos MC, Charpentier A, Epailly E, Levy F, Kretz JG, Eisenmann B, Haberey P. Endothelin participates in increased circulating atrial natriuretic peptide early after human heart transplantation. J Heart Lung Transplant 1998; 17:167–175.
27. Lerman A, Kub SH, Tschumperlin LK, Burnett JC. Plasma endothelin concentrations in humans with end stage heart failure and after heart transplantation. J Am Coll Cardiol 1992; 20:849–853.

Pathophysiology of Cardiovascular Disease,
edited by N.S. Dhalla, H. Rupp,
A. Angel & G.N. Pierce
Kluwer Academic Publishers, Boston, 2004

Activation of Mitogen-Activated Protein Kinases and Protein Kinase B/Akt Signaling by Oxidative Stress in Vascular Smooth Muscle Cells: Involvement in Vascular Pathophysiology

Ashok K. Srivastava, Nihar R. Pandey and Antoine Blanc

Research Center, Centre hospitalier de l'Université de Montréal, Hôtel-Dieu and Department of Medicine, Université de Montréal, Montreal (Quebec) H2W 1T7, Canada

Summary. Oxidative stress has been implicated in the pathogenesis of a host of vascular abnormalities such as atherosclerosis, hypertension and in restenosis followed by balloon angioplasty. However, the molecular mechanism by which oxidative stress causes these abnormalities remains poorly characterized. Recent studies have shown that exposure of vascular smooth muscle cells (VSMC) with H_2O_2, to mimic oxidative stress, activates components of growth promoting and proliferative signal transduction pathways. These components include mitogen-activated protein kinases (MAPKs) and protein kinase B (PKB/Akt), and are believed to be key players mediating growth, proliferation, hypertrophy, migration, survival and death of VSMC. In this article, we provide a brief overview of the effect of H_2O_2 on MAPKs and PKB/Akt signaling in VSMC in relation to their potential role in the pathogenesis of vascular abnormalities.

***Key words*:** VSMC, mitogen-activated protein kinases, protein kinase B, vascular pathophysiology, oxidative stress, H_2O_2.

INTRODUCTION

The past decade has witnessed a surge of interest in studies related to oxidative stress in biological systems. This is primarily due to the potential implication of increased oxidative stress in a variety of pathophysiological conditions including

Address for Correspondence: Ashok K. Srivastava, Ph.D., Research Center—CHUM, Hôtel-Dieu, 3840 St. Urbain Street, Rm 7-135, Montreal (Quebec) H2W 1T8, Canada. Tel.: (514) 890-8000 ext. 12917, Fax: (514) 412-7152, E-mail: ashok.srivastava@umontreal.ca.

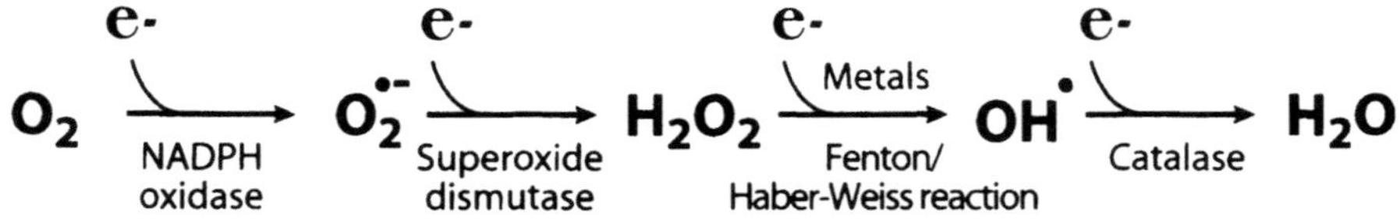

Figure 1. Simplified scheme showing key steps in the production of reactive oxygen species (ROS). ROS, such as $O^{\bullet-}_2$ (superoxide anion), H_2O_2 (hydrogen peroxide) and $OH^{\bullet}$ (hydroxyl ion) are generated by several pathways, some of which are shown here. NADPH oxidase catalyzes transfer of one electron from NADP to molecular oxygen to generate ($O^{\bullet-}_2$). $O^{\bullet-}_2$ is converted to H_2O_2 either by a protonation reaction or by the action of superoxide dismutase. $OH^{\bullet}$ is produced by Fenton or Haber-Weiss reaction in the presence of metal ions, and finally inert H_2O is formed by Catalase.

Table 1. Some of antioxidant enzymes and molecules involved in ROS scavanging

Conventional antioxidant enzymes
Superoxide dismutase
Catalase
Glutathione peroxidase
Small antioxidant molecules
α-tocopherol
Lipoic acid
Ascorbic acid

cardiovascular disorders (1–3). Oxidative stress is caused by an excessive accumulation of reactive oxygen species (ROS) as a result of a defective anti-oxidant defense system of the cell. Endogenously, the main source of ROS is believed to be mitochondria which converts about 1–2% of consumed molecular oxygen into superoxide anion ($O^{\bullet-}_2$) (4–5). Other sources for the generation of $O^{\bullet-}_2$ include NADPH oxidase, lipooxygenase, hypoxanthine/xanthine oxidase or cyclooxygenase (1). $O^{\bullet-}_2$ is a relatively short-lived ROS and is converted to H_2O_2 by superoxide dismutase (SOD), and H_2O_2 is degraded to produce water by several cellular enzymes such as catalase and glutathione (GSH) peroxidase. Under certain conditions, and in presence of metals H_2O_2 can generate extremely active hydroxyl radical ($OH^{\bullet}$) via Fenton and Haber-Weiss reactions (6–7) (Figure 1). Exogenous source of ROS includes ultraviolet light, ionizing radiation and environmental and chemical toxins (8). ROS generation is also believed to play an essential role in propagating the signals of various growth factors, peptide hormones and cytokines such as platelet-derived growth factor (PDGF) (9), epidermal growth factor (EGF) (10), angiotensin II (11) insulin (12) interleukin-1 and tumor necrosis factor-α (TNF-α) (13,14). Normally, the cells are able to detoxify and neutralize the ROS to produce inert substances by using a host of redox scavenging systems. These comprise of conventional antioxidant enzymes such as SOD, catalase, GSH peroxidase, and small antioxidant molecules such as α-tocopherol, lipioic acid, ascorbic acid, and vitamins A and E (Table 1). An aberration in anti-oxidant scavenging ability of cells leads to excessive accumulation of ROS, which by activating a series of growth promoting and proliferative signaling pathways contributes to the pathogenesis of vascular diseases (8). Among these pathways the activation of mitogen-activated protein kinases (MAPKs) and phosphatidylinositol-3-kinase (PI3-K)/protein kinase B (PKB/Akt) pathways appear to be key players in mediating growth promoting, proliferative, migratory, survival and death responses in the cell. Therefore, this article aims to provide a brief overview on the effect of oxidative stress, primarily by H_2O_2, on MAPKs, PI3-K/PKB/Akt signaling in vascular smooth muscle cell (VSMC) and their potential contribution in the pathogenesis of vascular diseases.

Mitogen-activated protein kinases

Mitogen-activated protein kinases (MAPKs) are serine/threonine protein kinases, which are activated in response to a variety of external stimuli, including growth factors, hormones and stress. MAPKs have been classified into several subfamilies: ERK 1/2 (extracellular signal-regulated kinases 1 and 2), $p38^{mapk}$, JNK/SAPK (c-Jun NH_2-terminal kinase/stress-activated protein kinase), ERK 3/4 and ERK 5 (reviewed in ref. 15,16). MAPKs are activated by dual phosphorylation on both tyrosine and threonine residues by dual specificity protein kinases known as MAPKK or MEK (**m**itogen **e**xtracellular signal-regulated kinase **k**inase) (15). The sequential upstream signaling molecules to MEK are MEK kinases (MEKKs), a serine/threonine kinase, and $p21^{ras}$ (16). $p21^{ras}$ belongs to family of low molecular weight GTP binding proteins which cycles between an active GTP-bound conformation and an inactive GDP-bound form. Activation of $p21^{ras}$ and other low molecular weight GTP-binding protein such as Rho, leads to sequential activation of several serine/threonine protein kinases which are involved in the activation of MAP kinase families (17). MAPKs phosphorylate downstream cytosolic regulatory proteins, such as $P90^{rsk}$, and many nuclear transcription factors, such as c-Jun, ATF-2, Elk-1, CHOP, CREB, and MEF-2 (15–25). $P90^{rsk}$ phosphorylates ribosomal proteins and participates in protein synthesis (25), whereas the phosphorylation of transcription factors by MAPKs leads to activation of several genes involved in growth and differentiation (16). Thus, activation of the MAPK pathway can potentially results in increased growth, hypertrophy, gene expression and proliferation of VSMC in response to oxidative stress. Therefore, several investigators have examined the effect of H_2O_2-induced oxidative stress on the activation of MAPKs in cultured VSMC (26–35). Although, in some earlier studies no significant activation of ERK1/ERK2 was demonstrated in response to H_2O_2 in VSMC (26,28,31) in subsequent studies a robust increase in ERK 1/2 phosphorylation was reported (27,29,30,32–35). Effect of H_2O_2 on the activation of $p38^{mapk}$, JNK/SAPK, and big mitogen activated protein kinase 1, (BMK1/ERK5), other members of MAPK family has also been investigated in rat aortic VSMC. Ushio-Fukio demonstrated that 200 µM H_2O_2 caused a rapid (within 2 minutes) increase in $p38^{mapk}$ phosphorylation which was sustained for up to 15 minutes (31). Under these conditions, however, no ERK 1/2 phosphorylation was detected (31). Abe et al. were also unable to observe any significant enhancement in ERK 1/2 phosphorylation in VSMC treated with 200 µM

H_2O_2 (28), but in subsequent studies $100\,\mu M$ H_2O_2 was shown to stimulate ERK 1/2 phosphorylation which peaked at 1 mM (32). These investigators also demonstrated a significantly enhanced phosphorylation of $p38^{mapk}$ and JNK/SAPK (32). A strong phosphorylation of BMK1 (ERK 5) in response to $200\,\mu M$ H_2O_2 has also been demonstrated (28). Interestingly, at this concentration of H_2O_2 no increased phosphorylation of ERK 1/2 was observed. Studies from our laboratory have also demonstrated that as low as $100\,\mu M$ H_2O_2 enhanced the phosphorylation of ERK 1/2 and $p38^{mapk}$ in VSMC from rat embryonal thoracic aorta (A-10 cells) (Figure 2). The H_2O_2-induced phosphorylation of both ERK 1/2 and $p38^{mapk}$ was time-dependent and could be detected within 5 minutes of H_2O_2 treatment of cells (Figure 3). However, the time course of the activation was different for 2 MAPKs, ERK 1/2 phosphorylation remained sustained for up to 30 min and then declined slightly during the next 120 min (Figure 3), whereas phosphorylation of $p38^{mapk}$ showed a biphasic response, with peaks at 10 min and 120 min (Figure 3). Thus it appears that the sensitivity of phosphorylation of different MAPKs in response to H_2O_2 is dependent on the source ofVSMC as well as on the type of MAPK being studied. Rao reported that H_2O_2-induced ERK 1/2 activation was associated with enhanced tyrosine phosphorylation of EGF receptor as well as guanine nucleotide exchange factors and adapter of protein Grb2 and SOS (27). In these studies a direct activation of $p21^{ras}$ was also demonstrated in VSMC stimulated with $200\,\mu M$ H_2O_2 for 30 minutes (27). Moreover, in pulmonary artery airway smooth muscle cells, H_2O_2 induced activation of ERK 1/2 was associated with stimulation of MEK-1 and c-raf activities demonstrating that all the upstream components of ERK-signaling cascade, i.e. ras/raf/MEK participate in this process (28–30). Since MAPKs can phosphorylate and activate several regulatory proteins which are intimately linked to gene expression, growth promotion and hypertrophy, activation of this signaling component has been implicated in mediating some of the physiological responses of H_2O_2 in VSMC (36). ERK 1/2 and $p38^{mapk}$ might also play a role in smooth muscle cell migration and vascular remodeling by regulating the transcription of matrix metalloproteinase-9 (MMP-9) (37).

Protein kinase B/Akt

Protein kinase B (PKB), also known as Akt (a product of akt protooncogene) a serine/threonine protein kinase (38) has received much attention due to its potential role in a myriad of physiological processes which include protein synthesis, glucose transport and cell survival (39). Three isozymic forms of PKB/Akt have been identified, and all of which are activated by dual phosphorylation on threonine (Thr^{308}) and serine (Ser^{473}) (40). Phosphorylation of PKB/Akt in thr^{308} is catalyzed by a phosphatidylinositol 3,4,5 tris phosphate (PIP3) dependent protein kinase-1 (PDK-1) (41). The protein kinase responsible for catalyzing ser^{473} phosphorylation remains uncharacterized at present but is tentatively referred to as PDK-2 (41). PIP3 is the product of the reaction catalyzed by phosphatidylinositol 3-kinase (PI3-K). PI3-Ks are heterodimeric enzymes composed of a 85 kDa (p85) regulatory subunit and a 110 kDa (p110) catalytic subunit. The p85 subunit contains the src

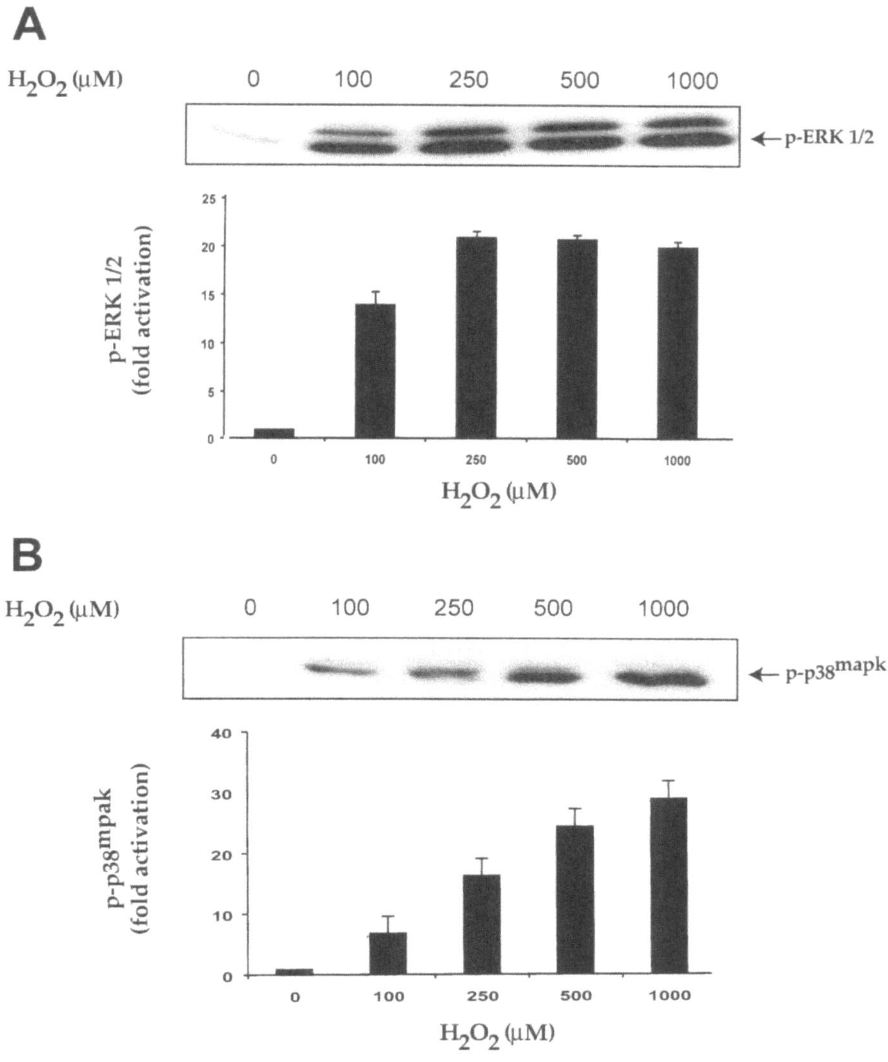

Figure 2. H_2O_2-induced phosphorylation of ERK 1/2 and $p38^{mapk}$. Serum-starved quiescent VSMCs (A10) were treated with different concentration of H_2O_2 for 10 min. Cell lysates were prepared, and an equal amount of proteins were separated on 12% SDS-PAGE. Phosphorylation of ERK 1/2 and $p38^{mapk}$ was detected by immunoblot analysis with phospho-specific-ERK and $p38^{mapk}$ antibodies (59). **(A)** ERK 1/2 phosphorylation. **(B)** $p38^{mapk}$ phosphorylation. Bottom panels represent average data quantified by densitometric scanning of immunoblots expressed as fold increase in phosphorylation. Values are the mean of ±SE for 3 independent experiments.

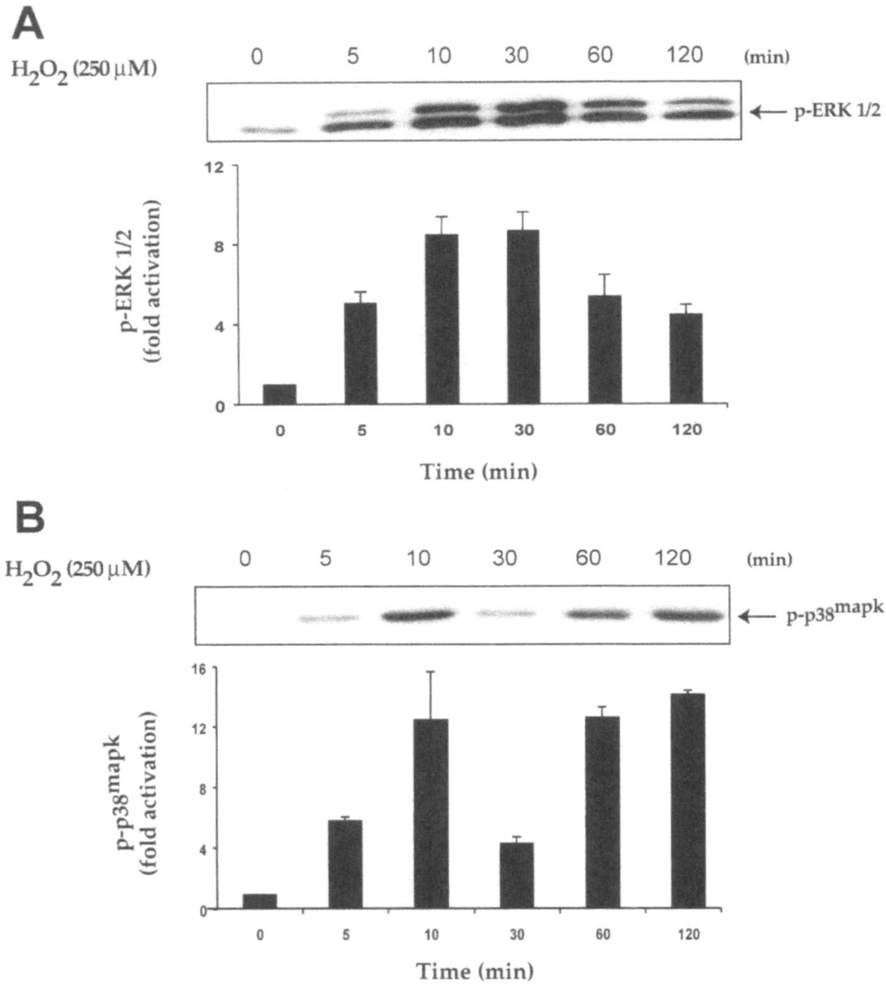

Figure 3. Time course of H_2O_2-induced phosphorylation of ERK 1/2 and p38^{mapk}. Serum-starved quiescent VSMCs (A10) were incubated with 250 μM H_2O_2 for the indicated time. Cell lysates were prepared, and an equal amount of proteins were separated on 12% SDS-PAGE. Phosphorylation of ERK 1/2 and p38^{mapk} was detected by immunoblot analysis with phospho-specific-ERK and p38^{mapk} antibodies (59). **(A)** ERK 1/2 phosphorylation. **(B)** p38^{mapk} phosphorylation. Bottom panels represent average data quantified by densitometric scanning of immunoblots expressed as fold increase in phosphorylation. Values are the mean of ±SE for 3 independent experiments.

homology-2 (SH-2) domain and is able to interact with phosphorylated tyrosine residues on receptor or other docking proteins. This interaction eventually stimulates the activity of the 110 kDa catalytic subunit. The catalytic subunit catalyzes the phosphorylation of phosphatidylinositol (PI), PI4 phosphate and PI4,5 biphosphate in the 3′ position of the inositol ring to generate PI3 phosphate and PI3,4 biphosphate, and PI3,4,5 triphosphate, respectively (42). PI3-Ks are activated in response to growth factors and other agents, including vasoactive peptides, and have been suggested to regulate a variety of cellular responses such as proliferation, differentiation, growth and cell survival (42,43). Given the importance of PI3K/PKB/Akt signaling in various cellular processes, effect of H_2O_2 on PI3K and PKB activation in VSMC has also been investigated. In VSMC from adult rat thoracic aortas exogenous H_2O_2 was shown to increase the phosphorylation of PKB/Akt in ser^{473} (44). The phosphorylation was completely blocked by wortmannin and LY294002; two specific inhibitors of PI3K (44,45), implicating PI3K as an upstream mediator of H_2O_2 response. Consistent with this observation, a direct activation of PI3K in rat A-10 VSMC treated with 200 μM H_2O_2 has been reported (35). In addition, similar to the studies of Ushio-Fukai, we have also shown that H_2O_2 in a dose and time-dependent fashion stimulated the Ser^{473} phosphorylation of PKB/Akt in A-10 VSMC (Figure 4). The phosphorylation was evident even at the lowest concentration of H_2O_2 (100 μM) tested and was further enhanced by increasing the concentration of H_2O_2 up to 1 mM which resulted in about 12-fold increase in PKB phosphorylation (Figure 3A). The PKB phosphorylation in response to 250 μM H_2O_2 was transient, reaching a peak value at 10 min and then quickly declining within the next 60 min (Figure 3A). Since PKB/Akt has emerged as key effectors of survival, growth and death responses (39,43), it is possible that oxidative stress induced activation of this important signaling pathway contributes to various abnormalities associated with vascular pathophysiology.

Potential mechanism of MAPKs/PKB/Akt activation by H_2O_2

The underlying mechanisms responsible for MAPKs or PKB/Akt activation in response to H_2O_2 in VSMC have not been fully characterized. However, there are reports demonstrating an increased transactivation of EGF (27) and PDGF-receptor PTK (46), as well as Src family of PTKs such as c-src, and fyn in cells treated with H_2O_2 (47,48). H_2O_2 has also been shown to inhibit the activities of protein tyrosine phosphatases (PTPases) such as PTP1B (49), and SHP-2 (50). Inhibition of PTPase is accomplished by oxidation of catalytically essential cysteine residue in the active site of PTPase (51) (Figure 5). Inhibition of cellular PTPases shifts the equilibrium of phosphorylation-dephosphorylation cycle by promoting a burst in PTK activity resulting in an increase in the tyrosyl phosphorylation of substrate proteins. These tyrosyl phosphorylated proteins promote the assembly of signaling modules responsible for activating various components of MAPKs and PI3-K signaling pathway (Figure 5). The activated MAPKs phosphorylate downstream regulatory target proteins such as p90^{rsk} and several transcription factors (Figure 5) leading to increased protein synthesis and gene expression. Similarly, PIP_3-induced activation

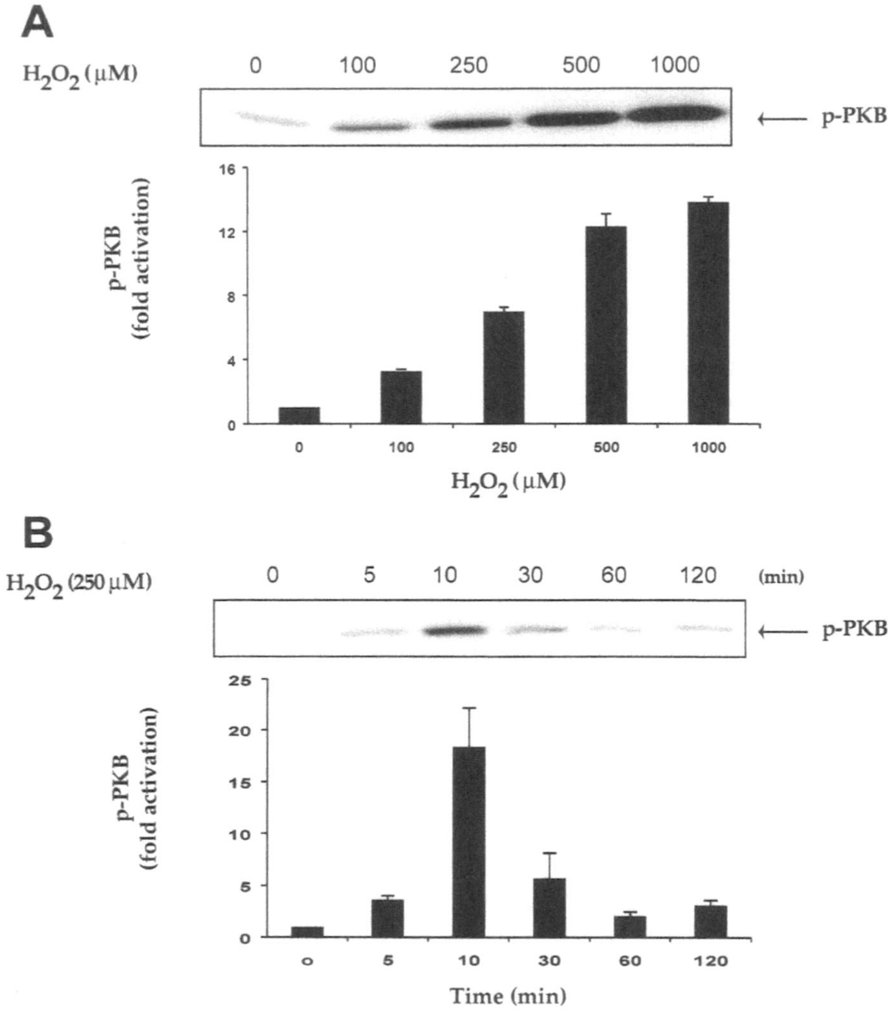

Figure 4. Dose and time course of H_2O_2-induced phosphorylation of PKB/Akt. Serum-starved quiescent VSMCs (A10) were treated with different concentration of H_2O_2 for 10 min **(A)** or incubated with 250μM H_2O_2 for the indicated time **(B)**. Cell lysates were prepared, and an equal amount of proteins were separated on 12% SDS-PAGE. Phosphorylated PKB/Akt was detected by immunoblot analysis with phospho-ser^{473}-specific-PKB/Akt antibody (59). Bottom panels represent average data quantified by densitometric scanning of immunoblots expressed as fold increase in phosphorylation. Values are the mean of ±SE for 3 independent experiments.

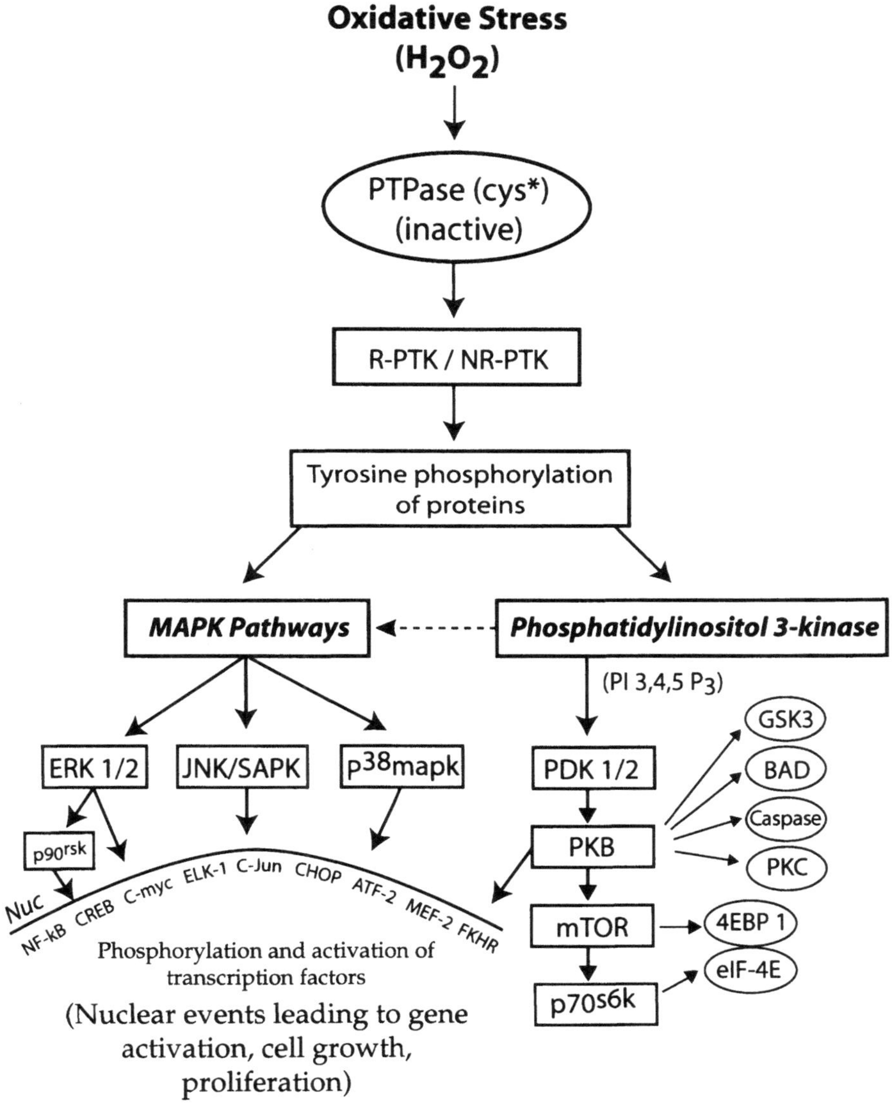

Figure 5. Schematic model showing potential steps in H_2O_2-induced activation of MAPK and PKB/Akt signaling. The transcription factors are shown inside the nucleus (Nuc). Further details are given in the text. Abbreviations are as follows: PTPase, protein tyrosine phosphatase; cys*, oxidised cysteine in the catalytic site of PTPase; R/NR-PTK, receptor protein tyrosine kinase; MAPK pathways, mitogen-activated protein kinase pathways include low molecular weight GTP-binding proteins (ras, rho), mitogen-activated protein kinase kinase (MAPKK such as c-raf) and MAPKK (MEK); ERK, extracellular signal-regulated kinase; JNK, c-Jun N-terminal kinase; SAPK, stress-activated protein kinase; NF-κB, nuclear factor κB; PI3,4,5,P_3, phosphatidylinositol 3,4,5-trisphosphate; PDK 1/2, phosphatidylinositol-dependent kinase 1/2; PKC, protein kinase C-ζ; PKB, protein kinase B; mTOR, mammalian target of rapamycin; $p70^{rsk}$, 70-kDa ribosomal S-6-kinase; $p90^{rsk}$, 90-kDa ribosomal S-6-kinase; GSK-3, glycogen synthase kinase 3; BAD, **B**CL-2 BCL-XL **a**ntagonist causing cell **d**eath; eIF-4E, eukaryotic translation initiation factor 4E; 4EBP 1, eIF-4E binding protein 1.

of downstream targets of PKB such as mTOR, $p70^{s6k}$, GSK-3, FKHR, BAD, caspase and some forms of PKC (Figure 5) are critical effectors of cellular growth responses including cell cycle entry, cell division, growth, gene expression and survival (43,52). In addition a potential cross-talk between PI3-K and MAPK pathway has also been suggested (35). A possible role of Ca^{2+}, in mediating H_2O_2-induced response has also been suggested because H_2O_2 has been shown to increase Ca^{2+} levels in several cell types including VSMC (53–58). H_2O_2-stimulated changes in Ca^{2+} are believed to be mediated at least in part through a PTK-dependent (56–58) and PI3-K-dependent pathway involving the activation of phospholipase C gamma (PLCγ) (56,58). However, the precise contribution of Ca^{2+} in stimulating MAPKs and PKB/Akt pathways in VSMC remains to be defined.

CONCLUSIONS

The results from the studies summarized in this article clearly demonstrate that H_2O_2 stimulates two key signaling pathways—MAPKs and PKB/Akt in VSMC. Inhibition of PTP activities by H_2O_2 appears to be a critical step in this activation response. Since several downstream targets of MAPKs and PKB/Akt are intimately involved in the regulations of growth, proliferation, hypertrophy, migration, survival and death of VSMC, the activation of these two signaling pathways may have an important role in vascular pathophysiology due to oxidative stress.

ACKNOWLEDGEMENTS

This study was supported by a grant from Heart and Stroke Foundation of Quebec to A.K.S. We thank Susanne Bordeleau-Chénier for preparing the manuscript and illustrations. We also thank Int. J. Mol. Med. for permission to reproduce our work published in Int. J. Mol. Med. 2003, 11:229–234.

REFERENCES

1. Droge W. 2002. Free radicals in the physiological control of cell function. Physiol Rev 82:47–95.
2. Dhalla NS, Temsah RW, Netticaden T. 2000. Role of oxidative stress in cardiovascular disease. J Hypertens 18:655–673.
3. Singal PK, Khaper N, Palace V, Kumar D. 1998. The role of oxidative stress in the genesis of heart disease. Cardiovas Res 40:426–432.
4. Turrens JF. 1997. Superoxide production by the mitochondrial respiratory chain. Bio Sc Rep 17:3–8.
5. Boveris A, Chance B. 1973. The mitochondrial generation of hydrogen peroxide: General properties and effect of hyperbasic oxygen. Biochem J 134:707–716.
6. Imalay JA, Chin SM, Lin S. 1988. Toxic DNA damage by hydrogen peroxide through the Fenton reaction in vivo and in vitro. Science 240:640–642.
7. Haber F, Weiss JJ. 1934. The catalytic decomposition of hydrogen peroxide by iron salts. Proc R Soc Lond Ser A 147:332–335.
8. Abe JI, Berk BC. 1998. Reactive oxygen species as mediators of signal transduction in cardiovascular disease. Trends in Cardiovas Med 8:59–64.
9. Sundaresan M, Yu ZX, Ferrari VJ, Ironi K, Finkel T. 1995. Requirement for the generation of H_2O_2 for platelet-derived growth factor signal transduction. Science 270:296–299.
10. Bae VS, Kang SW, Seo MS, Baines IC, Tekle E, Chock PB, Rhee SG. 1997. Epidermal growth factor (EGF)-induced generation of hydrogen peroxide. Role in EGF receptor-mediated tyrosine phosphorylation. J Biol Chem 272:217–221.

11. Griendling KK, Minieri CA, Ollerenshaw JD, Alexander RW. 1994. Angiotensin II stimulates NADH and NADPH oxidase activity in cultured vascular smooth muscle cells. Circulation 74:1141–1148.
12. Mahadev K, Wu X, Zilbering A, Zhu L, Lawrence JT, Goldstein BJ. 2001. Hydrogen peroxide generated during cellular insulin signaling cascade in 3T3-L1 adipocytes. J Biol Chem 276:48662–48669.
13. Lo YYC, Cruz TF. 1995. Involvement of reactive oxygen species in cytokines and growth factor induction of c-fos expression in chondrocytes. J Biol Chem 270:11727–11730.
14. Lo YYC, Wong JMS, Cruz TF. 1996. Reactive oxygen species mediate cytokine activation of c-Jun NH2-terminal kinases. J Biol Chem 271:15703–15707.
15. Seger R, Krebs EG. 1995. The MAPK signaling cascade. FASEB J 9:726–735.
16. Widman C, Gibson S, Jarpe MB, Johnson GL. 1999. Mitogen-activated protein kinase: conservation of a 3-kinase module from yeast to human. Physiol Rev 79:143–180.
17. Ip YT, Davis RJ. 1998. Signal transduction by c-Jun N-terminal kinase (JNK) from inflammation to development. Curr Opin Cell Biol 10:205–219.
18. Force T, Bonventre JV. 1998. Growth factors and mitogen-activated protein kinases. Hypertension 31 (Part II):152–161.
19. Davis RJ. 1993. The mitogen-activated protein kinase signal transduction pathway. J Biol Chem 268:14553–14556.
20. Denton RM, Tavare JM. 1995. Dose mitogen-activated protein kinases have a role in insulin action: The cases for and against. Eur J Biochem 227:597–611.
21. Tan Y, Rouse J, Zhang A, Cariati S, Cohen P, Comb MJ. 1996. FGF and stress regulate CREB and ATF-1 via a pathway involving p38 MAP kinase and MAPKAP kinase-2. EMBO J 15:4629–4642.
22. Wang X, Rao D. 1996. Stress-induced phosphorylation and activation of the transcription factor CHOP (GADD 153) by p38 MAP kinase. Science 272:1347.
23. Gupta S, Campbell D, Derijand B, Davis RJ. 1995. Transcription factor AFT-2 regulation by the JNK signal transduction pathway. Science 267:389–393.
24. Zinck R, Cahill MA, Kracht M, Sachsenmaier C, Hipskind RA, Nordheim A. 1995. Protein synthesis inhibitor reveals differential regulation of mitogen-activated protein kinase and stress-activated protein kinase pathways that converge on Elk-1. Mol Cell Biol 15:4930–4939.
25. Frodin M, Gammeltoft S. 1999. Role and regulation of 90kDa ribosomal S6 kinase (RSK) in signal transduction. Mol Cell Endocrinol 151:65–77.
26. Bass AS, Berk BC. 1995. Differential activation of mitogen-activated protein kinases by H_2O_2 and $O^{\bullet-}_2$ in vascular smooth muscle cells. Cir Res 77:29–36.
27. Rao GN. 1996. Hydrogen peroxide induces complex formation of SHC-Grb2-SOS with receptor tyrosine kinase and activates Ras and extracellular signal-regulated protein kinases group of mitogen-activated protein kinase. Oncogene 13:713–719.
28. Abe J, Kusuhara M, Ulevitch RJ, Berk BC, Lee J-D. 1996. Big mitogen-activated protein kinase 1 (BMK1) is a redox-sensitive kinase. J Biol Chem 271:16586–16590.
29. Rao GN. 1997. Protein tyrosine kinase activity is required for oxidant-induced extracellular signal-regulated protein kinase activation and c-fos and c-Jun expression. Cell Signal 9:181–187.
30. Zhang J, Jin N, Liu Y, Rhoades RA. 1998. Hydrogen peroxide stimulates extra-cellular signal-regulated protein kinase in pulmonary arterial smooth muscle cells. Am J Respir Cell Mol Biol 19: 324–332.
31. Ushio-Fukai, Alexander RW, Akers M, Griendling KK. 1998. p38 mitogen-activated protein kinase is a critical component of the redox-sensitive signaling pathways by angiotensin II: role in vascular smooth muscle cell hypertrophy. J Biol Chem 273:15022–15029.
32. Yoshizumi M, Abe J, Haendeler J, Huang Q, Berk BC. 2000. Src and Cas mediate JNK activation but not ERK1/2 and p38 kinases by reactive oxygen species. J Biol Chem 275:11706–11712.
33. Guyton KZ, Liv Y, Gorospe M, Xu Q, Holbrook NJ. 1996. Activation of mitogen-activated protein kinase by H_2O_2. Role in cell survival following oxidant injury. J Biol Chem 271:4138–4142.
34. Taher MM, Mahgouh MA, Abd-Elfattah ASA. 1998. Redox regulation of signal transduction in smooth muscle cells: distinct effects of PKC-down regulation and PKC inhibition on oxidant-induced MAP kinase. J Recept & Signal Trend Res 18:167–185.
35. Srivastava AK, Pandey SK. 2000. Stimulation of mitogen-activated protein kinases ERK-1 and ERK-2 by H_2O_2 in vascular smooth muscle cells. *In*: The Hypertrophied Heart, Takeda N, Nagao M and Dhalla NS (Eds). Boston, Kluwer Academic Publishers, pp. 197–206.
36. Griendling KK, Sorescu D, Lassegue B, Ushio-Fukai M. 2000. Modulation of protein activity and gene expression by reactive oxygen species and their role in vascular physiology and pathology. Arterioscler Thromb Vas Biol 20:2175–2183.

37. Chao A, Graves J, Reidy MA. 2000. Mitogen activated protein kinases mediated matrix metalloproteinase-9 expression in vascular smooth muscle cells. Arterioscler Thromb Vas Biol 20:2527–2532.
38. Bellacosa A, Testa JR, Staal SP, Tsichlis PN. 1991. A retroviral oncogene, akt, encoding a serine-threonine kinases containing an SH-2 like region. Science 254:274–277.
39. Datta SR, Brunet A, Greenberg ME. 1999. Cellular survival: a play in three Akts. Genes Dev 13: 2905–2927.
40. Alessi DR, Andjelkovic M, Caudwell B, Cron P, Morrice N, Cohen P, Hemmings BA. 1996. Mechanism of activation of protein kinase B by insulin and IGF-1. EMBO J 15:5541–6551.
41. Alessi DR, James SR, Downes CP, Holmes AB, Gaffeney PR, Reese CB, Cohen P. 1997. Characterization of a 3-phoshpoinositide-dependent protein kinase which phosphorylates and activates protein kinase B alpha. Curr Biol 7:261–269.
42. Fruman DA, Meyers RE, Cantley LC. 1998. Phosphoinositide kinase. Annu Rev Biochem 67: 481–507.
43. Kozma SC, Thomas G. 2002. Regulation of cell size in growth, development and human disease: PI3K, PKB and S6K. Bio Essays 24:65–71.
44. Ushio-Fukai M, Alexander W, Akers M, Yin QQ, Fujio Y, Walsh K, Griendling KK. 1999. Reactive oxygen species mediate the activation of Akt/Protein kinase B by angiotensin II in vascular smooth muscle cells. J Biol Chem 274:22699–22764.
45. Srivastava AK. 1998. Use of pharmacological inhibitors in elucidating the mechanism of insulin action. Trends Pharmacol Sci 19:205–209.
46. Jin N, Hatton ND, Harrington MA, Xia X, Larsen SH, Rhoades RA. 2000. H_2O_2-induced EGR-1, FRA-1 and c-Jun gene expression is mediated by tyrosine kinase in aortic smooth muscle cells. Free Rad Biol Med 29:736–746.
47. Abe J, Berk BC. 1999. Fyn and JAK2 mediate Ras activation by reactive oxygen species. J Biol Chem 274:21003–21010.
48. Abe J, Okuda M, Huang Q, Yoshizumi M, Berk BC. 2000. Reactive oxygen species activate p90 ribosomal S6 kinase via Fyn and Ras. J Biol Chem 275:1739–1748.
49. Lee SR, Kwon KS, Kim SR, Rhee SG. 1998. Reversible inactivation of protein tyrosine phosphatase 1B in A431 cells stimulated with epidermal growth factor. J Biol Chem 273:15366–15372.
50. Meng T-C, Fukada, Tonks NK. 2002. Reversible oxidation and inactivation of protein tyrosine phosphates in vivo. Mol Cell 9:387–399.
51. Guam KC, Dixon JE. 1991. Evidence for protein tyrosine phosphatase catalysis via a cysteine-phosphate intermediate. J Biol Chem 266:17026–17030.
52. Brazil DP, Hemming BA. 2001. Ten years of protein kinase B signaling: a hard Akt to follow. Trends Biochem Sci 26:657–664.
53. Roveri A, Coassin M, Maiorino M, Zamburlini A, Van Amesterdam FT, Ralti F. 1992. Effect of hydrogen peroxide on calcium homeostasis in smooth muscle cells. Arch Biochem Biophys 297: 265–270.
54. Yang ZW, Zheng T, Wang J, Zhang A, Altura BT, Altura BM. 1999. Hydrogen peroxide induces contraction and raises $[Ca^{2+}]_i$ in canine cerebral arterial smooth muscle: Participation of cellular signaling pathways. Naunyn Schmiedebergs Arch Pharmacol 360:646–653.
55. Meyer TN, Gloy J, Hug MJ, greger R, Schollmeyer P, Pavenstadt H. 1996. Hydrogen peroxide increases the intracellular calcium activity in rat mesanglial cells in primary culture. Kidney Int 49:388–395.
56. Qiu S, Stadman ER, Chock PB. 2000. Regulation of oxidative stress-induced calcium release by phosphatidylinositol 3-kinase and Bruton's tyrosine kinase in B cells. Proc Natl Acad Sci 97: 7118–7123.
57. Qin S, Chock PB. 2001. Bruton's tyrosine kinase is essential for peroxide-induced calcium signaling. Biochemistry 40:8085–8091.
58. González-Pacheco FR, Caramelo C, Castilla MA, Deudero JJP, Arias J, Yagüe S, Jiménez, Bragado R, Álvarez-Arroyo MV. 2002. Mechanism of vascular smooth muscle cells activation by hydrogen peroxide: role of phospholipase C gamma. Nephrol Dial Transplant 17:392–398.
59. Pandey SK, Théberge J-F, Bernier M, Srivastava AK. 1999. Phosphatidylinositol 3-kinase requirement in activation of ras/C-raf-1/MEK/ERK and $p70^{s6k}$ signaling cascade by the insulinomimetic agent vanadyl sulfate. Biochemistry 38:14667–14675.

Pathophysiology of Cardiovascular Disease,
edited by N.S. Dhalla, H. Rupp,
A. Angel & G.N. Pierce
Kluwer Academic Publishers, Boston, 2004

Preconditioning and Myocardial Angiogenesis

Nilanjana Maulik

Molecular Cardiology Laboratory, University of Connecticut School of Medicine, Farmington, CT 06030-1110, U.S.A.

Summary. In the Western World ischemic coronary disease is the leading cause of morbidity and mortality. Therapeutic approaches mostly aim to restore flow to a localized segment by angioplasty or bypass surgery. In order to develop better and more effective therapeutic strategies using the powerful concept of inducing new vessel growth or by employing vascular growth factors, it is essential to further our understanding of the molecular mechanisms and the chain of events underlying the fascinating process of **angiogenesis**. Among the various triggers of angiogenesis, tissue hypoxia/ischemia has been identified as being a particularly important stimulus for the induction of new vessel growth. Occlusion of a main coronary depletes the blood supply to the myocardium and subsequently reduces cardiac function, which ultimately leads to heart failure. Progressive, chronic coronary artery occlusion has been shown to induce development of collateral arteries to reestablish and maintain blood flow to the myocardium at risk via the growth of new capillary vessels or angiogenesis. Studies from our laboratory as well as from others have already confirmed the protective role of collaterals against myocardial ischemia and cell death. We have successfully demonstrated in adult rat myocardium (LV) effects of hypoxia/reoxygenation on significant upregulation of the protein expression profiles of vascular endothelial growth factor (VEGF) and its tyrosine kinase receptors (Flk-1 and Flt-1) as well as other angiogenic factors such as Ang-1, Ang-2 and their receptor Tie-2. Also, we were able to demonstrate increased capillary/arteriolar density, increased capillary to myocyte cross-sectional area (after 1–4 weeks and after 2 months) and decreased collagen volume fraction by hypoxic/ischemic preconditioning in a rat model of chronic myocardial infarction (MI) model.

Correspondence: Nilanjana Maulik, Ph.D., Molecular Cardiology Laboratory, Department of Surgery, L-2096, University of Connecticut, School of Medicine, 263 Farmington Avenue, Farmington, CT 06030-1110, U.S.A. Tel: (860) 679-2857, Fax: (860) 679-4606, E-mail: nmaulik@neuron.uchc.edu.

Key words: Hypoxic preconditioning, ischemic preconditioning, myocardial infarction, angiogenesis, VEGF, apoptosis, PI-3 kinase, Bcl-2, Survivin.

1. INTRODUCTION

Coronary heart disease is the primary cause of cardiovascular death. After myocardial infarction (MI), there is a progressive myocardial remodeling characterized by left ventricular (LV) dilation, contractile dysfunction, myocyte hypertrophy and increased matrix protein formation. Ischemic preconditioning provides the most powerful form of endogenous protection against lethal ischemic injury. The classical preconditioning can be induced by a variety of stimuli other than ischemia. Hypoxia (1–3), calcium (4), adenosine agonists (5) α_1-adrenergic agents (6), muscarinic agonists (7), and stretch (8) have been used as preconditioning stimulus to induce tolerance of heart to the subsequent ischemic episode. Hypoxia, an element of ischemia, plays an important role in the cardiovascular system. First described by Neely and Grotyohann in 1984 (9), hypoxic preconditioning, like ischemic preconditioning, can attenuate stunning caused by repeated coronary artery occlusions (1), and enhance postischemic recovery of myocardial function (2). We have recently showed that hypoxic preconditioning can exert potent cardioprotective effect by upregulating antioxidant reserve of the heart (10). Hypoxic preconditioning like ischemic preconditioning not only reduces the extent of infarction but also causes the salvaged myocardium to have better mechanical function. During the last several years, our laboratory has studied extensively the molecular mechanisms of preconditioning-mediated signal transduction. It was demonstrated that preconditioning-mediated signal transduction cascade triggered by tyrosine-kinase and coupled to phospholipase D leads to the activation of MAP kinases (11). Additionally, we documented that translocation and activation of the nuclear transcription factor of p38 MAP kinase in ischemic preconditioning resulting in the activation of MAPKAP kinase 2 (12). A modern experimental strategy for treating myocardial ischemia is to induce neovascularization of the heart by use of angiogens, mediators that induce the formation of blood vessels, or angiogenesis. The process of angiogenesis is regulated by the above mentioned signal transduction cascade triggered by the signals obtained from the transmembrane receptor tyrosine kinases (RTKs) and non-receptor tyrosine kinases (Src family) of endothelial cells.

Angiogenesis is known to be the body's natural healing process in which new blood vessels grow in response to injury. Therefore it is extremely important to develop the body's natural angiogenic process in order to create collateral circulation in areas where blocked coronary arteries deprive the heart muscle of sufficient blood flow e.g. in the settings of chronic myocardial ischemia. Development of coronary collateral circulation is the heart's own bypass mechanism by which it retains the blood supply to the myocardium at risk (13–16).

Angiogenic therapy for the human heart is currently being vigorously pursued. In the past ten years, alternative revascularization/angiogenesis strategies have progressed from bench to bedside, focussing on the capillary sprouting and/or growth of new vessels to replace the old. However most of the strategies involves the delivery of growth factors. Very little success with these strategies has been demonstrated so far for various reasons. Very recently, we have demonstrated that both hypoxic as well as ischemic preconditioning can stimulate myocardial angiogenesis to an extent sufficient to exert significant cardioprotection in a rat model of myocardial infarction progressing to heart failure as evidenced by increased capillary/arteriolar density and enhanced ventricular contractile functional reserve.

2. PRECONDITIONING MEDIATED ANGIOGENIC RESPONSE IN THE HEART

A. Ischemic preconditioning

Ischemic preconditioning mediated angiogenesis and growth factors/receptors stimulation for cardioprotection is a very novel approach and potentially very exciting therapeutic strategy. Preconditioning of heart by repeated ischemia and reperfusion has been found to delay the onset of subsequent irreversible ischemic injury (17). It is well known that preconditioning provides cardioprotection by reducing subsequent post-ischemic ventricular dysfunction, decreasing incidence of arrhythmias and infarct size. Such myocardial preservation by repeated short-term reversible ischemia leads to the development of the concept of stress adaptation (18). After myocardial infarction (MI), development of left ventricular (LV) enlargement occurs and this process is known as postinfarction ventricular remodeling. Ventricular remodeling is known to be affected by several factors. One major factor is the infarct size, which is limited by the presence of collateral vessels. Capillary density becomes lower in the border zone than in the remote areas of the infarcted ventricle (19). Ischemia or coronary artery occlusion has been shown to induce VEGF mRNA in rat hearts (20). VEGF is a well-known endothelial cell-specific angiogenic factor and also a critical regulator of angiogenesis that stimulates proliferation, migration, and proteolytic activity of the endothelial cells (21). Additional report suggested VEGF-induced expression of Bcl-2, which eventually functions to enhance the survival of endothelial cells in the toxic, oxygen-deficient environment (21). Substantial evidence exists to support the notion that oxygen-derived free radicals are generated during the reperfusion of ischemic myocardium resulting in the development of oxidative stress (22). Ischemia was found to induce angiopoietin/Tie receptor system in a focal cerebral ischemia model (23). Myocardial adaptation to ischemic stress in stunned pig myocardium demonstrated the induction of c-jun, c-fos, Egr-1 and jun-B that may be involved in repair process of angiogenesis (24). Adenosine is known to limit the degree of vascular injury during ischemia and reperfusion by inhibition of oxygen free radical release which prevents endothelial cell damage and that might help to preserve endothelial cell function and microvascular perfusion (25). In our ischemic preconditioned rat myocardial infarction model, we were able to induce angiogenesis after preconditioning. Recently our study demonstrated that in vivo brief repetitive cycles of coronary artery occlusion (5 min) followed by short duration of reperfusion (10 min) triggered myocardial angiogenesis at the capillary and arteriolar levels, which nicely corroborated with the improved myocardial contractile function (26).

B. Hypoxic preconditioning

There are numerous reports on the effects of environmental hypoxic exposure on cardiac pathophysiology. Hypoxia is characterized by inadequate oxygen delivery to the tissue such as myocardium with a resulting imbalance between oxygen demand and energy supply. The possibility that such hyoxic exposure can potentially act as a preconditioning stimulus, however, has not been adequately considered. A strong resemblance exists between the patterns of acute stress response induced by hypoxia/reoxygenation, ischemia/reperfusion or any means of generating ROS.

Hypoxia has been found to be the strongest inducer both in vitro and in vivo, of vascular endothelial growth factor (VEGF) which serves as a major angiogen in normal cardiac development (27,28). Tissue hypoxia exerts a proangiogenic action through various angiogenic factors, the most notable being vascular endothelial growth factor (VEGF). VEGF is mainly associated with initiating the process of angiogenesis through the recruitment and proliferation of endothelial cells VEGF, a protein coded by a 7-exon gene localized on chromosome 6, serves as a major angiogen in normal cardiac development (29). We found that non-lethal moderate hypoxic challenge is capable of increasing protein levels of important angiogenic factors and their receptors in the adult rat myocardium. Immunohistochemical analysis of VEGF revealed a diffuse pattern of distribution throughout the ventricular myocardium with strong localization around the coronary arterial wall where both coronary endothelium as well as vascular smooth muscle appeared to stain positive for VEGF (30). Hearts obtained from rats, which had been subjected to hypoxia followed by a 24 hour period of reoxygenation, displayed a progressive increase in intensity of staining for VEGF with increasing durations of hypoxia. Although higher in intensity as compared to control, the distribution pattern remained diffuse and there were no observable areas of localization around capillaries. However, VEGF remained strongly localized around the coronary arteries. VEGF migrated in SDS-PAGE as a dimer of approximately 40 KDa and a monomer of 20 kDa only (30). The expression patterns of the two VEGF forms seem similar although the level of the expression pattern seems significantly higher in the case of VEGF dimer. Significantly increased VEGF expression (about 50% compared to the control) was observed within 30 min of hypoxic challenge which remained upregulated at the same level even after 4 hrs of hypoxia. VEGF receptors, Flk-1 and Flt-1 were also found to be upregulated by hypoxic preconditioning (30). Recently we have demonstrated for the first time that hypoxic preconditioning induced by whole body hypoxia/reoxygenation induces myocardial angiogenesis in a rat model of myocardial infarction as evidenced by increased capillary/arteriolar density and blood flow (31). Our finding also documented increased VEGF expression along with decreased endothelial apoptosis during hypoxic adaptation. Several recent studies have shown that hypoxic preconditioning, like ischemic preconditioning, can attenuate post-ischemic ventricular dysfunction caused by stunning (31,32). Although the exact mechanism of preconditioning is under considerable debate, there is a general agreement that myocardial preservation is achieved through biphasic intrinsic mechanism.

The early effect is most likely to occur through the modulation of intracellular signaling mechanism, and the late effect is believed to be mediated by the gene expression (33) and protein synthesis.

Cellular injury occurs both during hypoxia and after reoxygenation. It was observed that brief hypoxia followed by reperfusion causes more damage at the cellular level than with more prolonged hypoxia alone. A major component of hypoxia induced death was due to apoptosis, as observed by DNA laddering (34). Hypoxic preconditioning induced VEGF and increased capillary density in the hypoxic preconditioned group even before MI (HS). The relative time course of protein expression in response to hypoxic preconditioning, as indicated from our previous experiment (30), seems to suggest the involvement of the VEGF system as well as Ang-Tie system in the early angiogenesis. Examination of non-MI left ventricle (border zone tissue) by anti-CD31 revealed significant increase in the capillary density in the hypoxic preconditioned group (HS) (1814 ± 56 vs. 1642 ± 43 counts/mm^2) compared to the control non-hypoxic group (CS) confirming that modulation of angiogenic factors and their receptors by hypoxia/reoxygenation was able to stimulate capillary proliferation evan in the non-MI animals. The increased capillary density was maintained in hypoxic preconditioned rat hearts subjected to LAD occlusion (HMI) even after 2 months of post-operation when compared to the corresponding sham-operated and non-hypoxic preconditioned controls subjected to LAD occlusion (CMI) groups (Figure 1 and Figure 2). Capillary density was determined by counting (superimposed grid overlay technique) CD-31 positive structure. The capillary density to myocyte ratio (C/M) was found to be increased significantly even after 2 months in the non-ischemic (area at risk) zone (Figure 3). The increase in C/M ratio increases the oxygenation of the surviving tissue. Thus this phenomenon seems beneficial to the living tissue. Arteriolar counts (purple color) for each field in the same region was averaged to yield the value of the particular density measurement. The arteriolar density was significantly elevated in the HMI group after 1 week post-op (2.23 ± 0.12 HMI vs. 1.53 ± 0.02 counts/mm^2 in CMI) and also after 2 and 3 weeks post-op in the HMI group compared to the CMI group (Figure 4). The increased arteriolar level remained significantly high in HMI compared to the CMI group even after 2 months (Figure 5). In our experimental set up, increased VEGF in HMI (33% after 1 week and 64% after 3 weeks of hypoxic preconditioning) probably potentiate a survival signal to reduce the extent of apoptosis associated with myocardial infarction when compared to the non-preconditioned CMI group of animals (31). This phenomenon perhaps accelerated the capillary and arteriolar density and increased regional myocardial blood flow in the hypoxic preconditioned myocardium after coronary occlusion. The extent of infarction was also reduced with 24 hrs after coronary occlusion and the amount of viable myocardium was more abundant in HMI group compared to the CMI group. Possible mechanisms to salvage infarcted myocardium in the HMI group compared to the CMI group may be due to increased VEGF expression which increases flow through the artery, opening of latent collateral vessels by its vasodilating effect, induction of vascular growth and direct cellular

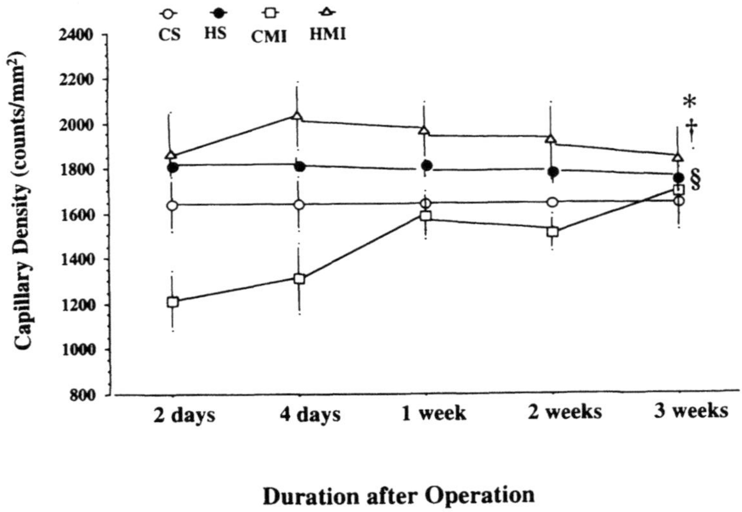

Figure 1. Left Ventricular Endocardial Capillary Density. Tissue sections from 2, 4 days and 1, 2, 3 weeks were processed for CD31 staining and 8 non-overlapping random fields were selected from endocardial regions of non-infracted of the left ventricle of two sections from each hearts (16 fields per region per heart, 64 fields per region per group, magnification of 400× were used, n = 6). Images were captured and stored in digital tiff file format for image analysis. Rats randomly assigned to various experimental groups (n = 12) were subjected to LAD ligation following different exposure protocols. CMI group: Normoxia + LAD occlusion. HMI group: Hypoxia/Reoxygenation + LAD occlusion; Rats which were randomly assigned to next two groups served as the respective sham-operated controls for groups CMI and HMI, respectively. CS group: Normoxia + Sham surgery; HS group: Hypoxia/Reoxygenation + Sham surgery. Where *$P < 0.01$, compared with Sham Operation; †$P < 0.01$, compared with CMI; §$p < 0.05$, compared to CS. (Data from Sasaki et al. [31] with permission).

protection against ischemia as we have observed by reduced endothelial apoptosis. All these parameters, no doubt significantly protects the cardiac function in the HMI group compared to the CMI group of animals. Thus, our result suggests that a relationship exists between programmed cell death (endothelial) and blood flow, and such relationship is dependent on the level of survival factor VEGF (31). Existing evidence indicates expression of a number of growth factors by hypoxia that influence vascular endothelial cell viability (35). Furthermore, it was suggested that in the abnormal regression of retinal capillaries (due to hypoxia), VEGF acts as a survival factor and can prevent the apoptosis associated with regression (36,37). Thus VEGF isoforms are considered as the prime candidates for circulating survival stimuli.

Pharmacological cardiac stress testing with dobutamine proved successful in effectively revealing the ability of hypoxic preconditioning to exert a pronounced and significant long term effect in maintaining LV contractile reserve even after 2 months (Figure 6). Hypoxic preconditioning was also able to reduce the extent of myocar-

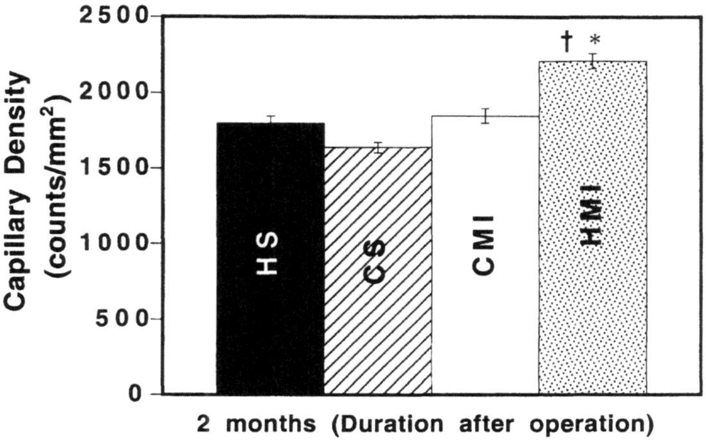

Figure 2. Left ventricular capillary density (Endocardial). The increased capillary density was maintained in HMI group even after 2 months of post-operation when compared to the corresponding sham-operated and CMI groups. Sections processed for CD31. *p < 0.05 compared to CMI, †p < 0.05 compared to sham.

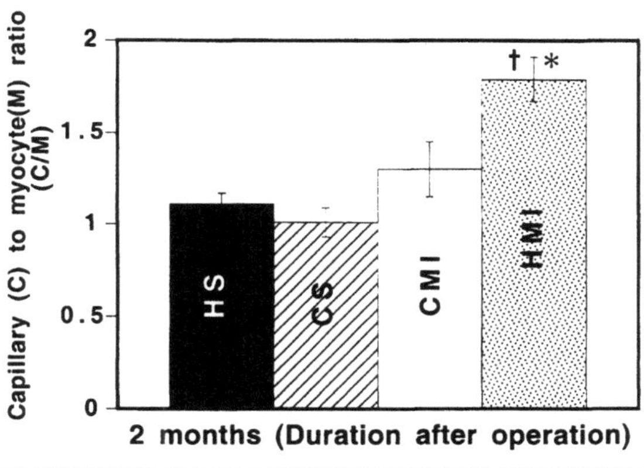

Figure 3. Capillary to myocyte ratio of endocardium (area ar risk) subjected to hypoxia and normoxia followed by LAD occlusion in rat. Where HS = Hypoxia followed by sham operated; CS = Control, sham operated; Normoxic control, LAD occluded, HMI = hypoxic, LAD occluded, *p < 0.05 compared to sham; †p < 0.05 compared to CMI.

dial infarction significantly in the permanent LAD occluded group and also demonstrated a significantly enhanced level of contractile function and increased blood flow. These results, also demonstrate that hypoxic preconditioning was not only able to increase capillary and arteriolar density but it also helped them to remain more functional than the corresponding CMI group.

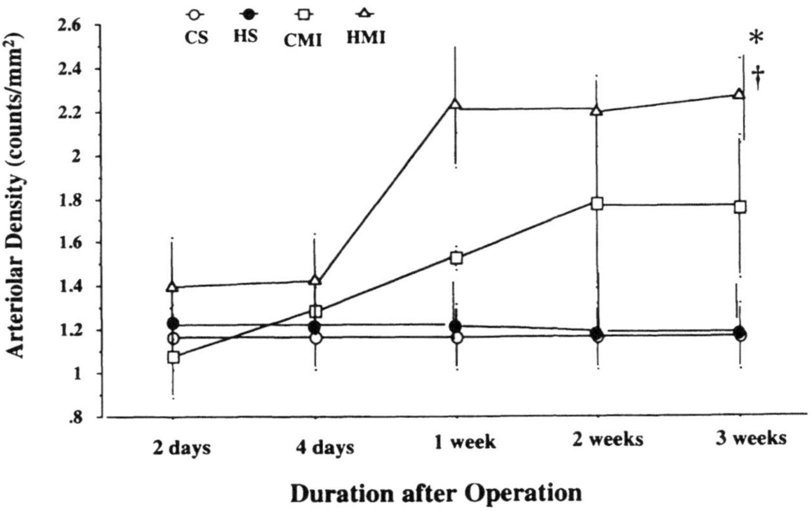

Figure 4. Left ventricular endocardial Arteriolar Density. Left ventricular tissue sections from 2, 4 days and 1–3 weeks were labeled using monoclonal anti-smooth muscle actin and 8 non-overlapping random fields were selected from endocardial region of the left ventricle. Where *P < 0.01, compared with Sham Operation; †P < 0.01, compared with CMI. (Data from Sasaki et al. [31] with permission).

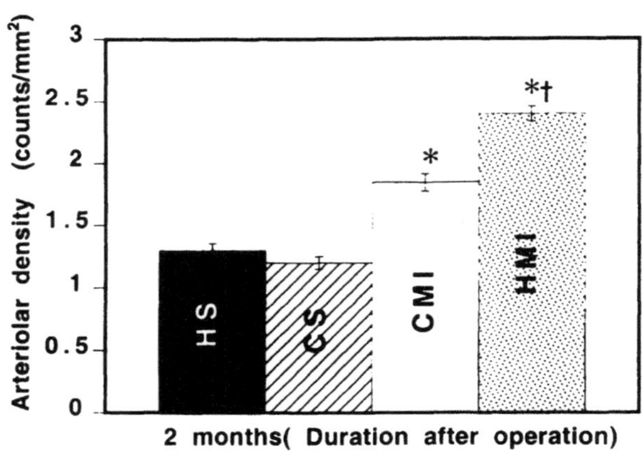

Figure 5. Left ventricular endocardial Arteriolar Density after 2 months. The arteriolar density remained significant in HMI group compared to the CMI group. *p < 0.05 compared to sham; †p < 0.05 compared to CMI.

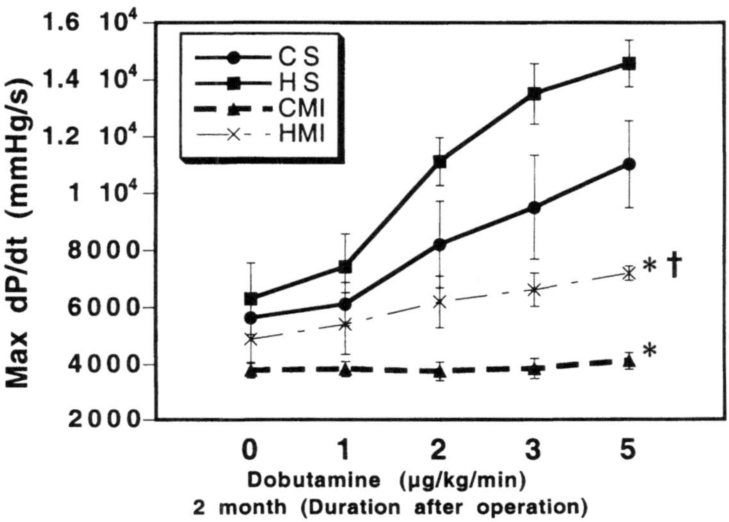

Figure 6. Dose Response Curves of dp/dt max during Dobutamine Stress Test. Changes in dp/dtmax in mm of mercury per second (mmHg/sec) from baseline with increasing dobutamine boluses (0–5 μg/kg/min) in 2 months MI hearts. The dose response curve for dobutamine for the various doses were tested for significant differences by repeated measure ANOVA. Difference were considered significant at $p < 0.05$. Where *$p < 0.01$ compared to Sham, †$p < 0.01$ compared to CMI.

3. INVOLVEMENT OF ANGIOGENIC FACTORS

A. Vascular endothelial growth factor (VEGF) and its tyrosine kinase receptors

The process of angiogenesis is regulated by the signals obtained from the transmembrane receptor tyrosine kinases (RTKs) and non-receptor tyrosine kinases (Src family) of endothelial cells. Flk-1/KDR and Flt-1 are two such RTKs. which together with their ligand VEGF, have been shown to control blood vessel development during embryogenesis (28,38). This receptor/ligand system has been shown to augment neovascularization (39,40). VEGF is not only an endothelial cell-specific angiogenic factor but also a critical regulator of angiogenesis that stimulates proliferation, migration, and proteolytic activity of endothelial cells (20). Yet the signaling pathways that modulate the mitogenic effects of VEGF in vascular endothelial cells are still ill defined (41). A recent study demonstrated VEGF localization and expression in the embryonic/fetal heart and remained high during the early postnatal period when capillary proliferation is high (42). It is now well established that alternate exon splicing of a single VEGF gene results in the generation of four different molecular species, having 121, 165, 189, and 206 amino acids following signal sequence cleavage. $VEGF_{165}$ is the predominant molecular species produced by a variety of normal and transformed cells. Transcripts encoding $VEGF_{121}$ and

$VEGF_{189}$ are detected in the majority of cells and tissues expressing the VEGF gene (43). Among the mechanisms that have been proposed to participate in the regulation of VEGF gene expression, oxygen tension is a particularly important mediator, both in vitro and in vivo. VEGF mRNA expression is rapidly and reversibly induced by exposure to low pO2 in a variety of normal and transformed cultured cell types (44). Also, ischemia caused by occlusion of the left anterior descending coronary artery results in a dramatic increase in VEGF mRNA levels in the pig and rat myocardium, suggesting the possibility that VEGF may mediate the spontaneous revascularization that follows myocardial ischemia (45).

VEGF is a potent vascular endothelial cell specific mitogen that stimulates endothelial cell proliferation, microvascular permeability, vasodilatation, and angiogenesis (46,47). VEGF is the only growth factor proven to be specific and critical for blood vessel formation (48). VEGF has also been shown to improve endothelial cell function and survival in vitro and vascular reactivity in vivo. (49).

Flt-1 (VEGFR-1) and Flk-1/KDR (VEGFR-2) are the endothelial specific tyrosine kinase receptors of VEGF through which its effects are primarily mediated (50,51). Two other angiogenic factors, the angiopoietins 1 and 2 (Ang-1 and Ang-2), have been found to regulate the maturation of new blood vessels from the proliferated endothelial cells (52). Tie-1 and Tie-2 comprise another family of endothelial specific receptor tyrosine kinases, Ang-1 and Ang-2 being the specific ligands for Tie-2.

It has been well established that expression of VEGF, Flt-1 and Flk-1 is upregulated in response to hypoxia in vitro and in vivo (53) and to ischemia in vivo (54).

Tissue hypoxia has been identified as being a particularly important stimulus for triggering angiogenesis. We demonstrated early effects of hypoxia/reoxygenation (H/R) on the protein expression profiles and localization patterns of the VEGF and Angiopoietin-Tie systems in adult rat myocardium. Western blot as well as immunohistochemical analyses were performed on hearts obtained from rats exposed to various duration's of in vivo systemic hypoxemic hypoxia followed by 24 hr reoxygenation. The relative time course of protein expression in response to increasing duration's of hypoxia, as indicated from our experiments, seems to suggest the involvement of the VEGF system and the Ang-Tie system in early angiogenesis. An apparent relationship between the expression profiles of Flk-1 and Ang-2 was observed. The most significant and interesting relationship which came to light was the surprisingly coincident yet opposite temporal trends between Ang-1 and Ang-2 protein levels. In the 1 hr hypoxia group, there was significant induction of Ang-2 expression (31.3% compared to its baseline control) in contrast to relatively mild Ang-1 expression (23.8% compared to its baseline control) (30). Thereafter Ang-1 displayed a progressive increase in expression, parallel to a progressive decrease in Ang-2 expression, becoming most pronounced in the 4 hr hypoxia group (Ang-1 50% and Ang-2 12.6% compared to respective baseline control values). This suggests that despite their being antagonists at the receptor level, regulation of Ang-1 and Ang-2 protein levels in response to hypoxia runs much deeper and seems to indicate modulatory control at the transcriptional and/or translational level. Two additional groups of rats were sacrificed 7 days after 4 hr hypoxia +24 hr reoxygenation, or after a 28 hr period of time-matched normoxia. Left ventricular tissue sections were used to determine capillary density (CD) by using anti-CD31 immunohistochemistry and computer-assisted morphometry. CD was significantly increased in the 4 hr hypoxia group compared to control (1814 ± 56 vs. 1642 ± 43 counts/mm^2) confirming that modulation of angiogenic factors and their receptors by H/R is capable of stimulating capillary proliferation in the myocardium. Our study presents the first evidence for the Ang-Tie system's involvement in early stages of myocardial angiogenesis along with the VEGF-Flk-1-Flt-1 system. The stimulation of myocardial angiogenesis by H/R may constitute a potential basis for a possible more long-lived adaptive response to stress afforded by preconditioning stimuli.

B. Transcription factors (NFκB, AP-1 and SP-1)

Several transcription factors, including HIF-1 (55), AP-1 (56), and NFκB(57), are known to be regulated by hypoxia. The activated form of NFκB is heterodimer and which consists of two proteins, a p65 subunit and a p50 subunit. In normal cells NFκB is maintained in the cytoplasm by protein-protein interaction with inhibitor IkappaB. Recently it was demonstrated in mice model that NFκB activation is obligatory for retinal angiogenesis and it was also documented that the administration of pyrrolidine dithiocarbamate (PDTC) suppressed retinal neovascularization (58). In another study, it was documented that hypoxia/reoxygenation, and not hypoxia alone, can cause formation of reactive oxygen species (ROS) and the activation of the NFκB both of which were inhibited by ROS-scavengers, and was accompanied by inhibition of tube formation in angiogenesis. Therefore, in clinical setting of hypoxia/reoxygenation during ischemic pre-conditioning, the activation of ROS-dependent intracellular signaling may accelerate the rate of neovascularization also in vivo. (59). Hypoxia has been shown to induce NFκB activation and increased IL-8 as well as VEGF gene expression in glial cells in vitro. Furthermore, PDTC, a very specific inhibitor of NFκB, prevented the induction of IL-8 gene expression, but had no effect on the VEGF gene in in vitro study. This finding suggested that IL-8 gene is induced by hypoxia and mediated by NFκB may contribute to the pathogenesis of intraoccular neovascularization (60).

The AP-1 binding complex consists of either Jun-Fos heterodimers or Jun-Jun homodimers (61). Several studies have shown that AP-1 and NFκB are differentially activated by oxygen tension. Several potential binding sites for the transcription factors AP-1, AP-2, and SP-1 are localized in the VEGF gene promoter (62). TNFα or bFGF appears to stimulate expression of the VEGF gene through SP-1 on its promoter (37). Among eight human glioma cell lines, cellular mRNA levels of transcription factors SP-1 and AP-1 were found to be closely correlated with those of VEGF (63). In vitro studies already documented the involvement of transcription factors in tubular morphogenesis of human microvascular endothelial cells by oxidative stress (64). The H_2O_2 induced tubular morphogenesis in HUVEC was blocked by the coadministration of NFκB anti-sense oligonucleotides. In another study,

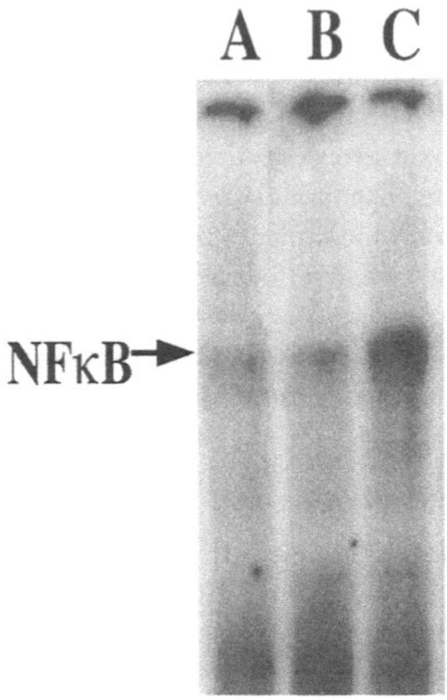

Figure 7. Gel shift analysis for NFκB after normoxia (A) , 4 hr hypoxia (B) and after hypoxia (4 hr) and reoxygenation (2 hrs) (C). Significant increase in NFκB DNA binding was observed in the hypoxia/reoxygenated rat myocardium (B) compared to normoxic (A) and hypoxic hearts (B).

NFκB antisense oligonucleotides reduced cell adhesion and tumorigenicity resulting in tumor regression in the mouse model (65). In our rat model, the hypoxia/reoxygenation induced myocardial angiogenesis, as measured by CD-31 labeling, was inhibited by the intraperitoneal injection of PDTC prior to hypoxia (66). It is already documented that hypoxia/reoxygenation, but not hypoxia alone, caused the production of ROS and thereby activated NFκB (67,68). In consistent with these previous reports, DNA binding activity of NFκB remained almost at the baseline level when rats were subjected to hypoxia only. In contrast, when the rats were subjected to hypoxia followed by reoxygenation, significant amount of NFκB DNA binding activity was observed in the rat myocardium (Figure 7). This study provides evidence for direct involvement of ROS and ROS mediated signaling via NFκB in vivo in myocardial angiogenesis. NFκB inhibitor, PDTC was found to have no influence on the upregulation of VEGF protein expression by hypoxia/reoxygenation (Figure 8). This documents that VEGF is upstream of NFκB and may therefore involve in the activation of NFκB.

C. Involvement of nitric oxide

Nitric oxide (NO) plays a significant role in intracellular signaling process in cardiovascular as well as in other systems. NO is an unique messenger in that it is produced in one cell and diffuses into adjacent target cells to activate cytosolic

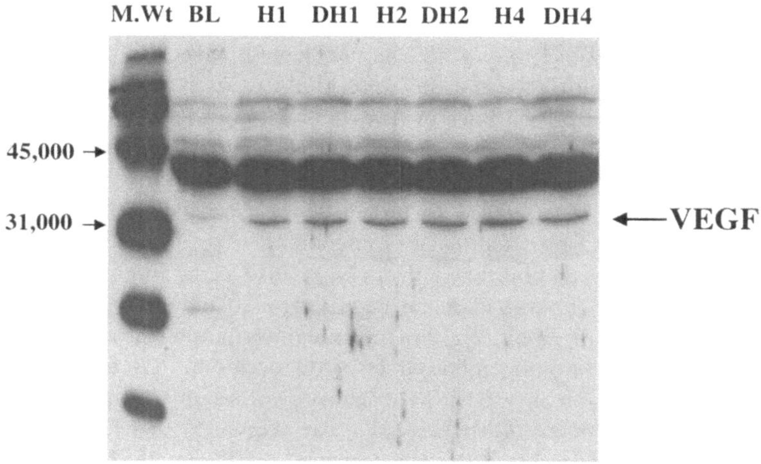

Figure 8. Western blot analysis for VEGF in the hypoxic preconditioned (various duration) hearts followed by reoxygenation. Diethyldithiocarbamate (DDTC), NFκB inhibitor was injected to the rats before hypoxia/reoxygenation. The VEGF protein expression was significant compared to the normoxic control (BL). DDTC was found to have no influence on the VEGF protein expression. Where BL = Normoxic control, H1 = 1 hr hypoxia/24 hrs reoxygenation, H2 = 2 hrs hypoxia/24 hrs reoxygenation, DH2 = DDTC + 2 hrs H/24 hrs reoxygenation, H4 = 4 hrs hypoxia/24 hrs reoxygenation, DH4 = DDTC + 4 hrs H/24 hrs reoxygenation.

guanylatecyclase bound heme to generate the NO-heme adduct of guanylate cyclase. NO can readily react with other cellular hemoproteins such as hemoglobin and myoglobin to produce corresponding NO-heme adducts which can rapidly activate guanylate cyclase (69).

Several in vitro studies directly established the role of NO in angiogenesis (70). Treatment of HepG2 cells with NO donor SNAP was found to increase VEGF mRNA expression. Guanylate cyclase is likely to be important for NO-mediated VEGF activation (71). There is also a considerable evidence that NO downregulates the expression of VEGF gene (72). In spite of several negative observations, activation of angiogenesis in mammalian (human) monocytes is believed to be NO-dependent (73). Indeed, several studies have documented that NO-generating compounds stimulate angiogenesis in human glastoma and hepatoma cells (71). A positive correlation was found between nitric oxide synthase, cGMP levels, and tumor angiogenesis in head and neck cancer (74). However, the role of NO in angiogenesis is still controversial. For example, NO donors were found to inhibit angiogenic activity in the chick chorioallantoic membrane (75) and the growth and metastatic properties of the Lewis lung tumor in mice (76). In addition, NO donors were found to inhibit VEGF expression in the arterial wall in response to balloon angioplasty (77), and in rat lungs during acute and chronic hypoxia (78). Transfer of eNOS (79) and iNOS (80) genes resulted in the inhibition of restenosis after balloon angioplasty. In contrast, a recent study showed L-arginine supplementation after balloon angioplasty of rabbit iliac arteries was beneficial for healing of endothelium demonstrating positive role of NO in angiogenesis (81). Human colon cancer cell lines transfected with a NOS-gene were found to grow faster and more vascularized than the normal cell lines in vivo (82). NO was also found to enhance the proliferation and migration of endothelial cells in vitro (70). Thus, a significant number of studies demonstrated that NO may stimulate the proliferation of endothelial cell while others failed to prove such a mechanism and suggested an inhibitory effect of NO on endothelial proliferation.

Recent studies have demonstrated that eNOS/NO plays an important role in many VEGF-induced actions. VEGF has been shown to induce the production of nitric oxide (NO) from rabbit, pig, bovine, and human vascular endothelial cells (83). Hypoxic preconditioning followed by LAD occlusion was found to induce eNOS protein expression in rat left ventricular myocardium (HMI) compared to non-hypoxic control MI rat (CMI) after one day (Figure 9) that lead to increased capillary and arteriolar density in the left ventricle (see Figure 1–5). The Inhibition of NO production by eNOS inhibitors significantly inhibited VEGF-induced mitogenic and angiogenic effects (84). Endothelial NOS/NO has been implicated as one of the important mediators for VEGF-induced hemodynamic changes and microvascular permeability. Although eNOS was originally described as a constitutive

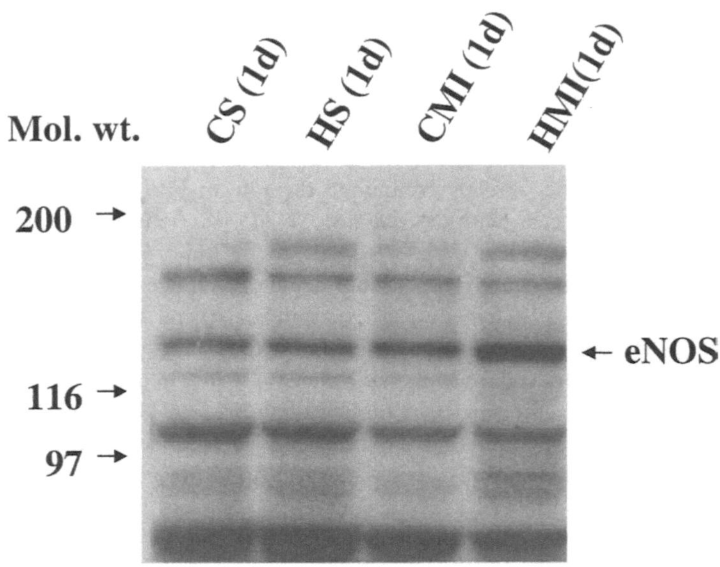

Figure 9. Western blot analysis of eNOS (endothelial nitric oxide synthase) expression at day 1 of LAD occlusion. HMI group demonstrated significant increase in the eNOS expression compared to the CS, HS and CMI groups.

enzyme. Recent studies indicated that a variety of stimuli including hypoxia, shear stress, inflammatory cytokines, high glucose, and injury could modulate eNOS expression and activity (85). In vitro experiments have shown that activation of KDR induces proliferation and migration of endothelial cells as well as expression of eNOS and iNOS (86). So far, the role of VEGF receptors in the release of NO remains controversial. In human placental trophoblast it has been suggested that Flt-1 can induce the release of NO (87). The precise signaling pathways that mediate those responses have not yet been elucidated; however, phospholipase D and nitric oxide synthase have been identified as key enzymes activated by VEGF. These 2 enzymes have been shown to depend on protein kinase c (PKC) and it's down stream mitogen activated protein kinase (MAPK) activation (88). Many studies showed that the effects of preconditioning are mediated by PKC, MAPKs and redox regulated transcription factors such as NFκB and STAT 1 activation. Among 10 PKC isoforms, PKC α, δ, ε and zeta isoforms are expressed in rat myocardium. Several groups have demonstrated that either PKC delta or PKC epsilon is required for the protective effect of preconditioning in rat myocardium. Inhibition of PKC activity with the chemical inhibitor calphostin prevents VEGF induced vascular growth (89) but the role of specific isoenzymes in the processes that constitute angiogenesis is unknown, because the specificity of most PKC inhibitors for individual isoenzymes has not been established.

4. INVOLVEMENT OF SURVIVAL PATHWAY

A. Phosphatidylinositol 3 OH-kinase (PI-3 Kinase) and AKT

Formation and regression of new vessels or angiogenesis are highly regulated processes in which the growth and death of endothelial cells play critical role. Recent studies documented the involvement of VEGF in cell-survival signaling (90). Experimental evidence has suggested that regulation of endothelial cell apoptosis may be an essential mechanism to maintain angiogenesis in vivo. Exposure of endothelial cells to TNF-α induces apoptosis. However, when the same culture was treated simultaneously with VEGF resulted in a dose dependent inhibition of apoptosis (91). Therefore it seems that VEGF, in addition to its action as an endothelial cell mitogen, also exerts a protective effect, inhibiting apoptosis in vascular endothelial cells. In vivo, VEGF functions as a survival factor for microvascular endothelial cells, specifically when apoptosis is induced by hyperoxia (36). Failure of angiogenesis has been directly linked to the advent of endothelial cell apoptosis.

These report points out that enhanced level of VEGF may have some role in the inhibition of endothelial cell apoptosis. Another very recent investigation was that VEGF a potent promoter of angiogenesis, upregulates the expression of the intracellular adhesion molecule-1 (ICAM-1) through a novel pathway that includes phosphatidylinositol 3 OH-Kinase (PI3-K) and AKT resulting in the migration of brain microvascular endothelial cells. It was found that in vitro VEGF treatment phosphorylates AKT in a PI3-K-dependent manner (92). The PI3-K/AKT pathway appears to be a general mediator of cytokine induced survival and anti-apoptotic

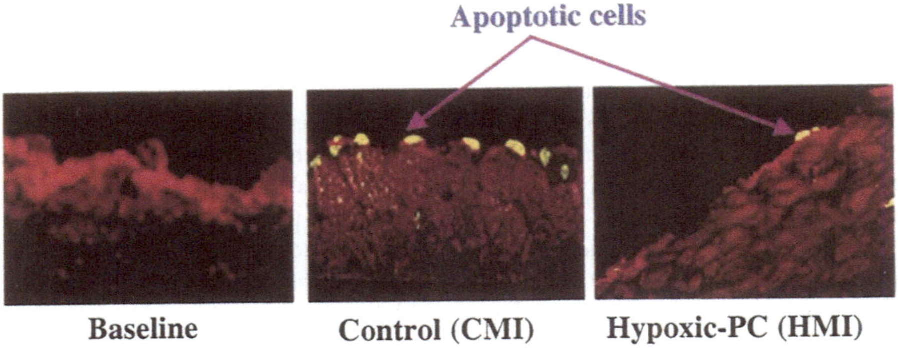

Figure 10. The sections (4 μm) were assayed for in-situ terminal transferase labeling (TUNEL) using Apop Tag® Plus (Oncor Inc., Gaithersburg, MD). Negative control slides were processed with the TdT enzyme excluded. The slides were first stained with a peroxidase-conjugated sheep polyclonal anti-digoxigenin antibodies and diaminobenzidine and with methyl green as a counterstain. For the detection of endothelial cell apoptosis, the serial sections were first stained by TUNEL with fluorescein isothiocyanate (FITC) and then incubated with polyclonal rabbit anti-von Willebrand factor antibodies (Dako Japan, Tokyo Japan) followed by incubation with tetrarhodamine isothiocyanate (TRITC)-conjugated goat anti-rabbit IgG (Dako Japan). FITC and TRITC fluorescence were viewed with a confocal laser microscopy (Fluo View, Olympus Co, and Tokyo, Japan). The number of TUNEL-positive endothelial cells was counted on 60 high power fields (magnification ×600) from the endocardium through the epicardium of the left ventricular free wall. TUNEL assay for apoptotic endothelial cells were performed on confocal laser microscopy. Representative photomicrographs of two week after LAD occlusion, showing immunohistochemical staining of extended DNA. Where Baseline = Normoxic control; Control = Non-hypoxic hearts subjected to LAD occlusion (CMI); Hypoxic PC (HMI) = rats were exposed to Hypoxia/reoxygenation (4 hrs hypoxia followed by 24 hrs of reoxygenation) followed by LAD occlusion. Note that apoptotic endothelial cells are lying on the luminal surface of the coronary vessels. Magnification ×1200.

signals. Recently, pro-apoptotic factor, BAD, was reported to be phosphorylated by activated AKT on a serine residue causing BAD to dissociate from BCL-X_L. No in vivo study have been done so far to investigate the involvement of PI3K/AKT pathway and endothelial cell survival. Another recent study reported inhibition of endothelial cell apoptosis by Ang-1 via the Akt/survivin pathway, which contributed Ang-1 mediated stabilization of vascular structures during angiogenesis (93). Recently we have shown that hypoxic preconditioning accelerates tubular morphogenesis along with the activation of reactive oxygen species-inducible NFκB, PI3-kinase and broad spectrum anti-apoptotic protein survivin in human coronary arteriolar endothelial cells (HCAEC). The formation of tubular morphogenesis was inhibited by using the PI3-kinase and NFκB antagonists LY294002 and SN50 respectively along with increased apoptosis in HCAEC. These data demonstrate a crucial role of PI3-kinase/Akt/NFκB signaling in tubular morphogenesis of HCAEC triggered by hypoxic preconditioning (94). In vivo study from our laboratory also demonstrated significant inhibition of endothelial cell apoptosis in hypoxic preconditioned hearts (HMI) when compared to the non-hypoxic control rat MI hearts (CMI) (Figure 10) (31). It is also reported that activation of the MAPK pathway together with inhibition of SAPK/JNK activity by VEGF appears to be a key event in determining whether an endothelial cell is going to survive or will undergo programmed cell death.

B. Anti-death gene: Bcl-2 and Survivin

Recent report suggested VEGF-induced expression of Bcl-2, which eventually functions to enhance the survival of endothelial cells in the toxic, oxygen-deficient environment (21). VEGF induced angiogenesis has been found to be associated with enhanced cell survival in human umbilical vein endothelial cells in vitro (36). This effect is thought to be due to a reduction in apoptotic endothelial cell death increased expression of the anti-apoptotic protein Bcl-2 and reduced expression of the pro-apoptotic protein caspase 3 are possible biochemical mediators of such effect of VEGF. Furthermore, it has been suggested that in the abnormal regression of retinal capillaries, VEGF acts as a survival factor and can prevent candidates for circulating survival stimuli (37,42). Most potent angiogenic activity ofVEGF was found by its ability to induce the expression of Bcl-2 in endothelial cells. This over expression of anti-death protein Bcl-2 might potentiate the ability of endothelial cells to differentiate into functional blood vessels. The most crucial information about the role of Bcl-2 in angiogenesis was obtained from the implantation of HDMEC-Bcl-2 (human dermal microvascular endothelial cells overexpressing Bcl-2) in SCID mice.

The results of our study also documented (unpublished) that ischemic preconditioning potentiates angiogenesis by increasing both capillary and arteriolar density through the upregulation of angiogenic factor VEGF and anti-apoptotic protein Bcl-2. It, thus appear that myocardial adaptation by intermittent ischemia is a highly promising approach to induce angiogenesis in a rat model of myocardial infarction.

Another anti-death protein, Survivin, a recently described inhibitor of apoptosis, also plays a significant role in the process of cell proliferation in human hepatocellular carcinoma (95). Survivin is the smallest member of the IAP gene family. It is expressed in several apoptosis-regulated fetal tissues, including lung, liver, heart, and gastrointestinal tract (95). Activation of survivin was found to inhibit caspase-3 activity (95). The inhibition of endothelial cell apoptosis during angiogenesis may occur simultaneously through diverse pathways, which might involve mitochondrial integrity by bcl-2 and suppression of caspase activity by survivin (96–98).

It was demonstrated that angiopoietin-1 (Ang-1) presumably acting via Tie-2 tyrosine kinase receptor, is capable of preventing endothelial cell apoptosis by activating Akt, a critical survival messenger. In addition, Ang-1 also can up-regulate the broad-spectrum apoptosis inhibitor, survivin (93). Thus, inhibition of endothelial cell apoptosis by survivin expression may have significant clinical significance in the realm of angiogenesis by improving endothelial cell viability.

5. CONCLUSION

A moderate hypoxic/ischemic exposure is capable of imparting a state of heightened myocardial preparedness to better enable it to mount a more aggressive salvage angiogenic response in the face of a later more severe hypoxic or ischemic challenge, is an obvious question. Further studies are necessary utilizing hypoxic/ischemic preconditioning to promote myocardial angiogenesis in experimental models of chronic myocardial infarction and heart failure that will help in the characterization and identification of promising target candidates for gene therapy to achieve clinically relevant myocardial angiogenesis.

ACKNOWLEDGEMENT

The study was supported by NIH HL 56803.

REFERENCES

1. Shizukuda Y, Iwamoto T, Mallet RT, Downey HF. 1993. Hypoxic preconditioning attenuates stunning caused by repeated coronary artery occlusions in dog heart. Cardiovasc Res 27:559–564.
2. Lasley RD, Anderson GM, Mentzer RM. 1993. Ischemic and hypoxic preconditioning enhance postischemic recovery of function in the rat heart. Cardiovasc Res 27:565–570.
3. Zhai X, Lawson CS, Cave AC, Hearse DJ. 1993. Preconditioning and postischemic contractile dysfunction: the role of impaired oxygen delivery vs. extracellular metabolite accumulation. J Mol Cell Cardiol 25:847–857.
4. Ashraf M, Suleiman J, Ahmad M. 1994. Ca^{2+} preconditioning elicits a unique protection against the ca^{2+} paradox injury in rat heart. Circ Res 74:360–367.
5. Thornton JD, Liu GS, Olsson RA, Downey JM. 1992. Intravenous pretreatment with A1-selective adenosine analogues protects the heart against infarction. Circulation 85:659–665.
6. Banerjee A, Locke-Winter C, Rogers KB, Mitchell MB, Brew EC, Cairns CB, Bensard DD, Harken AH. 1993. Preconditioning against myocardial dysfunction after ischemia and reperfusion by an A1-adrenergic mechanism. Circ Res 73:656–670.
7. Thornton JD, Downey JM. 1993. Pretreatment with pertussis toxin blocks the protective effects of preconditioning: evidence for a Gi-protein mechanism. J Mol Cell Cardiol 25:311–320.
8. Ovize M, Kloner RA, Przyklenk K. 1994. Stretch preconditions canine myocardium. AM J Physiol 266(1Pt 2):H137–H146.
9. Neely JR, Grotyohann LW. 1984. Role of glycolytic products in damage to ischemic myocardium. Dissociation of adenosine triphosphate levels and recovery of function of reperfused ischemic hearts. Circ Res 55:816–824.
10. Engelman DT, Watanabe M, Engelman RM, Rousou JA, Kisin E, Kagan VE, Maulik N, Das DK. 1995. Hypoxic preconditioning preserves antioxidant reserve in the working rat heart. Cardiovas Res 29:133–140.
11. Maulik N, Watanabe M, Zu Y-L, Huang C-K, Cordis GA, Schely JA, Das DK. 1996. Ischemic preconditioning triggers the activation of MAP kinases and MAPKAP kinase 2 in rat hearts. FEBS Lett 396:233–237, 396.
12. Maulik N, Yoshida T, Zu Y-L, Sato M, Banerjee A, Das DK. 1998. Ischemic preconditioning triggers tyrosine kinase signaling: a potential role for MAPKAP kinase 2. Am J Physiol 275(44): H1857–H1864.
13. Risau W, Ekblom P. 1986. Growth factors and the embrynic kidney. In Serino D (ed): "Progress in clinical and Biological Research. Hormonal Control of Embryonic and cellular differentiation." New York: Liss, pp. 221–232.
14. Angus JA, Ward JE, Smolich JJ, McPherson GA. 1991. Reactivity of canine isolated epicardial collateral coronary arteries. Circ Res 69:1340–1352.
15. Iruela-Arise ML, Diglio CA, Sage EH. 1991. Modulation of extracellular matrix proteins by endothelial cells undergoing angiogenesis in vitro. Arterioscl Thromb 11:805–815.

16. Cohen MV, Yang X-M, Neumann T, Heusch G, Downey JM. 2000. Favorable remodeling enhances recovery of regional myocardial function in the weeks after infarction in ischemically preconditioned hearts. Circulation 102:579.
17. Murry N, Jennings RB, Reimer KA. 1986. Preconditioning with ischemia: a delay of lethal cell injury in ischemic myocardium. Circulation 74:1124–1136.
18. Das DK, Moraru II, Maulik N, Engelman R. 1994. Gene expression during myocardial adaptation to ischemia and reperfusion. Ann NY Acad Sci 723:292–307.
19. Epstein S, Williams DO, Amsterdam EA, Miller RR. 1988. Functional significance of coronary collateral vessels in patients with acute myocardial infarction: Relation to pump performance, cardiogenic shock and survival. Am J Cardiol 61:345–351.
20. Maruyama K, Mori Y, Murasawa S, Masaki H, Takahashi N, Tsutusmi Y, Moriguchi Y, Shibazaki Y, Tanaka Y, Shibuya M, Inada M, Matsubara H, Iwasaka T. 1999. Interleukin-1 beta upregulates cardiac expression of vascular endothelial growth factor and its receptors KDA/FLK-1 via activation of protein tyrosine kinases. J Mol Cell Cardiol 31:607–617.
21. Nor JE, Christensen J, Mooney DJ, Polverini PJ. 1999. Vascular endothelial growth factor (VEGF)-mediated angiogenesis is associated with enhanced endothelial cell survival and induction of Bcl-2 expression. Am J Pathol 154:375–384.
22. Maulik N, Watanabe D, Engelman D, Engelman RM, Das DK. 1995. Oxidative stress adaptation improves postischemic ventricular recovery. Mol Cell Biochem 144:67–74.
23. Liu TN, Nian GM, Chen SE, Cheung WM, Chang C, Lin WC, Hsu CY. 2001. Induction of Tie-1 and Tie-2 receptor protein expression after cerebral ischemia-reperfusion. J Cerebral Blood flow and Metabolism 21:690–701.
24. Brand T, Sharma HS, Fleischmann KE, Duncker DJ, McFalls EO, Verdue PD, Schaper W. 1992. Proto-oncogene expression in porcine myocardium subjected to ischemia and reperfusion. Circ Res 71:1351–1360.
25. Ely SW, Berne RM. 1992. Protective effects of adenosine in myocardial ischemia. Circulation 85:893–904.
26. Fukuda S, Zhu Li, Engelman RM, Das DK, Maulik N. 2001. Ischemic preconditioning (PC) triggers myocardial angiogenesis by upregulating vascular endothelial growth factor (VEGF). Surgical Forum LII:65–67.
27. Senger DR, Connolly DT, L Van de W, Feder J, Dvorak HF. 1990. Purification and NH2-terminal amino acid sequence of guinea pig tumor-secreted vascular permeability factor. Cancer Res 50:1774–1778.
28. Fong G-H, Rossant J, Gerstenstein M, Breitman ML. 1995. Role of the Flt-1 receptor tyrosine kinase in regulating the assambly of vascular endothelium. Nature 376:66–70.
29. Takeshita S, Zheng LP, Brogi E, Kearney M, Pu LQ. 1994. Therapeutic angiogenesis. A single intra arterial bolus of vascular endothelial growth factor augments revascularization in a rabbit ischemic hind limb model. J Clin Invest 93:662–670.
30. Ray PS, Estrada-Hernandez T, Sasaki H, Zhu Li, Maulik N. 2000. Early effects of hypoxia/reoxygenation on VEGF, Ang-1, Ang-2 and their receptors in the rat myocardium: Implications for myocardial angiogenesis. Mol Cell Biochem 213:145–153.
31. Sasaki H, Fukuda S, Otani H, Zhu Li, Yamaura G, Engelman RM, Das DK, Maulik N. 2002. Hypoxic preconditioning triggers myocardial angiogenesis: a novel approach to enhance contractile function reserve in rat with myocardial infarction. J Mol Cell Cardiol 34:335–348.
32. Shizukuda Y, Iwamoto T, Mallet RT, Downey HF. 1993. Hypoxic preconditioning attenuates stunning caused by repeated coronary artery occlusion in dog heart. Cardiovascular Res 27:559–5564.
33. Das DK. 1993. Ischemic preconditioning and myocardial adaptation to ischemia. Cardiovasc Res 27:2077–2079.
34. Gerber H-P, Dixit V, Ferrara N. 1998. Vascular endothelial growth factor induces expression of the antiapoptotic proteins Bcl-2 and A1 in vascular endothelial cells. J Biol Chem 273:13313–13316.
35. Meeson A, Palmer M, Calfon M, Lang R. 1996. A relationship between apoptosis and flow during programmed capillary regression is revealed by vital analysis. Develop 122:3929–3938.
36. Alon T, Hemo I, Itin A, Pe'er J, Stone J, Keshet E. 1995. Vascular endothelial growth factor acts as a survival factor for newly formed retinal vessels and has implications for retinopathy of prematuarity. Nat Med 1:1024–1028.
37. Shweiki D, Itin A, Soffer C, Keshet E. 1992. Vascular endothelial growth factor induced by hypoxia may mediate hypoxia-initiated angiogenesis. Nature 359:843–845.

38. Takeshita S, Zheng LP, Brogi E, Kearney M, Pu LQ. 1994. Therapeutic angiogenesis. A single intra arterial bolus of vascular endothelial growth factor augments revascularization in a rabbit ischemic hind limb model. J Clin Invest 93:662–670.
39. Asahara T, Chen D, Takahashi T, Fujikawa K, Kearney M, Magner M, Yancopoulos GD, Isner JM. 1998. Tie-2 receptor ligands, angiopoietin-1 and angiopoietin-2, modulate VEGF-induced postnatal neovascularization. Circ Res 83:342–343.
40. Isner JM, Pieczek A, Schainfed R, Blair R, Haley L, Asahara T, Rosenfield K, Razvi S, Walsh K, Symes J. 1996. Clinical evidence of angiogenesis is after arterial gene transfer of ph VEGF165 in patient with ischemic limb. Lancet 348:370–374.
41. Rousseau S, Houle F, Landry J, Hout J. 1997. p38 MAP kinase activation by vascular endothelial growth factor mediates actin reorganization and cell migration in human endothelial cells. Oncogene 15:2169–2177.
42. Tomanek RJ, Ratajska A, Kitten GT, Yue X, Sandra A. 1999. Vascular endothelial growth factor expression coincides with coronary vasculogenesis and angiogenesis. Dev Dyn 215:54–61.
43. Houck KA, Ferrara N, Winer J, Cachianes G, Li B, Leung DW. 1991. The vascular endothelial growth factor family: identification of a fourth molecular species and characterization of alternative splicing of RNA. Mol Endocrinol 5:1806–1814.
44. Minchenko A, Bauer T, Salceda S, Caro J. 1994. Hypoxic stimulation of vascular endothelial growth factor expression in vivo and in vitro. Lab Invest 71:374–379.
45. Hasimoto E, Ogita T, Nakaoka T, Matsuoka R, Takao A, Kira Y. 1994. Rapid induction of vascular endothelial growth factor expression by transient ischemia in rat heart. Am J Physiol 267: H1948–H1954.
46. Neufeld G, Cohen T, Gengrinovitch S, Poltorak Z. 1999. Vascular endothelial growth factor (VEGF) and its receptors. FASEB J 13:9–22.
47. Ferrara N, Davis-Smyth T. 1997. The biology of vascular endothelial growth factor. Endocr Rev 18:4–25.
48. Ferrara N. 1999. Vascular endothelial growth factor: molecular and biological aspects. Curr Top Microbiol Immunol 237:1–30.
49. Spyridopoulos I, Brogi E, Kearney M, Sullivan AB, Cetrulo C, Isner JM, Losardo DW. 1997. Vascular endothelial growth factor inhibits endothelial cell apoptosis induced by tumor necrosis factor alpha balance between growth and death signals. J Mol Cell Cardiol 29:1321–1330.
50. DeVries C, Escobedo JA, Veno H, Houck K, Ferrara N, Williams LT. 1992. The fms-like tyrosine kinase, a receptor for vascular endothelial growth factor. Science 255:989–991.
51. Millauer B, Shawver LK, Plate KH, Risau W, Ullrich A. 1994. Glioblastoma growth inhibited in vivo by a dominant-negative Flk-1 mutant. Nature 367:576–579.
52. Witzenbichler B, Maisonpierre PC, Jones P, Yancopoulos GD, Isner JM. 1998. Chemotactic properties of angiopoietin-1 and-2, ligands for the endothelial-specific receptor tyrosine kinase Tie2. J Biol Chem 273:18514–18521.
53. Kuo N, Benhayon D, Przybylski RJ, Martin RJ, LaManna JC. 1999. Prolonged hypoxia increases vascular endothelial growth factor mRNA and protein in adult mouse brain. J Appl Physiol 86:260–264.
54. Banai S, Shweiki D, Pinson A, Chandra M, Lazarovici G, Keshet E. 1994. Upregulation of vascular endothelial growth factor expression induced by myocardial ischemia: implications for coronary angiogenesis. Cardiovasc Res 28:1176–1179.
55. Maxwell PH, Pugh CW, Ratcliffe PJ. 1993. Inducible operation of the erythropoitin 3′ enhancer in multiple cell lines: evidence for a widespread oxygen—sensing mechanism. PNAS 90: 2423–2427.
56. Bandyopadhyay RS, Phelan M, Faller DV. 1995. Hypoxia induces AP-1 regulated genes and AP-1 transcription factor binding in human endothelial and other cell types. Biochem Biophys 1264: 72–78.
57. Yao KS, Xanthoudakis S, Curran T, O'Dwyer PJ. 1995. Involvement of NFκB in the induction of NAD(P)H quinone oxidoreductase by hypoxia, oltipraz and mitomycin C. Biochem Pharmacol 49:275–282.
58. Yoshida A, Yoshida S, Ishibashi T, Kuwano M, Inomata H. 1999. Suppression of retinal neovascularization by the NF-κB inhibitor pyrrolidine dithiocarbamate in mice. Invest Ophthalmol Vis Sci 40:162–1629.
59. Lelkes PI, Hahn KL, Sukovich DA, Karmiol S, Schmidt DH. 1998. On the possible role of reactive oxygen species in angiogenesis. Adv Exp Med Biol 454:295–310.

60. Yoshida A, Yoshida S, Khalil AK, Ishibashi T, Inomata H. 1998. Role of NF-κB mediated interleukin-8 expression in intraocular neovascularization. Invest Ophthalmol Vis Sci 39:1097–1106.
61. Smeal T, Angel P, Meek J, Karin M. 1989. Different requirements for formation of Jun-Jun and Jun-Fos complexes. Genes Dev 3:2091–2100.
62. Tischer E, Mitchell R, Hartman T, Silva M, Gospodarowicz D, Fiddes JC, Abraham JA. 1991. The human gene for vascular endothelial growth factor. Multiple protein forms are encoded through alternative exon slicing. J Biol Chem 266:11947–11954.
63. Ryuto M, Ono M, Izumi H, Yoshida S, Weich HA, Kohno K, Kuwano M. 1996. Induction of vascular endothelial growth factor by tumor necrosis factor a in human glioma cells. J Biol Chem 271:28220–28228.
64. Shono T, Ono M, Izumi H, Jimi Sei-Ichiro, Matsushima K, Okamoto T, Kohno K, Kuwano M. 1996. Involvement of the transcription factor NF-κB in tubular morphogenesis of human microvascular endothelial cells by oxidative stress. Mol Cell Biol 16:4231–4239.
65. Higgins K, Perez JR, Colman TA, Dorshkind K, McComas WA, Sarmiento UM, Rosen CA, Narayanan R. 1993. Antisense inhibition of the p65 subunit of NFκB blocks tumorigenicity and causes tumor regression. Proc Natl Acad Sci USA 90:9901–9905.
66. Sasaki H, Ray PS, Zhu Li, Hajime O, Takayuki A, Maulik N. 2001. Hypoxia/Reoxygenation promotes myocardial angiogenesis via an NF-κB-dependent mechanism in a rat model of chronic myocardial infarction. J Mol Cell Cardiol 33:283–294.
67. Lelkes PI, Hahn KA, Karmiol S, Schmidt DH. 1998. Hypoxia/reoxygenation enhances tube formation of cultured human micro-vascular endothelial cells: the role of reactive oxygen species. In: ME Maragoudakis, ed. Angiogenesis: Models, Modulators, and Clinical Applications. New York: Plenum Press, pp. 321–336.
68. Stoltz RA, Abraham NG, Laniado-Schwartzman ML. 1996. The role of NFκB in the angiogenenic response of coronary microvessel endothelial cells. Proc Natl Acad Sci 93:2832–2837.
69. Maulik N, Engelman DT, Watanabe M, Engelman RM, Rousou JA, Flack JE, Deaton DW, Gorbunov NV, Elsayed NM, Kagan VE, Das DK. 1996. Nitric oxide/Carbon monoxide: A molecular switch for myocardial preservation during ischemia. Circulation 94:II-398–II-406.
70. Ziche M, Morbidelli L, Masini E, Amerini S, Granger HJ, Maggi CA, Geppetti P, Ledda F. 1994. Nitric oxide mediates angiogenesis in vivo and endothelial cell growth and migration in vitro promoted by substance P. J Clin Invest 94:2036–2044.
71. Chin K, Kurashima Y, Ogura T, Tajiri H, Yoshida S, Esumi H. 1997. Induction of vascular endothelial growth factor by nitric oxide in human glioblastoma and hepatocellular carcinoma cells. Oncogene 15:437–442.
72. Tsurumi Y, Murohara T, Krasinski K, Chen D, Witzenbichler B, Kearney M, Couffinhal T, Isner JM. 1997. Reciprocal relation between VEGF and NO in the regulation of endothelial integrity. Nat Med 3:879–886.
73. Leibovich SJ, Polverini PJ, Fong TW, Harlow LA, Koch A. 1994. Production of angiogenic activity by human monocytes requires an L-arginine/nitric oxide-synthase-dependent effector mechanism. Proc Natl Acad Sci USA 91: 4190–4194.
74. Gallo O, Masini E, Morbidelli L, Franchi A, Finistorchi I, Vergari WA, Ziche M. 1998. Role of nitric oxide in angiogenesis and tumor progression in head and neck cancer. J Natl Cancer Inst 90:587–596.
75. Pipili-Synetos E, Sakkoula E, Haralabopoulos G, Andriopoulou P, Peristeris P, Maragoudakis ME. 1994. Evidence that nitric oxide is an endogenous antiangiogenic mediator. Br J Pharmacol 111:894–902.
76. Pipili-Synetos E, Papa Georglou A, Sakkoula E, Sotiropoulou G, Fotsis T, Karakiulakis G, Maragoudakis ME. 1995. Inhibition of angiogenesis, tumour growth and metastasis by the NO-releasing vasodilators, isosorbide mononitrate and dinitrate. Br J Pharmacol 116:1829–1834.
77. Tsurumi Y, Murohara T, Krasinski K, Chen D, Witzenbichler B, Kearney M, Couffinhal T, Isner JM. 1997. Reciprocal relation between VEGF and NO in the regulation of endothelial integrity. Nat Med 3:879–886.
78. Tuder RM, Voelkel NF, Flook BE. 1995. Increased gene expression for VEGF and the VEGF receptors KDR/Flk and Flt in lungs exposed to acute or to chronic hypoxia. Modulation of gene expression by nitric oxide. J Clin Invest 95:1798–1807.
79. Von der Leyen HE, Gibbons GH, Morishita R, Lewis NP, Zhang L, Nakajima M, Kaneda Y, Cooke JP, Dzau VJ. 1995. Gene therapy inhibiting neointimal vascular lesion: in vivo transfer of endothelial cell nitric oxide synthase gene. Proc Natl Acad Sci USA 92:1137–1141.

80. Tzeng E, Shears LL, Robbins PD, Pitt Br, Geller DA, Watkins SC, Simmons Rl, Billiar TR. 1996. Vascular gene transfer of the human inducible nitric oxide synthase characterization of activity and effects of myointimal hyperplasia. Mol Med 2:211–225.
81. Schwarzacher S, Lim TT, Wang BY, Kernoff RS, Niebauer J, Cooke JP, Yeung AC. 1997. Local intramural delivery of L-arginine enhances nitric oxide generation and inhibits lesion formation after vascular injury. Circulation 95:1863–1869.
82. Jenkins DC, Charles IG, Thomsen LL, Moss DW, Holmes LS, Baylis SA, Rhodes P, Westmore K, Emson PC, Moncada S. 1995. Roles of nitric oxide in tumor growth. Proc Natl Acad Sci USA 92: 4392–4396.
83. Van der Zee R, Murohara T, Luo Z, Zollmann F, Passeri J, Lekutat C, Isner JM. 1997. Vascular endothelial growth factor/vascular permeability factor augments nitric oxide release from quiescent rabbit and human vascular endothelium. Circulation 95:1030–1037.
84. Papapetropoulos A, Garcia-Cardena G, Madri JA, Sessa WC. 1997. Nitric oxide production contributes to the angiogenic properties of vascular endothelial growth factor in human endothelial cells. J Clin Invest 100:3131–3139.
85. Sase K, Michel T. 1997. Expression and regulation of endothelial nitric oxide synthase. Trends Cardiovasc Med 7:28–37.
86. Waltenberger J, Claesson-Welsh L, Siegbahn A, Shibuya M, Heldin CH. 1994. Different signal transduction properties of KDR and Flt-1, two receptors for vascular endothelial growth factor. J Biol Chem 269:26988–26995.
87. Bussolati B, Dunk C, Mason M, Kontos CD, Ahmed A. 2001. Vascular endothelial growth factor receptor-1 modulates vascular endothelial growth factor mediated angiogenesis via nitric oxide. Am J Pathol 159:993–1008.
88. Seymour LW, Shoaibi MA, Martin A, Ahmed A, Elvin P, Kerr DJ, Wakelam MJO. 1996. Vascular endothelial growth factor stimulates protein kinase C-dependent phospholipase D activity in endothelial cells. Lab Invest 75:427–437.
89. Friedlander M, Brooks PC, Shaffer RW, Kincaid CM, Varner JA, Cheresh DA. 1992. Definition of two angiogenic pathways by distinct α v integrins. Science 270:607–614.
90. Spyridopoulos I, Brogi E, Kearney M, Sullivan AB, Cetrulo C, Isner JM, Losardo DW. 1997. Vascular endothelial growth factor inhibits endothelial cell apoptosis induced by tumor necrosis factor-alpha: balance between growth and death signals. J Mol Cell Cardiol 29:1321–1330.
91. Watanabe Y, Dvorak HF. 1997. Vascular permeability factor/vascular endothelial growth factor inhibits anchorage disruption induced apoptosis in microvessel endothelial cells by inducing scaffold formation. Exp Cell Res 233:340–349.
92. Gerber H-P, McMurtrey A, Kowalski J, Yan M, Keyt BA, Dixit V, Ferrara N. 1998. Vascular endothelial growth factor regulates endothelial cell survival through the phosphatidylinositol 3′-kinase/AKT signal transduction pathway. J Biol Chem 273:30336–30343.
93. Papapetropoulos A, Fulton D, Mahboubi K, Kalb RG, O'Connor DS, Li F, Altieri DC, Sessa WC. 2000. Angiopoietin-1 inhibits endothelial cell apoptosis via the AKT/survivin pathway. J Biol Chem 275:9102–9105.
94. Li Zhu, Fukuda S, Cordis G, Das DK, Maulik N. 2001. Anti-apoptotic survivin plays a significant role in tubular morphogenesis of human coronary arteriolar endothelial cells by hypoxic preconditioning. FEBS Lett 508:369–374.
95. Ito T, Shirak IK, Sugimoto K, Yamanaka T, Fujikawa K, Ito M, Takase K, Moriyama M, Kawano H, Hayashida M, Nakano T, Suzuki A. 2000. Survivin promotes cell proliferation in human hepatocellular carcinoma. Hapatology 31:1080–1085.
96. O'Connor DS, Schechner JS, Adida C, Mesri M, Rothermel AL, Li F, Nath AK, Pober JS, Altieri DC. 2000. Control of apoptosis during angiogenesis by survivin expression in endothelial cells. Am J Pathol 156:393–398.
97. Adams JM, Cory S. 1998. The Bcl-2 Protein Family: Arbiters of cell survival. Science 281:1322–1326.
98. Deveraux QL, Reed JC. 1999. IAP family proteins-suppressors of apoptosis. Genes Dev 13:239–252.

Pathophysiology of Cardiovascular Disease,
edited by N.S. Dhalla, H. Rupp,
A. Angel & G.N. Pierce
Kluwer Academic Publishers, Boston, 2004

The Arachidonic Acid Cascade: Thromboxane A_2 Antagonism and Synthetase Inhibition in Experimental Myocardial Infarction

Shiva D. Seth,[1] Uma Singh,[2] and Sandeep Seth[3]

[1] *Indian Council of Medical Research, Ansari Nagar, New Delhi—110 029; Departments of* [2] *Pharmacology and* [3] *Cardiology, All India Institute of Medical Sciences, New Delhi—110 029, India*

Summary. The pathogenesis of myocardial ischemia depends on the factors that upset the balance between myocardial oxygen supply and demand. Arachidonic acid accumulation in the ischemic myocardial tissue, its release and subsequent metabolisation into several classes of eicosanoids and lipid peroxides occurs simultaneously with the development of myocardial ischemic injury. Eicosanoid release in acute ischemia is primarily deleterious and enhances inflammatory-type reactions and functional disturbances of the myocardium. Endoperoxide intermediates (i.e. prostaglandin H_2), thromboxane A_2 (TXA_2) and other eicosanoids having vasoconstrictor and proaggregatory properties are thought to contribute to the pathophysiology of acute myocardial ischemia. Consequently, inhibition of the formation and/or action of TXA_2 and other deleterious eicosanoids have been shown to reduce the severity of myocardial ischemia in models of coronary occlusion. In the present study the effects of the specific thromboxane synthetase inhibitor UK 38485 (dazmegrel) and the TXA_2/prostaglandin endoperoxide receptor antagonist SQ 29548 on myocardial infarct size, cardiometabolic correlates and hemodynamics were determined in a canine model of 24 h acute myocardial infarction. Anesthetised open-chest dogs were subjected to left anterior descending coronary artery ligation. Twenty minutes post-ligation the dogs were given (i) intravenous saline (0.9% NaCl solution) ($n = 12$), (ii) UK 38485 (3 mg/kg i.v.) ($n = 10$) or SQ 29548 (0.2 mg/kg i.v. loading dose with 0.2 mg/kg/h i.v. for 4 hours) ($n = 10$). UK 38485 and SQ 29548 treatment resulted in a significant ($p < 0.01$) reduction in infarct size

Corresponding Author: Professor Shiva D. Seth, National Professor & Chair in Clinical Pharmacology, Indian Council of Medical Research, Ansari Nagar, New Delhi-110 029, India. Tel: 91-11-6565781, Fax: 91-11-6857791, E-mail: drsdseth@yahoo.com.

as compared to the controls. UK 38485 salvaged 51% and SQ 29548 salvaged 45% of the left ventricular area at risk of infarction. Both drugs significantly reduced the accumulation of lactate and restored the activity of myocardial enzymes creatine phosphokinase and lactate dehydrogenase in the ischemic zone of the myocardium. Heart rate, systolic blood pressure and the pressure-rate index were not markedly affected by these drugs. The sharp rise in the left ventricular end diastolic pressure in the saline-treated animals was significantly lowered by UK 38485 and SQ 29548 treatment. The ischemia-induced lowering of the maximal rate of rise of left ventricular pressure (LV dP/dt_{max}) observed in the saline-treated animals was corrected, *albeit* non-significantly, by SQ 29548 treatment. These results showing the beneficial effects of UK 38485 and SQ 29548 indicate that thromboxane A_2 may be involved in the pathogenesis of myocardial ischemia. The study suggests that attenuation of TXA_2 synthesis and action may have potential as a mode of therapy for ischemic heart disease.

Key words: Acute myocardial infarction, Arachidonic acid cascade, Thromboxane synthetase inhibitor, Thromboxane receptor antagonist, Cardioprotection, Dog.

INTRODUCTION

Myocardial ischemia results when the demand for oxygen and substrate exceeds the supply such that muscle cell metabolism is shifted to an anaerobic state, with resultant impairment of contractile function (1). The molecular events or series of events that cause myocytes to die in severe ischemia remain unknown. According to reperfusion studies in the open-chest anesthetized dog heart, severely ischemic myocytes are irreversibly injured after 40–60 minutes. The most likely explanation for the demise of these myocytes is believed to be due to the development of breaks in the sarcolemma of the dead myocyte (2).

Progression of cell necrosis and salvage of the ischemic myocardium

Myocardial ischemia, as a result of coronary occlusion, initiates a sequence of cellular changes which are initially reversible. Uninterrupted, these changes will progress to irreversible alterations, culminating in cellular death (3). Cell death occurs when the energy level of the cell is reduced sufficiently to render it incapable of integrated function by the combined effects of metabolic derangements, autolysis, advanced protein denaturation, and an increased permeability of the sarcolemma permitting excessive loss of coenzymes, enzyme molecules, and electrolytes. Eventually, the biochemical processes of energy production cannot maintain cellular integrity and cellular death occurs (3–5).

Major efforts have been devoted to the development of techniques designed to interfere with the sequence of events summarized above. Most patients with myocardial ischemia have a reduction in myocardial oxygen supply, usually owing to a 70% or greater intraluminal narrowing in a major coronary artery or one of its branches. It has been reported that the infarct size correlates well with the mortality rate (6,7). Thus, an intervention capable of reducing myocardial infarct (MI) size might be expected to reduce the incidence of postinfarction death. Hence, considerable attention has been directed to achieving timely reperfusion of occluded coronary arteries to interrupt the evolving myocardial infarction by restraining the process while

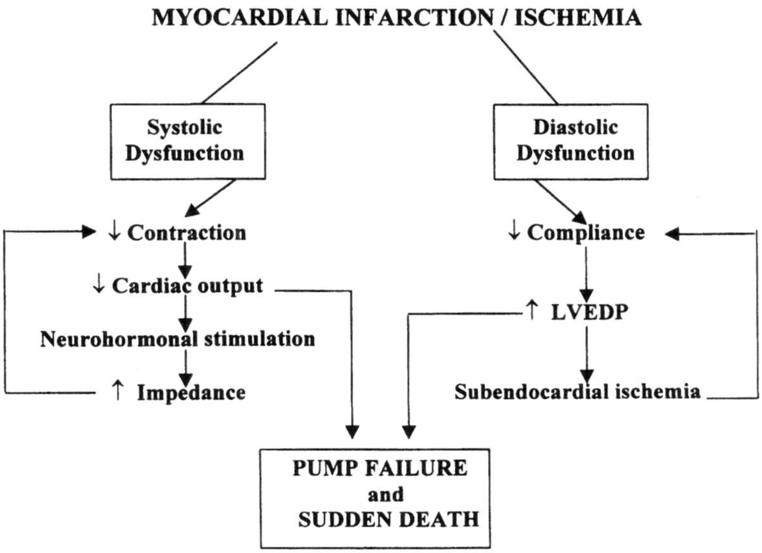

Figure 1. Dynamic processes that may contribute to left ventricular failure after acute myocardial infarction.

it is still in a reversible phase, thus, eventually reducing infarct size (8). The significance of reducing the extent of myocardial infarction relates to the possible therapeutic benefits of preventing cardiac pump failure and mortality that result from myocardial ischemic damage (Figure 1).

The studies on myocardial infarction and the salvage of ischemic myocardium are extensive, nevertheless, despite such enormous documentation, there still are substantial lacunae in our knowledge of the pathobiology of myocardial ischemic injury. Results obtained using animal models of acute myocardial infarction strongly suggest that interventions which can favorably influence the oxygen supply/demand balance in the coronary occluded myocardium can influence the ultimate size of the myocardial infarct, thus affecting the prognosis (9). Subsequent studies have shown that a decrease in infarct size is a definite possibility. Furthermore, animal experiments indicate that the irreversible cell damage and cell death can be delayed, and that one may limit the ultimate extent of necrosis by appropriate pharmacologic interventions during the initial period of ischemia. With the understanding of the "wavefront phenomenon" of myocardial cell death it would be correct to infer that the preservation of ischemic myocardium would largely depend on the time of intervention. Hence, a number of studies are directed towards determining the effect of the time interval between coronary artery occlusion and the application of an intervention (10,11). However, the results obtained in anesthetized or conscious laboratory animals cannot be easily extrapolated to humans. In humans, the duration of the wavefront phenomenon may be greater than six hours because of the presence of collateral vessels that develop in response to the progressive development of atherosclerotic lesions. These collaterals may alter the balance of oxygen supply and demand during ischemia and slow the wavefront of necrosis, thus, allowing more time to introduce an intervention aimed at reducing infarct size.

With the realization of the potential salvage of myocardium at risk with reperfusion, the introduction of reperfusion in the clinical setting with thrombolytic agents or other procedures becomes highly desirable. However, thrombolytic therapy only lyses thrombi and does not affect the underlying causes of the coronary artery occlusion. Therefore, therapy to reduce the chances of reinfarction and death must also be initiated. Since thrombus formation has occurred at least once in patients who survive an attack of myocardial infarction, it is probable that the conditions for thrombus formation still exist. Therefore, use of antiplatelet aggregating drugs, such as aspirin, and inhibitors of deleterious prostanoids would seem to be an appropriate prophylactic and treatment regimen. This concept has created interest to explore new and more effective therapeutic approaches to limit the extension of the myocardial infarct to delay cell death.

However, at present it is not possible to identify the specific interventions most likely to be effective in reducing infarct size and influencing favourably the subsequent size and shape of the left ventricle (12,13). Existing therapies are effective but have profound side effects. There is a need for drugs that have selective anti-ischemic potency. The main cellular effects of ischemia which could be addressed include retardation of high energy phosphate metabolism and prevention of several ischemic factors such as the formation of free radicals, activation of phospholipases and subsequent cascade of the arachidonic acid metabolites (14).

In general consideration, the basis of interventions aimed at reducing myocardial ischemic injury are:

1. Interventions that reduce myocardial oxygen demand (beta-adrenergic blocking agents e.g. propanolol, metoprolol etc. and selected calcium channel blockers (CCBs) e.g. gallopamil.
2. Interventions that increase substrate availability (carnitine, hyaluronidase etc.)
3. Interventions that increase coronary blood flow and certain modulators of the arachidonic acid metabolism.
4. Interventions that relieve vascular obstructions (modulators of arachidonic acid metabolism).

Arachidonic acid cascade

Enzyme oxidation of arachidonic acid leads to a multitude of biochemically important products. Among them are prostaglandins (PGs), thromboxanes (TXs), prostacyclin and leukotrienes (LTs). These substances, referred to collectively as eicosanoids, constitute what has become known as the *arachidonic acid cascade*, a term coined by workers in the field to describe an ever-increasing stream of compounds derived from arachidonic acid (15). These compounds are called eicosanoids because they are derived from 20-carbon essential fatty acids that contain three, four or five double bonds: eicosatrienoic acid, eicosatetraenoic acid and eicosapentaenoic acid respectively. In man, arachidonic acid is the most abundant precursor, and it is either derived from dietary linoleic acid or is ingested as a constituent of meat. Arachidonate is then esterified as a component of the phospholipids of cell membranes or is found in ester linkage in other complex lipids. The biosynthesis of the eicosanoids depends primarily upon the release of arachidonic acid from cellular stores by various acyl hydrolases (16). The enhanced biosynthesis of the eicosanoids is generally believed to involve activation of these enzymes, caused by increase in the intracellular concentration of calcium. Calmodulin, acting in concert with calcium, is important in the activation of the acyl hydrolases and the membrane-bound forms of both phospholipase A_2 and C are probably involved in the formation of precursor arachidonic acid. Once released from the membrane phospholipids, arachidonic acid and its congeners are rapidly metabolized to oxygenated products by several distinct enzyme systems. Products that contain ring structures (prostaglandins, thromboxanes and prostacyclin) result from the initial action of cyclooxygenase while the hydrolated derivatives of straight-chain fatty acids (e.g. leukotrienes) result from the action of various lipoxygenases. The term "eicosanoids" is applied to all the 20 carbon derivatives, whereas, "prostanoids" refer only to those with a ring contained in their structure (15).

Cyclooxygenase pathway

The cyclooxygenase enzyme forms the cyclic endoperoxide PGG_2. This is converted to PGH_2, which then converts enzymically or non-enzymically to the stable substances PGE_2, $PGF_{2\alpha}$ and PGD_2. The cyclic endoperoxides (PGG_2, PGH_2) are at crossroads of arachidonic acid metabolism, for they are precursors of substances with opposing biological properties (Figure 2).

Prostaglandin I_2 (PGI_2, prostacyclin) and thromboxane A_2 (TXA_2) currently appear to be the most active and major metabolites of PGH_2 associated with the cardiovascular system. Prostacyclin synthetase, a membrane-bound enzyme in vascular tissue, synthesizes PGI_2 (17) which possesses potent vasodilator and platelet-aggregation inhibitory activity that is associated with increased intracellular levels of cyclic AMP (17–19). The inactivation of PGI_2 occurs with a half-life of 5 min and involves a hydrolysis reaction to 6-keto-$PGF_{1\alpha}$, a relatively inactive metabolite. TXA_2 is synthesized in platelets by the membrane-bound thromboxane synthetase, an enzyme with a secondary catalytic activity that results in the decomposition of PGH_2 into malondialdehyde (MDA) and 17-carbon hydroxy acid (HTT). The decomposition reaction is also catalyzed by Fe^{2+}. Contraction of smooth muscle along with platelet aggregation and the release reaction are induced by TXA_2 at a 10-fold lower concentration than PGH_2. TXA_2 is inactivated by a hydrolysis reaction, with a half-life of about 30 sec, to the relatively inactive metabolite thromboxane B_2 (TXB_2) (20). Each substance has opposing effects on cAMP concentrations; thereby, affecting a balanced control mechanism which influences thrombus and hemostatic plug formation.

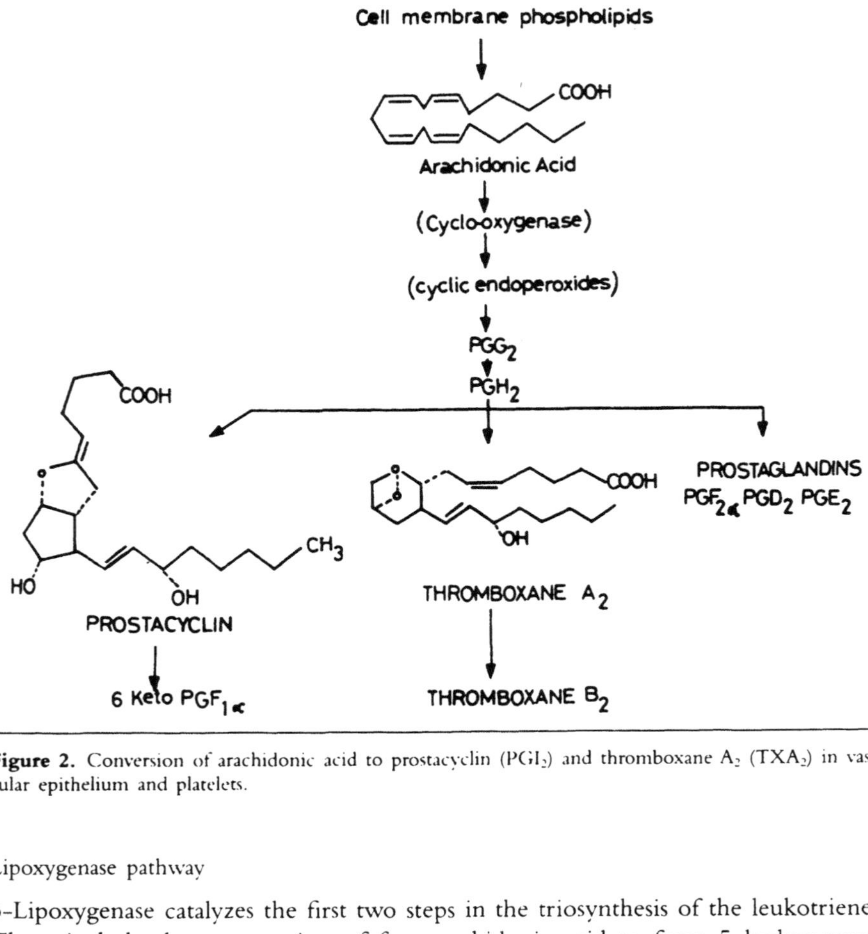

Figure 2. Conversion of arachidonic acid to prostacyclin (PGI_2) and thromboxane A_2 (TXA_2) in vascular epithelium and platelets.

Lipoxygenase pathway

5-Lipoxygenase catalyzes the first two steps in the triosynthesis of the leukotrienes. These include the oxygenation of free arachidonic acid to form 5 hydroperoxy-6,8,11,14-eicosatetraenoic acid (5 HPETE) followed by the subsequent conversion of 5 HPETE to 5,6-oxido-7,9,11,14-eicosatetraenoic acid, known as leukotriene A_4 (LTA_4). The LTA_4 is in turn converted to a variety of products including leukotriene B_4 and the peptidoleukotrienes LTC_4, LTD_4 and LTE_4. In addition, the enzyme is involved in the biosynthesis of the lipoxins through the oxidation of 15 HPETE. Because of their biological activities, leukotrienes (LTs) and lipoxins are believed to have an important role in the pathophysiology of a variety of inflammatory responses including those of myocardial ischemia (21). LTs have potent actions on various types of smooth muscles and are able to alter the calibre of blood vessels. Usually, LTC_4 or D_4 increase vascular resistance (21). It has also been indicated that leukotrienes may cause oedema of the vessel wall which would contribute to narrowing of the lumen. The LTs also cause reduction in ventricular developed tension which appears to be secondary to the impairment of coronary flow. Thus, LTs play an important role in the myocardial ischemic events. It is noteworthy that LTs C_4 and D_4 have differing profiles of action in the heart. Leukotriene C_4 is more potent than LTD_4 in reducing coronary flow and cardiac developed tension. Further, the reduction in cardiac developed tension induced by LTC_4 is of slower onset and longer duration than that due to LTD_4 (22).

Altered arachidonic acid metabolism in acute myocardial ischemia

The physiological and clinical importance of the eicosanoids is now widely accepted (15,23). Furthermore, it has been emphasized that cardiac eicosanoids which physiologically act as modulators of cell activity and are primarily beneficial, may become a major detrimental factor under conditions of cell injury i.e. myocardial ischemia and infarction. Arachidonic acid release occurs simultaneously with development of severe cellular and sarcolemmal damage (24). It has been proposed (15) that vascular homeostasis is determined by a balance between the production of thromboxane A_2 and PGI_2 and that the occurrence of thrombotic disorders might be influenced by factors that alter this balance. If this balance is shifted in favor of platelet aggregation and manifestation of atherosclerosis there is a profound decrease in PGI_2 formation by the coronary vascular endothelium. Age, an increase in low-density lipoproteins (LDLs), nicotine, diabetes and hereditary factors may also be associated with impaired PGI_2 production. Cholesterol, diabetes mellitus, catecholamines, mental stress and vascular trauma increase thromboxane synthetase activity. An increase in coronary sinus TXB_2 and platelet aggregation has been found in patients with myocardial ischemia brought on by an increase in myocardial oxygen demand (25). The occurrence of spontaneous episodes of myocardial ischemia in patients with unstable angina has been attributed to an imbalance between PGI_2 and TXA_2 production and increased thromboxane generation (26). There is evidence for increased neutrophil aggregation in patients with acute myocardial infarction and enhanced chemotaxis and LTB_2 formation in unstable angina. Although earlier reports suggested a decrease in PGI_2 formation by atherosclerotic vessels, current evidence suggests that arachidonic acid release and PGI_2 generation may be actually greater in these compared to normal vessels. Infarcted tissue from experimental animals has been shown to generate much greater amounts of TXA_2 and leukotrienes than the adjacent normal tissue. Release of these arachidonic acid derivatives in the ischemic myocardium may be responsible for reduction in coronary blood flow and subsequent extension of the region of infarct.

Modification of arachidonic acid cascade through phospholipase A_2-dependent mechanisms

As mentioned above the biosynthesis of the eicosanoids, *namely*, prostaglandins thromboxanes, prostacyclins, hydroxyeicosatetraenoic acids (HETEs), leukotrienes and lipoxins, whether synthesized by the cyclooxygenase pathway or the lipoxygenase pathway, all depend on the availability of free arachidonic acid. The control step for the production of that free arachidonic acid is believed to somehow involve a membrane-receptor event which involves phospholipase A_2 (27). Phospholipase has been suggested to play a role in the precipitation and consequences of myocardial ischemic injury and reduction of infarct size in a canine model of coronary artery occlusion by the administration of phospholipase inhibitor quinacrine, an antimalarial drug has been reported (28). Antiinflammatory and platelet antiaggregant activity of phospholipase A_2 inhibitors has been elucidated (29,30). Furthermore, the effect of arachidonic acid and phospholipases has been shown on the cardiac function, evidencing the correction of arachidonic acid-induced post-ischemic dysfunction brought about by co-administration of oxygen free radical scavengers (30).

Several potential sources of free radicals exist within the membrane of the cell. The metabolism of arachidonic acid involves the formation of intermediate peroxy compounds as well as hydroxy radicals (14,30). Furthermore, it has been suggested that not only is the arachidonic acid cascade an important source of toxic oxidant generation, it can itself be initiated by free radicals from other sources (31).

Isoprostanes: Bioactive products of lipid peroxidation

A decade ago, isoprostanes were discovered as a novel family of prostaglandins that are produced from arachidonic acid independent of the cyclooxygenases; through the free radical—catalyzed lipid peroxidation (32). The presence of representative isoprostane, 8—iso prostaglandin $F_{2\alpha}$ in human coronary atheroma evidences the involvement of isoprostanes in ischemic heart diseases (33). Furthermore, transcardiac 8-isoprostaglandin $F_{2\alpha}$ has been found to be a reliable marker for the size of the myocardium exposed to oxidative stress (34).

TXA_2 release during myocardial ischemia and the potential role of thromboxane inhibitors in preventing ischemic injury

Several animal studies have documented release of TXA_2 during myocardial ischemia. TXB_2 concentrations have been shown (35,36) to increase markedly 5 to 10 minutes after coronary occlusion and returned to baseline at 30 minutes despite continuing occlusion. This indicates a transient release of TXA_2 soon after coronary occlusion. It has been observed that animals with the highest level of TXB_2 had the most arrhythimas while there was a negative correlation between the rise in 6-keto-$PGF_{1\alpha}$ and the number of ventricular ectopic beats (35).

TXA_2 is also released during formation of platelet-rich thrombi in the narrowed coronary arteries of dog with endothelial injury (37). A reduction in PGI_2 release resulting in an "imbalance" between TXA_2 and PGI_2 may be a cause of cyclical flow reductions in these dogs (38). Furthermore, TXA_2 release has been documented in a variety of myocardial ischemic syndromes in humans as well. Increased TXA_2 release (and diminished PGI_2 synthesis) has been reported during exercise in patients with stable angina pectoris (39). Myocardial ischemia induced by atrial pacing is associated with release of TXB_2 in the coronary venous blood (40). Furthermore, elevation of TXB_2 in the peripheral and the coronary venous blood plasma (40,41) and of 2,3 diner-TXB_2 in urine (42) has been demonstrated in patients with unstable angina pectoris. Interestingly, enhanced TXB_2 synthesis has been observed just prior to acute coronary occlusion (43) and soon after acute myocardial infarction (36). Increased TXA_2 biosynthesis has been shown during thrombolysis (44) indicating that platelet activation and TXA_2 release may modulate the response to thrombolytic agents.

In most of these studies release of TXA_2 has been shown to be associated with the presence of myocardial ischemia and ischemia-related arrhythmias. Obviously, it is difficult to ascertain if TXA_2 release is the cause of myocardial ischemia or is a consequence of the ischemic process.

Thromboxane inhibitors in myocardial ischemia

In view of the preceding evidence of TXA_2 release during myocardial ischemia, a variety of agents with TXA_2 inhibitory activity have been developed over the past two decades for potential clinical use. Since the development of these agents, the potential roles of the PG endoperoxides and TXA_2 in platelet aggregation and in cardiovascular disease have been studied extensively (45). It was anticipated that this group of compounds could provide new therapeutic approaches for the treatment of a variety of cardiovascular diseases (45). These agents have undergone extensive trials in various animal models of myocardial ischemia. The information from these studies has provided new insights into the mechanisms of disease and possible new avenues for therapy.

For the purpose of reviewing, these agents can be divided into two broad categories (a) TXA_2 synthetase inhibitors and (b) TXA_2/endoperoxide receptor antagonists.

(a) Thromboxane A_2 (TXA_2) synthetase inhibitors

Imidazoles selectively block thromboxane synthetase but do not interfere with other enzymes of the prostanoid cascade (46,47). Thus, in contrast to cyclooxygenase blockers, thromboxane synthetase inhibitors (TXSIs) preserve formation of prostacyclin, the natural antagonist to TXA_2. Diversion of endoperoxides away from the blocked thromboxane pathway may enhance the formation of prostacyclin and prostaglandins that are beneficial in the setting of cardiovascular diseases. This shunting of cyclic endoperoxides evident in vitro as well as in vivo (48–50), has been termed "endoperoxide steal". It seems to be quantitatively significant only in the presence of thromboxane synthetase inhibition (51) and may be an important mechanism for these agents to reduce TXA_2 formation, increase PGI_2 formation, inhibit platelet aggregation and thrombus formation.

Furthermore, TXA_2 synthetase inhibitor-induced increase in PGI_2 has been shown to be responsible for the reduction in neutrophil superoxide radical generation when neutrophils are incubated with platelets in the presence of a selective TXA_2 synthetase inhibitor U63, 557A (52). Superoxide radicals have been incriminated in the extension of myocardial injury and a reduction in their synthesis may relate to the decrease in infarct size in animals treated with selective TXA_2 synthetase inhibitors evidenced by the encouraging results of some workers (53–61).

UK 38485, 3-(1H-imidazole 1-yl-methyl)-2-methyl-1H-indole-1-propanoic acid; (Dazmegrel), a specific inhibitor of thromboxane A_2 synthetase is an advanced imidazole compound of higher potency and longer action in vitro and in vivo (62).

(b) TXA_2/Cyclic endoperoxide receptor antagonists

Thromboxane A_2/endoperoxide receptor antagonists block platelet aggregation in response to a variety of stimuli in very low concentrations. This enhanced efficacy of TXA_2/endoperoxide receptor antagonists is probably on account of their ability to block the pro-platelet aggregatory activities of TXA_2 and cyclic endoperoxides. These drugs do not inhibit TXA_2 synthesis and do not augment PGI_2 generation. It is possible that excessive synthesis of PGI_2 shown to occur during use of TXA_2 synthetase inhibitors (48–51), may induce myocardial ischemia by shunting blood away from the ischemic zone. This possible deleterious effect is unlikely to occur with TXA_2/endoperoxide receptor antagonists such as SQ 29548 (IS – (1, 2β (5Z), 3β, 4)-7 (3-[(2-((phenylamino) – carbonyl) hydrazino] methyl-7-oxabicyclo (2.2.1)-hept-2-yl)-5-heptenoic acid). This agent has been shown to be a potent, selective antagonist of thromboxane A_2 and endoperoxide analogue induced platelet aggregation (63). In addition, SQ 29548 antagonizes CTA_2 (thromboxane agonist)—induced coronary vasoconstriction in vitro and prevented platelet aggregation and sudden death in vivo in rabbits given exogenous arachidonic acid, indicating significant antagonism of a variety of thromboxane-mediated biological actions (64). This agent is, therefore, a potent antagonist of both vascular and platelet thromboxane receptors (63).

TXA_2 inhibitors in animal models of myocardial ischemia

As reviewed above a variety of TXA_2 synthetase inhibitors such as dazoxiben, U63, 557A and UK 38485 have been shown to reduce infarct size in animal models of myocardial infarction (53–61). TXA_2/prostaglandin endoperoxide receptor antagonists also reduce myocardial infarct size (56,65–71). It has been observed that the degree of reduction in infarct size with the use of TXA_2 synthetase inhibitor or receptor antagonists is almost similar (56). Reduction in myocardial leukocyte infiltration and activation could be a factor in myocardial infarct size reduction by TXA_2 inhibitors, since aspirin, which also inhibits TXA_2 formation, does not limit infarct size and does not affect neutrophil infiltration. Both TXA_2 synthetase inhibitors or receptor antagonists have been shown to markedly reduce myocardial leukocyte infiltration and peripheral leucocytosis (55–57). BM 13505 a specific TXA_2/endoperoxide receptor antagonist has been shown (70) to reduce atherogenesis and the severity of ischemic damage resulting from total coronary occlusion in cholesterol-fed rabbits. BM 13505 was shown to protect the ischemic rabbit myocardium by three different methods: i) maintenance of mycoardial tissue creatine kinase (CK) activity; (ii) reduced loss of free amino nitrogen-containing compounds from the myocardium; and (iii) blunting the rise of plasma CK activity (70).

Furthermore, in animal models of platelet aggregation occurring in the narrowed coronary arteries, TXA_2 synthetase inhibitors have been found to be effective in decreasing the frequency and magnitude of blood flow reduction (72). As stated above, the increase in endogenous PGI_2 upon administration of TXA_2 synthetase inhibitor may be important in preventing coronary artery thrombosis. Administration of TXA_2/endoperoxide receptor antagonists SQ 29548 or ONO 3708 also decreases intracoronary platelet deposition in experimentally stenosed coronary arteries (73,74) evidencing that a local deficiency of PGI_2 does not play a significant role in platelet aggregation—related cyclical flow changes; since, TXA_2/endoperoxide receptor antagonists do not increase PGI_2 concentrations.

Coronary arterial reocclusion after initial thrombosis was found to be inhibited with the use of TXA_2 synthetase inhibitor CGS-13080 (75) or the TXA_2 receptor antagonists L636, 499 and AH-23848 (44). OKY 1581, a TXA_2 synthetase inhibitor was found to inhibit the incidence of thrombotic occlusion in a canine model of electrically-induced occlusive coronary thrombus (76). The TXA_2/prostaglandin endoperoxide receptor antagonists L636, 499, SQ29548 and ONO 3708 have also been reported to inhibit formation of occlusive thrombus in a similar model (74,77).

TXA_2 synthetase inhibitors as well as TXA_2/endoperoxide receptor antagonists reduce arrhythmias which have been shown to occur during the early phase of reperfusion following coronary occlusion (78–80). However, another study (81) failed to show an important contribution of TXA_2 to arrhythmogenesis during evolving myocardial infarction.

Clinical use of TXA_2 inhibitors

Because of the many documented beneficial effects of TXA_2 inhibitors in animals, several trials of TXA_2 inhibitors have been conducted in patients with ischemic heart disease. Since TXA_2 formation is clearly increased in unstable angina pectoris and in the early stages of acute myocardial infarction, it is possible that TXA_2 inhibitors may be effective in vasoocclusive syndromes. In a study (82) TXA_2 synthetase inhibitor given within 6 hours of the onset of acute myocardial infarction reduced TXA_2 synthesis and caused a modest (approximately 25%) decrease in cumulative creatine kinase levels. Furthermore, when used in conjunction with thrombolytic therapy in humans, these agents may prove to be beneficial, as reocclusion after initial restoration of coronary blood flow in acute myocardial infarction continues to be a major problem. Coronary bypass surgery and angioplasty are associated with intense platelet activation at the site of endothelial tear; hence, TXA_2 inhibitors are currently being assessed in these conditions. However, long-term dosing studies with these agents have not been reported in patients with primary cardiac disease. Pretreatment with dazoxiben has been shown to affect a modest correction of lactate accumulation and ST depression during pacing-induced angina (83).

In conclusion, evidence has been provided in the above account that TXA_2 is released during myocardial ischemia, and when platelets interact with stenotic coronary artery with injured endothelium. It is clear that TXA_2 is an important mediator of platelet aggregation and coronary arteriolar constriction. It is also evident that TXA_2 inhibitors reduce myocardial injury, prevent reperfusion arrhythmias, and decrease formation of intracoronary thrombus in a variety of animal models. Many of these salutary effects of TXA_2 inhibitors are mediated in part by reduction in leukocyte infiltration.

In the clinical setting of ischemic heart disease, trials thus far with TXA_2 inhibitors, have not shown much promise in stable or unstable ischemic syndromes. Clearly, there is a discrepancy between the efficacy of TXA_2 synthetase inhibitors and TXA_2 receptor antagonists in short-term animal preparation of thrombosis and their lack of effect in clinical studies. Modifications of either drug structure or delivery will be necessary to prolong the duration of maximal thromboxane inhibition throughout the dosing interval. Furthermore, combination of the inhibitors of TXA_2 synthesis with antagonists of TXA_2/prostaglandin endoperoxide receptors may be necessary to reveal the full therapeutic potential of these agents (84).

It has been shown that PGH_2 itself is active at thromboxane A_2/prostaglandin endoperoxide receptors, and accumulates when thromboxane synthetase is inhibited (56,63–67,85). Therefore, a novel class of agents that act as antagonists at the thromboxane A_2/endoperoxide receptors, as well as inhibiting thromboxane synthetase, has been developed (86–89). Considering the great therapeutic interest in dual activity compounds; that is, thromboxane A_2 receptor antagonists/thromboxane A_2 synthetase inhibitors (TXRAs/TXSIs), further in vivo experiments are warranted to confirm their efficacy in the treatment of cardiovascular diseases.

The objective of the present study was: (i) to define and compare the effects of a selective TXA_2 synthetase inhibitor and a specific TXA_2 receptor antagonist on myocardial infarct size in the canine acute coronary artery ligation model, (ii) to determine the metabolic and hemodynamic basis of such effects and (iii) to elucidate the role of TXA_2 in the pathophysiology of acute myocardial ischemia.

METHODS

The All India Institute of Medical Sciences (AIIMS), New Delhi Guidelines for care and use of laboratory animals were followed in this study.

Surgical procedures and experimental measurements

Thirty seven adult mongrel dogs of either sex, weighing 15–20 kg were studied during a 2 day protocol. On day 1 after fasting overnight, the animals were anesthetized with sodium pentobarbital (35 mg/kg, intraperitoneally), intubated and ventilated by positive pressure ventilation with room air by a respiration pump (INCO, Ambala, India). Polyethylene catheters filled with normal saline were inserted into the right femoral artery for measuring systolic arterial blood pressure (SAP) through a P23 ID Gould Statham pressure transducer and into the right femoral vein for drug and/or saline administration. A left thoracotomy was performed through the fifth intercostal space, the pericardium was opened and the heart was suspended in a pericardial cradle. The left anterior descending coronary artery (LAD) was gently dissected free from the surrounding tissue for a few millimeters, about 2 cm from its origin, just proximal to its first large diagonal branch. Myocardial ischemia was induced by ligation of LAD coronary artery in two stages according to the method of Harris (90). A sterile 18 G stainless steel needle was introduced into the left ventricular cavity through the apex of the heart to measure the left ventricular pressure using a P23 ID Statham pressure transducer. This signal was recorded at full scale and high gain for determination of left ventricular end diastolic pressure (LVEDP) and electronically differentiated to obtain the maximal rate of change of left ventricular pressure ($LVdP/dt_{max}$). Heart rate (HR) was calculated from the standard limb lead II electrocardiogram (ECG). The pressure-rate index (PRI) was calculated as the product of SAP and HR divided by 100 and was used as an approximation of myocardial oxygen consumption (91). After surgical preparation and stabilization, all pressure measurements and electrocardiographic signals were monitored continuously during the experiment, starting 10 min before LAD coronary artery ligation, on a Grass Polygraph Model 7D (Grass Instruments, Quincy, Massachusetts, USA); and recorded at preset time points (i.e. 5, 20, 30, 40, 80 min and 24 h after LAD ligation), on Day 1 of the experiment as well as 24 h after coronary ligation i.e. Day 2 of the experiment. The time course and trend of the response of hemodynamic variables to LAD ligation and the drug treatments in the different experimental groups at 80 min and 24 h post-ligation is calculated in terms of percent change from the initial pre-ligation (baseline) value (a decrease is designated by a minus sign (–), the other % change (unmarked) is an increase.

Experimental protocol

After ligation of LAD, the animals were randomly assigned to (i) the saline-treated (control) group (n = 12) including animals administered 10 ml of sterile 0.9% NaCl solution intravenously, 20 min post-ligation of the coronary artery; (ii) the UK 38485—treated group (n = 10) including animals given 3 mg/kg i.v. slow bolus 20 min post-ligation of the coronary artery (UK 38485 was dissolved in 3 ml of 0.1 N sodium hydroxide and made up to 10 ml with 0.9% NaCl solution) or (iii) the SQ 29548-treated group (n = 10) including animals given a loading dose of 0.2 mg/kg i.v. slow bolus 20 min post-ligation of the coronary artery followed by an infusion of 0.2 mg/kg/h for 4 hours). The total dose of the drug administered was 1 mg/kg. SQ 29548 was dissolved in 10% ethanol and made up to 10 ml with 0.9% NaCl solution.

The drug solutions were prepared immediately prior to each experiment. Ethanol did not alter the infarct size, or the hemodynamic variables that we measured. All animals received 1 million units of penicillin G intramuscularly for prophylaxis of infection.

On day 2, approximately 24 hours after ligation of LAD, all dogs were reanesthetized and the chest reopened at the site of the previous incision. All hemodynamic variables were measured as on day 1 of the experiment. Ten minutes before euthanasia, monastral blue dye (1.5 ml/kg) was injected into the left atrium to delineate the in vivo area at risk. All animals were euthanized with a sodium pentobarbital overdose. Thereafter, the heart was excised rapidly. The left ventricle was separated, frozen in saline solution (−10°C) and sliced into 5 mm thick transverse sections. The blue-stained normally perfused zone was separated from the unstained portion depicting the area at risk (AR) and weighed. Both the portions were incubated separately in buffered (1%) triphenyl tetrazolium chloride (TTC) solution for 20 minutes at 37°C (92). This differential staining distinguished the unstained area of necrosis (AN) from the normally perfused viable tissue (VT) and reversibly injured area at risk, stained red. Different zones of the myocardium are expressed in terms of percentage of the total LV mass and AR. The per cent reduction of the left ventricular area at risk of necrosis was calculated as follows:

$$[1-\{(AN_t/AR_t)/(AN_c/AR_c)\}]\times 100$$

Where the subscripts t and c represent the treated and control groups respectively.

Transmural myocardial tissue samples were rapidly taken from the normally perfused, ischemic and necrotic zones of the myocardium and fixed in accordance to the biochemical parameters to be estimated. Tissue samples immediately frozen in liquid nitrogen were weighed and homogenised at 4°C in 6% perchloric acid with a tissue homogenizer. These heart extracts were neutralized with 5 M K_2CO_3 and centrifuged, the supernatant was stored until further analysis of lactate (93). For estimation of myocardial enzymes tissue samples taken from different zones of the ventricular myocardium were processed as described previously (94,95). Creatine phosphokinase (CPK) was estimated according to the method of Foster et al. (96) and lactate dehydrogenase (LDH) by the method of Bergmeyer and Bernt (97). The total amount of protein was estimated in the supernatant fluid of the homogenate obtained for estimation of myocardial enzymes, by the method of Lowry et al. (98). Biochemical analysis were performed on Beckman DU 640 spectrophotometer. The weight of the tissue samples taken for biochemical estimations was added to the weight of the respective zones of the ventricular mass for quantification of infarct size.

Statistical analysis

The data were analysed by use of Student's *t* test and analysis of variance (ANOVA) to determine significance between groups. A p value less than 0.05 was considered significant. Data are expressed as mean ± SEM.

RESULTS

A total of 37 dogs were anesthetized for use of this study. Five dogs succumbed to fatal ventricular fibrillation shortly after coronary artery ligation. Transient arrhythmias occurred most commonly after the first ligation of the coronary artery, but subsided within 20–30 minutes after the second ligation. These arrhythmias never persisted in the drug-treated groups.

Infarct size

The effects of saline, UK 38485 and SQ 29548 treatment on experimental infarct size are shown in Table 1 and Figure 3. Twenty four hours after LAD ligation, infarc-

Table 1. Effect of thromboxane A_2 synthetase inhibitor, UK 38485 and thromboxane A_2/prostaglandin endoperoxide receptor antagonist, SQ 29548 on various zones of the myocardium in coronary artery ligated dogs

Groups	LV mass	Area at risk (AR)		Area of necrosis (AN)			Viable tissue (VT) within risk region			AN/AR	Per cent reduction of AR destined to necrosis
	g	g	% of LV mass	g	% of LV mass	% of AR	g	% of LV mass	% of AR		
Saline-treated (n = 12)	66 ± 6.0	22 ± 2.0	34 ± 1.8	17 ± 2.3	26 ± 1.8	77 ± 3.1	5 ± 0.6	8 ± 1.0	22 ± 3.1	0.8 ± 0.03	
UK 38485-treated (n = 10)	64 ± 3.4	19 ± 1.2	30 ± 1.5	7 ± 1.0	11 ± 0.9**	38 ± 2.7**	12 ± 1	19 ± 1.3**	62 ± 2.7**	0.4 ± 0.05	51.0
SQ 29548-treated (n = 10)	71 ± 1.8	18 ± 1.9	25 ± 2.1	7 ± 0.6	10 ± 0.8**	42 ± 1.5**	11 ± 1.3	15 ± 1.5*	58 ± 1.4**	0.4 ± 0.01	45.0

LV mass = left ventricular mass; g = gram; n = number of animals in each group;
AR = area at risk (hypoperfused zone lying distal to occluded left anterior descending coronary artery (LAD) i.e. in vivo tissue unstained with monastral blue dye);
AN = infarct zone, (i.e. in vitro tissue unstained with triphenyl tetrazolium chloride (TTC)).
All values are mean ± SEM.
* = $p < 0.05$; ** = $p < 0.01$ as compared to the same zone in the control group.

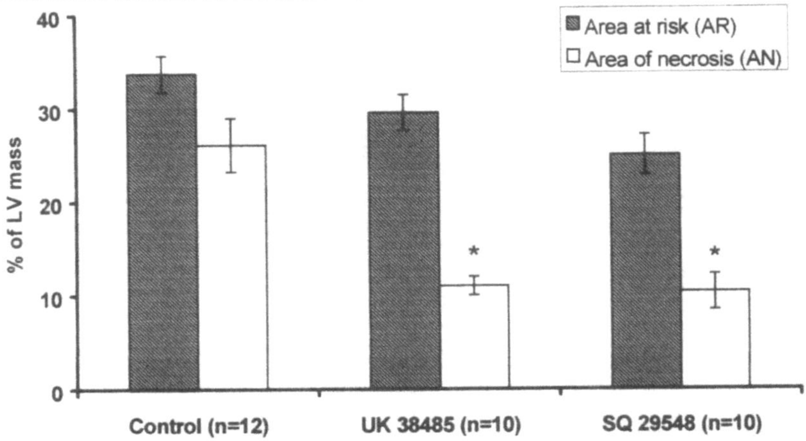

Figure 3. Left ventricular area at risk (AR) and area of necrosis (AN) in the controls and drug-treated groups, expressed as percentage of the left ventricular mass. In UK 38485-treated (n = 10) and SQ 29548-treated (n = 10) groups AN is significantly smaller as compared to the saline–treated control group (n = 12). Height of bars represent means; brackets represent ± SEM for respective values in each group.
*p < 0.01 as compared with the similar myocardial region in the control group.

tion was always transmural. The mean area at risk (AR) expressed as a per cent of total left ventricular (LV) mass was not significantly different in the saline-and drug-treated groups. However, the mean area of necrosis (AN) expressed as per cent of the total LV mass and AR was significantly ($p < 0.01$) less in the drug-treated groups as compared to the same area of the myocardium in the saline-treated group. Per cent reduction of AR destined to necrosis, was 51% in the UK 38485-treated group and 45% in the SQ 29548- treated group as compared with the saline-treated control group.

Biochemical parameters

The data regarding the effect of UK 38485 and SQ 29548 on cardiac metabolism following coronary artery ligation in dogs are shown in Figure 4. Myocardial ischemia resulted in depletion of the myocardial high energy phosphates (HEPs) (data not shown here) and a significant accumulation of lactate, the deleterious myocardial metabolite in the three differentiated zones of the left ventricular myocardium i.e. normally perfused, ischemic, and necrotic; *albeit*, more markedly in the ischemic zone of the myocardium in saline-treated control group. The lactate content μmole/g^{-1} wet weight of tissue of the ischemic zone was 10.19 ± 0.53 in the saline-treated group. These metabolic anomalies in the control group were found to occur in concert with the decreased activity of the myocardial enzymes creatine phosphokinase (CPK), and lactate dehydrogenase (LDH). Myocardial CPK activity was 5.10 ± 0.92 U/mg of protein and the myocardial LDH activity was found to be 1.55 ± 0.45 U/mg of protein in the ischemic zone of the myocardium.

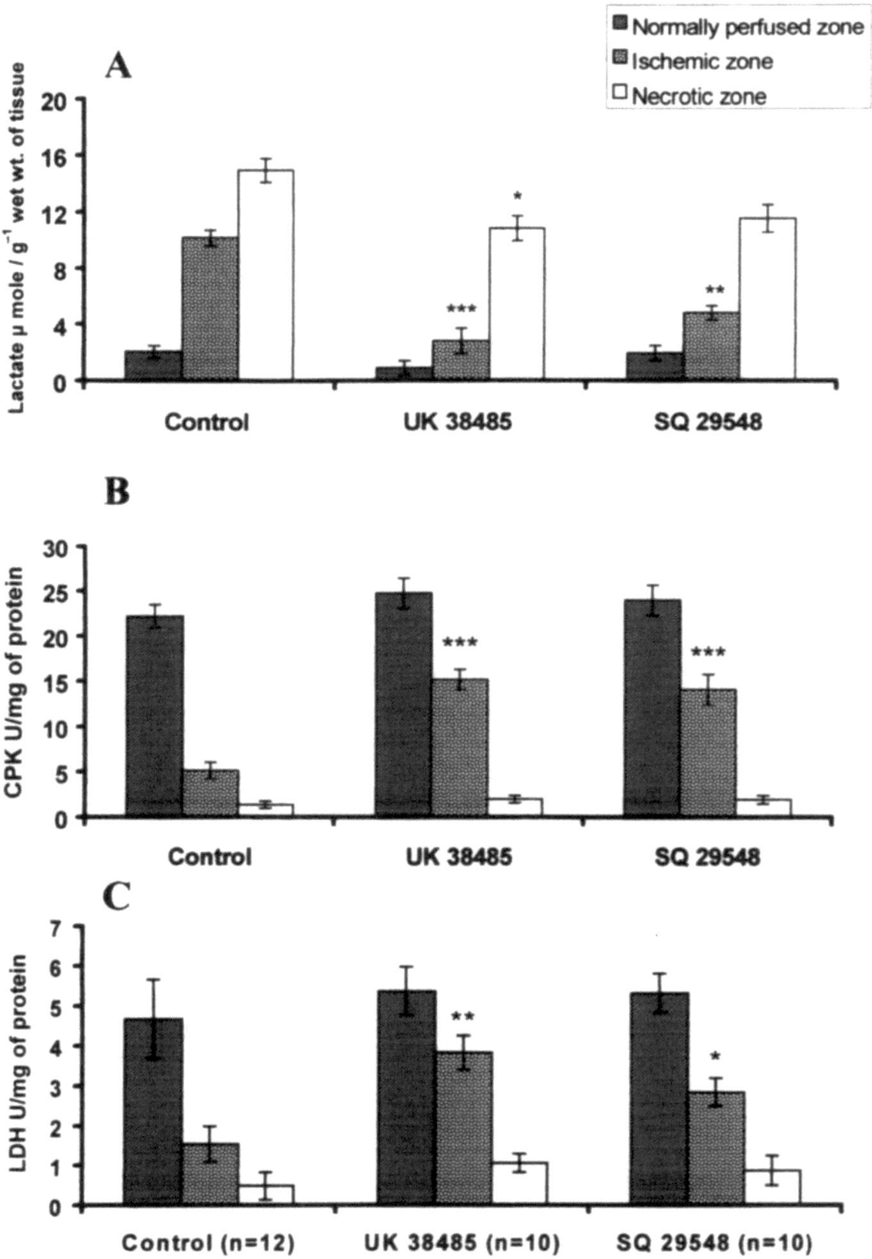

Figure 4. Effect of UK 38485 and SQ 29548 on **A.** myocardial lactate content; **B.** activity of the enzyme creatine phosphokinase (CPK) and **C.** activity of the enzyme lactate dehydrogenase (LDH) in the three differentiated zones of the myocardium, *viz.* normally perfused, ischemic and necrotic. Height of bars represent means; brackets represent ± SEM for respective values in each group. $^{*}p < 0.05$, $^{**}p < 0.01$; $^{***}p < 0.001$ as compared with the similar zone of the myocardium in the control group.

The lactate content of the ischemic zone of left ventricular myocardium was significantly ($p < 0.001$) lowered in the UK 38485-treated group, being 2.81 ± 0.89 μmole/g^{-1} wet weight of tissue. UK 38485 treatment restored the myocardial CPK activity to 15.16 ± 1.09 U/mg of protein in the ischemic zone which was a highly significant ($P < 0.001$) increase as compared to the CPK activity of the same zone in the controls. The level of the activity of the enzyme LDH in the ischemic zone of the myocardium in the UK 38485-treated group was 3.85 ± 0.42 U/mg of protein, which was significantly ($p < 0.01$) more compared to the activity of this enzyme in the ischemic zone of myocardium in the controls.

SQ 29548 treatment was effective enough to reduce accumulation of lactate in the ischemic zone of the myocardium. The lactate levels in this zone being 4.80 ± 0.51 μmole/g^{-1} wet weight of tissue which was significanlty ($p < 0.01$) less as compared to the lactate content of the same zone in the saline-treated control group. Furthermore, SQ 29548 significantly increased the activity of the myocardial enzymes; the level of CPK in the ischemic zone was 14.07 ± 1.65 U/mg of protein ($p < 0.001$) and that of LDH was 2.86 ± 0.35 U/mg of protein ($p < 0.05$) as compared to the ischemic zone of the ventricular myocardium in the controls.

Hemodynamics

Tables 2 and 3 show the hemodynamic data for saline-, UK38485- and SQ 29548-treated animals at preset time points during coronary artery ligation. There was no significant difference in the heart rate (HR, beats/min) between the saline-, UK 38485- and SQ 29548-treated groups for upto 80 min after coronary artery ligation. At 24 h post-ligation, heart rate was significantly ($p < 0.05$) higher in the saline-treated group as compared to the baseline pre-ligation value of the same group. However, at 24 h post-ligation, the heart rate was significantly ($p < 0.05$) less in the SQ 29548-treated group as compared to the value of this variable at the same time period in the control group. There was a significant ($p < 0.05$) fall in systolic arterial blood pressure (SAP, mmHg) in the saline-treated group 80 min post-ligation, this effect was sustained 24 h post-ligation. UK 38485 and SQ 29548 were effective in preventing this decrease in SAP. The pressure-rate index (PRI, mmHg beats/min/100) calculated as the product of SAP and HR is a crude measure of cardiac oxygen consumption. The PRI decreased in all the groups; the fall being significant ($p < 0.05$) in SQ 29548-treated group 80 min post-ligation of the coronary artery.

A steep, rapid rise in the left ventricular end diastolic pressure (LVEDP, mmHg) was observed following coronary artery ligation in the saline-, UK 38485- and SQ 29548-treated groups. In the saline-treated animals the sharp rise in LVEDP increased consistently with the duration of ischemia. The increase was significant at 20, 30 and 80 min post-ligation ($p < 0.01$) and highly significant ($p < 0.001$) at 40 min and 24 h post occlusion as compared to the baseline, pre-ligation value. Treatment with UK 38485 and SQ 29548 corrected the LVEDP to near pre-ligation values and affected a significant decrease in the values of this hemodynamic variable as compared to the values of the same time periods in the saline-treated group.

Table 2. Time course of changes in systolic arterial blood pressure, heart rate and pressure-rate product in saline, UK 38485- and SQ 29548-treated coronary artery ligated dogs

		Time after LAD ligation							
	Initial value before LAD	5 min	20 min	30 min	40 min	80 min		24 hrs	
Groups	ligation	ABS	ABS	ABS	ABS	ABS	%	ABS	%
1. Heart-rate (HR) (beats/min):									
Sal	165 ± 4.9	169 ± 5.0	171 ± 5.2	175 ± 5.24	176 ± 5.5	180 ± 5.2	9.4 ± 2.0	185 ± 5.6	12.4 ± 2.6*
UK	166 ± 3.9	168 ± 4.2	170 ± 4.5	165 ± 3.7	164 ± 4.3	164 ± 4.0	−0.6 ± 1.2	168 ± 3.5	11.4 ± 9.3
SQ	164 ± 6.8	170 ± 4.6	172 ± 5.1	150 ± 7.3	148 ± 7.2	149 ± 7.1	−9.2 ± 7.1	168 ± 5.9	2.5 ± 0.8#
2. Systolic arterial blood pressure (SAP) (mmHg):									
Sal	173 ± 10.0	164 ± 25.7	162 ± 25.4	157 ± 24.7	153 ± 24.4	151 ± 24.0	−12.7 ± 2.1*	150 ± 25	−13.9 ± 2.5*
UK	158 ± 8.7	154 ± 7.9	153 ± 7.8	151 ± 7.3	149 ± 8.0	149 ± 8.1	−5.0 ± 0.8	147 ± 8.1	−5.9 ± 1.0
SQ	157 ± 4.2	153 ± 4.2	152 ± 4.1	152 ± 5.0	152 ± 4.1	151 ± 4.0	−3.9 ± 0.6	151 ± 4.2	−3.8 ± 0.7
3. Pressure-rate Index (PRI) (mmHg beats/min/100):									
Sal	289 ± 22.3	288 ± 21.6	289 ± 22.2	286 ± 23	287 ± 24.2	286 ± 24.5	−1.2 ± 2.7	291 ± 29.5	0.2 ± 3.7
UK	259 ± 14.9	257 ± 12.8	260 ± 13.3	250 ± 11.1	244 ± 13.8	243 ± 13.3	−5.7 ± 1.3	246 ± 12.3	−4.5 ± 1.5
SQ	258 ± 12.1	262 ± 9.8	262 ± 10.3	229 ± 13.8	267 ± 13.5	226 ± 13.0	−12.9 ± 2.8*	255 ± 11.4	−1.1 ± 1.2

LAD = Left anterior descending coronary artery,
ABS = absolute value, % = percent change from initial value before coronary artery ligation (− = decrease; other % change in values = increase),
Sal = saline-treated group (n = 12); UK = UK 38485-treated group (n = 10); SQ = SQ 29548-treated group (n = 10); Administration of drug or saline.
20 minutes post-ligation of LAD coronary artery.
Values are given as mean ± SEM.
* = $p < 0.05$, as compared to the initial value before ligation.
# = $p < 0.05$, as compared to the value at the same time point in the saline-treated control group.

Table 3. Time course of changes in left ventricular dynamics in saline, UK 38485- and SQ 29548-treated coronary artery ligated dogs

		Time after LAD ligation							
	Initial value before LAD	5 min	20 min	30 min	40 min	80 min		24 hrs	
Groups	ligation	ABS	ABS	ABS	ABS	ABS	%	ABS	%
1. Left ventricular end diastolic pressure (LVEDP) (mm Hg):									
Sal	6 ± 0.3	8 ± 0.4	9 ± 0.4**	9 ± 0.4**	10 ± 0.5***	10 ± 0.5	44 ± 6.6**	12 ± 0.8	88 ± 12.8***
UK	6 ± 0.3	7 ± 0.3*	7 ± 0.3*	6 ± 0.2#	6 ± 0.2##	6 ± 0.2	6 ± 4.9##	6 ± 0.3	5 ± 6.5###
SQ	6 ± 0.4	8 ± 0.5*	7 ± 0.5*	6 ± 0.5#	6 ± 0.5##	7 ± 0.5	11 ± 3.6#	7 ± 0.5	20 ± 5.2*##
2. Maximal rate of rise of left ventricular pressure (LV dP/dt max) (mmHg/sec):									
Sal	2470 ± 101.6	2192 ± 88.8	2158 ± 86.0	2073 ± 69.7*	1933 ± 61.5*	1867 ± 78.4	−24 ± 1.5**	1575 ± 107.6	−37 ± 2.2**
UK	2557 ± 721.8	2268 ± 143.1	2050 ± 132.3**	2177 ± 118.3*	2170 ± 122.4*	2163 ± 132.4	−14 ± 3.4*	1968 ± 108.3	−22 ± 3.7**
SQ	2383 ± 698.0	2195 ± 118.2	2018 ± 103.2*	2210 ± 78.2	2164 ± 101.7*	2173 ± 112.3	−8 ± 4.2*#	2195 ± 123.5	−7 ± 4.5##

LAD = Left anterior descending coronary artery, OCCL = occlusion,
ABS = absolute value, % = percent change from initial value before coronary artery ligation (− = decrease; other % change in values = increase).
Sal = saline-treated group (n = 12), UK = UK 38485-treated group (n = 10),
SQ = SQ 29548-treated group (n = 10). Administration of drug or saline, 20 minutes post ligation of LAD coronary artery.
Values are mean ± SEM.
* = $p < 0.05$; ** = $p < 0.01$; *** = $p < 0.001$ as compared to the initial value before ligation.
# = $p < 0.05$; ## = $p < 0.01$ as compared to the value at the same time point in the saline-treated group.

Following coronary artery ligation a steady significant decrease was noted in the time course of response of the maximal rate of change of left ventricular pressure ($LVdP/dt_{max}$, mmHg/sec) in the saline-, UK 38485 and SQ 29548-treated groups. UK 38485-treatment did not exert any corrective effect on $LVdP/dt_{max}$. However, SQ 29548 treatment influenced $LVdP/dt_{max}$, which was significantly less depressed as compared to the saline-treated group at 80 min ($p < 0.05$) and 24 h ($p < 0.01$) after coronary artery ligation, as compared to the value of this variable at the same time point in the saline-treated control group.

DISCUSSION

The reasons for attempting to salvage myocardium threatened by ischemia during temporary or permanent occlusion of one or more coronary arteries is based on the observation that patient's survival depends largely on the extent of myocardial damage and the level of metabolic activity. The enormous complexity inherent in the deprivation of oxygen and substrates from the myocardium, accumulation of metabolic wastes and hemodynamic alterations during myocardial ischemia, and approaches addressing the attenuation of thromboxane A_2 synthesis and/or action during ischemia, are the focus of the present study.

Recent discoveries in the molecular biology of cardiac eicosanoids promise a new understanding of the pathobiology of myocardial infarction. Considerable interest centres on the possible role of TXA_2 release during myocardial ischemia (44). Hence, this discussion is confined to a consideration of the process of the acute phase of experimental myocardial infarction with emphasis on attenuation of TXA_2 release during ischemia in an attempt to reduce infarct size and to ameliorate the metabolic and functional consequences of the ischemic insult. TXA_2 release has been demonstrated during myocardial ischemia both in animals and humans (99–102). TXA_2 may cause or propagate myocardial ischemia by causing vasoconstriction and platelet aggregation (99,100). The actions of TXA_2 are countered by PGI_2. It has been proposed that an imbalance between TXA_2 and PGI_2 can cause cyclic flow reductions (103).

Selective TXA_2 synthetase inhibition has been demonstrated to have beneficial effects in animals with total and subtotal coronary occlusion. However, inhibition of TXA_2 synthesis leads to accumulation of precursor cyclic endoperoxides which can also stimulate TXA_2 platelet receptors and cause platelet aggregation. These cyclic endoperoxides are shunted to alternate metabolic pathways leading to formation of other eicosanoids like PGD_2 and $PGF_{2\alpha}$ which can also cause platelet aggregation and coronary vasoconstriction (39,40,104).

It has been evidenced that short and long term prognosis of patients with acute myocardial infarction mainly depends on the extent of infarcted myocardium and the resulting impairment of left ventricular function (6,105). Regional myocardial dysfunction is an inevitable consequence of acute myocardial infarction and the severity of dysfunction is determined by the extent of myocardial injury. Therefore, accurate estimation of infarct size is of major importance to assess clinical outcome. Furthermore, to determine the effectiveness of therapeutic strategies designed to salvage jeopardized myocardium, accurate determination of infarct size is of paramount value (105).

It has been reported (106) that acute and sudden interruption of blood flow for more than 15 to 20 minutes results in myocardial necrosis which progresses from the subendocardium toward the epicardium. In experimental animals, approximately 60% of the myocardium at risk undergoes necrosis when total cessation of flow continues for 3 hours and, following 6 hours of flow deprivation, 70% to 80% of the myocardium at risk undergoes irreversible damage (106). Similar to these reports, it was observed in the present study that irreversible injury within the risk region progressed transmurally with the duration of ischemia. There was no cell death at 20 minutes after coronary occlusion. Necrosis was limited to the subendocardial region in all dogs with 3 hours coronary artery occlusion. Prolongation of coronary occlusion beyond 6 hours duration in anesthetised dogs is associated with evolution of myocardial infarction evidenced by the development of tissue necrosis, rapid changes in myocardial metabolism and impairment in cardiac function. Similar findings have been reported earlier from our laboratory (95,107). Furthermore, the results of the present study are consistent with the concept that subendocardium is the region of the ventricular wall that is most vulunerable to ischemia and subsequent necrosis (108). The presence of subendocardial myocardial infarct involving 28–38% of the transmural myocardium has been reported (106), evidencing that after progressively longer periods of ischemia, greater proportion of the region at risk becomes infarcted. Thus a "wave front" (106) occurs, extending from the subendocardium to subepicardium, which is dependent on the duration of ischemia.

In the saline-treated control group in the present study, myocardial infarction was found to be transmural at 24 hours post-occlusion of coronary artery. In the canine myocardium, the final amount of myocardial necrosis is mainly dependent upon the size of the area at risk of the occluded coronary artery and upon the degree of collateral flow to the ischemic area. However, regional myocardial blood flow level was not measured in the present study. Furthermore, it is generally accepted that quantification of infarct size is only relevant when the size of the myocardium which is excluded from perfusion is also considered (109). This was done in the present study by calculating the ratios of the area of infarct to the area at risk. These ratios were significantly smaller in the treated groups compared to the controls. By definition, this indicated a decrease in infarct size due to the specific drugs employed in this study. Sham-operated controls with placement of the double ligature under the coronary artery using a small aneurysm needle, though evaluated, were not used for this study. Unpublished observations suggest that sham operated animals do not spontaneously develop occlusive thrombi, myocardial infarction or other functional impairment as a result of instrumentation.

Changes in cellular energy metabolism and other biochemical indices have long been regarded as sensitive markers of ischemic injury (110–112), and many studies have used the tissue content of high energy phosphates (HEPs) adenosine triphosphate (ATP), creatine phosphate (CP) and lactate in the assessment of tissue injury and protection (110–112).

Previous studies (1–7) indicate that CPK enzyme activity is a dynamic index of tissue necrosis. In the present study, following LAD coronary artery ligation, there was marked depletion in the tissue content of HEPs (data not shown here) and the activity of the myocardial enzymes creatine phosphokinase (CPK) and lactate dehydrogenase (LDH). This was accompanied by an increase in the myocardial lactate content. Similar findings have been reported by other workers (2,5). During ischemia, instead of aerobic respiration, anaerobic glycolysis becomes the significant source of production of adenosine triphosphate.

Initially the rate of anaerobic glycolysis in ischemia is high. Within 60–90 seconds, thereafter, anaerobic glycosis slows due to inhibition of glyceraldehyde-3-phosphate dehydrogenase by the high cytosolic reduced nicotinamide dinucleotide-nicotinamide adenine dinucleotide ratio. Moreover, anaerobic glycolysis is tightly coupled to deleterious lactate production. Increased intracellular accumulation of lactate and protons (H^+) as metabolic end products causes increased acidosis and exacerbate myocardial injury. Acidosis subsequently inhibits the glycolytic flux at the level of glyceraldehyde-3-phosphate and enzyme phosphofructokinase (1,110–112).

In the present study, following LAD coronary artery ligation, there was a gradual change in different hemodynamic variables, which intensified with time. A marked increase ($p < 0.05$) in heart rate (HR) was observed 24 hours after coronary artery ligation compared with the initial value before ligation. This was accompanied by a fall in systolic arterial blood pressure (SAP). These findings corroborate with those of other investigators (56,65) and previous studies from our laboratory (71,107). In the present study, there was marked depression in myocardial contractility and diastolic function as evidenced by a fall in left ventricular dP/dt in concert with a significant elevation of left ventricular end diastolic pressure (LVEDP); this ischemia-induced effect is well documented (113,114). The pressure-rate product which is related to the myocardial oxygen consumption was also depressed on account of the ischemic insult. Similar findings have been reported earlier (65–67).

The magnitude of reduction in myocardial infarct size by UK 38485 in the present study has not been reported earlier using animals with total coronary artery ligation. Further evidence of the protective effect of UK 38485 administration was obtained from metabolic studies. Treatment with UK 38485 resulted in a significant decrease in lactate content of the ischemic myocardium indicative of reduction in tissue acidosis. Similar results have been reported by other workers (61). In the present study, UK 38485 significantly preserved the myocardial CPK activity, which is a direct indication of its tissue protective and cardiometabolic potential. This is in corroboration with an earlier study which examined the effects of U 63557A, a selective TXA_2 synthetase inhibitor, in LAD coronary artery ligated rats (56). Myocardial LDH activity, the depletion of which depicts the profile of the later stages of progression of ischemia, was also restored in the ischemic zone of the myocardium by the administration of UK 38485. The preservation of the enzymatic activity in myocardial tissue is indicative of the cellular and subcellular membrane stabilizing properties of UK 38485. Furthermore, the hemodynamic changes induced by coronary occlusion were attenuated by UK 38485. There was a slight decrease in heart rate and blood pressure following administration of this drug, and LVEDP was significantly lowered. A similar finding has been reported earlier (113). The decreased LVEDP maintained the perfusion pressure despite of a fall in blood pressure, since perfusion pressure is a product of the difference between diastolic pressure in the stenosed coronary artery and LVEDP (114,115). Furthermore, this drug was effective enough to bring the value of dP/dt to a less depressed state as compared to the controls. These results indicate that the cardioprotective effect of UK 38485 evidenced by the significant salvage of the myocardial area at risk of infarction is not exerted through hemodynamic changes, but more on account of metabolic changes at the cellular level. This may encompass endoperoxide shunting towards the synthesis of PGD_2, PGE_2 and PGI_2, all of which, in turn, can inhibit neutrophil activation, and their accumulation in the ischemic heart. It is suggested (55) that this action of the thromboxane synthetase inhibitor UK 38485, may result in preventing neutrophil-mediated tissue injury.

Furthermore, previous studies with platelets indicate that stimulated platelets release PGH_2 in the presence of a thromboxane synthetase inhibitor, and that the vessel can use this source of PGH_2 to synthesize PGI_2. This transfer of PGH_2 has been termed "endoperoxide steal"; and as mentioned above, seems to be quantitatively significant only in the presence of thromboxane synthetase inhibition (48–50). This mechansim has been reported to reduce TXA_2 formation and increase PGI_2 formation (48–51). Thus, these drugs may inhibit the development of myocardial ischemia and damage (53–57,59,60).

Selective TXA_2 synthetase inhibition has been demonstrated to have beneficial effects in animals with total and subtotal coronary occlusion. However, TXA_2 synthesis inhibition leads to accumulation of precursor cyclic endoperoxides which can stimulate TXA_2 platelet receptors and cause platelet aggregation. These cyclic endoperoxides are also shunted to alternate metabolic pathways leading to formation of other eicosanoids like PGD_2 and $PGF_{2\alpha}$ which can also cause platelet aggregation and coronary vasoconstriction (63–67,85).

TXA_2 receptor antagonists, such as SQ 29548, offer unique advantages of inhibiting platelet aggregation in very small concentrations without affecting TXA_2, PGI_2 or PGD_2 formation. Several investigators have shown that TXA_2 receptor angatonists have a beneficial role in myocardial ischemia.

In the present study SQ 29548 reduced infarct size appreciably, leaving a large amount of viable tissue within the area at risk. This cardioprotective effect of SQ 29548 is consistent with previous reports (56,65–71). In these studies SQ 29548 was shown to reduce several indices of experimental myocardial ischemia in models of total coronary occlusion without reperfusion. The physiological mechanism for the mitigation of the ischemic injury with TXA_2 blockade via SQ 29548 is presently unknown. It may be related to alleviation of TXA_2—mediated vascular tone or inhibition of platelet aggregation.

In the present study SQ 29548 prevented the loss of myocardial enzyme CPK indicating an actual preservation of the ischemic area of the myocardium. This effect of SQ 29548 is consistent with previously reported findings using SQ 29548, other thromboxane antagonists or antagonist of vasoconstrictor eiosanoids (56,65–67).

In the present study following administration of SQ 29548 the heart rate was markedly reduced; whereas, there was no such effect on blood pressure (SAP). However, this agent was capable of blunting the acute hypotension associated with coronary ligation-induced tachycardia. The effect of SQ 29548 on myocardial O_2 consumption or demand is another potential mechanism of action of this compound. There was a significant ($p < 0.05$) decrease in the pressure-rate index, 80 min post occlusion as compared to the value of this variable at the same time period in the saline-treated animals; but this effect was not sustained 24 h post-occlusion of coronary artery. This effect of SQ 29548 on this derived hemodynamic variable, as a marker of cardiac oxygen demand, suggests that the beneficial effect of this drug involves a reduction in myocardial oxygen consumption. No alteration in this parameter has been reported by other workers (65,66). It has, however, been noted that the indirect indices of myocardial oxygen consumption should be interpreted with caution (116,117).

Furthermore, SQ 29548 significantly decreased LVEDP, prevented the fall in the value of the $LVdP/dt_{max}$ at 80 min post-occlusion i.e. 60 min after administration of the loading dose of the drug. This suggested that the beneficial effect of the drug manifests slowly in concert with metabolic corrections afforded by this agent. One of the marked and significant effects of SQ 29548 in our experiments was a reduction in LVEDP, accompanied by a slight increase in myocardial contractility. Reduction in LVEDP results in better myocardial flow through the inner layers of the myocardium. It has been shown that interventions that decrease cardiac size, LVEDP, and wall tension improve myocardial blood flow in the deeper (endocardial) regions (114,115). Thus, SQ 29548 prolonged the diastolic interval and maintained optimum perfusion pressure within the myocardial area at risk, thus improving the ultimate blood supply to the ischemic zone and affecting the myocardial infarct size.

The present study shows that SQ 29548 treatment during ischemia resulted in a significant protection of myocardial tissue and infarct size reduction. The mechanism of the protective effect appears to be related to antagonism of the cytolytic actions of TXA_2. Improved cardiac function may also contribute to its antiischemic action.

In conclusion, the present investigation elucidates the role of TXA_2 inhibitors in experimental myocardial infarction. The results of the study provide further support to the hypothesis that endogenously released thromboxane A_2 may play an active role in contributing to the pathophysiology of myocardial ischemia and tissue necrosis and mitigation of TXA_2—induced reactions in ischemic tissue seem to result in some degree of myocardial protection.

The present investigation further demonstrates the beneficial role of the drugs (that were employed in the present study) modulating the arachidonic acid metabolism in protecting the myocardium by limitation of infarct size and correction of the ischemia-induced deranged cardiometabolic correlates. This cardioprotection was achieved in concert with modest changes in myocardial oxygen supply-demand variables and other marked hemodynamic parameters.

Furthermore, thromboxane synthetase inhibition was found to be superior to thromboxane receptor blockade in terms of improved inotropy and salvage of post-ischemic myocardial injury. This may be on account of the substantially evidenced greater endoperoxide shunting to prostacyclin generation by thromboxane synthetase inhibition. This is indicative of the fact that the balance between thromboxane A_2 and prostacyclin is of considerable importance in the pathophysiology of myocardial ischemia/infarction (118).

These observations appear to furnish sufficient basis for further investigation on the effectiveness of TXA_2 inhibitors; that is thromboxane synthetase inhibitors and thromboxane A_2 / prostaglandin endoperoxide receptor blocking drugs in myocardial ischemia, other ischemic states and in clinical situations; with a view to further understanding ischemic heart disease and improving its therapy.

ACKNOWLEDGEMENTS

This study was supported by a research grant from the Indian Council of Medical Research (ICMR), New Delhi, India. The authors gratefully acknowledge the generous gifts of UK 38485 from Pfizer Research Laboratories, Sandwich, Kent, UK and SQ 29548 from Squibb Pharmaceuticals, New Jersey, USA.

REFERENCES

1. Opie LH. 1997. Oxygen lack: ischemia and angina. In: The Heart: Physiology, from Cell to Circulation, 515–541. New York: Lippincott-Raven Publishers.
2. Jennings RB, Murry CE, Steenbergen C Jr, Reimer KA. 1990. Acute myocardial ischemia: Development of cell injury in sustained ischemia. Circulation 82(II):II2–II12.
3. Reimer KA 1996. The slowing of ischemic energy demand in preconditioned myocardium. In: Myocardial Preservation, Preconditioning and Adaptation. Eds. DK Das, RM Engelman and KM Cherian, 13–26. New York: New York Academy of Sciences.
4. Piper HM. 1990. Irreversible ischemic injury, definition of the problem. In: Pathophysiology of Severe Ischemic Myocardial Injury. Ed. HM Piper, 3–14. London: Kluwer Academic Publishers.
5. Spieckermann PG. 1990. The critical ATP threshold hypothesis. In: Pathophysiology of Severe Ischemic Myocardial Injury. Ed. HM Piper, 27–39. London: Kluwer Academic Publishers.
6. Sobel BE, Bresnahan GF, Shell WE. 1972. Estimation of infarct size in man and its relation to prognosis. Circulation 46:640–648.
7. Thompson PL, Fletcher E, Katavatis V. 1979. Enzymatic indices of myocardial necrosis: Influence of short and long-term prognosis after myocardial infarction. Circulation 59:113–119.
8. Maroko PR, Kjekshus JK, Sobel BE, Watanabe T, Covell JW, Ross J Jr., Braunward E. 1971. Factors influencing infarct size following experimental coronary artery occlusions. Circulation 43:67–82.
9. Braunwald E, Sobel BE. 1984. Coronary blood flow and myocardial ischemia. In: Heart Disease. Ed. E Braunwald, 1235–1261. Philadelphia: WB Saunders Company.
10. Rude RE, Muller JE, Braunwald E. 1981. Efforts to limit the size of myocardial infarcts. Ann Int Med 95:736–761.
11. Lange LG, Sobel BE. 1982. Pharmacological salvage of myocardium. Annu Rev Pharmacol 22:115–143.
12. Pfeffer MA, Braunwald E. 1990. Ventricular remodeling after myocardial infarction. Circulation 81:1161–1172.
13. Jugdutt BI. 1993. Prevention of ventricular remodeling post myocardial infarction: Timing and duration of therapy. Can J Cardiol 9:103–114.
14. Karmazyn M, Dhalla NS. 1983. Physiological and pathological aspects of cardiac prostaglandins. Can J Physiol Pharmacol 61:1207–1225.

15. Moncada S, Vane JR. 1979. Pharmacology and endogenous roles of prostaglandin endoperoxides, thromboxane and prostacyclin. Pharmacol Rev 30:271–331.
16. Hamberg M, Svensson J, Samuelsson B. 1974. Prostaglandin endoperoxides: a new concept concerning the mode of action and release of prostaglandins. Proc Natl Acad Sci USA 71:3824–3828.
17. Moncada S, Gryglewski R, Bunting S. 1976. An enzyme isolated from arteries transforms prostaglandin endoperoxides to an unstable substance that inhibits platelet aggregation. Nature 263:663–665.
18. Moncada S, Vane JR. 1979. Arachidonic acid metabolites and the interactions between platelets and blood-vessel walls. N Engl J Med 300:1142–1147.
19. Vane JR, Botting RM. 1995. Prostaglandin profile of prostacyclin. Am J Cardiol 75:3A–10A.
20. Hamberg M, Svensson J, Samuelsson B. 1975. Thromboxanes: a new group of biologically active compounds derived from prostaglandin endoperoxides. Proc Natl Acad Sci USA 72:2994–3024.
21. Piper PJ. 1984. Formation and actions of leukotrienes. Physiol Rev 64:744–761.
22. Piper PJ, Sampson AP, Yaacob H bin, McLeod JM. 1990. Leukotrienes in the cardiovascular system. In: Advances in Prostaglandin, Thromboxane and Leukotriene Research. Eds. B Samuelsson, S-E Dahlen, J Fritsch and P Hedqvist, 146–152. New York: Raven Press.
23. Oates JA, FitzGerald GA, Branch RA, Jackson EK, Knapp HR, Roberts LJ. 1988. Clinical implications of prostaglandin and thromboxane A_2 formation (2 parts). N Engl J Med 319:689–698; 761–767.
24. Chien KR, Hau A, Sen A, Buja LM, Willerson JT. 1984. Accumulation of unesterified arachidonic acid in ischemic canine myocardium: relationship to a phosphatidylcholine deacylation reacylation cycle and depletion of membrane phospholipids. Circ Res 54:313–322.
25. Hirsch PD, Hillis LD, Campbell WB, Firth BG, Willerson JT. 1981. Release of prostaglandins and thromboxane into the coronary circulation in patients with ischemic heart disease. New Engl J Med 304:685–691.
26. Lorenz R. 1990. Antiplatelet therapy in unstable angina: rationale, effectiveness and dosage. In: Unstable Angina. Eds. W Bliefeld, CW Hamm and E Braunwald, 186–202. New York: Springer-Verlag.
27. Dennis EA. 1990. Modification of the arachidonic acid cascade through phospholipase A_2 dependent mechanisms. In: Advances in Prostaglandin, Thromboxane and Leukotriene Research. Eds. B Samuelsson, S-E Dahlen, J Fritsch and P Hedqvist, 217–223. New York: Raven Press.
28. Chiariello M, Ambrosio G, Cappelli-Bigazzi M, Perrone Filardi P, Tritto I, Nevota E, Golino P. 1990. Reduction in infarct size by the phospholipase inhibitor quinacrine in dogs with coronary artery occlusion. Am Heart J 120:801–807.
29. Vallee E, Gougat J, Navarro J, Delahayes JF. 1979. Anti-inflammatory and platelet antiaggregant activity of phospholipase A_2 inhibitors. J Pharm Pharmacol 31:588–592.
30. Pieper GM. 1990. Arachidonic acid causes postischemic dysfunction in control but not diabetic hearts. Am J Physiol 258:H923–H930.
31. Reilly RM, Schiller HJ, Bulkley GB. 1991. Pharmacologic approach to tissue injury mediated by free radicals and other reactive oxygen metabolites. Am J Surg 161:488–503.
32. Morrow JD, Roberts LJ. 1996. The isoprostanes. Current knowledge and directions for future research. Biochem Pharmacol 51:1–9.
33. Mehrabi MR, Ekmekcioglu C, Tatzber F, Oguogho A, Ullrich R, Morgan A, Tamaddon F, Grimm M, Glogar HD, Sinzinger H. 1999. The isoprostane, 8-epi-PGF 2 alpha, is accumulated in coronary arteries isolated from patients with coronary heart disease. Cardiovasc Res 43:492–499.
34. Kijima Y, Hashimura K, Matsu-ura Y, Kato Y, Yasuda T, Ueda T, Orita Y, Fukunaga M. 2001. Transcardiac 8-iso-prostaglandin $F_{2\alpha}$ generation from acute myocardial infarction heart: insight into abrupt reperfusion and oxidant stress. Prostaglandins Leukot Essen Fatty Acids 64:161–166.
35. Coker SJ, Ledingham I McA, Parratt JR, Zeitlin IJ. 1981. Aspirin inhibits the early myocardial release of thromboxane B_2 and ventricular ectopic activity following acute coronary artery occlusion in dogs. Br J Pharmacol 72:593–595.
36. Walinsky P, Smith JB, Lefer AM, Lebenthal M, Urban P, Greenspon A, Goldberg S. 1984. Thromboxane A_2 in acute myocardial infarction. Am Heart J 108:868–872.
37. Ashton J, Schmidtz JM, Campbell WB, Ogletree ML, Raheja S, Taylor AL, FitzGenald C, Buja LM, Willerson JT. 1986. Inhibition of cyclic flow variations in stenosed canine coronary arteries by thromboxane A_2/prostaglandin H_2 receptor agonists. Cir Res 59:568–578.
38. Tada M, Esumi K, Yamagishi M, Kuzuya T, Matsuda H. 1984. Reduction in prostacyclin synthesis as a possible cause of transient flow reduction in partially constricted canine coronary artery. J Mol Cell Cardiol 16:1137–1149.

39. Mehta JL, Roberts A. 1983. Human vascular tissues produce thromboxane as well as prostacyclin. Am J Physiol 244:R839–R844.
40. Mehta JL, Mehta P, Lopez LM, Ostrowski N, Aguila E. 1984. Platelet function and biosynthesis of prostacyclin and thromboxane A_2 in whole blood upon aspirin administration in man. J Am Coll Cardiol 4:806–811.
41. Tada M, Kuzuya T, Inoue M, Kodama K, Mishima M, Yumada M, Inui M, Abe H. 1981. Elevation of thromboxane B_2 levels in patients with classic and variant angina pectoris. Circulation 64:1107–1115.
42. FitzGerald DJ, Roy L, Catella F, FitzGerald GA. 1986. Platelet activation in unstable coronary disease. New Engl J Med 315:983–990.
43. Mehta J, Mehta P, Feldman RL. 1982. Severe intracoronary thromboxane release preceding acute artery occlusion. Prostaglandins Leukotrienes Med 8:599–607.
44. FitzGerald DJ, Wright F, FitzGerald GA. 1989. Increased thromboxane biosynthesis during coronary thrombolysis: evidence that platelet activation and thromboxane A_2 modulate the response to tissue-type plasminogen activator in vivo. Cir Res 65:83–94.
45. Mitchell JRA. 1983. Clinical aspects of the arachidonic acid-thromboxane pathway. Br Med Bull 39:289–295.
46. Needleman P, Raz A, Ferrendelli JA, Minkes M. 1977. Application of imidazole as a selective inhibitor of thromboxane synthetase in human platelets. Proc Natl Acad Sci USA 74:1716–1720.
47. Tai H, Lee N, Tai C. 1980. Inhibition of thromboxane synthesis and platelet aggregation by pyridine and its derivatives. In: Advances in Prostaglandin and Thromboxane Research. Eds. B Samuelsson, PW Ramwell and R Paoletti, 447–452. New York: Raven Press.
48. FitzGerald GA, Brash AR, Oates JA, Pederson AK. 1983. Endogenous prostacyclin biosynthesis and platelet function during selective inhibition of thromboxane synthetase in man. J Clin Invest 71:1336–1343.
49. Mehta JL, Mehta P, Lawson DL, Ostrowski N, Brigmon L. 1985. Influence of selective thromboxane synthetase blocker CGS-13080 on thromboxane and prostacyclin by leukocytes from platelet-derived endoperoxides. J Lab Clin Med 106:246–252.
50. Marcus AJ, Weksler BB, Jaffe EA, Broekman MJ. 1980. Synthesis of prostacyclin from platelet-derived endoperoxides by cultured human endothelial cells. J Clin Invest 66:979–986.
51. Defreyn G, Deckmyn H, Vermylen J. 1982. A thromboxane synthetase inhibitor reorients endoperoxide metabolism in whole blood towards prostacyclin and prostaglandin E_2. Thromb Res 26:389–400.
52. Mehta JL, Lawson DL, Mehta P. 1988. Modulation of human neutrophil superoxide production by selective thromboxane synthetase inhibitor U63, 557A. Life Sciences 43:923–928.
53. Burke SE, Lefer AM, Smith GM, Smith JB. 1983. Prevention of extension of ischemic damage following acute myocardial ischemia by dazoxiben, a new thromboxane synthetase inhibitor. Br J Clin Pharmacol 15:97s–101s.
54. Farber NE, Pieper GM, Gross GJ. 1988. Lack of involvement of thromboxane A_2 in post-ischemic recovery of stunned canine myocardium. Circulation 78:450–461.
55. Mullane KM, Fornabio D. 1988. Thromboxane synthetase inhibitors reduce infarct size by a platelet-dependent, aspirin-sensitive mechanism. Circ Res 62:668–678.
56. Nichols WW, Mehta J, Wargovich TJ, Franzini D, Lawson D. 1989. Reduced myocardial neutrophil accumulation and infarct size following thromboxane synthetase inhibitor or receptor antagonist. Angiology 40:209–221.
57. Wargovich TJ, Mehta J, Nichols WW, Ward MB, Lawson DL, Franzini D, Conti CR. 1987. Reduction in myocardial neutrophil accumulation and infarct size following administration of thromboxane inhibitor U63, 557A. Am Heart J 114:1078–1085.
58. Austin JC, Berrizbeitia LD, Schoen FJ, Kauffman RP, Hechtman HB, Cohn LH. 1988. Thromboxane synthetase inhibition reduces ventricular irritability after coronary occlusion and reperfusion. Am Heart J 115:505–509.
59. Hock CE, Phillips GR III, Lefer AM. 1985. Protective action of thromboxane synthetase inhibitor in preventing extension of infarct size in acute myocardial infarction. Prostaglandins Leukotrienes Med 17:339–346.
60. Hock CE, Lefer AM. 1986. CGS-12970, a thromboxane synthetase inhibitor, limits ischemic damage following coronary artery occlusion. Res Comm Pathol Pharmacol 52:285–294.
61. Hoeft A, Karob H, Bock J, Wolpers HG, Wober W, Hellige G. 1986. Preservation of myocardium in transient ischemia by the thromboxane synthetase inhibitor UK 38485. Res Exp 186:35–46.

62. Rebec MV, Skrinska VA. 1989. Pharmacology of UK 38485 (Dazmegral), a specific inhibitor of thromboxane A_2 synthetase. Prostaglandins Leukot Essen Fatty Acids 38:207–212.
63. Ogletree ML, Harris DN, Greenberg R, Haslanger MF, Nakane M. 1985. Pharmacological actions of SQ 29548, a novel selective thromboxane antagonist. J Pharmacol Exp Ther 234:435–441.
64. Darius H, Smith JB, Lefer AM. 1985. Beneficial effects of a new potent and specific thromboxane receptor antagonist (SQ 29548) in vitro and in vivo. J Pharmacol Exp Ther 235:274–281.
65. Brezinski ME, Yanagisawa A, Darius H, Lefer AM. 1985. Antiischemic actions of a new thromboxane receptor antagonist during acute myocardial ischemia in cats. Am Heart J 110:1161–1167.
66. Hock CE, Brezinski ME, Lefer AM. 1986. Anti-ischemic actions of a new thromboxane receptor antagonist SQ 29548 in acute myocardial ischemia. Eur J Pharmacol 122:213–219.
67. Brezinski ME, Yanagisawa A, Lefer AM. 1987. Cardioprotective actions of specific thromboxane receptor antagonist in acute myocardial ischemia. J Cardiovasc Pharmacol 9:65–71.
68. Grover GJ, Schumacher WA. 1988. Effect of thromboxane receptor antagonist SQ 29548 on myocardial infarct size in dogs. J Cardiovasc Pharmacol 11:29–35.
69. Thiemermann C, Ney P, Schror K. 1988. The thromboxane receptor antagonist daltrobin, protects the myocardium from ischemic injury resulting in suppression of leukocytosis. Eur J Pharmacol 155:57–67.
70. Osborne JA, Lefer AM. 1988. Cardioprotective actions of thromboxane receptor antagonism in ischemic atherosclerotic rabbits. Am J Physiol 255:H318–H324.
71. Singh Uma, Seth SD, Manchanda SC, Seth Sandeep. 1997. Protective actions of a thromboxane receptor antagonist SQ 29548 in the ischemic myocardium morphologic and hemodynamic effects. Prostaglandins Leukot Essen Fatty Acids 56:105–110.
72. Aiken JW, Shebuski RJ, Miller OV, Gorman RR. 1981. Endogenous prostacyclin contributes to the efficacy of a thromboxane synthetase inhibitor for preventing coronary artery thrombosis. J Pharmacol Exp Ther 219:299–308.
73. Golino P, Buja LM, Ashton JH, Kulkarni P, Taylor AL, Willerson JT. 1988. Effect of thromboxane and serotonin receptor antagonists on intracoronary platelet deposition in dogs with experimentally stenosed coronary arteries. Circulation 78:701–711.
74. Kondo K, Seo RM, Naka M, Kitagawa T, Wakitani K, Sakata M, Kira H, Okegawa T, Kawasaki A. 1989. Effects of ONO-3708, an antagonist of the thromboxane A_2/prostaglandin endoperoxide receptor, on platelet aggregation and thrombosis. Eur J Pharmacol 163:253–261.
75. Mickelson JK, Simpson PJ, Gallas MT, Lucchesi BR. 1987. Thromboxane synthetase inhibition with CGS 13080 improves coronary blood flow after streptokinase-induced thrombolysis. Am Heart J 113:1345–1352.
76. Shea MJ, Driscoll EM, Romson JL, Pitt B, Lucchesi BR. 1984. Effects of OKY-1581, a thromboxane synthetase inhibitor, on coronary thrombosis in the conscious dog. Eur J Pharmacol 105:285–291.
77. FitzGerald DJ, Doran J, Jackson E, FitzGerald GA. 1986. Coronary vascular occlusion mediated via thromboxane A_2-prostaglandin endoperoxide receptor activation in vivo. J Clin Invest 77:496–502.
78. Coker SJ, Parratt JR. 1985. AH-23848, a thromboxane antagonist, suppresses ischaemia and reperfusion-induced arrhythmias in anaesthetized greyhounds. Br J of Pharmacol 86:259–264.
79. Wainwright C, Parratt J. 1987. Antiarrhythmic effects of the thromboxane antagonist BM 13.177. Eur J Pharmacol 133:257–264.
80. Mehta JL, Nichols WW, Schofield R, Donnelly WH, Chandan VK. 1990. TXA2 inhibition and ischemia-induced loss of myocardial function and reactive hyperemia. Am J Physiol 25:H1402–H1408.
81. Kramer J, Davis A, Dean R McCluskey ER, Needleman P, Corr PB. 1985. Thromboxane A_2 does not contribute to arrhythmogenesis during evolving myocardial infarction. J Cardiovasc Pharmacol 7:1069–1076.
82. Tada M, Hoshida S, Kuzuya Y, Inoue M, Minamino T, Abe H. 1985. Augmented thromboxane A_2 generation and efficacy of its blockade in acute myocardial infarction. Int J Cardiol 8:301–312.
83. Thaulow E, Dale J, Myhre E. 1984. Effects of a selective thromboxane synthetase inhibitor, dazoxiben, and of acetylsalicylic acid in myocardial ischemia in patients with coronary artery diseases. Am J Cardiol 53:1255–1258.
84. Fitzgerald GA, Reilly IAG, Pedersen AK. 1985. The biochemical pharmacology of thromboxane synthase inhibition in man. Circulation 72:1194–1201.
85. Saussy DL, Mais DE, Knapp DR, Halushka PV. 1985. Thromboxane A_2 and prostaglandin endoperoxide receptors in platelets and vascular smooth muscle. Circulation 72:1202–1207.

86. Gresele P, Deckmyn H, Nenci GG, Vermylen J. 1991. Thromboxane synthase inhibitors, thromboxane receptor antagonists and dual blockers in thrombotic disorders. Trends Pharmacol Sci 12:158–163.
87. Quest DW, Wilson TW. 1998. Effects of ridogrel, a thromboxane synthetase inhibitor and receptor antagonist, on blood pressure in spontaneously hypertensive rat. Jpn J Pharmacol 78:479–486.
88. De Clerck F, Beetens J, de Chaffoy de Courcelles F, Freyne E, Janssen PAJ. 1989. R68070: Thromboxane A_2 synthase inhibition and thromboxane A_2/prostaglandin endoperoxide receptor blockade combined in one molecule I. Biochemical profile in vitro. Thromb Haemostatis 61:35–42.
89. Rolin S, Dogne JM, Michaux C, Delarge J, Masereel B. 2001. Activity of a novel dual thromboxane A_2 receptor antagonist and thromboxane synthase inhibitor (BM-573) on platelet function and isolated smooth muscle. Prostaglandins Leukot Essen Fatty Acids 65:67–72.
90. Harris AS. 1950. Delayed development of ventricular ectopic rhythms following experimental coronary oclusion. Circulation 1:1318–1328.
91. Merillon JP, Morgant C, Zygelman M, Abadie E, Charlier P, Gourgon R. 1980. Comparison of the hemodynamic and coronary effects of three vasodilator drugs: dipyridamole, nifedipine and glyceryl nitrate. In: 4th International Adalat Symposium. Eds. P Puech and R. Krebs, 41–54. Amsterdam: Excerpta Medica.
92. Lie JT, Pairolero PC, Holley KE. 1975. Macroscopic enzyme mapping verification of large, homogenous experimental myocardial infarcts of predictable size and location in dogs. J Thorac Cardiovasc Surg 69:599–605.
93. Gaweher K, Bergmeyer HU. 1974. Lactate determination. In: Methods of Enzymatic Analysis. Ed. HU Bergmeyer, 1492–1495. New York: Academic Press.
94. Gupta MP, Seth SD, Manchanda SC, Singh Uma. 1987. Effect of nifedipine and hyaluronidase alone and in combination on myocardial preservation in experimental myocardial infarction—A morphological and biochemical profile. In: Myocardial Ischemia. Eds. NS Dhalla, IR Innes and RE Beamish, 311–323. Boston: Martinus Nijhoff Publishing.
95. Seth SD, Gupta MP, Manchanda SC, Singh Uma. 1988. Effect of oxyfedrine on cardiac metabolism and infarct size in experimental myocardial infarction. Curr Ther Res 44:278–286.
96. Foster G, Bernt E, Bergmeyer HU. 1974. Creatine kinase determination with creatine phosphate as substrate. In: Methods of Enzymatic Analysis. Ed. HU Bergmeyer, 789–793.
97. Bergmeyer HU, Bernt E. 1974. Lactate dehydrogenase, UV assay with pyruvate and NADH. In: Methods of Enzymatic Analysis. Ed. HU Bergmeyer, 574–579. New York: Academic Press.
98. Lowry OH, Rosenbrough NJ, Farr AL, Randall RJ. 1951. Protein measurement with Folin phenol reagent. J Biol Chem 193:265–372.
99. Evers AS, Murphee S, Saffitz JE, Jakschik BA, Needleman P. 1985. Effects of endogenously produced leukotrienes, thromboxanes, and prostaglandins on coronary vascular resistance in rabbit myocardial infarction. J Clin Invest 75:992–999.
100. Schmitz JM, Apprill PG, Buja LM, Willerson JT, Campbell WB. 1985. Vascular prostaglandin and thromboxane production in a canine model of myocardial ischemia. Cir Res 57:223–231.
101. Lewy RI, Wiener L, Walinsky P, Lefer AM, Silver MJ, Smith JB. 1980. Thromboxane release during pacing induced angina pectoris: Possible vasoconstrictor influence on the coronary vasculature. Circulation 61:1165–1177.
102. Hirsh PD, Hillis LD, Campbell WB, Firth BG, Willerson JT. 1981. Release of prostaglandins and thromboxane into the coronary circulation in patients with ischemic heart disease. N Engl J Med 304:685–691.
103. Mehta J, Mehta P, Horalek C. 1983. The significance of platelet- vessel wall prostaglandin equilibrium during exercise-induced stress. Am Heart J 105:895–900.
104. Bertele V, Falanga A, Tomasiak M, Chiabrando C, Cerletti C, deGastano G. 1984. Pharmacologic inhibition of thromboxane synthetase and platelet aggregation: modulating role of cyclooxygenase products. Blood 63:1460–1466.
105. Holman ER, van Jonbergen H-PW, van Dijkman PRM, van der Laarse A, deRoos A, van der Wall EE. 1993. Comparison of magnetic resonance imaging studies with enzymatic indexes of myocardial necrosis for quantification of myocardial infarct size. Am J Cardiol 71:1036–1040.
106. Reimer KA, Jennings RB. 1979. The "wavefront phenomenon" of myocardial ischemic cell death: II. Transmural progression of necrosis within the framework of ischemic bed size (myocardium at risk) and collateral flow. Lab Invest 40:633–644.
107. Seth SD, Gupta MP, Manchanda SC, Singh Uma. 1987. Cardioprotective effects of oxyfedrine in experimental myocardial infarction: A morphologic and hemodynamic study. Curr Ther Res 42:846–856.

108. Forman R, Cho S, Factor SM, Kirk ES. 1983. Acute myocardial infarct extension into a previously preserved subendocardial region at risk in dogs and patients. Circulation 67:117–124.
109. Bishop PB. 1984. How well can we measure coronary flow, risk zones and infarct size? In: Therapeutic Approaches to Myocardial Infarct Size Limitation. Eds. J Hearse and DM Yellon, 139–148. New York: Raven Press.
110. Neely JR, Feuvray D. 1981. Metabolic products and myocardial ischemia. Am J Pathol 102:282–291.
111. Opie LH. 1990. Myocardial ischemia—metabolic pathways and implications of increased glycolysis. Cardiovasc Drug Ther 4:777–790.
112. Cross HR, Clarke K, Opie LH, Radda GK. 1995. Is lactate—induced ischemic injury due to decreased lactate / H^+ co-transport or inhibition of glycolysis? J Mol Cell Cardiol 27:1369–1381.
113. Coker SJ. 1984. Further evidence that thromboxane exacerbates arrhythmias: Effects of UK 38485 during coronary artery occlusion and reperfusion in anesthetized greyhounds. J Mol Cell Cardiol 16:633–641.
114. Lee SJK, Sung YK, Zaragoza AJ. 1970. Effects of nitroglycerin on left ventricular volume and wall tension in patients with ischaemic heart disease. Br Heart J 32:790–794.
115. Moore EM, Parratt JR. 1972. Effects of oxyfedrine on local myocardial blood flow, myocardial heat production, contractility and wall tension. In: Wirkugsweise von oxyfedrine. Eds. E Gerlach and K Moser, 181–191. Stuttgart, New York: Schattauer Verlage.
116. Grover GJ, Parham CS. 1989. Role of thromboxane A2 in the control of myocardial O_2 supply consumption balance and severity of ischemia during pacing induced ischemia. Cr Res 64:575–582.
117. Baller D, Bretschneider HJ, Hellige G. 1981. A critical look at currently used indices of myocardial oxygen consumption. Basic Res Cardiol 76:181–189.
118. Golino P, Ambrosio G, Villari B, Ragni M, Focaccia A, Pace L, de Clerk F, Condorlli M, Chiariello M. 1993. Endogenous prostaglandin endoperoxides may alter infarct size in presence of thromboxane synthase inhibition: studies in a rabbit model of coronary artery occlusion—reperfusion. J Am Coll Cardiol 21:493–501.

IV. Novel Therapeutic Approaches to Cardiovascular Disease

Pathophysiology of Cardiovascular Disease,
edited by N.S. Dhalla, H. Rupp,
A. Angel & G.N. Pierce
Kluwer Academic Publishers, Boston, 2004

Receptors and Channels of Nuclear Envelope Membranes as a New Target for Drug Action

Ghassan Bkaily, Danielle Jacques,[1] and Pedro D'orléans Juste

Department of Anatomy and Cell Biology,
[1] Department of Pharmacology, Faculty of Medicine, Université de Sherbrooke, Sherbrooke, Quebec, Canada J1H 5N4

Summary. The use of confocal microscopy and fluorescent dyes permitted us to show that the nucleus plays an important role in excitation-contraction and secretion coupling of several excitable and non excitable cells. In addition, our group suggested that, like the sarcolemmal membrane, the nuclear envelope membrane possesses receptors and channels which may play an important role in modulating Ca^{2+} influx and cytosolic Ca^{2+} buffering capacity of the nucleus.

Our results showed that the nucleus plays a role as a cytosolic Ca^{2+} buffer during spontaneous contraction of heart cells as well as during stimulation of cell surface membrane receptors of vascular endothelial and smooth muscle cells.

Using isolated nuclei of cardiomyocytes, vascular endothelial and smooth muscle cells, our results showed that receptors and channels are present at the nuclear envelope membranes. Our results also showed that receptors, such as Ang II and ET-1 receptors, undergo internalization and nuclear translocation and that the nucleus could be implicated in protein synthesis of these receptors. These results demonstrate that like the sarcolemmal membrane, the nuclear envelope membranes possesses channels and G-protein-coupled receptors that could be a new target for drug action.

Key words: Nucleus, Ca^{2+}, R-type Ca^{2+} channel, receptors, heart, vascular endothelial cells, smooth muscle cells, Ang II, ET-1, bradykinin.

Corresponding Author: Dr G. Bkaily, Dept. of Anatomy & Cell Biology, Faculty of Medicine—Univ. de Sherbrooke, 3001—12th Avenue North, Sherbrooke, Qc, Canada J1H 5N4. Tel: (819) 564-5303, Tel: (819) 820-6807, E-mail: gbkail01@courrier.usherb.ca.

INTRODUCTION

Drug development has been focused on regulation of cytosolic calcium level during excitation-contraction and secretion coupling by targeting sarcolemmal ion channels, exchangers, pumps and receptors. However, little attention was given to possible contribution of the nucleus in controlling agonist-induced increase of cytosolic Ca^{2+} (1). In addition growing evidence showed that nuclear Ca^{2+} levels constitute an important signal regulating many vital events in a cell such as cell growth (2), gene expression (3,4) activation of nuclear kinases and phosphatases (5) as well as DNA repair, break-down of nuclear envelope (6) and apoptosis (7–9). Although several studies have made use of isolated nuclei and nuclear membrane vesicles, these studies did not take into consideration the crosstalk between the nucleus and the content of the surrounding cytosolic milieu as well as the possible presence of receptors and channels at the nuclear membranes levels. In addition, most of the work using isolated nuclei was focused on nuclear pore transport of proteins and macromolecules (10) despite the few reports suggesting the presence of Ang II (11) and ET-1 (12) receptors in the nuclei of liver cells. It wasn't until the recent development of ionic probes, mainly Fluo-3 Ca^{2+} dye and fluorescent specific antibodies and peptides directed against specific receptors coupled to the use of 3-D confocal microscopy, that it was possible to rediscover the nucleus in living working cells. These techniques allow us to begin to realize the possible contribution of the nucleus not only as a locus for DNA synthesis and cell differentiation, but also as an important organelle that behaves as a cell within a cell, thus regulating its own ionic homeostasis as well as contributing to maintain cytosolic Ca^{2+} and Na^{+} homeostasis (13–17). Thus, the recent use of confocal microscopy in working cells allowed scientists to revisit the nucleus and visualize as well as quantify several mechanisms and pharmacological events occurring through the sarcolemma and the nuclear membrane. Figure 1 summarizes these findings (1,13–28). The receptors for several agonists such as those for PDGF (29), Ang II (14,30), vasopressin (30,31), bradykinin (15), PGE_2 (26) as well as ET-1, PAF and IL-1 (15) have all been shown to be present in the nuclear membranes and may directly modulate Ca^{2+} influx through the steady-state R-type Ca^{2+} channel (1,13–17) or via an IP_3 calcium pathway (27,32).

In the present study, we attempt to emphasize the nuclear contribution to excitation-contraction and secretion coupling. We further demonstrate that sarcolemma receptor-mediated calcium signals are transferred to the nucleus via the cytosol and that receptors as well as channels exist in the nuclear membranes and do participate in mediating signaling of an agonist. Therefore, nuclear membranes receptors blockade could be a new target for drug development research.

MATERIALS AND METHODS

All methods used for cell isolation and culture, confocal microscopy settings, Fluo-3 loading, Syto 11 staining, ligands fluorescent probe and analysis, solutions and sources of all compounds (where it is not indicated) as well as statistical analysis are the same as those reported recently (1,14–17).

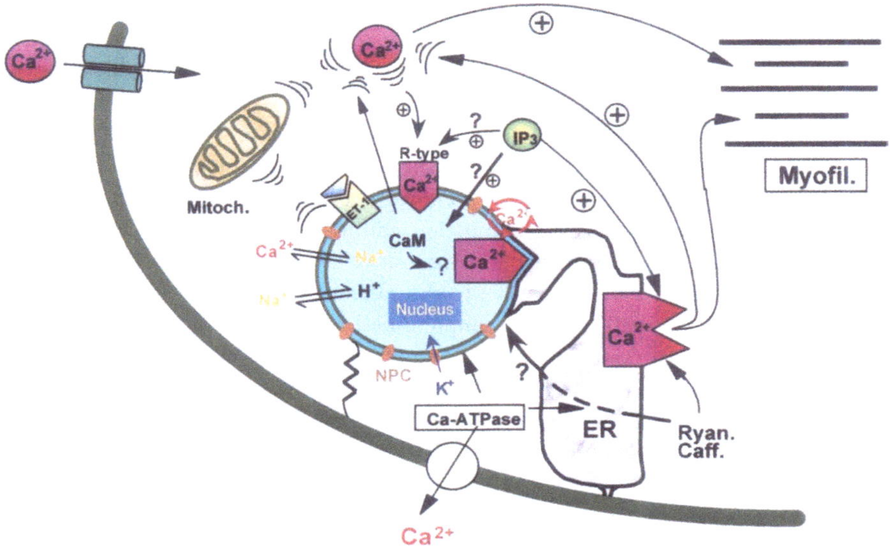

Figure 1. Diagrammatic representation of various possible components modulating Ca^{2+} waves. NPC is nuclear pore complex, CaM is calmodulin, r is ryanodine and caffeine-sensitive pools (from ref. 15).

For immunofluorescence studies the cells were fixed for 10 min in ice cold 4% paraformaldehyde. After washing with phosphate buffer solution (PBS), cells were incubated for 10 min with PBS containing sodium borohydride (2 mg/ml), then permeabilized and blocked with 0.1% Triton X-100, 7% normal goat serum (NGS) and 5% nonfat dry milk (NFDM) for 30 min. Finally, the cells were washed twice in PBS and incubated overnight at 4°C with rabbit anti-AT_1 and goat anti-AT_2 primary antibodies (Research Diagnostics Inc., Flanders, NJ, U.S.A.) in PBS containing 1.4% NGS, 1% NFDM and 0.1% Triton X-100. After two PBS washes, the cells were incubated for 1 hour at room temperature with rabbit and goat anti-IgG-FITC secondary antibodies (SIGMA-ALDRICH, Oakville, ON, Canada) for AT_1 and AT_2 immuno-fluorescence, respectively. The cells were examined using both a Nikon Eclipse 1000 fluorescence microscope and 3-D imaging confocal microscopy. The PBS is composed of 137 mM NaCl, 5.4 mM KCl, 1.0 mM $MgCl_2$, 1.5 mM $CaCl_2$, 5.5 mM D-glucose, 0.01 mM ethylenediamine-tetraacetic acid (EDTA) and 20.0 mM HEPES.

RESULTS

1. **Presence and implication of the steady-state voltage-dependent R-type Ca^{2+} channel in regulation of cytosolic Ca^{2+} by the nucleus during excitation contraction coupling;** recent work, using a fluorescent antagonist of the R-type Ca^{2+} and confocal microscopy techniques, showed the presence of this

type of Ca^{2+} channel at both the sarcolemmal and nuclear membranes of heart as well as human vascular endothelial and smooth muscle cells (1). Figure 2 shows fluorescent labeling of the R-type Ca^{2+} channel in human aortic vascular smooth muscle cells. However, using a fluorescent specific L-type calcium channel blocker did only show presence of this type of channel at the sarcolemma level. As can be seen in this figure (Figure 2), the R-type calcium channel is distributed at both the sarcolemmal and nuclear membranes in a cluster like fashion. Using isolated nuclei from human aortic vascular smooth muscles, administration of ET-1 induced an increase of nuclear free calcium that was insensitive to the specific L-type calcium channel blocker, nifedipine, but sensitive to the calcium chelator EGTA and R-type calcium channel blocker (Bkaily et al., unpublished results). Also, using sarcolemma perforated single human vascular smooth muscle cells, cytosolic administration of ET-1 induced increase of nuclear calcium that was blocked by chelating cytosolic calcium using EGTA (14). These results demonstrate that calcium influx through the nuclear envelope membranes is taking place via the opening of the steady-state R-type calcium channel.

Recent work using ion-sensitive dyes and confocal microscopy suggested possible crosstalk between cytosolic free calcium ($[Ca]_c$) and the nucleus (1,7,13–15,19,25). Using fluo-3 confocal microscopy and Syto 11 nuclear staining techniques (13) in cultured spontaneously contracting single cells from 10-day-old embryonic chick hearts, we were able to follow the simultaneous changes of Ca^{2+} entry through the sarcolemma and its propagation through the cytosol as well as its uptake by the nucleus. As can be seen in Figure 3, upon depolarization of the cell membrane, Ca^{2+} enters through the sarcolemma at a fixed region and then this calcium spreads into the cytosol and it is then taken by the nucleus. As can also be seen, relaxation takes place with decrease of the cytosolic Ca^{2+}, probably by calcium outflow through the sarcolemma Ca^{2+} pumps as well as the Na^+-Ca^{2+} exchanger. The decrease of cytosolic calcium is followed immediately by a slow release of the Ca^{2+} that was buffered by the nucleus during the period of cytosolic Ca^{2+} increase. During excitation-contraction coupling, the pattern of increase of transient cytosolic Ca^{2+} follows a fast increase of cytosolic and nuclear free calcium. This fast period of Ca^{2+} increase was followed by a slower period of decrease of $[Ca]_c$ and $[Ca]_n$ that characterizes relaxation of the cardiomyocyte. Blockade of Ca^{2+} entry through the sarcolemma membrane by L-type Ca^{2+} channel antagonists completely blocked the increase of cytosolic and nuclear Ca^{2+} (13,14). Several hormones that act at their sarcolemmal respective receptors, such as Ang II, ET-1, epinephrine and bradykinin, affect the kinetics of excitation-contraction coupling as well as the level of $[Ca]_c$ and $[Ca]_n$ (14,15).

Presence of functional G-protein coupled receptors at the nuclear membranes; recent development of ligands coupled to the fluorescent dyes permitted fluorescent labeling of several types of receptors in working single cells (14). Moreover, using specific receptor antibodies coupled to fluorescent probes as well as 3-D confocal microscopy allowed the visualization and quantification of ligands and receptors all through the single cells. As can be seen in Figure 4, simultaneous triple

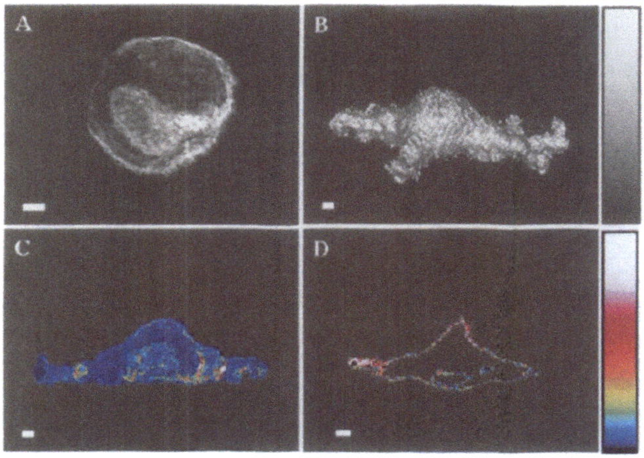

Figure 2. Single cultured VSM cells from bovine aorta labeled with the fluorescent R-type calcium channel probe. Three-dimensional reconstructions were generated from 37 serial optical sections (0.18 μm step-size) performed 20 minutes after addition of R-type probe, as described in the Materials and Methods. Panel (A) shows a rounded cell 30 minutes after attachment to the poly-L-lysine coated cover slip using depth-coded surface extraction. The cell remains rounded with only the nucleic and surface membranes respectively labeled. Panel (B) shows a cell 1 hour after attachment to the poly-L-lysine coated coverslip; the cell is now elongated with cytoplasmic projections extending over the surface of the coverslip. This reconstruction was performed using surface shading and emphasizes surface topography. Panel (C) shows a pseudocolored lateral cross-section of the cell in panel C. The middle portion of the cell has been sectioned exposing the nucleus, with the remainder of the cell intact. As with endothelial cells, labeling is associated with the outer cell membrane and the nucleus. As can be seen in Panel (D), a cross-section of the entire cell demonstrates that labeling is exclusively associated with the nucleus and plasma membrane. Surface extraction was used to visualize labeling associated with membrane surfaces. All images were Gaussian-filtered. Linear fluorescence intensity scale from 0 to 255 are shown at the right for both black and white and pseudocolored images (Bkaily et al. unpublished results).

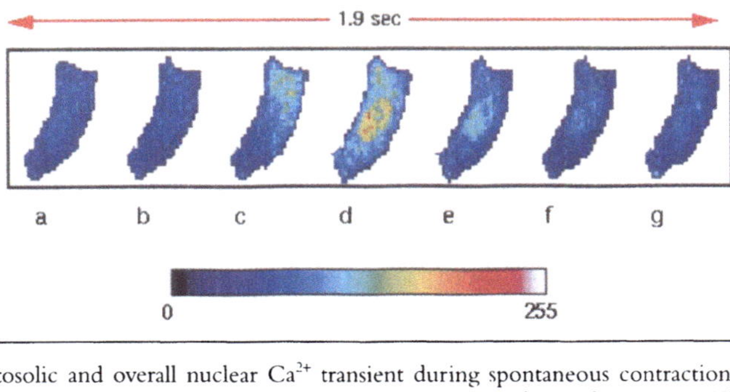

Figure 3. Cytosolic and overall nuclear Ca^{2+} transient during spontaneous contraction of single heart cell. Panels (a–g) are of spontaneously contracting human ventricular cardiomyocyte. The cell was scanned continuously at 3.5 second intervals. A, b, f and g are of quiescent periods, as witnessed by the low intensity of the stain. Panel C shows cytosolic Ca^{2+} increase and panel d shows spreading of the increase of cytosolic Ca^{2+} and its uptake by the nucleus. Panel e, shows decreasing of nuclear Ca^{2+} during relaxation of the single cell. Images are shown as pseudocolored representations according to intensity of fluorescence scale from 0–255 nm or calcium calibration scale from 0 to 40 μM.

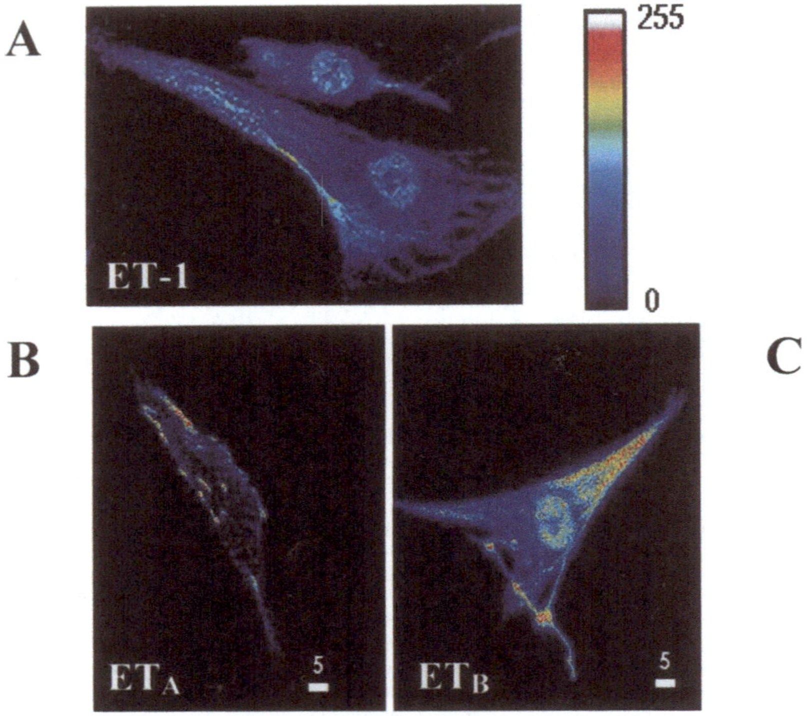

Figure 4. Fluorescence labeling of ET-1, ET_A and ET_B receptors in human aortic vascular smooth muscle cells. A: Labeling of ET-1. B: labeling of ET_A receptors. C: Labeling of ET_B receptors. Images are shown as pseudocolored representations according to an intensity of fluorescence scale from 0 to 255. White bar is in μm. Panels A, B and C are different cells.

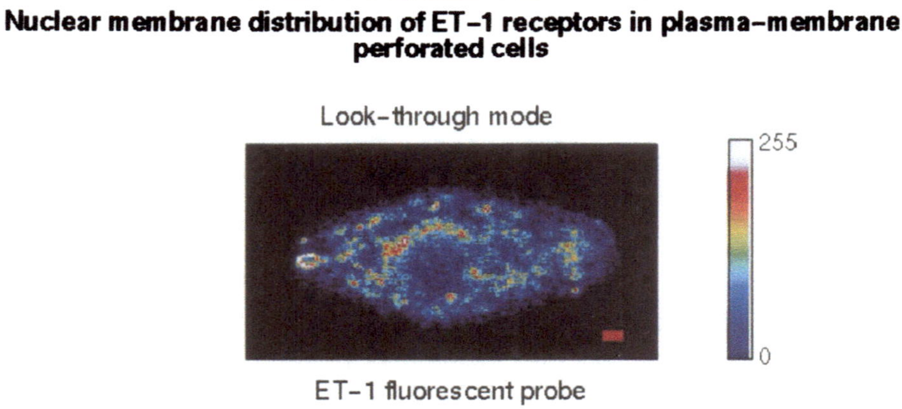

Figure 5. Three-dimensional reconstruction of ET-1 receptor distribution in sarcolemma-perforated chick heart cells. Use of the fluorescent ET-1 probe revealed the presence of ET-1 receptors in the nuclear envelope membrane (E). The colour scale represents the pseudocolor fluorescence intensity level (scale 0–255). The scale bar is in μm (modified from ref. 12).

labeling of ET-1, ET_A and ET_B receptors showed clearly that a ligand such as ET-1 is also present in the nuclear envelope space and within the nucleus. The ET_A receptors seem to be present only at the sarcolemma and in the cytosol. However, ET_B receptors seem to be present all through the cell including at the nuclear membranes, in the nuclear envelope space and in the nucleopasm. These results clearly show that some types of receptors seem to be mainly localized at the sarcolemma membrane; however, some other types of receptors could be also present at the nuclear envelope membranes. These results also suggest that a hormone as well its receptor could be present within the nucleus. These ligands and their respective receptors could also be synthesized within the nucleus (18). The presence of a specific type of receptor at the nuclear membranes may suggest that these receptors may indirectly regulate Ca^{2+} entry through the nuclear envelope and/or release of Ca^{2+} from the nucleoplasm (14). The presence of hormones such as ET-1 in the nucleoplasm and in the perinuclear plasma level may suggest that the peptides are able to activate their receptors at inner and outer membranes of the nucleus.

Recent work from our laboratory showed that cytosolic administration of ligands such as ET-1, directly activates ET-1 receptors (most probably the ET_B receptor) at the nuclear membrane level and induces an increase of nuclear Ca^{2+} (Figure 5) that was insensitive to known L-type Ca^{2+} channel blockers, but sensitive to cytosolic application of R-type Ca^{2+} channel antagonists and the calcium chelator, EGTA (9, 14, 17, Bkaily et al., unpublished results). This information suggests that G-protein-coupled receptors are present and functional at the nuclear envelope membranes. The stimulation of these receptors modulates nuclear Ca^{2+} entry through nuclear membranes. The stimulation of these receptors seems to regulate the minimum and maximum Ca^{2+} buffering capacity of the nucleus (9,13–15), thus preventing nuclear Ca^{2+} overload and cell death (8,9). As can be seen in Figure 6, once the sarcolemmal membrane of human aortic vascular smooth muscle cells was perforated using ionomycin or triton (9,13–15,17), cytosolic free Ca^{2+} concentration gradually changed from 100 to 2000 nM following the administration of ET-1 (100 nM) in the cytosol. Sequential addition of increasing concentrations of extranuclear free Ca^{2+} induced sequential increase of nuclear free Ca^{2+}. This increase started at 200 nM of cytosolic free calcium. Nuclear uptake of cytosolic calcium reached a steady-state at a cytosolic Ca^{2+} level near 1600 nM. In presence of 1600 or 2000 nM extranuclear free Ca^{2+}, addition of extranuclear ET-1 (100 nM) did not induce further increase in $[Ca]_n$. However, in presence of 100 nM cytosolic Ca^{2+}, superfusion with 100 nM ET-1 induced a significant increase of $[Ca]_n$. In presence of cytosolic ET-1, superfusion with sequential increasing concentrations of extranuclear free Ca^{2+} from 200 to 2000 nM did not further increase ET-1 induced sustained elevation of $[Ca]_n$ (Figure 6).

DISCUSSION

Using fluorescence confocal microscopy and fluo-3 Ca^{2+} measurements as well as ligand-fluorescent probes, this study, together with other recently published reports

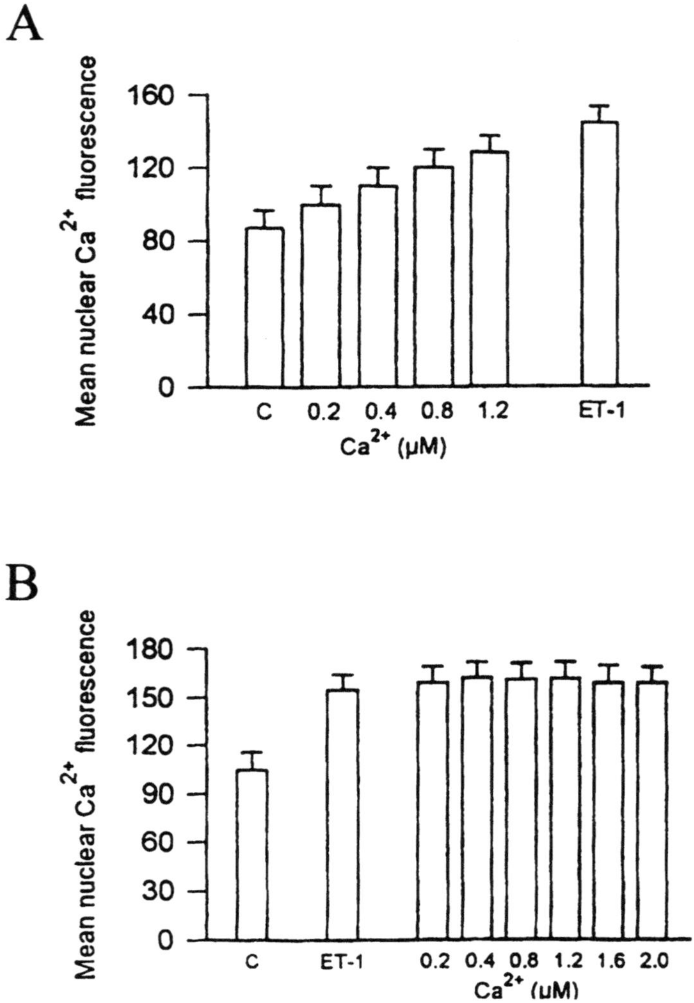

Figure 6. (A, B) Effect of cytosolic ET-1 on 3D measurements of intra-nuclear free Ca^{2+} fluorescence intensity of plasma membrane perforated human aortic vascular smooth muscle cells before and after sequential addition of increasing concentrations of Ca^{2+} (200–2000 nM) in the extranuclear medium. Mean fluorescence intensity values were calculated from the nuclei of 3D-reconstructed cells as described in Bkaily et al. (1996, 1997a). As can be seen, a gradual increase of free Ca^{2+}-fluorescence intensity is observed in nuclei of 3D-reconstructed cells. Steady-state increase of nuclear Ca^{2+} was achieved at 1200 nM $[Ca]_c$ and addition of (10^{-7} M) ET-1 to the cytosolic medium further increased $[Ca]_n$. (B) in the presence of 100 nM $[Ca]_c$, superfusion of the cytosol with 10^{-7} M ET-1 increased $[Ca]_n$ and no further increase could be observed when cytosolic Ca^{2+} was increased up to 2000 nM. The results are expressed in terms of absolute Ca^{2+} fluorescence intensity and C is control nuclear free Ca^{2+} in presence of zero cytosolic Ca^{2+} that did not change upon superfusion with 100 nM $[Ca]_c$. Values are expressed as means ± SEM. The number of experiments in A was 4 and in B, 14. Measurements were taken when the effect reached a steady-state level (modified from ref. 15).

by our group, clearly demonstrates the presence and contribution of the nifedipine-insensitive, isradipine-sensitive steady-state voltage dependent R-type Ca^{2+} channel to regulation of calcium transport through the surface cell membrane and nuclear membranes of excitable and non excitable cell types, including heart, vascular endothelial and smooth muscle cells of human and animal origin. In addition, our results described here and those recently reported by several laboratories, including ours, demonstrate that the nucleus plays an important role in excitation-contraction and secretion coupling by controlling cytosolic Ca^{2+} levels and waves. Our results highly suggest that the resting cytosolic and nuclear free Ca^{2+}, which determines the resting tension and/or secretion, is largely due to Ca^{2+} influx through the sarcolemmal and nuclear membrane R-type Ca^{2+} channels. However, the transient increase of $[Ca]_c$ and $[Ca]_n$ during excitation-contraction coupling is mainly due to Ca^{2+} influx through the sarcolemmal T and/or L-type Ca^{2+} channels (13–15). These two types of Ca^{2+} channels seem to act as "turbos" to bring a relatively fast increase of cytosolic Ca^{2+}. However, the R-type Ca^{2+} channel will maintain the Ca^{2+} influx necessary for the maintaining of the tension which compensates Ca^{2+} outflow through the Ca^{2+} pump and the Na^+—Ca^{2+} exchanger. This R-type Ca^{2+} channel seems to be distributed in a non-homogenous fashion, at both the surface cell membrane and the nuclear envelope membranes. This type of channel possesses no inactivation gate which would allow steady-state Ca^{2+} influx at rest, and upon sustained depolarization or receptor stimulation via a PTX and CTX sensitive G-protein coupled to the channel.

During excitation-contraction coupling, the R-type Ca^{2+} channel is implicated in allowing Ca^{2+} wave initiation by Ca^{2+} influx through the opening of the T- and L-type Ca^{2+} channels and the subsequent large Ca^{2+} release from the SR in order to attenuate the cytosolic Ca^{2+} wave amplitude by allowing Ca^{2+} influx through the nuclear membrane, thus permitting a smooth contraction. The subsequent release of the Ca^{2+} taken up permits the maintenance of Ca^{2+} waves and slow relaxation and propagation of the waves to the neighboring cells, most likely through gap-junctions, allowing in this manner synchronization of the contraction of ventricular cells (10). The fact that cytosolic Ca^{2+} waves cannot be completely absorbed by the nucleus is due to the maximum Ca^{2+} buffering capacity of the nucleus, which is shielded from variations in cytosolic Ca^{2+}, perhaps by gating mechanisms in the perinuclear envelope (Figure 1) once its maximum capacity is reached.

Other types of ionic channels were reported to be present at the nuclear membranes such as the calcium dependent potassium channel and chloride channel (26, for review see ref. 28). However, the role of these types of ionic channels in regulating nuclear function is not yet known.

Our results, as well as growing evidence in the literature, show that some G-protein coupled receptors such as ET_B, AT_1 and E_2-EP_3 receptors are also present at the nuclear membranes levels. The activation of these receptors by their respective agonists seems to increase nucleoplasmic free Ca^{2+}. EP_3 receptors were also reported to induce increase of eNOS RNA in cell-free isolated nuclear systems (26). Some of these receptors also participate in controlling the basal resting nucleoplasmic free calcium. Activation of these nuclear receptors seems to shield the nucleoplasm from sustained Ca^{2+} overload. Such a nuclear Ca^{2+} overload may highly contribute to premature contraction of heart cells. It is possible that cytosolic Ca^{2+} overload is not only the product of abnormal functioning of sarcolemmal receptors and/or channels and/or sarcoplasmic or endoplasmic reticulum, but could also be largely due to abnormal functioning of nuclear membranes receptors and channels. Thus, since receptors and channels seem to be present at the nuclear membranes and seem to directly affect nuclear Ca^{2+} transport and function, and since the nucleus seems to be tightly contributing to excitation-contraction and secretion coupling, these receptors and channels (more particularly the R-type Ca^{2+} channel) could be a new target for developing a new generation of therapeutical drugs.

ACKNOWLEDGEMENTS

This work was supported by a research grant from the Canadian Institutes of Health Research (CIHR), INSERC and the Quebec Heart Foundation. P. D'Orléans-Juste and D. Jacques are scholars of the Fonds de la recherche en santé du Québec (FRSQ). We thank Mss Chritiane Ducharme for secretarial assistance.

REFERENCES

1. Bkaily G. 1994. The possible role of Ca^{2+} and K^{+} channels in VSM pathophysiology. In: Ionic Channels in Vascular Smooth Muscle. Ed G. Bkaily, 103–113. Austin, TX: Molecular Biology Intelligence Unit, R.G. Landes.
2. Means AR. 1994. Calcium, calmodulin and cell regulation. FEBS Lett 347:1–4.
3. Peunova N, Enikolopov G. 1993. Amplicifaction of calcium-induced gene transcription by nitric oxide in neuronal cells. Nature 364:450–453.
4. Hardingham GE, Chawala S, Johnson CM, Bading H. 1997. Distinct function of nuclear and cytoplasmic calcium in the control of gene expression. Nature 385:260–265.
5. Czubryt MP, Ramjiawan B, Gilchrist JSC, Massaeli H, Pierce GN. 1996. The presence and partitioning of calcium binding proteins in hepatic and cardiac nuclei. J Mol Cell Cardiol 28:455–465.
6. Steinhardt RA, Alderton J. 1998. Intracellular free calcium rise triggers nuclear envelope breakdown in the sea urchin embryo. Nature 332:364–366.
7. Nicotera P, Rossi AD. 1994. Nuclear Ca^{2+}: Physiological regulation and role in apoptosis. Mol Cel Biochem 135:89–98.
8. Filippatos GS, Gangopadhyay N, Lalude O, Parameswaran N, Said SI, Spielman W, Uhal BD. 2001. Regulation of apoptosis by vasocactive peptides. Am J Physiol 281:L749–L761.
9. Bkaily G, Choufani S, Hassan G, El-Bizri N, Jacques D, D'Orléans-Juste P. 2000. Presence of functional endothelin-1 receptors in nuclear membranes of human aortic vascular smooth muscle cells. J Cardiovasc Pharmacol 36(Suppl. 1):S414–S417.
10. Gilchrist JSC, Czubryt MP, Pierce GN. 1994. Calcium and calcium-binding proteins in the nucleus. Mol Cell Biochem 135:79–88.
11. Bkaily G, Pothier P, D'Orléans-Juste P, Jacques D, Simaan M, Jaalouk D, Belzile F, Hassan G, Boutin F, Haddad G, Neugebauer W. 1997. The use of confocal microscopy in the investigation of cell structure and function in heart, vascular endothelium and smooth muscle cells. Mol Cell Biochem 172:171–194.
12. Bkaily G, Massaad D, Choufani S, Jacques D, D'Orléans-Juste P. 2002. Role of endothelin-1 receptors in the sarcolemma membrane and the nuclear membranes in the modulation of basal cytosolic and nuclear calcium levels in heart cells. Clinical Science 103:141–146.
13. Bkaily G, Gros-Louis N, Naik R, Jaalouk D, Pothier P. 1996. Implication of the nucleus in excitation contraction coupling of heart cells. Mol Cell Biochem 154:113–121.
14. Bkaily G, D'Orléans-Juste P, Pothier P, Calixto JB, Yuens R. 1997. Nuclear membrane receptors and channel potential therapeutical targets for drug action. Drug Devel Res 42:211–222.

15. Bkaily G, Jacques D, Pothier P 1999. Use of confocal microscopy to investigate cell structure and function. Methods Enzymol 307:119–135.
16. Claing A, Shbaklo H, Plante M, Bkaily G, D'Orléans-Juste P. 2002. Comparison of the contractile and calcium increasing properties of platelet-activating factor and endothelin-1 in the rat mesenteric artery and vein. Br J Pharmacol 135:433–443.
17. Bkaily G, Jacques D, Hassan G, Chouffani S, D'Orléans-Juste P. 2001. Using confocal imaging to measure changes in intracellular ions. In: Receptors: Structure and Function. Ed C Stanford and R Horton, 209–231. New York, Oxford University Press.
18. Iborra FJ, Jackson DA, Cook PR. 2001. Coupled transcription and translation within nuclei of mammalian cells. Science 293:1139–1142.
19. Al-Mohanna FA, Caddy KWT, Bolsover SR. 1994. The nucleus is insulated from large cytosolic calcium ion changes. Nature 367:745–750.
20. Himpsens B, De Smedt H, Casteels R. 1994. Relationship between $[Ca^{2+}]_i$ changes in nucleus and cytosol. Cell Calcium 16:239–246.
21. Jacques D, Sader S, Choufani S, D'Orléans-Juste P, Charest D. 2000. Endothlein-1 regulates cytosolic and nuclear Ca^{2+} in human endocardial endothelium. J Cardiovasc Pharmacol 36:S397–S400.
22. Cannel MB, Cheng H, Lederer WJ. 1994. Spatial non-uniformities in $[Ca]_i$ during excitation contraction coupling in cardiac myocytes. Biophys J 67:1942–1958.
23. O'Malley DM. 1994. Calcium permeability of the neuronal nuclear envelope: Evaluation using confocal volumes and intracellular perfusion. J Neurosci 14:7541–7558.
24. Santelle L. 1996. The cell nucleus: An eldorado to future calcium research? J Membr Biol 153:83–92.
25. Haller H, Lindschau C, Quass P, Distler A, Luft FC. 1994. Nuclear calcium signaling is initiated by cytosolic calcium surges in vascular smooth muscle cells. Kidney Int 46:1653–1662.
26. Gobeil F, Dumont I, Marrache AM, Vasquez-Tello A, Bernier SG, Arban D, Hou X, Beauchamp MH, Quiniou C, Bouayad A, Choufani S, Bhattacharya M, Molochnikoff S, Ribeiro-da-Silva A, Varma DR, Bkaily G, Chemtob S. Regulation of eNOS expression in brain endothelial cells by perinuclear EP_3 receptor. Cir Res 90:682–689.
27. Abrenica B, Gilchrist JSC. 2000. Nucleoplasmic Ca^{2+} loading is regulated by mobilization of perinuclear Ca^{2+}. Cell Calcium 28:127–136.
28. Mazzanti M, Bustamante JO, Oberleithner H. 2001. Electrical dimension of nuclear envelope. Pharmacol Rev 81:1–19.
29. Diliberto PA, Krishna S, Kwon S, Herman B. 1994. Isofor-specific induction of nuclear free calcium oscillations by platelet-derived growth factor. J Biol Chem 269:26349–26357.
30. Burnier M, Centeno G, Burki E, Brunner HR. 1994. Confocal microscopy to analyze cytosolic and nuclear calcium in cultured vascular cells. Am J Physiol 266:C1118–C1127.
31. Lin C, Hajnoczky G, Thomas AP. 1994. Propagation of cytosolic calcium waves into the nuclei of hepatocytes. Cell Calcium 16:247–258.
32. Carafoli E, Genazzani A, Guerini D. 1999. Calcium signallinig in the cell nucleus. Cell calcium 22:313–319.

Pathophysiology of Cardiovascular Disease,
edited by N.S. Dhalla, H. Rupp,
A. Angel & G.N. Pierce
Kluwer Academic Publishers, Boston, 2004

Smad Cofactors/Corepressors in the Fibrosed Post-MI Heart: Possible Therapeutic Targets

Julie C. Roth, Baiqiu Wang, Darren H. Freed, and Ian M.C. Dixon

Institute of Cardiovascular Sciences, St. Boniface General Hospital Research Centre and the Department of Physiology, University of Manitoba, Winnipeg, Canada

Summary. Several axioms that govern our understanding of the pathogenesis of post-MI congestive heart failure have recently been questioned or revisited in lieu of the relative flood of new data addressing the role of cardiac non-myocytes in this disease state. Among these is the assumption that the cardiac matrix necessarily serves a secondary role to initial progressive dysfunction of cardiac myocytes. It is now well established that remodeling of the cardiac interstitium is a separate and distinct contributor to the progression to heart failure, and that the turnover of extracellular matrix is rapid enough to be relevant to progressive heart disease. As cardiac fibroblasts and myofibroblasts are the main source of matrix and regulate cardiac wound healing, the study of stimulating ligands and their signaling pathways may reveal novel sites for therapeutic intervention. This chapter provides background on the cardiac interstitial components and their turnover as well as providing support for the argument for the need to address post-receptor components of classical TGF-β signaling in the regulation of cardiac (myo)fibroblast function, with an emphasis on Smad proteins and their cofactors/corepressors.

Key words: heart failure, myocardial infarction, wound healing, myofibroblasts, Smad proteins.

EXTRACELLULAR MATRIX AND THE HEART

The extracellular matrix (ECM or matrix) of all mammalian organ systems is comprised of a large variety of polysaccharides, proteins and proteoglycans, which form an intricate, continuous network between parenchymal cells (1). The majority of

Corresponding Author: Ian M.C. Dixon, PhD, Associate Professor of Physiology, Institute of Cardiovascular Sciences, Room 3038, St. Boniface General Hospital Research Centre, 351 Tache Avenue, Winnipeg, Manitoba, Canada R2H 2A6. Tel: 204 235 3419, Fax: 204 233 6723, E-mail: iand@sbrc.ca.

these macromolecules are secreted locally by fibroblasts and their derivative cell types that are found throughout the body as well as in abundance within the myocardium (2). The matrix was once thought to serve essentially only as an inert scaffolding, stabilizing tissue structure. It has been shown, however, that matrix plays an active role in the regulation of the behavior of those cells it contacts, influencing development, migration and proliferation (3). Within the myocardium, approximately 75% of total tissue volume is made up of cardiac myocytes. This, however, only makes up about one third of the total cell number while the remaining two thirds are non-myocytes existing within the interstitium. These non-myocytes include fibroblasts, myofibroblasts, endothelial cells and vascular smooth muscle cells. Cardiac fibroblasts and myofibroblasts form the largest group among these non-myocyte cells (4).

Components of the extracellular matrix

There are two main classes of extracellular macromolecules that make up the matrix. The first class is composed of polysaccharide chains known as glycosaminoglycans (GAGs) which are frequently linked by covalent bonds to proteins forming proteoglycans (5). GAGs, proteoglycans and hyaluronic acid work to hydrate the ECM and thereby facilitate cardiac contractility and optimize the pumping action in healthy heart tissue (6,7). The second class which comprise the fibrillar proteins that impart two main functions in the interstitium i.e., adhesion (laminin and fibronectin) and tensile strength (elastin and collagen) (4).

Laminin

Laminin is the first ECM protein to be synthesized during embryonic development. It is a large (~850 kDa) flexible complex consisting of three long polypeptide chains held together by disulfide bonds forming an asymmetric cross (8). Laminin contains four or more different structural domains that allow adherence to various molecules including those for collagen type IV, heparin, entactin, and laminin receptor proteins (9). In adult cardiac tissue, laminin is located along the length of the basement membrane and is concentrated in areas of morphological specialization including the sarcomeric Z bands (10). Laminin is known to mediate cell adhesion, migration, growth and differentiation and plays an intrinsic role in the survival of cardiac myocytes in culture (11,12).

Fibronectin

Fibronectin is a large glycoprotein with multiple sub-types but is essentially a dimer composed of two subunits joined together by a disulfide bond near their carboxy-termini. Each subunit contains complex folded rod-like domains that contain binding sites for heparin, collagen and cell adhesion. Fibronectin therefore serves as a liaison for cardiac myocytes to the interstitial collagen knit (13). Plasma fibronectin is a soluble form of the protein and participates in blood clotting, wound healing and phagocytosis. The remaining isoforms are all highly insoluble and assemble upon the surface of cells. These are then deposited into the ECM as fibronectin filaments. These various isoforms are, however, encoded by a single large gene containing approximately 50 exons. The diverse array of fibronectin isoforms is obtained by the method of alternative splicing in three regions of the primary RNA transcript. The predominant isoform expressed seems to rely upon the specific cell type and developmental stage of the tissue (14).

Elastin

The cardiac interstitium contains small amounts of elastin. The networks of elastic fibers allow for the resilience required to recoil after a transient stretch and interwoven inelastic fibers prevent tearing (15). Elastin is the main element of elastic fibers and is a highly hydrophobic protein. It is secreted into the extracellular space and the completed elastic fibers remain near the plasma membrane becoming highly cross-linked and form an web of fibers and sheets (15).

Collagen

Cardiac fibrillar collagens are characterized by the long, stiff, triple-stranded helical structure of each molecule, and collagen is an important structural component of the heart. In the myocardium, collagen fibers are laid out in an intricate network consisting of the epimysium, the perimysium and the endomysium, and are continuous with chordae tendineae and valve leaflets (16). Studies have shown that its tensile strength, which is proportional to collagen strut thickness, alignment, configuration and location confer myocardial stiffness and maintenance of myocyte alignment on a beat-to-beat basis (17). Collagen proteins also have other roles within the myocardium, including force transduction, prevention of muscle fiber slippage, protection from myocytic over-stretching, and storage of potential energy during cardiac systole (18,19).

Synthesis of collagen

Cardiac fibroblasts synthesize, then translate fibrillar collagen mRNAs on membrane bound ribosomes. These individual polypeptide chains contain a short amino-terminal peptide sequence signal and additional amino acids called propeptides at both ends. They are then directed into the lumen of the endoplasmic reticulum (ER) as large precursors called pro-α chains (20). Within the ER, the nascent pro-α chains undergo hydroxylation of selected proline and lysine amino acid residues. Further modifications are made such as glycosylation of certain hydroxylated prolines that enable the subsequent spontaneous assembly of three pro-α chains into mature procollagen proteins (2). The procollagen molecules are then transported to the cellular plasma membrane via a secretory vesicle and, within the extracellular space, the secreted procollagens are converted into mature collagen molecules by the removal of the propeptides by proteolytic enzymes. The characteristic triple helix structure is formed from three collagen molecules, organized individually as α-helices, wound together in a ropelike superhelix (21). All collagens are proline and glycine rich, this feature being a fundamental structural requirement. The amino acid proline contains a ring structure that acts to stabilize the helical conformation in each α-helix. Glycine, the simplest and smallest amino acid, is located in regular intervals, every third residue, along the central region of each α-chain. The significance of this arrangement of glycines is that it allows the three α-helices to coil tightly together to form the final superhelix (22). This superhelix is composed of either three identical α-chains (homotrimers) or different chains (heterotrimers) with the exception of the collagen isoform IV, which is a homodimer (23).

The propeptides serve two functions; the first role is to guide the procollagen to the ER previously mentioned, while the second lies in the prevention of self-assembling into large intracellular collagen fibrils. This could create unfavorable conditions in the cell, and would lead to the death of the synthetic source cell. Collagen fibrils are composed of many collagen molecules assembled into a 10–300 nm structure. These fibrils form close to the cell surface and aggregate together creating a collagen fiber (2).

Collagen sub-types

To date there have been about 25 distinct collagen α chains identified, each encoded by a separate gene. Different combinations of these genes are expressed in different tissues (22), although only 15 types have been isolated and characterized to date. The main isoforms found in heart tissues are types I, III, IV, V and VI although it is the fibrillar collagens (types I and III) which account for 90% of the collagens found within the myocardium (23,24). Type I collagen has the tensile strength approximating steel and accounts for nearly 75% of the total collagen makeup in the myocardium (25). This type of collagen forms aggregates of relatively thick fibers and has the molecular formula $[\alpha1(I)]_2\alpha2(I)$ (2). Type III collagen is much more distensible than type I and forms aggregates of finer fibers. Other than heart tissue, type III collagen is found in the skin and blood vessels and has a molecular formula of $[\alpha1(III)]_3$ (26). The remaining cardiac collagen isoforms (IV, V, and VI) only amount to about 10% of the total collagen complement. Collagen type IV is a basement membrane protein found in myocytes, myofibroblasts and most other cardiac cell types where it plays a crucial role in cell adhesion and molecular transport (23,24). The type V collagen isoform co-exists in the basement membranes in association with type IV collagen as well as with types I and III in the interstitium (24). Collagen type VI intermingles in the interstitium where it appears to coat the surface of the fibrillar collagens (24).

Cardiovascular disease and matrix remodeling

In consideration of stromal and parenchymal left ventricle, global remodeling is characterized by structural changes involving ventricular chamber size, wall thickness, and cellular composition (27). These changes are regulated by mechanical, neurohormonal, and genetic factors. Cardiac remodeling can be either physiological (adaptive) as occurs during normal growth or intensive atheletic training, or pathological (maladaptive) attending surviving myocardial tissue post-MI, or in cardiomyopathy (CMP), hypertension, and chronic valvular heart disease (28).

Myocardial infarction

Congestive heart failure results from diverse etiologies via multiple pathophysiological mechanisms. Common endpoints are chamber dilatation, contractile dysfunction, and stiffening of the myocardium. The most prevalent etiology leading to CHF is transmural MI (29). Myocardial infarction is defined as the death of a segment of the heart muscle following the interruption of blood flow (30). This occurs within minutes of occlusion and reduces ventricular pump performance in proportion to the amount of ischemic tissue (31). This loss of functioning myocardium provokes the redistribution of work load to the remaining viable heart muscle to maintain physiological cardiac output and blood flow (32). Cessation of blood flow to an area of the myocardium may be due to coronary thrombosis or vasospasm or embolization of one of the coronary arteries. The coronary arteries, particularly the left coronary artery, can also be blocked by atherosclerotic plaque, confining the majority of infarction to the left ventricle. A patient suffering from an acute MI presents such symptoms as sudden severe chest pain radiating to the arm and throat, arrhythmias, dyspnea, syncope, and either high or low blood pressure (33). MI involving coronary stenosis due to the presence of atherosclerotic plaques usually occurs in men aged 40 and above and the risk in women following menopause is equalized to the risk found in men (33). MI of this nature can also occur at a younger age due to genetic predisposition. Patients suffering from MI with normal coronary arteries are usually younger than those with coronary artery disease and are cigarette smokers or cocaine users (34).

Extracellular matrix remodeling

The process of ventricular remodeling is not well understood. It is defined as a repair process after an ischemic, toxic or inflammatory event in the heart or following hemodynamic overload, commonly resulting in fibrosis (35). This change in shape and composition is thought to compensate in acute or chronic alterations from normal demand on the myocardium (35). The term "slippage of myofilaments" has been coined referring to the destruction of connective tissue and may represent the initial remodeling event. Net removal or shuffling of matrix components is known to diminish the functional capacity of the heart (36). Myofibrillar slippage will result in reduced systolic performance leading to increased diastolic stiffness as a consequence of the changes of interstitial cell matrix composition. Cells of the myocardium also endure alterations such as hypertrophy or hyperplasia and may undergo a process known as programmed cell death or apoptosis (37). At a subcellular level, gene expression "switches" to a fetal phenotype in cardiomyocytes evidenced by elaboration of atrial natriuretic peptide and skeletal α-actin. In cardiac fibroblasts fibrillar collagen is synthesized at abnormal levels (38). Cardiovascular remodeling is affected by various factors including mechanical, neurohormonal and genetic influences.

Mechanical factors

The heart is constantly experiencing mechanical disturbances including shear and tensile stresses. During embryonic development, cells of the cardiovascular system respond to these mechanical forces in a positive manner and it is used as a stimulus for synthesis of ECM via cardiac fibroblasts. The adult myocardium is also exposed to constant mechanical forces. Cyclical stresses during the normal cardiac cycle do not create pathology in the myocardium, however, changes in either blood pressure or volume will stimulate cardiac remodeling (39). Therefore, it has been suggested that the mechanical environment of the heart is a key modulator of cellular function. ECM remodeling occurs in blood vessels and in the heart in response to increased mechanical load originating from such causes as elevated blood pressure or decreased cardiac output. There is strong evidence supporting the idea that increased mechanical forces have a direct effect on myocyte hypertrophy, however, the regulation of collagen deposition appears to involve additional players (40). Studies have shown that a change in mechanical load in the heart activates the release of pro-fibrotic growth factors such as transforming growth factor-β (39) and angiotensin II (26). These factors play a significant role in the increased production of total protein and fibrillar collagens type I and III (41). Volume and pressure overload will stimulate collagen synthesis, however, in the case of volume overload or the early phases of pressure overload, collagen deposition is matched by myocyte hypertrophy. Remodeling in this manner results in an enlarged ventricle with a normal collagen composition. With chronic pressure overload the myocardium remodels differently as collagen synthesis exceeds the hypertrophic response resulting in fibrosis (40).

Factors that require genetic regulation

Genes important to cell signaling via phosphorylation and dephosphorylation of specific target proteins are implicated in cardiac fibrosis and the pathogenesis of heart failure. Members of the mitogen activated protein kinase (MAPK) family including extracellular signal-regulated kinase (ERK) and c-*jun* NH2-terminal kinase (JNK) play key roles in the regulation of cell growth, apoptosis and gene expression (42). Although the role of MAPKs in cell growth and regulation may be crucial, specific mechanisms are not well defined. Recent studies of the activities of these kinases in cardiac hypertrophic or hypertensive rats have shown that JNK activity is chronically enhanced. This induction of JNK activity is followed by the upregulation of activator protein-1 (AP-1) activity. AP-1, a heterodimer of c-Fos and c-Jun, will then bind to a specific sequence in target promoters (43). Presently, data have shown that AP-1 transcriptional activation is mediated by TGF-β and involves an interaction with another important family of transcription factors important in cardiac remodeling, the Smad proteins (44). As ERK and JNK activity are elevated in hypertensive rats when compared to normotensive age-matched controls, these MAPKs may be a useful target for controlling cardiac remodeling (43).

Another genetic factor involved in cardiac remodeling is apoptosis, or programmed myocyte dropout (45). The process of apoptosis involves discrete genetic and molecular programs, de novo protein synthesis, and expression of a unique cellular phenotype. Certain myocytes within the myocardium undergo apoptosis following such traumas as acute MI, and ischemic and non-ischemic heart failure (46). Programmed cell death is potently stimulated by oxygen radicals, cytokines (FAS/TNFα receptor signaling), stress conditions (chemical and physical) and hormones (Angiotensin) (46). Apoptosis of ventricular myocytes will result in progressive pump failure, arrhythmias, and remodeling. It is thought that MAPKs are a potential target of preventative therapies (46).

Neurohormonal factors

Ventricular remodeling involves the activation of both the adrenergic and RAAS systems following increased wall stress. Studies indicate that myocardial contractility is vastly decreased during CHF, partially due to alterations in β-adrenergic receptor-mediated signal transduction (47). These changes in signaling result in a decreased production of cyclic AMP (cAMP) which, in turn, decreases intracellular calcium concentration. The reduction of intracellular calcium contributes to the attenuation of contractile force (48). In the failing heart, mRNA message levels for both β-1 and β-2 receptors are decreased as well as protein expression levels. However, α-1 receptor density remains unchanged or is slightly increased during heart failure and has been associated with myocyte growth and cardiac hypertrophy (49).

Considerable effort has been focused on the humoral mechanisms responsible for transducing increased hemodynamic load in cases of CHF into adaptive cardiac remodeling. The RAAS signaling system and its primary effector peptide, angiotensin II (angiotensin), contribute to cardiac remodeling (50). Reparative fibrosis, which is secondary to myocyte necrosis and reactive fibrosis contributes to this remodeling and angiotensin appears to play a key role in regulating this pathology (26). In the circulating RAAS, the liver derived precursor, angiotensinogen (A_0) is cleaved by the kidney derived protein renin to form angiotensin I. Angiotensin I is converted to the biologically active product, angiotensin, by either ACE, cathepsin G or Tonin (51). Angiotensin is a potent vasoconstrictor and promotes the production of aldosterone, an important player in renal function. In addition to this systematic angiotensin production, a local RAAS system exists in the cardiovascular system (52). Although the components of the RAAS are localized in cardiac cells, this system cannot function without the final effector protein, the angiotensin receptor. In human and rat myocardium, angiotensin receptors have been identified in human and rat atria, ventricles, myocytes and cardiac fibroblasts (50). Biochemical and pharmacological studies have characterized two angiotensin receptor subgroups, angiotensin type I (AT_1) and angiotensin type II (AT_2) (53). The AT_1 receptor is a seven transmembrane domain protein and is further subdivided into AT_{1A} and AT_{1B} classes (54). The effect of short-term AT_1 receptor blockade has been previously studied and this therapy has been associated with the normalization of collagen turnover in hearts affected by either MI or cardiomyopathy (55). Thus, antagonism of the AT_1 receptor is another method of reducing or reversing remodeling.

Until recently, it was generally accepted that angiotensin signaling was mediated mainly through the AT_1 receptors in the myocardium. Ironically, it had been observed that increased mechanical load causes the upregulation of both AT_1 and AT_2 mRNA and protein expression (56). New evidence however, has demonstrated that the re-expressed AT_2 receptor, located on cardiac fibroblast surfaces, exert an inhibitory effect on angiotensin-induced mitogen signals (57) thus introducing a novel mode of remodeling control.

Data from our lab acquired using cardiac fibroblasts have suggested that specific growth factors, primarily TGF-β_1 may indirectly mediate the stimulatory effects of angiotensin (114). In addition to increasing TGF-β_1 secretion, angiotensin may play a role in stimulating the conversion of latent TGF-β_1 propeptide into its active form (58). Also, simultaneous treatment with angiotensin and a TGF-β_1 neutralizing peptide in cultured cardiac fibroblasts reduces collagen type I and III mRNA expression. Together these data suggest that angiotensin may act upon fibroblasts indirectly through an autocrine/paracrine loop via TGF-β_1 to induce fibrotic protein synthesis (59).

The TGF-β/Smad signaling pathway

Two novel concepts or paradigms are emerging to explain the mechanism by which cells process cytokine signaling. First, it is becoming clear that cytokine signaling pathways are not discrete but that they act via a complex network with other factors through multiple cross-talk and feedback interactions. Secondly, although it has long been clear that a specific response to a pleuripotent factor is dependent on cell type, it is now becoming evident that this variable response may rely upon the specific expression of a specific subset of transcription cofactors within a given cell type (60). Thus cofactor specificity confers a characteristic cellular response, wherein that response is variable.

The TGF-β superfamily of cytokines consists of many structurally related peptides that function as either growth or differentiation factors. Members of the TGF-β superfamily are produced by numerous cell types and include the TGF-βs, the bone morphogenic proteins (BMP), the activins, and anti-Müllerian hormones (61). Each member of this family is capable of regulating a vast array of cellular processes including cell proliferation, lineage determination, motility, adhesion and cell death (62). Several of these modes of regulation stem from changes in the expression of key target genes; hence, understanding transcriptional control via the TGF-β family is of great importance (63).

All members of the TGF-β superfamily are synthesized as latent precursors containing a C-terminal domain that is cleaved upon secretion from the cell (64). This propeptide, known as the latency associated peptide (LAP), remains covalently bound to the secreted ligand and keeps it in a latent form by obstructing its binding to receptors (65). A large glycoprotein is the third component of the latent TGF-β complex. Latent TGF-β-binding protein (LTBP) is linked by a disulfide bond to LAP and while not essential for latency, plays a role in secretion and storage in the ECM, and for the eventual activation of the ligand (66).

TGF-β isoforms and function

Mature TGF-β proteins are 25 kDa homodimers, with each monomer consisting of 112 amino acids, 9 of which are cysteine residues. The three main isoforms are called TGF-β_1, -β_2 and -β_3. All TGF-β isoforms contain 72% to 80% sequence conservation, with the N-terminal region being the most divergent (67). TGF-β_1, which has been isolated in both cardiac myocytes and fibroblasts, has been implicated in fibrotic disorders (68,69). Increased expression of TGF-β_1 was observed using ELISA and Northern blot techniques within the bordering and scar regions of the myocardium following MI, suggesting a role for this cytokine in post-MI repair (70).

TGF-β receptors and ligand interaction

TGF-β and its related factors signal through the transmembrane serine/threonine receptors referred to as the TGF-β receptor family. The members of this family are divided into two sub-populations, type I (TβR-I) and type II (TβR-II) receptors, based on their structural and functional properties (62). However, the overall structure of all these receptors are the same and include an N-terminal signal sequence, a short cysteine-rich extracellular region, a single hydrophobic membrane spanning domain, and a kinase containing cytoplasmic region (71).

The relatively short (~150 amino acids) extracellular domain of these receptors contains 10 or more cysteine residues that determine the exterior folding pattern. Three of these residues are characteristically clustered near the transmembrane domain while spacing varies for the others (72). Unlike the other domains, the transmembrane domain and the cytoplasmic juxtamembrane region themselves bear no singular structural feature. However this region of TβR-II (Ser 213) is phosphorylated by a kinase in a ligand-independent manner and is required for signal propagation (73). TβR-II phosphorylates the Ser165 in the juxtamembrane region of TβR-I in a ligand-dependent fashion. It is the latter phosphorylation event that appears to selectively modulate the intensity of the different TGF-β responses (74).

A highly conserved 30 amino acid region immediately preceding the protein kinase domain is a feature unique to the TGF-β type I receptors. It is termed the "GS region" as it contains characteristically an "SGSGSG" domain (75). Directly after this sequence lies a leucine-proline motif found on all members of the type I receptor family. It has been shown that mutation of the amino acid in the second to last position of the GS domain, which is always a glutamine or threonine, elevates receptor activity in vitro, allowing constitutive signal activity (76). Therefore the GS domain is a key regulatory region which controls catalytic activity of the type I receptor kinase or substrate interaction (62).

Both type I and type II TGF-β receptors contain a serine-threonine kinase domain (71). Type I receptors have been observed to phosphorylate their only known substrates, the Smad proteins, on their serine residues while type II receptors phosphorylate themselves on serine and threonine moieties (77). Another difference between the receptor subtypes is that only the TβR-II contains a short C-terminal extension following the kinase domain. This extension may be phosphorylated, however, its deletion does not impair signaling (78). This is in vast contrast to the significant role the C-terminal extension plays in tyrosine kinase signal transduction (79).

TGF-β activates signaling by binding to and causing dimerization of TβR-I and TβR-II receptors. There are two general modes by which ligand binding have been observed: sequential binding and cooperative binding (62). Sequential binding involves TGF-β binding first to TβR-II and this complex then associating with TβR-I. The type I receptors can recognize the difference between free and bound ligand, complexing only with receptor bound ligand. The second binding mode is not typical of TGF-β binding but that of BMP and its receptors. These receptors bind with high affinity to the BMP ligand when expressed together but weakly when expressed separately (80).

Smad proteins as mediators of TGF-β signaling

The Smad family of proteins are substrates for what is becoming known as classical transduction of TGF-β receptor signaling. These proteins play a crucial role in the propagation of TGF-β receptor signals to target genes within the cellular nucleus (62). The first member of the Smad family was identified in *Drosophila* as the gene product of *Mad (mothers against decapentaplegic)* (81). Its discovery led to the identification of related genes in nematodes and vertebrates. Three homologous proteins were identified in *C. elegans* and were named Sma-2, Sma-3 and Sma-4 as their mutation caused a small body size (82). Following this discovery, many related genes were also identified in humans and were aptly named Smads meaning Sma/Mad related (83). It is now clear that Smad proteins are the downstream effectors of the TGF-β receptors.

Structure of smads: MH-1 and MH-2 domains

Smad proteins associated with TGF-β signaling are divided into three distinct subclasses (Figure 1A), the pathway specific or receptor-mediated Smads (R-Smads, Smad 2 and 3), the common mediator Smads (Co-Smads, Smad 4) and the inhibitory Smads (I-Smads, Smad 6 and 7) (62). The Smad family of proteins contain two highly conserved regions at their N- and C-terminal that are referred to as Mad homology region 1 (MH-1) and MH-2, respectively (Figure 1B). These regions are linked together by a non-conserved "linker" region of variable length (84). The MH-1 domain consists of approximately 130 amino acid residues and is only highly

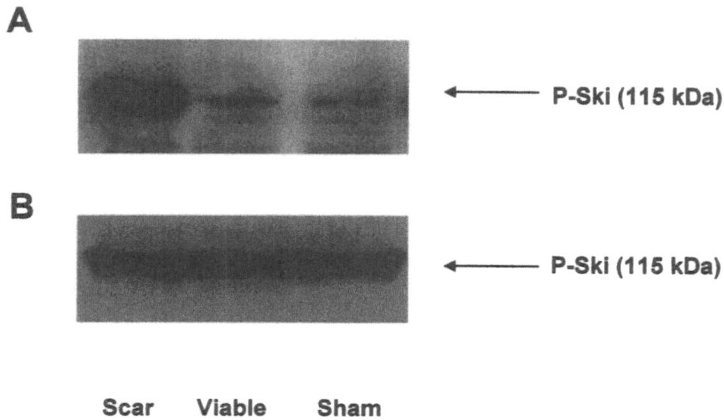

Figure 1. Expression of phosphorylated Ski (P-Ski) in 8 week post-MI cardiac tissues. **Panel A:** Cytosolic fraction of the tissue samples. **Panel B:** Total protein from tissue samples. A preponderance of localized P-Ski exists in the cytosol of infarct scar cells (mainly myofibroblasts at 8 weeks post-MI).

conserved in R- and Co-Smads but not the inhibitory Smads (62). The MH-1 domain of an R-Smad protein contains a β-hairpin loop which is responsible for DNA recognition upon activation. The MH-1 domain is also known to inhibit MH-2 binding to TβR-I (85). The MH-2 domain is approximately 200 residues in length and is responsible for Smad/receptor interaction, R-Smad/Co-Smad interaction, DNA binding protein interaction and transcriptional activation (62). These functions are conferred by the presence of receptor phosphorylation sites required for R-Smad activation (i.e., Co-Smad binding and subsequent translocation). Overexpression of the MH-1 domain has been demonstrated to inhibit R-Smad/Co-Smad association. Specificity of receptor binding is conferred by a sequence of amino acids located within the L3 loop of the MH2 domain (86). The highly variable (i.e. not conserved) MH-1/MH-2 linker region likely subserves R-Smads as its domain contains MAPK phosphorylation sites, which when activated by MAPK signaling, inhibit Smad nuclear translocation (119).

The Smad protein signaling pathway

R-Smads are cytoplasmic proteins that become activated upon association with the C-terminal kinase domain of the type I receptors (87). These Smads contain a highly conserved "SSXS" motif at their carboxy-termini with the last two serine residues becoming phosphorylated upon activation (77,87). Blocking R-Smad phosphorylation prevents two subsequent signal-dependent actions, association with Co-Smads and translocation to the nucleus (88). The protein SARA contains a lipid-binding FYVE domain and functions to recruit unphosphorylated R-Smads to the transmembrane spanning TGF-β receptors (89). Association of R-Smads with the type I receptor is transient and upon phosphorylation of the two C-terminal serine residues the R-Smad protein is released from both the TGF-β receptor and SARA protein (77,89,90).

Following its release from the type I receptor, R-Smads form a heterodimer with the common mediator Smad, Smad 4 (88). The Smad 4 chaperone protein, TGF-β receptor associated protein 1 (TRAP1), mediates the association of Smad 2 and Smad 4 (91). TRAP1 binds to inactive TβR-II and dissociates upon receptor activation, altering its conformation to form a transient complex with Smad 4. This TRAP1/Smad 4 complex is observed only in the absence of Smad 2 suggesting that it is a transient complex visualized only in the absence of an acceptor R-Smad (91). Upon signal activation, cytosolic R-Smads form heterodimers with Co-Smads and regulated translocation of this complex to the cellular nucleus is a key event leading to the induction of target gene transcription (84,92,93). The Co-Smads are structurally similar to R-Smads but do not contain the SSXS motif at their C-terminal end as their R-Smad counterparts. Lacking this SSXS-carboxy region results in the inability to associate with type I receptors, implicating a novel function for this Smad subtype (77,90,94). To date, Smad 4 is the only known vertebrate common-mediator protein (95,96).

Translocation is recognized as a receptor activation dependent phenomenon, as mutations within the sites for type I receptor phosphorylation prevent this event (88). It has also been observed that although phosphorylation triggers both Co-Smad association and translocation, dimerization with a Co-Smad is not required for R-Smad nuclear entry (88). How these Smad proteins propagate the signal to target genes is still relatively unclear. Smad 2, in contrast to Smad 3, contains an extra amino acid sequence at its N-terminal domain that makes it unable to bind to DNA with great affinity. In such cases it is Smad 4 that is responsible for DNA interaction (97). This, along with the recent discovery of various response elements within target genes, suggests that transcriptional co-factors are required for competent gene transcriptional regulation (98).

The third subclass of Smad proteins, the inhibitory Smads, contain the MH2 domain of the other Smad types, but similar to Co-Smads lack the SSXS motif. The I-Smads are unique, however, as they do not share any homology in their N-terminals (99, 100). Similar to R-Smads, Smad 6 and 7 directly interact with the type I receptor. Since these Smads do not contain the C-terminal phosphorylation sites, the association with the receptor is more stable and thus impedes signal transduction by competitive inhibition with R-Smads (99,100,101). Smad 7 is localized within the nucleus prior to TGF-β stimulation. Translocation to the cytosol is observed following this stimulation (102). Western analysis of the cytosolic fraction of adult rat primary cardiac fibroblasts following TGF-β_1 stimulation indicate an increase in Smad 7 expression after 60 minutes and a return to basal levels at 120 minutes (103). These proteins may therefore act as an auto-regulatory negative feedback system in TGF-β signal transduction. The two vertebrate I-Smads function as inhibitors differently. Smad 7 is a general inhibitor of R-Smad activation whereas Smad 6 selectively inhibits Smad 2 but not Smad 3 (99,100).

Other TGF-β downstream factors

In addition to Smad proteins, G proteins and MAPKs have been thought to function as downstream effectors of the TGF-β signaling pathway (104). There are several classes of G proteins including the small G proteins and the heterotrimeric signal transducing proteins (105,106). Of the latter class, G_s and G_i modulate the activity of adenylyl cyclase and G protein function may be modified by bacterial exotoxins (105). These toxins can also affect some of the biological activities of TGF-β (107). Experiments have shown that stable transfection of $G_{i\alpha 1}$ into NIH 3T3 fibroblasts, a cell line exhibiting little responsiveness to TGF-β, restores TGF-β responsiveness to control levels (108). Many studies have shown that the small G proteins (Ras, Rac, Rho and Cdc42) play a central role in the downstream signaling pathway of receptors, such as angiotensin type I receptor (AT_1 receptor). The association of the small G protein, Ras, and tyrosine receptors requires protein intermediates such as Grb2 (109). Other proteins are required to activate the membrane bound Ras which leads to the activation of members of the MAPK family (ERKs, p38-MAPK and JNKs) (109,110).

Small G proteins and the MAPK cascade are also involved in serine/threonine kinase receptor signaling. This was demonstrated using a TGF-β sensitive cell line and showing small G proteins are activated by TGF-β (111). TGF-β has been shown to activate all three of the MAPK cascades (112). Collectively these results indicate that signal transduction by the TGF-β family can be influenced by the small G proteins and the MAPKs (112). Smad proteins are also known to interact with the MAPK cascade (113). Whether or not the integration of actions between the Smad and MAPK proteins are a general feature in TGF-β signaling remain unknown. However, it has been suggested that the specificity of ligand response is determined not only by activated TβRI/TβRII complexes but also by the combination of other downstream effectors present in the cell. This is a concept that is supported by evidence for cross-talk between the angiotensin/TGF-β signaling pathways (104,114).

Transcriptional co-factors

Eukaryotic transcription is highly regulated by two classes of transcription factors. General transcription factors (GTFs) bind to an unspecific core promoter sequence in close proximity to the transcription initiation site. Together with an RNA polymerase they form the pre-initiation complex (115). The second class includes sequence-specific transcription factors that bind to regulatory DNA elements located at various distances upstream from the transcription start site. These DNA regulatory elements contain a specific DNA sequence necessary for transcription factor binding. Factors bound to this region will interact with other downstream factors to either activate or inhibit gene transcription (116). Both of these classes of transcription factors contact RNA polymerases, either directly or via mediators in an initiation complex (117).

DNA binding motifs

DNA-protein interactions are highly specific and among the tightest molecular interactions found in nature (118). X-ray crystallography and NMR spectroscopy techniques of approximately 20 gene regulatory proteins found to complex with DNA were used to examine the structure of these proteins. It was discovered that these proteins contain one of a small subset of structural motifs and interact with the major groove of DNA using either α-helices or β-sheets (119).

The helix-turn-helix motif

The helix-turn helix (HTH) was the first DNA binding motif to be discovered and has been found to be the most common form (120). This motif is constructed of two α-helices linked together connected by a short sequence of amino acids making up the "turn" of this structure (121). The two helices are maintained at a fixed angle, via interactions of the helices, to allow for accurate interaction with the DNA binding sites. Interaction with DNA by this type of transcription factor occurs through the carboxy-terminal helix known as the *'recognition'* helix and the varying amino acid side chains of each protein (122). The remaining protein structure varies greatly allowing each protein to present itself to the major groove uniquely. The polypeptide outside of the motif also helps to fine tune the protein/DNA interaction (123).

Homeodomain transcription factors are a separate class of HTHs. It was noted that several sequences of homeotic selector genes contained a conserved 60 amino acid residue sequence known as the homeodomain (124). The structure of a homeodomain protein bound to a specific DNA sequence differs from the previously described HTHs. The HTH motif of homeodomain proteins is always surrounded by the same structure and therefore will present itself to the DNA sequence in the same fashion (125).

The zinc finger motif

The zinc finger motif is structurally more complex than the HTH, which is comprised solely of amino acid residues. This second class of binding motif utilizes one or more atoms of zinc as a structural component, lending to its name (2). Previously, due to a schematic depiction of these proteins, these motifs were known as zinc fingers. Subsequent structural analyses have shown that these proteins are structurally different and can be placed into various classes (126). One subtype of zinc finger is a relatively simple structure consisting of an α-helix and a β-sheet that is held together by a zinc molecule. These proteins are usually found clustered together in a manner that does not allow the α-helices to contact the major groove. This long repeated series of zinc fingers allows for strong and specific DNA-protein interaction (127).

Another class of zinc finger is found to be a member of a large family of intracellular receptor proteins forming a structure similar to the HTH motif. This structure consists of two α-helices held together by two zinc molecules. This subclass of zinc fingers form dimers that allow one of the two α-helices of each subunit to interact with the DNA major groove. It is good to note that both of these zinc finger subtypes use zinc as a structural element and that both use an α-helix to recognize DNA (128).

The leucine zipper motif

Many of the regulatory proteins function as dimers as it is a simple way to achieve strong and specific interactions with DNA. Often, these regulatory proteins are segmented into two areas, one that is responsible for dimerization and one that makes contact with DNA. A third separate motif known as the leucine zipper combines both of these functions (129). The leucine zipper is composed of two HTH α-helix monomers joined together to form a short coiled-coil. These helices are held together by hydrophobic interactions of amino acid side chains which often contain leucine residues (130). Just past the dimerization interface, the two α-helices separate forming a Y shaped structure that contacts DNA in a clothespin-like manner.

Leucine zippers can either form homodimers or heterodimers. Heterodimers usually form between two proteins that possess distinct DNA-binding specificities. This vastly increases the number of DNA-binding sequences that these proteins can recognize, an example of *'combinational control'*. Combinational control refers to the process by which combinations of proteins, rather than individual proteins, control cellular processes (128). As there are many types of leucine zipper proteins it is thought that the majority form homodimers, otherwise the amount of crosstalk possible between the gene regulatory circuits could create havoc in the cell (131).

The winged helix motif

As previously described, transcription factors are usually classified according to the structural motif involved in DNA binding or, in some cases, the oligomerization domain. Within the last ten years, a structurally unique motif was identified in the *Drosophila* fork head gene product and in rat hepatocyte nuclear factor 3 (HNF3) (132). The *Drosophila* fork head gene is responsible for proper terminal structure formation during embryonic development. In situ studies revealed the early presence of both forkhead mRNA and protein in endodermal tissues, the protein being localized mainly within the nucleus. This nuclear localization suggested that it might function as a transcriptional regulator (133). Following the discovery of forkhead, a small family of hepatocyte enriched DNA binding transcription factors were identified in rodents, the HNF-3 family (134). Upon inspection of the HNF-3 110 amino acid DNA binding domain it was noted that the sequence did not bear any resemblance to any known transcription factor motifs (135). When compared to the *Drosophila* forkhead protein DNA binding sequence, it was discovered that there was a high degree of sequence homology. Thus forkhead and HNF-3 are considered to be the founding members of the family of transcription factors, for structural reasons, named winged helix proteins (WH) (134,136). Members of the WH family are expressed throughout a wide range of tissues during various stages of development and act to control the regulation of cell determination, cell proliferation and differentiation (137,138).

The topological structure of the winged helix motif is a compact α/β structure consisting of two wings, (W1 and W2), three α-helices (H1, H2 and H3) and three β-sheets (S1, S2, and S3) (139). The N-terminal domain contains a largely helical structure while the C-terminal domain is composed of two of the three β-strands arranged in a twisted anti-parallel β-sheet motif and the two characteristic loops or "wings", W1 and W2 (140). WH proteins are members of the HTH superfamily but they differ in the length of the "turn" connecting helices H2 and H3. This structural variation eliminates stereochemical-binding restrictions found in canonical HTH proteins. The angle between the H2 and H3 helix ranges in typical WH proteins from 100° in the biotin operator repressor protein BirA, to 150° in the transcription factor DP2. In contrast, 120° is the typical angle found in the HTH proteins (134). The majority of WH proteins present the H3 helix, known as the recognition helix, to the major groove of target DNA sequences. Fourteen protein-DNA contacts occur, five within the length of the recognition helix, four involving the wing W2 and the minor groove of DNA, and five others distributed throughout the length of the polypeptide chain (141). Binding of this transcription factor causes a deformation in the DNA structure, inducing a 13° bend by narrowing the major groove to allow it to embrace the H3 helix (142). Frequently WH proteins exhibit a length of rosette-like hydrophobic residue side-chains. These side-chains are thought to mediate protein-protein interactions in conjunction with DNA binding (141).

Smad transcriptional co-activators and co-repressors

Although both R-Smads and Co-Smads can bind to the promoter regions of various genes at a specific Smad binding element (SBE), regulation of transcription usually occurs via a partnership with additional proteins (143). Many studies have been performed to identify how these Smad binding proteins modulate the TGF-β signal. Positive regulators of the TGF-β pathway signal include both upstream accessory regulators such as SARA, and downstream effectors that function as either general or tissue-specific transcriptional regulators (144). In addition to positive regulators there have been numerous proteins identified that will attenuate TGF-β signaling by interfering with Smad function, for example anti-Smads. These proteins are required for the prevention of inappropriate activation of TGF-β or play a role in a negative feedback loop (144).

Positive modulators and effectors

AP-1 and TFE3

The members of the AP-1 family, c-Jun and c-Fos, have now been shown to physically interact with Smads. These proteins have long been thought to play a role in

TGF-β target gene regulation due to the identification of intact AP-1 binding sites within the promoter regions of several TGF-β responsive genes such as PAI-1 (145). c-Jun binds directly to amino acid residues within the variable linker region of Smad 3 while c-Fos binds to the MH2 domain. This ligand induced complex, which also includes Smad 4, binds to overlapping or adjacent AP-1 sites and SBEs in target gene promoters. Activation of transcription by this complex has been demonstrated by utilizing the classical TGF-β responsive promoter, PAI-1 (146).

Another HTH transcription factor known to induce the transcription of PAI-1 in response to TGF-β signaling is TFE3. TFE3 has been shown to associate with Smads 3 and 4, and the three proteins bind to specific DNA sequences within the PAI-1 promoter and contribute to maximizing TGF-β target gene activation (147). Recently, data has been obtained to suggest that Smad 3 binds directly to TFE3 and that this interaction is enhanced after the C-terminal serines of Smad 3 are phosphorylated (148). The DNA binding sites used by TFE3 and Smads differ from those employed by the AP-1 family of factors, indicating distinct mechanisms of generating specificity at Smad-responsive promoters. It has not been reported if TFE3 interacts with AP-1 family members (144).

CREB binding protein/p300

CRE binding protein (CBP/CREB) and p300 are well characterized transcriptional co-activators known to enhance gene activation via a host of unrelated proteins through two different mechanisms (149). First, CREB and p300 help to mediate the interaction of sequence-specific transcription factor complexes to the basal transcription machinery required for transcriptional activation (150). The second transcription enhancing feature these proteins possess is that they have histone acetyl transferase (HAT) activity (151). Acetylation of the core histone amino-terminal ends alters nucleosomal structure by "unwinding" the chromatin, thereby increasing the accessibility to the transcriptional machinery (152).

Many studies have shown that CBP/p300 will bind directly to the phosphorylated MH2 domain of the R-Smads, Smad 2 and Smad 3 (153). These studies used a variety of TGF-β-responsive reporter gene complexes in cell culture including p3TP-Lux, SBE-Luc and PAI1-Luc (154). Definitively, the specificity and ultimate requirement of CBP/p300 in these assays was confirmed using the adenoviral protein E1A. E1A protein is an inhibitor of CBP/p300 enhancement and it was observed that inhibition of CBP/p300 blocks the increase of Smad-dependant transactivation by this complex (155).

Vitamin D receptor

Another transcriptional co-activator that specifically interacts with R-Smad, Smad 3, is the vitamin D receptor (VDR). VDR is a vitamin D regulated nuclear receptor which functions with co-activators of the steroid receptor co-activator 1/transcriptional intermediary factor 2 family (144). The VDR-Smad 3 interaction is unique, as the VDR appears to bind to the MH1 domain of Smad 3, not the typical protein-protein interacting domain, MH2 (156). Additionally, activation of TGF-β signaling appears to result in Smad 3 mediated enhancement of VDR dependant transcription. Therefore Smad 3 transcriptional effects are not limited to the activation of transcription of TGF-β target genes, but are also involved in co-operative enhancement of other signal related pathways (157).

Forkhead transcription factors

Currently there are approximately 90 members of the Forkhead superfamily with orthologues expressed in a vast array of species ranging from yeast to man (158). All members exhibit a large degree of sequence homology within their "winged-helix" DNA binding domain. Forkhead activin signal transducer 1 (FAST-1, also known as FOX-H1) is a WH protein identified by its ability to mediate transcriptional induction of activin, a member of the TGF-β superfamily (98). This transcriptional induction by TGF-β and activin has been shown to involve the TGF-β downstream effector proteins, the Smads. In fact, this discovery not only confirmed a role for Smads inside the nucleus, but was also the first indication that the majority of direct Smad binding proteins are transcriptional regulators (144). FAST-1 interacts directly with R-Smad, Smad 2, to form a transcriptionally active complex on the promoter region of the *Xenopus mix.2* gene at a site known as the activin response element (159,160). Subsequently, Smad 3 and Smad 4 were shown to be present within the complex. Smad 2 and 3 associate with FAST-1 however this interaction is not known to occur with Smad 4, as its role resides in DNA binding specificity (160). It has yet to be determined whether or not FAST-1 alone has any transcriptional activity.

Several FAST-1 homologues have been identified in various species including mouse and human. Studies have defined the conserved DNA binding sequence for these transcription factors (TGT G/TT/G ATT), which has also been identified in the activin response element (ARE) of the *Xenopus Mix.2* gene. Additionally, the importance of Smad binding elements adjacent to the FAST-1 binding sequence has been noted (161). Cloning and further analysis of FAST2, a murine FAST-1 homologue, has indicated that its transcriptional regulation is elegant and complex. In a similar fashion to FAST-1 DNA binding, FAST2 complexes with Smad 2 on the *goosecoid (gsc)* promoter to activate target gene transcription. However the FAST2/Smad 3 complex works to inhibit transcription from the same promoter. This is the first case that suggests that Smad 2 and Smad 3 may have opposing roles within selected systems (97).

Negative modulators and effectors

TGIF

TGIF (tumor growth interacting factor), a ubiquitously expressed homeodomain binding protein belonging to the TALE (three amino acid loop extension) family of proteins, has been shown to repress TGF-β-induced transcriptional activity by associating with histone deacetylases (HDAc) (162). TGIF has also been shown to interact with both Smad 2 and Smad 3 (162,163). For these genes, the expression level of TGIF appears to set a maximal response to TGF-β signaling by competing with CBP/p300 for binding to the Smad complex (63). TGIF can bind to the TGF-β target gene promoter element independently of the Smad complex. It is, however, unclear whether this DNA-binding function is involved in the context of Smad-dependant transcription (164).

Ski and SnoN

Ski protein was originally discovered as the gene product of a retroviral oncogene (v-*ski)* that is responsible for transformation of chick embryo fibroblasts and muscle hypertrophy in mice (63). Recently, the cellular counterpart, c-Ski and the related protein SnoN, was discovered to function as a co-repressor that recruits HDACs via an adaptor protein, N-CoR (165). Studies have shown that both c-Ski and SnoN interact with Smad 3 and Smad 4 proteins and act as Smad2/3 co-repressors (166). Unlike TGIF, which is induced by TGF-β stimulation, the interaction of Smads with the Ski and SnoN proteins has been observed under basal conditions and decreases to non-detectable levels after the first hours of TGF-β stimulation (167). This effect is cell type dependant, and may be mediated by TGF-β induced, proteasome-mediated degradation of these two proteins (63).

Possible roles for Ski and SnoN in cardiac post-MI remodeling

The argument has been made that TGF-β regulation of transcription seems to depend on the ability of Smad proteins to recruit and complex together with proteins possessing different chromatin-remodeling activities. TGIF functions as a negative regulator of TGF-β signaling while Ski and SnoN serve to protect against agonist-independent gene activation by Smad proteins. It is thought that a wave of SnoN expression may function as a negative feedback mechanism, terminating TGF-β responses (167). This change is also thought to occur in concert with TGIF, thereby moderating TGF-β signaling (164). Therefore, the relative expression levels of regulator proteins that interact with Smads can be auto-regulated and cross-regulated, resulting in the negative regulation of TGF-β signaling (63).

Enhanced Ski expression in the infarct scar of 8-week post-MI animals

It could be argued that if a cell type is responsive to TGF-β1, enhanced expression of negative modulators may inhibit the effect of the ligand. Could there be an anti-fibrotic effect associated with Ski or Sno-N expression in the chronically healed post-MI scar? To begin to address this issue, we have recently begun investigating the function of Ski in cardiac tissues in post-MI heart. One of our initial observations are depicted in Figure 1, and we describe a marked increase in phosphorylated Ski protein in the cytosolic fraction of cardiac myofibroblasts of the 8 week post-MI rat heart. The rats used in this experiment suffered surgical coronary ligation and were allowed to recover without any form of pharmacologic intervention. After 8 weeks, the animals of all groups (age-matched sham-operated control and MI group) were sacrificed and infarct scar, remnant (surviving) left ventricular myocardial tissues, as well as myocardium from sham-operated hearts were extracted for total proteins as well as the cytosolic protein fraction. The hemodynamic status of the animals was tested and they were found to be in decompensated heart failure with markedly increased LVEDP values *vs* controls. Western analysis (see Figure 1) was employed to detect immunoreactive Ski using a specific commercial primary antibody (Upstate, Lake Placid, NY) and equal protein loading in gel lanes was tested using the Ponceau S staining technique. In healed myocardial infarct scar tissues (populated almost exclusively by myofibroblasts) we noted a marked increase in the appearance of Ski expression in the cytosolic fraction when compared to fractions taken from viable cardiac left ventricle (remote to the infarct scar) and uninfarcted myocardium. No differences were observed between groups when total cardiac extracts were compared among the same groups. The cytosolic localization of Ski overexpression within the cell raises many questions about its function, i.e., is the enhanced bioavailability of Ski in place to provide rapid modulation ofTGF-β mediated transcriptional events when needed? What is preventing Ski from moving to the nucleus to modify histones? These questions require further investigation to be resolved. Precisely how enhanced Ski expression is linked to modulation of myofibroblast activation begets experimentation that addresses wound healing at various time points in the evolution of cardiac infarct scar formation. An understanding of the role of negative regulators and effectors of R-Smads may provide the rationale for a novel mode of therapy that may provide a means to target progressive cardiac fibrosis and the development of congestive heart failure after MI.

ACKNOWLEDGEMENTS

This study was supported by a grant from the Canadian Institutes of Health Research and the Heart and Stroke Foundation of Manitoba. BW and DHF are supported by PDF awards from the Manitoba Health Research Council. IMCD is a CIHR group scientist (CIHR group in Experimental Cardiology).

REFERENCES

1. Dell'Orbo C. 2000. Structural molecules of connective tissue matrices. Ital J Anat Embryol 105:51–96.
2. Alberts B. Molecular Biology of the Cell third ed. Bray D, Lewis J, Raff M, Roberts K, Watson JD. 1994. New York, Garland Publishing Inc.
3. Bendeck MP, Zempo N, Clowes AW, Galardy RE, Reidy MA. 1994. Smooth muscle cell migration and matrix metalloproteinase expression after arterial injury in the rat. Circ Res 75:539–545.
4. 1991. Effect of enalapril on survival in patients with reduced left ventricular ejection fractions and congestive heart failure. The SOLVD Investigators. N Engl J Med 325:293–302.
5. Scott JE. 1992. Supramolecular organization of extracellular matrix glycosaminoglycans, in vitro and in the tissues. FASEB J 6:2639–2645.
6. Honda A, Murota SI, Mori Y. 1982. Comparative study on glycosaminoglycan-sulfotransferases in rat costal cartilage and chick embryo cartilage. Comp Biochem Physiol B 71:41–47.
7. Ruoslahti E, Yamaguchi Y. 1991. Proteoglycans as modulators of growth factor activities. Cell 64: 867–869.
8. Martin GR, Timpl R. 1987. Laminin and other basement membrane components. Annu Rev Cell Biol 3:57–85.

9. Tryggvason K. 1993. The laminin family. Curr Opin Cell Biol 5:877–882.
10. Kreis T. Guidebook to the extracellular matrix and adhesion proteins. Vale R. 1993. London, Oxford University Press.
11. Lundgren E, Gullberg D, Rubin K, Borg TK, Terracio MJ, Terracio L. 1988. In vitro studies on adult cardiac myocytes: attachment and biosynthesis of collagen type IV and laminin. J Cell Physiol 136:43–53.
12. Lundgren E, Terracio L, Mardh S, Borg TK. 1985. Extracellular matrix components influence the survival of adult cardiac myocytes in vitro. Exp Cell Res 158:371–381.
13. Hynes RO. 1986. Fibronectins. Sci Am 254:42–51.
14. Schwarzbauer JE. 1991. Alternative splicing of fibronectin: three variants, three functions. Bioessays 13:527–533.
15. Cleary EG, Gibson MA. 1983. Elastin-associated microfibrils and microfibrillar proteins. Int Rev Connect Tissue Res 10:97–209.
16. Robinson TF. 1983. The skeletal framework of mammalian heart muscle: arrangement of inter- and pericellular connective tissue structures. Cohen-Gould L, Factor SM. Lab Invest 49:482–487.
17. Borg TK. 1982. Functional arrangement of connective tissue in striated muscle with emphasis on cardiac muscle. Sullivan T, Ivy J. SEM 4:1775–1784.
18. Ju H, Dixon IMC. 1996. Extracellular matrix and cardiovascular diseases. Can J Cardiol 12: 1259–1267.
19. Weber KT, Sun Y, Tyagi SC, Cleutjens JP. 1994. Collagen network of the myocardium: function, structural remodeling and regulatory mechanisms. J Mol Cell Cardiol 26:279–292.
20. Bornstein P, Horlein D, McPherson J. 1984. Regulation of collagen synthesis. Prog Clin Biol Res 154:61–80.
21. Eghbali M, Weber KT. 1990. Collagen and the myocardium: fibrillar structure, biosynthesis and degradation in relation to hypertrophy and its regression. Mol Cell Biochem 96:1–14.
22. Burgeson RE. 1988. New collagens, new concepts. Annu Rev Cell Biol 4:551–577.
23. Pelouch V, Dixon IMC, Golfman L, Beamish RE, Dhalla NS. 1993. Role of extracellular matrix proteins in heart function. Mol Cell Biochem 129:101–120.
24. Bishop JE, Laurent GJ. 1995. Collagen turnover and its regulation in the normal and hypertrophying heart. Eur Heart J 16 Suppl C:38–44.
25. Weber KT. 1989. Cardiac interstitium in health and disease: the fibrillar collagen network. J Am Coll Cardiol 13:1637–1652.
26. Weber KT, Brilla CG. 1991. Pathological hypertrophy and cardiac interstitium. Fibrosis and renin-angiotensin-aldosterone system. Circulation 83:1849–1865.
27. Janicki JS. Molecular and subcellular cardiology: effects of structure and function. Brower G, Henegar J, Wang L. 1995. New York, Plenum Press.
28. Sutton MG, Sharpe N. 2000. Left ventricular remodeling after myocardial infarction: pathophysiology and therapy. Circulation 101:2981–2988.
29. Jaffe R. Analytical and quantitative cardiology. Flugelman MY, Halon DA, Lewis RB. 257–266. 1997. New York, Plenum Press.
30. Weber KT, Sun Y, Katwa LC. 1996. Wound healing following myocardial infarction. Clin Cardiol 19:447–455.
31. Anversa P, Sonnenblick EH. 1990. Ischemic cardiomyopathy: pathophysiologic mechanisms. Prog Cardiovasc Dis 33:49–70.
32. Olivetti G, Quaini F, Lagrasta C, Cigola E, Ricci R, Maestri R, Anversa P. 1995. Cellular basis of ventricular remodeling after myocardial infarction in rats. Cardioscience 6:101–106.
33. Berne RM, Physiology, Levy MN, Koeppen BM, Stanton BA. 2001. St Louis, MI, Mosby Inc.
34. Tun A, Khan IA. 2001. Myocardial infarction with normal coronary arteries: the pathologic and clinical perspectives. Angiology 52:299–304.
35. Maisch B. 1996. Ventricular remodeling. Cardiology 87 Suppl 1:2–10.
36. McKay RG, Pfeffer MA, Pasternak RC, Markis JE, Come PC, Nakao S, Alderman JD, Ferguson JJ, Safian RD, Grossman W. 1986. Left ventricular remodeling after myocardial infarction: a corollary to infarct expansion. Circulation 74:693–702.
37. Gerdes AM, Capasso JM. 1995. Structural remodeling and mechanical dysfunction of cardiac myocytes in heart failure. J Mol Cell Cardiol 27:849–856.
38. Dietz KJ. Ventricular remodeling after acute myocardial infarction. Osterziel R, Gulba DC, Harsdorf RV. Thromb Haemost 82:73–75. 1999.

39. Villarreal FJ, Dillmann WH. 1992. Cardiac hypertrophy-induced changes in mRNA levels for TGF-beta 1, fibronectin, and collagen. Am J Physiol 262:H1861–H1866.
40. Bishop JE, Lindahl G. 1999. Regulation of cardiovascular collagen synthesis by mechanical load. Cardiovasc Res 42:27–44.
41. Bishop JE, Rhodes S, Laurent GJ, Low RB, Stirewalt WS. 1994. Increased collagen synthesis and decreased collagen degradation in right ventricular hypertrophy induced by pressure overload. Cardiovasc Res 28:1581–1585.
42. Cobb MH. 1999. MAP kinase pathways. Prog Biophys Mol Biol 71:479–500.
43. Kim S, Iwao H. 1999. Activation of mitogen-activated protein kinases in cardiovascular hypertrophy and remodeling. Jpn J Pharmacol 80:97–102.
44. Liberati NT, Datto MB, Frederick JP, Shen X, Wong C, Rougier-Chapman EM, Wang XF. 1999. Smads bind directly to the Jun family of AP-1 transcription factors. Proc Natl Acad Sci USA 96:4844–4849.
45. Narula J, Hajjar RJ, Dec GW. 1998. Apoptosis in the failing heart. Cardiol Clin 16:691–710, ix.
46. Feuerstein GZ, Young PR. 2000. Apoptosis in cardiac diseases: stress- and mitogen-activated signaling pathways. Cardiovasc Res 45:560–569.
47. Schwartz K, Mercadier JJ. 1996. Molecular and cellular biology of heart failure. Curr Opin Cardiol 11:227–236.
48. Bristow MR, Ginsburg R, Umans V, Fowler M, Minobe W, Rasmussen R, Zera P, Menlove R, Shah P, Jamieson S, et al. 1986. Beta 1- and beta 2-adrenergic-receptor subpopulations in nonfailing and failing human ventricular myocardium: coupling of both receptor subtypes to muscle contraction and selective beta 1-receptor down-regulation in heart failure. Circ Res 59: 297–309.
49. Lombardi WL, Gilbert EM. 2000. The effects of neurohormonal antagonism on pathologic left ventricular remodeling in heart failure. Curr Cardiol Rep 2:90–98.
50. Dostal DE. 2000. The cardiac renin-angiotensin system: novel signaling mechanisms related to cardiac growth and function. Regul Pept 91:1–11.
51. Rosendorff C. 1996. The renin-angiotensin system and vascular hypertrophy. J Am Coll Cardiol 28:803–812.
52. Danser AH. 1996. Local renin-angiotensin systems. Mol Cell Biochem 157:211–216.
53. Wong PC, Hart SD, Zaspel AM, Chiu AT, Ardecky RJ, Smith RD, Timmermans PB. 1990. Functional studies of nonpeptide angiotensin II receptor subtype-specific ligands: DuP 753 (AII-1) and PD123177 (AII-2). J Pharmacol Exp Ther 255:584–592.
54. Goodfriend TL, Elliott ME, Catt KJ. 1996. Angiotensin receptors and their antagonists. N Engl J Med 334:1649–1654.
55. Dixon IMC, Ju H, Reid NL, Scammell-La Fleur T, Werner JP, Jasmin G. 1997. Cardiac collagen remodeling in the cardiomyopathic Syrian hamster and the effect of losartan. J Mol Cell Cardiol 29:1837–1850.
56. Kijima K, Matsubara H, Murasawa S, Maruyama K, Mori Y, Ohkubo N, Komuro I, Yazaki Y, Iwasaka T, Inada M. 1996. Mechanical stretch induces enhanced expression of angiotensin II receptor subtypes in neonatal rat cardiac myocytes. Circ Res 79:887–897.
57. Tsutsumi Y, Matsubara H, Ohkubo N, Mori Y, Nozawa Y, Murasawa S, Kijima K, Maruyama K, Masaki H, Moriguchi Y, et al. 1998. Angiotensin II type 2 receptor is upregulated in human heart with interstitial fibrosis, and cardiac fibroblasts are the major cell type for its expression. Circ Res 83:1035–1046.
58. Campbell SE, Katwa LC. 1997. Angiotensin II stimulated expression of transforming growth factor-beta1 in cardiac fibroblasts and myofibroblasts. J Mol Cell Cardiol 29:1947–1958.
59. Omura T, Kim S, Takeuchi K, Iwao H, Takeda T. 1994. Transforming growth factor beta 1 and extracellular matrix gene expression in isoprenaline induced cardiac hypertrophy: effects of inhibition of the renin-angiotensin system. Cardiovasc Res 28:1835–1842.
60. Massague J. 2000. How cells read TGF-beta signals. Nat Rev Mol Cell Biol 1:169–178.
61. Massague J. 1990. The transforming growth factor-beta family. Annu Rev Cell Biol 6:597–641.
62. Massague J. 1998. TGF-beta signal transduction. Annu Rev Biochem 67:753–791.
63. Massague J, Wotton D. 2000. Transcriptional control by the TGF-beta/Smad signaling system. EMBO J 19:1745–1754.
64. Derynck R, Jarrett JA, Chen EY, Eaton DH, Bell JR, Assoian RK, Roberts AB, Sporn MB, Goeddel DV. 1985. Human transforming growth factor-beta complementary DNA sequence and expression in normal and transformed cells. Nature 316:701–705.

65. Gentry LE, Webb NR, Lim GJ, Brunner AM, Ranchalis JE, Twardzik DR, Lioubin MN, Marquardt H, Purchio AF. 1987. Type 1 transforming growth factor beta: amplified expression and secretion of mature and precursor polypeptides in Chinese hamster ovary cells. Mol Cell Biol 7:3418–3427.
66. Miyazono K, Ichijo H, Heldin CH. 1993. Transforming growth factor-beta: latent forms, binding proteins and receptors. Growth Factors 8:11–22.
67. Lawrence DA. 1995. Transforming growth factor-beta: an overview. Kidney Int Suppl 49:S19–S23.
68. Eghbali M. 1989. Cellular origin and distribution of transforming growth factor-beta in the normal rat myocardium. Cell Tissue Res 256:553–558.
69. Border WA, Noble NA. 1994. Transforming growth factor beta in tissue fibrosis. N Engl J Med 331:1286–1292.
70. Hao J, Ju H, Zhao S, Junaid A, Scammell-La Fleur T, Dixon IMC. 1999. Elevation of expression of Smads 2, 3, and 4, decorin and TGF-beta in the chronic phase of myocardial infarct scar healing. J Mol Cell Cardiol 31:667–678.
71. Mathews LS, Vale WW. 1991. Expression cloning of an activin receptor, a predicted transmembrane serine kinase. Cell 65:973–982.
72. Wrana JL, Tran H, Attisano L, Arora K, Childs SR, Massague J, O'Connor MB. 1994. Two distinct transmembrane serine/threonine kinases from Drosophila melanogaster form an activin receptor complex. Mol Cell Biol 14:944–950.
73. Luo K, Lodish HF. 1997. Positive and negative regulation of type II TGF-beta receptor signal transduction by autophosphorylation on multiple serine residues. EMBO J 16:1970–1981.
74. Souchelnytskyi S, ten Dijke P, Miyazono K, Heldin CH. 1996. Phosphorylation of Ser165 in TGF-beta type I receptor modulates TGF-beta1-induced cellular responses. EMBO J 15:6231–6240.
75. Josso N, Cate RL, Picard JY, Vigier B, di Clemente N, Wilson C, Imbeaud S, Pepinsky RB, Guerrier D, Boussin L, et al. 1993. Anti-mullerian hormone: the Jost factor. Recent Prog Horm Res 48:1–59.
76. Wieser R, Wrana JL, Massague J. 1995. GS domain mutations that constitutively activate T beta R-I, the downstream signaling component in the TGF-beta receptor complex. EMBO J 14:2199–2208.
77. Macias-Silva M, Abdollah S, Hoodless PA, Pirone R, Attisano L, Wrana JL. 1996. MADR2 is a substrate of the TGFbeta receptor and its phosphorylation is required for nuclear accumulation and signaling. Cell 87:1215–1224.
78. Wieser R, Attisano L, Wrana JL, Massague J. 1993. Signaling activity of transforming growth factor beta type II receptors lacking specific domains in the cytoplasmic region. Mol Cell Biol 13:7239–7247.
79. Ullrich A, Schlessinger J. 1990. Signal transduction by receptors with tyrosine kinase activity. Cell 61:203–212.
80. Liu F, Ventura F, Doody J, Massague J. 1995. Human type II receptor for bone morphogenic proteins (BMPs): extension of the two-kinase receptor model to the BMPs. Mol Cell Biol 15:3479–3486.
81. Sekelsky JJ, Newfeld SJ, Raftery LA, Chartoff EH, Gelbart WM. 1995. Genetic characterization and cloning of mothers against dpp, a gene required for decapentaplegic function in Drosophila melanogaster. Genetics 139:1347–1358.
82. Savage C, Das P, Finelli AL, Townsend SR, Sun CY, Baird SE, Padgett RW. 1996. Caenorhabditis elegans genes sma-2, sma-3, and sma-4 define a conserved family of transforming growth factor beta pathway components. Proc Natl Acad Sci USA 93:790–794.
83. Hahn SA, Schutte M, Hoque AT, Moskaluk CA, da Costa LT, Rozenblum E, Weinstein CL, Fischer A, Yeo CJ, Hruban RH, et al. 1996. DPC4, a candidate tumor suppressor gene at human chromosome 18q21.1. Science 271:350–353.
84. Raftery LA, Sutherland DJ. 1999. TGF-beta family signal transduction in Drosophila development: from Mad to Smads. Dev Biol 210:251–268.
85. Hata A, Lo RS, Wotton D, Lagna G, Massague J. 1997. Mutations increasing autoinhibition inactivate tumour suppressors Smad2 and Smad4. Nature 388:82–87.
86. Chen YG, Hata A, Lo RS, Wotton D, Shi Y, Pavletich N, Massague J. 1998. Determinants of specificity in TGF-beta signal transduction. Genes Dev 12:2144–2152.
87. Kretzschmar M, Liu F, Hata A, Doody J, Massague J. 1997. The TGF-beta family mediator Smad1 is phosphorylated directly and activated functionally by the BMP receptor kinase. Genes Dev 11:984–995.

88. Wisotzkey RG, Mehra A, Sutherland DJ, Dobens LL, Liu X, Dohrmann C, Attisano L, Raftery LA. 1998. Medea is a Drosophila Smad4 homolog that is differentially required to potentiate DPP responses. Development 125:1433–1445.
89. Tsukazaki T, Chiang TA, Davison AF, Attisano L, Wrana JL. 1998. SARA, a FYVE domain protein that recruits Smad2 to the TGFbeta receptor. Cell 95:779–791.
90. Nakao A, Imamura T, Souchelnytskyi S, Kawabata M, Ishisaki A, Oeda E, Tamaki K, Hanai J, Heldin CH, Miyazono K, et al. 1997. TGF-beta receptor-mediated signalling through Smad2, Smad3 and Smad4. EMBO J 16:5353–5362.
91. Wurthner JU, Frank DB, Felici A, Green HM, Cao Z, Schneider MD, McNally JG, Lechleider RJ, Roberts AB. 2001. Transforming growth factor-beta receptor-associated protein 1 is a Smad4 chaperone. J Biol Chem 276:19495–19502.
92. Liu F, Hata A, Baker JC, Doody J, Carcamo J, Harland RM, Massague J. 1996. A human Mad protein acting as a BMP-regulated transcriptional activator. Nature 381:620–623.
93. Hoodless PA, Haerry T, Abdollah S, Stapleton M, O'Connor MB, Attisano L, Wrana JL. 1996. MADR1, a MAD-related protein that functions in BMP2 signaling pathways. Cell 85:489–500.
94. Zhang Y, Feng X, We R, Derynck R. 1996. Receptor-associated Mad homologues synergize as effectors of the TGF-beta response. Nature 383:168–172.
95. Das P, Maduzia LL, Wang H, Finelli AL, Cho SH, Smith MM, Padgett RW. 1998. The Drosophila gene Medea demonstrates the requirement for different classes of Smads in dpp signaling. Development 125:1519–1528.
96. Attisano L, Silvestri C, Izzi L, Labbe E. 2001. The transcriptional role of Smads and FAST (FoxH1) in TGFbeta and activin signalling. Mol Cell Endocrinol 180:3–11.
97. Labbe E, Silvestri C, Hoodless PA, Wrana JL, Attisano L. 1998. Smad2 and Smad3 positively and negatively regulate TGF beta-dependent transcription through the forkhead DNA-binding protein FAST2. Mol Cell 2:109–120.
98. Chen X, Rubock MJ, Whitman M. 1996. A transcriptional partner for MAD proteins in TGF-beta signalling. Nature 383:691–696.
99. Nakao A, Afrakhte M, Moren A, Nakayama T, Christian JL, Heuchel R, Itoh S, Kawabata M, Heldin NE, Heldin CH, et al. 1997. Identification of Smad7, a TGFbeta-inducible antagonist of TGF-beta signalling. Nature 389:631–635.
100. Imamura T, Takase M, Nishihara A, Oeda E, Hanai J, Kawabata M, Miyazono K. 1997. Smad6 inhibits signalling by the TGF-beta superfamily. Nature 389:622–626.
101. Hayashi H, Abdollah S, Qiu Y, Cai J, Xu YY, Grinnell BW, Richardson MA, Topper JN, Gimbrone MA, Jr, Wrana JL, et al. 1997. The MAD-related protein Smad7 associates with the TGFbeta receptor and functions as an antagonist of TGFbeta signaling. Cell 89:1165–1173.
102. Itoh S, Landstrom M, Hermansson A, Itoh F, Heldin CH, Heldin NE, ten Dijke P. 1998. Transforming growth factor beta1 induces nuclear export of inhibitory Smad7. J Biol Chem 273: 29195–29201.
103. Wang B, Hao J, Jones SC, Yee M, Roth JC, Dixon IMC. 2002. Overexpression of Smad 7 attenuates collagen synthesis in cardiac fibroblasts; Smad 7 in cardiac fibroblasts. Am J Physiol (Heart and Circ Physiol), in press.
104. Visser JA, Themmen AP. 1998. Downstream factors in transforming growth factor-beta family signaling. Mol Cell Endocrinol 146:7–17.
105. Gilman AG. 1987. G proteins: transducers of receptor-generated signals. Annu Rev Biochem 56:615–649.
106. Bourne HR, Sanders DA, McCormick F. 1990. The GTPase superfamily: a conserved switch for diverse cell functions. Nature 348:125–132.
107. Howe PH, Leof EB. 1989. Transforming growth factor beta 1 treatment of AKR-2B cells is coupled through a pertussis-toxin-sensitive G-protein(s). Biochem J 261:879–888.
108. Kataoka R, Sherlock J, Lanier SM. 1993. Signaling events initiated by transforming growth factor-beta 1 that require Gi alpha 1. J Biol Chem 268:19851–19857.
109. McCormick F. 1993. Signal transduction. How receptors turn Ras on. Nature 363:15–16.
110. Egan SE, Weinberg RA. 1993. The pathway to signal achievement. Nature 365:781–783.
111. Mulder KM, Morris SL. 1992. Activation of p21ras by transforming growth factor beta in epithelial cells. J Biol Chem 267:5029–5031.
112. Frey RS, Mulder KM. 1997. Involvement of extracellular signal-regulated kinase 2 and stress-activated protein kinase/Jun N-terminal kinase activation by transforming growth factor beta in the negative growth control of breast cancer cells. Cancer Res 57:628–633.

113. Atfi A, Djelloul S, Chastre E, Davis R, Gespach C. 1997. Evidence for a role of Rho-like GTPases and stress-activated protein kinase/c-Jun N-terminal kinase (SAPK/JNK) in transforming growth factor beta-mediated signaling. J Biol Chem 272:1429–1432.
114. Hao J, Wang B, Jones SC, Jassal DS, Dixon IM. 2000. Interaction between angiotensin II and Smad proteins in fibroblasts in failing heart and in vitro. Am J Physiol Heart Circ Physiol 279: H3020–H3030.
115. Green MR. 2000. TBP-associated factors (TAFIIs): multiple, selective transcriptional mediators in common complexes. Trends Biochem Sci 25:59–63.
116. Muller CW. 2001. Transcription factors: global and detailed views. Curr Opin Struct Biol 11:26–32.
117. Cramer P, Bushnell DA, Fu J, Gnatt AL, Maier-Davis B, Thompson NE, Burgess RR, Edwards AM, David PR, Kornberg RD. 2000. Architecture of RNA polymerase II and implications for the transcription mechanism. Science 288:640–649.
118. Harrison SC. 1991. A structural taxonomy of DNA-binding domains. Nature 353:715–719.
119. Pabo CO, Sauer RT. 1992. Transcription factors: structural families and principles of DNA recognition. Annu Rev Biochem 61:1053–1095.
120. Murre C, McCaw PS, Baltimore D. 1989. A new DNA binding and dimerization motif in immunoglobulin enhancer binding, daughterless, MyoD, and myc proteins. Cell 56:777–783.
121. Harrison SC, Aggarwal AK. 1990. DNA recognition by proteins with the helix-turn-helix motif. Annu Rev Biochem 59:933–969.
122. Laughon A, Scott MP. 1984. Sequence of a Drosophila segmentation gene: protein structure homology with DNA-binding proteins. Nature 310:25–31.
123. Sauer RT, Yocum RR, Doolittle RF, Lewis M, Pabo CO. 1982. Homology among DNA-binding proteins suggests use of a conserved super-secondary structure. Nature 298:447–451.
124. Scott MP, Tamkun JW, Hartzell GW, III. 1989. The structure and function of the homeodomain. Biochim Biophys Acta 989:25–48.
125. Wuthrich K. Transcriptional regulation by homeodomain proteins: structural, functional and genetic aspects. Transcriptional Regulation. 1992. Cold Springs Harbor, New York, Cold Springs Harbor Laboratory Press.
126. Laity JH, Lee BM, Wright PE. 2001. Zinc finger proteins: new insights into structural and functional diversity. Curr Opin Struct Biol 11:39–46.
127. Rhodes D, Klug A. 1993. Zinc fingers. Sci Am 268:56–65.
128. Lamb P, McKnight SL. 1991. Diversity and specificity in transcriptional regulation: the benefits of heterotypic dimerization. Trends Biochem Sci 16:417–422.
129. Landschulz WH, Johnson PF, McKnight SL. 1988. The leucine zipper: a hypothetical structure common to a new class of DNA binding proteins. Science 240:1759–1764.
130. O'Shea EK, Rutkowski R, Kim PS. 1989. Evidence that the leucine zipper is a coiled coil. Science 243:538–542.
131. McKnight SL. 1991. Molecular zippers in gene regulation. Sci Am 264:54–64.
132. Kaufmann E, Knochel W. 1996. Five years on the wings of fork head. Mech Dev 57:3–20.
133. Weigel D, Jurgens G, Kuttner F, Seifert E, Jackle H. 1989. The homeotic gene fork head encodes a nuclear protein and is expressed in the terminal regions of the Drosophila embryo. Cell 57:645–658
134. Lai E, Clark KL, Burley SK, Darnell JE, Jr. 1993. Hepatocyte nuclear factor 3/fork head or "winged helix" proteins: a family of transcription factors of diverse biologic function. Proc Natl Acad Sci USA 90:10421–10423.
135. Lai E, Prezioso VR, Smith E, Litvin O, Costa RH, Darnell JE, Jr. 1990. HNF-3A, a hepatocyte-enriched transcription factor of novel structure is regulated transcriptionally. Genes Dev 4: 1427–1436.
136. Weigel D, Jackle H. 1990. The fork head domain: a novel DNA binding motif of eukaryotic transcription factors? Cell 63:455–456.
137. Winnier GE, Hargett L, Hogan BL. 1997. The winged helix transcription factor MFH1 is required for proliferation and patterning of paraxial mesoderm in the mouse embryo. Genes Dev 11: 926–940.
138. Xuan S, Baptista CA, Balas G, Tao W, Soares VC, Lai E. 1995. Winged helix transcription factor BF-1 is essential for the development of the cerebral hemispheres. Neuron 14:1141–1152.
139. Clark KL, Halay ED, Lai E, Burley SK. 1993. Co-crystal structure of the HNF-3/fork head DNA-recognition motif resembles histone H5. Nature 364:412–420.

140. Marsden I, Chen Y, Jin C, Liao X. 1997. Evidence that the DNA binding specificity of winged helix proteins is mediated by a structural change in the amino acid sequence adjacent to the principal DNA binding helix. Biochemistry 36:13248–13255.
141. Gajiwala KS, Burley SK. 2000. Winged helix proteins. Curr Opin Struct Biol 10:110–116.
142. Jin C, Liao X. 1999. Backbone dynamics of a winged helix protein and its DNA complex at different temperatures: changes of internal motions in genesis upon binding to DNA. J Mol Biol 292:641–651.
143. Zawel L, Dai JL, Buckhaults P, Zhou S, Kinzler KW, Vogelstein B, Kern SE. 1998. Human Smad3 and Smad4 are sequence-specific transcription activators. Mol Cell 1:611–617.
144. Zimmerman CM, Padgett RW. 2000. Transforming growth factor beta signaling mediators and modulators. Gene 249:17–30.
145. Zhang Y, Feng XH, Derynck R. 1998. Smad3 and Smad4 cooperate with c-Jun/c-Fos to mediate TGF-beta-induced transcription. Nature 394:909–913.
146. Wong C, Rougier-Chapman EM, Frederick JP, Datto MB, Liberati NT, Li JM, Wang XF. 1999. Smad3-Smad4 and AP-1 complexes synergize in transcriptional activation of the c-Jun promoter by transforming growth factor beta. Mol Cell Biol 19:1821–1830.
147. Hua X, Liu X, Ansari DO, Lodish HF. 1998. Synergistic cooperation of TFE3 and smad proteins in TGF-beta-induced transcription of the plasminogen activator inhibitor-1 gene. Genes Dev 12: 3084–3095.
148. Hua X, Miller ZA, Wu G, Shi Y, Lodish HF. 1999. Specificity in transforming growth factor beta-induced transcription of the plasminogen activator inhibitor-1 gene: interactions of promoter DNA, transcription factor muE3, and Smad proteins. Proc Natl Acad Sci USA 96:13130–13135.
149. Giordano A, Avantaggiati ML. 1999. p300 and CBP: partners for life and death. J Cell Physiol 181: 218–230.
150. Feng XH, Zhang Y, Wu RY, Derynck R. 1998. The tumor suppressor Smad4/DPC4 and transcriptional adaptor CBP/p300 are coactivators for smad3 in TGF-beta-induced transcriptional activation. Genes Dev 12:2153–2163.
151. Ogryzko VV, Schiltz RL, Russanova V, Howard BH, Nakatani Y. 1996. The transcriptional coactivators p300 and CBP are histone acetyltransferases. Cell 87:953–959.
152. Lee DY, Hayes JJ, Pruss D, Wolffe AP. 1993. A positive role for histone acetylation in transcription factor access to nucleosomal DNA. Cell 72:73–84.
153. Janknecht R, Wells NJ, Hunter T. 1998. TGF-beta-stimulated cooperation of smad proteins with the coactivators CBP/p300. Genes Dev 12:2114–2119.
154. Pouponnot C, Jayaraman L, Massague J. 1998. Physical and functional interaction of SMADs and p300/CBP. J Biol Chem 273:22865–22868.
155. Nishihara A, Hanai J, Imamura T, Miyazono K, Kawabata M. 1999. E1A inhibits transforming growth factor-beta signaling through binding to Smad proteins. J Biol Chem 274: 28716–28723.
156. Yanagi Y, Suzawa M, Kawabata M, Miyazono K, Yanagisawa J, Kato S. 1999. Positive and negative modulation of vitamin D receptor function by transforming growth factor-beta signaling through smad proteins. J Biol Chem 274:12971–12974.
157. Yanagisawa J, Yanagi Y, Masuhiro Y, Suzawa M, Watanabe M, Kashiwagi K, Toriyabe T, Kawabata M, Miyazono K, Kato S. 1999. Convergence of transforming growth factor-beta and vitamin D signaling pathways on SMAD transcriptional coactivators. Science 283:1317–1321.
158. Kops GJ, Burgering BM. 1999. Forkhead transcription factors: new insights into protein kinase B (c-akt) signaling. J Mol Med 77:656–665.
159. Chen X, Weisberg E, Fridmacher V, Watanabe M, Naco G, Whitman M. 1997. Smad4 and FAST-1 in the assembly of activin-responsive factor. Nature 389:85–89.
160. Liu F, Pouponnot C, Massague J. 1997. Dual role of the Smad4/DPC4 tumor suppressor in TGFbeta-inducible transcriptional complexes. Genes Dev 11:3157–3167.
161. Zhou S, Zawel L, Lengauer C, Kinzler KW, Vogelstein B. 1998. Characterization of human FAST-1, a TGF beta and activin signal transducer. Mol Cell 2:121–127.
162. Wotton D, Lo RS, Lee S, Massague J. 1999. A Smad transcriptional corepressor. Cell 97:29–39.
163. Bertolino E, Reimund B, Wildt-Perinic D, Clerc RG. 1995. A novel homeobox protein which recognizes a TGT core and functionally interferes with a retinoid-responsive motif. J Biol Chem 270: 31178–31188.
164. Wotton D, Lo RS, Swaby LA, Massague J. 1999. Multiple modes of repression by the Smad transcriptional corepressor TGIF. J Biol Chem 274:37105–37110.

165. Luo K, Stroschein SL, Wang W, Chen D, Martens E, Zhou S, Zhou Q. 1999. The Ski oncoprotein interacts with the Smad proteins to repress TGFbeta signaling. Genes Dev 13:2196–2206.
166. Akiyoshi S, Inoue H, Hanai J, Kusanagi K, Nemoto N, Miyazono K, Kawabata M. 1999. c-Ski acts as a transcriptional co-repressor in transforming growth factor-beta signaling through interaction with smads. J Biol Chem 274:35269–35277.
167. Stroschein SL, Wang W, Zhou S, Zhou Q, Luo K. 1999. Negative feedback regulation of TGF-beta signaling by the SnoN oncoprotein. Science 286:771–774.

Pathophysiology of Cardiovascular Disease,
edited by N.S. Dhalla, H. Rupp,
A. Angel & G.N. Pierce
Kluwer Academic Publishers, Boston, 2004

Phytoestrogens and Cardiovascular Disorders

Ricky Y.K. Man, Susan W.S. Leung, Hwee Teoh, Adrian Quan, Wendy Keung, and Mary Y.K. Lee

Department of Pharmacology
Faculty of Medicine
University of Hong Kong
Level 2, Laboratory Block
21 Sassoon Road
Hong Kong

Summary. Dietary intake of phytoestrogens has been associated with a reduction in risk of cardiovascular disorders including coronary heart disease, hypertension and atherosclerosis. Phytoestrogens have long been found to be structurally similar to the female sex hormone, 17β-estradiol, and have a wide range of estrogenic effects. Like 17β-estradiol, phytoestrogens such as genistein have been suggested to exert its cardioprotective actions through favourable alteration of serum lipoprotein levels. There is also evidence indicating the role of isoflavones as tyrosine kinase inhibitors, calcium channel antagonists as well as antioxidants. Since genistein has a higher affinity for estrogen receptor β than for estrogen receptor α, it is predicted to have a preferential protective effect on the cardiovascular system without significant actions on the reproductive system. Therefore, genistein and related compounds may have superiority over 17β-estradiol as potential therapeutic agent.

Key words: Estrogen receptors, Genistein, Lipoproteins, Tyrosine kinase, Vasodilatation.

INTRODUCTION

Cardiovascular disease (CVD) is one of the leading disorders in the world, resulting in severe morbidity and mortality (1). Epidemiological studies have indicated a lower incidence of coronary heart disease in women compared to age-matched man (2). As this female superiority disappears after the menopause (2–4), the major female

Corresponding author: Ricky Y. K. Man, Department of Pharmacology, Faculty of Medicine, University of Hong Kong, Level 2, Laboratory Block, 21 Sassoon Road, Hong Kong. Phone: (852) 2819 9250, Fax: (852) 2817 0859, E-mail: rykman@hkucc.hku.hk.

sex hormones, 17β-estradiol, is believed to be cardioprotective. This postulation was further supported by the findings that estrogen replacement therapy reduced CAD risk and mortality in postmenopausal women (5–7).

Unfortunately, the therapeutic potential of 17β-estradiol is hampered by its actions on the reproductive systems, which, in turn, result in adverse effects including endometrial hyperplasia and breast cancer (8,9). Recently, the novel estrogen receptor (ER) β was identified (10,11) and was later found to be the predominant subtype in human smooth muscle cells (12), whereas ERα appeared to be predominantly expressed in the uterus and the mammary glands (13). As such, ligands with selectivity for ERβ over ERα may have the advantage of producing cardiovascular protective effects with minimal risks. There have been reports that the naturally occurring phytoestrogens bind preferentially to ERβ (14,15). Among this group of compound, genistein is found to have an ERβ binding affinity close to that of 17β-estradiol, and the most pronounced selectivity for ERβ compared to ERα (15,16).

PHYTOESTROGENS: BACKGROUND

Phytoestrogens are a group of plant-derived nonsteroidal compounds that possess estrogenic activity (17,18). There are three main types of phytoestrogens: isoflavones, coumestans and lignans. They are all characterized by the presence of diphenolic rings and are structurally similar to 17β-estradiol (Figure 1). The most commonly studied phytoestrogen is genistein, partially because it shows the highest estrogenic properties. Genistein is the major component of the dietary isoflavones, which are found extensively in soy proteins. Interestingly, soy-rich diets have been associated with a lower incidence of CVD in Asia compared to Western countries (19,20). It is, therefore, believed that genistein has cardioprotective effects, and these effects may correlate with its estrogenic activity.

BIOLOGICAL EFFECTS OF GENISTEIN

Genistein has been demonstrated to have diverse biological actions. In 1987, Akiyama et al. (21) reported that genistein was a potent inhibitor of the enzyme tyrosine kinase. Genistein has been shown to inhibit several tyrosine kinase-dependent events such as angiotensin II-induced Ras activation and fibroblast growth factor 2 expression (22,23). It also elicits similar effects to another structurally different selective tyrosine kinase inhibitor, tyrphostin (24): e.g., both of them inhibited calcium currents in isolated rabbit ear arterial smooth muscle cells (25), reduced noradrenaline-induced contraction in isolated endothelium-denuded rat aorta (26), and decreased the potency and efficacy of 5-hydroxytryptamine-induced contraction in isolated endothelium-denuded rat carotid arterial strips (27). In contrast, genistein, unlike tyrphostin, failed to inhibit the tyrosine autophosphorylation of the epidermal growth factor receptor in LNCaP prostate cancer cells and MCF-7 breast cancer cells (28,29). Instead of directly inhibiting tyrosine kinase activity, the effect of genistein on this enzyme may occur at the transcriptional level, result-

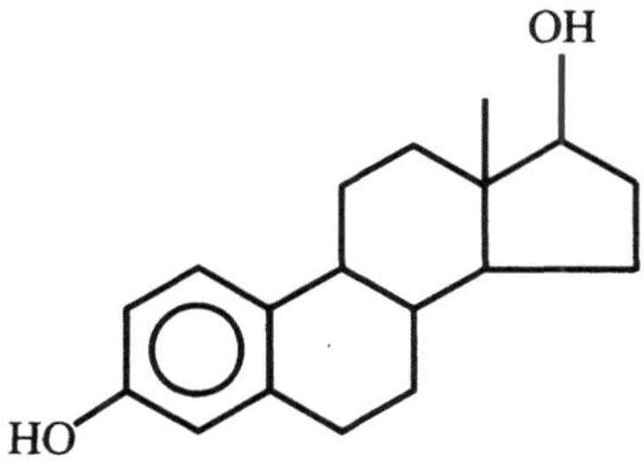

17β-estrodiol
(estrogen)

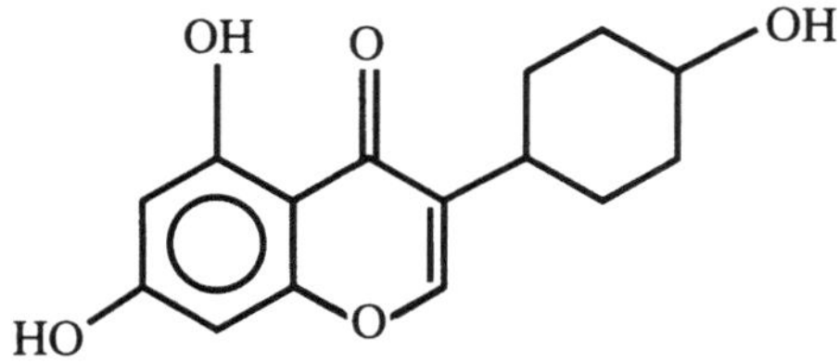

Genistein
(Isoflavones)

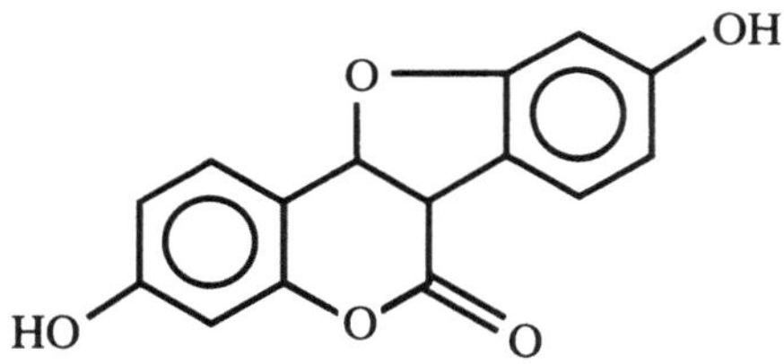

Coumestrol
(Coumestan)

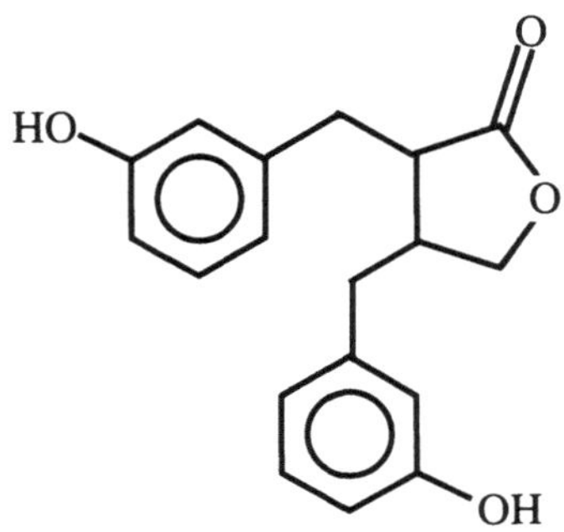

Enterolactone
(Lignan)

Figure 1. A structural comparison of estrogen and different subgroups of phytoestrogens. Representatives of the three main types of phytoestrogens, isoflavones, coumestans and lignans are shown on the right. They are all characterized by the presence of diphenolic rings and are structurally similar to 17β-estradiol, the most abundant type of estrogen.

ing in the down regulation of tyrosine kinase expression (30). Moreover, the inhibition of tyrosine kinase by genistein occurs only at pharmacological concentrations (>10 μM), and so genistein is unlikely to cause tyrosine kinase inhibition under physiological conditions.

While the role of genistein as tyrosine kinase inhibitor is not clear, there were accumulative evidence indicating a calcium channel antagonistic activity of genistein. Genistein may inhibit calcium entry through voltage-dependent L-type calcium channels via the inhibition of tyrosine kinase (23,31–33). However, a tyrosine kinase-independent pathway is also likely, since both genistein and daidzein, another isoflavone which is devoid of tyrosine kinase inhibitor activity (34), were shown to be equally effective in inhibiting L-type calcium currents in cardiac myocytes (35,36). In addition to L-type calcium channel, Morikawa et al. (37) reported that genistein blocked N-type and T-type calcium channel activities in NG108-15 cells. These effects appears to be both dependent and independent of the tyrosine kinase inhibitory action, since other tyrosine kinases and daidzein were capable of suppressing N-type and T-type calcium currents, albeit to a smaller extent compared to genistein.

That genistein has a diphenolic ring structure suggests that it may have antioxidant properties. Indeed, superoxide anion generation by xanthine/xanthine oxidase was inhibited by genistein (38). Genistein was also reported to increase the resistance of low-density lipoprotein (LDL) to oxidation (39). Despite findings from in vitro studies suggesting that the concentration of genistein required to produce antioxidant action exceeds the physiological range (40,41), dietary administration of 250 ppm genistein for 30 days enhanced the activity of antioxidant enzymes in the skin and small intestine in mice (38). Furthermore, LDL fractions from healthy volunteers consuming 36 mg genistein and 21 mg daidzein per day for 2 weeks were found to contain significant amount of genistein and daidzein, and their lag time to oxidation was prolonged (42). Therefore, the antioxidant property of genistein appears to be physiologically relevant.

PHYTOESTROGENS AND CARDIOVASCULAR RISK FACTORS

At present, there is a general consensus that dietary intake of phytoestrogens from soy proteins is related with a reduction of cardiovascular risks (19,20). One of the well-established cardiovascular risk factors is increased serum cholesterol concentration; more specifically, a high LDL to high-density lipoprotein (HDL) ratio, since LDL is the primary carrier of cholesterol in the circulatory system (43) whereas HDL is responsible for carrying cholesterol to the liver to be broken down and excreted (44). The oxidative modification of LDL facilitates the cellular deposition of cholesterol hence the formation of atherosclerotic lesions (45,46). Oxidized LDL also causes attenuation of fibrinolytic process (47) and impairment of vascular reactivity (48–50), both of which contribute to the pathogenesis of hypertension and atherosclerosis. The cardiovascular protective actions of phytoestrogens and the proposed mechanisms underlying these effects are summarized in Table 1.

Table 1. Effects and possible mechanisms of soy phytoestrogens in cardiovascular protection

	Actions of Phytoestrogens	Possible Mechanisms
Lipoprotein Profiles	↓ Cholesterol (51–53) ↓ LDL Cholesterol (51,52,56) ↑ HDL Cholesterol (56,57)	↑ Thyroxine Level (51,58) ↑ Bile Acids Level (59) ↑ LDL Removal (62,63)
Oxidation Stress	↑ Resistance to LDL Oxidation (64,65)	Scavenging Free Radicals (39) ↑ Antioxidant Enzyme Activity (38,66)
Coagulation	↓ Thrombosis (67,68)	Tyrosine Kinase Inhibition (67,69) Thromboxane Receptor Antagonism (70)
Vascular Reactivity	↓ Smooth Muscle Contraction (26,84–86) ↑ Smooth Muscle Relaxation (82,83) ↑ Endothelium-dependent Relaxation (72,73,76,77,78)	Endothelium Protection (65) ↑ NOS activity (76–78) Phosphodiesterase Activation (81,82) K_{Ca} Channel Activation (81,83)

LDL, low-density lipoprotein; HDL, high-density lipoprotein; NOS, nitric oxide synthase; K_{Ca} channel, calcium-activated potassium channel.

Lowering of cholesterol level

The cardioprotective effect of phytoestrogen has been attributed to the favourable alteration of the lipoprotein profiles. Dietary consumption of phytoestrogens has been documented to reduce total serum cholesterol level in animals and humans (51–53). However, studies from Shorey et al. (54) and Hodgson et al. (55) suggested that intake of soy diets and isoflavone pill, respectively, did not affect the total cholesterol concentration in men with normal cholesterol level. In view of the findings that men with more severe hypercholesterolemia showed a greater cholesterol lowering response to soy diets, it is believed that the hypocholesterolemic effect of phytoestrogens is related to the baseline cholesterol level (54,56). However, total serum cholesterol may not be the best indicator for cardiovascular risks because it includes both LDL-cholesterol and HDL-cholesterol. Phytoestrogen diets have also been shown to decrease LDL-cholesterol level and increase HDL-cholesterol level (52,56,57). The improvement of lipoprotein profiles by soy diets is accompanied with the reduction of atherosclerotic lesions and blood pressure (52,53).

Several mechanisms have been proposed for the actions of phytoestrogens on lipoprotein profiles. It has been reported that feeding soy proteins to animals resulted in elevated plasma thyroxine level, which may, in turn, lead to decreases in plasma cholesterol level (51,58). Soy proteins have also been reported to increase the excretion of bile acids in rabbits and rats (59). Given that bile acids enhance the excretion of cholesterol, alteration of bile acid balance is another potential mechanism underlying the hypocholesterolemic effect of phytoestrogen. This postulation was argued against by the findings that soy proteins decreased cholesterol level without changes in bile acid excretion in humans (60,61). Study from Sirtori's group suggested that soy phytoestrogens lower cholesterol concentration by augmenting the degradation of LDL (62). This action appears to be resulted from increases in the LDL receptor activity, since the hypocholesterolemic effect of soy phytoestrogens was only observed in wild-type but not in LDL receptor-deficient mice (63).

Inhibition of LDL oxidation

As discussed previously, oxidative modification of LDL by macrophages or endothelial cells is one of the processes involved in the pathogenesis of atherosclerosis (45–50). Therefore, the antioxidant property of phytoestrogens may be partially responsible for its ability to reduce cardiovascular risks. It has been shown that soy phytoestrogens increase the resistance of LDL to oxidation by macrophages and endothelial cells (64,65). Ruiz-Larrea et al. (39) suggested that isoflavones inhibited LDL oxidation by scavenging free radicals. On the other hand, genistein has been shown to increase the activity of antioxidant enzymes and this action may contribute to the reduction of the oxidative stress on LDL (38,66).

Augmentation of thrombosis

That oxidized LDL attenuates the process of fibrinolysis suggests that phytoestrogens may inhibit thrombus formation. Indeed, a thrombotic activity of isoflavones has been demonstrated and this is not limited to oxidized LDL-mediated thrombosis (67,68). Tyrosine kinase is a key enzyme involved in thrombus formation, and so the prothrombotic property of phytoestrogens may be a consequence to its inhibitory action on tyrosine kinase (67,69). Conversely, McNicol (70) demonstrated that thromboxane receptor antagonism, but not tyrosine kinase inhibition, was the mechanism through which genistein decreased platelet aggregation hence thrombus formation. As a result, it remains unclear how phytoestrogen exerts its thrombotic actions.

Improvement of vascular reactivity

Impairment of endothelium-dependent vasodilatation has been implicated in several CVD such as atherosclerosis (71). Preclinical studies have shown that chronic dietary intake of isoflavones improves acetylcholine-elicited coronary arterial dilation in female rhesus monkeys with atherosclerosis (72,73). It is plausible that this vasculoprotective action of phytoestrogens is related to the inhibition of oxidized LDL formation, since oxidized LDL is known to reduce endothelium-dependent relaxation (74,75). Kapiotis et al. (65) also reported that genistein was able to reduce endothelial damage by oxidized LDL.

On the other hand, phytoestrogens may also directly affect endothelium-dependent vasodilatation. In vitro studies showed that acute administration of genistein and daidzein concentration-dependently relaxed rat aortic and pulmonary arterial rings preconstricted with phenylephrine and this action was inhibited by endothelium removal (76). It is also unlikely that phytoestrogens inhibited tyrosine kinase to induce vasodilatation, because both genistein and daidzein, isoflavones with and without tyrosine kinase inhibitory activity, respectively, were equally effective in eliciting endothelium-dependent relaxation. Instead, nitric oxide synthase activation appears to be the downstream pathway that account for the vasodilatory action of phytoestrogens, according to the observation that genistein- and daidzein-induced relaxations were attenuated by the inhibition of nitric oxide synthase activity (76).

Similarly, through activation of nitric oxide synthase, genistein restored endothelium-dependent relaxation to carbachol in hypoxic rat pulmonary arterial rings (77). Nitric oxide also appears to play a role in the improvement of endothelium dysfunction by genistein supplementation in ovariectomized rats (78). A beneficial action of phytoestrogen on vascular reactivity was also confirmed in clinical studies, in which isoflavones were found to improve arterial compliance in menopausal women (79,80).

In addition to the nitric oxide pathway, other mechanisms have been proposed for the ability of genistein to enhance vasodilatation. The potentiating effect of genistein on isoproterenol-induced relaxation in rat aortic rings appeared to involve inhibition of phosphodiesterase, as well as activation of calcium-activated potassium channels and cytochrome P-450 enzymes (81). Whereas Satake et al. (82) proposed that genistein enhanced nitric oxide-mediated endothelium-independent relaxation by inhibiting cyclic AMP-dependent phosphodiesterase, the findings from Nevala et al. (83) indicated that genistein and daidzein directly relaxed rat mesenteric arterial rings through the activation of calcium-activated potassium channels.

Genistein has also been shown to directly inhibit the contractile responses of porcine coronary artery and rat aorta to angiotensin II, phenylephrine and noradrenaline (26,84,85). This direct vascular effect has been attributed to the tyrosine kinase inhibitory activity of genistein, since other tyrosine kinase inhibitors similarly reduced agonist-induced contraction in these tissues. On the contrary, Nevala et al. (86) reported that both genistein and daidzein, active and inactive isoflavone tyrosine kinase inhibitors, respectively, effectively inhibited potassium chloride- and noradrenaline-elicited contraction in isolated rat aortic and mesenteric arterial rings. As such, phytoestrogen-induced inhibition of contraction may also be mediated through a tyrosine kinase-independent pathway; one possible mechanism is the antagonism of calcium channels on smooth muscle, as phytoestrogens have been shown to directly inhibit L-type calcium currents (35,87).

PHYTOESTROGENS AND ESTROGEN RECEPTORS

In view of the structural similarity between 17β-estradiol and phytoestrogens (Figure 1), it is possible that phytoestrogens act on estrogen receptors to exert its cardioprotective actions. Indeed, estrogen has been documented to reduce cardiovascular risk in postmenopausal female (5–7). The cardioprotective effects of estrogen are attributable to favourable alterations of serum lipoprotein levels (5,7,88), antioxidant protection of LDL (89–91) and favourable vascular actions (92–95), all of which are also elicited by phytoestrogens and have accounted for the beneficial cardiovascular effects of phytoestrogens. Currently, two ER subtypes have been identified: the classical ERα and the more recently identified ERβ (10,96). Despite the fact that both ERα and ERβ are expressed in the vascular system (12,97,98), ERβ appears to be the predominate subtype in human smooth muscle cells (12). Interestingly, 17β-estradiol was found to produce similar vascular protection in ERα knock-out mice as in wild-type counterparts (99). There were also reports indicating an upregulation of the expression of ERβ mRNA in rat arteries following vascular injury, without significant changes in ERα expression (100,101). Therefore, the cardiovascular benefits of estrogen may be mediated through ERβ.

While 17β-estradiol binds to both ERα and ERβ with equal affinity, phytoestrogens appear to have differential binding affinities for the two receptors. Kuiper et al. (15) demonstrated that phytoestrogens binds only weakly to ERα, e.g., the most potent one, genistein, has a binding affinity that is 25 times lower than that of 17β-estradiol. On the other hand, the binding affinity of genistein for ERβ is 87% compared to 17β-estradiol (15). Therefore, there is about 20-fold difference in the binding affinity of genistein for ERα and ERβ. In comparison, the binding affinity of daidzein for ERβ is 5-fold greater than that for ERα, and is 200-fold less than that of 17β-estradiol. The preferential binding of phytoestrogens to ERβ may have physiological importance, since the distribution of ERα and ERβ varies in different tissues. ERβ is expressed mainly in non-reproductive tissues (10) and is capable of mediating the cardiovascular protective effects. As such, the ability of isoflavones, particularly genistein, to bind preferentially to ERβ may result in a relatively selective vasculoprotective action with fewer undesired effects. This postulation was indeed supported by the work of Makela et al. (101).

CONCLUDING REMARKS

Presently, there is substantial evidence for the protective effects of phytoestrogens against cardiovascular risks. However, data concerning the dose regime required to produce optimal beneficial actions is lacking. As well, phytoestrogens are a group of diverse plant-derived compounds with estrogenic activity, and little is known relating the type of phytoestrogens to be recommended for cardiovascular protection. More importantly, the potential adverse effects of chronic phytoestrogens supplementation should be considered with caution. Although to date there is no definite evidence suggesting that the consumption of phytoestrogens may be harmful, it is still possible that cumulative intake of phytoestrogens at large quantities may lead to adverse effects like those caused by unopposed estrogen treatment. Therefore, further studies are needed before phytoestrogens can be recommended for therapeutic uses.

ACKNOWLEDGEMENTS

The authors wish to thank the Areas of Excellence Scheme established under the University Grants Committee of the Hong Kong Special Administrative Region, China.

REFERENCES

1. Kannel DW. 1998. Overview of atherosclerosis. Clin Ther 20(Suppl B):B2–B17.
2. Barrett-Conner E. 1997. Sex differences I coronary heart disease. Why are women so superior? The 1995. Ancel Keys Lecture. Circulation 95:252–264.
3. Saaman SA, Crawfore MH. 1995. Estrogen and cardiovascular function after menopause. J Am Coll Cardiol 26:1403–1410.
4. Van der Schouw YT, Van der Graaf Y, Steyerberg EW, Eijkemans MJC, Banga JD. 1996. Age at menopause as a risk factor for cardiovascular mortality. Lancet 347:714–718.

5. Dallongeville J, Marecaux N, Isorez D, Zylberberg G, Fruchart J-C, Amouyel P. 1995. Multiple coronary heart disease risk factors are associated with menopause and influenced by substitutive hormonal therapy in a cohort of French women. Athersclerosis 118:123–133.
6. Stampfer MJ, Colditz GA, Willett WC, Manson JE, Rosner B, Speizer FE, Hennekens CH. 1991. Postmenopausal estrogen therapy and cardiovascular disease. N Engl J Med 325:756–762.
7. Ettinger N, Friedman GC, Bush T, Quesenberry CP Jr. 1996. Reduced mortality associated with long-term postmenopausal estrogen therapy. Obstet Gynecol 87:6–12.
8. Harlap S. 1992. The benefits and risks of hormone replacement therapy: an epidemiologic overview. Am J Obstet Gynecol 166:1986–1992.
9. Colditz GA, Hankinson SE, Hunter DJ, Willett WC, Manson JE, Stampfer MJ, Hennekens CH, Rosner B, Speizer FE. 1995. The use of estrogens and progestins and the risk of breast cancer in postmenopausal women. N Engl J Med 332:1589–1593.
10. Kuiper GG, Enmark E, Pelto-Huikko M, Nilsson S, Gustafsson JA. 1996. Cloning of a novel receptor expressed in rat prostate and ovary. Proc Natl Acad Sci USA 93:5925–5930.
11. Mosselman S, Polman H, Dijkema R. 1996. ER beta: identification and characterization of a novel human estrogen receptor. FEBS Lett 392:49–53.
12. Hodges YK, Tung L, Yan XD, Graham JD, Horwitz KB, Horwitz LD. 2000. Estrogen receptor α and β—prevalence of estrogen receptor β mRNA in human vascular smooth muscle and transcriptional effects. Circulation 101:1792–1798.
13. Gustafsson JA. 2000. Novel aspects of estrogen action. J Soc Gynecol Invest 7(1 Suppl):S8–S9.
14. Barkhem T, Carlsson B, Nilsson Y, Enmark E, Gustafsson JA, Nilsson S. 1998. Differential response of estrogen receptor alpha and estrogen receptor beta to partial estrogen agonists/antagonists. Mol Pharmacol 54:105–112.
15. Kuiper GG, Lemmen JG, Carlsson B, Corton JC, Safe SH, van der Saag PT, van der Burg B, Gustafsson JA. 1998. Interaction of estrogenic chemicals and phytoestrogens with estrogen receptor β. Endocrinology 139:4252–4263.
16. Kuiper GG, Carlsson B, Grandien K, Enmark E, Haggblad J, Nilsson S, Gustafsson JA. 1997. Comparison of the ligand binding specificity and transcript tissue distribution of estrogen receptors alpha and beta. Endocrinology 138:863–870.
17. Martin PM, Horwitz KB, Ryan DS, McGuire WL. 1978. Phytoestrogen interaction with estrogen receptors in human breast cancer cells. Endocrinology 103:1860–1867.
18. Murkies AL, Wilcox G, Davis SR. 1998. Phytoestrogens. J Clin Endocrin Metab 83:297–303.
19. Adlercreutz H, Mazur W. 1995. Phyto-oestrogens and western diseases. Ann Med 29:95–120.
20. Clarkson TB, Anthony MS, Hugher CL Jr. 1995. Estrogenic soybeans isoflavones and chronic disease. Risks and benefits. Trends Endocrinol Metab 6:11–16.
21. Akiyama T, Ishida J, Nakagawa S, Ogawara H, Watanabe S, Itoh N, Shibuya M, Fukami Y. 1987. Genistein, a specific inhibitor of tyrosine-specific protein kinases. J Biol Chem 262:5592–5595.
22. Takahashi T, Kawahara Y, Okuda M, Ueno H, Takeshita A, Yokoyama M. 1997. Angiotensin II stimulates mitogen-activated protein kinases and protein synthesis by a Ras-independent pathway in vascular smooth muscle cells. J Biol Chem 272:16018–16022.
23. Peifley KAA, Winkles JA. Angiotensin II and endothelin-1 increase fibroblast growth factor-2 mRNA expression in vascular smooth muscle cells. Biochem Biophys Res Commun 242:202–208.
24. Gazit A, Yaish P, Gilon C, Levitzki A. 1989. Tyrphostins I: synthesis and biological activity of protein tyrosine kinase inhibitors. J Med Chem 32:2344–2352.
25. Wijetunge S, Aalkjaer C, Schachter M, Hughes AD. 1992. Tyrosine kinase inhibitors block calcium channel currents in vascular smooth muscle cells. Biochem Biophys Res Commun 189:1620–1623.
26. Abebe W, Agrawal DK. 1995. Role of tyrosine kinases in norepinephrine-induced contraction of vascular smooth muscle. J Cardioivasc Pharmacol 26:153–159.
27. Watts SW, Yeum CH, Campbell G, Webb RC. 1996. Serotonin stimulates protein tyrosyl phosphorylation and vascular contraction via tyrosine kinase. J Vasc Res 33:288–298.
28. Peterson TG, Barnes S. 1993. Isoflavones inhibit the growth of human prostate cancer cell lines without inhibiting epidermal growth factor receptor autophosphorylation. Prostate 22:335–345.
29. Peterson TG, Barnes S. 1996. Genistein inhibits both estrogen and growth factor stimulated proliferation of human breast cancer cell. Cell Growth Diff 7:1345–1351.
30. Dalu A, Haskel IJF, Coward L, Lamartiniere CA. 1998. Genistein, a component of soy, inhibits the expression of the EGF and erbB2/Neu receptors in the rat dorsolateral prostate. Prostate 37:36–43.
31. Low AM. 1996. Role of tyrosine kinase on Ca^{2+} entry and refilling of agonist-sensitive Ca^{2+} stores in vascular smooth muscles. Can J Physiol Pharmacol 74:298–304.

32. Strauss O, Mergler S, Wiederholt M. 1997. Regulation of L-type calcium channels by protein tyrosine kinase and protein kinase C in cultured rat and human retinal pigment epithelial cells. FASEB J 11:859–867.
33. Liu H, Li K, Sperelakis N. 1997. Tyrosine kinase inhibitor, genistein, inhibits macroscopic L-type calcium current in rat portal vein smooth muscle cells. Can J Physiol Pharmacol 75:1058–1062.
34. Sargeant P, Farndale RW, Sage SO. 1993. ADP- and thapsigargin-evoked Ca^{2+} entry and protein-tyrosine phosphorylation are inhibited by the tyrosine kinase inhibitors genistein and methyl-2,5-dihydroxycinnamate in fura-2-loaded human platelets. J Biol Chem 268:18151–18156.
35. Chiang CE, Chen SA, Chang MS, Lin CI, Luk HN. 1996. Genistein directly inhibits L-type calcium currents but potentiates cAMP-dependent chloride currents in cardiomyocytes. Biochem Biophys Res Commun 223:598–603.
36. Yokoshiki H, Sumii K, Sperelakis N. 1996. Inhibition of L-type calcium current in rat ventricular cells by the tyrosine kinase inhibitor, genistein and its inactive analog, daidzein. J Mol Cell Cardiol 28:807–817.
37. Morikawa H, Fukuda K, Mima H, Shoda T, Kato S, Mori K. 1998. Tyrosine kinase inhibitors suppress N-type and T-type Ca^{2+} channel currents in NG108-15 cells. Pflugers Archiv—Eur J Physiol 436:127–132.
38. Wei H, Bowen R, Cai Q, Barnes S, Wang Y. 1995. Antioxidant and antipromotional effects of the soybean isoflavone genistein. Proc Soc Exp Biol Med 208:124–130.
39. Ruiz-Larrea MB, Mohan AR, Paganga G, Miller NJ, Bolwell GP, Rice-Evans CA. 1997. Antioxidant activity of phytoestrogenic isoflavones. Free Radical Res 26:63–70.
40. Kerry N, Abbey M. 1998. The isoflavone genistein inhibits copper and peroxyl radical medicated low density lipoprotein oxidation in vitro. Atherosclerosis 140:341–347.
41. Mitchell JH, Gardner PT, McPhail DB, Morrice PC, Collins AR, Duthie GG. 1998. Antioxidant efficacy of phytoestrogens in chemical and biological model systems. Arch Biochem Biophys 360:142–148.
42. Tikkanen MJ, Wahala K, Ojala S, Vihma V, Adlercreutz H. 1998. Effect of soybean phytoestrogen intake on low density lipoprotein oxidation resistance. Proc Natl Acad Sci USA 95:3106–3110.
43. Brown MS, Goldstein JL. 1976. Receptor-mediated control of cholesterol metabolism. Science 191:150–154.
44. Miller GJ, Miller NE. 1975. Plasma-high-density-lipoprotein concentration and development of ischaemic heart-disease. Lancet 1:16–9.
45. Steinberg D, Parthasarathy S, Carew TE, Khoo JC, Witzum JL. 1989. Beyond cholesterol, modifications of low density lipoprotein that increase its atherogenecity. N Engl J Med 320:915–924.
46. Witztum JL, Steinberg D. 1991. Role of oxidized low density lipoprotein in atherogenesis. J Clin Invest 88:1785–1792.
47. Ren S, Man RYK, Angel A, Shen GX. 1997. Oxidative modification enhances lipoprotein(a)-induced overproduction of plasminogen activator inhibitor-1 in cultured vascular endothelial cells. Atherosclerosis 128:1–10.
48. Simon BC, Cunningham LD, Cohen RA. 1990. Oxidized low-density lipoproteins cause contraction and inhibit endothelium-dependent relaxation in the pig coronary artery. J Clin Invest 86:75–79.
49. Mangin EL Jr, Kugiyama K, Nguy JH, Kerns SA, Henry PD. 1993. Effects of lysolipids and oxidatively modified low density lipoprotein on endothelium-dependent relaxation of rabbit aorta. Circ Res 72:161–166.
50. Abebe W, Mustafa SJ. 1997. Effect of low density lipoprotein on adenosine receptor-mediated coronary vasorelaxation in vitro. J Pharmacol Exp Ther 282:851–857.
51. Balmir F, Staack R, Jeffrey E, Jimerez MDB, Wand L, Potter SM. 1996. An extract of soy flour influences serum cholesterol and thyroid hormones in rats and hamsters. J Nutr 126:3046–3053.
52. Anthony MS, Clarkson TB, Bullock BC, Wagner JD. 1997. Soy protein versus soy phytoestrogens in prevention of diet-induced coronary artery atherosclerosis of male cynomolgus monkeys. Atheriosler Thromb Vasc Biol 17:2524–2531.
53. Washburn S, Burke GL, Morgan T, Anthony M. 1999. Effect of soy protein supplementation on serum lipoproteins, blood pressure and menopausal symptoms in perimenopausal women. Menopause. 6:7–13.
54. Shorey RL, Bazan B, Lo GD, Steinke FH. 1981. Determinants of hypocholesterolemic response to soy and animal protein based diets. Am J Clin Nutr 34:1769–1778.

55. Hodgson JM, Puddey IB, Beilin LJ, Mori TA, Croft KD. 1998. Supplementation with isoflavonoid phytoestrogens does not alter serum lipid concentrations: a randomized controlled trial in humans. J Nutr 128:728–732.
56. Anderson JW, Johnstone BM, Cooke-Newell ME. 1995. Meta-analysis of the effects of soy protein intake on serum lipids. N Engl J Med 333:272–286.
57. Clarkson TB, Anthony MS. 1998. Phytoestrogens and coronary heart disease. Baillieres Clin Endocrinol Metab 12:589–604.
58. Forsythe WA III. 1995. Soy protein, thyroid regulation and cholesterol metabolism. J Nutr 125:619S–623S.
59. Potter SM. 1996. Soy protein and serum lipids. Curr Opin Lipidol 7:260–264.
60. Fumagalli R, Soleri L, Farina R, Musanti R, Mantero O, Noseda G, Gatti E, Sirtori CR. 1982. Facal cholesterol excretion studies in type II hypercholesterolemic patients treated with the soybean protein diets. Atherosclerosis 43:341–353.
61. Grundy SM, Abrams JJ. 1983. Comparison of actions of soy protein and casein on metabolism of plasma lipoproteins and cholesterol in humans. Am J Clin Nutr 38:245–252.
62. Sirtori CR, Lovati MR, Manzoni C, Monetti M, Pazzucconi F, Gatti E. 1995. Soy and cholesterol reduction: clinical experience. J Nutr 125(suppl):599S-605S.
63. Kirk EA, Sutherland P, Wang SA, Chait A, LeBoeuf RC. 1998. Dietary isoflavones reduce plasma cholesterol and atherosclerosis in C57BL/6 mice but not LDL receptor-deficient mice. J Nutr 128:954–959.
64. de Whalley CV, Rankin SM, Hoult JRS, Jessup W, Leake DS. 1990. Flavonoids inhibit the oxidative modification of low density lipoproteins by macrophages. Biochem Pharm 39:1743–1750.
65. Kapiotis S, Hermann M, Held I, Seelos C, Ehringer H, Gmeiner BMK. 1997. Genistein, the dietary-derived angiogenesis inhibitor, prevents LDL oxidation and protects endothelial cells from damage by atherogenic LDL. Arterioscler Thromb Vasc Biol 17:2868–2874.
66. Cai Q, Wei H. 1996. Effect of dietary genistein on antioxidant enzyme activities in SENCAR mice. Nutr Cancer 25:1–7.
67. Sargeant P, Farndale RW, Sage SO. 1993. The tyrosine kinase inhibitors methyl 2,5-dihydroxycinnimate and genistein reduce thrombin-evoked tyrosine phophorylation and Ca^{2+} entry in human platelets. FEBS Lett 315:242–246.
68. Wilcox JN, Blumenthal BF. 1995. Thrombotic mechanisms in atherosclerosis: potential impact of soy protein. J Nutr 125:631S–638S.
69. Nakashima S, Koike T, Nozawa Y. 1991. Genistein, a protein tyrosine kinase inhibitor, inhibits thromboxane A2-mediated human platelet responses. Mol Pharm 39:475–480.
70. McNicol A. 1992. The effects of genistein on platelet function are due to thromboxane receptor antagonism rather than inhibition of tyrosine kinase. Prostaglandins, Leukotrienes, Essential Fatty Acids 48:379–384.
71. Vogel RA. 1997. Coronary risk factors, endothelial function, and atherosclerosis: a review. Clin Cardiol 20:426–432.
72. Honore EK, Williams JK, Anthony MS, Clarkson TB. 1997. Soy isoflavones enhance coronary vascular reactivity in atherosclerotic female macaques. Fertil Steril 67:148–154.
73. Williams JK, Clarkson TB. 1998. Dietary soy isoflavones inhibit in-vivo constrictor responses of coronary arteries to collagen-induced platelet activation. Coronary Art Disease 9:759–764.
74. Kugiyama K, Kerns SA, Morrisett JD, Roberts R, Henry PD. 1990. Impairment of endothelium-dependent arterial relaxation by lysolecithin in modified low-density lipoproteins. Nature 344: 160–162.
75. Tanner FC, Noll G, Boulanger CM, Lüscher TF. 1991. Oxidized low density lipoproteins inhibit relaxations of porcine coronary arteries: role of scavenger receptor and endothelium-derived nitric oxide. Circulation 83:2012–2020.
76. Mishra SK, Abbot SE, Choudhury Z, Cheng M, Khatab N, Maycock NJ, Zavery A, Aaronson PI. 2000. Endothelium-dependent relaxation of rat aorta and main pulmonary artery by the phytoestrogens genistein and daidzein. Cardiovasc Res 43:539–546.
77. Karamsetty MR, Klinger JR, Hill NS. 2001. Phytoestrogens restore nitric oxide-mediated relaxation in isolated pulmonary arteries from chronically hypoxic rats. J Pharmacol Exp Ther 297:968–974.
78. Squadrito F, Altavilla D, Squadrito G, Saitta A, Cucinotta D, Minutoli L, Deodato B, Ferlito M, Campo GM, Bova A, Caputi AP. 2000. Genistein supplementation and estrogen replacement therapy improve endothelial dysfunction induced by ovariectomy in rats. Cardiovasc Res 45:454–462.

79. Nestel PJ, Yamashita T, Sasahara T, Pomeroy S, Dart A, Lomesaroff P, Owen A, Abbey M. 1997. Soy isoflavones improve systemic arterial compliance but not plasma lipids in menopausal and perimenopausal women. Arterioscler Thromb Vasc Biol 17:3392–3398.
80. Nestel PJ, Pomeroy S, Kay S, Komesaroff P, Behrsing J, Cameron JD, West L. 1999. Isoflavones from red clover improve systemic arterial compliance but not plasma lipids in menopausal women. J Clin Endocrinol Metab 84:895–898.
81. Satake N, Shibata S. 1999. The potentiating effect of genistein on the relaxation induced by isoproterenol in rat aortic rings. Gen Pharmacol 33:221–227.
82. Satake N, Imanishi M, Shibata S. 1999. Increased nitroglycerin-induced relaxation by genistein in rat aortic rings. Eur J Pharmacol 377:193–197.
83. Nevala R, Paukku K, Korpela R, Vapaatalo H. 2001. Calcium-sensitive potassium channel inhibitors antagonize genistein- and daidzein-induced arterial relaxation in vitro. Life Sci 69:1407–1417.
84. Laniyonu AA, Saifeddine M, Yang SG, Hollenberg MD. 1994. Tyrosine kinase inhibitors and the contractile action of G-protein-linked vascular agonists. Can J Physiol Pharmacol 72:1075–1085.
85. Filipeanu CM, Brailoiu E, Huhurez G, Slatineanu S, Baltatu O, Branisteanu DD. 1995. Multiple effects of tyrosine kinase inhibitors on vascular smooth muscle contraction. Eur J Pharmacol 281:29–35.
86. Nevala R, Korpela R, Vapaatalo H. 1998. Plant derived estrogens relax rat mesenteric artery in vitro. Life Sci. 63:PL95–PL100.
87. Figtree GA, Griffiths H, Lu YQ, Webb CM, MacLeod K, Collins P. 2000. Plant-derived estrogens relax coronary arteries in vitro by a calcium antagonistic mechanism. J Am Coll Cardiol 35:1977–1985.
88. Sack MN, Rader DJ, Cannon RO III. 1994. Oestrogen and inhibition oxidationof low-density lipoproteins in postmenopausal women. Lancet 343:267–270.
89. Keaney Jr JF, Shwaery GT, Xu A-M, Nicolosi RJ, Loscalzo J, Foxall TL, Vita JA. 1994. 17β-estradiol preserves endothelial vasodilator function and limits low-density lipoprotein oxidation in hypercholesterolemic swine. Circulation 89:2251–2259.
90. Shwaery GT, Vita JA, Keaney JF Jr. 1995. Antioxidant protection of LDL by physiological concentrations of 17β-estradiol. Requirement for estradiol moification. Circulation 95:1378–1385.
91. Huang M, Li J, Teoh H, Man RYK. 1999. Low concentrations of 17β-estradiol reduce oxidative modification of low-density lipoproteins in the presence of vitamin C and vitamin E. Free Radical Biol Med 27:438–441.
92. Herrington DM, Braden GA, Williams JK, Morgan TM. 1994. Endothelial-dependent coronary vasomotor responsiveness in postmenopausal women with and without estrogen replacement therapy. Am J Cardiol 73:951–952.
93. Gilligan DM, Quyyumi AA, Cannon RO. 1994. Effects of physiological levels of estrogen on coronary vasomotor function in postmenopausal women. Circulation 89:2545–2551.
94. Han S-Z, Karaki H, Ouchi Y, Akishita M, Orimo H. 1995. 17β-Estradiol inhibits Ca^{2+} influx and Ca^{2+} release induced by thromboxane A_2 in porcine coronary artery. Circulation 91:2619–2626.
95. Teoh H, Leung SWS, Man RYK. 1999. Short-term exposure to physiological levels of 17β-estradiol enhances endothelium-independent relaxation in porcine coronary artery. Cardiovasc Res 42:224–231.
96. Greene GL, Gilna P, Waterfield M, Baker A, Hort Y, Shine J. 1986. Sequence and expression of human estrogen receptor complementary DNA. Science 231:1150–1154.
97. Inoue S, Hoshino S, Miyoshi H, Akishita M, Hosoi T, Orimo H, Ouchi Y. 1996. Identification of a novel isoform of estrogen receptor, a potential inhibitor of estrogen action, in vascular smooth muscle cells. Biochem Biophys Res Commun 219:766–772.
98. Register TC, Adams MR. 1998. Coronary artery and cultured aortic smooth smucle cells express mRNA for both the classical estrogen receptor and the newly described estrogen receptor β. J Steroid Biochem Mol Biol 64:187–191.
99. Iafrati MD, Karas RH, Aronovitz M, Kim S, Sullivan TR, Lubahn DB, O'Donnell TF, Korach KS, Mendelsohn ME. 1997. Estrogen inhibits the vascular injury response in estrogen receptor α-deficient mice. Nature Med 3:545–548.
100. Lindner V, Kim SK, Karas RH, Kuiper GG, Gustafsson JA, Mendelsohn ME. 1998. Increased expression of estrogen receptor-β in male blood vessels after vascular injury. Circ Res 83:224–229.
101. Makela S, Savolainen H, Aavik E, Myllarniemi M, Strauss L, Taskinen E, Gustafsson JA, Hayry P. 1999. Differentiation between vasculoprotective and uterotrophic effects of ligands with different binding affinities to estrogen receptors α and β. Proc Natl Acad Sci USA 96:7077–7082.

Pathophysiology of Cardiovascular Disease,
edited by N.S. Dhalla, H. Rupp,
A. Angel & G.N. Pierce
Kluwer Academic Publishers, Boston, 2004

Non-Conventional Long-Lasting Cardioprotection Induced by Chronic Exposure to Ambient Heat

Michal Horowitz,[1] Omer Cohen,[1] Yonathan Hasin,[2]
Alina Maloyan,[1] and Einat Levi[1]

[1] *Division of Physiology, Faculty of Dental medicine, The Hebrew University, and*
[2] *Dept of Cardiology, Hadassah University Hospital, Jerusalem 91120, Israel*

Summary. Heat acclimation produces concerted favorable adaptations in the mechanical and metabolic performance of the heart. Under normoxic condition this is displayed primarily by greater pressure generation with lower oxygen consumption. Altered expression of calcium regulatory proteins, EC coupling proteins and redistribution of the myosin isoenzyme to predominancy of the V3 isoform contribute to the changes observed. An important acclimatory response is the finding that heat acclimation confers protection against ischemic/reperfusion insults. This is due to numerous factors among which greater constitutive HSP 72 kDA (Δ200%) reserves and energy sparing mechanisms such as redistribution of ATPase isoforms and altered glycolytic flux leading to better intracellular pH concomitantly with continuous ATP production, are likely candidates. Collectively, our data indicate that both transcriptional and post-transcriptional processes are involved and underline the integrative response.

Key words: Ca^{2+} regulatory proteins, cardiac myosin isoenzymes, cardioprotection, ischemia, reperfusion, heat shock proteins, heat acclimation.

1. INTRODUCTION

Cells have many mechanisms for adapting and surviving stress. Some of these coping mechanisms are constitutive, others are inducible and therefore require time for their expression. An example of an inducible response to stress occurs in the heart, where protection is conveyed to limit the undesirable effects of ischemic and anoxic stress

Corresponding author: Prof Michal Horowitz, Division of Physiology, Faculty of Dental Medicine, The Hebrew University, POB 12272, Jerusalem 91120, Israel. Tel: 972 2 6757588, Fax: 972 2 6439736, E-mail: horowitz@cc.huji.ac.il.

by repetitive short ischemic episodes (1,2). This preconditioning effect is considered to be part of a rapid adaptive mechanism induced by many physiological and pharmacological stressors, and shared by many cells and organs. The preconditioning effect is of limited duration and represents the cellular attempt to protect both structure and function from stress-induced damage.

In the heart, ischemic or thermal preconditioning involves early, rapid (1–2 hrs) and delayed (24–72 hrs) phases of protection (1).

In contrast to the relatively short duration preconditioning effect, we have substantiated a long-lasting (2–3 weeks) cardioprotection conferred by heat acclimation via the "cross-tolerance" mechanism (3,4,5). This mini-review gives us the opportunity to delve into the mechanisms of the cross-tolerance phenomenon. In the first part of this chapter the characteristics of the heat acclimated phenotype will be delineated. In the second part, questions pertaining to the heat acclimation induced cardioprotection will be discussed.

2. WHAT ARE HEAT ACCLIMATION AND CROSS-TOLERANCE?

Heat acclimation is achieved via chronic exposure to moderate ambient heat. It enhances heat tolerance in terms of duration and extremes of toleration to heat stress, and delays heat injuries. The development of the acclimated phenotype is biphasic. A short transient phase (2–5 days), during which a variety of acute processes are recruited to alleviate the initial stress, is followed by a long-term phase (>3 weeks) during which long-lasting mechanisms are developed, to establish the acclimatory homeostasis (6). This involves an orchestrated reprogramming of gene expression and post-translational variations (7). The signaling pathways and molecules targeted to acclimatory responses include G proteins, cytoprotection (via heat shock protein HSP72 kDa and antioxidants (8,9)), several transcription factors, metabolic enzymes and contractile proteins, collectively leading to enhanced metabolic efficiency and cellular performance (10).

Cross-tolerance is achieved when adaptation to the one stress reinforces protection against an additional type of stress. We have substantiated the development of cross-tolerance between heat acclimation and O_2 deprivation/reoxygenation in the heart. Our findings lend support the hypothesis that heat acclimation predisposes protective signaling pathways utilized by both stressors (4,5).

3. HEAT ACCLIMATION AND THE HEART UNDER NORMOXIC CONDITIONS

Heat acclimation produces concerted favorable adaptations in the mechanical and metabolic performance of the heart. At the integrative level, in the intact individual, this is displayed by a lowered heart rate and greater stroke volume, collectively leading to increase in cardiac work efficiency. Previous studies on human subjects related the enhanced stroke volume occurring upon heat acclimation to a greater venous return, resulting from the heat-acclimation-induced plasma volume expansion (11,12). It was debatable whether intrinsic changes occur in the cardiac muscle to increase contractile force. Using the Langendorff perfused system we have shown

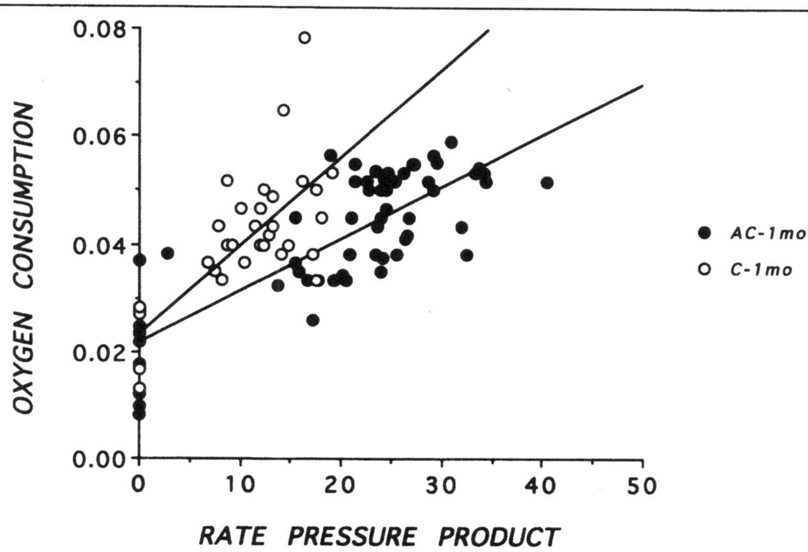

Figure 1. Oxygen consumption: cardiac work (expressed as rate-pressure product) relationship in hearts of sedentary nonacclimated (C) and 1-mo heat-acclimated (AC) rats. Marked shift to the left of the regression line of heat-acclimated hearts compared with that of nonacclimated hearts ($P < 0.0007$) implies significantly increased work efficiency of heat-acclimated heart. From Horowitz M, 1998 (6), with permission of the Am. Physiol. Soc.

in the isolated heart that left ventricular compliance and systolic pressure are increased, while oxygen consumption is lowered (5,13). This suggests that the acclimated heart becomes more energetically efficient. This is depicted in Figure 1, which presents the O_2-cardiac work relationships before and after heat acclimation. Studies on other rodent species confirm these observations (Horowitz, Meiri, Bromberg, unpublished observations). We can conclude that cardiac adaptations *per se* take place to match changes in peripheral blood flow and venous return, producing greater stroke volume and cardiac output at lower energy cost.

It is evident from Table 1 that intrinsic molecular and biochemical adaptive modalities in the contractile machinery of the heart are associated with the enhanced pressure and work efficiency of the heat-acclimated phenotype. The redistribution of cardiac myosin isoezymes, from the fast V_1 isoform with high ATPase activity to the predominance of the slow V_3 form with low ATPase activity (14), underlies the increased contractile efficiency and velocity of pressure generation of the heat-acclimated heart.

Adaptations at the level of the sarcoplasmic reticulum (SR) were also found. Eynan et al., 1999 (15) reported on 49% increase in the Ca^{2+} regulatory protein phospholamban. Concomitantly, the SR-Ca^{2+}ATPase decreased by 25%. These adaptations, lead to an increased phospholamban/Ca^{2+}-ATPase ratio and produce a negative lusitropic effect. In the sarcolemma, the protein level of L-type Ca^{2+} channels

Table 1. Qualitative presentation of heat acclimation effect on the contractile machinery of the heart

Feature	Acclimatory response	Reference
Cardiac mechanical performance		
Pressure	↑	Horowitz et al. (13)
Velocity of contraction	↓	Levi et al., Levy et al. (4,5)
Velocity of relaxation	↓	Eynan et al. (15)
Noradrenaline inotropic effect	↑	Eynan et al. (15)
Biochemical & molecular		
Predoninant myosin isoform	V_3	Horowitz et al. (14)
β-Adrenergic receptors affinity	↓	Eynan et al. (15)
Ca^{2+} Transient	↑	Cohen et al. (3)
Phospholamban	↑	Eynan et al. (15)
Ca^{2+} ATPase	↑	Eynan et al. (15)
Phospholamban/Ca^{2+}-ATPase	↑	Eynan et al. (15)
Ryanodine/Dihydropyridine	Altered	Cohen et al. (3)

was elevated by 70% (3) and the β-adrenergic receptors—decreased their affinity by 63% with no change in their density (16). We suggest that the described adaptations are part of a well architectured reprogramming of gene expression resulting in altered Ca^{2+} turnover to yield greater Ca^{2+} transients, and greater pressure generation.

Phospholamban, in its phosphorylated state, activates the sarcoplasmic reticulum (SR) Ca^{2+}-ATPase to re-uptake Ca^{2+} to SR pools for subsequent contractions. We hypothesize that the greater phospholamban mass available for phosphorylation (by noradrenaline stimulation) enhances the positive inotropic response in these hearts (7,15).

A large body of evidence links alterations in the velocities of contraction and relaxation of the cardiac contractile proteins to the influence of thyroid hormones, via an effect on the expression of genes coding for cardiac contractile elements (e.g., 17,18,19). Heat acclimation leads to a sustained drop in thyroid hormone levels (14,15). Subjection of rats to heat acclimation with maintenance of euthyroid state blunts the phenomenology of the heat-acclimated cardiac phenotype (14,15). We therefore suggest that the sustained drop in plasma thyroid hormones, occurring upon acclimation, is responsible for the observed adaptations in cardiac mechanical performance.

4. HEAT ACCLIMATION AND THE HEART UNDER ISCHEMIC CONDITIONS

Levy et al. (1993) (5) have shown that in the heart heat acclimation induces favorable adaptations allowing improved coping with mismatching between the oxygen supply and oxygen demand. This is displayed by improved mechanical and metabolic performance upon ischemia/reperfusion insults and by decreased ischemic injury. The infarct size developed in hearts of non-acclimated and heat-acclimated hearts subjected to global ischemia/reperfusion insults (Figure 2) clearly demonstrates this phenomenon.

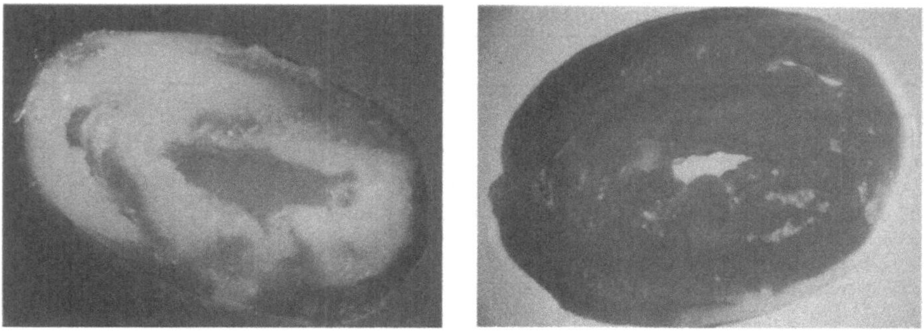

Non-acclimated **Heat acclimated**

Figure 2. Infarct size in heat acclimated and non-acclimated hearts subjected to 30 min of total ischemia and subsequent 30 min of reperfusion. Immediately after, the hearts, mounted on a Langendorff perfusion system, were stained with 10% (wt/vol) solution of triphenyltetrazolium in phosphate buffer. Normoxic perfusion pressure was 100 cm H_2O.

Additional features characterizing the beneficial effects of heat acclimation on the heart include a decreased number of cases of ischemic contracture (IC) in the acclimated heart population, delayed onset of IC by almost 100%, and better recovery of systolic and diastolic pressures upon reperfusion (Table 2). Some of these beneficial features, e.g. decreased number of cases of IC contraction within a group or recovery of pulse pressure upon reperfusion, could also be obtained via exercise training, however, features such as delayed onset of IC and enhanced recovery of the intraventricular pressure upon reperfusion seem to be heat acclimation-specific (Table 2).

What, then, makes the heat-acclimated heart so well protected against O_2 deprivation/reperfusion insults? Species that routinely experience long periods of oxygen shortage in their life history (20,21) have acquired constitutive adaptations to endure aggressive hypoxia/reoxygenation stress. These can be grouped into two categories: 1) those leading to metabolic depression and, in turn, energy sparing, and 2) those upregulating the energetic efficiency of ATP-producing pathways. There are a host of additional physiological, biochemical and molecular strategies, including alterations in membrane activity and enhanced cytoprotection, which complement the above-mentioned adaptations. The acclimatory responses seem to recapitulate the evolutionary adaptations (6). It is therefore likely that enhancement of protective signaling pathways, similar to those that have emerged through many generations of a species' history under shortage of oxygen, underlies the cardioprotection achieved upon heat acclimation. Currently, evidence for metabolic and cytoprotective adaptation has already been substantiated. This will be discussed below (4.1 and 4.2).

4.1. Metabolic adaptations and cardioprotection in the heat-acclimated heart

During ischemia/hypoxia, energy conservation can be achieved either by matching energy requirements to a greatly curtailed supply of ATP, or by increasing energy

Table 2. Evidence for heat-acclimation/ischemic-reperfusion cross-tolerance. Data were derived from studies on Langendorff perfused heart. Bold lettering denotes significant acclimation effect

	Non-acclimated	Heat acclimated	Reference
IC within a group			
(Sedentary 1 mo)			
50% decrease in PP	**46%**	**15%**	Levi et al. (4)
75% decrease in PP	85%	85%	
IC within a group			
(Exercise 1 mo)			
50% decrease in PP	no IC	9%	Levy et al. (5)
75% decrease in PP	40%	50%	
Time to onset of IC (min)			
S	**6.0 ± 1.3**	**8.0 ± 2**	Levi et al. (4)
EX	**8.7 ± 2**	**10.87 ± 2.5**	Levy et al. (5)
% recovery of pulse pressure			
S	20%	60%	Levi et al. (4)
Ex	80%	90%	Levy et al. (5)
Recovery of intraventricular pressure			
S	**35%**	**80%**	Levi et al. (4)
EX	35%	40%	Levy et al. (5)

PP—perfusion pressure.
IC—ischemic contracture.
S—acclimation under sedentary conditions.
EX—Exercise training.
Bold letters denote significant heat acclimation effect.

production efficiency. When enhanced glycolysis is the operating mechanism for increasing energy, intercellular pH must be protected. Levi et al. (1993, [4]) and Levy et al. (1996, [5]) showed that during total ischemia (TI) heat-acclimated hearts (as well as hearts from heat-acclimated trained rats) display better preservation of global myocardial high-energy compounds and a shorter period of IC and acidosis During TI the heat acclimated hearts show slowed ATP depletion and after 30 min of the TI ~ 25% of the initial ATP pool was preserved, whereas no ATP was detectable (by P NMR) in the non-acclimated hearts. During reperfusion the recovery of phosphocreatine (PCr) and the sum of PCr and ATP, was higher (4). Improvement in each of the above-mentioned parameters is known to preserve myocardial integrity and to enhance mechanical recovery during reperfusion. Improved energy status and delayed acidosis during ischemia decrease sodium accumulation via Na^+/H^+ exchange and therefore decrease the Na^+-mediated Ca^{2+} gain and subsequent Ca^{2+}-mediated rise in diastolic pressure and mitochondrial Ca^{2+} overload during reperfusion. Concomitantly, the shorter IC may reduce possible mechanical distortion and, together with the faster high-energy phosphate compounds, enhance mechanical performance upon reperfusion. We have indications that the delayed fall in ATP in the heat-acclimated ischemic heart is the result of changes in 1. the properties of the cellular ATPases, e.g. myosin V_3 predominance (vs myosin V_1 predominance before acclimation), or decreased Na^+/Ka^+ pump affinity for its

Table 3. Effects of global ischemia on cardiac ATPase properties in sarcolemmal enriched fraction of control and heat-acclimated hearts

	ΣATPase (37°)		ΣATPase (GI)		Na/K ATPase (37°)		Na/K ATPase (GI)	
	Vmax	EC50	Vmax	EC50	Vmax	EC50	Vmaxc	EC50
C	0.323	430.0	0.403	487.0	0.103	518.3	0.098	479.0
	0.01	12.6	0.01	16.7	0.003	8.08	0.003	8.9
AC	0.413	422.0	0.425	596.0	0.116	777.3	0.141	914.0
	0.02	18.9	0.015	11.6	0.014	2.22	0.003	5.52

substrate during ischemia (in the acclimated heart only; Table 3) 2. metabolic enzymes. These, lead to energy sparing and sustained, but slower ATP production via altered glycolytic flux (10,22).

On-line measurements of glucose metabolites, using ^{13}C-NMR spectroscopy, showed that in non-acclimated hearts glycolysis is rapid, but apparently self-limiting, due to a rapid fall in intracellular pH (pHi). These findings were confirmed by our findings that a longer time was required to inhibit glycolysis (by sodium iodoacetate) in acclimated hearts compared to the non-acclimated (30 min for AC hearts vs 15 min for C hearts) (22). Features such as greater pre-ischemic glycogen stores (10,22), enhanced glucose uptake and elevated GLUT4 (the cardiac insulin-dependent glucose transporter) levels, may suggest glucose is the preferred fuel (elevated ATP/mole fuel), emphasize the metabolic advantage of the acclimated heart, and provide new perspectives for our understanding of the induced metabolic adaptations.

4.2. Cytoprotection in the heat-acclimated heart

Another aspect of the heat-acclimation/ischemia cross-tolerance and molecular response is the role of heat shock proteins (HSP), among which the HSP 72 kDa has been the most extensively studied (23,24,25). Heat-acclimated hearts constitutively express a markedly higher level of the inducible HSP 72 kDa (Δ 200%), and show faster HSP induction upon heat or ischemic stress. For example, following heat stress peak HSP 72 kDa level in the heat acclimated rats hearts is attained following 1 hr compared to 4 hrs in the non-acclimated ones (23,24). Teleologically, the advantage of larger HSP reserves is immediate protection without a need for time-consuming *de novo* protein synthesis. This conclusion is compatible with the finding that genetic manipulation to produce overexpression of HSP renders ischemic tolerance in both cardiomyocytes and whole hearts (26,27) and with our observations that the index for the severity of the stress, infarct size, measured subsequent to coronary occlusion, is markedly smaller in the heat-acclimated heart. An additional acclimatory phenomenon is the faster rate of increased hsp mRNA expression, and the higher oxygen threshold required for HSP production during ischemia. This dynamic response may produce an outcome different from that obtained by constitutively high HSP expression.

5. FUTURE PERSPECTIVES

Our data provide evidence that the process of heat acclimation confers long-term intrinsic adaptations. The manifestations of this adaptive response are multifaceted: in normoxic conditions, it is displayed by improved mechanical and metabolic performance. Upon ischemia/reperfusion insults, cross-tolerance is shown, involving ionic (not shown), metabolic and cytoprotective adaptations. The data indicate that both transcriptional and post-transcriptional processes are involved. Understanding of this natural adaptation may constitute novel therapeutic targets.

ACKNOWLEDGEMENT

Most of the studies reported were supported, over the years, by the USA-Israel Binational Science Fundation and was carried out in collaboration with G. Gerstenblith, M. Stern and Y. Hasin.

REFERENCES

1. Sommerschild Hilchen T, Kirkeboen KA. 2002. Preconditioning endogenous defence mechanisms of the heart. Acta Anaesthesiol Scand 46:123–137.
2. Xi L, Tekin D, Bhargava P, Kukreja RC. 2001. Whole body hyperthermia and preconditioning of the heart: basic concepts, complexity, and potential mechanisms. Int J Hyperthermia 17:439–455.
3. Cohen O, Stern M, Horowitz M. 2001. Heat acclimation improves cardiac contractility and ischemic tolerance: Is heat acclimation memorized? J Mol Cell Cardiol 33:22A.
4. Levi E, Vivi A, Hasin Y, Tassini M, Navon G, Horowitz M. 1993. Heat acclimation improves cardiac mechanics and metabolic performance during ischemia and reperfusion. J Appl Physiol 75:833–839.
5. Levy E, Navon G, Hasin Y, Horowitz M. 1997. Chronic heat improves mechanical and metabolic performance of trained rats heart upon ischemia and reperfusion. Am J Physiol 272:H2085–H2094.
6. Horowitz M. 1998. Do cellular heat acclimatory responses modulate central thermoregulatory activity? NIPS 13:218–225.
7. Horowitz M. 2002. From Molecular and Cellular to Integrative Heat Defense during Exposure to Chronic Heat. Comp Biochem Physiol A Mol Integr Physiol 131(3):475–483.
8. Maloyan A, Palmon A, Horowitz M. 1999. Heat acclimation increases basal HSP72 level and alters its production dynamics during heat stress. Am J Physiol 276:R1506–R1515.
9. Beit-Yannai E, Kohen R, Horowitz M, Trembovler V, Shohami E. 1997. Changes of biological reducing activity in rat brain following closed head injury: a cyclic voltammetry study in normal and heat acclimated rats. J Cerb BF Metab 17:273–279.
10. Eynan M, Knubuvetz T, Meiri U, Navon G, Gerstenblith G, Bromberg Z, Hasin Y, Horowitz M. ••. Elevated glycogen and enhanced glycolysis underlie ischemic cardioprotection following heat acclimation (submitted).
11. Wyndham CH, Rogers GG, Senay LC, Mitchell D. 1976. Acclimatization ration in a hot, humid environment: cardiovascular adjustments. J-Appl Physiol 40:779–785.
12. Senay LC, Mitchell D, Wyndham CH. 1976. Acclimatization in a hot, humid environment: body fluid adjustments. J-Appl Physiol 40:786–796.
13. Horowitz M, Shimoni Y, Parnes S, Gottsman MS, Hasin Y. 1986a. Heat acclimation: cardiac performance of isolated rat heart. J Appl Physiol 60:9–13.
14. Horowitz M, Peiser MJ, Muhlrad A. 1986b. Alterations in cardiac myosin distribution as an adaptation to chronic environmental heat stress. J Mol Cell Cardiol 18:511–515.
15. Eynan M, Palmon A, Hasin Y, Horowitz M. 1999. Heat acclimation induces changes in cardiac mechanical performance: the role of thyroid hormone. Am J Physiol 276:R550–R558.

16. Eynan M, Gross M, Hasin Y, Palmon A, Horowitz M. 2000. Changes in cardiac mechanics with heat acclimation: adrenergic signaling and SR-Ca regulatory proteins. Am J Physiol 279:R77–R85.
17. Chizzonite RA, Zak R. 1984. Regulation of myosin isoenzyme composition in fetal and neonatal rat ventricle by endogenous thyroid hormones. J Biol Chem 259:12628–12632.
18. Kiss E, Jakab G, Kranias EG, Edes I. 1994. Thyroid hormone-induced alterations in phospholamban protein expression. Circ Res 75:245–251.
19. Koss KL, Kranias EG. 1996. Phospholamban: A prominent regulator of myocardial contractility. Circ Res 79:1059–1063.
20. Hochachka PW, Lutz PL. 2001. Mechanism origin, and evolution of anoxia tolerance in animals Comp-Biochem-Physiol B. Biochem-Mol-Biol 130:435–459.
21. Hochachka PW. 2000. Oxygen homeostasis, and metabolic regulation. Adv-Exp-Med-Biol 475: 311–335.
22. Eynan M, Knubovetz T, Meiri U, Navon G, Hasin Y, Horowitz M. 1995. Beneficial effects of heat acclimation during ischemia and reperfusion: metabolic adaptations. J Cel Mol Cardiol 27:A166.
23. Horowitz M, Maloyan A, Shlaier J. 1997. HSP 70 kDa dynamics in animals undergoing heat stress superimposed on heat acclimation. Ann NY Acad Sci 813:617–619.
24. Maloyan A, Palmon A, Horowitz M. 1999. Heat acclimation increases basal HSP72 level and alters its production dynamics during heat stress. Am J Physiol 276:R1506–R1515.
25. Maloyan A, Horowitz M. 2002. β-Adrenergic signaling and thyroid hormones cross-talk affects HSP72 expression during heat acclimation J Appl Physiol 93:107–116.
26. Marber MS, Mestril R, Chi SH, Sayen MR, Yellon DM, Dillmann WH. 1995. Overexpression of the rat inducible 70-kD heat stress protein in a transgenic mouse increases the resistance of the heart to ischemic injury. J-Clin-Invest 95:1446–1456.
27. Hutter JJ, Mestril R, Tam EK, Sievers RE, Dillmann WH, Wolfe CL. 1996. Overexpression of heat shock protein 72 in transgenic mice decreases infarct size in vivo. Circulation 94:1408–1411.

Pathophysiology of Cardiovascular Disease,
edited by N.S. Dhalla, H. Rupp,
A. Angel & G.N. Pierce
Kluwer Academic Publishers, Boston, 2004

Fetal and Neonatal Cardiomyocyte Transplantation for the Treatment of Myocardial Infarction

Jochen Müller-Ehmsen,[1] Jonathan Leor,[2] Larry Kedes,[3] Kirk L. Peterson,[4] and Robert A. Kloner[5,6]

[1] *Department of Internal Medicine III, University of Cologne, Germany,*
[2] *Neufeld Cardiac Research Institute, Sheba Medical Center, Tel-Hashomer, Israel,*
[3] *Institute for Genetic Medicine, Keck School of Medicine, University of Southern California, Los Angeles, CA,*
[4] *Division of Cardiology, Department of Medicine, University of California, San Diego, CA,* [5] *Department of Medicine, Keck School of Medicine, University of Southern California, and* [6] *Heart Institute, Good Samaritan Hospital, Los Angeles, CA*

Summary. In the past few years cardiac cell transplantation has emerged as a novel therapeutic strategy to treat myocardial infarction and heart failure. Several clinical studies are already underway, and multipotent stem cells show promise to become the favored cell type to replace lost myocardium. Nevertheless, important data providing insight into the basic mechanisms of this technique and supporting the beneficial effects of a cell transfer therapy have already been obtained in studies involving fetal or neonatal cardiomyocytes. In this chapter we summarize the results from investigations on the potential of fetal and neonatal rat cardiomyocytes to survive, to mature and to structurally and functionally replace host myocardium after implantation in infarcted and non-infarcted rat hearts.

In initial experiments human and rat fetal cardiomyocytes were injected into infarcted hearts of adult Sprague-Dawley rats. The expression of smooth muscle α-actin was followed as a marker of the fetal cell grafts. The cellular grafts survived for at least 60 days. On the other hand, the continuous presence of fetal α-actin also demonstrated a lack of complete maturation of these cells. Immunohistochemical analysis with connexin-43 antibody and the presence of surrounding collagen revealed that areas of transplanted cells were isolated from

Address for Correspondence: Robert A. Kloner, MD, PhD, Heart Institute, Good Samaritan Hospital, 1225 Wilshire Boulevard, Los Angeles, CA 90017. 213-977-4050, 213-977-4107 (fax), E-mail: rkloner@goodsam.org.

host myocardium. Nevertheless, transplanted fetal cardiomyocytes attenuated the deterioration of cardiac function after myocardial infarction by echocardiographic analysis, albeit they were not able to reverse defective function.

One of the factors that likely determines the effective support of cardiac function, will be the survival of transplanted cells. To assess the number of surviving cells, we used the Y chromosomes of male neonatal cardiomyocytes which were transplanted into female adult rats. PCR analysis revealed that the number of surviving cells decreased within 24 hours after transplantation to about 15% of those injected, but was quite stable thereafter for at least 12 weeks. When injected into the infarcts of rat hearts (ligation of left coronary artery), neonatal cardiomyocytes increased the thickness of the infarcts and improved LV ejection fraction as measured by angiography. Echocardiography revealed a trend towards a smaller internal diameter and an increased fractional shortening. The grafted neonatal cells developed organized sarcomeric structures and aligned parallel to the circumference of the LV. However, similar to the observation for fetal cells, the neonatal cells did not achieve the maturity of adult rat cardiomyocytes.

In summary, our data demonstrate that fetal and neonatal cardiomyocytes can survive long term after injection into normal or infarcted adult hearts. They undergo a certain degree of maturation over time, but during the observed periods they do not reach the maturation status of adult cells. Fetal and neonatal cells have beneficial effects on cardiac function of hearts after myocardial infarction, however, they do not completely reverse the contractile deficit of these hearts. Nevertheless, the data are encouraging that in the not too distant future cell therapy may be a new hope for patients who suffer from ischemic heart disease.

Key words: myocardial infarction, therapy, cells, cardiomyocytes, transplantation, cardiomyoplasty, rat, fetal, neonatal.

INTRODUCTION

Cardiac tissue is terminally differentiated and myocardial cells only rarely undergo mitoses. Accordingly, regeneration of myocardium after injury is quite limited (1). Cell transplantation is a new strategy for provision of surrogate cardiomyocytes and for treatment of multiple diseases including myocardial infarction and heart failure. Currently, several studies in humans are already underway using different cell types like skeletal myoblasts (2) or hematopoietic stem cells (3,4) to replace the lost cardiac muscle. Although it is still unclear which cell type is most suitable for this task, it is likely that some kind of stem cell—whether embryonic, mesenchymal or hematopoietic—will be advantageous over other approaches. The reasons include their proliferative activity, their potential to differentiate into different types of cells and the possibility of using autologous cells that do not require immunosuppresive therapy. Despite these gaps in our knowledge that need to be filled in order to make cardiac cell transplantation a valuable therapy for human heart disease, it is also important to assess the validity of the concept and the therapeutic potential of this therapy in simple models. Therefore, in our laboratory we focused on the questions whether transplanted cells can survive in an hypoxic environment of a myocardial infarction, and whether they can contribute actively or passively to cardiac function. To address these questions we used fetal and neonatal cardiomyocytes as supposedly optimal cellular grafts because they have the potential to survive after cell transplantation (5) and they already possess the properties that are characteristic of myocardial cells.

SURVIVAL OF FETAL CARDIOMYOCYTES AFTER TRANSPLANTATION IN NORMAL AND INFARCTED ADULT HEARTS

We initially studied fetal human and rat myocardial fragments that were injected into the scar tissue of rat hearts, 7 to 24 days after ligation of the left coronary artery (6). The grafted cells were detected by α-actin immunohistochemistry and electron microscopy. α-actin is a smooth muscle cell isoform of actin expressed in fetal, but not in adult, cardiomyocytes. Engrafted human tissue was detected in 6 rats (55%) at 7 days and in 4 rats (100%) 14 days after injection. Engrafted rat fetal tissue was found 8 days (1 of 2), 14 (1 of 2), 32 (1 of 1) and 65 days (2 of 2) after transplantation. Overall graft survival was 71%, and was independent of the time of implantation after myocardial infarction. This high percentage of survival was somewhat surprising given the adverse environment of an infarcted region. However, from this data it is unclear how many or what fraction of the injected cells did survive. Ultrastructural analysis revealed the persistence of the fetal phenotype of cells until the latest time point studied (65 days). The regions of engraftment also stained positive for smooth muscle α-actin, indicating the survival of the grafted cell clusters, but also the incomplete maturation of the grafted cells. This data is supported by other investigations that found that fetal rat cardiomyocytes survived one month after transplantation in infarcted myocardium of adult rats (7), and they survived for as long as 20 weeks after transplantation into cryoinfarcted myocardium (8).

FUNCTIONAL IMPACT OF FETAL CARDIOMYOCYTES AFTER TRANSPLANTATION IN INFARCTED HEARTS

The most important objective that needs to be met by cell transplantation is the improvement of function in the state of cardiac disease. To test for this, we performed echocardiographic studies in rats that had infarcted left ventricles half of which received an injection of 1.5×10^6 fetal ventricular cardiomyocytes (9). The histologic analysis revealed that cells were successfully grafted and survived for 8 weeks in 10 of 11 cases. However, no gap junctions were found between grafted cells and host myocardium by anti-connexin 43 staining, and the grafted cells frequently were surrounded by fibrous tissue that separated them from the host myocardium. Nevertheless, rats that received fetal cardiomyocyte transplantation showed a reduced deterioration of cardiac function assessed by echocardiography compared to untreated animals. The reduction in fractional shortening and the increase in LV diastolic diameter was greater in untreated than in cell treated animals. This beneficial effect on cardiac function also translated in an—albeit non-significant—reduction in mortality: No rat died in the cell treated group ($n = 11$) while 4 of 16 rats in the control group died within 8 weeks ($P = 0.12$).

Other investigators also found potential beneficial effects of fetal cardiomyocytes when transplanted into diseased hearts. In a comparative study, fetal cardiomyocytes were found to have a better effect on cardiac function after myocardial infarction than did either smooth muscle cells or fibroblasts (10). In Langendorff preparations, cell-treated infarcted hearts were able to develop higher pressures than non-treated hearts (11). It was even found that echocardiographic function was improved in vivo not only in rats after myocardial infarction (7), but also in rats with doxorubicin-induced heart failure that had received fetal cardiomyocyte grafts (12). However, in this study the successful grafting of the cells could not be demonstrated.

Besides their effects on cardiac function, transplanted fetal cardiomyocytes can also enhance angiogenesis. In our study (9) there was a non-significant trend towards a higher density of capillaries or arterioles ($P = 0.13$), but Li et al. (8) found significant angiogenesis in the animals that were treated with cells.

SURVIVAL AND DEVELOPMENT OF NEONATAL CARDIOMYOCYTES AFTER TRANSPLANTATION INTO NORMAL AND INFARCTED RAT HEARTS

Acute myocardial infarction is followed by an early inflammatory response which results in degradation of necrotic tissue and in scar formation. For cell transplantation it is still not completely clear what time after the acute event would be optimal for both survival of cells and therapeutic effect. Zhang et al. (13) addressed the first question by using neonatal cardiomyocytes, and found that cell death is most pronounced if cells are injected into the acute myocardial infarct. Recent data from other investigators support this finding, in that fetal cardiomyocytes that were injected 2 or 4 weeks after myocardial infarct, i.e. after the acute inflammatory phase, but before scar expansion, had the most beneficial effects on cardiac function after myocardial infarction (14). Not surprisingly the death of grafted neonatal cardiomyocytes occurs largely during the first few days after cell transplantation (13). In our study we sought to establish a quantitative time course of survival of male neonatal cardiomyocytes transplanted into the heart of female healthy adult syngeneic rats (15). Cell survival was followed by quantitative real-time PCR of a Y chromosome marker. We found that more than 75% of the injected cells were no longer detectable at 24 hours. At 12 weeks we found that 15% of the grafted cells had survived. Thus, after a rapid drop in cell number, the number of surviving cells remains relatively stable for several months, which is encouraging with regard to a long term therapeutic effect of the surviving cells. Interestingly, even immediately after injection only 57% of injected cells could be identified in the host myocardium, indicating that technical issues of cells grafting play a role in cell loss. In our study, unlike others using neonatal cardiomyocytes (13), skeletal myoblasts (16) or dopaminergic neurons (17), cell survival was not improved by anti-apoptotic interventions. The grafted cells matured over time, but even at 12 weeks they did not reach the length, the width and the nuclear size of adult rat cardiomyocytes (15). When male neonatal cells were injected into infarcted female

A

B

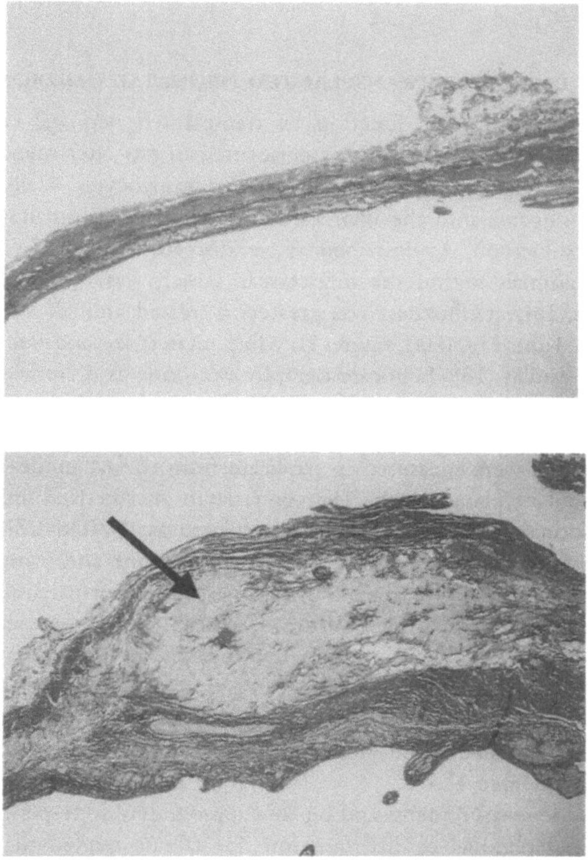

Figure 1. Bulked-up myocardial scar after cell transplantation.
(A) Picrosirius-red stained infarct scar at 6 months of an untreated (media only) rat. The infarct scar is collagenous and appears dark grey. The infarct is transmural, thin, and devoid of myocytes. The visceral pericardium appears thickened (picrosirius red, X4 objective).
(B) Transplanted neonatal cardiomyocytes survived for 6 months in the infarct and formed a bulky muscle patch (arrow) within the scar. Viable myocytes are pale. The cell implant is a discrete lump of cells found in the midmyocardium and subepicardium of the scar. We did not observe similar structures in media-treated hearts (picrosirius red, X4 objective).

hearts, the number of surviving cells, as assessed by PCR of a Y chromosomal region, was significantly higher: After 6 months we found that 62% of grafted cells survived (n = 6) (18). Although this number likely overestimates the number of surviving cardiomyocytes (since about 10% of injected cells were non-cardiomyocytes), histology revealed large clusters of transplanted cardiomyocytes that survived to "bulk-up" the thin-walled scar of the infarct (Figure 1). Whether these cells would suffice to improve function of the infarcted hearts was tested in our subsequent experiments.

FUNCTIONAL IMPACT OF TRANSPLANTED NEONATAL CARDIOMYOCYTES

To assess support of cardiac function by transplanted neonatal cells, echocardiographic and angiographic analysis was performed in rats after myocardial infarction and after the transplantation of neonatal rat cardiomyocytes. Cells ($3–5 \times 10^6$) or medium were injected into the infarct zone of syngeneic animals 1 week after left coronary artery ligation. As described above, discrete clusters of cells were present in all treated animals within the infarct scar, which was not observed in media-injected hearts. Infarct thickness was greater in treated animals vs. controls (909 ± 97 vs. 619 ± 43 µm, $P < 0.02$, figure 1), while infarct size and left ventricular (LV) volumes were similar (18). Echocardiography was done as described previously (19) at baseline (1 wk after infarct, before cell injection) and 1, 2, 3 and 6 months after cardiomyocyte transplantation. Left ventricular diastolic (LVIDd) and systolic (LVIDs) diameters were measured in triplicate from the M-mode of the short axis view (mid-papillary plane) of the left ventricle in anesthetized animals (methoxyflurane). Fractional shortening (FS) was calculated as (LVIDd-LVIDs)/ LVIDd. In addition, anterior wall thickening was assessed at baseline and 6 months after treatment (0 = akinesis or dyskinesis, 1 = hypokinesis, 2 = normal wall motion). Two-dimensional LV areas in diastole (LVAd), in systole (LVAs) were calculated as well as area EF, at baseline and at 6 months. At 6 months, we found that treated animals ($n = 17$) had a smaller LVIDs than untreated ($n = 17$, $P < 0.05$) and there was a trend for reduced LVIDs ($P = 0.07$) (Figure 2, Table 1). Also thickening of the antero-septal wall tended to be greater in animals treated with neonatal cardiomyocytes ($P = 0.10$, Table 1).

Hemodynamic measurements and biplane angiography were performed to further assess global and regional cardiac function. Heart rate tended to be lower in the

Table 1. M-mode and two-dimensional echocardiography 6 months after cell transplantation or medium injection

	Untreated (n = 17)	Cell transplantation (n = 17)	P-value
LVIDd (cm)	0.72 ± 0.03	0.66 ± 0.02	0.066
LVIDs (cm)	0.52 ± 0.03	0.44 ± 0.02	***0.046***
FS (%)	28.7 ± 3.1	33.1 ± 2.8	0.31
Area EF (%)	27.6 ± 3.5	31.4 ± 3.1	0.44
Anterior wall Thickening	0.35 ± 0.15	0.71 ± 0.17	0.10

LVIDd/ LVIDs = left ventricular internal diameter in diastole/ systole; FS = fractional shortening calculated as (LVIDd − LVIDs)/ LVIDd; area EF = area ejection fraction calculated as (diastolic LV area—systolic LV area)/ diastolic LV area; n = number of animals.

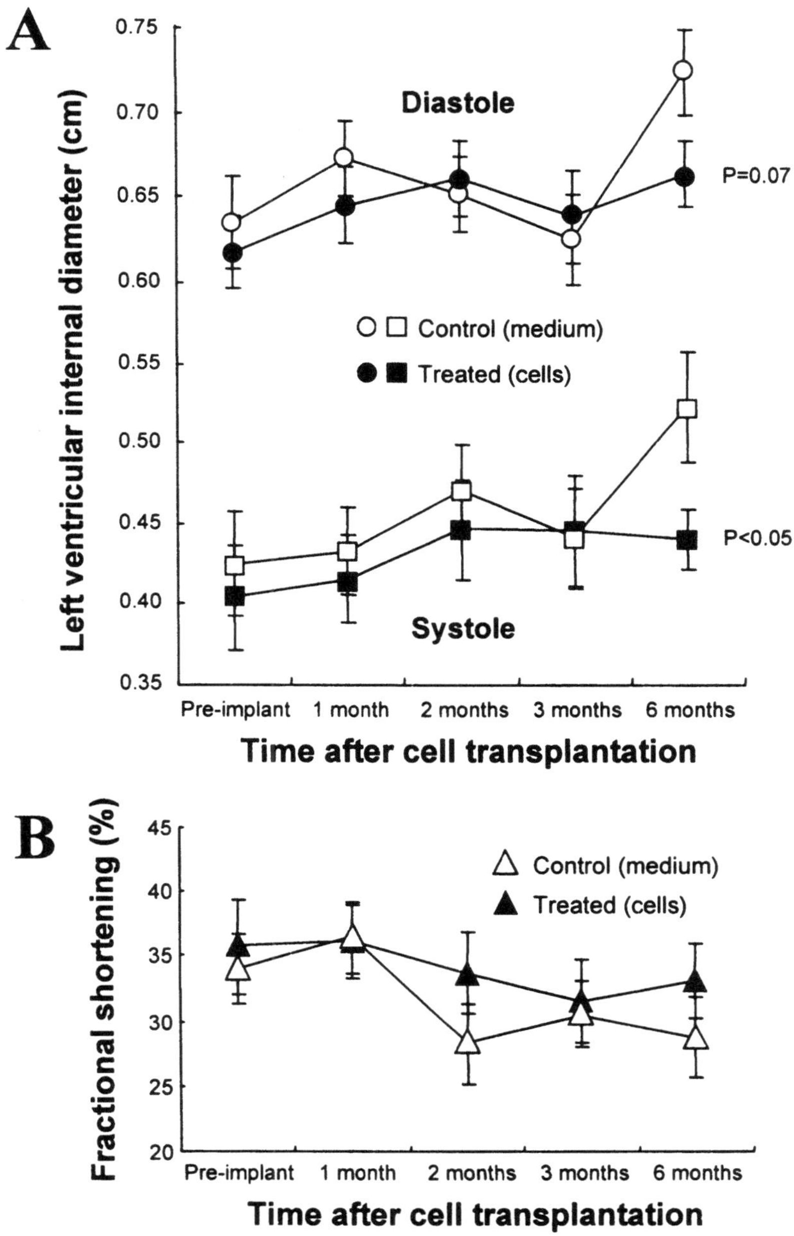

Figure 2. LV dimensions (A) and fractional shortening (B) in M-mode echocardiography, before and after cell transplantation. At 6 months (n = 17) fractional shortening was the same in the cell transplantation group (solid symbols) as in control rats (open symbols), but there was a trend for smaller LV internal diameter in diastole, while systolic LV internal diameter was significantly smaller than in control rat hearts.

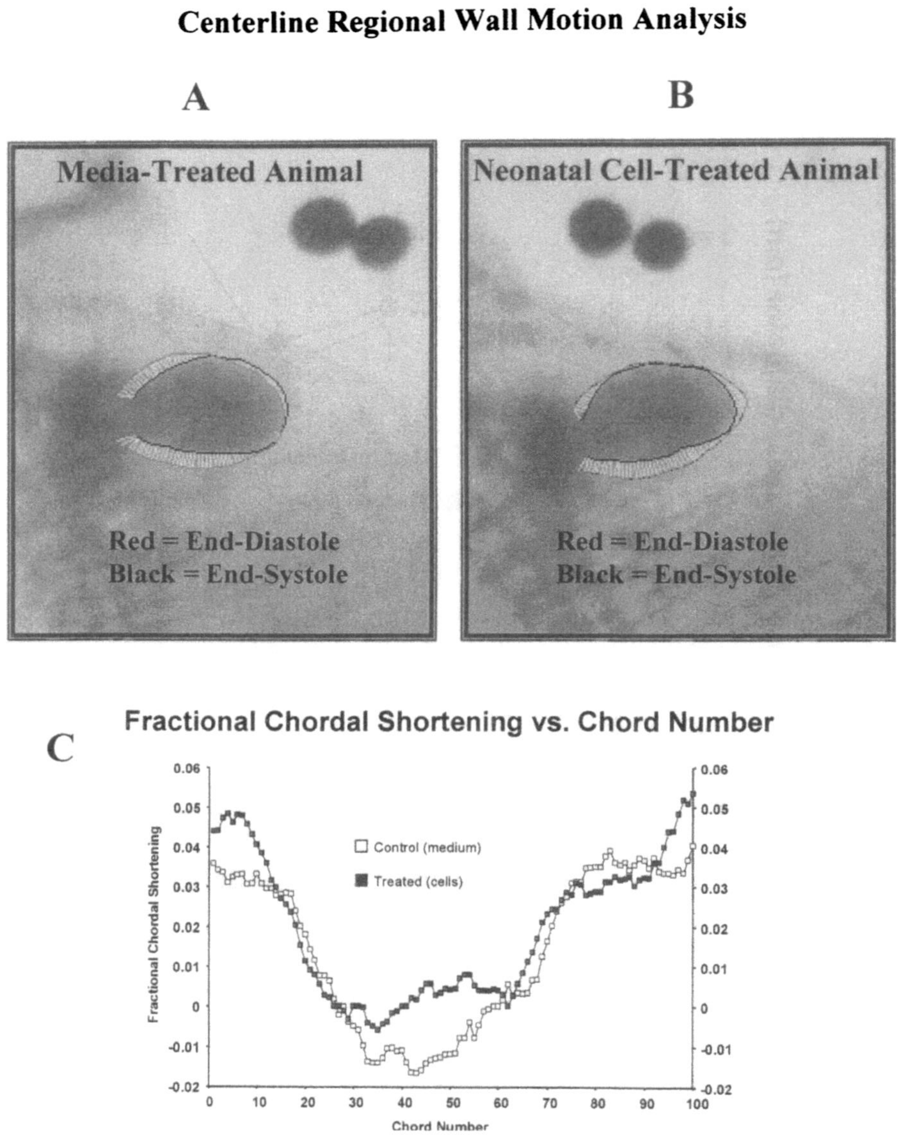

Figure 3. Right lateral projection videoangiography of rats 6 months after treatment with either media (A) or neonatal cardiomyocytes (B), injected into the anteroapical region of the infarct. The end-diastolic perimeter is shown in light grey and the end-systolic perimeter in black. Chords from regional wall motion analysis by the centerline technique are shown in white. Cardiomyocyte treated animals were found to have hypokinesis or akinesis of the infarct, as opposed to dyskinesis (paradoxic systolic bulging). (C) Chordal shortening versus chord number around the left ventricular perimeter of the same animals as in (A) and (B) in lateral projection. The media-treated animal is shown in open squares and the neonatal cardiomyocyte-treated animal is shown in solid squares. The size of the myocardial infarct zone (fractional shortening <0.005) was similar in both animals, but the paradoxic systolic bulging of the untreated animal was not seen in the treated rat.

treated group, but LV pressures, peak positive dP/dt and peak negative dP/dt were similar in both groups. In contrast, LV ejection fraction at 6 months was greater in treated animals versus untreated (0.36 ± 0.03 vs. 0.25 ± 0.02, $P < 0.01$), and they had significantly less infarct zone dyskinesis (0.30 ± 0.08 vs. 0.55 ± 0.07, $P = 0.035$, lateral projection, Figure 3) (18).

Thus, both echocardiography and LV angiography support the hypothesis that transplantation of neonatal cardiomyocytes can improve cardiac function after myocardial infarction. Whether this is due to active contribution of grafted cells to contractility or secondary to passive thickening the wall of the infarct is unclear. In either case, ejection fraction was enhanced and paradoxical systolic bulging of the infarct reduced.

SUMMARY AND CONCLUSIONS

Regenerating damaged myocardium by the injection or transplantation of new cells is a challenge currently being met by many cardiovascular researchers and clinicians. Apart from choosing a suitable cell type that, ideally, should be autologous, it is important to understand the basic mechanisms that are initiated by this procedure. Evidence must be accumulated to confirm the hypothesis that new myocardial cells can not only survive in an environment that has killed or damaged the endogenous cells, but that they can also support myocardial function long-term. For this, immature cardiomoycytes appear superior to other cell types because, unlike adult cells, they are known to endure the mechanical stress of transplantation (5), and they are ready to resume contractile activity without differentiation or even trans-differentiation needed. Studies to date indicate that fetal and neonatal cardiac cell transplants exhibit long-term survival when injected into myocardial infarct scars of small animals. They show maturation to some extent, but even after 6 months they remain easily distinguishable from adult cells by their smaller size and smaller and rounder nuclei. Nevertheless, both fetal and neonatal cardiomyocytes improve function or at least attenuate deterioration of function as assessed by echocardiography and LV angiography. For neonatal cardiomyocytes, a decreased dilatation of the left ventricle, less paradoxical bulging of the scar and larger ejection fraction was found in comparison to untreated controls. The goal of complete replacement of damaged myocardium and full restoration of cardiac function has not yet been met by the transplantation of immature cardiomyocytes. However, some long-term improvement of cardiac function was observed that serves as a proof of principle and as a compelling basis for further research in this field.

ACKNOWLEDGEMENTS

We thank Sharon Hale for statistical analyses, Seda Dzhandzhapanyan for assistance in histology, and Dr. Yusu Gu for assistance in cardiac angiography. Supported by NIH/NHLBI (R01 HL61488-01 RAK, R0152771-06 LK), the Deutsche Forschungsgemeinschaft (DFG Mu 1469/1-1 JME), and the San Diego Foundation for Cardiovascular Research and Education (KLP).

REFERENCES

1. Kajstura J, Leri A, Finato N, Di Loreto C, Beltrami CA, Anversa P. 1998. Myocyte proliferation in end-stage cardiac failure in humans. Proc Natl Acad Sci U S A 95:8801–8805.
2. Menasche P, Hagege AA, Scorsin M, Pouzet B, Desnos M, Duboc D, Schwartz K, Vilquin JT, Marolleau JP. 2001. Myoblast transplantation for heart failure. Lancet 357:279–280.
3. Strauer BE, Brehm M, Zeus T, Gattermann N, Hernandez A, Sorg RV, Kogler G, Wernet P. 2001. [Intracoronary, human autologous stem cell transplantation for myocardial regeneration following myocardial infarction]. Dtsch Med Wochenschr 126:932–938.
4. Hamano K, Nishida M, Hirata K, Mikamo A, Li TS, Harada M, Miura T, Matsuzaki M, Esato K. 2001. Local implantation of autologous bone marrow cells for therapeutic angiogenesis in patients with ischemic heart disease: clinical trial and preliminary results. Jpn Circ J 65:845–847.
5. Reinecke H, Zhang M, Bartosek T, Murry CE. 1999. Survival, integration, and differentiation of cardiomyocyte grafts: a study in normal and injured rat hearts. Circulation 100:193–202.
6. Leor J, Patterson M, Quinones MJ, Kedes LH, Kloner RA. 1996. Transplantation of fetal myocardial tissue into the infarcted myocardium of rat. A potential method for repair of infarcted myocardium? Circulation 94:II332–336.
7. Scorsin M, Hagege AA, Marotte F, Mirochnik N, Copin H, Barnoux M, Sabri A, Samuel JL, Rappaport L, Menasche P. 1997. Does transplantation of cardiomyocytes improve function of infarcted myocardium? Circulation 96:II188–193.
8. Li RK, Mickle DA, Weisel RD, Mohabeer MK, Zhang J, Rao V, Li G, Merante F, Jia ZQ. 1997. Natural history of fetal rat cardiomyocytes transplanted into adult rat myocardial scar tissue. Circulation 96:II179–186.
9. Etzion S, Battler A, Barbash IM, Cagnano E, Zarin P, Granot Y, Kedes LH, Kloner RA, Leor J. 2001. Influence of embryonic cardiomyocyte transplantation on the progression of heart failure in a rat model of extensive myocardial infarction. J Mol Cell Cardiol 33:1321–1330.
10. Sakai T, Li RK, Weisel RD, Mickle DA, Jia ZQ, Tomita S, Kim EJ, Yau TM. 1999. Fetal cell transplantation: a comparison of three cell types. J Thorac Cardiovasc Surg 118:715–724.
11. Li RK, Jia ZQ, Weisel RD, Mickle DA, Zhang J, Mohabeer MK, Rao V, Ivanov J. 1996. Cardiomyocyte transplantation improves heart function. Ann Thorac Surg 62:654–660.
12. Scorsin M, Hagege AA, Dolizy I, Marotte F, Mirochnik N, Copin H, Barnoux M, le Bert M, Samuel JL, Rappaport L, Menasche P. 1998. Can cellular transplantation improve function in doxorubicin-induced heart failure? Circulation 98:II151–II155.
13. Zhang M, Methot D, Poppa V, Fujio Y, Walsh K, Murry CE. 2001. Cardiomyocyte grafting for cardiac repair: graft cell death and anti-death strategies. J Mol Cell Cardiol 33:907–921.
14. Li RK, Mickle DA, Weisel RD, Rao V, Jia ZQ. 2001. Optimal time for cardiomyocyte transplantation to maximize myocardial function after left ventricular injury. Ann Thorac Surg 72:1957–1963.
15. Müller-Ehmsen J, Whittaker P, Kloner RA, Dow JS, Sakoda T, Long TI, Laird PW, Kedes L. 2002. Survival and development of neonatal rat cardiomyocytes transplanted into adult myocardium. J Mol Cell Cardiol 34:107–116.
16. Suzuki K, Smolenski RT, Jayakumar J, Murtuza B, Brand NJ, Yacoub MH. 2000. Heat shock treatment enhances graft cell survival in skeletal myoblast transplantation to the heart. Circulation 102: III216–III221.
17. Schierle GS, Hansson O, Leist M, Nicotera P, Widner H, Brundin P. 1999. Caspase inhibition reduces apoptosis and increases survival of nigral transplants. Nat Med 5:97–100.
18. Müller-Ehmsen J, Peterson KL, Kedes L, Whittaker P, Dow JS, Long TI, Laird PW, Kloner RA. 2002. Rebuilding a damaged heart: Long-term survival of transplanted neonatal rat cardiomyocytes after myocardial infarction and effect on cardiac function. Circulation 105:1720–1726.
19. Schwarz ER, Pollick C, Dow J, Patterson M, Birnbaum Y, Kloner RA. 1998. A small animal model of non-ischemic cardiomyopathy and its evaluation by transthoracic echocardiography. Cardiovasc Res 39:216–223.

Pathophysiology of Cardiovascular Disease,
edited by N.S. Dhalla, H. Rupp,
A. Angel & G.N. Pierce
Kluwer Academic Publishers, Boston, 2004

Cellular Cardiomyoplasty: The Role of Bone Marrow Stromal Cells in Myocardial Tissue Regeneration

Derek J. MacDonald and Ray C.-J. Chiu

Division of Cardiac Surgery, McGill University Health Center, Montreal, Quebec, Canada

Summary. Bone marrow stromal cells (MSCs) are multipotent adult stem-cells capable of regenerating a variety of adult tissues. This review examines the use of MSCs as a source of myocardial tissue regeneration, concentrating on the recent history of experiments examining the pathophysiological role of MSCs following myocardial infarction and their therapeutic use in regenerative cardiology. It provides an up-to-date review of current trials examining the functional impact of MSC implantation into an injured myocardium and discusses the future directions and potential pitfalls of this active area of research.

Key words: Marrow stromal cells, cellular cardiomyoplasty, myocardial regeneration, heart failure.

INTRODUCTION

The burden of cardiovascular disease in Western society has been well documented, with approximately 550,000 people in the United States suffering from morbidity related to myocardial ischemia each year (1). The final common pathophysiological pathway in the development of the symptoms of congestive heart failure (CHF) in these patients is the irreversible loss of terminally differentiated cardiomyocytes. While medical and surgical therapies for CHF exist, they target the symptoms of the disease and are ineffective against the underlying pathophysiological process. It is clear that novel therapeutic strategies are needed to confront the growing problem of CHF-related morbidity.

Corresponding Author: Ray C.-J. Chiu, M.D., Ph.D., Division of Cardiac Surgery, The Montreal General Hospital, MUHC, 1650 Cedar Avenue, Suite C9-169, Montreal, Quebec, H3G 1A4, Canada. Tel: (514) 937-6011 ext. 42839, Fax: (514) 934-8289, E-mail: *rchiu@po-box.mcgill.ca*.

CELLULAR THERAPY

The concept of cellular cardiomyoplasty (CCM), the transplantation of cardiomyocyte precursors or functioning cardiomyocytes into infarcted tissue, has been described as early as 1989 (2). The term itself was first used by Chiu et al. 1995 in reference to skeletal muscle satellite cell implantation (3). Although many types of donor cells have been described including skeletal myoblasts (4), fetal (5,6) and adult cardiomyocytes and embryonic stem cells (7), the most actively pursued area of research has been adult stem cells (8). Adult bone marrow is known to contain an autologous source of these stem cells, often referred to as marrow stromal cells (MSCs). The availability, plasticity and relative ease of culture of these cells are enormous advantages, in addition to the avoidance of the ethical considerations and issues of immune tolerance associated fetal tissue and embryonic stem cells.

The initial observations by Friedenstein and colleagues in the 1970s led many to believe that the bone marrow contained a subpopulation of cells capable of differentiating into fibroblasts and other mesenchymal cells (9). Since this initial work, many investigators have reported evidence of the existence of multipotential stems cells in adult bone marrow capable of differentiation into a variety of different cells types, including those of nonmesenchymal origin (10,11). This led to use of the term *adult stem cells* as opposed to mesenchymal stem cells.

THE PATHOPHYSIOLOGICAL ROLE OF MSCS FOLLOWING MYOCARDIAL INFARCTION

Cardiomyocyte necrosis following myocardial infarction (MI) triggers a cascade of events ultimately leading to decreased left ventricular performance. Scar formation, angiogenesis and ventricular remodeling occur in the weeks to months following an acute MI. Until recently, this process was thought only to involve local myofibroblasts and the cellular mediators of inflammation. Emerging data, however, suggests that the MSC may also play a role in the reparative and regenerative processes.

Bittira et al. (12) provided convincing evidence of the pathophysiological role of these cells following an acute MI. Following systemic injection of autologous MSCs labeled with the LacZ reporter gene into Lewis rats, it was demonstrated that these cells were able to "home" into the bone marrow compartment. Furthermore, following creation of an anterior myocardial infarction by LAD ligation, it was shown that labeled MSCs subsequently engrafted the injured area of myocardium. A proportion of these cells underwent morphologic changes representative of cardiomyogenic differentiation. The presence of connexin 43, the main constituent protein of gap junctions, was evident at 4-weeks. Lebeled cells did not, however, stain positively for sarcomeric myosin heavy chain. Additionally, some labeled cells stained positively for alpha smooth muscle actin and incorporated into the endothelium of small arterioles and capillaries, suggesting a role in angiogenesis.

These findings suggest that MSCs play a role in the normal biological repair process following acute myocardial infarction. The therapeutic implications of this discovery are obvious if a method can be developed to deliver a meaningful number of cardiomyocyte progenitors to the injured ventricle.

INDUCING MSCS TO FORM CARDIOMYOCYTES IN VITRO

The ability to promote high proportions of MSCs to form cardiomyocyte progenitors in vitro is desirable if cell-based therapy is to become clinical reality. Wakatani and colleagues (13) demonstrated in 1995 that rat MSCs, following exposure to the demethylating agent 5-azacytidine, transdifferentiated into multinucleated myotubes that stained positively for skeletal muscle-specific myosin. Additionally, these cells demonstrated contractile properties following exposure to acetylcholine.

Makino et al. (14) demonstrated that adult MSC's can be induced to differentiate into cardiomyocytes in vitro following 5-azacytidine exposure. In approximately 30% of the initial cell mass, cellular morphology gradually changed from fibroblast-like appearance to myotube-like structures over a three-week period. Surprisingly, spontaneous contractility was demonstrated after two weeks and synchronous contraction after three weeks. Electrophysiologic studies demonstrated that these cells developed node-like action potentials at 3-weeks and subsequently ventricular myocyte-like action potentials after 4-weeks. Cardiomyocyte characteristics were confirmed by a number of techniques: The cells expressed atrial natriuretic peptide, myosin heavy and light chains and were interconnected by intercalated disks.

The exact mechanism of action of 5-azacytidine has not yet been elucidated. Konieczny et al. proposed that embryonic cells contain a myogenic determination locus in a methylated state with a transcriptionally inactive phase. It is hypothesized that this locus becomes demethylated and transcriptionally active following the addition of 5-azacytidine, causing myogenic differentiation to occur (15).

The establishment of a cardiomyogenic cell line from adult bone marrow cells yielded a powerful model for the study cardiomyocyte differentiation that has since been used by many investigators. Additionally, this study raised the possibility of using 5-azacytidine as an adjunct to induce cardiomyocyte differentiation in MSCs used for cellular cardiomyoplasty.

EXPERIMENTAL EVIDENCE FOR CELLULAR CARDIOMYOPLASTY AS A TREATMENT FOR HEART FAILURE

The first in vivo study to definitively demonstrate that adult MSCs were capable of differentiation into cardiomyogenic cells was conducted by Tomita et al. (16). A BrdU labeled population of rat MSCs pretreated with 5-azacytidine was injected directly into the scar area of the rat left ventricle 3-weeks post-cryoinjury. At 5-weeks post-transplantation, labeled cells were identified in the scar area of animals in the study group. MSCs were found to contribute to angiogenesis, and transplanted hearts were found to contain higher capillary density than controls. Additionally, infarct size was smaller in transplanted animals. Although functional improvement was demonstrated using the Langendorff model, as evidenced by significantly higher peak systolic and developed pressures in animals who received cell therapy, the study was criticized because left-ventricular end-diastolic volumes in the study group were lower relative to controls. The predictive value of the Langendorff model when used in hearts of differing sizes is poor, since peak systolic pressures may be falsely elevated in smaller hearts.

The role of milieu-dependent differentiation has been examined by Wang and colleagues (17,18). These investigators injected autologous MSCs into the left-ventricular free wall of isogenic Lewis rats and sacrificed the animals in a serial manner. At 1 month, labeled MSCs acquired a phenotype similar to normal host cardiomyocytes and stained positively for the presence of sarcomeric myosin heavy chain and the cardiomyocyte-specific protein connexin 43. Thus the ultimate fate of implanted MSCs appeared to be determined by the cardiac milieu they were placed in, rather than by exogenous stimulation by a chemical agent such as 5-azacytidine. Wang continued his work in a second series of experiments where autologous MSCs were again injected into the left ventricle. This time, however, cells remaining in the needle tract (surrounded by fibrous scar) were compared with those that had migrated into surrounding normal myocardium. Cells surrounded by fibrous scar did not undergo cardiomyogenic differentiation and appeared morphologically similar to fibroblasts. In findings consistent with the initial experiment, MSCs that were in close proximity to normal host cariomyocytes acquired a cardiomyocyte phenotype and again stained positively for sarcomeric myosin heavy chain and connexin 43. The findings suggested that even without the use of exogenous synthetic agents, local stimuli from the microenvironment are sufficient to guide the differentiation of MSCs into functioning cardiac tissue.

These observations support the growing body of evidence that MSCs are in fact multipotent adult stem cells, capable of differentiating into functioning cardiomyocytes with and without the use of exogenous agents. Many questions remained unanswered, however, and a positive effect of MSC implantation on myocardial function had not yet been definitively shown. Additionally, it appeared that following implantation, MSCs only underwent cardiomyogenic differentiation if they were surrounded by normal host cardiocytes. The ability to regenerate functioning myocardium from mature scar remained a significant hurdle on the path to true cellular cardiomyoplasty.

AUTOLOGOUS MSC IMPLANTATION AS A SOURCE OF MYOCARDIAL TISSUE REGENERATION: RESULTS OF FUNCTIONAL STUDIES

Recently, very promising studies have been conducted demonstrating that MSCs are able to promote meaningful repair of the left ventricle in ischemia/reperfusion injury models using mice and rats (19,20). Orlic et al. (19) sorted lineage-negative (Lin^-) bone marrow cells from transgenic mice expressing enhanced green fluorescent protein by fluorescence-activated cell sorting on the basis of c-kit expression. They then created a myocardial infarction model by ligating the proximal LAD, and shortly afterward Lin^- $c\text{-}kit^{POS}$ cells were injected in the contracting wall bordering the infarct (the peri-infarct zone of the left ventricle). Reconstitution of the myocardium was found to occur in 12 of 30 animals (40%) with the failure rate being attributed to the technical difficulty of direct implantation into a left ventricle contracting at approximately 600 beats per minute. Newly formed myocardium occupied 68% of the infarcted portion of the ventricle 9 days after transplanting bone marrow cells. The percentage of new myocytes, endothelial cells and smooth muscle cells expressing the GFP marker was 53 ± 9%, 44 ± 6% and 49 ± 7%, respectively, and the values were consistent with the fraction of transplanted Lin^{-} $c\text{-}kit^{POS}$ bone marrow cells that expressed GFP. Furthermore, the authors were able to demonstrate that all nuclei of labeled cells expressed myocyte enhancer factor-2 (MEF2) and the cardiac specific transcription factor GATA-4. Only 40 ± 9% expressed the early marker for myocyte development Csx/Nkx2. These findings were consistent with the belief that newly formed myocytes represented maturing cells aimed at functional competence. Connexin 43 was apparent in the cell cytoplasm and at the surface of closely aligned differentiating cells, a result consistent with expected functional competence.

To determine whether developing myocytes derived from MSCs actually had an impact on function, haemodynamic parameters were obtained prior to sacrifice. In MSC-treated animals, left ventricular end-diastolic pressure (LVEDP) was lower and left ventricular developed pressure (LVDP) was higher (36% and 32%, respectively). These findings are significant, as this was the first study to definitively demonstrate that a subpopulation of primitive bone marrow cells regenerate myocardium in vivo with an improvement in left ventricular performance.

In a follow-up to their original study, Orlic and colleagues hypothesized that MSCs mobilized by stem cell factor (SCF) and granulocyte-colony stimulating factor (G-CSF) would target the infarcted region of myocardium, replicate, differentiate and ultimately promote myocardial repair (20). The authors again used C57BL/6 male mice which were splenectomized at 2 months of age. Two weeks later these animals were injected with recombinant rat SCF (200 mg/kg/day) and recombinant human G-CSF (50 mg/kg/day) daily for 5 days, resulting in a 250-fold increase in Lin^{-} $c\text{-}kit^{POS}$ circulating cells. A significant myocardial infarction was then created by proximal LAD ligation. BrdUrd (50 mg/kg) was given once a day for 13 days before the mice were killed. Echocardiography and hemodynamics were performed and the mice were sacrificed at 27 days.

The results were dramatic. The combination of SCF and G-CSF administration with the consequent MSC mobilization (and presumably translocation to the infarcted portion of the heart) led to significant myocardial repair. Infarct size was reduced by 40%, cavitary dilation by 26% and diastolic stress by 70% in treated animals. From the echocardiographic data, ejection fraction was found to progressively increase and hemodynamics were significantly improved. Perhaps the most notable result from the study was the 68% improvement in mortality that was noted in the treatment group. Eleven of fifteen treated animals survived to the 27 day endpoint compared to 9 of 52 untreated animals (83% mortality). These results should be interpreted with great caution, however, as the mortality in control animals is significantly higher than would be expected from LAD ligation in a rodent model.

Perhaps the most significant finding of this study is the suggestion anatomical and functional improvement can be accomplished with the use of a noninvasive procedure. In the previous landmark study, and in other studies assessing myocardial function following cell therapy, MSCs were delivered directly to the infarct region by local infiltration. The development of an effective, noninvasive therapy for myocardial regeneration following acute myocardial infarction has obvious implications for the treatment of ischemic heart disease in humans.

DELIVERY METHODS: AN ONGOING DEBATE

The optimum method of delivery of MSCs to an injured myocardium has not yet been established. Direct intramyocardial injection of MSCs has already been established as an effective means of augmenting ventricular performance in an animal model of myocardial injury (19), however the procedure is highly invasive and may not be amenable to many clinical situations. The recruitment of host MSCs using intravenous cytokines such as stem cell factor or G-CSF is promising (20), however studies examining this therapy have utilized pretreatment with these agents prior to infarct creation.

Intracoronary administration of MSCs as a potential method of delivery to an injured myocardium has been examined, and is considerably less invasive than direct injection since it can be accomplished by catheter-based techniques. Wang et al. (21) examined the pathophysiologic and therapeutic implications of this delivery technique. Two weeks after LAD ligation in Lewis rats, LacZ labeled autologous MSCs were injected directly into the aortic root after cross-clamping. The hearts were harvested immediately or at 4 weeks following injection. LacZ labeled MSCs were seen trapped within the coronary capillaries immediately following infusion. After 4 weeks, these cells were detected within the myocardial scar or, in some instances, outside the infarction area. Cells trapped within the newly formed scar had a fibroblastic phenotype, however cells that had developed in close contact with normal host cardiomyocytes acquired morphologic features consistent with cardiac muscle cells. These cells stained positively for connexin 43 and integrated with surrounding normal myocardium. Some cells were also incorporated into the endocardium and capillary endothelium. Although this study did not have a functional component, it did demonstrate that coronary delivery of MSCs is a potential therapeutic option for cellular cardiomyoplasty.

Intravenous delivery of MSCs would obviously represent the least invasive form of therapy. A study focusing on the neoangiogenic potential of marrow-derived cells by Kocher and colleagues (22) was performed which demonstrated that systemically administered $CD34^+$ human bone marrow cells are also capable of supporting repair after myocardial infarction. Following proximal LAD ligation in a rat model, 2 million freshly-obtained, labeled human $CD34^+$ MSCs were injected intravenously. Histologic examination at two weeks post-infarction revealed that injection of $CD34^+$ cells was accompanied by a significant increase in infarct zone microvascularity, cellularity and numbers of factor-VIII^+ angioblasts and capillaries, and was also accompanied by reduction in matrix deposition and fibrosis in comparison to controls. Labeled cells supported healing by contributing to approximately 25% of the newly developed myocardial capillary vasculature. At 15-weeks post-infarction, mean cardiac index in rats injected with $CD34^+$ cells was only reduced by $26 \pm 8\%$ relative to normal rats, whereas in the other groups it was reduced by 48–59% ($P < 0.001$). In a follow-up experiment, the authors demonstrated that the infarct zone vasculogenesis and peri-infarct angiogenesis induced by cell injection prevented a radially extending pro-apoptotic process that was evident in controls. This phenomenon enabled survival of hypertrophied myocytes within the peri-infarct zone and improved myocardial function.

This study demonstrates that the marrow-derived cells do not necessarily need to be injected directly into the site of injury or even into the coronary circulation to provide a beneficial effect. Additionally, the therapeutic effect of these cells could, at least in part, be due to induction of endothelial growth and neovascularization of the myocardium.

MECHANISMS OF FUNCTIONAL IMPROVEMENT: MYOGENESIS, ANGIOGENESIS, OR SOMETHING MORE?

It is important to note that the mechanisms of the beneficial effects resulting in improved cardiac functionality following MSC or cytokine administration are not yet clear. The remarkable results achieved in some studies are probably due to more than just recruitment of cells to become myocytes. The positive effects on cardiac performance may be due to myogenesis, angiogenesis or prevention of left ventricular dilatation and remodeling (12).

The case for myogenic precursor cells integrating into damaged myocardial tissue and contributing to meaningful, synchronous contractile function has already been made, and the evidence continues to mount. As already discussed, a growing body of literature suggests that MSCs are able to migrate to the left ventricle following an acute myocardial infarction (12), or they can be delivered directly into the area of injury by local injection (19). These cells phenotypically resemble cardiac myocytes, stain positively for cardiomyocyte specific markers and form connections with surrounding myocytes via gap junctions (13,14,16,19). Data derived from in vitro experiments demonstrates that bone-marrow derived myocardial precursors are induced to express beta-adrenergic and muscarinic receptors after they had been induced to express the cardiomyocyte phenotype. These newly-formed receptors had functional signal-transduction pathways and were shown to be capable of modulating cell function (23). Clearly, for regenerated cardiomyocytes to be useful for transplantation, they must not only differentiate and integrate into the host myocardium but they must express functional adrenergic and muscarinic receptors capable of responding to in vivo stimuli. From a synthesis of the evidence at hand, this appears to be the case.

Angiogenesis also clearly plays a part in the healing/remodeling process following an ischemic insult to the myocardium. Although myogenesis was not examined in the recent study by Kocher et al. (22), significant angiogenesis and attenuation of myocyte apoptosis at the infarct border-zone was demonstrated, and the functional improvement in the hearts of treated animals relative to controls was significant. In fact, optimum cellular therapy for the failing myocardium may ultimately be myocardial progenitor cells induced to express angiogenic growth factors such as vascular endothelial growth factor (VEGF). Studies examining this potential "combination" therapy are currently underway.

In a letter to the editor of The Journal of Thoracic and Cardiovascular Surgery, it was hypothesized that strategies aimed at preventing myocardial impairment resulting from left ventricular remodeling would be superior to ones that propose to restore ventricular function by the introduction of newly formed cardiomyocytes in an attempt to replace those lost as the result of an infarction (24). The author points out that both myocytes and intercellular collagen matrix are destroyed following myocardial infarction, and the random implantation of myogenic precursors into an infarct in the absence of a viable collagen network to "harness them together" are unlikely to provide effective contractile force. It was suggested that the beneficial effects attributed to cellular cardiomyoplasty seen in recent studies are partially attributable to alteration in infarct material properties and prevention of infarct expansion, thereby limiting the myopathic effects of postinfarction left ventricular remodeling. Despite the enthusiasm generated by the favorable results in animal studies, a clearer understanding of the mechanisms of functional improvement following cellular cardiomyoplasty with MSCs is required before the works pushes forward to human trials.

"STEM CELLS THAT AREN'T?": A FINAL CAVEAT

Much evidence has accumulated suggesting that bone-marrow derived precursor cells are capable of in vivo differentiation to many host tissues (4,6,25,26), presumably through milieu-dependent differentiation. A recent study published in Nature, however, casts some doubt on the current body of evidence and raises questions whether or not transdifferentiation of MSCs actually occurs at all (27). In an experiment designed to determine if embryonic stem cells can influence MSCs to transdifferentiate into embryonic-like cells, Terada et al. co-cultured these two cell types. Although transdifferentiation of adult stem cells into embryonic-like cells did not occur, the authors found that cell fusion occurred between the ES cells and MSCs. The hypothesis has been put forth that fusion is perhaps being misinterpreted as transdifferentiation.

Fusion was assumed to be a rare event, with an estimated frequency of 1 in 10^4 to 1 in 5×10^5. Since MSCs used in these experiments represent a nonhomogeneous population, it is conceivable that perhaps only a small proportion of these cells are capable of fusion. Additionally, cell fusion has so far only been shown to occur in vitro, while transdifferentiation has been demonstrated both in vitro and in vivo. While it is much too early to conclude that the phenomenon of transdifferentiation exhibited by MSCs is not a real one, it is clear that further in vivo studies must include detailed genetic analysis in an effort to answer this question.

MSCS FOR CELLULAR CARDIOMYOPLASTY: THE ROAD TO HUMAN CLINICAL TRIALS

This chapter provides a broad overview of the topic of MSCs as a source of myocardial tissue regeneration. Although much has been accomplished to date, it is apparent that an enormous amount of work remains to be completed before human clinical trials can be effectively undertaken. While the new issue of trandifferentiation versus cell fusion remains to be sorted out, other questions such as the optimum subpopulation of MSCs for implantation, timing of therapy, mode of cell delivery and mechanism of functional improvement must be answered before effective and safe therapy can be designed. The roles of cytokine therapy and angiogenic therapy also need to be defined. With the phenomenal amount of interest in this active area of research worldwide, the enormous strides made so far are certain to continue at an accelerated pace, laying the groundwork for the first therapy to target the basic underlying pathophysiologic process of heart failure.

REFERENCES

1. American Heart Association. Heart and Stroke Statistical Update. 2000.
2. Kao RL, Rohwedel J, Magovern GJ. 1989. Satellite cells for myocardial regeneration. Physiologist 32:220.
3. Chiu RCJ, Zibaitis A, Kao RL. 1995. Cellular cardiomyoplasty: myocardial regeneration with satellite cell implantation. Ann Thor Surg 60:12–18.
4. Marelli D, Desrosiers C, El-Alfy M, Kao RL, Chiu RCJ. 1992. Cell transplantation for myocardial repair: an experimental approach. Cell Transplant 1:383–390.
5. Muller M, Fleishmann B, Selbert S, Ji G, Endl E, Middeler G, Muller O. 2000. Selection of ventricular-like cardiomyocytes from ES cells in vitro. FASEB 14:254–2548.
6. Li RK, Mickle DA, Weisel RD, Zhang J, Mohabeer MK. 1996. In vivo survival and function of transplanted rat cardiomyocytes. Circ Res 78:283–288.
7. Maltsev VA, Rohwedel J, Hescheler J, Wobus AM. 1993. Embryonic stem cells differentiate in vitro into cardiomyocytes representing sinusnodal, atrial and ventricular cell types. Mech Dev 44:41–50.
8. Wang JS, Shum-Tim D, Galipeau J, Chedrawy E, Eliopoulos N, Chiu RCJ. 2000. Marrow stromal cells for cellular cardiomyoplasty: feasability and potential clinical advantages. J Thorac Cardiovasc Surg 120:999–1006.
9. Friedenstein AJ, Gorskoja V, Kulagina NN. 1976. Fibroblast precursors in normal and irradiated mouse hematopoietic organs. Exp Hematol 4:267–274.
10. Peterson BE, Bowen WC, Patrene KD, Mars WM, Sullivan AK, Murase N, Boggs SS, Greenberger JS, Goff JP. 1999. Bone marrow as a potential source of hepatic oval cells. Science 284:1168–1170.
11. Woodbury D, Schwarz EJ, Prockop DJ, Black IB. 2000. Adult rat and human bone marrow stromal cells differentiate into neurons. J Neurosci Res 61:364–370.
12. Bittira B, Kuang JQ, Piquer S, Shum-Tim D, Al-Khaldi A, Chiu RCJ. 2002. The Pathophysiological Roles of Bone Marrow Stromal Cells (MSCs) In Myocardial Infarction. Circulation 104(Suppl II):II–523.
13. Wakitani S, Saito T, Caplan A. 1995. Myogenic cells derived from rat bone marrow mesenchymal stem cells exposed to 5-azacytidine. Muscle Nerve 18:1417–1426.
14. Makino S, Fukudi K, Miyoshi S, Konishi F, Kodama H, Pan J, Sano M, Takahashi T, Hori S, Abe H, Hata J, Umezawa A, Ogawa S. 1999. Cardiomyocytes can be generated from marrow stromal cells in vitro. J Clin Invest 103:697–705.
15. Konieczny SF, Emerson CP. 1984. 5-azacytidine induction of stable mesodermal stem cell lineages from 10T1/2 cells: evidence for regulatory genes controlling determination. Cell 38:791–800.
16. Tomita S, Li RK, Weisel RD. 1999. Autologous transplantation of bone marrow stromal cells improves damaged heart function. Circulation 100(Suppl II):247–256.
17. Wang JS, Shum-Tim D, Galipeau J, Chedrawy E, Eliopoulos N, Chiu RCJ. 2000. Marrow stromal cells for cellular cardiomyoplasty: feasability and potential clinical advantages. J Thorac Cardiovasc Surg 20:999–1006.
18. Wang JS, Shum-Tim D, Chedrawy E, Eliopoulos N, Galipeau J, Chiu RCJ. 2000. Marrow stromal cells for cellular cardiomyoplasty: The importance of microenvironment for milieu dependent differentiation. Presented at the annual meeting of the American Heart Association.
19. Orlic D, Kajstura J, Chimenti S, Jakoniuk I, Anderson SM, Li B, Pickel J, McKay R, Nadel-Ginard B, Bodine DM, Leri A, Anversa P. 2001. Bone marrow cells regenerate infarcted myocardium. Nature 410:701–705.

20. Orlic D, Kajstura J, Chimenti S, Jakoniuk I, Anderson SM, Li B, Pickel J, McKay R, Nadel-Ginard B, Bodine DM, Leri A, Anversa P. 2001. Mobilized bone marrow cells repair the infarcted heart, improving function and survival. Proc Nat Acad Sci 98:10344–10349.
21. Wang JS, Shum-Tim D, Chedrawy E, Chiu RCJ. 2001. The coronary delivery of marrow stromal cells for myocardial regeneration: Pathophysiologic and therapeutic implications. J Thorac Cardiovasc Surg 121:1–7.
22. Kocher AA, Schuster MD, Szabolcs MJ, Takuma S, Burkhoff D, Wang J, Homma S, Edwards NM, Itescu S. 2001. Neovascularization of ischemic myocardium by human bone marrow derived angioblasts prevents cardiomyocyte apoptosis, reduces remodeling and improves cardiac function. Nat Med 7:430–436.
23. Hakuno D, Fukuda K, Makino S, Konishi F, Tomita Y, Manabe T, Suzuki Y, Umezawa A, Ogawa S. 2002. Bone marrow-derived regenerated cardiomyocytes (CMG cells) express functional adrenergic and muscarinic receptors. Circulation 105:380–386.
24. Gorman R. 2002. Cellular myoplasty: What are we really trying to achieve? J Thorac Cardiovasc Surg 123:582–583.
25. Brazelton TR, Rossi FM, Keshet GI, Blau HM. 2000. From marrow to brain: expression of neuronal phenotypes in adult mice. Science 290:1779.
26. Bruder SP, Fink DJ, Caplan AI. 1994. Mesenchymal stem cells in bone development, bone repair, and skeletal regeneration therapy. J Cell Biochem 56:283–294.
27. Terada N, Hamazaki T, Oka M, Hoki M, Mastalerz DM, Nakano Y, Meyer EM, Morel L, Petersen BE. 2001. Bone marrow cells adopt the phenotype of other cells by spontaneous cell fusion. Nature 416:542–545.

Pathophysiology of Cardiovascular Disease,
edited by N.S. Dhalla, H. Rupp,
A. Angel & G.N. Pierce
Kluwer Academic Publishers, Boston, 2004

Xenotransplantation: An Overview

Rizwan A. Manji,[1,2] Jacqueline S. Manji,[2]
Ray V. Rajotte,[2] and Arvind Koshal[1]

[1] *The Division of Cardiac Surgery, Department of Surgery,*
and [2] *the Surgical Medical Research Institute, University of Alberta,*
Edmonton, Alberta, Canada

Summary. Interest in xenotransplantation has increased significantly as a possible response to the critical organ shortage crisis that exists in clinical transplantation today. The pig has been suggested to be the most suitable animal for human xenotransplantation, however, many barriers still exist to the clinical application of xenotransplantation.

This review takes a brief look at the history of clinical xenotransplantation, the choice of organ donor, the obstacles to xenotransplantation and attempts to overcome them. Xenografted organs will likely be used as a bridge to allotransplantation before being used as full blown replacements for human organs. There is much progress that has been made, but still much to do. With perseverance and good science, solutions are likely to be found.

Key words: xenotransplantation, rejection, alpha-Gal antibody, pig.

1. INTRODUCTION

The field of allotransplantation (transplantation within a species) has had considerable success due in part to the important contributions of individuals like Carrel (1,2)—vascular anastomosis; Medawar & Burnett (3–5)—immunological tolerance; and Calne (6)—cyclosporine. There have been over a million organ and tissue transplants performed world wide to date. In 1999, worldwide (~29 countries comprising ~790.5 million people), there were over 40,000 cadaveric solid organ transplants performed, with most being performed in the USA (7). Further scientific advances will undoubtedly increase the application of allotransplantion. No matter

Corresponding Author: Dr. Rizwan A. Manji, Surgical Medical Research Institute, 1074 Dentistry/Pharmacy Centre, University of Alberta, Edmonton, Alberta, Canada, T6G 2N8. Phone: (780) 984-1724, Fax: (780) 492-1627, E-mail: rmanji@ualberta.ca.

how advanced the science becomes, allotransplantation will be limited by the shortage of human tissues and organs. In 1999, only about 30% of people who needed an organ in the USA were transplanted. It is estimated, that by the year 2010, the number of people on the waiting list will have increased by about 200% (i.e. ~225,000 people); whereas the number of organ donors in the USA will likely stay static around 20,000 (8). This means that instead of the current 30% of people who get transplanted, only about 10% of people on the waiting list will get transplanted.

Investigators are trying to solve this organ shortage crisis in many ways including but not limited to: increasing awareness of both the general public and physicians (9), artificial organs (10,11), cloning technology (12), and xenotransplantation. Xenotransplantation has many advantages such as an endless supply of organs, ability to genetically engineer organs, and optimizing technical factors (ex. decreasing ischemia time or eliminating "middle of the night" operations by being able to plan ahead of time). However, many obstacles still exist such as xenograft rejection, possible physiological mismatch between humans and animals, and xenozoonoses.

2. HISTORY OF XENOTRANSPLANTATION

Ideas of xenotransplantation date back centuries as can be seen in stories of mythological beings that were more powerful since they had the advantages of being part animal as well as part human (8). The first surgical attempt at xenotransplantation was reported in 1804 when Boronio described the failure of a cow and horse skin graft exchange (13). Similar failures were reported by Bert (13) who tried to transplant guinea pig skin to rats. In the early 1920s, Russian surgeon Serge Voronoff transplanted 51 older human males with baboon testicles in an attempt to "rejuvenate their physical, mental, and sexual condition" (8). Voronoff writes of how successful his procedure was; however, with today's knowledge, we know that these grafts probably did not survive very long and most of the beneficial effects were probably placebo effects.

2.1. Solid organ xenotransplants

The first vascularized renal xenotransplant in a human was described by Jaboulay in 1905 (13). Jaboulay transplanted two patients—one was pig-to-human and the second was goat-to-human. Both of these grafts failed and pathology showed vessel thrombosis. There were a few other failed trials early in the 20th century and then the interest in xenotransplantation subsided. Interest picked up again in the 1960's when Reemtsma (14) did xenotransplantation using immunosuppressive protocols that were known to blunt allograft rejection. Reemtsma performed 12 chimpanzee-to-human renal transplants in the early 1960's using an immunosuppressive protocol of steroids, actinomycin C and local irradiation of the graft. Most of the xenografts rejected within 2 months; however, one functioned for greater than 9 months and at the patient's death (from infection), there was little evidence of rejection in the graft. Starzl (15) and Milliard (13) did 6 baboon-to-human kidney transplants in the early 1960's as well. Even though they used similar immunosuppression as Reemstma had used, the grafts functioned only a mean of 36 days (range 10–60 days). In their studies there was no evidence of hyperacute rejection but there were more frequent and more severe rejection episodes. Pathology showed hemorrhagic infarcts compatible with vascular lesions characteristic of humoral rejection (15). The baboon is more phylogenetically disparate from humans than chimpanzees and the more phylogenetically disparate the combination, the more humoral the rejection response. This may help explain why Starzl and Millard had shorter survivals with their transplants as compared to Reemstma.

In 1964, Hardy (16) performed the first heart xenotransplant using a chimpanzee donor. The recipient died on the operating room table and the cause of death was said to be an undersized graft and perhaps hyperacute rejection contributing to cardiac failure. In 1984, Bailey (17) transplanted a baboon heart into an infant ("Baby Fae") with hypoplastic left heart syndrome under cyclosporine immunosuppression. The graft functioned well for about 2 weeks with subsequent problems leading to cardiac failure and death at 20 days postoperatively. The pathology was very educational. There was only minimal cellular infiltrate, suggesting that cyclosporine had been effective; however, there was evidence of interstitial hemorrhage and coagulative necrosis implying that humoral rejection was predominant. This latter finding was confirmed by demonstrating antibody and complement deposition on the vasculature (17).

Realizing the importance of the humoral arm of rejection in xenotransplantation, Starlz's group preformed two liver transplants from baboons to humans between June 1992 and January 1993 (16). This was done under immunosuppressive coverage to target both the humoral and cellular arms of the rejection process with tacrolimus, steroids, cyclophosphamide and prostaglandin E1. One recipient died at 25 days and the other at day 70, both from infection. Pathology did not show any humoral or cellular rejection in either case, but there was evidence of complement activation. Starzl concluded that xenotransplantation may be achievable once complement activation was overcome.

The next important xenotransplant performed was in 1993 by Makowka (18). A porcine liver was used as a bridge to support a young woman with hepatic encephalopathy until a liver allograft became available. The graft did function for a period of time, but the patient deteriorated, being declared brain dead prior to getting allotransplanted. Makowka deduced that the graft failed because of a rapid return of xenoreactive natural antibodies against the graft following transfusion of the patient with human blood products. This deduction was indirectly confirmed by pathology consistent with hyperacute rejection (19). From 1906 to 1992, there have been over 40 attempted human solid organ xenotransplants. Though all have failed; they all have provided useful information about the xenorejection process such as the importance of phylogenetic distance, antibody mediated rejection, and the problems with infection secondary to heavy immunosuppression.

2.2. Extracorporeal xenograft perfusion

Because of problems with solid organ xenotransplantation, some have investigated extracorporeal perfusion. Extracorporeal perfusion is done outside the body and involves passing the recipients blood through the donor organ or through a system of tubes which has the donors' cells lining them. An advantage of being outside the body is when the xenograft rejects, another xenograft can more easily be attached. Most perfusions have been done with pig livers in patients with hepatic failure and improvements were seen in many of the patients "buying them time" until a human liver was available for transplantation leading to their survival (20). Kidney perfusions have also been done, with less success (21). Currently, researchers are looking into further development of extracorporeal xenograft perfusions (8,20).

2.3. Cellular xenotransplants

There have also been a number of cellular xenotransplants performed particularly in the last half of the 1990's (Table 1). The clinical results from cellular transplantation are not overwhelmingly encouraging or discouraging but they suggest that further studies are needed. A reason that some of the results are not as encouraging as hoped may simply be methodological issues such as incorrect number of cells injected, improper processing of cells for implantation, need for xenogenic growth factor injection in addition to cell transplantation, and need for larger sample sizes (21). Hopefully, as further studies in cellular xenotransplantation continue, the results will be more positive.

Advantages of the use of xenogenic cellular transplants as opposed to vascularized whole organ transplants are many. Firstly, as cellular xenotransplants are not immediately vascularized, they do not undergo hyperacute rejection though they can still undergo acute vascular rejection (22,23). Secondly, cells can be more easily manipulated compared to organs. Cells can be encapsulated (24,25) or treated to remove certain epitopes, genetically engineered (26–28) to express certain factors (e.g. human complement regulatory proteins or anti-apoptotic factors), or injected into "immunopriviledged sites" like the brain. These factors help to decrease the likelihood of rejection of xenogenic cells as compared to solid organs (8,29). Thirdly, cellular transplants are technically easier and physically less demanding on the

Table 1. Clinical cellular xenotransplantation

Cell Type	Application (reference)
Bovine chromaffin cells	Pain relief (146)
Fetal hamster cells	Amyotrophic lateral sclerosis (147)
Fetal porcine dopamine neurons	Parkinson's disease (148)
Fetal porcine GABA neurons	Huntington's chorea (148)
Fetal porcine islets	Diabetes mellitus (149)
Baboon bone marrow	Advanced HIV (150)

patients than a major organ operation. The last theoretical advantage of cellular transplants is that they are less likely (than large and solid organs) to be an infection hazard because the number of cells (and thus potential infectious agents) transplanted is less than would be transplanted if a solid organ were used (12). This latter reason is why the FDA has allowed cellular xenotransplant clinical studies to continue while banning any further solid organ xenotransplant studies (12).

3. CHOICE OF ORGAN DONOR

Animal species that are evolutionarily closer together would likely be best to transplant as antigen similarity between the two would hopefully avoid very severe rejection. The best animals to choose are the great apes which include chimpanzees and gorillas (30). Many of the early xenotransplants done in humans actually used organs from the great apes (such as the chimpanzee heart that was transplanted into "Baby Fae"). These nonhuman primate (NHP)-to-human organs did not undergo hyperacute rejection but underwent a less severe form of antibody mediated rejection called acute vascular rejection (16,17).

It would seem that with the recent advances in immunosuppression and genetic engineering, primate organs would still be the most suitable. Unfortunately, there are many problems with the use of great apes, and in fact any NHP. One reason for abandoning the use of NHP is the moral dilemma as to whether it is right to breed a higher animal species specifically for the purpose of killing them for human transplantation. A second major concern is the potential transmission of zoonoses between NHP and humans (12,31–33). In general, the closer two animal species are to each other, the more likely that an infectious organism in one will infect the other and the more likely that this infection will cause problems. A third reason for rejecting NHP as donors is that the numbers of primates that could be bred does not come close to the numbers needed, due to long gestation times and small litter sizes (8).

Other species that may be used for xenotransplantation are some of the lower mammals. Pigs have been looked at very favorably for many reasons. First, ethical concerns are less volatile with pigs since they are slaughtered (over 90 million in the United States each year) for food on a daily basis (8). Second, having an unlimited supply of pig organs and tissues should be obtainable almost immediately because of the breeding characteristics of pigs. Sows have large litters (5–12 per litter on average), they have short (114 days) and frequent gestational periods, have frequent estrous cycles, reach sexual maturity by ~4–8 months of age and reach adult/transplantable size by about 6 months of age (8). Third, because of these breeding characteristics, genetic engineering would also be much easier in pigs. Fourth, though there is a risk of infection transmission to humans with pigs, this is less of a concern than with NHP. Finally, anatomically, pig organs (especially the heart) would be of an appropriate size for humans. Pig organs (such as kidneys and heart) have been transplanted into baboons and have been able to support the baboons for significant periods of time before rejection caused organ failure (8,34,35). At present, the pig is felt to be the ideal organ/tissue donor for human xenotransplantation.

4. OBSTACLES TO XENOTRANSPLANTATION

The potential benefits of transplanting pig organs into humans are great, however, numerous obstacles remain. The biggest obstacle at present is rejection of the donor organ by the recipient. Other obstacles are physiological differences between the donor and recipient and the potential spread of infectious agents from the donor to recipient and perhaps those in contact with the recipient.

4.1. Xenograft rejection

There are two types of xenografts, concordant and discordant (36). Concordant xenografts are grafts between phylogenetically close species such as mouse-to-rat or chimpanzee-to-human. These combinations undergo a form of rejection termed acute vascular rejection/delayed xenograft rejection (AVR) (37). This form of rejection develops within a few days of transplantation and induced antibody (usually IgG) is important in its development. Discordant grafts, on the other hand, are between phylogenetically disparate combinations (such as pig-to-human), where one has naturally occurring antibodies, usually of the IgM isotype, against the other. These combinations undergo a very rapid (minutes to hours) form of rejection termed hyperacute rejection (HAR) (38–40). If HAR is overcome, discordant grafts undergo AVR.

There are some common features of the rejection process in HAR and AVR These include the role of antibodies and endothelial cell activation.

4.1.1. Xenoantibodies of interest in the pig-to-human combination

There are a number of porcine endothelial cell (PEC) (41) glycoproteins recognized by human antibodies. All of these glycoproteins have not been fully characterized, but it is known that they are members of the integrin family (41). The primary entity on PEC that is recognized by human xenoantibodies (Xabs) and is the cause of hyperacute rejection is the Gal α 1–3 Gal (α Gal) epitope (42–49).

The anti-α-gal antibody was first discovered around 1983 by Uri Galili who was studying mechanisms of destruction of abnormal human red blood cells (RBC) (sickle cell, β-thalassemia) (50,51). As was learned later, abnormal red blood cells and old senescent red cells are bound by anti-α-gal antibody and can be removed from the circulation in this way (50–54). When Galili was first doing his studies, he did hemagglutination inhibition experiments using various sugars and found that very specific sugars, and thus very specific antibodies were inhibiting hemagglutination. He found that anti-α-gal antibody was very important and on quantifying the antibody, he found that a very high proportion of antibody, around 1% of circulating IgG was this anti-α-gal antibody (55). It was quite interesting that there was so much antibody of a specific type in humans. Galili subsequently looked at other animal species and found that many did not have this antibody but in fact, these species expressed the corresponding antigen, sometimes in very high amounts. After extensively studying many animal species, Galili determined that all mammals including new world monkeys (e.g. spider monkey, owl monkey) express the α-gal antigen except for humans, apes (e.g. gorillas, chimpanzees, and orangutans) and old world monkeys (e.g. baboons, rhesus, and cynomologous monkeys) (56–60). Humans, apes and old world monkeys have naturally occurring antibodies against the α-gal antigen. Though all humans have anti-α-gal antibodies, there is variability in titres and cytotoxicity (61,62). Non-mammals (e.g. birds, fish, reptiles) do not express the α-gal antigen (60).

It was felt that anti-α-gal antibodies developed due to exposure of the host to certain pathogens. Many studies have been done showing that many bacteria (e.g. *Escherichia coli*) and parasites express the α-gal antigen (63,64). Galili (57) proposed that there was an evolutionary force (such as infectious agent) which contained the Gal α 1–3 Gal epitope that eventually lead to the inactivation of the α1,3 galactosyltransferase (α1,3 GT) gene in apes, Old World monkeys and humans, thus preventing the expression of αGal on these species' tissues. Indeed, the above animal species do have a α1,3 GT pseudogene that is nonfunctional due to a frameshift mutation (57). This allowed these species to make antibodies to the pathogens to help destroy them and protect the host.

A seminal experiment which first identified the anti-α-gal antibody as the important antibody for pig-to-human xenotransplantation was done by Good et al. (42). This group perfused 4 pig hearts and 4 pig kidneys with human plasma and subsequently eluted the antibodies from the organs. They tested these antibodies against various carbohydrate antigens of known specificity and they found the strongest reactions to the following antigens: Gal α1,3 gal β 1,4 Glc, Gal α1,3 gal β1,4 GlcNAc and Gal α1,3 gal. An important experiment confirming that α-gal antigen was the important antigen was done by Vaughan et al. (65) who demonstrated that transfection of COS cells (a galα1,3gal negative cell line of old world monkey origin) with cDNA for α1,3-galactosyltransferase caused *de novo* expression of α-gal antigen and rendered these cells sensitive to human serum mediated cytotoxicity. In additon, if human serum was adsorbed against the α-gal expressing COS cells, then this signficantly decreased toxicity of the serum to porcine endothelial cells. A further experiment showing that it was the α-gal antigen that was important was done by Collins et al. (46). They found that hearts from new world monkeys (express α-Gal antigen) were hyperacutely rejected by old world monkeys (do not express α-Gal antigen but have anti-α-gal antibody).

Over the past 10 years, much has been learned about the α-gal antigen and the anti-α-gal antibody. The following summarizes some of the important points.

1. The anti-α-gal antibody recognizes the galα1,3 gal terminal sugar (i.e. gal α1,3gal; gal α1,3gal β1,4Glc-R; gal α1,3gal β1,4 GlcNAc-R (R = glycoprotein or glycolipid)). It is very specific for the α1,3 linkage (42–49).
2. Anti-α-gal antibody is present in humans, apes, and old world monkeys. Anti-α-Gal antibody is not present in other mammals but instead, these other mammals (e.g. pigs) express the α-gal antigen (56–60).
3. The α-gal antigen is quite heavily expressed. It is estimated that there are $\sim 10^6$–10^7 α-gal epitopes per porcine endothelial cell (60). Like the A and B blood

group antigens on human cells, the α-gal antigen on other mammal's cells is probably involved with functions such as cell signaling, and cell-cell interactions (60).

4. The α-gal antigen is also known to be expressed by human tissues in disease states like cancer or autoimmune disease (e.g. Grave's). (66–68) and in hemolytic diseases (50). The α-gal antigen is expressed on senesent red blood cells (54,69,70).
5. Anti-α-Gal antibody is estimated to comprise 4–8% of total IgM (71) and ~1% of total IgG in humans (60). The physiologic purpose of anti-α-gal antibody is believed to be for protection against pathogens such as gut bacteria and parasites (63,64) for protection against cancer (68) and to aid in the identification and removal of sensescent or abnormal RBC (54).
6. The α-gal antigen is variably distributed on pig tissues. McKenzie et al. (72) used an α-gal specific biotinylated lectin ($BSIB_4$) and adult pig tissues to determine α-gal antigen expression. Endothelial cells in all small vessels such as capillaries, arterioles and venules stained uniformly and diffusely. Larger vessels like the aorta stained but were less reactive than small vessels. The highest concentrations of staining in organs was in liver and kidney. Liver parenchyma stained uniformly. The brush borders of the proximal convoluted tubules in kidneys stained strongly, the distal convoluted tubules in kidneys stained moderately and there was no staining of collecting ducts or glomeruli (except for the endothelium). Heart muscle was non-reactive although the endothelium on capillaries and other vessels stained strongly. Lung alveoli and bronchioles stained moderately.
7. The α-gal antigen is very similar in structure to the blood group B antigen (60).
8. The anti-α-gal antibody (mostly naturally occurring IgM) is felt to be the most important antibody leading to HAR in the pig-to-human combination (73). The anti-α-gal antibody (mostly induced IgG) as well as other induced anti-pig antibodies are felt to be important in AVR (73).

4.1.2. Endothelial cell (EC) activation

EC cell activation is a very important part of xenograft rejection (74). EC normally provide a barrier function and provide an anticoagulant environment. EC activation contributes to rejection by causing the loss of these barrier and anticoagulant properties (75), and by secreting and expressing various proteins to recruit inflammatory cells. EC can be activated in numerous ways. Human Xab alone deliver signals to PEC to induce some, but not all, of the genes associated with full EC activation (76). Adding complement and Xab (by adding human serum) onto PEC will also lead to up-regulation of genes for EC activation that are not induced by Xab alone (76). Nuclear factor κB (NF-κB) is a transcription factor that is essential for the up-regulation of a number of genes associated with EC activation (77). Human natural killer cells (78) and macrophages (in vitro) (79), can also activate and lyse PEC (this effect is augmented by adding IgG) (80).

Once the EC has been stimulated, there are a number of events that occur. The occurrence of these events can be broken down into two phases. Phase I (or type I activation) occurs very early does not require protein synthesis (77) and is important in HAR. This phase involves the following: alteration of normal defense mechanisms (such as ADPase, regulators of complement activation (RCA), thrombomodulin, tissue factor pathway inhibitor and heparin sulfate); retraction of the EC; expression of P-selectin/E-selectin; expression of different inflammatory mediators (including C3a, C5a, histamine, PAF thrombin and leukotrienes); leukocyte recruitment; platelet aggregation; and fibrin deposition (81,82). Phase II (type II EC activation) occurs later, requires protein synthesis (77), and is more important in AVR. In phase II, there is macrophage and natural killer cell recruitment. There is also involvement of tissue factor, lectins, and cytokines (IL-8 and MCP-1) in this phase (81,83,84). These two phases of EC activation play a crucial role in xenograft rejection.

4.1.3. Hyperacute rejection

Hyperacute rejection (HAR) occurs in minutes to hours of transplantation of a vascularized solid organ in discordant xenografting. There is binding of natural antibody (mostly IgM, but also some IgG) to endothelial cell antigens and this leads to activation of the complement cascade, endothelial cell (EC) activation, and possibly EC death (43,44). There are two pathways in the complement cascade, the classical and alternate pathways. In the porcine-to-primate combination, HAR appears to primarily involve the classical complement pathway (85–87). Activation of the EC causes it to undergo numerous changes, which leads to cell injury as well as thrombus formation in the graft vasculature causing graft ischemia and ultimate organ failure (82,83). Pathologic examination of tissues having undergone HAR show deposition of antibody and complement on the graft vasculature as well as evidence of interstitial hemorrhage, and intravascular coagulation (88,89).

Knowledge about the mechanisms of HAR and the importance of Xabs, complement, and EC activation have led to many studies investigating methods to overcome HAR. Methods have included: depletion or blocking the Xabs, inhibition of complement, and EC based strategies.

Xab depletion can be accomplished via plasmaphoresis, ex vivo perfusion of the recipient's blood through another of the donor's organs or via immunosuppressive drug therapy (85). Xab depletion/blockade is also possible using specific carbohydrate-based immunoadsorptive techniques (90–94). Plasmaphoresis is a method of removing antibodies. Unfortunately, it has the undesirable effect of removing other proteins nonspecifically as well as antibodies. A more specific removal of just the antibodies against the donor organ can be achieved using ex vivo donor organ perfusion. Investigators (95) perfused rhesus monkey blood or plasma through a porcine liver and obtained removal of 60–70% of the monkey's Xab against pig antigens. Also, perfusion of a pig kidney with baboon blood demonstrated a decrease in IgM by 68% (96). Perfusing recipient blood through a donor organ (e.g. kidney) prior to transplanting another of the donor's organs (e.g. heart) into the recipient does increase survival of the transplanted organ; however the initial perfusion through the organ (e.g. kidney) causes EC activation of the perfused organ as well as other changes. The activated EC release various mediators into the recipient's blood, thus when the second organ (e.g. heart) is transplanted, the blood perfusing it is already "hostile". A solution to this problem is to use immunoadsorbant columns that can pull antibodies out without having the added problems of EC activation. Various immunoadsorbants have been shown to be effective in depleting the α-Gal Xab levels (97–100). Pascher et al. (98) found that using columns to deplete human xenoantibodies increased survival of ex-vivo perfused porcine livers. Levanthal et al. (99) and Taniguchi et al. (100) found that using immunoadsorption with splenectomy and immunosuppressive drug regimens prevented HAR in porcine-to-baboon xenografts and pig kidney cells respectively. Manji et al. (101) found that specifically depleting the α-gal antibody lead to prolongation of cardiac xenografts in a porcine-to-human working heart model.

Another method of avoiding hyperacute rejection is to inhibit complement (102). There are different methods to do this including: giving an agent that inhibits complement such as cobra venom factor (CVF) (103), employing an agent that binds complement (sCR1) (104), and through the use of organs from transgenic pigs expressing human complement regulatory proteins such as human DAF and human CD59 (105–107). All of these methods have been shown to prevent HAR, however inhibiting complement can cause severe immunosuppression, leading to infection.

Another way of preventing HAR includes EC based strategies. Investigators have produced knockout mice in which α-gal expression on EC cells has been eliminated and have shown that HAR can be prevented in this model (108,109). It has been more difficult to produce α-gal knockout pigs. Recently, 2 groups have reported producing live pigs in which one allele of the α-1,3-galalctosyltransferase locus was knocked out (110). At this point, studies on the ability of organs from these pigs to resist rejection caused by α-gal antibodies have yet to be performed. Because of the difficulties in developing the α-gal knockout pig model, investigators have also studied the effects of competitive glycosylation by creating transgenic pigs which have increased expression of the fucosyl transferase enzyme thereby producing more H antigen (blood group O antigens) and thus fewer α-gal antigens (111). Use of these transgenic pigs has shown some benefit but no significantly prolonged xenograft survival.

Modification of EC activation and death has been attempted as another method of preventing xenograft rejection. As described earlier, the ubiquitous transcription factor NF-κB is felt to be central in activating the genes involved in EC activation (112). It has been found that the agent pyrrolidine dithiocarbamate (PDTC) inhibits NF-κB activation and inhibits E-selectin, IL-8 and tissue factor expression (77,113). Another method of inhibiting NFκB is to over-express IκBα (a natural inhibitor of NFκB) in the cytoplasm (114). The use of a recombinant IκBα adenoviral vector to transduce PEC suppressed the transcription of various EC activated genes (113). These methods have been tried in xenotransplantation with limited success.

Another method of addressing the EC in xenograft rejection is to focus more peripherally on the adhesion molecules. The use of NPC 15669—a drug which inhibits neutrophil adhesion to the EC of the donor has been used combined with complement inhibition in small animal models of discordant heart transplantation with prolonged xenograft survival (115). P-selectin is an important adhesion molecule for platelets and neutrophils, and experiments using antibodies against P-selectin as well as a PAF antagonist, led to prolonged xenograft survival in a rat model (116). Investigators have also used monoclonal antibodies against porcine VCAM to address the EC issue in xenograft rejection.

Directly addressing the issues of thrombosis may help prevent the consequences of EC activation. The use of GPI 562 (a gpIIb/IIIa antagonist) has been used to inhibit platelet aggregation with some results (117); and 5D2 MTH 958 (a thrombin inhibitor) has been used in an isolated perfused heart model (118).

An alternative method of dealing with EC death (apoptosis) is in regards to the alteration of expression of various genes for endothelial cell survival. There are genes which if expressed by the EC are "apoptotic" and these include Bad, Bax, and BcL-X_S (119). There are also genes, which if expressed are "protective" or "anti-apoptotic" and these include A20, Bcl, Bcl-X_L and hemoxygenase (119). All of the above endothelial cell strategies have potential but more research is needed.

4.1.4. Acute vascular rejection

Once HAR (rejection within minutes to hours) has been prevented, discordant xenografts are subject over the subsequent days to weeks to the next phase of the rejection process termed acute vascular rejection (AVR) by Platt (120) or delayed xenograft rejection by Bach (121). AVR is characterized by intravascular coagulation, endothelial thickening, ischemia, and interstitial inflammation. It can be observed in concordant xenografts which do not undergo HAR. This phase of rejection is initiated by xenoantibody binding to signal molecules on EC leading to activation (26). Complement is generally thought not to be involved in AVR as in most cases the assembly of terminal complement components is inhibited by therapies that prevent HAR (26). Other mechanisms may be important as well, including the elaboration of tissue factor by macrophages, cytotoxcity and activation of EC by natural killer cells, activation of EC and oxidant mediated injury by neutrophils, secretion of vasoactive factors by platelets and thrombomodulin incompatibility with recipient proteins due to molecular incompatibility (26).

Methods to prevent AVR are similar to those dealing with HAR. Since xenoantibodies trigger AVR, methods that deal with altering the interaction of the antibody with the organ by removing the antibodies or preventing their synthesis (ex. immunoadsorption, and immune suppression by splenectomy and immunosuppressive therapy) (122,123) or altering the α-gal antigen on EC (124) may be beneficial. Inducing immunological tolerance would also be beneficial. Tolerance induction involves teaching the recipient's immune system that the donor's antigens are self antigens. If this can be done, the organ can be transplanted without the need for any immunosuppression (125,126). Sykes and colleagues in Boston are showing that suppressing the synthesis of antibodies via specific immunological tolerance induced by mixed chimerism is a promising approach (127,128). Also, like

HAR, developing lines of pigs that are low or non-expressers of α-gal could also be helpful.

Another approach being investigated is inducing accommodation. Accommodation is the ability of a vascularized xenograft to avoid humoral rejection in the face of normal circulating anti-graft antibodies and normal levels of complement. Evidence for the occurrence of accommodation of xenografts was initially based on anecdotal observations of studies in primate models (119). Accommodation has been seen in hamster hearts transplanted into rats and in pig-to-NHP transplants (77). Accommodation was also seen by Alexandre (129), who demonstrated that reduction of anti-AB blood group antibodies before renal allotransplantation and maintenance of decreased antibody levels for several days after grafting, prevented rejection by humoral mechanisms once antibody levels returned to normal. The mechanisms of accommodation may involve a change in antibody repertoire (to non-complement binding antibodies), change in antigen, or acquired resistance to EC injury by the expression of anti-apoptotic genes and may be induced in situations in which anti-donor antibodies have been removed and then allowed to return or where antibody mediated rejection is temporarily prevented (such as by complement inhibition) (26,130).

4.1.5. Cellular rejection

Cellular rejection may be seen in solid organ xenografts if HAR and AVR are prevented and is the first type of rejection seen in cellular and tissue xenografts (26). T-cell dependent immunity to xenografts varies strongly according to the type of graft (i.e. cellular versus vascularized). Destruction of the vascular endothelium by HAR and AVR may prevent study of a later cellular response. In pig-to-primate xenotransplantation, $CD4^+$ and $CD8^+$ T-lymphocytes can recognize pig antigens and do not require antigen-presenting cells. $CD4^+$ cells seem to have a critical role and methods of preventing cellular rejection will have to deal with this cell population. Natural killer cells may also be involved (131).

Approaches to preventing cellular rejection include the use of immunosuppressive drugs to modulate T-cells and natural killer cells and the induction of tolerance though the generation of mixed chimerism, transplantation of donor thymus, or manipulating mature T-cells through the generation of microchimerism so that xenoreactive cells are eliminated (131).

4.2. Physiologic concerns

While investigations have generally focused on the prevention of xenograft rejection, recent interest has developed in the question of whether pig xenografts will be able to physiologically replace the function of human organs. The answer to this question may depend on the organ. Humans are a species that stands upright and are generally active. The human heart is designed for an active lifestyle and to function for an upright posture (there are different pressure loads put upon the heart depending on posture). The pig has a horizontal posture and most pigs are quite sedentary. For all of these reasons, there are concerns that pig hearts will not be able to work in humans (132–134). Pigs which will be used for transplantation will likely have to have a very refined diet and may need to be exercised fairly aggressively and regularly (8). The above maneuvers will still not solve the posture issues. For this reason some investigators have been looking at the kangaroo as a potential donor (8). This may prove to be an important source; however for the time being because of expense, number, breeding issues, and other factors, the pig is still felt to be the animal to concentrate on.

There are other major concerns related to the differences in physiology between pigs and humans (8,30,132,133). Some of these differences are subtle but may become important. They include electrolyte and body temperature differences. It is known that enzymes are very sensitive to temperature changes. Whether enzymes in pig organs will work as efficiently in a human body is not known. Certain hormones (such as growth hormone) are very species specific and it is unlikely that pig organs will respond properly to some of these hormones (8). It is possible that humans may need to be injected with certain pig hormones for the pig organs to respond appropriately. This may be problematic if humans subsequently produce antibodies to these injected proteins. This hopefully will not be a problem as the humans will likely be immunosuppressed.

Just as pig organs may not respond to human proteins (such as hormones or growth factors), the proteins made by certain pig organs such as the liver may not function appropriately in a human. The liver is a very important biosynthetic organ making, among others, coagulation proteins, albumin (to which many other substances attach for transport purposes), and complement proteins. Studies (8) have shown that porcine albumin is quite different from human albumin and this may cause problems with other substances being able to bind to the albumin for transport. Another problem may be pig complement. One of the ways that an organism protects itself against its own complement proteins is via complement regulatory proteins (CRP) which are present on endothelium. Indeed, researchers have used the species-specific knowledge about CRP to successfully make transgenic pigs which express human CRP to prevent hyperacute rejection caused by perfusion of human blood (135). Pig livers will produce pig complement which may attack normal human tissue as normal human tissue will not express pig CRP (132).

Despite these concerns, recent studies have suggested that pig kidneys can maintain relatively normal fluid and electrolyte levels in NHP (136), porcine hearts orthotopically transplanted into baboons can keep them alive (137), pig lungs can perform normal gas exchange in baboons for a short time (138), and pig livers have been used as a extracorporeal device in humans for a limited time (139). Thus physiology issues will have to be kept in mind but should not deter from further research into pig-to-human xenotransplantation.

4.3. Xenozoonoses

A potential disadvantage of xenotransplantation that has garnered much publicity recently, is the possibility of infection transfer from animal organ to the human recipient and then possibly to contacts of the recipient. Researchers (140) have found that a porcine cell line having the ubiquitous porcine endogenous retrovirus (PERV) can infect human cell lines in the lab under certain simulated conditions. This is a concern; however, there have been no consequences documented from this infection. Researchers have also tested the blood of patients exposed to pig xenografts via PCR and have found no evidence of PERV infection (141). Many researchers feel that xenogenic infection will probably not be a major problem for reasons which include the following: firstly, infection spread is usually species-specific and as a result it is less likely that pig infectious agents will easily infect humans and cause problems; secondly, facilities used for breeding pigs will be highly regulated and controlled and only "clean" pigs will be bred thus making baseline infection in the pigs less likely; in fact, inbreeding of transgenic pigs in Boston has shown that some of the pigs have less or no PERV in their genetic makeup; and thirdly, humans have used pig products for many years (pig heart valves, porcine insulin, pig meat) without serious problems thus if there was something very serious, it probably would have manifested by now (12,141,142). Even if PERV does lead to infection in a human recipient, it is hypothesized that the problem caused will likely be a lymphoma-like illness which has a probable chance of being cured by current protocols (12,143). Because all infection issues are still not well known, regulatory agencies have been formed to make some basic rules that xeno-researchers would have to comply with before trying human clinical trials (12). Currently, solid organ clinical trials are banned by the FDA but cellular transplant clinical trials are ongoing.

5. THE FUTURE OF CLINICAL XENOTRANSPLANTATION

Considerable progress has been made in the quest to alleviate the critical organ shortage crisis through the use of xenogenic organs. Our group has been studying human ABO blood groups in porcine-to-human xenotransplantation using an ex-vivo porcine working heart model (144,145). Human ABO matching is important in human allotransplantation yet its importance in xenotranplantation has not been fully explored. The similarity of the α-gal antigen and the blood group A and B antigens would suggest a blood group dependence. Indeed we have found that the survival of porcine working hearts is severely shortened by perfusion with human blood group A and O plasma compared to blood group B, which gives survival times comparable to porcine controls. Blood group AB fell between the two extremes. While much of the focus in xenotransplantation research to date has been on finding the ideal donor, our studies show that there are characteristics of recipients such as blood group that need to be investigated. This may help the development of transplantation protocols uniquely suited to each recipient.

Clinical trials of cellular and tissue xenografts and extracorporeal animal organs are currently underway in a limited capacity and these techniques will likely be the first to come into mainstream use. Subsequently, xenografted organs will likely be used as a bridge to allotransplantation before being used as full blown replacements for human organs. There is much progress that has been made, but still much to do. With perseverance and good science, solutions are likely to be found.

REFERENCES

1. Carrel A. 1983. Landmark article, Nov 14, 1908: Results of the transplantation of blood vessels, organs and limbs. JAMA 250:944–953.
2. Carrel A. 1992. The preservation of tissues and its applications in surgery. 1912. Clin Orthop 2–8.
3. Gunn A, Medawar P. 1971. Studies on the induction of transplantation tolerance at low zones of dosage. Br J Surg 58:858.
4. Burnet FM. 1991. The Nobel Lectures in Immunology. The Nobel Prize for Physiology or Medicine, 1960. Immunologic recognition of self. Scand J Immunol 33:3–13.
5. Medawar P. 1981. The Nobel awards for medicine or physiology. Nature 289:345.
6. Merion RM, White DJ, Thiru S, Evans DB, Calne RY. 1984. Cyclosporine: five years' experience in cadaveric renal transplantation. N Engl J Med 310:148–154.
7. International Figures on Organ Donation and Transplantation (1999). 1999. (Online). Council of Europe. www.social.coe.int/en/qoflife/publi/steercom.htm.
8. Cooper DK, Lanza RP. 2000. Xeno: the promise of transplanting animal organs into humans. Ed. 1–274. New York: Oxford University Press.
9. Meeting the organ shortage (1999). 1999. (Online). Council of Europe. www.social.coe.int/en/qoflife/publi/donation.htm.
10. Mihaylov D, Verkerke GJ, Rakhorst G. 2000. Mechanical circulatory support systems—a review. Technol Health Care 8:251–266.
11. Mussivand T. 1999. Mechanical circulatory devices for the treatment of heart failure. J Card Surg 14:218–228.
12. Organ Farm—Frontline. 2001. (Online). PBS online—Frontline. www.pbs.org/wgbh/pages/frontline/shows/organfarm.
13. Markmann JF, Barker CF. 1994. Current Problems in Surgery. Ed. SAJ Wella, New York: Mosby.
14. Reemtsma K. 1964. Renal Heterotransplantation in Man. Annals of Surgery 160:384.
15. Starzl TE. 1964. Renal Heterotransplantation from Baboon to Man: Experience with 6 Cases. Transplantation 2:752.
16. Taniguchi H, Cooper DKC. 1997. Clinical xenotransplantation—A brief review of the world experience. In: Xenotransplantation: The Transplantation of Organs and Tissues Between Species. Ed. DKC Cooper, E Kemp, JL Platt, DJG White, 776–785. New York: Springer.
17. Bailey LL, Nehlsen-Cannarella SL, Concepcion W, Jolley WB. 1985. Baboon-to-human cardiac xenotransplantation in a neonate. JAMA 254:3321–3329.
18. Makowka L, Wu GD, Hoffman A, Podesta L, Sher L, Tusa PJ, Breda M, Chapman FA, Cosenza C, Yasunga C. 1994. Immunohistopathologic lesiona associated with the rejection of a pig-to-human liver xenograft. Transplant Proc 26:1074–1075.
19. Makowka L. 1995. The use of a pig liver xenograft for the temporary support of a patient with fulminanat hepatic failure. Transplantation 59:1654–1659.
20. Abouna GM. 1997. Extracorporeal xenogenic liver perfusion for the treatment of hepatic failure. In: Xenotransplantation: The Transplantation of Organs and Tissues Between Species. Ed. DKC Cooper, E Kemp, JL Platt, DJG White, 785–793. New York: Springer.
21. State of the art report on Xenotransplantation (2000)—Clinical Activities. 2000. (Online). Council of Europe. www.social.coe.int/en/qoflife/publi/artreport/clinact.htm.
22. Korbutt GS, Aspeslet L, Ao Z, Warnock GL, Ezekowitz J, Koshal A, Rajotte RV, Yatscoff RW. 1996. Porcine islet cell antigens are recognized by xenoreactive natural human antibodies of both IgG and IgM subtypes. Transplant Proc 28:837–838.
23. Rayat GR, Rajotte RV, Elliott JF, Korbutt GS. 1998. Gal alpha(1,3)Gal expression on neonatal porcine islet cells and susceptibility to human antibody/complement lysis. Transplant Proc 30:325.
24. Korbutt GS, Ao Z, Flashner M, Rajotte RV. 1997. Neonatal porcine islets as a possible source of tissue for humans and microencapsulation improves the metabolic response of islet graft posttransplantation. Ann N Y Acad Sci 831:294–303.
25. Rayat GR, Rajotte RV, Ao Z, Korbutt GS. 2000. Microencapsulation of neonatal porcine islets: protection from human antibody/complement-mediated cytolysis in vitro and long-term reversal of diabetes in nude mice. Transplantation 69:1084–1090.
26. Platt JL. 2001. A primer on xenotransplantation. In: Xenotransplantation. Ed. JL Platt, 3–28. Washington: ASM Press.

27. Thomas F. 1997. Isolated pancreatic islet xenografting. In: Xenotransplantation: The Transplantation of Organs and Tissues Between Species. Ed. DKC Cooper, E Kemp, JL Platt, DJG White, 545–565. New York: Springer.
28. Rabinovitch A, Suarez-Pinzon W, Strynadka K, Ju Q, Edelstein D, Brownlee M, Korbutt GS, Rajotte RV. 1999. Transfection of human pancreatic islets with an anti-apoptotic gene (bcl-2) protects beta-cells from cytokine-induced destruction. Diabetes 48:1223–1229.
29. Groth CG, Korsgren O, Tibell A, Reinholt F, Wennberg L, Satake M, Moller E, Rydberg L, Samuelsson B, Andersson A, Hellerstrom C. 1997. Clinical islet xenotransplantation. In: Xenotransplantation: The Transplantation of Organs and Tissues Between Species. Ed. DKC Cooper, E Kemp, JL Platt, DJG White, 812–821. New York: Springer.
30. Hammer C. 1997. Evolutionary obstacles to xenotransplantation. In: Xenotransplantation: The Transplantation of Organs and Tissues Between Species. Ed. DKC Cooper, E Kemp, JL Platt, DJG White, 716–736. New York: Springer.
31. Van der Riet FS, Human PA, Cooper DKC, Reichart B, Fincham JE, Kalter SS, Kanki, PJ, Essex M, Madden DL, Lai-Tung MT. 1987. Virological implications of the use of primates in xenotransplantation. Transplant Proc 19:4068–4069.
32. Chapman LE, Folks TM, Salomon DR, Patterson TE, Eggerman TE, Noguchi PD. 1995. Xenotransplantation and xenogenic infection. N Engl J Med 333:1498–1501.
33. Luo Y, Taniguchi S, Kobayashi T, Niekrasz M, Cooper DKC. 1995. Screening of baboon potential liver donors for humans. Xenotransplantation 2:244–252.
34. Sachs DH. 1994. The pig as a potential xenograft donor. Vet Immunol Immunopathol 43:185–191.
35. Sachs DH. 1994. The pig as a xenograft donor. Pathol Biol (Paris) 42:217–219.
36. Calne RY. 1970. Organ Transplantation Between Widely Disparate Species. Transplant Proc 2:553.
37. Platt JL. 1997. Hyperacute xenograft rejection. In: Xenotransplantation: The Transplantation of Organs and Tissues Between Species. Ed. DKC Cooper, E Kemp, JL Platt, DJG White, 8–17. New York: Springer.
38. Schilling A, Land W, Pratschke E, et al. 1976. Dominant role of complement in the hyperacute xenograft rejection reaction. Surgery, Gynecology & Obstetrics 142:29–32.
39. Auchincloss H. 1988. Xenogenic transplantation: a review. Transplantation 46:1–20.
40. Platt JL, Vercellotti GM, Dalmasso AP. 1990. Transplantation of discordant xenografts: a review of progress. Immunol Today 11:450–456.
41. Holtzknecht ZE, Platt JL. 1995. Identification of porcine endothelial cell membrane antigens recognized by human xenoreactive natural antibodies. Journal of Immunology 154:4565–4575.
42. Good AH, Cooper DK, Malcolm AJ, Ippolito RM, Koren E, Neethling FA, Ye Y, Zuhdi N, Lamontagne LR. 1992. Identification of carbohydrate structures that bind human antiporcine antibodies: implications for discordant xenografting in humans. Transplant Proc 24:559–562.
43. Sandrin MS, Vaughan HA, Dabkowski PL, McKenzie IF. 1993. Anti-pig IgM antibodies in human serum react predominantly with Gal(alpha 1–3)Gal epitopes. Proc Natl Acad Sci USA 90: 11391–11395.
44. Galili U. 1993. Interaction of the natural anti-Gal antibody with alpha-galactosyl epitopes: a major obstacle for xenotransplantation in humans. Immunol Today 14:480–482.
45. Sandrin MS, McKenzie IF. 1994. Gal alpha (1,3)Gal, the major xenoantigen(s) recognised in pigs by human natural antibodies. Immunol Rev 141:169–190.
46. Collins BH, Cotterell AH, McCurry KR, Alvarado CG, Magee JC, Parker W, Platt JL. 1995. Cardiac xenografts between primate species provide evidence for the importance of the alpha-galactosyl determinant in hyperacute rejection. J Immunol 154:5500–5510.
47. Lin SS, Kooyman DL, Daniels LJ, Daggett CW, Parker W, Lawson JH, Hoopes CW, Gullotto C, Li L, Birch P, Davis RD, Diamond LE, Logan JS, Platt JL. 1997. The role of natural anti-Gal alpha 1–3Gal antibodies in hyperacute rejection of pig-to-baboon cardiac xenotransplants. Transpl Immunol 5:212–218.
48. Cooper DK, Good AH, Koren E, Oriol R, Malcolm AJ, Ippolito RM, Neethling FA, Ye Y, Romano E, Zuhdi N. 1993. Identification of alpha-galactosyl and other carbohydrate epitopes that are bound by human anti-pig antibodies: relevance to discordant xenografting in man. Transpl Immunol 1: 198–205.
49. Cooper DK, Koren E, Oriol R. 1994. Oligosaccharides and discordant xenotransplantation. Immunol Rev 141:31–58.
50. Galili U, Korkesh A, Kahane I, Rachmilewitz EA. 1983. Demonstration of a natural antigalactosyl IgG antibody on thalassemic red blood cells. Blood 61:1258–1264.

51. Galili U, Clark MR, Shohet SB. 1986. Excessive binding of natural anti-alpha-galactosyl immunoglobin G to sickle erythrocytes may contribute to extravascular cell destruction. J Clin Invest 77: 27–33.
52. Galili U, Kobrin E, Macher BA, Shohet SB. 1989. Anti-Gal and human red cell aging. Prog Clin Biol Res 319:225–241.
53. Galili U, Flechner I, Rachmilewitz EA. 1985. A naturally occurring anti-alpha-galactosyl IgG recognizing senescent human red cells. Prog Clin Biol Res 195:263–278.
54. Galili U. 1988. The natural anti-Gal antibody, the B-like antigen, and human red cell aging. Blood Cells 14:205–228.
55. Galili U, Anaraki F, Thall A, Hill-Black C, Radic M. 1993. One percent of human circulating B lymphocytes are capable of producing the natural anti-Gal antibody. Blood 82:2485–2493.
56. Galili U, Shohet SB, Kobrin E, Stults CL, Macher BA. 1988. Man, apes, and Old World monkeys differ from other mammals in the expression of alpha-galactosyl epitopes on nucleated cells. J Biol Chem 263:17755–17762.
57. Galili U, Swanson K. 1991. Gene sequences suggest inactivation of alpha-1,3-galactosyltransferase in catarrhines after the divergence of apes from monkeys. Proc Natl Acad Sci USA 88:7401–7404.
58. Galili U. 1993. Evolution and pathophysiology of the human natural anti-alpha-galactosyl IgG (anti-Gal) antibody. Springer Semin Immunopathol 15:155–171.
59. Galili U, Clark MR, Shohet SB, Buehler J, Macher BA. 1987. Evolutionary relationship between the natural anti-Gal antibody and the Gal alpha 1–3 Gal epitope in primates. Proc Natl Acad Sci USA 84:1369–1373.
60. Galili U. 1999. Evolution of alpha 1,3galactosyltransferase and of the alpha-Gal epitope. Subcell Biochem 32:1–23.
61. Wang L, Anaraki F, Henion TR, Galili U. 1995. Variations in activity of the human natural anti-Gal antibody in young and elderly populations. J Gerontol A Biol Sci Med Sci 50:M227–M233.
62. Koren E, Neethling FA, Ye Y, Niekrasz M, Baker J, Martin M, Zuhdi N, Cooper DK. 1992. Heterogeneity of preformed human antipig xenogeneic antibodies. Transplant Proc 24:598–601.
63. Avila JL, Rojas M, Galili U. 1989. Immunogenic Gal alpha 1–3 Gal carbohydrate epitopes are present on pathogenic American Trypanosoma and Leishmania. J Immunol 142:2828–2834.
64. Galili U, Mandrell RE, Hamadeh RM, Shohet SB, Griffiss JM. 1988. Interaction between human natural anti-alpha-galactosyl immunoglobulin G and bacteria of the human flora. Infect Immun 56: 1730–1737.
65. Vaughan HA, Loveland BE, Sandrin MS. 1994. Gal alpha(1,3)Gal is the major xenoepitope expressed on pig endothelial cells recognized by naturally occurring cytotoxic human antibodies. Transplantation 58:879–882.
66. Galili U. 1989. Abnormal expression of alpha-galactosyl epitopes in man. A trigger for autoimmune processes? Lancet 2:358–361.
67. Galili U. 1999. Graves' disease as a model for anti-Gal involvement in autoimmune diseases. Subcell Biochem 32:339–360.
68. Galili U, Macher BA. 1989. Interaction between anti-Gal and human tumor cells: a natural defense mechanism? J Natl Cancer Inst 81:178–179.
69. Galili U, Flechner I, Knyszynski A, Danon D, Rachmilewitz EA. 1986. The natural anti-alpha-galactosyl IgG on human normal senescent red blood cells. Br J Haematol 62:317–324.
70. Lane PA, Galili U, Iarocci TA, Shew RL, Mentzer WC. 1988. Cellular dehydration and immunoglobulin binding in senescent neonatal erythrocytes. Pediatr Res 23:288–292.
71. McMorrow IM, Comrack CA, Sachs DH, DerSimonian H. 1997. Heterogeneity of human anti-pig natural antibodies cross-reactive with the Gal(alpha1,3)Galactose epitope. Transplantation 64: 501–510.
72. McKenzie IF, Xing PX, Vaughan HA, Prenzoska J, Dabkowski PL, Sandrin MS. 1994. Distribution of the major xenoantigen (gal (alpha 1–3)gal) for pig to human xenografts. Transpl Immunol 2: 81–86.
73. Kobayashi T, Cooper DKC. 1999. Anti-Gal, α-Gal epitopes and xenotransplantation. Subcell Biochem 32:229–259.
74. Blakely ML. 1994. Activation of intragraft endothelial and mononuclear cells during discordant xenograft rejection. Transplantation 58:1059–1066.
75. Platt JL. 1994. A perspective on xenograft rejection and accommodation. Immunol Rev 141:127–149.

76. Bach FH. 1994. Xenotransplantation: problems posed by endothelial cell activation. Transplant Proc 26:1020–1030.
77. Bach FH, Stuhlmeier KM, Vanhove B, Van der Werf WJ, Blakely ML, de Martin R, Hancock WW, Winkler H. 1994. Endothelial cells in xenotransplantation: Do they accomodate? Transplant Proc 26:1167–1169.
78. Malyguine AM, Saadi S, Platt JL. 1996. Human natural killer cells induce morphologic changes in porcine endothelial cell monolayers. Transplantation 61:161–164.
79. Hancock WW, Miyatake T, Koyamada N, Kut JP, Soares M, Russell ME, Bach FH, Sayegh MH. 1997. Effects of leflunomide and deoxyspergualin in the guinea pig→rat cardiac model of delayed xenograft rejection: suppression of B cell and C-C chemokine responses but not induction of macrophage lectin. Transplantation 64:696–704.
80. Goodman DJ, Von Albertini M, Wilson A, Millan MT, Bach FH. 1996. Direct activation of porcine endothelial cells by human natural killer cells. Transplantation 61:763–771.
81. Bach FH, Robson SC, Winkler H, Ferran C, Stuhlmeier KM, Wrighton CJ, Hancock WW. 1995. Barriers to xenotransplantation. Nat Med 1:869–873.
82. Palmetshofer A, Galili U, Dalmasso AP, Robson SC, Bach FH. 1998. Alpha-galactosyl epitope-mediated activation of porcine aortic endothelial cells: type I activation. Transplantation 65:844–853.
83. Palmetshofer A, Galili U, Dalmasso AP, Robson SC, Bach FH. 1998. Alpha-galactosyl epitope-mediated activation of porcine aortic endothelial cells: type II activation. Transplantation 65: 971–978.
84. Mantovani AF, Bussolino AF, Dejana E. 1992. Cytokine regulation of endothelial cell function. FASEB J 6:2591–2599.
85. Platt JL, Fischel RJ, Matas AJ, Reif SA, Bolman RM, Bach FH. 1991. Immunopathology of hyperacute xenograft rejection in a swine-to-primate model. Transplantation 52:214–220.
86. Lexer G, Cooper DK, Rose AG, Wicomb WN, Rees J, Keraan M, Du TE. 1986. Hyperacute rejection in a discordant (pig to baboon) cardiac xenograft model. J Heart Transplant 5:411–418.
87. Dalmasso AP. 1991. Inhibition of complement-mediated endothelial cell cytotoxicity by decay accelerating factor: potential for prevention of xenograft hyperacute rejection. Transplantation 52: 530–533.
88. Rose AG, Cooper DK, Human PA, Reichenspurner H, Reichart B. 1991. Histopathology of hyperacute rejection of the heart: experimental and clinical observations in allografts and xenografts. J Heart Lung Transplant 10:223–234.
89. Rose AG, Cooper DK. 1996. A histopathologic grading system of hyperacute (humoral, antibody-mediated) cardiac xenograft and allograft rejection. J Heart Lung Transplant 15:804–817.
90. Ye Y, Neethling FA, Niekrasz M, Koren E, Richards SV, Martin M, Kosanke S, Oriol R, Cooper DK. 1994. Evidence that intravenously administered alpha-galactosyl carbohydrates reduce baboon serum cytotoxicity to pig kidney cells (PK15) and transplanted pig hearts. Transplantation 58: 330–337.
91. Ye Y, Neethling FA, Niekrasz M, Richards SV, Koren G, Merhav H, Kosanke S, Oriol R, Cooper DK. 1994. Intravenous administration of alpha-galactosyl carbohydrates reduces in vivo baboon serum cytotoxicity to pig kidney cells and transplanted pig hearts. Transplant Proc 26: 1399.
92. Neethling FA, Koren E, Ye Y, Richards SV, Kujundzic M, Oriol R, Cooper DK. 1994. Protection of pig kidney (PK15) cells from the cytotoxic effect of anti-pig antibodies by alpha-galactosyl oligosaccharides. Transplantation 57:959–963.
93. Rieben R. 1995. Detection immunoadsorption and inhibition of cytotoxic activity of anti-α-gal antibodies using newly developed substances with synthetic Gal α 1,3 Gal disaccharide epitopes. Xenotransplantation 2:98–106.
94. Cairns T, Lee J, Goldberg L, Cook T, Simpson P, Spackman D, Palmer A, Taube D. 1995. Inhibition of the pig to human xenograft reaction, using soluble Gal alpha 1–3Gal and Gal alpha 1–3Gal beta 1–4GlcNAc. Transplantation 60:1202–1207.
95. Azimzadeh A, Watier H, Meyer C, Guillaumin JM, Beller JP, Kieny R, Zibolt P, Boudjema K, Jaeck D, Cinqualbre J, Wolf P. 1997. Rebound in complement-dependent cytotoxicity and antibody-dependent cell-mediated cytotoxic activity in primate serum after pig liver perfusion. Transplantation 29:2385–2386.
96. Cooper DK. 1996. Experimental progress and clinical perspectives in xenotransplantation. Transplant Proc 28:1153.

97. Aspeslet LJ, Chackowsky P, Sekhon H, Malcolm AJ, Mosleh Z, Koshal A, Yatscoff RW. 1996. Identification of porcine membrane antigens involved in the cytotoxic response mediated by human xenoreactive antibodies. Xenotransplantation 3:1–10.
98. Pascher A, Poehlein C, Stangl M, Hoebel G, Thiery J, Mueller-Derlich J, Hammer C. 1997. Application of immunoapheresis for delaying hyperacute rejection during isolated xenogeneic pig liver perfusion. Transplantation 63:867–875.
99. Leventhal J, Fryer JP, Witson JC, Derlich JM, Remiszewski J, Dalmasse AP, Matas AJ, Bolman RM. 1995. Removal of baboon and human anti-porcine IgG and IgM natural antibodies by immunoadsorption: Results of *in vitro* and *in vivo* studies. Transplantation 59:294–300.
100. Taniguchi S, Neethling FA, Korchagina EY, Bovin N, Ye Y, Kobayashi T, Niekrasz M, Li S, Koren E, Oriol R, Cooper DK. 1996. In vivo immunoadsorption of antipig antibodies in baboons using a specific Gal(alpha)1–3Gal column. Transplantation 62:1379–1384.
101. Manji RA, Manji JS, Rajotte RV, Afrouzian M, Yatscoff RW, Koshal A. 2000. Specific α-gal antibody removal and human ABO blood group in human-porcine cardiac xenograft survival. Can J Cardiol 16:211F.
102. Ryan US. 1995. Complement inhibitory therapeutics and xenotransplantation. Nat Med 1995: 967–968.
103. Kobayashi T, Taniguchi S, Neethling FA, Rose AG, Hancock WW, Ye Y, Niekrasz M, Kosanke S, Wright LJ, White DJ, Cooper DK. 1997. Delayed xenograft rejection of pig-to-baboon cardiac transplants after cobra venom factor therapy. Transplantation 64:1255–1261.
104. Pruitt SK, Kirk AD, Bollinger RR, Marsh HC, Collins BH, Levin JL, Mault JR, Heinle JS, Ibrahim S, Rudolph AR. 1994. The effect of soluble complement receptor type 1 on hyperacute rejection of porcine xenografts. Transplantation 57:363–370.
105. Cozzi E, White DJG. 1995. The generation of transgenic pigs as potential organ donors for humans. Nat Med 1:964–966.
106. van Denderen BJW, Pearse MJ, Katerelos M, Nottle MB, Du ZT, Aminian A, Adam WR, Shenoy-Scaria A, Lublin D, Shinkel TA, d'Apice AJ. 1996. Expression of a functional human decay-accelerating factor (CD55) in transgenic mice protects against human complement-mediated attack. Transplantation 61:582–588.
107. Fodor WL, Williams BL, Matis LA, Madri JA, Rollins SA, Knight JW, Velander W, Squinto SP. 1994. Expression of a functional human complement inhibitor in a transgenic pig as a model for the prevention of xenogenic hyperacute rejection. Proc Natl Acad Sci USA 91:11153–11157.
108. Chong A, Blinder L, Ma L, Yin D, Shen J, Williams JW, Byrne G, Schwarz A, Diamond LS, Logan JE. 2000. Anti-galactose-alpha(1,3) galactose antibody production in alpha1,3-galactosyltransferase gene knockout mice after xeno and allo transplantation. Transpl Immunol 8:129–137.
109. Cowan PJ, Chen CG, Shinkel TA, Fisicaro N, Salvaris E, Aminian A, Romanella M, Pearse MJ, d'Apice AJ. 1998. Knock out of alpha1,3-galactosyltransferase or expression of alpha1,2-fucosyltransferase further protects CD55- and CD59-expressing mouse hearts in an ex vivo model of xenograft rejection. Transplantation 65:1599–1604.
110. Kaiser J. 2002. Making pigs more like people. ScienceNOW 103:1–3.
111. Costa C, Zhao L, Burton WV, Bondioli KR, Williams BL, Hoagland TA, Ditullio PA, Ebert KM, Fodor WL. 1999. Expression of the human alpha1,2-fucosyltransferase in transgenic pigs modifies the cell surface carbohydrate phenotype and confers resistance to human serum-mediated cytolysis. FASEB J 13:1762–1773.
112. Cheng Q, Cant CA, Moll T, Hofer-Warbinek R, Wagner E, Birnstiel ML, Bach FH, de Martin R. 1994. NFκB subunit specific regulation of the IκBα promoter. J Biol Chem 269:13551.
113. Winkler H. 1995. Genetic engineering of endothelial cells. Xenotransplantation 2:197–200.
114. Bach F, Robson SC, Ferran C, Millan MT, Anrather J, Kopp C, Lesnikoski B, Goodman DJ, Hancock WW, Wrighton CJ. 1995. Xenotransplantation: Endothelial cell activation and beyond. Transplant Proc 27:77.
115. Zehr KJ, Herskowitz A, Lee PC, Kumar P, Gillinov AM, Baumgartner WA. 1994. Neutrophil adhesion and complement inhibition prolongs survival of cardiac xenografts in discordant species. Transplantation 57:900–906.
116. Coughlan AF, Berndt MC, Dunlop LC, Hancock WW. 1993. In vivo studies of P-selectin and platelet-activating factor during endotoxemia, accelerated graft rejection and discordant xenograft rejection. Transplant Proc 25:2930.
117. Lesnikoski BA, Candinas D, Hancock WW, Otsu I, Siegel J, Bach FH, Robson SC. 1996. Inhibition of platelet GPIIbIIIa prolongs survival of discordant cardiac xenografts. Transplant Proc 28:703.

118. Robson SC, Young VK, Cook NS, Metternich R, Kasper-Konig W, Lesnikoski BA, Pierson RN, Hancock WW, Candinas D, White DJ, Bach FH. 1996. Thrombin inhibition in an ex vivo model of porcine heart xenograft hyperacute rejection. Transplantation 61:862–868.
119. Mark W. 1997. Xenograft accomodation is accompanied by intragraft Th2 cytokines and vascular expression of protective genes. Xenotransplantation 4:154–160.
120. Lin SS, Weidner BC, Byrne GW, Diamond LE, Lawson JH, Hoopes CW, Daniels LJ, Daggett CW, Parker W, Harland RC, Davis RD, Bollinger RR, Logan JS, Platt JL. 1998. The role of antibodies in acute vascular rejection of pig-to-baboon cardiac transplants. J Clin Invest 101:1745–1756.
121. Candinas D, Bach FH, Hancock WW. 1996. Delayed xenograft rejection in complement-depleted T-cell-deficient rat recipients of guinea pig cardiac grafts. Transplant Proc 28:678.
122. McCurry KR, Parker W, Cotterell AH, Weidner BC, Lin SS, Daniels LJ, Holzknecht ZE, Byrne GW, Diamond LE, Logan JS, Platt JL. 1997. Humoral responses to pig-to-baboon cardiac transplantation: implications for the pathogenesis and treatment of acute vascular rejection and for accommodation. Hum Immunol 58:91–105.
123. Lin SS, Hanaway MJ, Gonzalez-Stawinski GV, Lau CL, Parker W, Davis RD, Byrne GW, Diamond LE, Logan JS, Platt JL. 2000. The role of anti-Galalpha1–3Gal antibodies in acute vascular rejection and accommodation of xenografts. Transplantation 70:1667–1674.
124. Lin SS, Platt JL. 1998. Genetic therapies for xenotransplantation. J Am Coll Surg 186:388–396.
125. Kirk AD. 1999. Transplantation tolerance: a look at the nonhuman primate literature in the light of modern tolerance theories. Crit Rev Immunol 19:349–388.
126. Sachs DH. 1998. Transplantation tolerance. Transplant Proc 30:1627–1629.
127. Yang YG, deGoma E, Ohdan H, Bracy JL, Xu Y, Iacomini J, Thall AD, Sykes M. 1998. Tolerization of anti-Galalpha1–3Gal natural antibody-forming B cells by induction of mixed chimerism. J Exp Med 187:1335–1342.
128. Bartholomew AM, Powelson J, Sachs DH, Bailin M, Boskovic S, Colvin R, Hong HZ, Johnson M, Kimikawa M, LeGuern A, Meehan S, Sablinski T, Wee SL, Cosimi AB. 1999. Tolerance in a concordant nonhuman primate model. Transplantation 68:1708–1716.
129. Alexandre GP, Squifflet JP, De Bruyere M, Latinne D, Reding R, Gianello P, Carlier M, Pirson Y. 1987. Present experience in a series of 26 ABO-incompatible living donor renal allografts. Transplant Proc 19:4538.
130. Soares MP, Lin Y, Anrather J, Csizmadia E, Takigami K, Sato K, Grey ST, Colvin RB, Choi AM, Poss KD, Bach FH. 1998. Expression of heme oxygenase-1 can determine cardiac xenograft survival. Nat Med 4:1073–1077.
131. Buhler L, Sachs DH, Alwayn IPJ, Cooper DKC. 2001. Therapeutic Strategies for Xenotransplantation. In: Xenotransplantation. Ed. JL Platt, 117–135. Washington: ASM Press.
132. Hammer C. 1998. Physiological obstacles after xenotransplantation. Ann N Y Acad Sci 862: 19–27.
133. Hammer C. 1994. Fundamental problems of xenotransplantation. Pathol Biol (Paris) 42:203–207.
134. Hammer C. 1991. Xenotransplantation: perspectives and limits. Blood Purif 19:322–328.
135. Buhler L, Yamada K, Alwayn I, Kitamura H, Basker M, Barth RN, Appel J, Awwad M, Thall A, White-Scharf ME, Sachs DH, Cooper DK. 2001. Miniature swine and hDAF pig kidney transplantation in baboons treated with a nonmyeloablative regimen and CD154 blockade. Transplant Proc 33:716.
136. Zaidi A, Schmoeckel M, Bhatti FN, Waterworth PD, Tolan M, Cozzi E, Chavez G, Langford G, Thiru S, Wallwork J, White D, Friend P. 1998. Life-supporting pig-to-primate renal xenotransplantation using genetically modified donors. Transplantation 65:1584–1590.
137. Schmoeckel M, Bhatti FN, Zaidi A, Cozzi E, Waterworth PD, Tolan MJ, Pino-Chavez G, Goddard M, Warner RG, Langford GA, Dunning JJ, Wallwork J, White DJ. 1998. Orthotopic heart transplantation in a transgenic pig-to-primate model. Transplantation 65:1570–1577.
138. Daggett CW, Yeatman MF, Lodge AJ, Chen EP, Gullotto C, Frank MM, Platt JL, Davis RD. 1998. Total respiratory support from swine lungs in primate recipients. J Thorac Cardiovasc Surg. 115: 19–27.
139. Chari RS, Collins BH, Magee JC, Kirk AD, Harland RC, McCann RL, Platt JL, Meyers WC. 1994. Treatment of hepatic failure with ex-vivo pig liver perfusion followed by liver transplantation. N Engl J Med 331:234–237.
140. Takeuchi Y, Patience C, Magre S, Weiss RA, Banerjee PT, Le Tissier P, Stoye JP. 1998. Host range and interference studies of three classes of pig endogenous retrovirus. J Virol 72:9986–9991.

141. Patience C, Patton GS, Takeuchi Y, Weiss RA, McClure MO, Rydberg L, Breimer ME. 1998. No evidence of pig DNA or retroviral infection in patients with short-term extracorporeal connection to pig kidneys. Lancet 352:699–701.
142. State of the Art Report on Serological Screening Methods for the Most Relevant Microbiological Disease. 2000. (Online). Council of Europe. www.social.coe.int/en/qoflife/publi/artreport/biosaf.htm.
143. Qari SH, Magre S, Garcia-Lerma JG, Hussain AI, Takeuchi Y, Patience C, Weiss RA, Heneine W. 2001. Susceptibility of the porcine endogenous retrovirus to reverse transcriptase and protease inhibitors. J Virol 75:1048–1053.
144. Manji RA, Manji JS, Rajotte RV, Korbutt GS, Koshal A. 2001. Critical factors in survival of a pig-human xenograft. Can J Cardiol 17:243c.
145. Manji RA, Manji JS, Koshal A, Korbutt GS, Rajotte RV. 2001. The ideal human ABO blood group and red blood cells for porcine-to-human cardiac xenotransplantation. Xenotransplantation 8:95.
146. Buchser E, Goddard M, Heyd B, Joseph JM, Favre J, de Tribolt N, Lysaght M, Aebischer P. 1996. Immunoisolated xenogenic chromaffine cells therapy for chronic pain. Anesthesiology 85:1005–1012.
147. Zurn AD, Henry H, Schluep M, Aubert V, Winkel L, Eilers B, Bachmann C, Aebischer P. 2000. Evaluation of an intrathecal immune response in amyotrophic lateral sclerosis patients implanted with encapsulated genetically engineered xenogeneic cells. Cell Transplant 9:471–484.
148. Fink JS, Schumacher JM, Ellias SL, Palmer EP, Saint-Hilaire M, Shannon K, Penn R, Starr P, VanHorne C, Kott HS, Dempsey PK, Fischman AJ, Raineri R, Manhart C, Dinsmore J, Isacson O. 2000. Porcine xenografts in Parkinson's disease and Huntington's disease patients: preliminary results. Cell Transplant 9:273–278.
149. Groth CG, Korsgren O, Tibell A, Tollemar J, Moller E, Bolinder J, Ostman J, Reinholt FP, Hellerstrom C, Andersson A. 1994. Transplantation of porcine fetal pancreas to diabetic patients. Lancet 344:1402–1404.
150. Taylor R. 1996. Baboon graft fails, but patient thrives. Nat Med 2:259.

Pathophysiology of Cardiovascular Disease,
edited by N.S. Dhalla, H. Rupp,
A. Angel & G.N. Pierce
Kluwer Academic Publishers, Boston, 2004

Assessment of Myocardial Injury Using Magnetic Resonance Imaging

Ganghong Tian,[1,2] John Mark,[1,2] Bo Xiang,[1]
Guangping Dai,[1] Gang Li,[1,2] Jiankang Sun,[1] Robert Corne,[3]
Normand Lazarow,[4] and Roxanne Deslauriers[1,2]

[1] *Institute for Biodiagnostics, National Research Council of Canada,*
[2] *Department of Physiology,* [3] *Department of Cardiology, University of Manitoba,*
[4] *Manitoba Clinic, Winnipeg, Canada*

Summary. Left ventricular dysfunction may be the result of reversible or irreversible myocardial injury. Because there is significant difference in prognosis and clinical management for patients with damaged but viable myocardium and for those with nonviable myocardium, distinguishing the status of the myocardium has significant clinical relevance. One avenue that has been explored is magnetic resonance (MR) imaging in conjunction with MR contrast agents. However, there are conflicting reports on the accuracy of the common MR techniques, T1- and T2*-weighted imaging, as indicators of myocardial viability. To improve the reliability, we have developed a new MR imaging method, interleaved T1–T2* imaging. This new approach monitors both T1 and T2* signal intensities simultaneously during the first pass of an MR contrast agent. The interleaved T1–T2* imaging separates viable from nonviable myocardium based on the contrast-induced changes in T1 and T2* signals, which is related to the integrity of the cell membrane. Using isolated pig hearts we have found that the signal-time courses of the interleaved-T1–T2* images obtained from normal myocardium are significantly different from those obtained from infarct rim and infarct core. The results of this study suggest that the integrity of cell membrane can be reliably assessed using the interleaved T1–T2* imaging in conjunction with an MR contrast agent. Moreover, use of the new MR approach differentiates viable from nonviable myocardial viability.

Key words: Myocardial infarction, magnetic resonance imaging, contrast reagent, pig heart.

Address for Correspondence: Ganghong Tian, MD., Ph.D., Institute for Biodiagnostics, National Research Council of Canada, 435 Ellice Ave. Winnipeg, Manitoba, Canada R3B 1Y6. Tel: (204)984-6654, Fax (204)984-7036, E-mail: Hong.Tian@nrc.ca.

INTRODUCTION

After a heart attack it is important to determine whether myocardium is viable or not because patient management, therapy, and prognosis are different for patients with injured but living heart muscle and for those with irreversibly injured muscle (1). Patients whose heart muscle is still alive, but dysfunctional following a heart attack are expected to benefit from heart surgery or a balloon procedure to open up the narrowed coronary artery (2,3). These procedures will not lead to any improvement of heart function in patients with dead heart muscle. Accurate assessment of heart damage is one of the major challenges in contemporary cardiology. Therefore, it is clinically significant to establish methods that can reliably differentiate injured but alive heart muscle from dead heart muscle.

Electrocardiography (ECG), echocardiography, contrast ventriculography, single photon emission computed tomography (SPECT), and positron emission tomography (PET) have all been used for post-infarction assessment of myocardial injury (4–9). The former three techniques listed above provide indirect information about myocardial injury. They do not provide accurate information on the size and severity of tissue damage. Viable and non-viable myocardium cannot be differentiated when these two types of tissue are mixed in a small region of tissue. SPECT and PET have been established as standard methods for assessing regional myocardial perfusion and viability. However, these modalities involve the use of radioactive tracers and do not have very high spatial resolution (4–9).

Magnetic resonance (MR) imaging and spectroscopy have the potential to become important approaches for distinguishing viable from non-viable myocardium (10,11). For example, the loss of cell membrane integrity in infarcted myocardium results in an increase in the distribution volume of MR contrast agent, leading to a greater increase in MR signal intensity in non-viable myocardium than in viable myocardium (12,13). In addition, irreversible injury-associated depletion of energy metabolites can be readily monitored using phosphorus-31 (^{31}P) MR spectroscopy (14–16). Akinesia of infarcted myocardium can be visualized using fast MR imaging (17–19). More recently, necrosis-specific (such as Gadophrin-2) and live cell-specific contrast agents (such as manganese-based contrast agent) have also been used for assessment of myocardial viability (20–25). Evaluation of myocardial ionic homeostasis may be another avenue for evaluating tissue injury (26). At present, however, these MR modalities are not routinely used in clinic.

It has been shown that T1-weighted MR imaging is more sensitive to the regional concentration of MR contrast agent while T2*-weighted imaging is more sensitive to the homogeneity of contrast distribution (27–29). An increase in regional contrast concentration due to loss of cell membrane intensity or expansion of the interstitial space (tissue edema) leads to enhancement of T1 signal intensities (12,13,30). As such, an increase in T1 signal intensity may not necessarily indicate myocardial infarction. It has been shown that T1 imaging overestimates the amount of infarcted myocardium (30). On the other hand, the susceptibility effect of an MR contrast agent depends on the homogeneity of contrast distribution as well as its regional concentration. Loss of cell membrane integrity in infarcted myocardium, increases the regional concentration of the contrast agent, resulting in a change in T2* signal intensity. On the other hand, loss of cell membrane integrity also makes the distribution of the contrast agent more homogeneous, which may reduce the T2* effect of a contrast agent. Therefore, the use of T2*-weighted imaging alone may not be sufficient to assess myocardial viability. We believe that regional concentration and distribution of contrast agent may be more reliably assessed by simultaneous monitoring the contrast-induced changes in T1 and T2* signal intensities. We propose that the combination of T1- and T2*-weighted imaging may provide an effective MR method for the assessment of myocardial injury.

The present study was designed to determine whether myocardial viability can be assessed by simultaneous monitoring of T1–T2* signal intensities during the first passage of a MR contrast agent. We found that infarcted myocardium in isolated pig hearts was accurately detected by use of our interleaved T1–T2* imaging sequence.

METHODS AND MATERIALS

Isolated pig heart preparation

All animals used in this project received humane care in compliance with the "Guide to the Care and Use of Experimental Animals" formulated by the Canadian Council on Animal Care. The protocols used in this study were approved by the Animal Care Committee of the National Research Council of Canada.

Domestic pigs weighing 45–50 kg were sedated with an intramuscular injection of atropine (0.5 mg/kg body wt), diazepam (0.4 mg/kg body wt) and ketamine (20 mg/kg body wt). Anesthesia was maintained with 1.5–2.0% isoflurane in a mixture of oxygen and medical air. The rate and volume of positive pressure ventilation were adjusted to keep the arterial blood gases within the normal physiological range. The brachiocephalic artery was cannulated at the level of the common carotid artery for arterial pressure monitoring, blood sampling, and injection of cardioplegic solution.

A sternotomy was performed and the brachiocephalic and subclavian arteries were dissected. A cannula for the infusion of cardioplegic solution was placed in the ascending aorta from the brachiocephalic artery. The pericardium was opened longitudinally along the midline. The descending aorta was isolated at its origin for cross-clamping. The animals were then heparinized by the injection of heparin (3000 I.U.) into the right atrium. Heparinized cold (~4°C) cardioplegic solution was delivered into the aortic root (10 ml/kg body wt) to arrest the heart. The heart was then excised from the animal. Pig blood in the chest was collected and mixed in 1:1 ratio with Krebs-Henseleit (K-H) solution. The mixture was oxygenated with 95% O_2 and 5% CO_2 to a final pH and oxygen partial pressure of 7.41 ± 0.02 pH units and 530 ± 35 mmHg, respectively.

Protocol

After removal from the animal, the pig heart was mounted on a Langendorff apparatus through the aorta. A hydraulic occluder was placed at the origin of the anterior descending artery (LAD). After 30 minutes of control perfusion, the LAD of the pig heart was occluded for 2 hours to create a myocardial infarct, followed by 50 minutes of reperfusion. During control perfusion, LAD occlusion, and reperfusion periods, myocardial energy metabolism in the myocardium supported by the LAD (LAD myocardium) was continuously assessed using ^{31}P MR spectroscopy. After reperfusion, the heart was then arrested with a hyperkalemic mixture of pig blood and K-H solution containing 16 mmol/L KCl and 16 mmol/L $MgCl_2$. The pig heart was perfused with this mixture throughout MRI studies to prevent artifacts of cardiac motion on T2* images. The hearts were kept at 37 ± 0.5°C throughout protocol.

T1 and T2* relaxation times were first measured without contrast agent to obtain their control values. To determine whether simultaneous monitoring of contrast-enhanced T1 and T2* signal intensities helps assess myocardial viability, one milliliter of an extracellular MR contrast agent (Gadolinium-diethyl triamine pentaacetic acid, Gd-DTPA) was injected as a bolus into the perfusion line at a total coronary flow rate of 300 ml/min. The first passage of the contrast agent through the hearts was monitored using an interleaved T1–T2* imaging sequence (described below). Perfusion medium containing the contrast agent was discarded and not allowed to return to the perfusion system. Once the contrast agent was completely washed from the heart, a second injection of the same amount of contrast agent was performed at a total coronary flow of 150 ml/min. The second injection of contrast agent at the different coronary flow was to determine whether the efficacy of simultaneous T1 and T2* for assessment of myocardial injury depends on blood flow. Likewise, the first passage of the second injection of contrast agent was followed using the interleaved T1–T2* imaging sequence. Unlike the first injection of contrast agent, the second one was allowed to re-circulate into the perfusion apparatus. Once the contrast agent reached a steady state, myocardial T1 and T2* relaxation times were measured.

At the end of the MRI studies, the heart was removed from the magnet and perfused at a flow rate of 300 ml/min. Colored microspheres were injected into the perfusion line to assess regional myocardial blood flow. The heart was then disconnected from the Langendorff system and sectioned for triphenyl tetrazolium chloride (TTC) staining and microsphere analysis. TTC stained sections were used as references for analysis of the MR images and selecting regions of interest as well as to identify regions for analysis of regional blood flow. The heart was sectioned according to a T1 scout image (long cardiac axis) to ensure correlation with the MR images.

^{31}P MR spectroscopy

^{31}P MR spectroscopy was performed in a 7 Tesla, 40-cm diameter, horizontal bore magnet equipped with a Bruker Spectrometer (Bruker Inc, Karlsruhe, Germany). A 1.5-cm diameter surface coil was positioned over the LAD region of the myocardium. Free induction decay (FID) signals were acquired with single hard pulses. The pulse length and repetition time were 75 µsec and 2.0 seconds, respectively. Sixty FID signals were accumulated for each spectrum. Thus, each ^{31}P spectrum was averaged over a 2-minute sample period. The observed phosphorus compounds included Pi, PCr, and three peaks of ATP. The areas under each phosphorus peak were used to assess the metabolite levels.

Magnetic resonance imaging

MR imaging was performed in the magnet mentioned above using a dual-ring Helmholtz coil. Simultaneous monitoring of T1 and T2* signal intensities was performed using an interleaved T1–T2* imaging sequence. The sequence consists of one gauss-shaped 90° pulse (3.1 ms) followed by read and phase spoiler gradients to eliminate residual magnetization from the T2*-weighted imaging portion of the sequence. A 1.5-sec delay was inserted to allow magnetization to relax back and then followed by a sech-shaped 180° pulse (13 ms) to invert the magnetization, after which spoiler gradients were applied in all three directions. To acquire both T1 and T2* images in a single set of scans a dynamically variable echo time was set to 4.2 ms for T1 and 15 ms for T2*-weighted imaging with repetition times of 6.75 and 18.15 ms, respectively. The preparation portion of the sequence consisting of the 90° pulse and spoiler gradients lasted 1524 ms. Each T1-weighted image was acquired in approximately 864 ms, T2*-weighted images were acquired in 2324 ms. The sequence was repeated 50 times for a total of 100 images over 235 seconds.

To measure T1 relaxation times, 14 images were acquired using a TurboFLASH sequence with 14 inversion recovery intervals (10, 50, 100, 200, 400, 800, 1000, 1400, 1800, 2200, 3000, 4000, 6000, and 10,000 ms) in a total of 152 seconds. The signal intensities of these images were then fitted using the following function:

$$SI = A[1 - 2\exp(-(t - q)/\mathrm{T1})]$$

Where SI is signal intensity, A is initial signal intensity (at time zero), t is the inversion recovery interval, and q is the time shift between the calculated T1 (inversion recovery) and preset T1.

To measure T_2* relaxation time, 4 MR images were acquired over 256 seconds using a FLASH sequence with different echo times (4.6, 12, 20, and 25 ms). T2* relaxation times were then calculated by fitting image intensities with the function:

$$SI = A[\exp(-t/\mathrm{T2}^{\star})]$$

Where *t* is echo time, SI is signal intensity, and A is signal intensity at time zero.

All imaging (T1–T2*, T1, T2*) was performed with a slice thickness of 3 mm, FOV set at 12 × 12 cm, with a matrix size of 128 × 128 pixels. To minimize imaging time, only one slice was observed for each heart. However, to that the slice contained a sizable infarct several slices were observed prior to injection of contrast agent.

Colored microsphere and TTC staining procedure

At the end of each experiment, approximately 5×10^6 to 8×10^6 non-radioactive red colored microspheres (15 ± 1.9 mm diameter, suspended in 2 ml of saline solution) were injected through the aorta via the perfusion line to measure regional myocardial blood flow. Arterial reference samples were withdrawn simultaneously from the ascending aorta at a constant rate of 5 ml/min for five minutes, starting one minute before the injection of microspheres. The heart was then removed and sectioned. Alternating sections were used for TTC staining with the opposing side used for analysis of the colored microspheres. A proper spatial correlation to MR images was ensured by using the T1-weighted, long axis, scout image as a reference. Sections used for microsphere analysis were divided into 6 pie-shaped regions and tissue was collected from the endocardium, midmyocardium, and epicardium. These sample regions were correlated to TTC-stained regions in the matched, opposing sections. The average weight of each sample was approximately 0.5 g.

After collection, tissue and blood samples were digested. The numbers of microspheres in the samples were counted using a spectrophotometer at a wavelength of 536 nm. Actual blood flow was obtained based on calculation of the microsphere count in the tissue and blood samples and the speed of blood collection.

Data analysis

Image data were processed using Marevisi and EvIdent 5.2 (NRC—Institute for Biodiagnostics, Winnipeg, Canada). Marevisi was used to determine T1 and T2* relaxation times using the fitting functions described previously. Round, hand-drawn regions of interest (ROI) were created by correlating the MR images to the TTC stained pictures. The sizes of the regions selected varied between hearts due to the variability in both heart size and size of infarction. However, the regions selected were approximately 400 pixels for normal, 300 pixels for reperfused, and 200 pixels for low-reflow infarct ROIs, while the total heart covered approximately 3500 pixels.

Statistical analyses were performed using Statistica (Statsoft Inc, Tulsa, Oklahoma, USA). One-way analysis of variance (ANOVA) with a Tukey's Honest Significant Difference test was used to compare R_1 (1/T1) and R_2^* (1/T2*) relaxation rates, percentage recovery of T2* at maximum T1 signal intensity, and regional blood flow between normal, reperfused infarcted, and low-flow infarcted myocardium. A significant difference was said to exist at a probability value of less than 0.05.

RESULTS

A representative heart section (TTC-stained, short axis) is shown in Figure 1. The heterogeneity in myocardial injury after 2 hours of occlusion of the LAD was quite clear. Apart from the stained viable tissue, there were regions of the myocardium that remained completely unstained (midmyocardium in the LAD region), which suggests myocardial infarction. Several hearts also showed regions in the subepicardium that contained TTC-stained and TTC-unstained tissue, indicating the

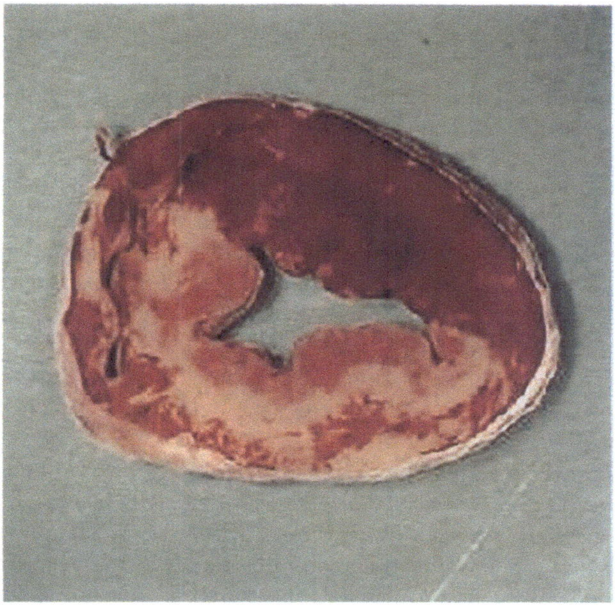

Figure 1. A representative TTC-stained section of a pig heart taken along the short cardiac axis. Red stained, lightly stained, and unstained regions identify normal, low-reflow infarcted, and reperfused infarcted myocardium, respectively. The subepicardium in the LAD region contained stained and unstained myocardium and is termed "mixed myocardium".

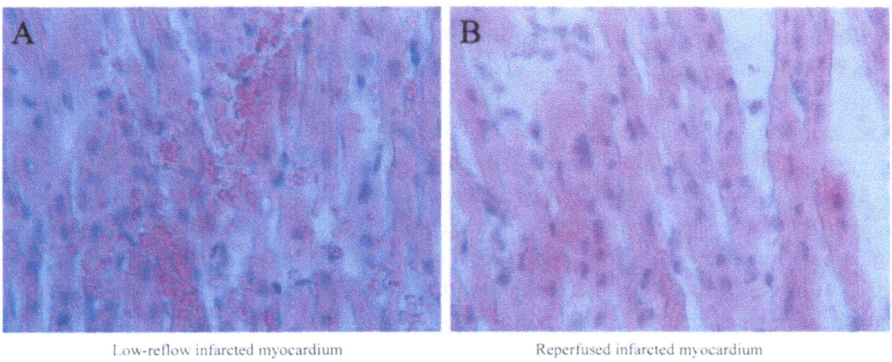

Figure 2. Tissue sections of a pig heart taken from the lightly TTC-stained areas (low-reflow infarcted myocardium, panel A) and TTC-unstained areas (reperfused infarcted myocardium, panel B). The tissue section sampled from the low-reflow infarcted myocardium shows sludging of red blood cells in the capillaries.

co-existence of viable and non-viable myocardium so-called "mixed region"(4 out of 8 hearts). Lastly, there were regions of lightly stained tissue (subendocardium in the LAD region).

The results obtained with colored microspheres show that regional blood flow in regions that remained unstained or were lightly stained with TTC was 1.20 ± 0.29 ml/g/min and 0.123 ± 0.19 ml/g/min, respectively. Histological sections of the pig hearts displayed severe blood cell sludging in the capillaries in the lightly stained myocardium (Figure 2). We believe that obstruction of the capillaries by red blood cells contributed to the severe reduction of blood flow in the lightly stained myocardium. The unstained and lightly stained myocardium are therefore designated as reperfused and low-reflow infarct, respectively. The non-LAD region, which stained brick red, is termed normal myocardium.

The average R_1 and R_2★ relaxation rates measured during equilibration of Gd-DTPA are shown in the top and bottom panels of Figure 3, respectively. Mean R_1 values were 17 ± 1.59 × $10^{-4}ms^{-1}$, 21.2 ± 2.05 × $10^{-4}ms^{-1}$, and 11.8 ± 2.08 × $10^{-4}ms^{-1}$ for normal, reperfused, and low-reflow infarct myocardium, respectively. Since not all hearts displayed a mixed region, R_1 and R_2★ relaxation rates for this area were not analyzed systematically. It is clear that the R_1 relaxation rate of normal myocardium is not significantly different from those of the reperfused infarcted myocardium and low-reflow infarcted myocardium, suggesting that contrast-enhanced T1 imaging may not be able to differentiate viable from non-viable myocardium.

Average R_2★ relaxation rates measured during equilibration of Gd-DTPA from normal, reperfused infarcted, and low-reflow infarcted myocardium were 7.23 ± 0.73 × $10^{-2}ms^{-1}$, 4.06 ± 0.25 × $10^{-2}ms^{-1}$, and 4.36 ± 0.37 × $10^{-2}ms^{-1}$, respectively. The average R_2★ relaxation rates in the reperfused infarcted and low-reflow infarcted myocardium were significantly shorter than that of normal myocardium. The difference in R_2★ between the reperfused infarcted and low-reflow infarcted myocardium was not statistically significant.

Representative signal-time courses of interleaved T1–T2★ images are shown in Figure 4. The four regions of the myocardium exhibited very distinct temporal relationships between T1 and T2★ signal intensities. It is apparent that T2★ signal intensity at the maximum T1 signal intensity was very different in the four types of tissue (normal, reperfused infarct, low-reflow infarct, and mixed myocardium). Figure 5 shows that low-reflow infarcted myocardium exhibited the greatest percentage recovery of T_2★ signal (93% of its initial value), followed by reperfused infarcted myocardium (71.3%) and mixed myocardium (38.3%). Normal myocardium displayed the least recovery of T2★ signal intensity at maximum T1 signal (19.3%). This suggests that simultaneous monitoring of T1 and T2★ signal intensities during the first passage of Gd-DTPA can differentiate viable from non-viable myocardium.

Figure 6 shows representative T1 maps obtained without (A) and with Gd-DTPA (B), a map of T2★ percentage recovery (C), and a section of heart (D). The map of percentage T2★ recovery was constructed using the interleaved T1–T2★ signal-time

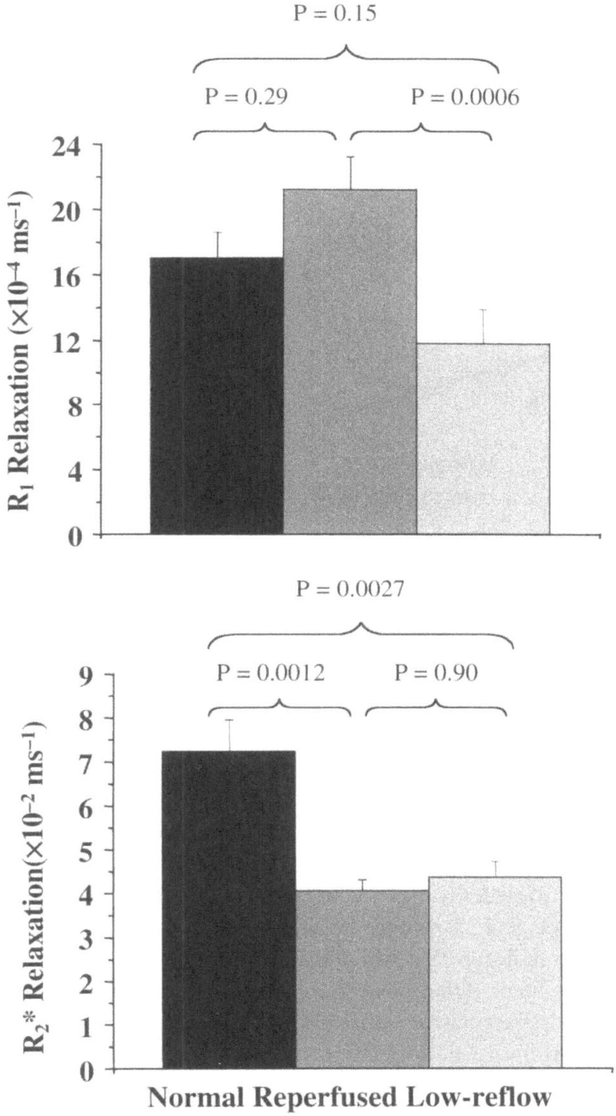

Figure 3. Contrast-enhanced R_1 (1/T1, top panel) and R_2* (1/T2*, bottom panel) relaxation rates measured in normal, reperfused infarcted, and low-reflow infarcted myocardium. During equilibration of Gd-DTPA, the R_1 relaxation rate in normal myocardium was not significantly different from those in reperfused infarcted and low-reflow infarcted myocardium. The R_2* relaxation rate in normal myocardium was significantly longer than those in the reperfused infarcted and low-reflow infarcted myocardium.

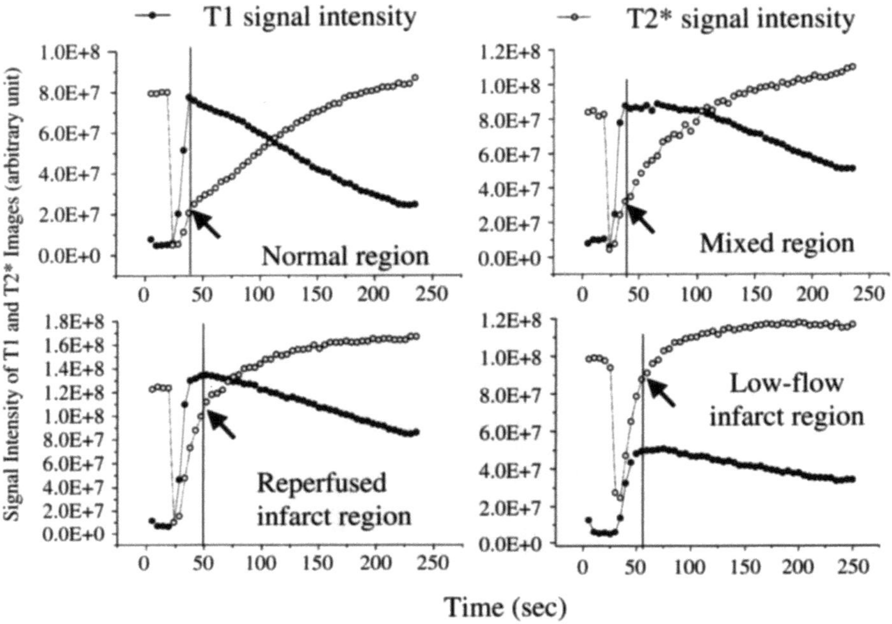

Figure 4. Representative signal-time courses of interleaved T1–T2* images reconstructed from normal, mixed, reperfused infarcted, and low-reflow infarcted regions of myocardium. The arrows indicate percentage recovery of T2* signal intensity when T1 signal intensity reached its maximum level.

courses. Bright areas on the map have a larger percentage of T2* recovery relative to the dark areas. It is clear that neither of the T1 maps, obtained with or without the contrast agent, provides a clear delineation of infarcted myocardium. The map showing the percentage T2* recovery correlates best with the TTC stained photograph. This further indicates that simultaneous monitoring of contrast-enhanced T1 and T2* intensities helps differentiate non-viable from viable myocardium.

Figure 7 shows representative signal-time courses of interleaved T1–T2* images obtained from a pig heart at total coronary flows of 300 ml/min and 150 ml/min. Variation in total coronary flow did not alter the percentage recovery of T2* at maximum T1 intensity during the first passage of Gd-DTPA. Significant differences in percentage T2* recovery between normal and reperfused infarcted myocardium were observed at both flow rates (Figure 8). Maps of percentage T2* recovery obtained at the two coronary flows and the corresponding sections of the same heart are shown in Figure 9. Both maps clearly delineate infarcted myocardium, suggesting that the efficacy of the interleaved T1–T2* imaging sequence is not affected by regional blood flow.

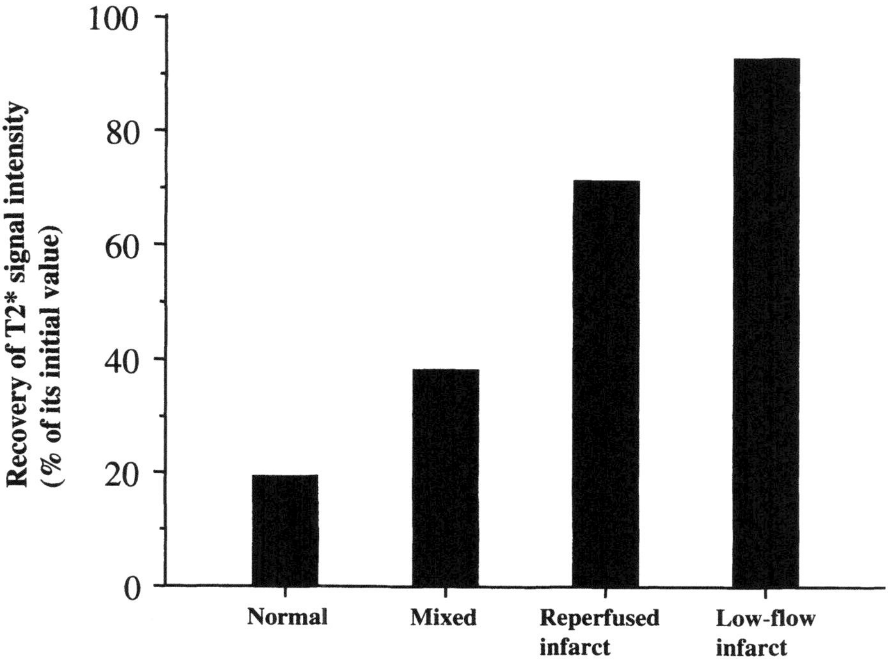

Figure 5. Percentage of T2* recovery at the maximum T1 signal intensity obtained from normal, mixed, reperfused infarcted, and low-reflow infarcted myocardium. Myocardial viability is closely related to the percentage recovery of T2*.

The time courses of PCr and Pi measured in the LAD supported region of the heart during control perfusion, LAD occlusion, and reperfusion are shown in Figure 10. As expected, occlusion of the LAD resulted in a rapid decrease in PCr with a corresponding increase in Pi. After one hour of occlusion, PCr was almost undetectable. Reperfusion did not result in any significant recovery of the PCr level, but it did wash out some Pi. This demonstrates that the areas of myocardium in the LAD region that had remained unstained or lightly stained by TTC had been irreversibly damaged.

DISCUSSIONS

The ability to differentiate viable from nonviable myocardium has important clinical implications for patient management and prognosis because patients with different degrees of myocardial injury may require different therapeutic management and may have different prognosis (1–3). Thus, methods that can accurately assess myocardial viability and further quantify the extent of infarcted tissue are still needed

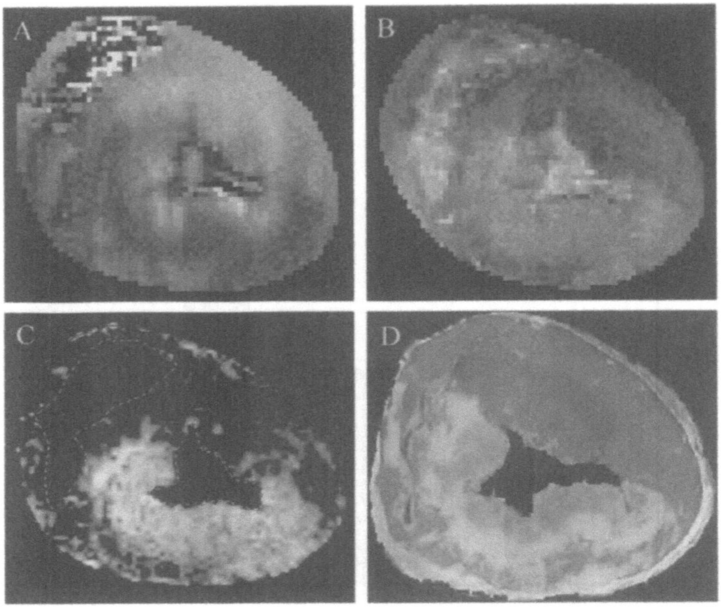

Figure 6. Panels A and B show T1 and T2★ maps acquired during equilibration of Gd-DTPA. A map of percentage T2★ recovery and a TTC-stained section of the same heart are shown in panels C and D, respectively.

clinically. The present study shows that simultaneous monitoring of T1 and T2★ signal intensities during the first passage of Gd-DTPA can differentiate viable from non-viable myocardium.

We observed that a 2-hour period of occlusion of the LAD resulted in heterogeneous myocardial injury in the mid-myocardium (no staining with TTC) and subendocardium (lightly staining with TTC). Analysis of the distribution of colored microspheres showed regional blood flow to be significantly higher in the mid-myocardium than in the subendocardium. Reduced nicotinamide adenine dinucleotide (NADH) was washed from the mid-myocardium, TTC was therefore not reduced to a brick red precipitate and as a result, the section of myocardium remained unstained. In addition, we found that the subendocardium in the LAD region was lightly stained by TTC. We believe that the staining was due to incomplete washout of NADH resulting from obstructed capillaries. As demonstrated by localized ^{31}P spectroscopy, this region of the myocardium was also irreversibly injured. Therefore, light TTC-staining in the subendocardium is not an indication of viable myocardium but a sign of more severe damage.

It has been shown that Gd-DTPA-enhanced R_1 relaxation rates depend mainly on the regional concentration of contrast agent, not its distribution (28,31). This is because protons exchange very rapidly between the intravascular, interstitial, and

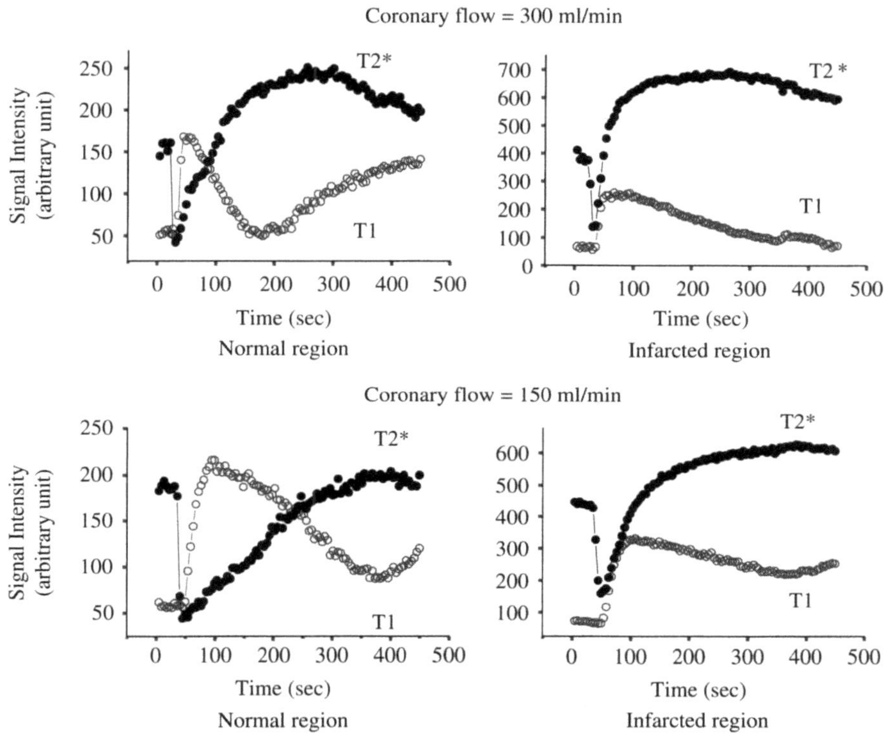

Figure 7. Representative signal-time courses of interleaved T1–T2* images acquired from normal and infarcted myocardium at flow rates of 300 ml/min and 150 ml/min.

intracellular spaces (32,33). In irreversible injury, the loss of cell membrane integrity increases the space accessible to Gd-DTPA. Obstruction of the capillaries (as observed in the subendocardium), however, limits delivery of contrast agent. Therefore, the combinative effect of membrane disruption and capillary obstruction may result in a shorter, longer, or similar R_1 in infarcted myocardium relative to normal myocardium. This study shows that R_1 relaxation rates measured during equilibration of Gd-DTPA in normal myocardium were indistinguishable from those in reperfused and low-reflow infarcted myocardium. This suggests that contrast-enhanced R_1 values may not be able to distinguish normal from infarcted myocardium. As mentioned above, a significant decrease of blood flow in the low-reflow infarcted myocardium limited delivery of the contrast agent. As a result, the regional concentration of contrast agent may not be significantly different in normal and low-reflow infarcted myocardium. At present, it is not completely clear why the R_1 values were not significant different in normal and reperfused infarcted myocardium. This may be due to non-linear signal enhancement from Gd-DTPA

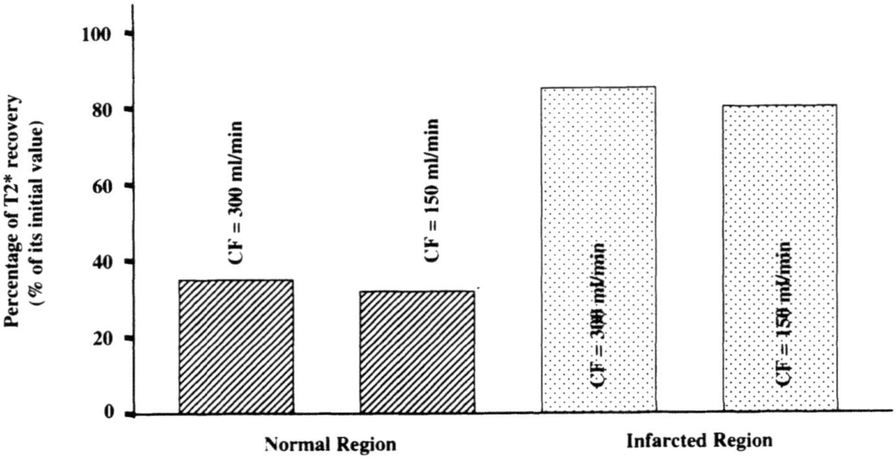

Figure 8. Comparison of percentage T2* recovery from normal and infarcted myocardium measured at flow rates of 300 ml/min and 150 ml/min.

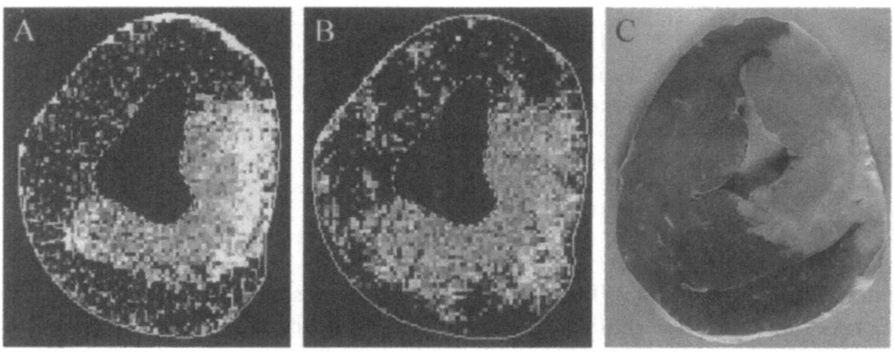

Figure 9. The maps of percentage T2* recovery obtained at flow rates of 300 ml/min (A) and 150 ml/min (B) and a TTC-stained section (C) of the same heart. Infarcted myocardium was delineated on both maps.

at high magnetic field or that the concentration of Gd-DTPA used in this study was not optimal. Nevertheless, the results suggest it is difficult to separate viable from non-viable heart tissue using R_1 values alone.

Following injection of contrast agent and equilibration, the reperfused and low-reflow myocardium displayed a significantly decreased R_2* rate compared to normal tissue. This is presumably due to a loss of membrane integrity in the infarcted tissues,

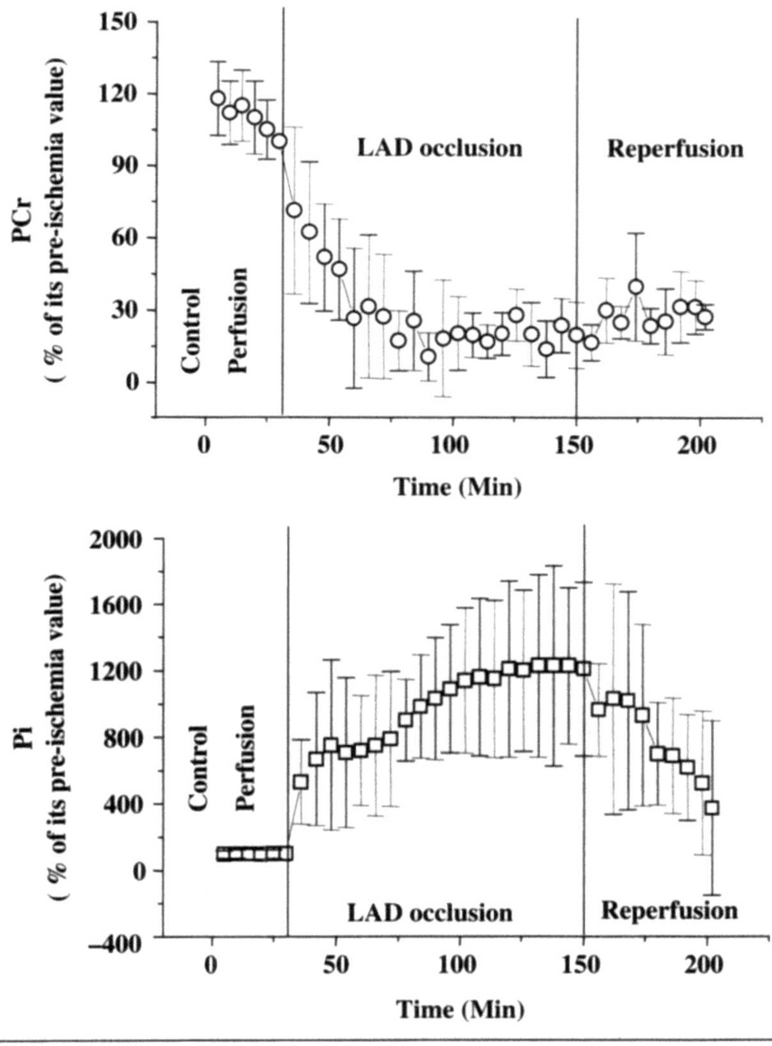

Figure 10. Time courses of phosphocreatine (PCr, top panel) and inorganic phosphate (Pi, bottom panel) acquired from the LAD region during control perfusion, LAD occlusion, and reperfusion. The lack of recovery of PCr during reperfusion indicated that LAD myocardium in the LAD region had been irreversibly damaged.

leading to a more homogeneous distribution of contrast agent in the infarcted areas than in normal areas of the myocardium. The results suggest that homogeneity of contrast distribution can be assessed using T2★-weighted imaging. Moreover, the contrast-enhanced R_2★ relaxation rate may be a more sensitive parameter for assessment of myocardial viability relative to the R_1 relaxation rate. Obstruction of the capillaries in the low-reflow infarcted myocardium does not appear to affect the sensitivity of R_2* to myocardial viability. However, it should be mentioned that this study was performed using a 7 Tesla magnet where the susceptibility effect of Gd-DTPA may have been magnified. In other words, it may not be possible to differentiate infarcted from normal myocardium using the R_2* values obtained with low-strength (0.5–3 Tesla) magnets such as those used in clinic.

It has also been shown that R_2* depends not only on the distribution of contrast agent, but also on its regional concentration (28,29). A relatively shorter R_2* relaxation rate could be due to a more homogeneous distribution of contrast agent (loss of cell membrane integrity) or a relatively lower concentration of contrast agent (intact cell membrane or obstruction of the capillaries). A relatively shorter R_2* value in a given region of myocardium may indicate a more homogeneous distribution of contrast agent only if the regional concentration of contrast agent is equal or higher in this region relative to remote areas of the myocardium. The regional concentration of contrast agent can be assessed in a relative manner with T1-weighted imaging. We believe that simultaneous monitoring of Gd-DTPA-induced changes in both T1 and T2* signal intensities provides more accurate assessment of myocardial viability than use of either T1 or T2* imaging alone.

During the first pass of Gd-DTPA, the peak level of T1 signal intensity indicates that the regional concentration of contrast agent has reached its maximum level. At this time, it is expected that the intravascular and interstitial spaces in normal tissue and all three compartments in infarcted myocardium will be completely filled with contrast agent. T2* signal intensities at this time are then expected dependent mainly on cell membrane integrity. This is the reason why we used T2* signal intensity at maximum T1 intensity to evaluate myocardial viability. However, the baseline levels of T2* intensity in normal and injured tissue (prior to contrast injection) may be different due to the differences in microstructure, regional blood flow, ratio of oxygenated to deoxygenated hemoglobin, or other factors. Conceivably, changes in T2* intensity relative to baseline values are expected to be more closely related to homogeneity of contrast distribution (or membrane integrity) than absolute T2*. As a result, we used the percentage recovery of T2* signal at maximum T1 intensity as a parameter for the assessment of myocardial viability.

We found that normal and infarcted myocardium have very distinct temporal relationships in T1 and T2* signal intensities. In normal myocardium there was a rapid decline of T2* signal intensity. This decrease in T2* signal intensity may be due to the heterogeneous distribution of the contrast agent as it was flushing into the capillaries. As the contrast agent diffused into the interstitial space, T1 signal intensity reached its peak level. This infiltration may at the same time lead to more homogeneous distribution of contrast agent, leading to recovery of T2* signal intensity. However, because Gd-DTPA cannot diffuse into the intracellular space in normal myocardium, the distribution was still relatively heterogeneous and the return of T2* signal in normal tissues was likely due to washout. Consequently, T2* signal intensity in normal myocardium displayed a small percentage recovery when correlated to peak T1. In the infarcted myocardium, Gd-DTPA may not only diffuse into the interstitial space, but also into the intracellular space (9). The distribution of the contrast agent became more homogeneous relative to that in normal myocardium by the time the regional concentration of contrast agent was maximal. This led to a greater percentage recovery of T2* signal intensity in the reperfused infarcted, low-reflow infarcted, and mixed myocardium than in normal myocardium.

In this study, we did not assess the homogeneity of Gd-DTPA distribution directly in various regions of the myocardium. Therefore, it is still a working hypothesis that the percentage recovery of T2* signal intensity depends on the severity of membrane injury. Nevertheless, our results suggest that simultaneous monitoring of contrast-enhanced T1 and T2* signal intensities during the first pass of Gd-DTPA delineates infarcted myocardium in a flow independent manner.

ACKNOWLEDGEMENTS

The authors would like to thank Mrs. Lori Gregorash, Shelley Germscheid, Rachelle Mariash, and Mr. Allen Turner for their assistance with surgery and heart perfusion experiments. This study was partially supported by the Heart and Stroke Foundation of Manitoba and the Canadian Institutes for Health Research.

REFERENCES

1. Baxter B. 1998. Cellular mechanisms of ischemic myocardial damage. In: Kaski J and Holt D. Eds. Myocardial Damage. Early Detection by Novel Biochemical Markers. Kluwer Academic Publishers, Boston. pp 1–16.
2. Agati L, Autore C, Iacoboni C, Castaldo M, Veneroso G, Voci P, Fedele F, Dagianti A. 1998. The complex relation between myocardial viability and functional recovery in chronic left ventricular dysfunction. Am J Cardiol 81(12A):33G–35G.
3. Huitink JM, Visser FC, Bax JJ, Visser CA. 1995. Detection of viability after myocardial infarction: available techniques and clinical relevance—a review. International J Cardiol 51:253–266.
4. Previtali M, Fetiveau R, Lanzarini L, Cavalotti C, Klersy C. 1998. Prognostic value of myocardial viability and ischemia detected by dobutamine stress echocardiography early after acute myocardial infarction treated with thrombolysis. J Am Coll Cardiol 32(2):380–386.
5. Pierard LA, Hoffer EP. 1998. Role of stress echocardiography in heart failure. Am J Cardiol 81(12A): 111G–114G.
6. Ajisaka R, Watanabe S, Masuoka T, Yamanouchi T, Saitoh T, Toyama M, Takeda T, Itai Y, Sugishita Y. 1998. Relationship between normalization of negative T waves on exercise ECG and residual myocardial viability in patients with previous myocardial infarction and no post-infarction angina. Jpn Circ J 62(3):153–159.
7. Casans-Tormo I, Gomez-Aldaravi R, Bodi-Peris V, Sanchis-Fores J, Ciudad-Platero J, Insa-Perez L, Manjon-Soriano J. 1998. Determination of ejection fraction and left ventricular fraction using isotopic ventriculography and bidimensional echocardiography. Comparison with contrast ventriculography. Rev Esp Cardiol 51(Suppl 1):10–18.
8. Flotats-Giralt A, Carrio-Gasset I, Estorch-Cabrera M, Berna-Roqueta L, Catafau-Alcantara AM, Mari-Aparici C, Ballester-Rodes M. 1998. Evaluation of myocardial viability using perfusion cardiac SPECT. 201-thallium rest/redistribution, 201-thallium rest/reinjection and technetium 99m tetrofosmin rest/postnitrates. Rev Esp Cardiol 51(Suppl 1):45–52.
9. Cloninger KG, Puey EG, Garcia EV, Roubin GS, Robbins WL, Nody A, De Pasquale EE, Berger HJ. 1988. Incomplete redistribution in delayed thallium-201 single photon emission computed tomographic (SPECT) images: an overestimation of myocardial scarring. J Am Coll Cardiol 12: 955–963.

10. Ryan T, Tarver RD, Duerck JL, Sawada SG, Hollenkamp NC. 1990. Distinguishing viable from infarcted myocardium after experimental ischemia and reperfusion by using nuclear magnetic resonance imaging. J Am Cardiol 15:1355–1364.
11. Iskandrian AS, Heo J, Schelbert HR. 1996. Myocardial viability: Methods of assessment and clinical relevance. Am Heart J 132:1226–1235.
12. Pereira RS, Prato FS, Wisenberg G, Sykes J. 1996. The determination of myocardial viability using Gd-DTPA in a canine model of acute myocardial ischemia and reperfusion. Magn Reson Med 36: 684–693.
13. Pereira RS, Prato FS, Sykes J, Wisenberg G. 1999. Assessment of myocardial viability using MRI during a constant infusion of Gd-DTPA: Further studies at early and late periods of reperfusion. Magn Reson Med 42:60–68.
14. Weiss RG, Bottomley PA, Hardy CJ, Gerstenblith G. 1990. Regional myocardial metabolism of high-energy phosphates during isometric exercise in patients with coronary artery disease. N Engl J Med 323:1593–1600.
15. Path G, Robitaille PM, Merkle H, Tristani M, Zhang J, Garwood M, From AHL, Bache RJ, Ugurbil K. 1990. Correlation between transmural high energy phosphate levels and myocardial blood flow in the presence of graded coronary stenosis. Circ Res 67:660–673.
16. Yabe T, Mitsunami K, Okada M, Morikawa S, Inubushi T, Kinoshita M. 1994. Detection of myocardial ischemia by 31P magnetic resonance spectroscopy during handgrip exercise. Circulation 89: 1709–1716.
17. Geskin G, Kramer CM, Rogers WJ, Theobald TM, Pakstis D, Hu YL, Reichek N. 1998. Quantitative assessment of myocardial viability after infarction by dobutamine magnetic resonance imaging. Circulation 98:217–223.
18. Croisille P, Moore CC, Judd RM, Lima JAC, Arai M, McVeigh ER, Becker LC, Zerhouni EA. 1999. Differentiation of viable and nonviable myocardium by the use of three-dimensional tagged MRI in 2-day-old reperfused canine infarcts. Circulation 99:284–291.
19. Lima JAC, Ferrari VA, Reichek N, Kramer CM, Palmon L, Llaneras MR, Tallant B, Young AA, Axel L. 1995. Segmental motion and deformation of transmurally infarcted myocardium in acute postinfarcted period. Am J Physiol 268(Heart Circ Physiol 37):H1304–H1323.
20. Pislaru SV, Ni Y, Pislaru C, Bosmans H, Miao Y, Bogaert J, Dymarkowski S, Semmler W, Marcha G, Van de Werf FJ. 1999. Noninvasive measurements of infarct size after thrombolysis with a necrosis-avid MRI contrast agent. Circulation 99:690–696.
21. Marchal G, Ni Y, Herijigers P, Flameng W, Petre C, Bosmans H, Yu J, Ebert W, Hilger CS, Semmler W, Baert AL. 1996. Paramagnetic metaloporphrins: infarct avid contrast agents for diagnosis of acute myocardial infarction by MRI. Eur Radiol 6:2–8.
22. Barkhausen J, Ebert W, Heyer C, Debatin JF, Weinmann HJ. Imaging of myocardial infarction: comparison of magnevist and gadophrin III in rabbits. Proceedings of the ISMRM April 2001, Galsgow, UK, pp 225.
23. Wyttenbach R, Saeed M, Wendland MF, Geschwind JF, Bremerich J, Arheden H, Higgins CB. 1999. Detection of acute myocardial ischemia using first-pass dynamics of MNDPDP on inversion recovery echoplanar imaging. J Magn Reson Imaging 9:209–214.
24. Elst LV, Colet JM, Muller RN. 1997. Spectroscopic and metabolic effects of MnCl2 and MnDPDP on the isolated and perfused rat heart. Invest Radiol 32:581–588.
25. Bremerich J, Saeed M, Arheden H, Higgins CB, Wendland MF. 2000. Normal and infarcted myocardium: differentiation with cellular uptake of manganese at MR imaging in a rat model. Radiology 216:524–530.
26. Kim RJ, Lima AC, Chen EL, Reeder SB, Klocke FJ, Zerhouni EA, Judd RM. 1997. Fast 23Na magnetic resonance imaging of acute reperfused myocardial infarction. Potential to assess myocardial viability. Circulation 95:1877–1885.
27. Mühler A. 1995. Assessment of myocardial perfusion using contrast-enhanced MR imaging: current status and future developments. MAGMA 2:21–33.
28. Mühler A. 1995. Assessment of myocardial perfusion using contrast-enhanced MR imaging: current status and future developments. MAGMA 3:21–33.
29. Keennan RR, Zhong J, Gore JC. 1994. Intravascular susceptibility contrast mechanisms in tissue. Magn Reson Med 31:9–21
30. Saeed M, Lund G, Wendland MF, Bremerich J, Weinmann H, Higgins CB. 2001. Magnetic resonance characterization of the peri-infarction zone of reperfused myocardial infarction with necrosis-specific and extracellular nonspecific contrast media. Circulation 103:871–876.

31. Edelman RR, Li W. 1994. Contrast-enhanced echo-planar MR imaging of myocardial perfusion: preliminary study in humans. Radiology Mar; 190(3):771–777
32. Donahue KM, Burstein D. 1993. Proton exchange rates in myocardial tissue with Gd-DTPA administration. SMRM Annual Meeting, New York.
33. Keenan RP, Zhong J, Gore JC. 1991. On the relative importance of paramagnetic relaxation and diffusion-mediated susceptibility losses in tissues. Magn Reson Med 22:197–203.

Pathophysiology of Cardiovascular Disease,
edited by N.S. Dhalla, H. Rupp,
A. Angel & G.N. Pierce
Kluwer Academic Publishers, Boston, 2004

Magnetic Resonance in the Assessment and Development of Surgical Strategies for Protecting the Brain During Cardiovascular Surgery

Jian Ye, Luojia Yang, and Roxanne Deslauriers

Institute for Biodiagnostics, National Research Council Canada, Department of Surgery, University of Manitoba, Winnipeg, Canada

Summary. Neurological and neuropsychological impairments are important sequelae of cardiac surgery with cardiopulmonary bypass, particularly aortic arch surgery requiring circulatory arrest. The nature of the underlying brain injuries remains poorly understood. Hypothermic circulatory arrest is commonly used for brain protection during aortic arch surgery, but the safe duration of this technique is limited. Other techniques used for brain protection have been proposed and used clinically. However, none of the techniques is completely satisfactory. Precise, and preferably quantitative, definition of the localization and nature of the underlying injuries is a precondition for rigorous evaluation of the efficacy of prophylactic measures. Magnetic resonance (MR) spectroscopy and imaging are unique, non-invasive tools that are amenable to repetitive monitoring of the anatomical, biochemical and physiological status of the brain. The ability of magnetic resonance spectroscopy and imaging to detect physiological and pathological changes in the brain is described. Results of serial experiments studying the biochemical and physiological sequelae of various selective brain perfusion techniques in an in vivo pig model are presented. MR imaging can provide non-invasive information on physiological and pathological changes and MR spectroscopy can follow changes in metabolism and intracellular pH in the brain in various body perfusion situations. The combined information can help understand the nature of underlying brain injuries and develop better surgical strategies for protecting the brain during cardiovascular surgery.

Key words: magnetic resonance, antegrade cerebral perfusion, retrograde cerebral perfusion, circulatory arrest, brain protection.

Correspondence: Jian Ye, MD, Institute for Biodiagnostics, National Research Council of Canada, 435 Ellice Ave., Winnipeg, Manitoba, Canada R3B 1Y6. Tel. (204) 984-7037, Fax (204) 984-7036, E-mail: jian.ye@nrc.ca.

INTRODUCTION

Cardiovascular disease is the leading cause of mortality in North American. Cardiac surgery is now performed on 400,000 patients each year in the United States and on 800,000 worldwide. Neurological complications account for the major adverse sequelae and are among the leading causes of morbidity and mortality in cardiovascular surgery. Complications are increasing because sicker and older patients with advanced atherosclerotic vascular disease now have surgery. The neurological complications are even more common in cardiac surgery requiring circulatory arrest (1,2). Operative repair of aortic aneurysms or aortic dissections extending into the aortic arch requires temporary interruption of blood flow through the cerebral vessels. Deep hypothermic circulatory arrest (HCA) has been the primary technique used for intraoperative protection against cerebral injury (1,3). However, neurological injury remains a serious consequence despite the protective effect of hypothermia. Periods of deep HCA longer than 40 minutes are associated with high incidences of brain injury, stroke and neurological sequelae (1,4). Various techniques, including antegrade cerebral perfusion (ACP, perfusing the brain through the carotid arteries) and retrograde cerebral perfusion (RCP, perfusing the brain through the superior vena cava), have been devised to provide better brain protection during circulatory arrest than deep hypothermia alone. However, the efficacy of these techniques remains to be fully determined, particularly when used for prolonged periods of circulatory arrest.

Magnetic resonance (MR) imaging and spectroscopy have been demonstrated to be unique non-invasive tools that allow continuous monitoring of anatomical, biochemical and pathophysiological changes in the brain under various conditions. We have developed and used these techniques to determine the efficacy of various brain protection techniques during prolonged HCA and to optimize/develop surgical techniques for brain protection during cardiac surgery. In this paper, we have reviewed our experimental studies on deep HCA, RCP, bilateral ACP, and unilateral ACP in an acute pig model.

ANIMAL MODEL

The pig has become an increasingly popular model for the study of cerebral protection during cardiothoracic surgery.

As described previously (5–8), pre-anesthesia was induced with midazolam (0.3 mg/kg), ketamine (20 mg/kg) and atropine (0.02 mg/kg) intramuscularly. After endotracheal intubation, the animal was ventilated mechanically with 60% oxygen and 40% air. The ventilator rate and tidal volume were adjusted to maintain the arterial carbon dioxide tension at 35–45 mm Hg. Anesthesia was maintained with 1.5–2.0% isoflurane. A temperature probe was placed in the esophagus to monitor the core temperature. Urine output was collected through a bladder catheter.

The chest was opened via a midline sternotomy. Heparin (500 IU/kg) was given intravenously. A cannula (20 F) was inserted into the ascending aorta and used for arterial blood return to the body during normal cardiopulmonary bypass (CPB).

A venous cannula (32 F, single stage) was placed into the right atrium just above the inferior vena cava. A small catheter was placed in the superior vena cava (SVC) through an internal mammary vein for monitoring central venous pressure. The lungs were not inflated during CPB or circulatory arrest. After preparation, the pig was placed supine in a specially designed cradle and transferred into a 7 Tesla MR instrument.

For RCP, a catheter was placed into the superior vena cava and a small pressure line inserted into the internal jugular vein for monitoring retrograde perfusion pressure. For ACP, the right axillary artery was isolated and a 12 F modified double lumen retrograde cannula was placed in the right axillary artery. The large lumen of the cannula was used for ACP, while the small lumen was used to monitor arterial and perfusion pressures. In the pig, both carotid arteries arise from the common carotid artery, which is a branch of the brachiocephalic artery. Therefore, during unilateral ACP, the left carotid artery was occluded.

The CPB circuit consisted of Cobe roller pumps (model C22.2), cardiotomy reservoir (Cobe HVRF 3700), a membrane oxygenator (Cobe Optima) with integral heat exchanger, arterial filter (40 µm, Dideco D733), and water bath (Laud MGW type RMSG). The circuit was primed with 700 ml lactated Ringer's solution, 500 ml Pentaspan (Du Pont Pharma, Canada, commonly used in clinic in Canada), 25 ml 1 M sodium bicarbonate and 5000 IU heparin. Sodium bicarbonate was given to maintain arterial blood pH within the normal range (7.35–7.45), when necessary. No correction was made for temperature during hypothermia, and the alpha-stat approach was used. Arterial blood gases were monitored and measured at 37°C with a blood gas analyzer (Stat 9, NOVA Biomedical, U.S.A). Blood electrolytes, glucose and osmolality were monitored and kept within the normal range.

After completing preparations for the MR studies (i.e., positioning the head and shimming the magnetic field), normothermic CPB was initiated in the pigs. During cooling and re-warming, the temperature gradient between the water bath and blood was less than 10°C. Hypothermic circulatory arrest with or without RCP or ACP was achieved when the body temperature reached the proposed level. During RCP, the perfusion pressure in the internal jugular vein was maintained at less than 25 mm Hg. During ACP, perfusion pressure was monitored continuously and maintained at around 60 mm Hg in the common carotid artery. No inotropic support was used during the experiments. The animals were not weaned from CPB, therefore no means of myocardial preservation (such as cardioplegia) was used.

HISTOPATHOLOGICAL EXAMINATION

The experimental details of the histopathological studies are described in a previous publication (6). At the end of each experiment, the brain was removed and fixed with formaldehyde solution. Tissue samples obtained from 10 different regions were cut into 5-µm-thick slices, which were stained with hematoxylin and eosin. Injury was graded (0 to 5) on the basis of the number of damaged neurons within each region, as follows: grade 0, normal; grade 1, <10%; grade 2, 10–25%; grade 3, 26–50%; grade 4, 51–75%; and grade 5, >75%.

MAGNETIC RESONANCE SPECTROSCOPY AND IMAGING

1. ^{31}P MR spectroscopy

Magnetic resonance spectroscopy allows non-invasive measurements of intracellular pH (pHi) based on the spectral shift of the inorganic phosphate (Pi) resonance, which is observed in most cells (9,10). MR spectroscopy can monitor steady-state levels of ATP and the kinetics of reactions responsible for the maintenance of ATP levels. Phosphocreatine (PCr), another high-energy compound, can readily be detected and its equilibrium with ATP investigated because both these compounds exist in high concentrations in brain tissue. Ischemia results in loss of high-energy phosphates and tissue acidosis. Under anaerobic conditions, ATP is initially generated with the breakdown of PCr and accumulation of Pi. Therefore, a decrease in PCr and an increase in Pi are very sensitive indicators of ischemic metabolism. The changes can be detected by MR spectroscopy within a few minutes following the onset of ischemia, but the change in ATP levels is more gradual (11). We have developed an experimental protocol for the in vivo study of brain energetics and intracellular pH by ^{31}P MR spectroscopy in large animals (5,6,12). Figure 1 shows typical ^{31}P MR spectra obtained in vivo from a normal pig brain. Both Pi and PCr have distinct resonance frequencies producing single peaks in a spectrum. ATP gives rise to three groups of peaks representing the three phosphate groups.

Quantification of peak areas (Pi, PCr and ATP) can be carried out using various computer programs (13). The β-ATP peak is commonly used to follow changes in ATP levels, since it is the least influenced by other merging resonances such as those that arise from adenosine diphosphate (ADP) and nicotinamide adenine dinucleotide phosphate (NADP). It is well established that pHi can be determined from the chemical shift of Pi (9,10).

2. MR perfusion imaging

Dynamic MR perfusion imaging has been developed to allow non-invasive assessment of tissue perfusion (7,8). The method uses fast T2*-sensitive imaging techniques, such as fast gradient-echo imaging sequences, to follow the change in signal intensity in a series of images acquired after a bolus injection of a paramagnetic contrast agent such as gadopentetate dimeglumine (Gd-DTPA, Berlex, Montreal, Canada). The amount and rate of signal change are related to blood flow. Since local magnetic field homogeneity is disturbed as a bolus of contrast agent passes through tissue vasculature, T2* image intensity decreases in the perfused area. The effect on local magnetic field homogeneity in an underperfused/non-perfused area is much less than in a normally perfused area. Therefore, image intensity decreases less in an underperfused area than in normal perfused tissue. Using this technique, underperfused tissue can be characterized by mapping regional cerebral blood volume (rCBV), mean transit time (MTT, the time required for blood (or contrast agent) to pass through each region of the brain), and regional cerebral blood flow (rCBF) (7,8,14). We have designed and constructed a 12 cm diameter dual-ring surface coil (transmit/receive) (15) and devised a method that is suitable for use in

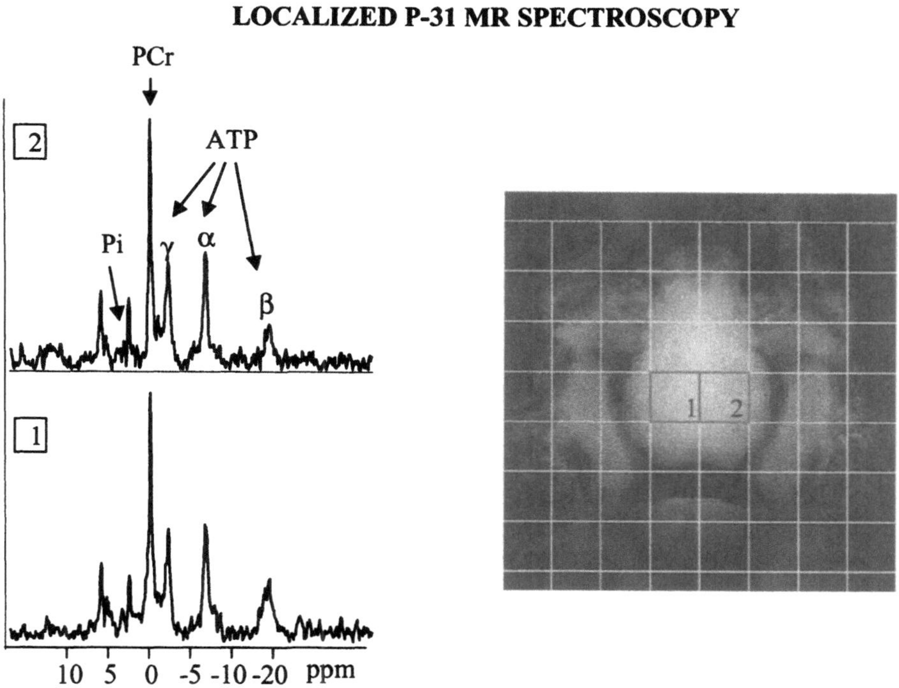

Figure 1. Localized ^{31}P MR spectra obtained from a normal pig brain in vivo.

vivo with various types of brain perfusion (7,8). Each image can be acquired in less than 1 second. The image intensity decreases significantly when Gd-DTPA is present in the vasculature of the brain and returns to precontrast levels as the Gd-DTPA washes out. A series of images can then be analyzed to produce individual maps of rCBV, MTT, and rCBF.

To obtain unambiguous information about the brain and avoid signal contamination from surrounding muscle, an experimental protocol for localized ^{31}P MR spectroscopy was developed (12). We also developed an experimental protocol for MR perfusion imaging that is repeatable within a very short period after each injection of the contrast agent (7,8), which allows acquisition of multiple MR perfusion images in the same animal.

DEEP HYPOTHERMIC CIRCULATORY ARREST

Deep HCA (15°C) is commonly used for brain protection during aortic arch surgery. The technique is simple and consists primarily of deep core cooling of the patient to temperatures in the range of 12 to 18°C, drainage of the patient's blood into a cardiotomy reservoir and total circulatory arrest. This allows precise and

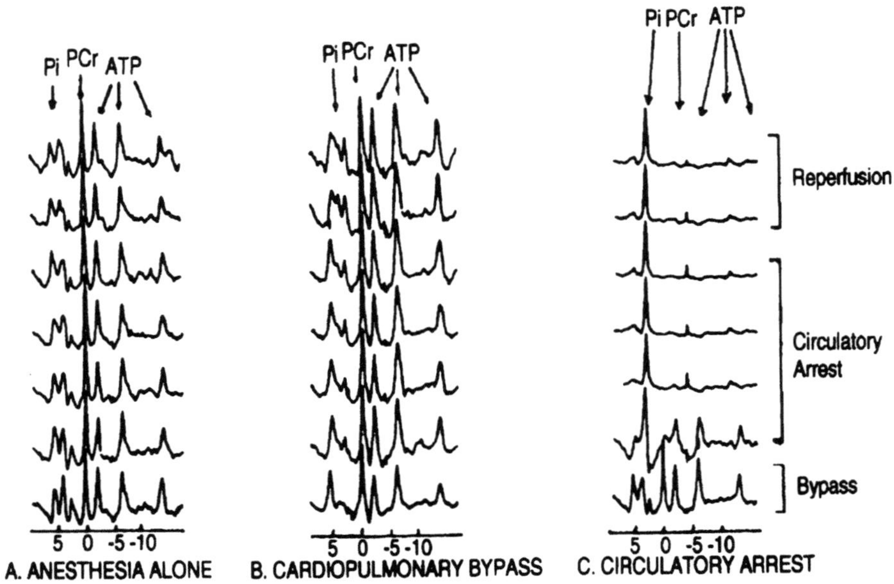

Figure 2. Representative MR spectra from the anesthesia control group (A), deep hypothermic cardiopulmonary bypass group (B), and deep hypothermic circulatory arrest group (C).

bloodless surgery on the heart and/or arch vessels without additional cannulation in a limited operation field. The "safe" period of HCA depends on the length of arrest and the core temperature achieved prior to circulatory arrest. Neurological complications increase significantly if HCA is prolonged (over 40–60 minutes) (16–18). The effects of hypothermia on ATP, PCr, and pHi in the brain during circulatory arrest and the degree of recovery of these parameters have been measured in a porcine model (5). Control animals (anesthesia alone or deep hypothermic CPB) showed no significant alterations in ATP, PCr, Pi (Figure 2A and 2B). In the test animals, circulatory arrest was performed at 15°C for 2 hours, followed by 60 minutes of reperfusion with normothermic CPB. The spectra showed a total loss of high energy phosphate metabolites after approximately 40 minutes of deep HCA with no, or incomplete, recovery of energy metabolites during reperfusion with normothermic CPB (Figure 2C). In the two control groups, that is, anesthesia and deep hypothermic CPB, the pHi was 7.1 ± 0.1 and 7.3 ± 0.1, respectively. There were no changes in pHi during the entire period of anesthesia or deep hypothermic CPB. A significant decrease in pHi (intracellular acidosis) was observed during deep HCA and during reperfusion with CPB in the HCA group (Figure 3). Significant brain injury following 120 minutes of deep HCA was confirmed by histopathological studies (Figure 4) (4). Our findings that ATP, PCr and pHi levels decrease to minimal levels after only 30 minutes of deep HCA are consistent with the clinical

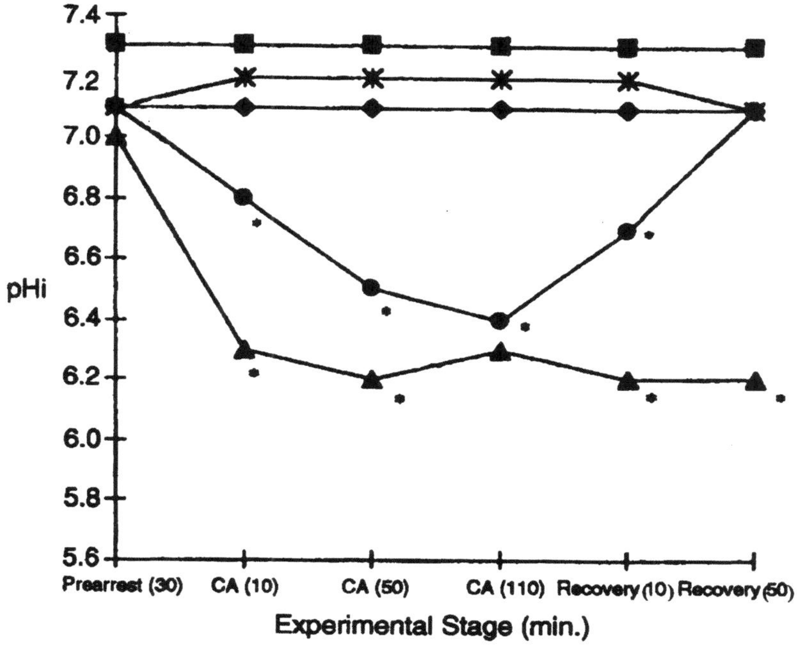

Figure 3. Intracellular pH (pHi) of the brain measured throughout the experimental protocol in the following groups: anesthesia (control) (-◆-), deep hypothermic cardiopulmonary bypass (control) (-■-), deep hypothermic circulatory arrest (-★-), antegrade cerebral perfusion (-▶-) and retrograde cerebral perfusion (-●-). CA = circulatory arrest. ★$P < 0.05$ compared with the anesthesia and deep hypothermic cardiopulmonary bypass control groups.

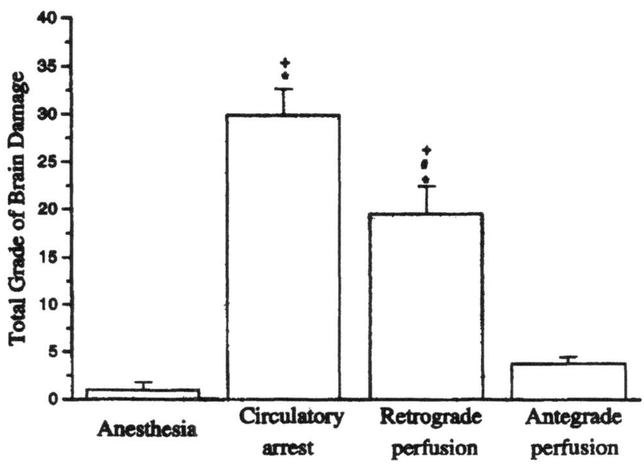

Figure 4. Total brain damage evaluated as the sum of the grades obtained from ten areas of the brain. All the values are mean ± standard error of the mean. ★$P < 0.05$ versus anesthesia group, +$P < 0.05$ versus antegrade perfusion group, #$P < 0.05$ versus circulatory arrest group.

observation that deep HCA alone for periods longer than 30–40 minutes significantly increases the risk of brain injury. Surgeons, therefore, "battle the clock" once the circulation is arrested and in some cases in which complication occurs, time runs out. Ergin et al. (19) recently conducted a clinical study comparing patients undergoing routine cardiac surgery with CPB to patients having surgery of the thoracic aorta with periods of deep HCA. It was demonstrated that prolonged deep HCA (>25 minutes) is a significant predictor of poor performance for both memory and fine motor domains at 6 weeks post-surgery. Prolonged deep HCA (>25 minutes) was also a determinant of prolonged hospital stay.

RETROGRADE CEREBRAL PERFUSION

Ueda and associates first introduced intermittent and later continuous RCP via the superior vena cava (SVC) as an adjunct to HCA. Retrograde cerebral perfusion has several potential advantages (20,21). The technique may provide uniform cerebral cooling, wash out air bubbles, embolic debris, neurological toxins and other byproducts of ischemic metabolism, and it may be able to deliver cerebral metabolic substrates (22–24). These factors may extend the "safe period" of circulatory arrest. Over the past decade, the technique has become popular and has been used by many surgeons (1,20,25–28) because of its simplicity and potential benefits. A number of clinical studies have suggested potential benefits of RCP (29–32) for brain protection and the reduction of stroke. Retrograde cerebral perfusion may "extend" the safe period of deep HCA (31,33). However, the maximum safe duration of RCP is still unknown. Many clinical studies suffer from serious shortcomings: (1) most trials are uncontrolled and anecdotal; (2) many reports include fewer than 50 patients; and (3) the majority of patients received RCP for only short intervals, which should have resulted in a good outcome with deep HCA alone (25,32–36). Therefore, the efficacy of RCP in brain protection is still undetermined. In particular, there are very limited data available regarding the protective effects of RCP during prolonged circulatory arrest. We have used MR imaging and spectroscopy to assess tissue perfusion, energy metabolism, and acid-base balance during prolonged RCP in a pig model (5–7).

Localized MR spectra were obtained during initial normothermic cardiopulmonary bypass, RCP, and reperfusion with CPB. Figure 5A shows spectra obtained during an experiment in which deep (15°C) hypothermic RCP was performed during deep HCA. A baseline, "normal", spectrum was recorded during initial normothermic CPB. There was a significant decrease in signals from the high-energy phosphate metabolites (ATP and PCr) and a significant increase in the signal from Pi during the first 40 minutes of deep hypothermic RCP. Complete loss of signals from ATP and PCr was observed after about 40 minutes of deep hypothermic RCP. In contrast to the results obtained from the pigs that received deep HCA alone (Figure 2C), the pigs treated with RCP during circulatory arrest showed recovery of all signals during reperfusion with normothermic CPB following RCP. After 60 minutes of reperfusion, the Pi, ATP, and PCr signal intensities returned to their baseline levels. Similarly, pHi also decreased continuously during RCP and then gradually recovered to its baseline after 60 minutes of reperfusion with normothermic CPB (Figure 3). In the pigs that received deep HCA alone, pHi remained low during 60 minutes of reperfusion with normothermic CPB. These results suggest that RCP does not provide a sufficient amount of oxygen to maintain normal energy metabolism during deep HCA, but provides additional brain protection relative to deep HCA alone as demonstrated by complete recovery of Pi, high-energy metabolites, and pHi during reperfusion following 120 minutes of RCP.

Many disavantages of deep hypothermia have been reported (1,37,38), including prolonged cardiopulmonary bypass time, which is a good predictor of postoperative death and stroke, prolonged rewarming which increases the risk of microembolism, disturbances of acid-base balance and the coagulation system, increase of blood viscosity, and decrease of oxygen dissociation. Although neuropsychological tests following cardiac surgery have shown better results in patients that were submitted to normothermic cardiopulmonary bypass (39), it is not known whether moderate hypothermia (28°C) can also be used effectively for RCP during circulatory arrest. To explore this question, pigs were subjected to 120 minutes of RCP at 28°C. Localized ^{31}P MR spectra were obtained every 30 minutes during the entire protocol (6). Representative ^{31}P MR spectra (Figure 5B) showed a significant

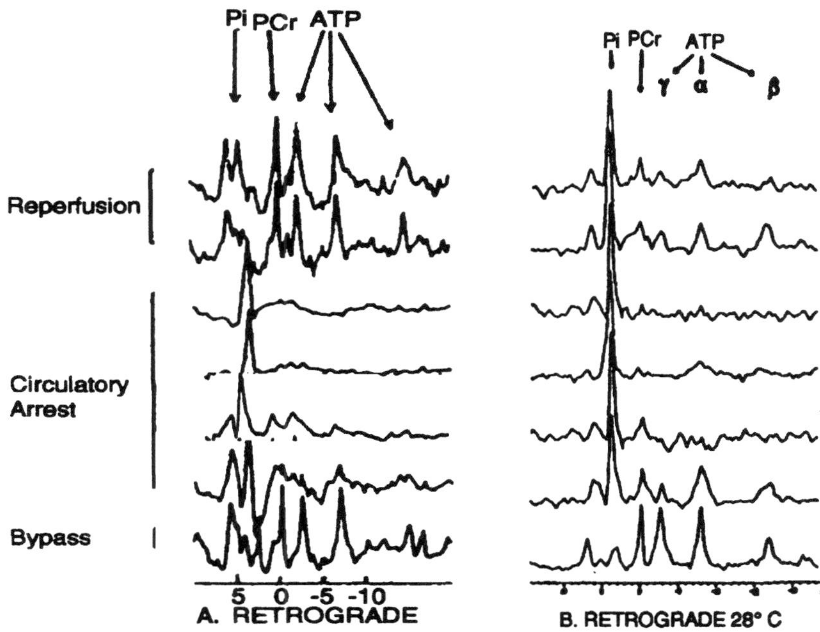

Figure 5. Representative MR spectra from a pig that received retrograde cerebral perfusion at 15°C (A) and 28°C (B).

increase in Pi signal, a decrease followed by complete loss of the ATP and PCr signals during moderate hypothermic RCP, which is similar to the spectra seen during deep hypothermic RCP. However, the Pi, ATP and PCr signals did not recover during normothermic reperfusion with CPB following moderate hypothermic RCP. In the pigs subjected to moderate hypothermic RCP, the pH_i was 7.1 ± 0.1 prior to RCP, decreased significantly to 6.4 ± 0.1 during the RCP, and remained at 6.4 ± 0.1 during the normothermic reperfusion period. These data suggest that RCP provides less brain protection during moderate hypothermia than during deep hypothermia, as indicated by the lack of recovery of energy metobolites and pHi during reperfusion. Deep hypothermia is necessary when RCP is used during HCA.

Measurements of cerebral blood flow during RCP range from 0 to 50% of baseline values (23–25,40–44). We used India ink to study cerebral perfusion in pigs and showed that only 10% of the capillaries are filled during RCP at a perfusion pressure of 20–25 mm Hg (without clamping the IVC) relative to the number with ACP (45). Magnetic resonance perfusion imaging in the same animal model showed very little blood flow to the brain during RCP and remarkably poor perfusion in many areas of the brain during reperfusion (Figure 6). The limited blood flow to the brain was mainly distributed in the midbrain. These findings help explain the MR spectroscopic findings that RCP causes ischemic metabolism and intracellular acidosis. Retrograde cerebral perfusion provides some blood to brain tissue and leads to relatively less brain injury relative to deep HCA alone. Complete recovery of Pi and PCr is seen during normothermic reperfusion after 2 hours of deep RCP, which is not observed after 2 hours of deep HCA alone. Our studies help explain the clinical findings that deep hypothermic RCP provides additional protection relative to deep HCA alone, but cannot extend the safe duration of deep HCA very significantly. This was also supported by histopathological studies. Prolonged RCP resulted in significant neuronal injury, but the injury was significantly less than that seen in the pigs that received HCA alone (Figure 4). Retrograde cerebral perfusion should be used with precaution when a prolonged period of HCA is needed during cardiac surgery.

BILATERAL ANTEGRADE CEREBRAL PERFUSION

Repair of aortic arch pathology was initially performed with antegrade perfusion of the head vessels (46), an approach that is still used (47–51). Clinical and experimental results show that conventional (bilateral) ACP through both carotid arteries can prevent ischemic brain damage during HCA and appears to be clinically useful when HCA times exceed 30–45 minutes (49,52). Kazui and co-worker suggest that it is probably safer to select ACP from the beginning, because it is often difficult to estimate pre-operatively the time required for obliteration of the distal false lumen, perform distal graft anastomosis and reconstruction of the aortic arch in patients with aortic arch dissection (53). ^{31}P MR spectroscopy and dynamic MR perfusion imaging techniques have been used to assess the effects of ACP on brain energetics/pHi and blood distribution, respectively, in an acute porcine model

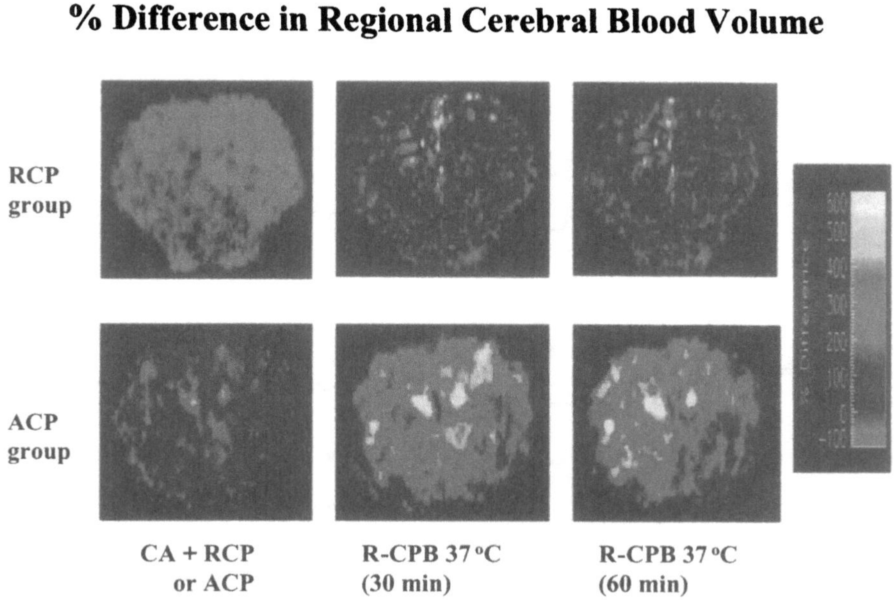

Figure 6. Representative regional cerebral blood volume (rCBV) maps showing the difference (%) between initial cardiopulmonary bypass (CPB) at 37°C and (left) circulatory arrest (CA) + retrograde cerebral perfusion (RCP) or CA + antegrade cerebral perfusion (ACP), (middle) 30 minutes of reperfusion with CPB (R-CPB) at 37°C, and (right) 60 minutes of reperfusion with CPB (R-CPB) at 37°C. A negative sign (–) indicates that rCBV is lower than during initial CPB; zero indicates that rCBV is the same as during initial CPB; a positive sign (+) indicates a rCBV that is higher than during initial CPB.

(1,2,5,7,54,55). In these studies, pigs were subjected to deep hypothermic bilateral ACP during 2 hours of deep HCA, followed by a one hour period of normothermic reperfusion with CPB.

Figure 7A consists of spectra obtained during an experiment in which deep (15°C) hypothermic bilateral ACP was performed during deep HCA. There was no significant difference between any of the spectra during the entire experimental protocol. In particular, there was no increase in the Pi signal or decrease in the signals from ATP and PCr during 120 minutes of ACP. Furthermore, pHi also remained at its baseline level during prolonged ACP (Figure 3). These findings suggest that the amount of oxygen supplied by ACP is sufficient to prevent ischemic metabolism during deep HCA.

Is deep hypothermia necessary when bilateral ACP is used during circulatory arrest? We used the same method to monitor energy metabolites and intracellular pH during moderate (28°C) hypothermic ACP in the pig model. Figure 7B shows representative spectra obtained from a pig. No ischemic events were observed during

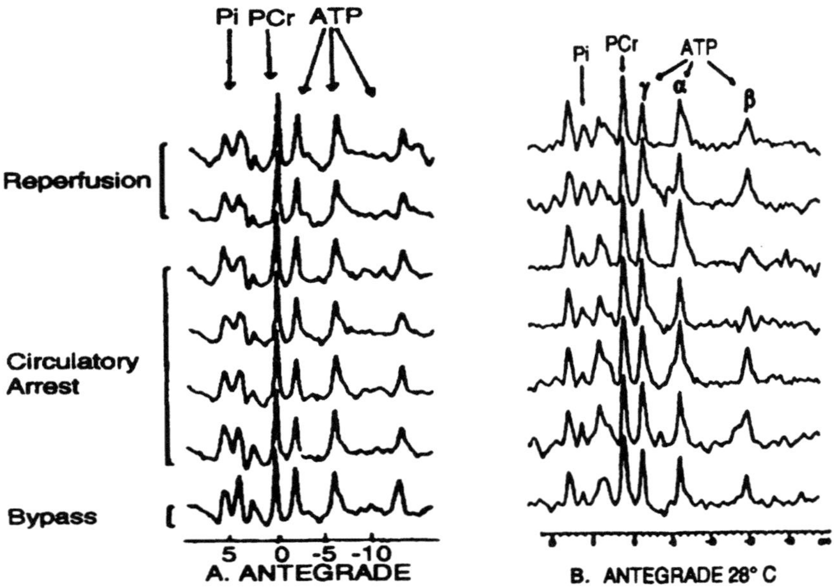

Figure 7. Representative MR spectra from pigs that received antegrade cerebral perfusion at 15°C (A) and 28°C (B).

120 minutes of moderate hypothermic ACP, as indicated by no increase in Pi and no decrease in ATP or PCr. Furthermore, there was no decrease in pHi during moderate hypothermic ACP. This suggests that deep hypothermia may not be necessary when bilateral ACP is used.

MR perfusion imaging was used in the pig to understand the blood distribution during ACP. The results showed that rCBV increased significantly in most regions of the brain during hypothermic bilateral ACP, and remained uniformly high during normothermic reperfusion. A few small regions showed slightly reduced reperfusion (Figure 6). These findings help explain how ACP maintains normal energy metabolites/pHi during either deep or moderate hypothermia.

Histopathological findings are consistent with the MR results (4,6). There was very minimal neuronal injury in pigs that received prolonged ACP at either 15°C or 28°C, which was similar to that observed in pigs subjected to anesthesia alone (Figure 4).

UNILATERAL ANTEGRADE CEREBRAL PERFUSION

Our studies have demonstrated that bilateral ACP through both carotid arteries provides uniform blood distribution and maintains normal energy metabolism in the entire brain, providing the best brain protection during heart surgery that requires circulatory arrest (4,5,7,55–57). However, ACP has decreased in popularity over the past decade. This may be due to the complexity of the technique, the fact that it is cumbersome, and the possible increased risk of cerebral embolism that results from direct cannulation of diseased cephalic arteries (49,58). Antegrade cerebral perfusion may be the optimal method of cerebral protection for aortic arch repair if its disadvantages can be eliminated. The axillary artery has been used successfully for normal CPB in patients with severe atherosclerotic disease in whom it is impossible or undesirable to place a cannula into the femoral artery (59). Several reports (59,60) have suggested that the axillary artery offers a safe and effective means of providing arterial inflow during cardiopulmonary bypass. In some centers, the axillary artery has replaced the femoral artery as the preferred alternate site for cannulation in patients with ascending aortic disease. We hypothesized that unilateral ACP through the right carotid artery via a cannula placed in the right axillary artery could provide uniform flow distribution to both hemispheres of the brain. Relative to bilateral ACP, unilateral ACP is simple (one cannulation outside the chest) and eliminates the need for direct cannulation of the arch vessels. We have used dynamic MR perfusion imaging to monitor the distribution of blood, regional cerebral blood volume (rCBV), and mean transit time (MTT) before, during and after unilateral ACP (8). In the study, pigs received 2 hours of deep (15°C) hypothermic ACP through the right carotid artery (unilateral ACP) during deep HCA, followed by a 1 hour period of normothermic reperfusion with CPB.

Uniform flow distribution to the entire brain was observed in all pigs during initial nomorthermic CPB. Regional cerebral blood volume increased in most regions of the brain during unilateral ACP, and remained high 60 minutes after normothermic reperfusion with CPB. Only a few small areas showed reduced rCBV (Figure 8). No significant difference in rCBV was observed between the left and right hemispheres of the brain during unilateral ACP or at 60 minutes of reperfusion with normal CPB. These data indicate that both hemispheres received the same amount of blood during unilateral ACP.

During nomorthermic CPB, the mean transit time was 7–10 seconds throughout the brain (Figure 8). The mean transit time was significantly prolonged in all brain regions during ACP at 15°C (2–3 times longer than the baseline level obtained during initial normothermic CPB at 37°C), indicating a significant decrease in the speed of blood passing through brain tissue. A similar prolongation of MTT was also observed during CPB at 15°C. Mean transit time recovered to its baseline level in both groups during normothermic reperfusion with CPB. There was no significant difference in MTT between the left and right hemispheres either during unilateral ACP or reperfusion with CPB.

The study clearly demonstrates that unilateral ACP through the right axillary artery provides equal and uniform flow distribution to both hemispheres of the brain. During unilateral ACP, the speed of blood passing through the brain tissue is also similar in both hemispheres of the brain. These findings indicate that during unilateral ACP, the left hemisphere receives an amount of blood similar to that of the right hemisphere through collateral circulation, and through the circle of Willis

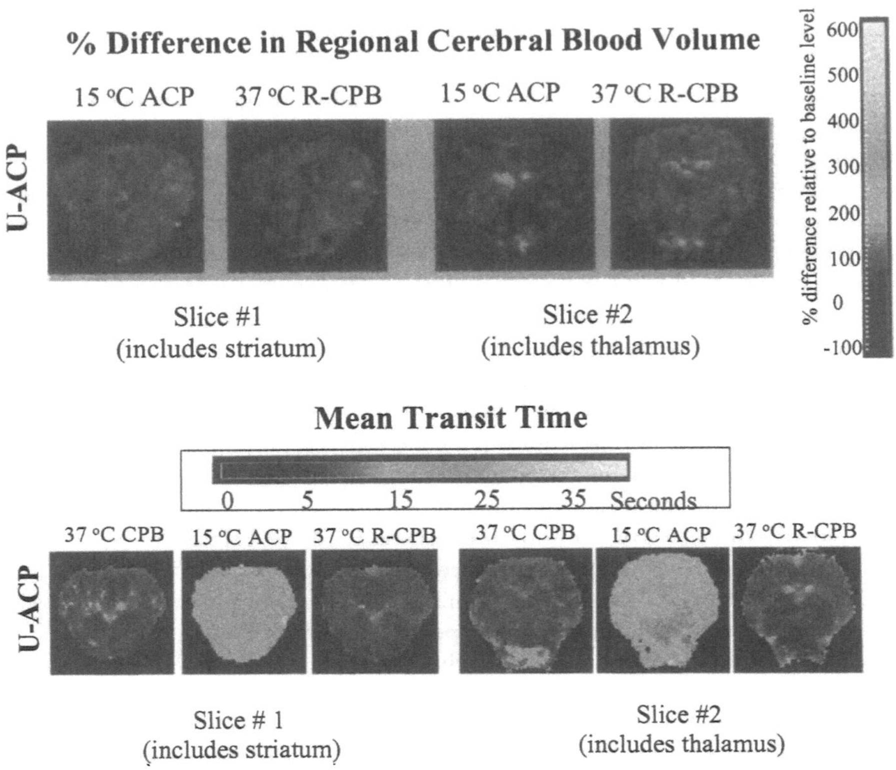

Figure 8. Upper panel: Representative regional cerebral blood volume (rCBV) maps showing the difference (%) between initial cardiopulmonary bypass (CPB) at 37°C and unilateral antegrade cerebral perfusion (U-ACP) at 15°C, and 60 minutes of reperfusion with CPB (R-CPB) at 37°C. A negative sign (–) indicates that rCBV is lower than during initial CPB; zero indicates that rCBV is the same as during initial CPB; a positive sign (+) indicates that rCBV is higher than during initial CPB. Lower panel: Representative mean transit time (MTT) maps obtained during initial normothermic cardiopulmonary bypass (37°C CPB), unilateral antegrade cerebral perfusion at 15°C (15°C ACP), and at 60 minutes of reperfusion with normothermic cardiopulmonary bypass (37°C R-CPB).

in particular. However, it was not known whether the amount of blood supplied to both hemispheres of the brain is sufficient to prevent ischemic events during prolonged deep hypothermic ACP. Therefore, we used localized ^{31}P MR spectroscopy to monitor energy metabolism and pHi in both hemispheres of the brain during deep (15°C) hypothermic unilateral ACP (61).

Representative localized MR spectra obtained from each hemisphere of a pig brain that received deep hypothermic unilateral ACP are shown in Figure 9. The spectra from the left and right hemispheres changed similarly throughout the protocol. As shown in Figure 10, all peak areas (PCr, and ATP) increased during CPB

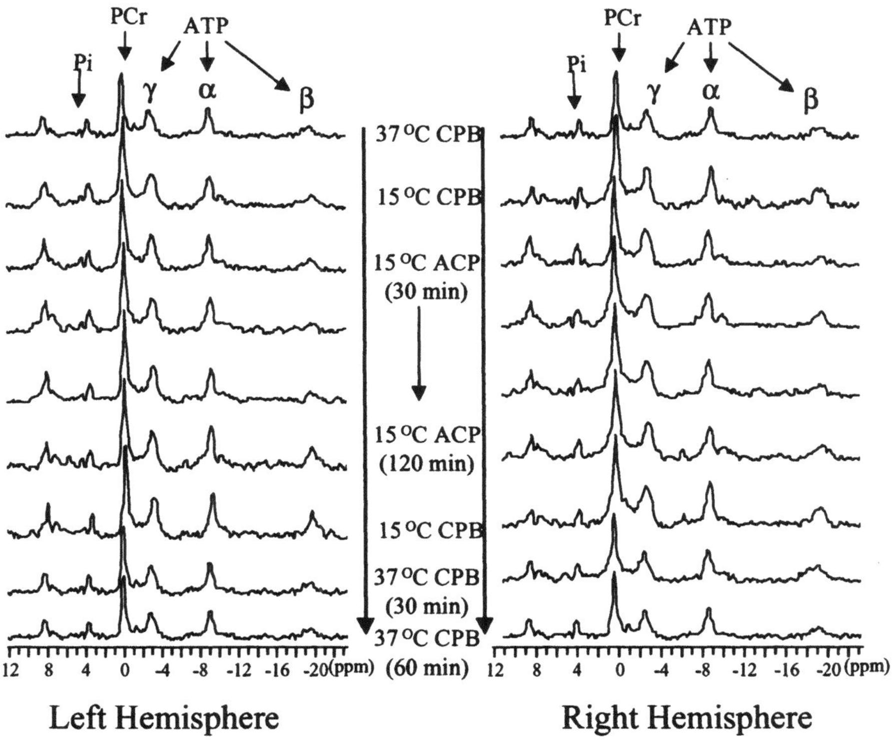

Figure 9. Representative spectra obtained from the left and right hemispheres of a pig brain at each stage of the protocol. Pi = inorganic phosphorus, PCr = phosphocreatine, ATP = adenosine triphosphate.

at 15°C relative to their baselines and remained constant during 120 minutes of unilateral ACP and reperfusion with CPB at 15°C. The peak areas returned to their baseline levels when the pig was rewarmed to 37°C. When the peak areas obtained during initial CPB at 15°C were used as baseline levels to eliminate possible effects of low temperature on the spectra (such as changes in RF coil loading), no significant decrease was observed in PCr or ATP in either hemisphere during 120 minutes of unilateral ACP, indicating no ischemic metabolism during unilateral ACP at deep hypothermia. No significant differences in PCr or ATP were observed between the left and right hemispheres during unilateral ACP and reperfusion. During the initial period of CPB at 37°C, the baseline pHi was 7.12 ± 0.14 and 7.08 ± 0.08 in the right and left hemispheres, respectively. A significant increase in intracellular pH was observed during deep hypothermic CPB and unilateral ACP. No significant decrease in pHi was seen in either hemisphere during 120 minutes of unilateral ACP relative to the pHi observed during initial CPB at 15°C, indicating no

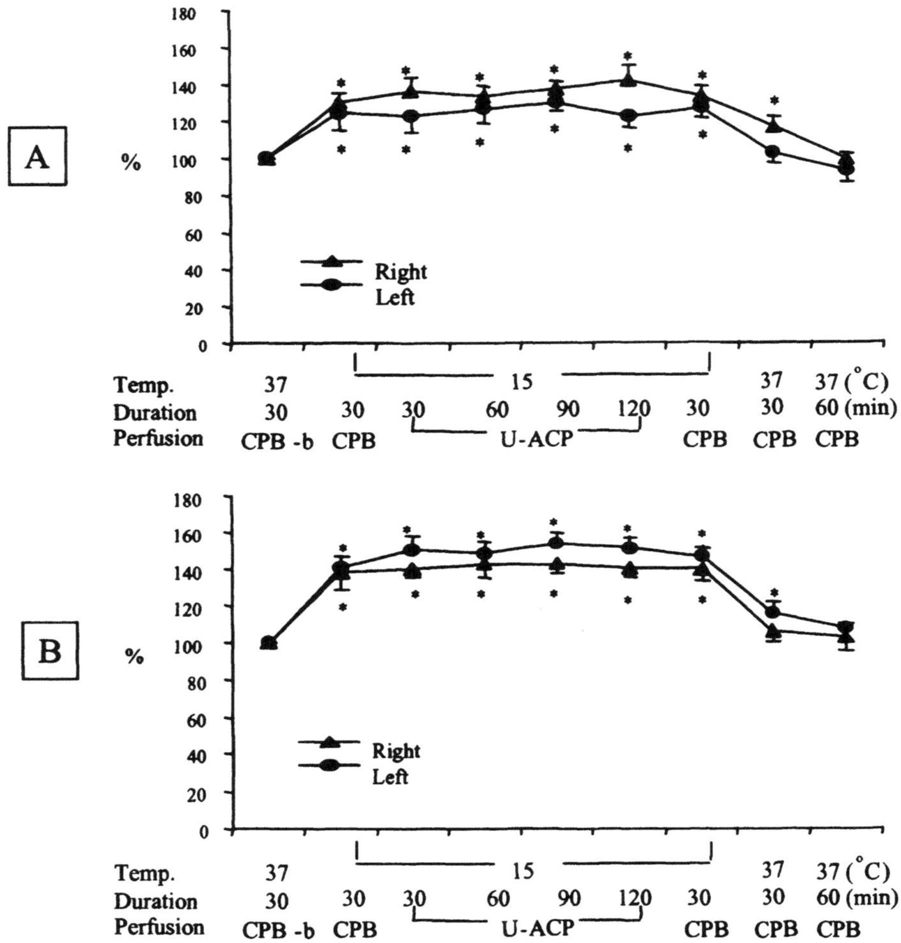

Figure 10. Changes in phosphocreatine (PCr) (A) and adenosine triphosphate (ATP) (B) in a pig brain that received unilateral antegrade cerebral perfusion at 15°C. The levels obtained during initial normothermic cardiopulmonary bypass (CPB) were used as the baseline (CPB-b = 100%). $^{\star}P < 0.05$ vs the baseline level (CPB-b). There was no statistically significant difference between the right and left hemispheres.

ischemia-induced intracellular acidosis. There were no significant differences in pHi between the right and left hemispheres during unilateral ACP and reperfusion with CPB (Figure 11).

Localized ^{31}P MR spectroscopic imaging also confirmed that there was no decrease in ATP or PCr, nor any increase in intracellular acidosis in either hemisphere of the brain during 120 minutes of moderate (28°C) hypothermic unilateral

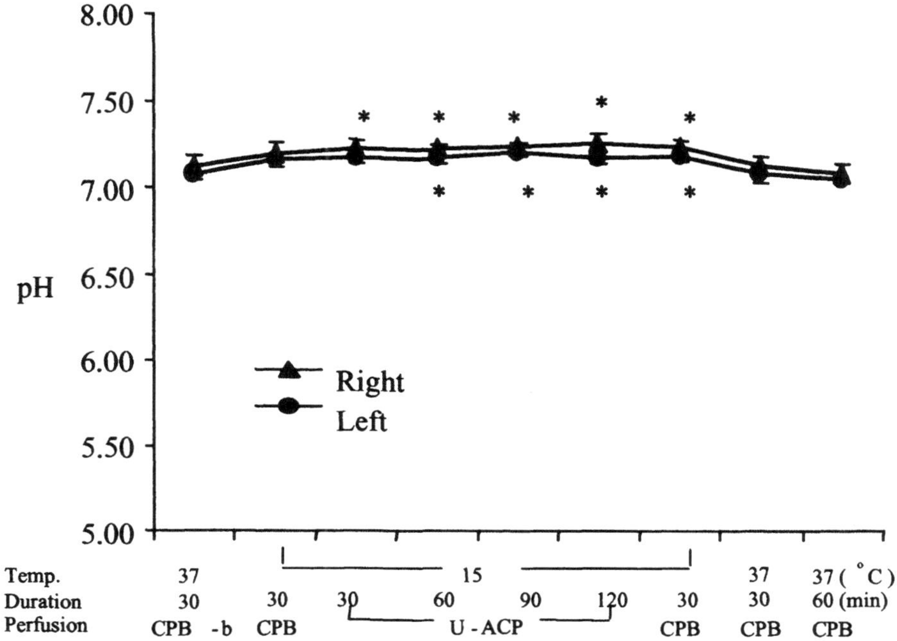

Figure 11. Changes in intracellular pH (pHi) in a pig brain that received unilateral antegrade cerebral perfusion at 15°C. *$P < 0.05$ vs the baseline level (CPB-b) obtained during initial normothermic cardiopulmonary bypass (CPB). There was no statistically significant difference between the right and left hemispheres.

ACP. There was no difference between the right and left hemispheres of the brain in spectra obtained during moderate hypothermic unilateral ACP (61). These findings are similar to those seen in pigs that received deep (15°) hypothermic unilateral ACP, indicating deep hypothermia may not be necessary when unilateral ACP is used.

Histopathological studies confirmed there were no significant morphological changes in the neurons of the right and left hemispheres of the brain subjected to either deep or moderate hypothermic unilateral ACP (61).

CONCLUSION

Magnetic resonance is a technology that allows repetitive, non-invasive in vivo measurements in both animals and humans. The combination of spectroscopic and imaging techniques provides comprehensive information on the biochemical and physiological status of the brain in large animal models of cardiac surgery. This allows better choices and design of methods to protect the brain during cardiac surgery. Our studies with MR imaging and spectroscopic imaging, combined with other techniques, have demonstrated that: (1) prolonged deep HCA alone (2 hours) results in severe brain injury; (2) RCP provides a limited amount of blood flow to the brain, does not provide complete brain protection during prolonged deep HCA (2 hours), but provides better brain protection than deep HCA alone; (3) bilateral ACP provides uniform flow distribution throughout the brain and prevents ischemic/hypoxic injury to the brain during 2 hours of deep or moderate HCA; (4) unilateral ACP through the right axillary artery provides a similar amount of blood to both the right and left hemispheres of the brain and prevents ischemic/hypoxic injury to the brain during prolonged HCA at either deep or moderate hypothermia; and (5) temperature is a critical factor when RCP is used during circulatory arrest. Therefore, we conclude that: (1) ACP provides the best brain protection; (2) RCP should be used cautiously during prolonged HCA; (3) deep hypothermia is necessary when RCP is used, but not when ACP is used; and (4) in certain clinical situations, unilateral ACP through the right axillary artery may be the optimal choice for brain protection during prolonged HCA. These findings need to be confirmed by clinical studies.

REFERENCES

1. Svensson LG, Crawford ES, Hess KR, Coselli JS, Raskin S, Shenaq SA, Safi HJ. 1993. Deep hypothermia with circulatory arrest. Determinants of stroke and early mortaliy in 656 patients. J Thorac Cardiovasc Surg 106:19–28.
2. Fessatidis IT, Thomas VL, Shore DF, Sedgwick ME, Hunt RH, Weller RO. 1993. Brain damage after profoundly hypothermic circulatory arrest: Correlation between neurophysiologic and neuropathologic findings. An experimental study in vertebrates. J Thorac Cardiovasc Surg 106:32–41.
3. Griepp EB, Griepp RB. 1992. Cerebral consequences of hypothermic circulatory arrest in adults. J Cardiac Surg 7:134–155.
4. Ye J, Yang L, Del Bigio MR, Filgueiras CL, Ede M, Summers R, Salerno TA. 1996. Neuronal damage after hypothermic circulatory arrest and retrograde cerebral perfusion in the pig. Ann Thorac Surg 61:1316–1322.
5. Filgueiras CL, Winsborrow B, Ye J, Scott J, Aronov A, Kozlowski P, Shabnavard L, Summers R, Saunders JK, Deslauriers R. 1995. A ^{31}P-magnetic resonance study of antegrade and retrograde cerebral perfusion during aortic arch surgery in pigs. J Thorac Cardiovasc Surg 110:55–62.
6. Filgueiras CL, Ryner L, Ye J, Yang L, Ede M, Sun J, Kozlowski P, Summers R, Saunders JK, Salerno TA, Deslauriers R. 1996. Cerebral protection during moderate hypothermic circulatory arrest: Histopathology and magnetic resonance spectroscopy of brain energetics and intracellular pH in pigs. J Thorac Cardiovasc Surg 112:1073–1080.
7. Ye J, Ryner L, Kozlowski P, Yang L, Del Bigio MR, Sun J, Donnelly M, Summers R, Salerno TA, Somorjai RL, Saunders JK, Deslauriers R. 1998. Retrograde cerebral perfusion results in flow distribution abnormalities and neuronal damage: a magnetic resonance imaging and histopathology study in pigs. Circulation 98:II-313–II-318.
8. Ye J, Dai G, Ryner LN, Kozlowski P, Yang L, Summers R, Sun J, Salerno TA, Somorjai RL, Deslauriers R. 1999. Unilateral antegrade cerebral perfusion through the right axillary artery provides uniform flow distribution to both hemispheres in the brain: A magnetic resonance and histopathological study in pigs. Circulation 100(suppl II):II-309–II-315.
9. Kost GJ. 1990. pH Standardization for phosphorus-31 magnetic resonance heart spectroscopy at different temperatures. Magn Reson Med 14:496–506.
10. Gadian DG. 1995. NMR and its applications to living systems, 42–45. New York: Oxford University Press.
11. Ye J, Sun JK, Shen J, Gregorash L, Summers R, Salerno TA, Deslauriers R. 1997. Does retrograde warm blood cardioplegia provide equal protection to both ventricles? A magnetic resonance spectroscopy study in pigs. Circulation 96(suppl II):II-210–II-215.

12. Winsborrow BG, Kozlowski P, Filgueiras CL, Shabnavard L, Ye J, Aronov A, Scott J, Salerno TA, Saunders JK, Deslauriers R. 1994. ^{31}P spectroscopic imaging of pig brain during circulatory arrest. Can J Appl Spectrosc 32:135–142.
13. Somorjai RL, Morrow R, Deslauriers R. 1991. Quantification of in vivo NMR spectra: A robust, interactive approach. Proc 10th Annual Meeting of the Society of Magnetic Resonance for Medicine 1219.
14. Rosen BR, Belliveau JW, Vevea JM, Brady TJ. 1990. Perfusion imaging with NMR contrast agents. Magn Reson Med 14:249–265.
15. Tomanek B, Ryner LN, Hoult DI, Kozlowski P, Saunders JK. 1997. Dual surface coil with high-B1 homogeneity for deep organ MR imaging. Magn Reson Imaging 15:1199–1204.
16. Mault JR, Ohtake S, Klingensmith ME, Heinle JS, Greeley WJ, Ungerleider RM. 1993. Cerebral metabolism and circulatory arrest: Effects of duration and strategies for protection. Ann Thorac Surg 55:57–64.
17. Fisk GC, Wright JS, Hicks RG, Anderson RM, Turner BB, Baker WC, Lawrence LC, Stacey RB, Lawrie GM, Kalnins I, Rose M. 1976. The influence of duration of circulatory arrest at 20°C on cerebral changes. Anaesth Intensive Care 4:126–134.
18. Griepp RB, Stinson EB, Hollingworth JF, Buehler D. 1975. Prosthetic replacement of the aortic arch. J Thorac Cardiovasc Surg 70:1051–1063.
19. Ergin MA, Uysal S, Reich DL, Apaydin A, Lansman SL, McCullough JN, Griepp RB. 1999. Temporary neurological dysfunction after deep hypothermic circulatory arrest: a clinical marker of long-term functional deficit. Ann Thorac Surg 67:1887–1890.
20. Lemole GM, Strong MD, Spagna PM, Karmilowicz NP. 1982. Improved results for dissecting aneurysms, intraluminal sutureless prosthesis. J Thorac Cardiovasc Surg 83:249–255.
21. Ueda Y, Miki S, Kusuhara K, Okita Y, Tahata T, Yamanaka K. 1990. Surgical treatment of aneurysm or dissection involving the ascending aorta and aortic arch, utilizing circulatory arrest and retrograde cerebral perfusion. J Cardiovasc Surg Torino 31:553–558.
22. Buket S, Alayunt A, Discigil B, Apaydin A, Yuksel M, Durmaz I. 1995. Continuous retrograde cerebral perfusion supplies substrates for brain metabolism during hypothermic circulatory arrest. Perfusion 10:237–244.
23. Usui A, Oohara K, Liu TL, Murase M, Tanaka M, Takeuchi E, Abe T. 1994. Determination of optimum retrograde cerebral perfusion conditons. J Thorac Cardiovasc Surg 107:300–308.
24. Usui A, Hotta T, Hiroura M, Murase M, Maeda M, Koyama T, Tanaka M, Takeuchi E, Yasuura K, Watanabe T. 1992. Retrograde cerebral perfusion through a superior vena caval cannula protects the brain. Ann Thorac Surg 53:47–53.
25. Deeb GM, Jenkins E, Bolling SF, Brunsting LA, Williams DM, Quint LE, Deeb ND. 1995. Retrograde cerebral perfusion during hypothermic circulatory arrest reduces neurologic morbidity. J Thorac Cardiovasc Surg 109:159–168.
26. Watanabe T, Iijima Y, Abe K, Abe H, Saito H, Naruke Y, Washio M. 1996. Retrograde brain perfusion beyond the venous valves. J Thorac Cardiovasc Surg 111:36–44.
27. Safi H, Brien HW, Winter JN, Thomas AC, Maulsby RL, Doerr HK, Svensson LG. 1993. Brain protection via cerebral retrograde perfusion during aortic arch aneurysm repair. Ann Thorac Surg 56: 270–276.
28. Ergin MA, Galla JD, Lansman SL, Quintana C, Griepp RB. 1994. Hypothermic circulatory arrest in operations on the thoracic aorta. Determinants of operative mortality and neurologic outcome. J Thorac Cardiovasc Surg 107:788–799.
29. Coselli JS, Lemaire SA. 1997. Experience with retrograde cerebral perfusion during proximal aortic surgery in 290 patients. J Card Surg 12:322–325.
30. Cheung AT, Bavaria JE, Weiss SJ, Patterson T, Stecker MM. 1998. Neurophysiologic effects of retrograde cerebral perfusion used for aortic reconstruction. J Cardiothorac Vasc Anesth 12:252–259.
31. Usui A, Abe T, Murase M, Takeuchi E, Ishihara T, Hoshino M, Ogawa Y, Seki A, Okamoto H, Moriya H. 1997. Early experience of retrograde cerebral perfusion. Cardiovasc Surg 5:510–515.
32. Kitamura M, Hashimoto A, Akimoto T, Tagusari O, Aomi S, Koyanagi H. 1995. Operation for type A dissection: introduction of retrograde cerebral perfusion. Ann Thorac Surg 59:1195–1199.
33. Sasaguri S, Yamamoto S, Hosoda Y. 1996. What is the safe time limit for retrograde cerebral perfusion with hypothermic circulatory arrest in aortic surgery? J Cardiovasc Surg 37:441–444.
34. Bavaria JE, Woo YJ, Hall RA, Carpenter JP, Gardner TJ. 1995. Retrograde cerebral and distal aortic perfusion during ascending and thoracoabdominal aortic operations. Ann Thorac Surg 60:345–353.

35. Lytle BW, McCarthy PM, Meaney KM, Stewart RW, Cosgrove KM. 1995. Systemic hypothermia and circulatory arrest combined with arterial perfusion of the superior vena cava: effective intraoperative cerebral protection. J Thorac Cardiovasc Surg 109:738–743.
36. Pagano D, Carey JA, Patel RL, Allen SM, Tsang GM, Hutton P, Lilley JP, Farogui MH, Bonser RS. 1995. Retrograde cerebral perfusion: clinical experience in emergency and elective aortic operations. Ann Thorac Surg 59:393–397.
37. Henriksen L, Hjelms E, Linderburgh T. 1983. Brain hyperperfusion during cardiac operations. J Thorac Cardiovasc Surg 86:202–208.
38. Hickey PR, Andersen NP. 1987. Deep hypothermic circulatory arrest: a review of pathophysiology and clinical experience as a basis for anesthetic management. J Cardiothorac Anesth 1:137–155.
39. Wong BES, McLean RF, Naylor CD, Snow WG, Harrington EM, Gawel MJ, Woods RB, Fremes SE. 1992. Central nervous system dysfunction after warm or hypothermic cardiopulmonary bypass. Lancet 339:1383.
40. Boeckxstaens CJ, Flameng WJ. 1995. Retrograde cerebral perfusion does not perfuse the brain in nonhuman primates. Ann Thorac Surg 60:319–328.
41. De Brux JL, Subayi JB, Pegis JD, Pillet J. 1995. Retrograde cerebral perfusion: Anatomic study of the distribution of blood to the brain. Ann Thorac Surg 60:1294–1298.
42. Nojima T, Magara T, Nakajima Y, Waterida S, Onoe M, Surgita T, Mori A. 1994. Optimal perfusion pressure for experimental retrograde cerebral perfusion. J Card Surg 9:548–559.
43. Oohara K, Usui A, Murase M, Tanaka M, Abe T. 1995. Regional cerebral tissue blood flow measured by the colored microsphere method during retrograde cerebral perfusion. J Thorac Cardiovasc Surg 109:772–779.
44. Lin PJ, Chang C, Tan PPC, Chang CN, Lee ST, Wang CC, Chang JP, Liu DW, Chu JJ, Tsai KT, Kao CL, Hsieh MJ, Hua MS. 1996. Prolonged circulatory arrest in moderate hypothermia with retrograde cerebral perfusion. Is brain ischemic? Circulation 94(suppl II):II-169–II-172.
45. Ye J, Yang L, Del Bigio MR, Summers R, Jackson K, Somorjai RL, Salerno TA, Deslauriers R. 1997. Retrograde cerebral perfusion provides limited distribution of blood to the brain: A study in pigs. J Thorac Cardiovasc Surg 114:660–665.
46. DeBakey ME, Cooley DA, Crawford ES, Morris GC Jr. 1957. Successful resection of fusiform aneurysm of aortic arch with replacement by hemograft. Surg Gynecol Obstet 105:656–664.
47. Tabayashi K, Niibori K, Iguchi A, Shoji Y, Ohmi M, Mohri H. 1993. Replacement of the transverse aortic arch for type A acute aortic dissection. Ann Thorac Surg 55:864–867.
48. Biglioli P, Spirito R, Agrifoglio M, Parolari A, Pompilia G, Alamanni F. 1993. Two cases of staged replacement of the thoracic aorta using the "elephant trunk" technique. Cardiovasc Surg 1:64–66.
49. Alamanni F, Agrifoglio M, Pompilio G, Spirito R, Sala A, Arena V, Roberto M, Biglioli P. 1995. Aortic arch surgery: pros and cons of selective cerebral perfusion. J Cardiovasc Surg 36:31–37.
50. Bachet J, Teodori G, Goudot B, Diaz F, el Kerdany A, Dubois C, Brodaty D, de Lentdecker D, Guilment D. 1988. Replacement of the transverse aortic arch dissection. Report of 26 cases. J Thorac Cardiovasc Surg 96:878–886.
51. Tabayashi K, Ohmi M, Togo T, Miura M, Yokoyama H, Akimoto H, Murata S, Ohsaka K, Mohri H. 1994. Aortic arch aneurysm repair using selective cerebral perfusion. Ann Thorac Surg 57:1305–1310.
52. Jezzard P, Heineman F, Taylor J, DesPres D, Wen H, Balaban RS, Turner R. 1994. Comparison of EPI gradient-echo contrast changes in cat brain caused by respiratory challenges with direct simultaneous evaluation of cerebral oxygenation via a cranial window. NMR Biomed 7:35–44.
53. Kazui T, Inoue N, Yamada O, Komatsu S. 1992. Selective cerebral perfusion during operation for aneurysms of the aortic arch: a reassessment. Ann Thorac Surg 53:109–114.
54. Deslauriers R, Filgueiras CL, Ye J, Yang L, Ede M, Summers R, Salerno TA, Ryner L, Kozlowski P, Winsborrow B, Sun JK, Saunders JK. 1995. Magnetic resonance in the development of better strategies for protecting the brain during surgery. Bull Magn Reson 17:79–86.
55. Filgueiras EL, Winsborrow B, Ye J, Aronov A, Kozlowski P, Saunders JK, Deslauriers R, Salerno TA. 1994. Brain protection during circulatory arrest. Surg Forum 45:263–265.
56. Ye J, Yang L, Del Bigio MR, Summers R, Salerno TA, Deslauriers R. 1997. The effect of circulatory arrest and retrograde cerebral perfusion on microtubule-associated protein 2: An immunohistochemcal study in pig hippocampus. Neurosci Lett 222:9–12.
57. Ye J, Yang L, Del Bigio MR, Summers R, Jackson D, Somorjai RL, Salerno TA, Deslauriers R. 1997. Retrograde cerebral perfusion provides limited distribution of blood to the brain: A study in pigs. J Thorac Cardiovasc Surg 114:660–665.

58. Bachet J, Goudot B, Teodori G, Brodaty D, Dubois C, De Lentdecker P, Guilmet D. 1990. Surgery of type A acute aortic dissection with Gelatine-Resorcine-Formol biological glue: a twelve-year experience. J Cardiovasc Surg 31:263–273.
59. Sabik JF, Lytle BW, McCarthy PM, Cosgrove DM. 1995. Axillary artery: an alternative site of arterial cannulation for patients with extensive aortic and peripheral vascular disease. J Thorac Cardiovasc Surg 109:885–891.
60. Baribeau YR, Westbrook BM, Charlesworth DC, Maloney CT. 1998. Arterial inflow via an axillary artery graft for the severely atheromatous aorta. Ann Thorac Surg 66:33–37.
61. Ye J, Dai G, Yang L, Sun J, Summers R, Deslauriers R. 2001. Is deep hypothermia necessary for unilateral antegrade cerebral perfusion during circulatory arrest? A magnetic resonance study in a pig model. Cardiovasc Surg 9:600–607.

Pathophysiology of Cardiovascular Disease,
edited by N.S. Dhalla, H. Rupp,
A. Angel & G.N. Pierce
Kluwer Academic Publishers, Boston, 2004

Heart Failure Recovery with Mechanical Circulatory Support

Tofy Mussivand,[1] Paul Heerdt,[2] Kenneth B. Margulies,[3]
O. Howard Frazier,[4] Gurmeet Singh,[5] and Hideo A. Baba[6]

[1] *University of Ottawa Heart Institute, Ottawa, Canada*
[2] *New York Presbyterian Hospital, New York, USA*
[3] *Temple University School of Medicine, Philadelphia, USA*
[4] *Texas Heart Institute, Houston, USA*
[5] *University of Manitoba Health Sciences Centre, Winnipeg, Canada*
[6] *Institute of Pathology, University of Essen, Essen, Germany*

Summary. An Expert Panel was convened in the Winnipeg 2001 Heart Congress to examine heart failure recovery mechanisms after mechanical circulatory support. The Expert Panel consisted of leading researchers, and clinicians knowledgeable in the use of mechanical circulatory support devices, and focused on heart failure recovery mechanisms at the gene, cell, organ and patient levels after prolonged mechanical circulatory support. Views of each panelist, about the circulatory support, was presented in this report. Clearly, the use of mechanical circulatory devices have not only saved many individual lives but the approach also seems to modulate or even reverse heart failure processes at all levels of body organization.

***Keys words*:** Bridge to Recovery, Reverse Remodeling, Mechanical Circulatory Support.

Tofy Mussivand (Director, Cardiovascular Devices Division, University of Ottawa Heart Institute, Ottawa, Canada) organized the Expert Panel and provided the topic introduction. Clinical experimentation with mechanical circulatory support (MCS) devices started in the 1960's, with the main objective being to determine if they were suitable for supporting and/or replacing a diseased natural heart. Since that time, they have been utilized quite successfully in over 11,000 reported clinical cases. Today mechanical circulatory support devices are used typically for short

Address for Correspondence: Tofy Mussivand, FRSC, Professor of Surgery & Engineering, University of Ottawa, & University of Ottawa Heart Institute, 40 Ruskin Street, Rm. H560B, Ottawa, Ontario, Canada K1Y 4W7. Tel: (613) 761-4545, Fax: (613) 761-5448, E-mail: tofym@ottawaheart.ca.

term post-cardiotomy or as a bridge to cardiac transplantation. In addition, their use as a definitive treatment in destination therapy and bridge to recovery applications is emerging as an important clinical option in the treatment of late stage heart failure.

The concept of the use of mechanical assist devices as a treatment for heart failure is relatively recent (1). In the late eighteenth century, healing of the heart was largely considered impossible. By the mid-twentieth century, bed rest was considered the best available remedy for heart failure (2). Today, evidence of improvement of the diseased natural heart after prolonged left ventricular unloading via mechanical circulatory support is clearly documented in the literature (3). Unfortunately the exact pathophysiology of this phenomenon is not clearly understood. The Expert Panel was convened to further elucidate these findings and was structured to examine heart failure recovery after mechanical circulatory support: 1) at the gene level, 2) at the myocyte level, 3) at the organ level, and 4) at the patient level. Mechanical circulatory systems currently available for use, as well as various future technologies were also reviewed.

RECOVERY AFTER MECHANICAL CIRCULATORY SUPPORT AT THE GENE LEVEL

Paul Heerdt (Anaesthesiology / Critical Care Medicine Director, Thoracic Anaesthesia Service, New York Presbyterian Hospital, New York, USA) spoke on work by the group at Columbia related to addressing the question, "Does improved intrinsic myocardial contractile function contribute to recovery after circulatory support ?"

While heart failure had previously been thought to be unrecoverable, reports emerged showing evidence of both echocardiographic and hemodynamic improvement after circulatory support (4,5). A study was designed to determine whether hemodynamic unloading leads to normalized expression of genes involved in intracellular calcium handling.

To allow matched controls, tissue samples from the apical core taken during implantation of a left ventricular assist device (LVAD) and those taken from the left ventricular free wall from the same patient at the time of subsequent transplantation were utilized. In these studies, mRNA expression involved in intracellular calcium cycling normalized after prolonged left ventricular unloading with the LVAD. Sarcoplasmic endoreticular calcium ATPase (SERCA2a), ryanodine-sensitive Ca^{2+} release channel (RyR) and Na:Ca exchanger (NCX) relative Ca^{2+} expressions all tended to improve after LVAD support. The SERCA2a expression also showed clear time dependence, with significant improvements leading to near normal levels (compared to non-failing hearts) occurring over the first 35–40 days of LVAD support. This suggested that the majority of the initial reverse remodelling may occur early during LVAD support.

While it was determined that mRNA expression was clearly increased for SERCA2a, RyR and NCX, western blot analysis for individual patients (pre and post LVAD) showed that the increase in expression only translated to a detectable change in proteins for SERCA2a, but not for RyR or NCX. To assess whether this also led to actual alterations in SERCA2a function, SERCA2a mediated uptake of calcium into sarcoplasmic reticular (SR) membranes was assessed. While it has been previously demonstrated that there is a depressed calcium uptake in heart failure patients, those on the LVAD in this study clearly showed normalized uptake levels. This demonstrates that in addition to an increased mRNA expression, and increased protein levels, that there is also an increased SERCA2a function in those patients supported by LVADs (6).

In regard to contractile function, it is well established that one of the markers of muscle dysfunction in heart failure is a negative force-frequency relationship (FFR). Left ventricular trabeculae from the free wall were taken from the same hearts (pre and post LVAD) and the FFR was examined. These results showed that prior to LVAD support the hearts had a negative FFR as typically seen in heart failure, whereas after LVAD support the FFR was positive, as is typically seen in healthy heart muscle. Since the FFR has been linked primarily to SERCA2a, this suggests the LVAD support has in fact resulted in stabilizing and returning to normal the SERCA2a function, as observed by the mRNA changes, protein changes, the uptake of calcium in the isolated SR and the altered FFR.

To assess whether the observed changes in gene expression were related to hemodynamic unloading or to changes in the neurohormonal milieu, properties of the unloaded left ventricle (LV) were compared to those of the continually loaded right ventricle (RV) in the same heart during LVAD support (7). Normalization of SERCA2a protein expression, FFR, and stimulation frequency on force development is observed only in the LV. However, the hormone drive (isoproterenol response) normalizes in both LV and RV. These results imply that both the reduction of hemodynamic load and the neurohormonal factors are linked to recovery after LVAD support.

The need to identify the pivotal molecular mechanisms that lead to these structural and functional improvements, now becomes even more important as they may play an important role in finding new therapeutic approaches to treating heart failure.

RECOVERY AFTER MECHANICAL CIRCULATORY SUPPORT AT THE MYOCYTE LEVEL

Kenneth Margulies (Cardiology Research Director, Temple University School of Medicine, Philadelphia, USA) presented on his group's work on the study of myocyte function and structure after circulatory support. Dr. Margulies first reviewed the phenotype of advanced dilated cardiomyopathy (DCM), outlining the manifestation at the myocyte level as shown in Figure 1.

Myocyte geometry is significantly altered in advanced heart failure both in size and shape. However after mechanical circulatory support, the cardiomyocyte size regresses towards the non-failing state in terms of volume, length, and width. In addition, the distorted shape of the failing myocytes also return to normal (8). As shown in Figure 2, post-LVAD myocytes are both less flattened with a length to height ratio of 14.2 (compared to 21.0 pre-LVAD) and a width to height ratio of

Manifestation Dilated Cardiomyopathy at the Myocyte Level

- Increased myocyte size
- Abnormal cell shape (elongation, flattening, branching)
- Impaired cellular shortening
- Slowed cellular relaxation
- Action potential prolongation
- Impaired β-adrenergic responsiveness
- Negative force-frequency relation
- Increased rates of apoptosis
- Deregulation of key signaling pathways
- Pathologic patterns of protein expression

(Impaired cellular shortening and Slowed cellular relaxation: *Linked to abnormal calcium handling*)

Figure 1. Phenotype of Advanced Dilated Cardiomyopathy.

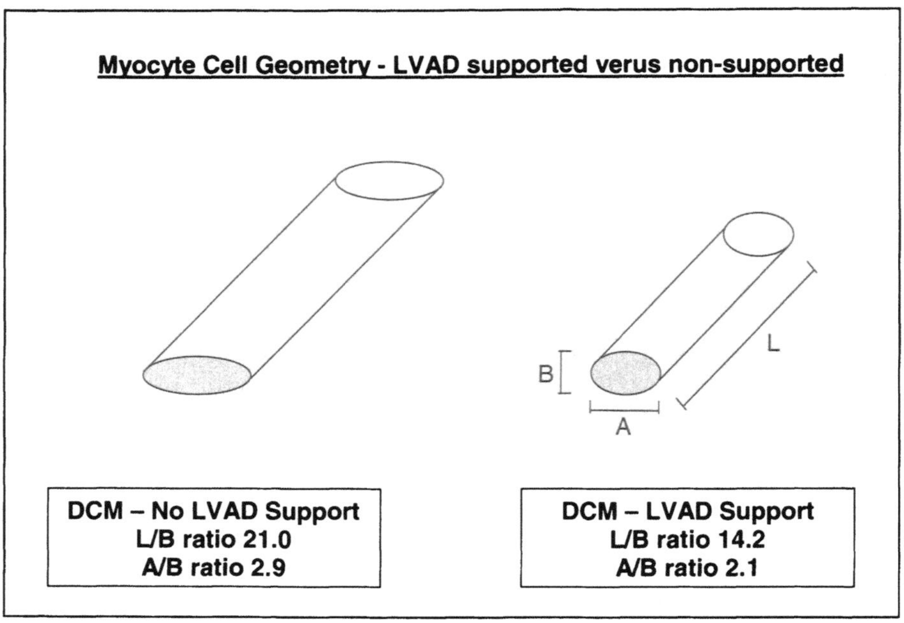

Figure 2. Changes in Myocyte Geometry of the Failing Heart.

2.1 (compared to 2.9 pre-LVAD) and tends to return to the more cylindrical shape of normal non-failing cell (8). At the same time, a decrease of the LV mass can be directly correlated to a decrease in myocyte diameter (9). These changes in cell geometry after prolonged left ventricular unloading raises the question of how these changes influences cell function.

Improvements in myocyte cell function after mechanical assist device support have also been demonstrated. Parameters of contractility, including rates and magnitudes of fractional shortening, time to peak contraction, and time to relaxation, all of which are impaired in the failing heart, have shown significant improvements after LVAD support (10). Studies were also conducted to assess responsiveness to β-adrenergic stimulation in individual myocytes which is substantially depressed during heart failure. These studies demonstrated a doubling of responsiveness to β-adrenergic stimulation in the LVAD supported myocytes (10).

Decreased rates of calcium binding in the SR membrane of cardiac cells is a biochemical characteristic of heart failure and believed to be linked to the contractility of the myocyte. Myocyte Ca^{2+} transients have been demonstrated to exhibit greater peak amplitude and faster decay rates after prolonged left ventricular unloading (10). In addition, calcium uptake increases significantly post-LVAD in isolated SR membranes, and calcium binding rates normalize (11). Cellular markers of hypertrophy and calcium regulatory protein abundance were measured in failing heart hearts (n = 12), non-failing hearts (n = 12) and LVAD supported hearts (n = 8). As had previously been established, in the failing hearts there were significant decreases in SERCA, and significant increases in both NCX and cell volume. After LVAD support, in contrast to previous reports, there was little improvement in SERCA protein abundance but a substantial decrease in NCX abundance.

In addition, studies utilizing a histochemical assay, myocyte apoptosis was measured pre and post LVAD (n = 12), and significant reductions in the incidence of myocyte apoptosis were seen in the LVAD supported group.

In summary, multi-level regression of the pathologic phenotype seen in the failing human myocardium is observed after LVAD support. The magnitude of the demonstrated changes observed indicates that reverse remodelling can occur even in the most advanced stages of cardiomyopathy. By defining the molecular correlations to the cellular changes, it may be possible to identify the pivotal molecular mechanisms involved. Ultimately however, the degree to which cellular recovery translates into organ and patient level recovery, will determine the viability of the use of mechanical circulatory support devices as a "bridge to recovery".

RECOVERY AFTER MECHANICAL CIRCULATORY SUPPORT AT THE ORGAN LEVEL

O. Howard Frazier (Chief Cardiopulmonary Transplantation, Director Cardiovascular Surgery Research, Texas Heart Institute, Houston, USA) spoke about the observed ventricular improvement seen during extended bridge to transplant studies, the need for approved weaning protocols in the US, and the future role of axial flow pumps in the field of bridge to recovery studies.

Dr. Frazier provided an overview of the history of clinical circulatory support beginning in the 1960's with the first clinical ventricular assist device and total artificial heart implants. The first patient in their experience, who demonstrated substantial "recovery" or ventricular improvement, was in 1991. This patient who subsequently died of stroke 505 days after implant, demonstrated during pump-off studies a significant reduction in Left Ventricular End Diastolic Dimension (LVEDD), an improvement in Left Venticular Ejection Fraction (LVEF), and a normalized Stroke Volume (SV).

The early clinical observations of myocardial improvement at the time of transplant in VAD supported patients, were subsequently correlated with histological studies using tissue samples from the apical core taken at time of device implant and subsequently after circulatory support at the time of transplant. These studies demonstrated significant reduction in the areas of attenuated fibers which seemed to indicate a degree of ventricular improvement after prolonged LVAD support (12). Subsequent studies, published in 1996 examined chronic heart failure patients (mean duration of prior heart failure 33 months) supported for extended times (mean support of 137 days). In this study, patients exhibited the following improvements after extended circulatory support: 1) Reduced cardiothoracic ratio, 2) Reduced LVEDD, 3) Increased LVEF, 4) Reduced Pulmonary Capillary Wedge Pressure (PCWP), and 5) Reduced Pulmonary Vascular Resistance (PVR). Moreover, the calcium handling (uptake and binding rates) of the SR membrane were also significantly improved after extended LVAD support in this study (11).

The primary cause of weaning patients off these pumps in the US, has been due to medical complications such as infections or defective devices, since there are no approvals for the bridge to recovery application. Case studies of substantial myocardial improvement after prolonged ventricular unloading, leading to pump removal, at the Texas Heart Institute were presented (13). The overall results for some of these long-term survivors showed substantial normalization with: 1) Decreased LVEDD, 2) Increased LVEF, and 3) Improved contractibility of the ventricles even years after the LVAD explant. It is important to note that while these patients improved significantly allowing weaning from the device, true clinical recovery (total normalization of cardiac function) was not achieved. Nevertheless, the need of a protocol to identify patients with sufficient myocardial recovery to consider device removal is urgently needed.

To identify potential patients for LVAD weaning, a protocol using dobutamine stress echocardiography (DSE) was developed at the Texas Heart Institute (14). Once patients have achieved stable functional capacity as measured by reaching NYHA Class I without inotropic support, and improvements in LV function (both LVDD and LVEF), the weaning protocol is initiated. The weaning protocol is dependent on the device type, however the general approach is to operate in a fixed mode decreasing the beat rate by 10 BPM weekly or as tolerated until a minimal threshold is achieved. Patients are returned to automatic mode if any of the following conditions are observed: 1) Shortness of Breath, 2) Fluid Overload, 3) Fatigue, or 4) Increased LVDD. The DSE study is performed with the pump off (patients are fully heparinized) and consists of conventional dobutamine infusion at increasing doses. Of the 16 patients studied, 9 patients were termed favourable responders and were explanted, while the remaining patients were transplanted. Six of the nine patients explanted survived for over 12 months, suggesting that DSE may be a useful tool for prospectively identifying patients for potential device removal. Furthermore, 2 of the 3 deaths in the favourable responders group were surgical deaths, with one of the major difficulties encountered being removing the pump after extended implant. This may suggest an important role for the newer generation of the pumps, such as the Jarvik 2000 axial flow pump. These devices which should be considered ventricular assist devices as opposed to ventricular replacement devices, are much smaller and therefore easier to both implant and explant, and thus may play an important role in the field of myocardial recovery (15).

RECOVERY AFTER MECHANICAL CIRCULATORY SUPPORT AT THE PATIENT LEVEL

Gurmeet Singh (Surgery Assistant Professor, University of Manitoba Health Sciences Centre, Winnipeg, Canada) gave an overview of mechanical circulatory support devices, the characteristics of the ideal device, and the future of the mechanical circulatory devices.

Mechanical assist devices can clearly provide substantial health benefits to heart failure patients, however, the patients need to clearly understand the possible complications such as thromboembolism and infections. Most recently, newer technologies have permitted the reduction in the size of the device consoles and the development of wearable devices allowing more mobility for the patients and the opportunity to return to normal life outside the hospital.

The different device types currently available are shown in Figure 3 and were reviewed. In the category of the Total Artificial Hearts there is the CardioWest and the recently utilized Abiocor system. Ventricular Assist Devices (VADs) can be broadly categorized into 3 areas: 1) Paracorporeal Devices such as the Thoratec VAD which can be used for bi-ventricular support and in smaller patients, but requires a large external console, 2) Extracorporeal Devices such as the Abiomed BVS-5000 which is relatively inexpensive, but limits patient mobility and is only designed for short-term use, 3) Implantable devices, of which there are two basic categories: a) Implantable Pulsatile Devices such as the Novacor and HeartMate, both of which offer the patient complete mobility outside of the hospital, but can not be utilized in some smaller patients due to the Body Surface Area (BSA) limitations, and b) Implantable Continuous Flow Devices which have just begun clinical application such as the MicroMed DeBakey, and Jarvik 2000, which are smaller than traditional pulsatile devices. On the issue of weaning patients, it was noted that with the Novacor LVAS there were 28 patients reported weaned. These patients were typically young patients (40 years of age) with cardiomyopathy and there was a 1 year survival of 79% in these patients, and 65% were deemed successfully weaned (i.e. alive and with no transplant at 1 year).

- ***Total Artificial Heart*** (CardioWest)
- ***Ventricular Assist Device***
  - Paracorporeal (Thoratec)
  - Extracorporeal (Abiomed)
  - Implantable
    - Pulsatile (Novacor, HeartMate)
    - Non-Pulsatile Continuous (MicroMed DeBakey, Jarvik 2000)

Figure 3. Device Classification.

The experience at the Cleveland Clinic with devices capable of discharge was presented. A total of 184 devices in this group were implanted (120 HeartMate VE, 8 HeartMate IP, 56 Novacor, 2 Debakey). No patients on the Debakey were discharged as both received early transplants, 30 of 56 (54%) of Novacor patients were discharged, and 55 of 120 (46%) of HeartMate VE patients were discharged. A total of 74 HeartMate IP devices were implanted, but initially were not capable for release until later in this experience, when 6 of 8 (75%) were discharged. The patients were either discharged to a rehab centre (54%) or home (46%). Re-admission was required in 24 patients (26%) due to neurological events, multiple infections, or pump malfunctions.

Based on this clinical experience, important aspects for the ideal device were determined. The ideal device should be: 1) durable, 2) reliable, 3) have infection and thromboembolic rates comparable to current prosthetic cardiac valves. It was noted that with the HeartMate system the inflow valve becomes incompetent at around 1 year, and this becomes a clinical concern for patients considered for chronic support. These valves can also develop endocarditis, in addition to the other drive-line and pump pocket infections experienced by these patients. Some devices also have a higher risk for thromboembolic events, sometimes seen in the conduits, which pre-disposes the patients to Cerebral Vascular Accidents (CVA).

In the future, the Total Artificial Heart (TAH) may prove to be a good alternative for a small percent of the patients. However at this time, implantable VADs may be more advantageous for the wider patient population, especially a totally implantable system that eliminates the needs for a percutaneous driveline and its associated infectious complications. In addition a booster or continuous flow system in which the left ventricular function is augmented rather that replaced may also prove to be beneficial in this area.

EFFECTS ON THE MYOCARDIUM AFTER LEFT VENTRICULAR ASSIST DEVICE SUPPORT

Hideo A. Baba (Institute of Pathology, University of Essen, Essen, Germany) spoke about the effects on the myocardium after prolonged left ventricular unloading. It was noted that in the US, over 27,000 LVAD candidates per year could potentially be bridged to recovery.

LVAD support has been established to improve the clinical status of heart failure patients. As demonstrated by x-ray the heart size and shape improves in a mechanism termed "reverse remodelling". Reverse modelling mechanisms are not completely understood, however reduction of myocyte diameter, decreased apoptosis, improved myocyte contractibility, decreased Tissue Necrosis Factor (TNF-α) tissue content and increased myocardial perfusion are all observed.

Heart tissue samples before and after LVAD support were studied to investigate the effects of mechanical assist devices on the myocardium. Three specific areas were examined: 1) Morphometric analysis, 2) Stress proteins (Metallothionein, Heme Oxygenase), and 3) Transcription factor (NF-κB).

At the organ level it has been well established that the diastolic diameter decreases and the wall thickness increases after LVAD support (16). At the cellular level studies were conducted to assess cardiomyocyte diameter, demonstrating significant reduction in cardiomyocyte diameter after LVAD support (17). To determine if there was a regional distribution of myocytes with reduced diameters, measurements were taken at the subendocardium, midmyocardium and sub-epicardium areas. While decreases in myocyte diameter were noted in each of these areas, the only area with significant changes was the subendocardial area. Studies to quantify the amount of fibrosis pre and post LVAD were also conducted. These studies utilized staining with Sirius Red for collagen I and III, followed by polarization microscopy, and image analysis for quantification. No significant difference in fibrosis was determined either at the interstitial or perivascular areas.

Stress proteins (Metallothionein—MT, and Heme Oxygenase—HO-1), were examined pre and post LVAD to assess immunoreactivity. MT is a stress protein with cell protection and detoxification properties, and has an association with impaired heart function. MT immunoreactivity was measured in 17 LVAD patients (15 Novacor LVAS, 2 HeartMate IP) pre and post LVAD using immunohistochemical methods (17). This study clearly demonstrated significant decreases for MT positive myocytes at both the subendocardial, and subepicardial regions after mechanical circulatory support (17). Heme Oxygenase (HO-1) another stress protein with cell protection and anti-apoptotic properties, was measured in LVAD patients pre and post implantation. Immunoreactivty for HO-1 was also found to be markedly reduced at both the subendocardial, and subepicardial regions after mechanical circulatory support. The reductions in immunoreactivity of stress proteins could be due to the relief of tissue hypoxia after LVAD support. To assess this hypothesis, cardiomyocytes from neonatal rats were assessed under hypoxic conditions and after re-oxygenation. HO-1 protein expression was assessed and showed a clear pattern of increased content during hypoxic conditions and a reduction after re-oxygenation.

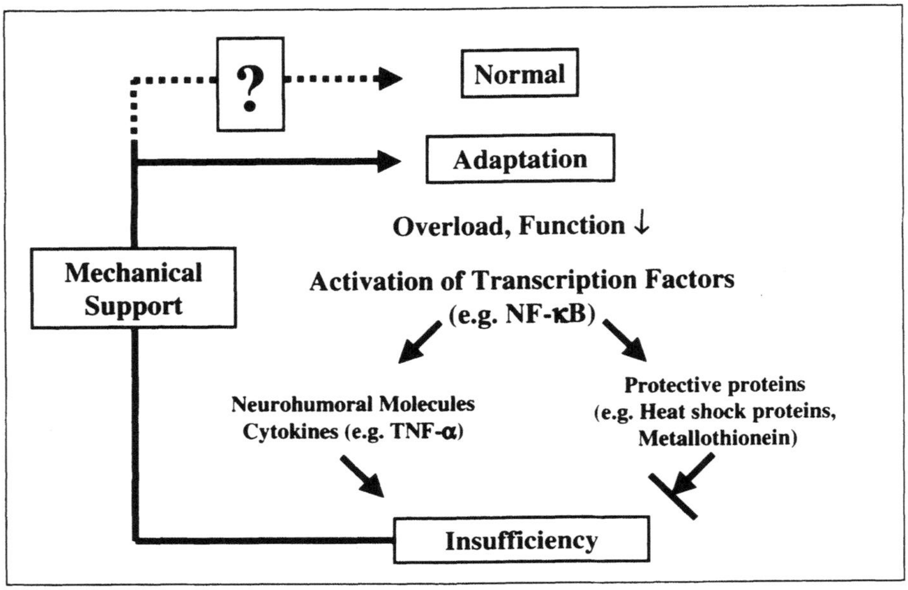

Figure 4. Schematic of Heart Failure Process supported by Mechanical Circulatory Support.

Next studies were conducted to assess if the transcription factor NF-κB was altered by mechanical circulatory support. Utilizing immunohistochemistry, active NF-κB was measured pre and post LVAD support, and a significant reduction was observed after LVAD support. Gel shift analysis was also conducted to examine NF-κB DNA-binding activity and again showed marked reduction after circulatory support.

These studies demonstrated that after LVAD support the following observations can be made: 1) The myocyte diameter decrease, mainly in the subendocardium, 2) No change in the content of fibrosis, 3) Reduction in stress proteins (MT, HO-1), and 4) Reduction in NF-κB activity. These findings suggest that the relief of tissue hypoxia during mechanical circulatory support may play an important role in the recovery process.

Based on these studies, a simplified schematic of the process of heart failure and the subsequent recovery after circulatory support was developed (Figure 4). Basically an event or process (cardiac insult) leads to adaptation, overload and decreasing cardiac function. Subsequently transcription factors such as NF-κB are activated. In addition, the neurohormonal cytokines such as TNF-α are released, leading to increased cardiac insufficiency. At the same time protective proteins such as MT and HO-1 are increased. After mechanical circulatory support, these various adaptations tend to improve, however the question remains, can these improvements lead to sustainable normalization ?

CONCLUSIONS

Not long ago healing of damaged heart muscle was largely believed to be impossible. However, clinical observations of advanced heart failure patients after mechanical circulatory support clearly showed a potential for significant improvements in cardiac structure and function. The ability to obtain substantial tissue samples at the time of device implant and at the subsequent transplant, has allowed in-depth investigation from basic neurohormonal assays to advanced DNA analysis. This new experimental laboratory for the study of heart failure has led to breakthroughs in knowledge of the heart failure process. The Expert Panel summarized an increasingly sizable body of evidence that now supports the previous clinical observations of cardiac recovery after circulatory support. This data demonstrates clearly reversal of heart failure mechanisms at the:

1) gene level
2) cell level
3) organ level
4) patient level

This new knowledge can be potentially utilized in the future to identify patients with substantial cardiac recovery, leading to increased use of mechanical circulatory support devices as a "bridge to recovery". Ultimately, mechanical circulatory devices may be remembered not for the many individual lives they have saved, but for providing a greater understanding of heart failure, leading to new innovative treatments for this epidemic.

TABLE OF ABBREVIATIONS

Abbreviation	Definition
BSA	Body Surface Area
CVA	Cerebral Vascular Accident
DCM	Dilated CardioMyopathy
DSE	Dobutamine Stress Echocardiography
FFR	Force Frequency Relationship
HO-1	Heme Oxygenase
LV	Left Ventricle
LVAD	Left Ventricular Assist Device
LVEDD	Left Ventricular End Diastolic Dimension
LVEF	Left Ventricular Ejection Fraction
MCS	Mechanical Circulatory Support
mRNA	messenger RNA
MT	Metallothionein (stress protein)
NCX	Sodium Calcium Exchanger
NF-kB	Nuclear Factor kB (transcription factor)
PVR	Pulmonary Vascular Resistance

PWCP	Pulmonary Capillary Wedge Pressure
RV	Right Ventricle
RyR	Ryanodine-sensitive Ca^{2+} Release
SERCA2a	Sarco/Endoplasmic Reticulum Calcium ATPase
SR	Sarcoplasmic Reticulum
SV	Stroke Volume
TAH	Total Artificial Heart
TNF	Tumor Necrosis Factor

ACKNOWLEDGMENTS

The organizers of this Expert Panel would like to thank the International Society of Heart Research Program Committee for providing time for the workshop. In addition, the authors would like to thank WorldHeart Corporation (Ottawa, Canada) for providing financial support. Finally, the authors would like to acknowledge the Cardiovascular Devices Division of the University of Ottawa Heart Institute & their staff for their efforts in both organizing the panel and compiling this manuscript. Dr. Singh would like to acknowledge the staff of the Cleveland Clinic Foundation for their assistance in providing materials for his presentation.

REFERENCES

1. Young JB. 2001. Healing the heart with ventricular assist device therapy: Mechanisms of cardiac recovery. Ann Thorac Surg 71:S210–S219.
2. McDonald CD, Burch GE, Walsh JJ. 1972. Prolonged bed rest in the treatment of idiopathic cardiomyopathy. Amer J Med 52(1):41–50.
3. Mussivand T, Hendry PJ, Masters RG, Holmes KS, Keon WJ. 1998. Circulatory support devices for bridge to recovery. Ann Thorac Surg 66:975–976.
4. Levin HR, Oz MC, Chen JM, Packer M, Rose EA, Burkhoff D. 1995. Reversal of chronic ventricular dilation in patients with end-stage cardiomyopathy by prolonged mechanical unloading. Circulation. 91:2717–2720.
5. Levin HR, Oz MC, Catanese KA, Rose EA, Burkhoff D. 1996. Transient normalization of systolic and diastolic function after support with a left ventricular assist device in a patient with dilated cardiomyopathy. J Heart Lung Transplant 15:840–842.
6. Heerdt PM, Holmes JW, Cai B, Barbone A, Madigan JD, Reiken S, Lee DL, Oz MC, Marks AR, Burkhoff D. 2000. Chronic unloading by left ventricular assist device reverses contractile dysfunction and alters gene expression in end-stage heart failure. Circulation 102:2713–2719.
7. Barbone A, Holmes JW, Heerdt PM, The' AH, Naka Y, Joshi N, Daines M, Marks AR, Oz MC, Burkhoff D. 2001. Comparison of right and left ventricular responses to left ventricular assist device support in patients with severe heart failure: A primary role of mechanical unloading underlying reverse remodeling. Circulation 104:670–675.
8. Zafeiridis A, Jeevanandam V, Houser SR, Margulies KB. 1998. Regression of cellular hypertrophy after left ventricular assist device support. Circulation 98:656–662.
9. Altemose GT, Gritsus V, Jeevanandam V, Goldman B, Margulies KBl. 1997. Altered myocardial phenotype after mechanical support in human beings with advanced cardiomyopathy. J Heart Lung Transplant 16:765–773.
10. Dipla K, Mattiello JA, Jeevanandam V, Houser SR, Margulies KB. 1998. Myocyte recovery after mechanical circulatory support in humans with end-stage heart failure. Circulation 97:2316–2322.
11. Frazier OH, Benedict CR, Radovancevic B, Bick RJ, Capek P, Springer WE, Macris MP, Delgado R, Buja LM. 1996. Improved left ventricular function after chronic left ventricular unloading. Ann Thorac Surg 62:675–682.

12. Scheinin SA, Capek P, Radovancevic B, Duncan JM, McAllister HA, Frazier OH. 1992. The effect of prolonged left ventricular support on myocardial histopathology in patients with end-stage cardiomyopathy. ASAIO J 38:M271-M274.
13. Frazier OH, Myers TJ. 1999. Left ventricular assist system as a bridge to myocardial recovery. Ann Thorac Surg 68:734–741.
14. Khan T, Okerberg K, Hernandez A, Miller K, Myers TJ, Radovancevic B, Delgado RM, Gregoric I, Frazier OH. 2001. Assessment of myocardial recovery using dobutamine stress echocardiography in LVAD patients. J Heart Lung Transplant 20(2):202–203.
15. Westaby S, Katsumata T, Houel R, Evans R, Pigott D, Frazier OH, Jarvik R. 1998. Jarvik 2000 Heart potential for bridge to myocyte recovery. Circulation 98:1568–1574.
16. Nakatani S, McCarthy PM, Kottke-Marchant K, Harasaki H, James KB, Savage RM, Thomas JD. 1996. Left ventricular echocardiographic and histologic changes: Impact of chronic unloading by an implantable ventricular assist device. J Am Coll Cardiol 27(4):894–901.
17. Baba HA, Grabellus F, August C, Plenz G, Takeda A, Tjan TD, Schmid C, Deng MC. 2000. Reversal of metallothionein expression is different throughout the human myocardium after prolonged left-ventricular mechanical support. J Heart Lung Transplant 19:668–674.

Index

D

MIX
Papier aus verantwortungsvollen Quellen
Paper from responsible sources
FSC® C105338

If you have any concerns about our products,
you can contact us on
ProductSafety@springernature.com

In case Publisher is established outside the EU,
the EU authorized representative is:
Springer Nature Customer Service Center GmbH
Europaplatz 3, 69115 Heidelberg, Germany

Printed by Libri Plureos GmbH
in Hamburg, Germany